INTERNATIONAL SERIES O
ON PHYSIC

International Series of Monographs on Physics

144. T. R. Field: *Electromagnetic scattering from random media*
143. W. Götze: *Complex dynamics of glass-forming liquids – a mode-coupling theory*
142. V. M. Agranovich: *Excitations in organic solids*
141. W. T. Grandy: *Entropy and the time evolution of macroscopic systems*
140. M. Alcubierre: *Introduction to 3 + 1 numerical relativity*
139. A. L. Ivanov, S. G. Tikhodeev: *Problems of condensed matter physics – quantum coherence phenomena in electron–hole and coupled matter–light systems*
138. I. M. Vardavas, F. W. Taylor: *Radiation and climate*
137. A. F. Borghesani: *Ions and electrons in liquid helium*
136. C. Kiefer: *Quantum gravity*, Second edition
135. V. Fortov, I. Iakubov, A. Khrapak: *Physics of strongly coupled plasma*
134. G. Fredrickson: *The equilibrium theory of inhomogeneous polymers*
133. H. Suhl: *Relaxation processes in micromagnetics*
132. J. Terning: *Modern supersymmetry*
131. M. Mariño: *Chern–Simons theory, matrix models, and topological strings*
130. V. Gantmakher: *Electrons and disorder in solids*
129. W. Barford: *Electronic and optical properties of conjugated polymers*
128. R. E. Raab, O. L. de Lange: *Multipole theory in electromagnetism*
127. A. Larkin, A. Varlamov: *Theory of fluctuations in superconductors*
126. P. Goldbart, N. Goldenfeld, D. Sherrington: *Stealing the gold*
125. S. Atzeni, J. Meyer-ter-Vehn: *The physics of inertial fusion*
123. T. Fujimoto: *Plasma spectroscopy*
122. K. Fujikawa, H. Suzuki: *Path integrals and quantum anomalies*
121. T. Giamarchi: *Quantum physics in one dimension*
120. M. Warner, E. Terentjev: *Liquid crystal elastomers*
119. L. Jacak, P. Sitko, K. Wieczorek, A. Wojs: *Quantum Hall systems*
118. J. Wesson: *Tokamaks*, Third edition
117. G. E. Volovik: *The universe in a helium droplet*
116. L. Pitaevskii, S. Stringari: *Bose–Einstein condensation*
115. G. Dissertori, I. G. Knowles, M. Schmelling: *Quantum chromodynamics*
114. B. DeWitt: *The global approach to quantum field theory*
113. J. Zinn-Justin: *Quantum field theory and critical phenomena*, Fourth edition
112. R. M. Mazo: *Brownian motion – fluctuations, dynamics, and applications*
111. H. Nishimori: *Statistical physics of spin glasses and information processing – an introduction*
110. N. B. Kopnin: *Theory of nonequilibrium superconductivity*
109. A. Aharoni: *Introduction to the theory of ferromagnetism*, Second edition
108. R. Dobbs: *Helium three*
107. R. Wigmans: *Calorimetry*
106. J. Kübler: *Theory of itinerant electron magnetism*
105. Y. Kuramoto, Y. Kitaoka: *Dynamics of heavy electrons*
104. D. Bardin, G. Passarino: *The standard model in the making*
103. G. C. Branco, L. Lavoura, J. P. Silva: *CP violation*
102. T. C. Choy: *Effective medium theory*
101. H. Araki: *Mathematical theory of quantum fields*
100. L. M. Pismen: *Vortices in nonlinear fields*
99. L. Mestel: *Stellar magnetism*
98. K. H. Bennemann: *Nonlinear optics in metals*
96. M. Brambilla: *Kinetic theory of plasma waves*
94. S. Chikazumi: *Physics of ferromagnetism*
91. R. A. Bertlmann: *Anomalies in quantum field theory*
90. P. K. Gosh: *Ion traps*
88. S. L. Adler: *Quaternionic quantum mechanics and quantum fields*
87. P. S. Joshi: *Global aspects in gravitation and cosmology*
86. E. R. Pike, S. Sarkar: *The quantum theory of radiation*
83. P. G. de Gennes, J. Prost: *The physics of liquid crystals*
73. M. Doi, S. F. Edwards: *The theory of polymer dynamics*
69. S. Chandrasekhar: *The mathematical theory of black holes*
51. C. Møller: *The theory of relativity*
46. H. E. Stanley: *Introduction to phase transitions and critical phenomena*
32. A. Abragam: *Principles of nuclear magnetism*
27. P. A. M. Dirac: *Principles of quantum mechanics*
23. R. E. Peierls: *Quantum theory of solids*

Physics of Ferromagnetism

SECOND EDITION

SÓSHIN CHIKAZUMI
Professor Emeritus,
University of Tokyo

English edition prepared with the assistance of
C. D. GRAHAM, JR
Professor Emeritus,
University of Pennsylvania

Great Clarendon Street, Oxford OX2 6DP

Oxford University Press is a department of the University of Oxford.
It furthers the University's objective of excellence in research, scholarship,
and education by publishing worldwide in

Oxford New York

Auckland Cape Town Dar es Salaam Hong Kong Karachi
Kuala Lumpur Madrid Melbourne Mexico City Nairobi
New Delhi Shanghai Taipei Toronto
With offices in
Argentina Austria Brazil Chile Czech Republic France Greece
Guatemala Hungary Italy Japan South Korea Poland Portugal
Singapore Switzerland Thailand Turkey Ukraine Vietnam

Oxford is a registered trade mark of Oxford University Press
in the UK and in certain other countries

Published in the United States
by Oxford University Press Inc., New York

Reprinted with corrections 2010

ISBN 978-0-19-956481-1

Printed in the United Kingdom by
Lightning Source UK Ltd., Milton Keynes

PREFACE

This book is intended as a textbook for students and investigators interested in the physical aspect of ferromagnetism. The level of presentation assumes only a basic knowledge of electromagnetic theory and atomic physics and a general familiarity with rather elementary mathematics. Throughout the book the emphasis is primarily on explanation of physical concepts rather than on rigorous theoretical treatments which require a background in quantum mechanics and high-level mathematics.

Ferromagnetism signifies in its wide sense the strong magnetism of attracting pieces of iron and has long been used for motors, generators, transformers, permanent magnets, magnetic tapes and disks. On the other hand, the physics of ferromagnetism is deeply concerned with quantum-mechanical aspects of materials, such as the exchange interaction and band structure of metals. Between these extreme limits, there is an intermediate field treating magnetic anisotropy, magnetostriction, domain structures and technical magnetization. In addition, in order to understand the magnetic behavior of magnetic materials, we need some knowledge of chemistry and crystallography.

The purpose of this book is to give a general view of these magnetic phenomena, focusing its main interest at the center of this broad field. The book is divided into eight parts. After an introductory description of magnetic phenomena and magnetic measurements in Part I, the magnetism of atoms including nuclear magnetism and microscopic experiments on magnetism, such as neutron diffraction and nuclear magnetic resonance (NMR), is treated in Part II. The origin and mechanism of para-, ferro- and ferrimagnetism are treated in Part III. Part IV is devoted to more material-oriented aspects of magnetism, such as magnetism of metals, oxides, compounds and amorphous materials. In Part V, we discuss magnetic anisotropy and magnetostriction, to which I have devoted most of my research life. Part VI describes domain structures, their observation technique and domain theory. Part VII is on magnetization processes, analyzed on the basis of domain theory. Part VIII is devoted to phenomena associated with magnetization such as magnetothermal, magnetoelectrical and magneto-optical effects, and to engineering applications of magnetism.

Throughout the book, the SI or MKSA system of units using the E–H analogy is used. As is well known, this system is very convenient for describing all electromagnetic phenomena without introducing troublesome coefficients. This system also uses practical units of electricity such as amperes, volts, ohms, coulombs and farads. This system is particularly convenient when we treat phenomena such as eddy currents and electromagnetic induction which relate magnetism to electricity. However, old-timers who are familiar with the old CGS magnetic units such as gauss, oersted, etc., must change their thinking from these old units to new units such as tesla, ampere per meter, etc. Once they become familiar with the new magnetic unit system, however, they may come to appreciate its convenience. To aid in the transition, a conversion table between MKSA and CGS units is given in Appendix 5.

In the previous edition I tried to refer to as many papers as possible. By the time of the revised edition, so many papers had become available that I was obliged to select only a small number of them to keep the text clear and simple. I have no doubt omitted many important papers for this reason, for which I apologize and beg their authors for tolerance and forgiveness. Many authors have kindly permitted me to use their beautiful photographs and unpublished data, for which I want to express my sincere thanks.

This book was originally published in Japanese by Shyokabo Publishing Company in Tokyo in 1959. The English version of that edition was published by John Wiley & Sons, Inc. in New York in 1964. The content of the English version was increased by about 55% from the Japanese version. At that time my English was polished by Dr Stanley H. Charap. A revised Japanese edition was published in two volumes in 1978 and 1984, respectively. The content was about 30% larger than the previous English edition. The preparation of the present English version of the revised edition was started in 1985 and took about ten years. This time my English was polished by Professor C. D. Graham, Jr using e-mail communication. The content has not been greatly increased, but has been renewed by introducing recent developments and omitting some old and less useful material.

Thanks are due to the staff of Oxford University Press who have helped and encouraged me throughout the period of translation.

Tokyo S.C.
March 1996

CONTENTS

Part I Classical Magnetism

Part II Magnetism of Atoms

Part III Magnetic Ordering

Part IV Magnetic Behavior and Structure of Materials

Part VI Domain Structures

Part VII Magnetization Processes

Part I
CLASSICAL MAGNETISM

Magnetism is one of the oldest phenomena in the history of natural science. It is said that magnetism was first discovered by a shepherd who noticed that the iron tip of his stick was attracted by a stone. This stone was found in Asia Minor, in the Magnesia district of Macedonia or in the city of Magnesia in Ionia. The word 'magnetism' is believed to originate from these names. Later in history, it was found that these natural magnets, if properly suspended from a thread or floated on a cork in water, would always align themselves in the same direction relative to the North Star. Thus they came to be known as 'lode stones' or 'loadstones', from a word meaning 'direction'.

In Part I, we introduce the phenomena of magnetostatics, which were investigated through the eighteenth and nineteenth centuries, following the first detailed studies of magnetism by William Gilbert early in the seventeenth century.

1
MAGNETOSTATIC PHENOMENA

1.1 MAGNETIC MOMENT

The most direct manifestation of magnetism is the force of attraction or repulsion between two magnets. This phenomenon can be described by assuming that there are 'free' magnetic poles on the ends on each magnet which exert forces to one another. These are called 'Coulomb forces' by analogy with the Coulomb forces between electrostatically charged bodies. Consider two magnetic poles with strengths of m_1 (Wb (weber)) and m_2 (Wb) respectively, separated by a distance r (m). The force F (N (newton)) exerted on one pole by the other is given by

$$F = \frac{m_1 m_2}{4\pi\mu_0 r^2}, \tag{1.1}$$

where μ_0 is called the *permeability of vacuum*, and has the value

$$\mu_0 = 4\pi \times 10^{-7} \qquad \text{henrys per meter } (\mathrm{H\,m^{-1}}). \tag{1.2}$$

It is also found that an electric current exerts a force on a magnetic pole. Generally a region of space in which a magnetic pole experiences an applied force is called a *magnetic field*. A magnetic field can be produced by other magnetic poles or by electric currents. A uniform magnetic field exists inside a long, thin solenoid carrying an electric current. When a current of i (A) flows in the winding of a solenoid having n turns per meter, the intensity of the field H at the center of the solenoid is defined by

$$H = ni. \tag{1.3}$$

The unit of the magnetic field thus defined is the ampere per meter or $\mathrm{A\,m^{-1}}$. ($1\,\mathrm{A\,m^{-1}} = 4\pi \times 10^{-3}\,\mathrm{Oe} = 0.0126\,\mathrm{Oe}$; $1\,\mathrm{Oe} = 79.6\,\mathrm{A\,m^{-1}}$, see Appendix 6.)

When a magnetic pole of strength m (Wb) is brought into a magnetic field of intensity H (A/m), the force F (N) acting on the magnetic pole is

$$F = mH. \tag{1.4}$$

(μ_0 is defined so as to avoid a coefficient in (1.4).) If a bar magnet of length l (m), which has poles m and $-m$ at its ends, is placed in a uniform magnetic field H, each pole is acted upon by a force as indicated by the arrows in Fig. 1.1, giving rise to a couple or torque, whose moment is

$$L = -mlH \sin\theta, \tag{1.5}$$

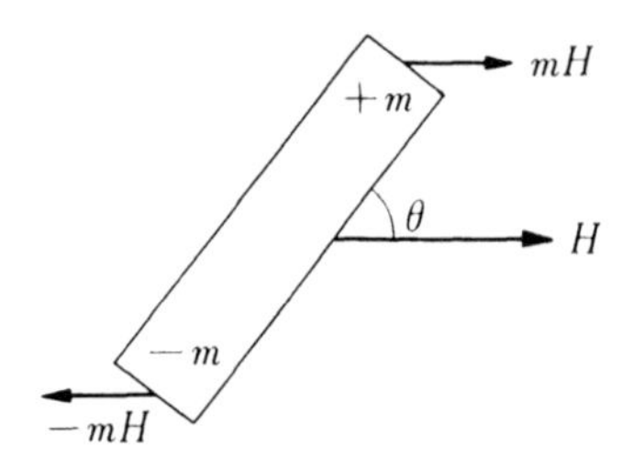

Fig. 1.1. A magnet under the action of a torque in a uniform magnetic field.

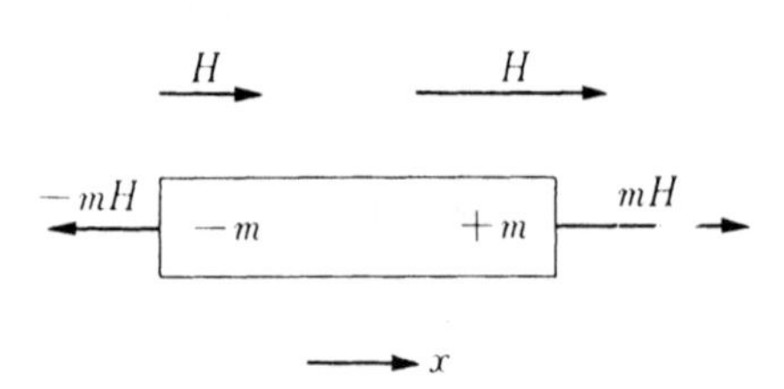

Fig. 1.2. A magnet under the action of a translational force in a gradient magnetic field.

where θ is the angle between the direction of the magnetic field H and the direction of the magnetization $(-m \to +m)$ of the magnet. Thus a uniform magnetic field exerts a torque on a magnet, but no translational force. A translational force acts on the magnet only if there is a gradient of the field $\partial H_x/\partial x$. The translational force is given by

$$F_x = ml \frac{\partial H}{\partial x} \tag{1.6}$$

in the x-direction (Fig. 1.2).

As seen in (1.5) and (1.6), any kind of force which acts on the magnet involves m and l in the form of the product ml. We call this product

$$M = ml \tag{1.7}$$

a *magnetic moment*; it has the unit of weber meter (Wb m)

$$(1 \text{ Wb m} = (1/4\pi) \times 10^{10} \text{ gauss cm}^3).$$

In terms of M, the torque exerted on a magnet in a uniform field H is given by

$$L = -MH \sin\theta, \tag{1.8}$$

irrespective of the shape of the magnet. If no frictional forces act on the magnet, the work done by the torque (1.8) is reversible, giving rise to a potential energy,

$$U = -MH \cos\theta. \tag{1.9}$$

The direction from the negative pole to the positive pole is defined as the direction of the magnetic moment, so that equations (1.8) and (1.9) can be expressed using the magnetic moment vector $\boldsymbol{M}$ as

$$\boldsymbol{L} = \boldsymbol{M} \times \boldsymbol{H}, \tag{1.10}$$

and

$$U = -\boldsymbol{M} \cdot \boldsymbol{H}. \tag{1.11}$$

Although the magnetic moment is defined here as (magnetic pole) × (distance between poles), in practice it is rather hard to define the position of magnetic poles

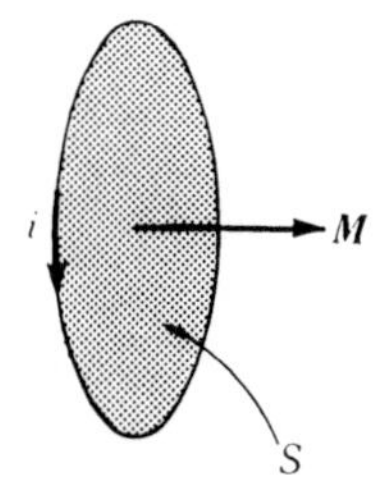

Fig. 1.3. A magnetic moment produced by a closed electric current.

accurately.* However, the moment of force or torque, L, is a measurable quantity, so that we can use equation (1.8) as the definition of the magnetic moment.

A magnetic moment can also be produced by a closed loop carrying an electric current. The magnetic moment produced by a current i (A) which flows in a closed circuit or loop enclosing an area S (m^2) is defined as

$$M = \mu_0 iS. \tag{1.12}$$

The direction of the magnetic moment is defined as the direction of movement of a right-hand screw which rotates in the same direction as the current in the closed circuit (Fig. 1.3).

Now let us consider how the magnetic moment M_1 of a magnet or a current loop placed at the origin O produces magnetic fields in space (Fig. 1.4). Consider the magnetic field at a point P whose position is given by (r, θ) in polar coordinates. For simplicity we assume the size of the magnet l or of the closed current loop $\sqrt{S}$ is negligibly small compared with r. A source of field meeting this condition is called a *magnetic dipole*. The components of the field at P are given by

$$\begin{aligned} H_r &= \frac{M_1}{4\pi\mu_0} \frac{2\cos\theta_1}{r^3} \\ H_\theta &= \frac{M_1}{4\pi\mu_0} \frac{\sin\theta_1}{r^3}. \end{aligned} \tag{1.13}$$

The distribution of magnetic fields in a space can be shown by *lines of force* running parallel to the direction of $\boldsymbol{H}$ at each point of the space. Figure 1.4 shows a computer drawing of the lines of force calculated from (1.13).

Now in addition to the moment M_1 at the origin, we place at P another magnetic dipole M_2, which makes an angle θ_2 with $\boldsymbol{r}$ (for simplicity, we assume that the two

* Some authors object to the concept of a 'magnetic pole' as unrealistic, because of the difficulty of realizing a 'point' pole. However, the same objection can be made to the 'element' of electric current which is used in the Biot–Savart law describing the magnetic field produced by an electric current. Both concepts are useful to calculate magnetic fields.

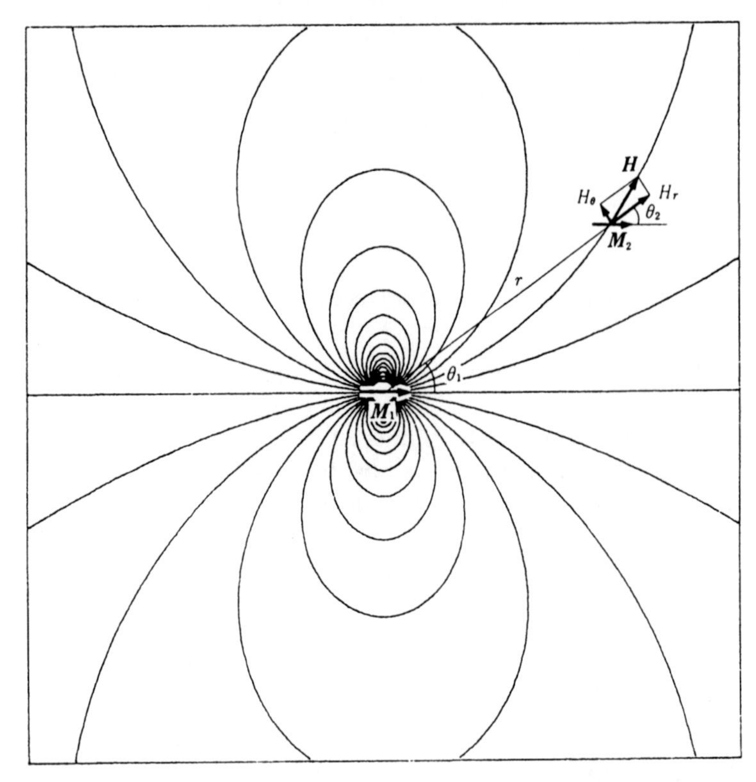

Fig. 1.4. Computer-drawn diagram of lines of magnetic force produced by a magnetic dipole.

dipoles lie in the same plane). Using (1.9), the potential energy of this system is given by

$$\begin{aligned} U &= -M_2 H_r \cos\theta_2 - M_2 H_\theta \sin\theta_2 \\ &= -\frac{M_1 M_2}{4\pi\mu_0 r^3}(2\cos\theta_1\cos\theta_2 - \sin\theta_1\sin\theta_2). \end{aligned} \tag{1.14}$$

If the two magnetic dipoles are equal in magnitude and parallel to one another, so that $M_1 = M_2 = M$, and $\theta_1 = \theta_2 = \theta$, as in the case of individual atomic dipoles in a ferromagnetic material, (1.14) becomes

$$U = -\frac{3M^2}{4\pi\mu_0 r_{12}^3}\left(\cos^2\theta - \tfrac{1}{3}\right). \tag{1.15}$$

This potential energy is minimum at $\theta = 0$, so that the configuration as shown in Fig. 1.5(a) is stable. Maximum energy occurs at $\theta = \pi/2$, so that the configuration as shown in Fig. 1.5(b) is unstable. This interaction between dipoles is called *dipole interaction.*

Fig. 1.5. Two arrangements of parallel dipoles: (a) stable, (b) unstable.

In Fig. 1.5(a), if we rotate M_1 by a small angle θ_1 while keeping the direction of M_2 fixed, we have, from (1.14),

$$U = -\frac{M_1 M_2}{2\pi\mu_0 r^3}\cos\theta_1. \tag{1.16}$$

Taking $M_1 = M_2 = M = 1.2 \times 10^{-29}$ (the moment of a single electron spin see (3.7)) and $r = 1\,\text{Å}$, the coefficient in (1.16) is

$$\frac{M_1 M_2}{2\pi\mu_0 r^3} = \frac{1.2^2 \times (10^{-29})^2}{8 \times 3.14^2 \times 10^{-7} \times 10^{-30}} = 1.8 \times 10^{-23}\,\text{J}$$
$$= 0.9\,\text{cm}^{-1} = 1.3\,\text{K} \tag{1.17}$$

(see Appendix 2 for energy conversions). If M_2 is antiparallel to M_1, as is the case for atomic dipoles in an antiferromagnetic material (see Chapter 7), the coefficient in (1.16) changes sign.

In (1.14), we assumed that M_1 and M_2 are both in the x–y plane. In the general case, the potential energy of the dipole interaction is given by

$$U = \frac{1}{4\pi\mu_0 r^3}\left\{\boldsymbol{M}_1 \cdot \boldsymbol{M}_2 - \frac{3}{r^2}(\boldsymbol{M}_1 \cdot \boldsymbol{r})(\boldsymbol{M}_2 \cdot \boldsymbol{r})\right\}. \tag{1.18}$$

1.2 MAGNETIC MATERIALS AND MAGNETIZATION

Magnetic materials are materials which are magnetized to some extent by a magnetic field. There are strongly magnetic materials that are attracted by a permanent magnet, and weakly magnetic materials whose magnetization can only be detected by sensitive instruments.

When a magnetic material is magnetized uniformly, the magnetic moment per unit volume is called the magnetic polarization or *intensity of magnetization*, usually denoted by I. If there are magnetic moments $\boldsymbol{M}_1, \boldsymbol{M}_2, \ldots, \boldsymbol{M}_n$ in a unit volume of a magnetic material, the intensity of magnetization is given by

$$\boldsymbol{I} = \sum_{i=1}^{n} \boldsymbol{M}_i. \tag{1.19}$$

If these moments have the same magnitude, M, and are aligned parallel to each other (Fig. 1.6), (1.19) simplifies to

$$I = NM, \tag{1.20}$$

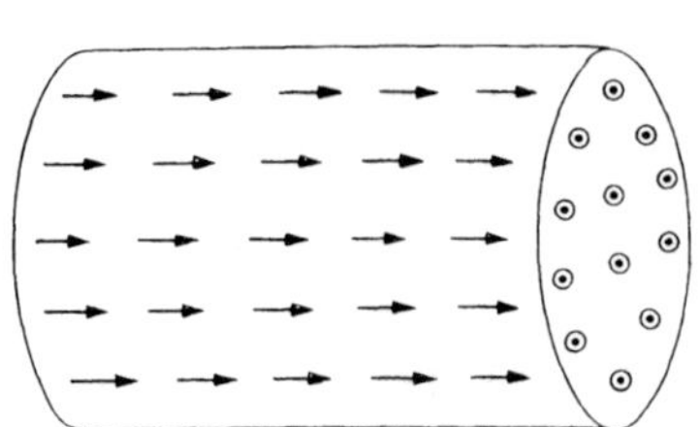

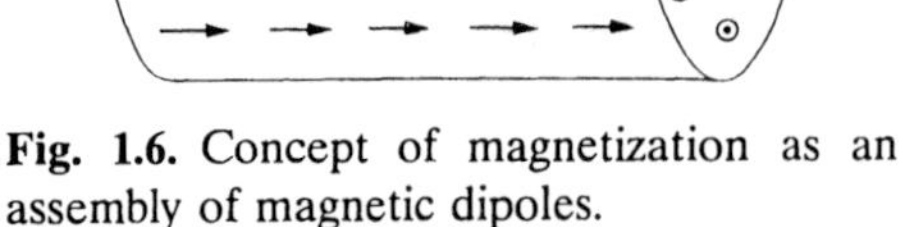

Fig. 1.6. Concept of magnetization as an assembly of magnetic dipoles.

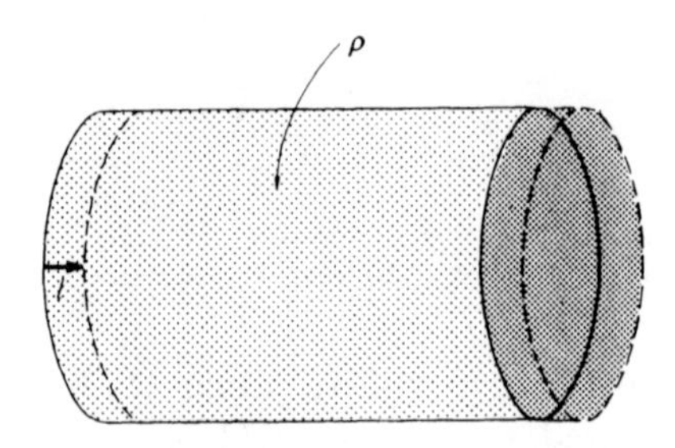

Fig. 1.7. Concept of magnetization as a displaced magnetic charge density.

where N is the total number of moments M in a unit volume. Since the unit of N is m^{-3}, we find from (1.20) that the unit of I is $\mathrm{Wb\,m}^{-2}$, which has the alternative and simpler name of tesla (T) (1 T of I $= 10^4/4\pi$ gauss $= 7.9 \times 10^2$ gauss). If we adopt the definition of equation (1.7) for the magnetic dipole, we have from (1.20)

$$I = Nml. \tag{1.21}$$

In this expression, Nm signifies the total quantity of magnetic poles existing in a unit volume of material, or the magnetic pole density, ρ ($\mathrm{Wb/m}^3$), so that we have

$$I = \rho l. \tag{1.22}$$

From this expression, the concept of magnetization is also interpreted as a displacement of magnetic pole density ρ relative to $-\rho$ by the distance l (Fig. 1.7). Consequently uncompensated magnetic poles of surface density

$$\omega = \rho l \tag{1.23}$$

will appear at the ends of the specimen. Comparing (1.22) with (1.23), we have

$$I = \omega. \tag{1.24}$$

Thus we can define the magnetization to be the number of magnetic poles displaced across a unit cross-section.

How can we connect the concept of elementary magnetic moments consisting of closed current loops with the concept of magnetization? Suppose that the magnetic material is filled with many elementary closed current loops as shown in Fig. 1.8. Since neighboring currents cancel one another, only the surface currents remain uncompensated. If we assume that there are n current layers per unit length along the direction of magnetization, and also that the cross-section of the elementary closed current is S (m^2), we have $1/S$ elementary current loops per unit area of the cross section, and accordingly n/S elementary current loops per unit volume of the magnetic material. Using equation (1.12), we have from (1.20)

$$I = \frac{n}{S}(\mu_0 iS) = \mu_0 ni. \tag{1.25}$$

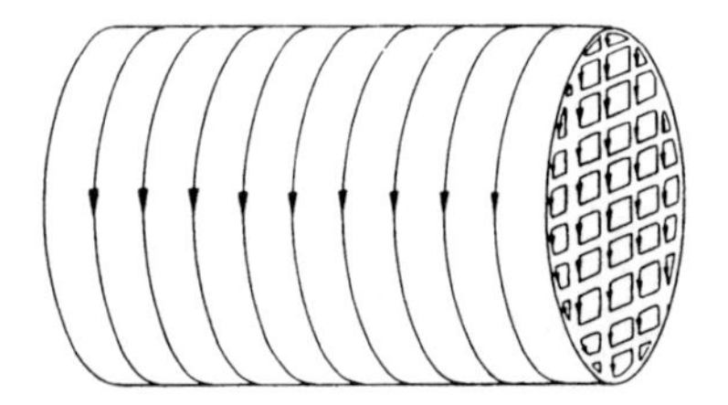

Fig. 1.8. Concept of magnetization as an assembly of small closed electric currents loops.

Comparing (1.25) with (1.3), we see that the magnetization is given by μ_0 times the magnetic field H' produced by the intrinsic current, or

$$I = \mu_0 H'. \tag{1.26}$$

Thus we have various concepts of 'magnetization': an ensemble of elementary magnetic moments, a displacement of magnetic poles, and an intrinsic current. When we calculate magnetic fields outside a magnetic material, we obtain the same result no matter which of these concepts we use. We will discuss the magnetic field inside magnetic materials in Section 1.3.

The intensity of magnetization can be determined by measuring the magnetic fields produced outside a magnetized specimen. Alternatively, we have another method utilizing the electromagnetic induction. Suppose that we have a search coil of cross-sectional area S (m^2) and N turns of wire placed as shown in Fig. 1.9(a). If we apply a magnetic field H perpendicular to the cross-section of the coil, the voltage

$$V = -NS\mu_0 \frac{\mathrm{d}H}{\mathrm{d}t} \tag{1.27}$$

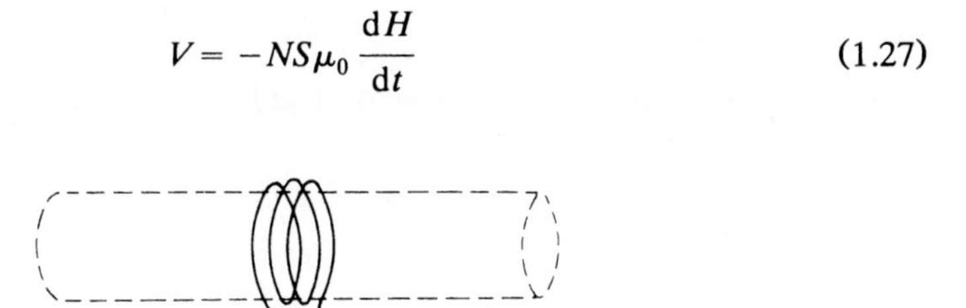

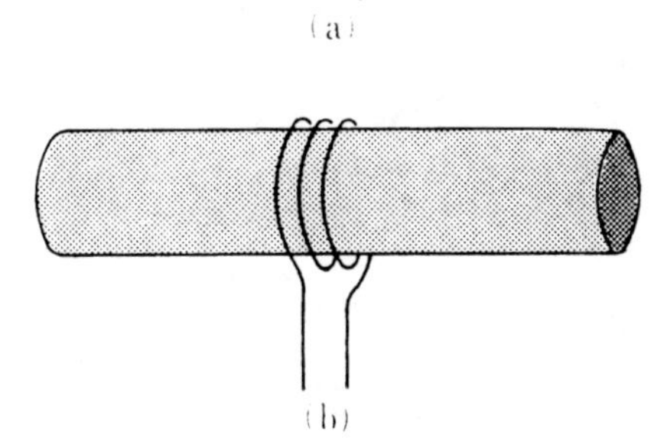

Fig. 1.9. (a) An air-core search coil; (b) a search coil wound on a magnetic rod.

is produced across the coil by the law of electromagnetic induction. If we insert a magnetic material in the coil (Fig. 1.9(b)), the voltage is increased to

$$V = -NS\frac{dB}{dt},^{*} \tag{1.28}$$

where B is the *magnetic flux density* or magnetic induction. This quantity B is defined by the relationship

$$B = I + \mu_0 H. \tag{1.29}$$

The unit of B is also T (1 T of $B = 10^4$ gauss).

If we use (1.26) in (1.29), we have

$$B = \mu_0(H' + H). \tag{1.30}$$

This relationship tells us that the reason that the voltage of electromagnetic induction is increased by the insertion of a magnetic material is that the magnetic field produced by the intrinsic current is added to the external magnetic field. From a different point of view, we can also interpret (1.30) to mean that B can be regarded as the sum of the magnetization of a magnetic material and that of vacuum.†

The product of B and the cross-section S is called the *magnetic flux*:

$$\Phi = BS. \tag{1.31}$$

The unit of flux is the weber. Using this quantity, (1.28) is written as

$$V = -n\frac{d\Phi}{dt}. \tag{1.32}$$

If the magnetization, I, is proportional to the magnetic field, H, we have

$$I = \chi H, \tag{1.33}$$

where the proportionality factor χ is called the *magnetic susceptibility*. In this case (1.29) is written as

$$B = (\chi + \mu_0)H$$
$$= \mu H, \tag{1.34}$$

where μ is called the *magnetic permeability*.

The units of χ and μ are both $\mathrm{H\,m^{-1}}$, which is the same unit as μ_0 (see (1.2)). Therefore we can also measure χ and μ in units of μ_0. We call these quantities *relative susceptibility* and *relative permeability*, respectively, and denote them $\bar{\chi}$ and $\bar{\mu}$. From the relationship in (1.34), we have

$$\bar{\mu} = \bar{\chi} + 1. \tag{1.35}$$

* In this case, we assume that the cross-section of the magnetic material is exactly the same as that of the search coil. We also ignore the effect of the demagnetizing field. For a more detailed discussion of a practical case see Equation (2.25) in Chapter 2.

† Some scientists feel that the magnetization of vacuum is a bad term, because vacuum is not matter. But we must remember that matter itself is an electromagnetic phenomenon, so that we cannot specify the nature of vacuum on the basis of a naive concept of matter. Comparing Equations (1.27) and (1.28), we must believe that the vacuum is magnetizable!

In the case of weakly magnetic materials, $\bar{\chi}$ and $\bar{\mu}$ are normally field-independent, while in the case of strongly magnetic materials, I is a complex function of H (see Section 1.3), so that (1.34) holds only approximately in a limited range of fields.

1.3 MAGNETIZATION OF FERROMAGNETIC MATERIALS AND DEMAGNETIZING FIELDS

Since ferromagnetic materials can be highly magnetized by a magnetic field, they have relatively large magnetic permeabilities $\bar{\mu}$, ranging from 10^2 to 10^6. Their magnetization is changed by a magnetic field in a complex way, which is described by a magnetization curve as shown in Fig. 1.10.

Starting from a demagnetized state ($I = H = 0$), the magnetization increases with increasing field along the curve OABC and finally reaches the *saturation magnetization*, which is normally denoted by I_s. In the region OA the process of magnetization is almost reversible; that is, the magnetization returns to zero upon removal of the field. The slope of the curve OA is called the *initial susceptibility* χ_a. Beyond this region the processes of magnetization are no longer reversible. If the field is decreased from its value at point B, the magnetization comes back, not along BAO, but along the *minor loop* BB′. The slope BB′ is called the *reversible susceptibility* χ_{rev} or the *incremental susceptibility*. The slope at any point on the initial magnetization curve OABC is called the *differential susceptibility* χ_{diff}, and the slope of the line which connects the origin O and any point on the initial magnetization curve is called the *total susceptibility* χ_{tot}. The maximum value of the total susceptibility, that is, the slope of the tangent line drawn from the origin O to the initial magnetization curve, is called the *maximum susceptibility* χ_{max}; it is a good measure of the average slope of the initial magnetization curve. Changes in χ_{rev}, χ_{diff}, and χ_{tot} along the initial magnetization curve are shown in Fig. 1.11. Starting from the value of χ_a, χ_{rev} decreases monotonically, while χ_{diff} has a sharp maximum, and χ_{tot} goes to its maximum value χ_{max} and drops off at $I = I_s$. The difference between χ_{diff} and χ_{rev} represents the susceptibility due to irreversible magnetization; it is called the *irreversible susceptibility* χ_{irr}; that is,

$$\chi_{diff} = \chi_{rev} + \chi_{irr}. \tag{1.36}$$

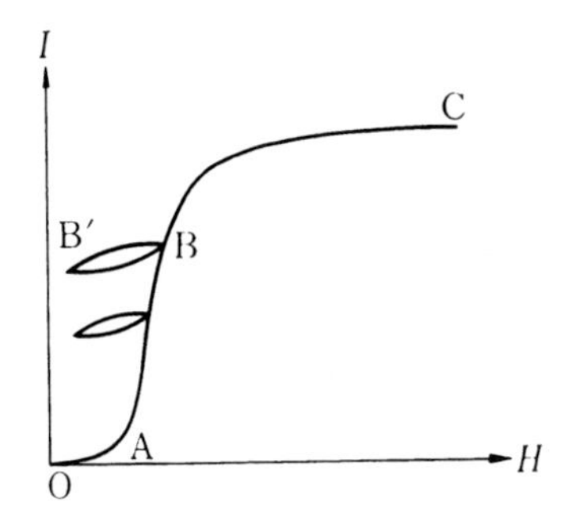

Fig. 1.10. Initial magnetization curve and minor loops.

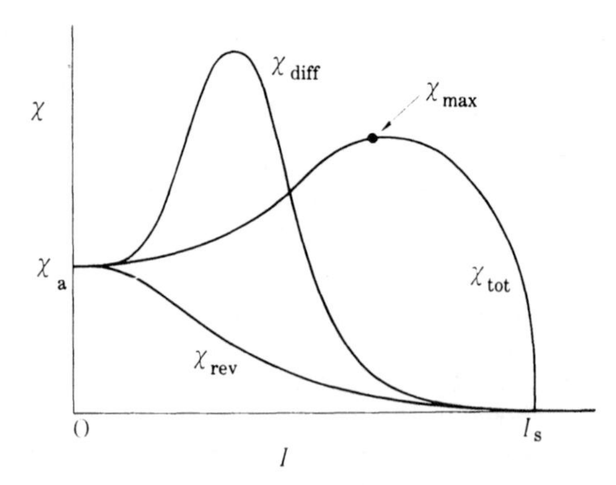

Fig. 1.11. Various kinds of magnetic susceptibilities as functions of the intensity of magnetization.

If the magnetic field is decreased from the saturated state C (Fig. 1.12), the magnetization I gradually decreases along CD, not along CBAO, and at $H=0$ it reaches the non-zero value I_r (= OD), which is called the *residual magnetization* or the remanence. Further increase of the magnetic field in a negative sense results in a continued decrease of the intensity of magnetization, which finally falls to zero. The absolute value of the field at this point is called the *coercive force* or the *coercive field* H_c (= OE). This portion, DE, of the magnetization curve is often referred to as the *demagnetizing curve*. Further increase of H in a negative sense results in an increase of the intensity of magnetization in a negative sense and finally to negative saturation magnetization. If the field is then reversed again to the positive sense, the magnetization will change along FGC. The closed loop CDEFGC is called the *hysteresis loop*.

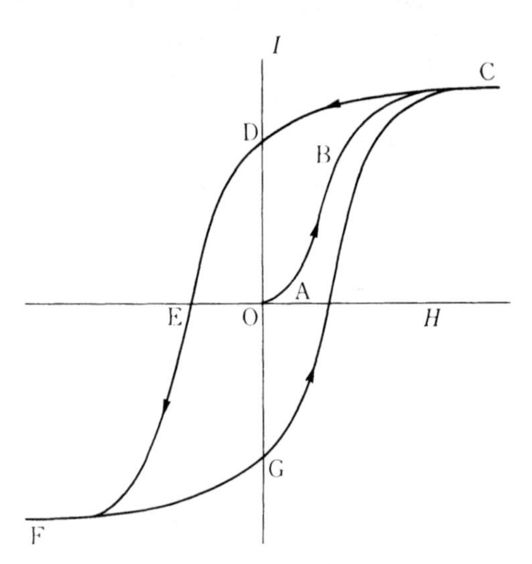

Fig. 1.12. Hysteresis loop.

The value of the saturation magnetization I_s does not exceed 2.5 T at room temperature for any material, but the value of the coercive field can vary over a wide range: it is about $8\,\mathrm{A\,m^{-1}}$ ($= 0.1$ Oe) or less for Permalloy and steel containing a few per cent of silicon, which are used for motor and transformer core materials, while it ranges from 50 to $200\,\mathrm{kA\,m^{-1}}$ (600 to 2500 Oe) or more for Alnico and Ba ferrites, which are used as permanent magnets. The value of relative maximum susceptibility also varies widely, from 10 to 10^5.

The apparent magnetization curve of a material depends not only on its magnetic susceptibility, but also on the shape of the specimen. When a specimen of finite size is magnetized by an external magnetic field, the free poles which appear on its ends will produce a magnetic field directed opposite to the magnetization (Fig. 1.13). This field is called the *demagnetizing field*. The intensity of the demagnetizing field H_d is proportional to the magnetic free pole density and therefore to the magnetization, so that we have

$$H_d = N\frac{I}{\mu_0}. \tag{1.37}$$

where N is called the *demagnetizing factor*, which depends only on the shape of the specimen. For instance, N approaches zero for an elongated thin specimen magnetized along its long axis, whereas it is large for a thick and short specimen.

Let us calculate the demagnetizing factor for a semi-infinite plate magnetized perpendicular to its surface (Fig. 1.14). If the intensity of magnetization is I, then the surface density of free poles on both sides of the plate is $\pm I$ (Wb/m^2) (see (1.24)). In order to calculate the demagnetizing field in this case, we use Gauss' theorem, which states that the surface integral of the normal component of the magnetic field H_n is equal to the magnetic free poles contained in the volume defined by the integral, divided by μ_0. That is,

$$\iint H_n \, dS = \frac{m}{\mu_0}. \tag{1.38}$$

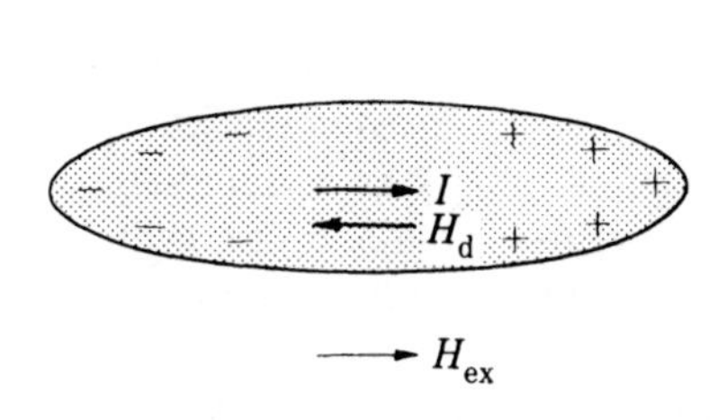

Fig. 1.13. Surface magnetic free poles and resulting demagnetizing field.

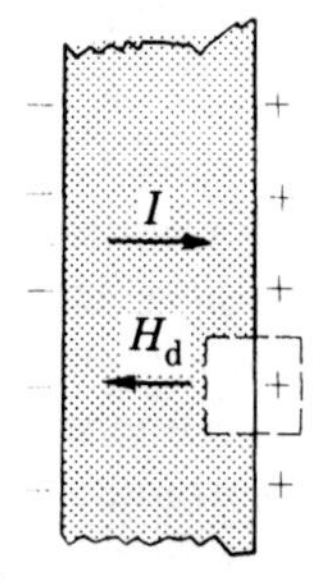

Fig. 1.14. Demagnetizing field produced by magnetization perpendicular to the surface of a magnetic plate.

Applying this theorem to the closed surface shown by the broken line in Fig. 1.14, and assuming that there is no field outside the plate other than the external field, we have

$$H_d = \frac{I}{\mu_0}. \tag{1.39}$$

The demagnetizing factor in this case is given by

$$N = 1. \tag{1.40}$$

When we magnetize the plate parallel to its surface, the effect of the free poles is negligible, provided that the plate is sufficiently wide and thin. Then

$$N = 0. \tag{1.41}$$

Thus the demagnetizing factor depends upon the direction of magnetization.

The calculation of the demagnetizing factor is generally not so simple. If the shape of the specimen is irregular, the demagnetizing field is not uniform but varies from place to place in the specimen. In such a case, we cannot define a single demagnetizing factor. The only shape for which the demagnetizing factor can be calculated exactly is the general ellipsoid. The general result is in the form of a complex integral, which can be simplified for special cases.[1] For example, for an elongated rotational ellipsoid (an ellipsoid with a circular cross-section) magnetized along its long axis, the demagnetizing factor is given by

$$N = \frac{1}{k^2 - 1}\left\{\frac{k}{\sqrt{k^2 - 1}} \ln\left(k + \sqrt{k^2 - 1}\right) - 1\right\}, \tag{1.42}$$

where k is the *aspect ratio* or the *dimensional ratio*, that is the ratio of length to diameter. In the special case of $k \gg 1$, (1.42) reduces to

$$N = \frac{1}{k^2}(\ln 2k - 1). \tag{1.43}$$

For a flat circular ellipsoid, approaching the shape of a disk, with the magnetization parallel to the plane of the surface, we have

$$N = \frac{1}{2}\left\{\frac{k^2}{(k^2 - 1)^{3/2}} \sin^{-1}\frac{\sqrt{k^2 - 1}}{k} - \frac{1}{k^2 - 1}\right\}, \tag{1.44}$$

where k is the ratio of diameter to thickness. In these calculations, the magnetization is assumed to be uniform throughout the material. Table 1.1 gives numerical values of the demagnetizing factor calculated from (1.43) and (1.44) as a function of the dimensional ratio k, together with experimental values obtained for cylinders of various k.[2] Experimental values of N show considerable scatter depending on the method of measurement and the type of material.

Assuming a uniform demagnetizing field in a general ellipsoid, we can prove that the demagnetizing factors for the three principal axes have a simple relationship

$$N_x + N_y + N_z = 1. \tag{1.45}$$

Table 1.1. Demagnetizing factors for rods and ellipsoids magnetized parallel to the long axis (after Bozorth[2])

Dimensional ratio k	Rod	Prolate ellipsoid	Oblate ellipsoid
1	0.27	0.3333	0.3333
2	0.14	0.1735	0.2364
5	0.040	0.0558	0.1248
10	0.0172	0.0203	0.0696
20	0.00617	0.00675	0.0369
50	0.00129	0.00144	0.01532
100	0.00036	0.000430	0.00776
200	0.000090	0.000125	0.00390
500	0.000014	0.0000236	0.001567
1000	0.0000036	0.0000066	0.000784
2000	0.0000009	0.0000019	0.000392

From this relationship we can easily obtain the demagnetizing factors for simple ellipsoids with high symmetry. For a sphere in which $N_x = N_y = N_z$, we have from (1.45)

$$N = \tfrac{1}{3}. \tag{1.46}$$

When a long cylinder is magnetized perpendicular to its long axis (which we take as the z-axis), $N_z = 0$, and $N_x = N_y$, so that we have

$$N = \tfrac{1}{2}. \tag{1.47}$$

When a semi-infinite plate is magnetized normal to its surface (which we take as the z-axis), $N_x = N_y = 0$, so that

$$N = 1, \tag{1.48}$$

in agreement with (1.40).

If we plot a magnetization curve as a function of the external field, the shape of the curve is sheared as shown by the broken line in Fig. 1.15 as compared with the true magnetization curve shown by the solid line. This is because the actual or true field acting in the material (usually called the *effective field** H_{eff}) is smaller than the applied external field H_{ex}. That is,

$$H_{eff} = H_{ex} - N\frac{I}{\mu_0}. \tag{1.49}$$

The correction for converting the sheared magnetization curve to the true curve is called the demagnetizing correction or the *shearing correction*. As a result of this correction, the magnetization curve takes on a more upright form, with increased susceptibility and remanence. Unless otherwise stated, published magnetization curves

* It may seem that the real effective field should include H' in (1.26). This is, however, not the case, because H' is produced by the magnetization itself, so that it produces only an internal force which has no effect on the magnetization.

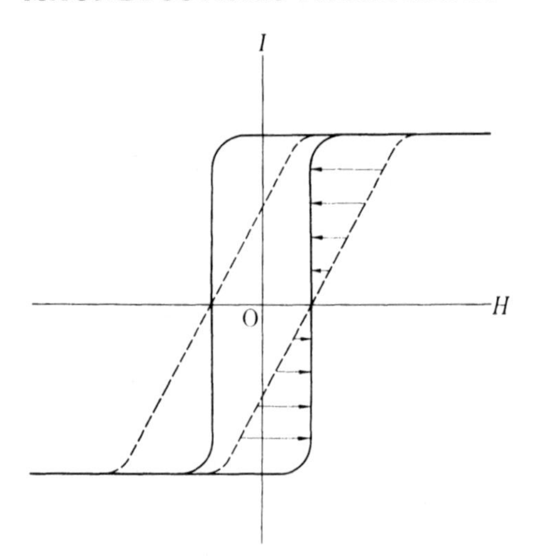

Fig. 1.15. Shearing correction of a magnetization curve.

have been corrected for demagnetizing effects and therefore represent the properties of the material independently of the shape of the sample used for the test.

In general we cannot ignore the effect of the demagnetizing field. A high field will be needed to magnetize a sample with a large demagnetizing factor, even if it has a high susceptibility. Suppose that we magnetize a Permalloy sphere with $H_c = 2\,\mathrm{A\,m^{-1}}$ ($= 0.025\,\mathrm{Oe}$) to the saturated state. Since the saturation magnetization of Permalloy is $1.16\,\mathrm{T}$ ($= 920\,\mathrm{gauss}$), the maximum demagnetizing field is given by

$$H_d = N\frac{I_s}{\mu_0} = \frac{1}{3} \times \frac{1.16}{4\pi \times 10^{-7}} = 3.08 \times 10^5\,\mathrm{A\,m^{-1}}\ (= 3860\,\mathrm{Oe}). \qquad (1.50)$$

In order to saturate this sphere, therefore, we must apply an external field which exceeds this value, which is 10^5 times larger than H_c. We must keep this fact in mind when we use magnetic materials with high susceptibility; otherwise the favorable properties of the material become useless.

Finally we consider the field inside a cavity in a material magnetized to an intensity I (Fig. 1.16). The free pole distribution on the surface of the cavity is the same as that on the surface of a solid body with the same shape as the cavity, and with the same magnetization as the material surrounding the cavity, except that the poles are of opposite sign. This must be true, because if we superpose the body and the hole, we have a uniformly magnetized solid without free poles. Thus the field produced by the surface free poles of the cavity is given by

$$H_{in} = N\frac{I}{\mu_0}, \qquad (1.51)$$

where N is the demagnetizing factor of a body with the same shape as the cavity. In the case of a spherical cavity, the field is given by

$$H_{in} = \frac{I}{3\mu_0}, \qquad (1.52)$$

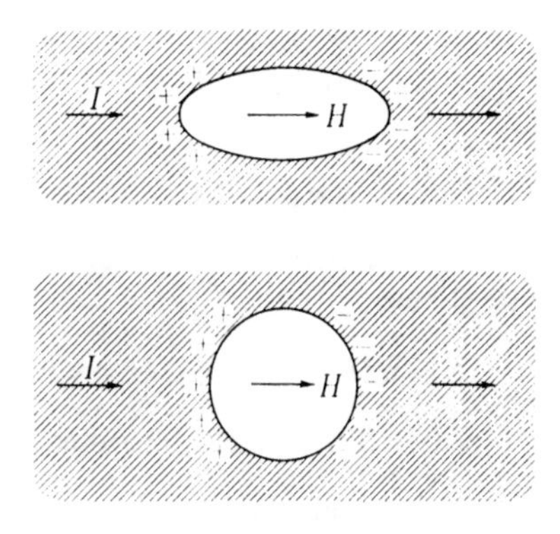

Fig. 1.16. Magnetic field inside holes in a magnetic body.

and its direction is the same as that of the magnetization. This field is called the *Lorentz field*.

If we bore an elongated hole parallel to the magnetization, the field inside H_{in} is the same as H_{eff}. If we make a thin flat cavity perpendicular to the magnetization, H_{in} is the same as $H' + H$, as given in (1.30).

1.4 MAGNETIC CIRCUIT

As mentioned in the preceding section, the demagnetizing field of a specimen of irregular shape is not uniform, and a complicated distribution of magnetization results.

The same situation exists when we try to find the distribution of free electric charges in a space containing a dielectric material of irregular shape. In the absence of true electric charge,

$$\operatorname{div} \boldsymbol{D} = 0, \tag{1.53}$$

where $\boldsymbol{D}$ is the electric displacement or electric flux density. On the other hand,

$$\boldsymbol{D} = \varepsilon \boldsymbol{E} = -\varepsilon \operatorname{grad} \phi, \tag{1.54}$$

where ϕ is the electric potential, so that (1.53) becomes

$$\operatorname{div}(\varepsilon \operatorname{grad} \phi) = 0. \tag{1.55}$$

In a uniform dielectric material we can put $\varepsilon = \text{constant}$, so that (1.55) becomes

$$\operatorname{div} \operatorname{grad} \phi = 0,$$

or

$$\Delta \phi = 0, \tag{1.56}$$

which is Laplace's equation. The problem of finding a distribution of electric potential is reduced to the problem of solving the Laplace equation under the given boundary condition. If, however, there are dielectric materials distributed in an irregular manner, we must solve (1.55) for the given spatial distribution of ε; this is a fairly complex problem.

In magnetostatics, where there are ferromagnetic materials of irregular shape, the situation is quite similar to the case just mentioned.

In magnetism, the relation

$$\operatorname{div} \boldsymbol{B} = 0 \tag{1.57}$$

is always valid. If we assume that the magnetization curve is given by a straight line without hysteresis, we can simply write

$$\boldsymbol{B} = \mu \boldsymbol{H} = -\mu \operatorname{grad} \phi_{\mathrm{m}}, \tag{1.58}$$

where ϕ_{m} is the magnetic potential. Then (1.57) becomes

$$\operatorname{div}(\mu \operatorname{grad} \phi_{\mathrm{m}}) = 0, \tag{1.59}$$

which has exactly the same form as (1.54). If, however, μ is non-uniformly distributed, the problem becomes very difficult.

A similar problem arises in finding the distribution of electric current density in a conducting medium. For a steady current,

$$\operatorname{div} \boldsymbol{i} = 0. \tag{1.60}$$

Since

$$\boldsymbol{i} = \sigma \boldsymbol{E} = -\sigma \operatorname{grad} \phi, \tag{1.61}$$

where σ is the electric conductivity, (1.60) becomes

$$\operatorname{div}(\sigma \operatorname{grad} \phi) = 0, \tag{1.62}$$

which is also of the same form as (1.55) or (1.59). Therefore if conductors of irregular shape with different conductivities are distributed in a medium with finite conductivity, the problem becomes very difficult, as in the two cases mentioned above.

If, however, the conductor is surrounded by insulators, the situation becomes very simple. For instance, if a conducting wire is placed in vacuum as shown in Fig. 1.17, the electric current is contained in the wire no matter how complicated the shape of the wire. This is, however, not true for dielectric materials; here the $\boldsymbol{D}$ vector leaks more or less into the vacuum, because the dielectric constant of vacuum is not zero ($\varepsilon_0 = 8.85 \times 10^{-12}\,\mathrm{F\,m^{-1}}$). For ferromagnetic materials the situation is intermediate between these two cases. Since the permeability of vacuum is not zero ($\mu_0 = 4\pi \times 10^{-7}\,\mathrm{H\,m^{-1}}$), the situation is formally similar to the case of electrostatics. But actually the permeability of ferromagnetic materials is 10^3–10^5 times larger than that of vacuum; hence the actual situation is quite similar to that of a steady current in a

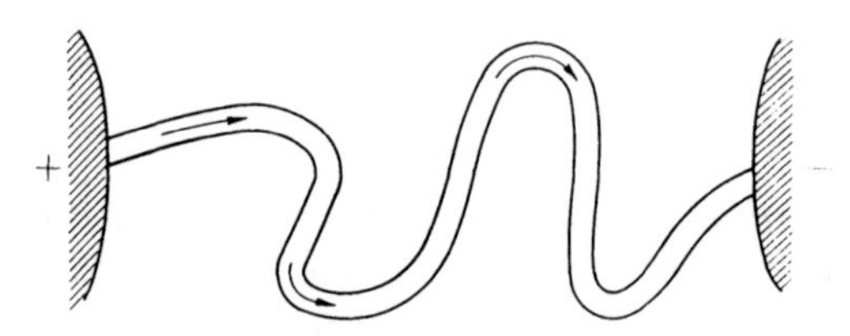

Fig. 1.17. Current distribution in a conductor of complex shape connecting two electrodes.

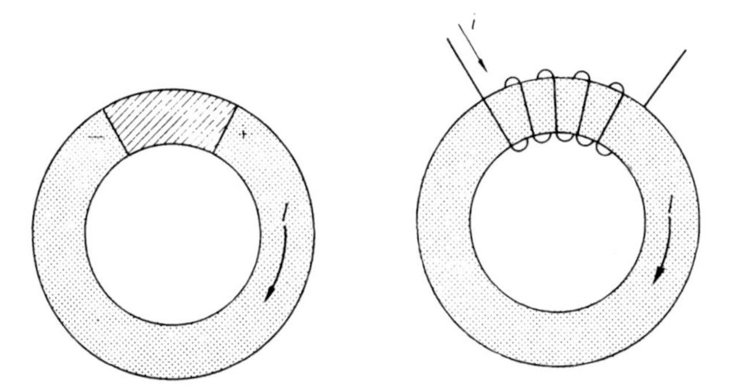

Fig. 1.18. Two kinds of magnetic circuit.

conductor. In other words, if a magnetic material, especially one of high permeability, forms a circuit, the magnetic flux density $\boldsymbol{B}$ is almost fully confined inside the magnetic material. Thus we can treat the magnetic circuit and electric circuit similarly.

Figure 1.18 shows two typical magnetic circuits. One is magnetized by a permanent magnet, and the other by a coil. The gray portions of the circuits are made from soft magnetic materials. The direction of magnetization should always be parallel to the surface of the magnetic material, because otherwise free poles will appear on the surface, producing strong demagnetizing fields that act to rotate the magnetization parallel to the surface again.

The flux density B in the circuit corresponds to the electric current density i of an electric circuit. The total flux in the circuit,

$$\Phi = \iint B_n \, \mathrm{d}S, \tag{1.63}$$

corresponds to the total electric current. If B is uniform throughout the cross-section of the magnetic circuit,

$$\Phi = BS. \tag{1.64}$$

When (1.58) and (1.61) are compared, the magnetic permeability μ is found to correspond to the electrical conductivity σ. Thus, corresponding to the resistance of the electric circuit,

$$R = \int \frac{\mathrm{d}s}{\sigma S}, \tag{1.65}$$

where S is the cross-section of the conducting wire, and $\mathrm{d}s$ is an element of length in the circuit, we can define the magnetic resistance or *reluctance* as

$$R_m = \int \frac{\mathrm{d}s}{\mu S}. \tag{1.66}$$

Corresponding to the electromotive force, we can define the *magnetomotive force* in the circuit as

$$V_m = \oint H_s \, \mathrm{d}s, \tag{1.67}$$

where the line integral is taken once around the circuit. When the magnetic circuit is magnetized by an electric current i which flows in a coil of N turns wound around the magnetic circuit, we have, from Ampère's theorem,

$$\oint H_s \,\mathrm{d}s = Ni, \tag{1.68}$$

so that the magnetomotive force of the coil is given by

$$V_m = Ni. \tag{1.69}$$

The unit of magnetomotive force is the ampere turn (AT).*

Using relations (1.58), (1.64), and (1.66), we have

$$V_m = Ni = \oint H_s \,\mathrm{d}s = \oint \frac{B}{\mu}\,\mathrm{d}s = \oint \frac{\Phi}{\mu S}\,\mathrm{d}s = \Phi \oint \frac{\mathrm{d}s}{\mu S} = \Phi R_m. \tag{1.70}$$

This relation corresponds to Kirchhoff's second law of electric circuits; that is

$$V = iR. \tag{1.71}$$

The calculation of the magnetomotive force for a permanent magnet is fairly complex. We can proceed as follows: when a magnet supplies the magnetomotive force in a magnetic circuit, the direction of the magnetic field H_p inside the permanent magnet is always opposite to that of magnetic flux B_p. The reason is that the demagnetizing field of the permanent magnet can be reduced in value by inserting it into a magnetic circuit, but can never be changed in direction unless another magnet or coil supplies an additional magnetomotive force. Thus the magnetic state of a permanent magnet is represented by a point, say P, in the second quadrant of the B–H plot (Fig. 1.19). If the magnetic field is changed, the point P will move reversibly on the line RQ whose slope $\mathrm{d}B/\mathrm{d}H$ is given by the reversible permeability μ_r. If we translate the origin O to S, the extrapolation, SR, of RQ to the abscissa can be regarded as a simple reversible magnetization curve, Thus a permanent magnet is equivalent to a magnetic material with permeability μ_r which can drive magnetic flux against a hypothetical field OS ($=H_r$). The magnetomotive force is given by integrating H_r over the length of the permanent magnet, so that

$$V_m = -H_r l_p = \frac{B_r l_p}{\mu_r}, \tag{1.72}$$

where l_p is the length of the permanent magnet and B_r is the remanent magnetic flux density (OR). It must be remarked, however, that the permanent magnet itself has a considerable reluctance given by

$$R'_m = \frac{l_p}{\mu_r S_p}. \tag{1.73}$$

* Since the number of turns is not a unit in the SI system, we can say that a coil of 10 turns carrying a current of 1 ampere is equivalent to a coil of 1 turn carrying a current of 10 amperes. The product of amperes × turns can thus be considered to have the units of amperes (A).

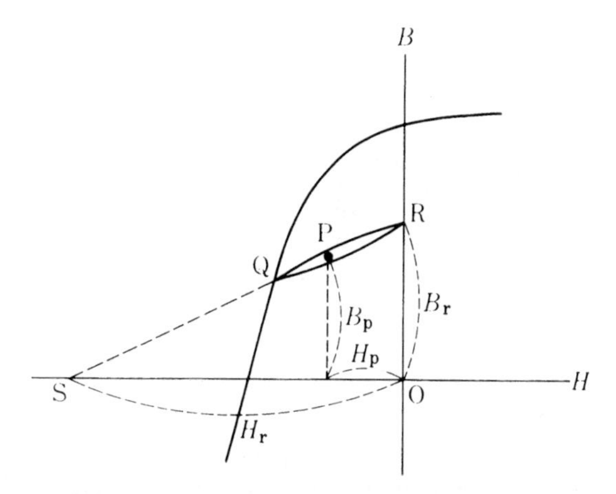

Fig. 1.19. Demagnetizing curve of a permanent magnet showing the operating point P.

Generally μ_r of a permanent magnet is very small, so that R'_m is fairly large. Thus a permanent magnet corresponds to a high-voltage battery with a fairly large internal resistance which supplies a constant current independent of the output impedance. For a magnetic circuit containing a permanent magnet,

$$\oint H_s \, \mathrm{d}s = 0,$$

because the electric current is zero. Then

$$-H_p l_p = \int_{\text{ex}} H_s \, \mathrm{d}s, \tag{1.74}$$

where H_p is the field inside the permanent magnet and the integral is taken along the circuit excluding the permanent magnet. Thus

$$\begin{aligned} V_m &= \frac{B_r l_p}{\mu_r} = -H_r l_p = -\left(H_p - \frac{B}{\mu_r}\right) l_p = -H_p l_p + \frac{B l_p}{\mu_r} \\ &= \int_{\text{ex}} H_s \, \mathrm{d}s + \Phi \frac{l_p}{\mu_r S_p} = \Phi\left(\int_{\text{ex}} \frac{\mathrm{d}s}{\mu S} + \frac{l_p}{\mu_r S_p}\right) \\ &= \Phi(R_m + R'_m). \end{aligned} \tag{1.75}$$

This relation corresponds to Kirchhoff's second law.

If a magnetic circuit has a shunt or shunts, we have Kirchhoff's first law,

$$\Phi_1 + \Phi_2 + \cdots = 0, \tag{1.76}$$

where $\Phi_1, \Phi_2, \ldots$ are the fluxes which leave the shunt point through the branches, $1, 2, \ldots$.

By using Kirchhoff's first and second laws, we can solve any kind of magnetic circuit, just as we can an electric circuit. It must be noted, however, that a magnetic circuit has more or less leakage of the magnetic flux, like an electric circuit dipped in an electrolytic solution. In extreme cases, the leakage is so large than the actual flux is only a small fraction of the calculated value. The non-linear relationship between B and H also makes the problem more complex. In particular, if a part of the circuit is magnetically saturated, the permeability becomes very small, so that the circuit is practically broken at this point. The presence of magnetic hysteresis also makes the situation more complicated. However, when the product of the coercive field times the length of the circuit is negligibly small compared to the magnetomotive force, we can neglect the effect of hysteresis.

1.5 MAGNETOSTATIC ENERGY

Let us discuss the energy involved in a magnetostatic system. Consider a system consisting of several permanent magnets (Fig. 1.20). Let the intensities of the free poles be $m_1, m_2, \ldots, m_i, \ldots, m_n$ (Wb) and the magnetic potential at the position of each pole be $\phi_1, \phi_2, \ldots, \phi_i, \ldots, \phi_n$ (A). Then the potential energy of the system is

$$U = \frac{1}{2} \sum_{i=1}^{n} m_i \phi_i \qquad \text{(J)}. \tag{1.77}$$

Since ϕ_i is the potential due to the Coulomb interaction of free poles other than m_i, it is expressed as

$$\phi_i = \sum_{j \neq i} \frac{m_j}{4\pi\mu_0 r_{ij}}, \tag{1.78}$$

where r_{ij} is the distance between the ith and the jth free poles. If the free poles are

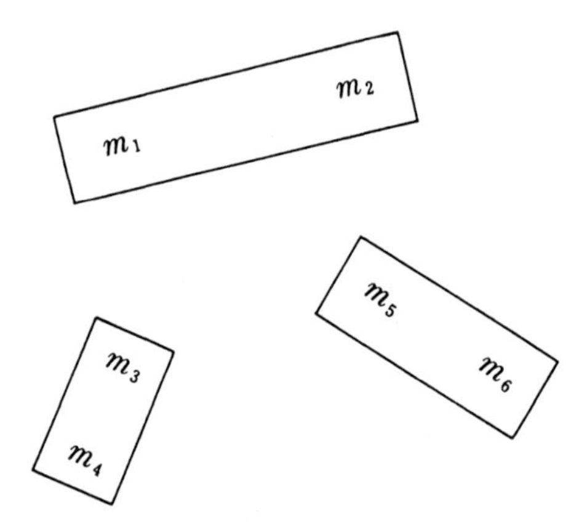

Fig. 1.20. Assembly of permanent magnets.

distributed over space with the density ρ_m, the energy is expressed by the volume integral

$$U=\tfrac{1}{2}\iiint \rho_m \phi \, dv, \tag{1.79}$$

where the potential ϕ is calculated from

$$\phi=\iiint \frac{\rho_m}{4\pi\mu_0 r}\, dv. \tag{1.80}$$

It can also be calculated by solving the Poisson equation,

$$\Delta\phi = -\frac{\rho_m}{\mu_0}, \tag{1.81}$$

under the given boundary conditions, or the Laplace equation,

$$\Delta\phi = 0, \tag{1.82}$$

at points where there are no magnetic free poles.

The free pole density ρ_m induced in a ferromagnetic medium is expressed in terms of the magnetization as

$$\operatorname{div} \boldsymbol{I}_s = -\rho_m. \tag{1.83}$$

If the material is homogeneous, the magnitude of $\boldsymbol{I}_s$ should be constant, so that div $\boldsymbol{I}_s$ can be expressed in terms of the direction cosines $(\alpha_1, \alpha_2, \alpha_3)$ of $\boldsymbol{I}_s$ as

$$\frac{\partial\alpha_1}{\partial x}+\frac{\partial\alpha_2}{\partial y}+\frac{\partial\alpha_3}{\partial z} = -\frac{\rho_m}{I_s} \tag{1.84}$$

or

$$\operatorname{div} \boldsymbol{\alpha} = -\frac{\rho_m}{I_s}, \tag{1.85}$$

where α is a unit vector parallel to $\boldsymbol{I}_s$.

For example, for a sphere which is radially magnetized as shown in Fig. 1.21, (1.85) becomes

$$\frac{1}{r^2}\frac{\partial}{\partial r}(r^2) = -\frac{\rho_m}{I_s}. \tag{1.86}$$

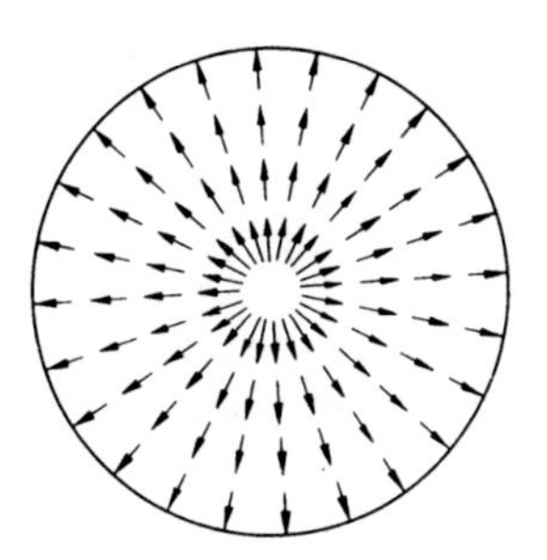

Fig. 1.21. Ferromagnetic sphere magnetized radially.

Solving this equation, we have

$$\rho_{\mathrm{m}} = -\frac{2I_{\mathrm{s}}}{r}. \tag{1.87}$$

Then (1.81) becomes

$$\frac{1}{r^2}\frac{\partial}{\partial r}\left(r^2\frac{\partial\phi}{\partial r}\right) = \frac{2I_{\mathrm{s}}}{\mu_0 r}. \tag{1.88}$$

Solving (1.88), we obtain

$$\phi = \frac{I_{\mathrm{s}}}{\mu_0}(r-R). \tag{1.89}$$

The magnetostatic energy of the system is then

$$U = \frac{-4\pi I_{\mathrm{s}}^2}{\mu_0}\int_0^R (r^2 - rR)\,\mathrm{d}r = \frac{I_{\mathrm{s}}^2}{\mu_0}\frac{2\pi}{3}R^3. \tag{1.90}$$

The magnetostatic energy can be expressed in terms of I and H instead of ρ_{m} and ϕ. Let the length of the kth magnet be l_k and the magnetic poles at its ends be m_k and $-m_k$. Then the magnetic moment is

$$M_k = m_k l_k. \tag{1.91}$$

The difference in the magnetic potential between the positions of the positive and the negative magnetic poles is

$$\phi_+ - \phi_- = -H_{\|} l_k, \tag{1.92}$$

where $H_{\|}$ is the component of the magnetic field parallel to the magnetic moment M_k. Then

$$m_k\phi_+ - m_k\phi_- = -\left(\frac{M_k}{l_k}\right)\times(H_{\|}l_k) = -\boldsymbol{M}_k\cdot\boldsymbol{H}_k, \tag{1.93}$$

and (1.77) becomes

$$U = -\frac{1}{2}\sum_{k=1}^{N}\boldsymbol{M}_k\cdot\boldsymbol{H}_k, \tag{1.94}$$

where N is the total number of magnets. When the magnetization $\boldsymbol{I}$ is distributed over space, the magnetostatic energy is

$$U = -\tfrac{1}{2}\iiint \boldsymbol{I}\cdot\boldsymbol{H}\,\mathrm{d}v. \tag{1.95}$$

By using this equation, we can calculate the magnetostatic energy of the sphere shown in Fig. 1.21. Since the intensity of the magnetic field inside the sphere is given by

$$H_{\mathrm{r}} = -\frac{\partial\phi}{\partial r} = -\frac{I_{\mathrm{s}}}{\mu_0}, \tag{1.96}$$

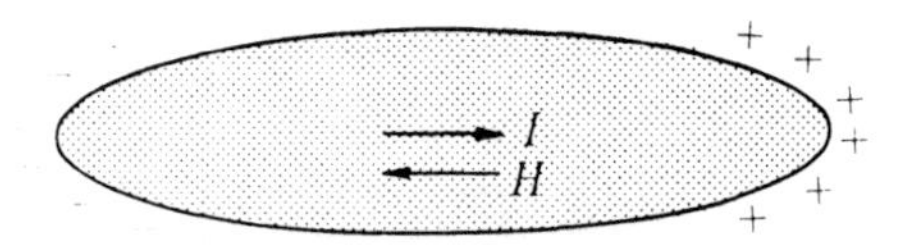

Fig. 1.22. A magnetized ferromagnetic body of finite size.

and is zero outside the sphere, the magnetostatic energy can be calculated from (1.94) as

$$U = \frac{1}{2}\frac{I_s^2}{\mu_0}\frac{4\pi R^3}{3} = \frac{I_s^2}{\mu_0}\frac{2\pi}{3}R^3, \tag{1.97}$$

which agrees exactly with (1.90).

Let us calculate the magnetostatic energy for another example. Consider a magnetic body with demagnetizing factor N magnetized to intensity I (Fig. 1.22). Since the demagnetizing field is

$$H = -\frac{N}{\mu_0}I, \tag{1.98}$$

the magnetostatic energy is

$$U = \frac{1}{2\mu_0}NI^2 v, \tag{1.99}$$

where v is the volume of the magnetic body.

It should be noted that (1.79) and (1.95) can also be used to calculate the magnetostatic energy even when soft magnetic materials are present with the permanent magnets. In this case ρ_m in (1.79) signifies the pole density produced by the permanent magnetization only, and I in (1.95) denotes the permanent magnetization. The effect of magnetization induced in the soft magnetic material comes into (1.79) and (1.95) only through the change in ϕ or H, respectively.

When some soft magnetic materials are present in the system, ϕ or H always decrease, so that the magnetostatic energy also decreases. For example, consider that a space in which permanent magnetic free poles are distributed with density ρ_m is filled with a soft magnetic material of relative permeability $\bar{\mu}$. In the absence of the soft magnetic material, the divergence of the field due to ρ_m is given by

$$\operatorname{div} \boldsymbol{H} = \frac{\rho_m}{\mu_0}. \tag{1.100}$$

When the soft magnetic material is placed in the space, it produces additional magnetization I', so that additional free poles ρ_m' are induced:

$$\operatorname{div} \boldsymbol{I}' = -\rho_m'. \tag{1.101}$$

As a result, the field is reduced to H' which is related to ρ_m and ρ_m' by

$$\operatorname{div} \boldsymbol{H}' = \frac{\rho_m}{\mu_0} + \frac{\rho_m'}{\mu_0}. \tag{1.102}$$

The magnetization I' is magnetized by this resultant field H', according to $I' = \chi H'$, so that, using (1.101), (1.102) becomes

$$\operatorname{div} \boldsymbol{H}' = \frac{\rho_m}{\mu_0} - \frac{1}{\mu_0} \operatorname{div} \boldsymbol{I}',$$
$$= \frac{\rho_m}{\mu_0} - \frac{\chi}{\mu_0} \operatorname{div} \boldsymbol{H}'$$

or

$$\operatorname{div} \boldsymbol{H}' = \frac{\rho_m}{\mu}. \tag{1.103}$$

Comparing (1.103) with (1.100), we find that the field H' is reduced by a factor $1/\bar{\mu}$ from the field H. Accordingly the magnetic potential ϕ is also reduced to $1/\bar{\mu}$ times its original value. Therefore, the magnetostatic energy of the system is also reduced by a factor $1/\bar{\mu}$. Since this factor is substantially less than unity for most soft magnetic materials, we cannot ignore this effect in the calculation of the magnetostatic energy. We shall discuss this problem once again in Chapter 17 as what is known as the μ^* *correction*.

The magnetostatic energy can be also expressed in terms of B and H:

$$U = \tfrac{1}{2} \iiint \boldsymbol{B} \cdot \boldsymbol{H} \, \mathrm{d}v. \tag{1.104}$$

The integration is carried out over the space where magnetic field $\boldsymbol{H}$ is present. It should be noted that $\boldsymbol{B}$ in the integrand is calculated by excluding the permanent magnetization.* When no soft magnetic material is present. $\boldsymbol{B}$ should equal $\mu_0 \boldsymbol{H}$, so that (1.104) becomes

$$U = \frac{\mu_0}{2} \iiint H^2 \, \mathrm{d}v. \tag{1.105}$$

If there is some soft magnetic material, (1.104) becomes

$$U = \frac{\mu}{2} \iiint_{\substack{\text{magnetic}\\ \text{material}}} H^2 \, \mathrm{d}v + \frac{\mu_0}{2} \iiint_{\substack{\text{other}\\ \text{space}}} H^2 \, \mathrm{d}v. \tag{1.106}$$

Considering again the example shown in Fig. 1.21, the magnetic field inside the sphere is given by (1.96) and is zero outside the sphere; hence (1.105) gives

$$U = \frac{I_s^2}{\mu_0} \frac{2\pi}{3} R^3, \tag{1.107}$$

which is again in agreement with (1.90) and (1.97).

As seen from (1.105), the magnetostatic energy is stored in any space where magnetic fields exist. The energy stored per unit volume is given by

$$E_m = \tfrac{1}{2} \mu_0 H^2 \qquad (\mathrm{J\,m^{-3}}). \tag{1.108}$$

* The reason is that (1.104) is deduced from (1.79) using the relation $\rho_m = \operatorname{div} \boldsymbol{B}$. If we include the permanent magnetization in $\boldsymbol{B}$, we have always $\operatorname{div} \boldsymbol{B} = 0$.

A line of magnetic force therefore tends to shrink so as to decrease the energy in (1.108). In other words, a tension given by

$$T_{\parallel} = \tfrac{1}{2}\mu_0 H^2 \qquad (\mathrm{N\,m^{-2}}) \tag{1.109}$$

acts parallel to the line of force. On the other hand, when the line of force expands perpendicular to the force, the energy density (1.108) again decreases, because a decrease in energy density caused by a decrease in the intensity of the field overcomes an increase in energy density caused by a decrease in the density of the lines of force. In other words, a pressure expressed by

$$T_{\perp} = -\tfrac{1}{2}\mu_0 H^2 \tag{1.110}$$

acts perpendicular to the line of force. The stresses given by (1.109) and (1.110) are called *Maxwell stresses*. The forces calculated by the interaction between magnetic poles and the magnetic field, or from the interaction between electric current and the flux density, must coincide with the result calculated from the Maxwell stresses.

1.6 MAGNETIC HYSTERESIS

In the preceding discussion, we assumed that the energy supplied to construct the system is conserved and stored as magnetostatic energy. This assumption is, however, not necessarily true for actual ferromagnetic materials.

Consider the work necessary to magnetize a ferromagnetic material. Suppose that the magnetization is increased from I to $I + \delta I$ under the action of a magnetic field H, parallel to I. If we consider a cylindrical section of magnetic material whose length is l (parallel to I) and whose cross-section is S, an increase of magnetization, δI, is accomplished by transporting a magnetic pole of magnitude IS through the distance l from the bottom to the top of the cylinder under the action of the force ISH. The work required for this transportation is $H\,\delta IS$. Since the volume of the cylinder is Sl, the work necessary to magnetize a unit volume of the magnetic material is given by

$$\delta W = H\,\delta I. \tag{1.111}$$

Then the work required to magnetize a unit volume from $I = I_1$ to I_2 is given by

$$W = \int_{I_1}^{I_2} H\,\mathrm{d}I \qquad [\mathrm{J\,m^{-3}}]. \tag{1.112}$$

For example, the work required to magnetize the volume from the demagnetized state to saturation, I_s, is given by (1.112), putting $I_1 = 0$ and $I_2 = I_s$. This is equal to the area enclosed by the ordinate axis, the line $I = I_s$, and the initial magnetization curve, as shown in Fig. 1.23. The energy supplied by this work is partially stored as potential energy and partly dissipated as heat which is generated in the material. After one full circuit of the hysteresis loop, the potential energy must return to its original value, so that the resultant work must appear as heat. This heat or energy is called the *hysteresis loss* and is given by

$$W_h = \oint H\,\mathrm{d}I, \tag{1.113}$$

which is equal to the area inside the hysteresis loop.

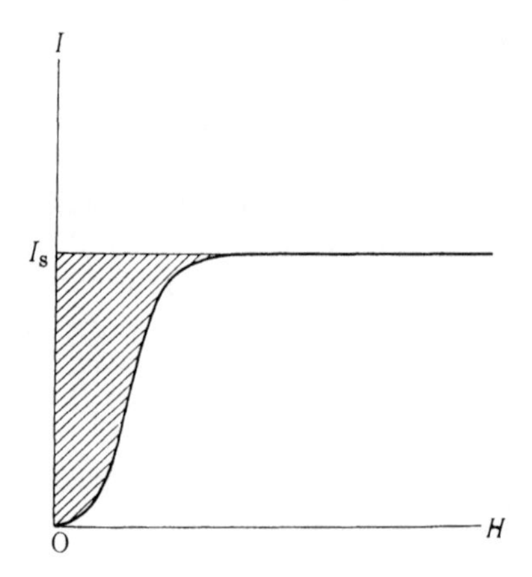

Fig. 1.23. Work required to saturate a unit volume of ferromagnetic material.

For weakly magnetic materials without hysteresis, whose magnetic behavior is described by $I = \chi H$, the potential energy stored in a unit volume is

$$E_m = -\tfrac{1}{2}\chi H^2, \tag{1.114}$$

which is obtained by putting $I_1 = 0$, and $I_2 = I$ in (1.112).

For engineering applications, ferromagnetic materials can be classified as 'soft' and 'hard'. Soft magnetic materials are normally used for the cores of transformers, motors, and generators; for these purposes high permeability, low coercive force, and small hysteresis loss are required. On the other hand, hard magnetic materials are used as permanent magnets for various kinds of electric meters, loudspeakers, and other apparatus for which high coercivity, high remanence, and large hysteresis loss are desirable. It is interesting that the main applications of ferromagnetic materials fall into two groups which require almost opposite properties. There is, however, a large and growing application for magnetic materials in magnetic recording where properties intermediate between the traditional hard and soft materials are needed. Magnetic materials have been developed with characteristics ranging from extremely soft to extremely hard.

At the beginning of the twentieth century, soft iron was almost the only available soft magnetic material. This material has a hysteresis loop which is very wide compared to that of Permalloy, a modern high-quality soft magnetic material, as shown in Fig. 1.24. Similarly, hardened carbon steel was the standard permanent magnet material until the beginning of the twentieth century. It has a hysteresis loop fairly narrow compared to a modern permanent magnet material such as MK steel (Fig. 1.25). Recently developed rare-earth magnets exhibit hysteresis loops about twenty times wider than that of the MK steel.

It should be noted that hysteresis is necessary for the observation of purely magnetostatic phenomena. If there are no permanent magnets and no electric current, so that all magnetic materials are soft magnetic materials which can be magnetized only by external magnetic fields, the only stable condition is $I = 0$ and

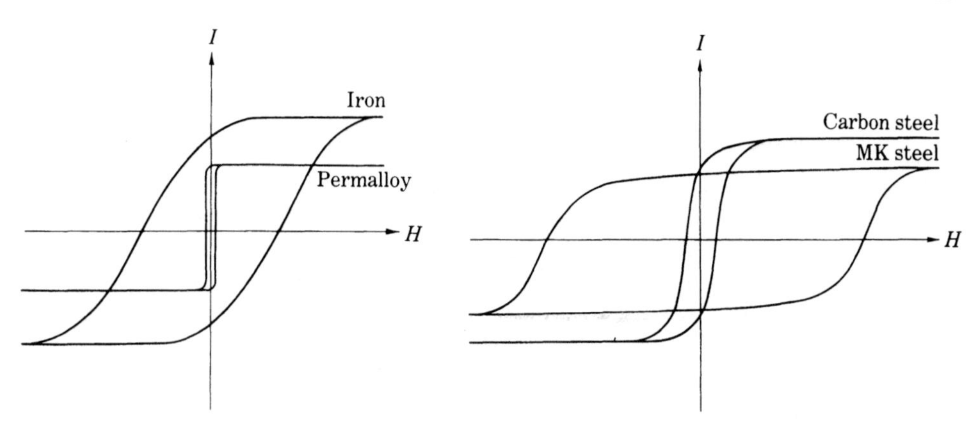

Fig. 1.24. Comparison of hysteresis loops of soft iron and Permalloy.

Fig. 1.25. Comparison of hysteresis loops of hardened carbon steel and MK steel.

$H = 0$. Thus no magnetostatic phenomena will be observed. Therefore we can stay that permanent magnets are the only source of magnetostatic energy.

Next we consider the performance of a permanent magnet as an energy source. Let us consider how a permanent magnet produces a magnetic field in the air gap of a magnetic circuit, as shown in Fig. 1.26, where P is the permanent magnet and S is soft magnetic material. Using (1.104), the energy stored in this system is given by

$$U = \tfrac{1}{2} \iiint \boldsymbol{B} \cdot \boldsymbol{H} \, \mathrm{d}v, \tag{1.115}$$

where $\boldsymbol{B}$ is the flux density of the magnetic circuit excluding the permanent magnetization (see the footnote concerning (1.104)). Therefore the integrand in (1.115) in the permanent magnet becomes $\mu_0 H^2$, which, however, does not produce a magnetic field in the air gap. An energy term that gives the effectiveness of the permanent magnet in producing a field in the air gap is given by the integration of (1.115) over

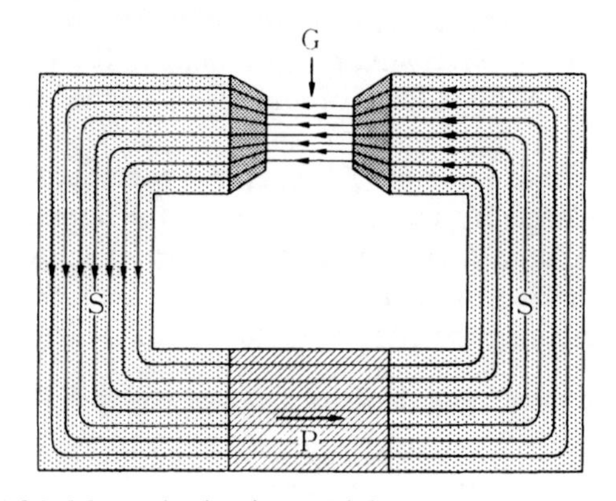

Fig. 1.26. Magnetic circuit containing a permanent magnet.

the magnetic circuit excluding the permanent magnet. Since the volume element $\mathrm{d}v$ is equal to the cross-section S times the line element $\mathrm{d}s$, (1.115) becomes

$$U = \tfrac{1}{2}\int_{\mathrm{G\&S}} BSH_s\,\mathrm{d}s = \tfrac{1}{2}\Phi\int_{\mathrm{G\&S}} H_s\,\mathrm{d}s. \tag{1.116}$$

Where G means the air gap and S means the magnetic material. For an ideal soft magnetic material, $\bar{\mu} = \infty$, so that $H = 0$ for a given B, and the integration in (1.116) is non-zero only in the air gap. Since the purpose of this magnetic circuit is to supply a magnetic field in the air gap, this is quite reasonable. Even if the permeability of the soft magnetic material is finite, so that the permanent magnet must supply additional energy to magnetize it, this contribution should count as a part of the effectiveness of the permanent magnet. In other words, the quantity U in (1.116) gives a good measure of the effectiveness of the permanent magnet.

Now considering that

$$\oint H_s\,\mathrm{d}s = 0 \tag{1.117}$$

for one circuit of the magnetic path, it follows that

$$\int_{\mathrm{G\&S}} H_s\,\mathrm{d}s = -\int_{\mathrm{P}} H_s\,\mathrm{d}s, \tag{1.118}$$

so that (1.116) becomes

$$U = -\tfrac{1}{2}\Phi\int_{\mathrm{P}} H\,\mathrm{d}s = -\tfrac{1}{2}\int_{\mathrm{P}} \boldsymbol{B}\cdot\boldsymbol{H}\,\mathrm{d}v, \tag{1.119}$$

where the superscript P means that the integral should be taken only over the permanent magnet. The final result (1.119) means that the permanent magnet has greater effectiveness for larger values of the quantity $-\boldsymbol{B}\cdot\boldsymbol{H}$ (note that $\boldsymbol{H} < 0$), and for larger volume.

Therefore, if a permanent magnet material has a demagnetizing curve as shown in Fig. 1.27, the magnet can be most effectively utilized by setting the working point at A where (BH) has its maximum value, rather than at A′ or A″ where (BH) is lower. Accordingly the quality of a permanent magnetic material can be expressed by the maximum value of (BH), which is usually denoted by $(BH)_{\max}$, and is called the *maximum BH product* or the *maximum energy product*.* If a permanent magnet has a high value of $(BH)_{\max}$, a small volume of material is needed to produce a given field in a given air gap.

In order to obtain a large value of $(BH)_{\max}$, it is essential to make the value of B_r $(=I_r)$ large, to have a large H_c, and finally to have the demagnetizing curve close to rectangular in shape. A large I_r can be achieved by making the value of I_s as large as possible, and also by making the remanence ratio I_r/I_s as close to unity as possible. This can be done by orienting the crystallites so that the magnetocrystaline anisotropy

* In spite of this name, the actual energy produced by a unit volume of the magnet is one-half of the quantity (BH), as shown by (1.119).

The value of $(BH)_{\max}$ does not necessarily correspond to the magnetomotive force V_m. The magnet with a small μ_r results in a large V_m, but cannot produce a large magnetic flux, because it has a large internal reluctance R'_m.

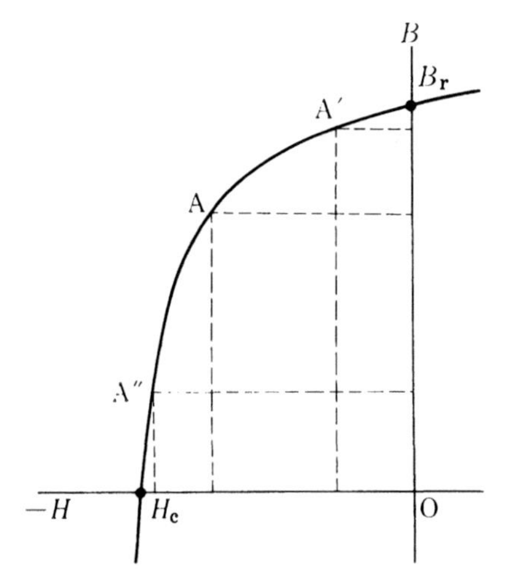

Fig. 1.27. Demagnetizing curve of a permanent magnet.

(see Chapter 12) aligns the magnetization parallel to the direction of easy magnetization, or by creating an induced magnetic anisotropy (see Chapter 13), so as to align the easy axes parallel to the axis of magnetization. Any change in the material that increases the remanence ratio will also tend to produce a rectangular demagnetizing curve. A high coercive field H_c can be achieved either by increasing internal stresses (caused by a crystallographic transformation, precipitation hardening, or superlattice formation); or by making the size of particles or crystallites less than the critical size for single domain behavior. These points will be considered in more detail later. It must be noted here that the important coercivity is not the field at which $I = 0$, or ${}_IH_c$, but the field at which $B = 0$, or ${}_BH_c$, as shown in Fig. 1.27. The value of ${}_BH_c$ is, however, limited by I_r, since

$$|{}_BH_c| = \frac{I_c}{\mu_0} < \frac{I_r}{\mu_0} = \frac{B_r}{\mu_0}, \tag{1.120}$$

where I_c is the value of I at $B = 0$. Thus ${}_BH_c$ is also limited by I_s/μ_0. We therefore see that high I_s is essential for obtaining a high-quality permanent magnet.

PROBLEMS

1.1 Calculate the magnetic moment of a sphere of radius R made from a magnetic material with magnetic susceptibility χ, when it is magnetized by an external magnetic field H. How is the value of the moment changed in the limit of $\chi \to \infty$?

1.2 Calculate the force needed to separate two semicircular permanent magnets of cross sectional area S, magnetized to intensity I, while keeping the air gaps equal, as shown in the figure. Discuss this problem in terms of the energy.

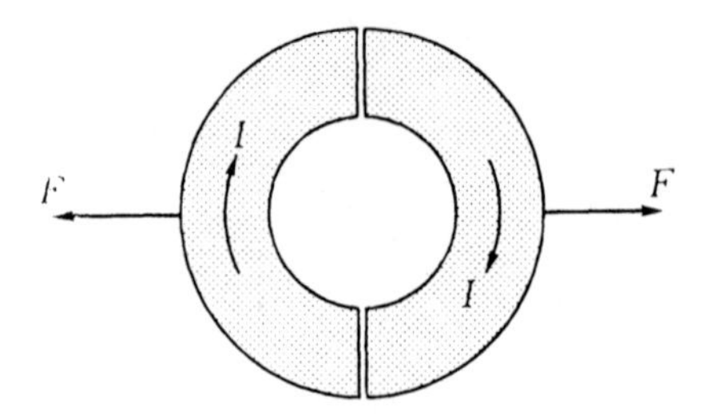

Fig. Prob. 1.2

1.3 Calculate the intensity of the magnetic field in the air gap of the magnetic circuit shown in the figure. Use the values $N = 200$, $i = 5\,\mathrm{A}$, $S_1 = 2.5 \times 10^{-3}\,\mathrm{m}^2$, $S_2 = 5 \times 10^{-4}\,\mathrm{m}^2$, $l_1 = 1\,\mathrm{m}$, $l_2 = 0.01\,\mathrm{m}$, $\bar{\mu} = 500$.

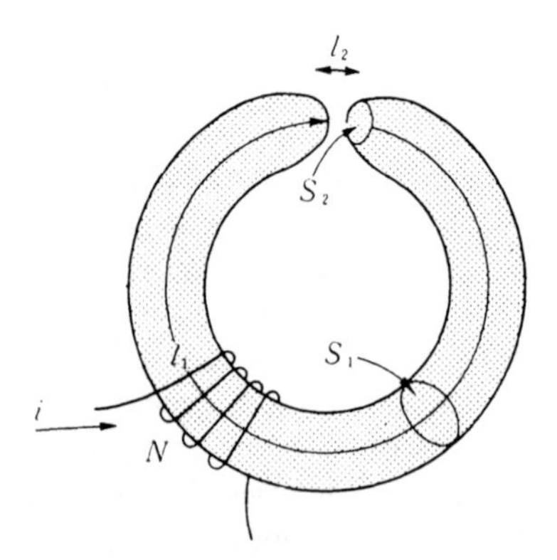

Fig. Prob. 1.3

1.4 Using a permanent magnet material with a maximum BH product at $B = 0.15\,\mathrm{T}$ and $H = 60\,\mathrm{kA\,m^{-1}}$, we want to produce a magnetic field of $250\,\mathrm{kA\,m^{-1}}$ in an air gap of cross-sectional area $12\,\mathrm{cm}^2$ and width 14 mm. Design the size and shape of the magnet, assuming that the magnetic circuit has no leakage of magnetic flux and no reluctance.

1.5 Calculate the magnetostatic energy per unit length of an infinitely long ferromagnetic rod of radius R which has residual magnetization I_r perpendicular to its long axis. How is the energy changed when the rod is dipped into a ferromagnetic liquid of relative permeability $\bar{\mu}$?

REFERENCES

1. R. Becker, *Theorie der Electrizitaet*, Vol. 1 (Verlag u. Druck von B. G. Tuebner, 1933).
2. R. M. Bozorth, *Ferromagnetism* (Van Nostrand, New York, 1951).

2

MAGNETIC MEASUREMENTS

2.1 PRODUCTION OF MAGNETIC FIELDS

There are various methods of producing magnetic fields. The appropriate method depends on the intensity, the volume, and the uniformity (in space and in time) of the field required. In this book, magnetic fields are measured in amperes per meter ($\mathrm{A\,m^{-1}}$), although for high magnetic fields the tesla (T) is a more convenient unit. For conversion factors between $\mathrm{A\,m^{-1}}$ and Oe or T, see Appendix 6.

The intensity of the magnetic field produced by an *air-core coil* (that is, a coil containing no iron or other strongly magnetic material) is proportional to the electric current which flows in the coil, or

$$H = Ci \qquad (\mathrm{A\,m^{-1}}), \tag{2.1}$$

where C is the *coil constant* which depends on the shape of the coils and on the number of turns in the windings. For an infinitely long coil or *solenoid*, C is given by

$$C = n, \tag{2.2}$$

where n is the number of turns per unit length (along the axis) of the solenoid. A *Helmholtz coil*, which consists of two identical thin coils located on a common axis and separated by a distance equal to the coil radius, has the coil constant

$$C = \frac{8N}{5\sqrt{5}\,R} = \frac{0.716}{R} N, \tag{2.3}$$

where N is the number of turns in the winding of each separate coil.

A Helmholtz coil is used to produce a highly uniform field in a particular space, for instance for cancelling the Earth's magnetic field. In general, for fields on the central axis of a single-layer solenoid of finite length, the coil constant is given by

$$C = \frac{n}{2}\left[\frac{l+z}{\sqrt{R^2+(l+z)^2}} + \frac{l-z}{\sqrt{R^2+(l-z)^2}}\right], \tag{2.4}$$

where $2l$ is the length of the solenoid, z is the distance from the center O, R is the radius of the solenoid, and n is the number of turns per unit length along the

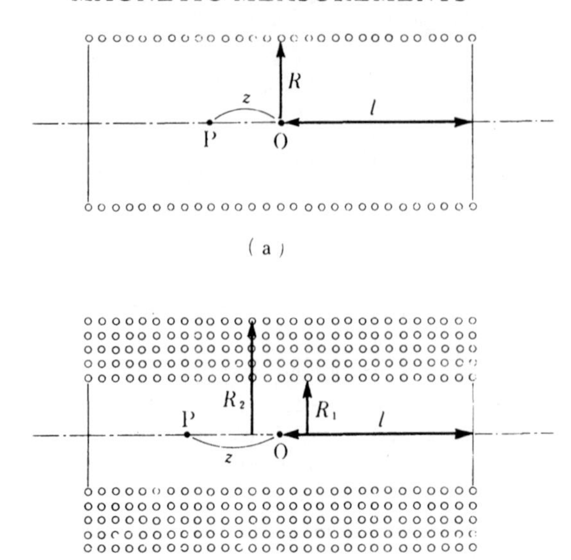

Fig. 2.1. (a) Single-layer solenoid; (b) multilayer solenoid.

solenoid axis (Fig. 2.1(a)). For a *multi-layer solenoid* of finite length, the coil constant is given by

$$C = \frac{n}{2(R_2 - R_1)}\left[(l+z)\ln\frac{R_2 + \sqrt{R_2^2 + (l+z)^2}}{R_1 + \sqrt{R_1^2 + (l+z)^2}} + (l-z)\ln\frac{R_2 + \sqrt{R_2^2 + (l-z)^2}}{R_1 + \sqrt{R_1^2 + (l-z)^2}}\right], \tag{2.5}$$

where R_1 and R_2 are the inner and the outer radii of the solenoid (Fig. 2.1(b)).

The maximum field is determined by the maximum current, which in turn is usually limited by the heat generated in the windings. In the case of a simple solenoid cooled only by natural convection, the maximum current density is about 1 $\mathrm{A\,mm^{-2}}$. Figure 2.2 shows various methods for cooling the windings. In Fig. 2.2(a), pancake-type windings of copper tape insulated with paper or polymer film are separated by thick copper disks which are water-cooled at their circumferences; in Fig. 2.2(b), thin copper disks provided with internal water-cooling tubes are used instead of the thick copper disks in (a); in Fig. 2.2(c) pancake-type windings are immersed in circulating water or oil; in Fig 2.2(d), pancake-type windings are insulated by spirally wound filaments instead of continuous films, and cooling water or oil is sent through the pancakes; in Fig. 2.2(e), wide helical copper windings provided with many holes are cooled by sending cooling water through the holes; in Fig. 2.2(f) the windings are made of rectangular hollow copper conductors which are cooled by sending purified

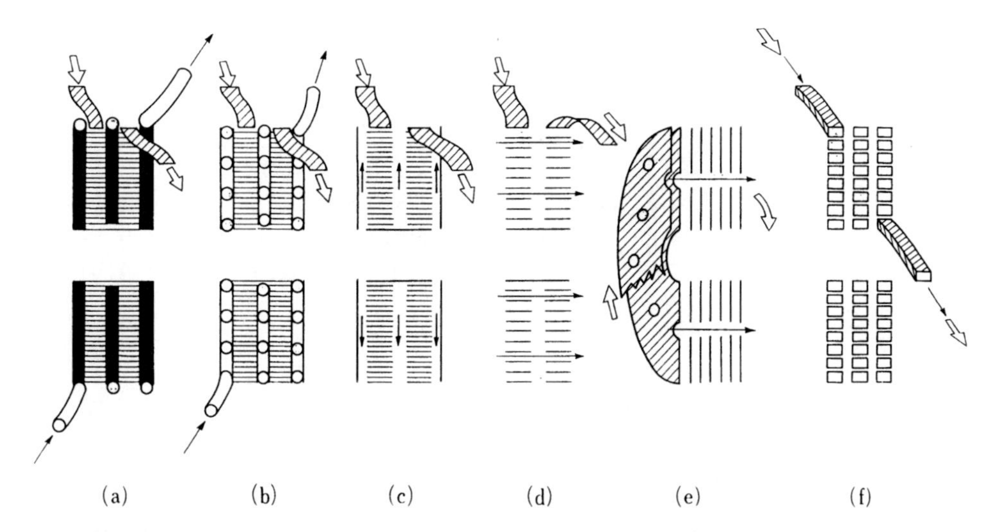

Fig. 2.2. Various types of coil windings. (Thin arrows show water flow, open arrows show electric current.)

water through the conductor. By these methods, current densities of 3–50 A mm^{-2} can be carried by the coils, and magnetic fields from 80 kA m^{-1} to 8 MA m^{-1} (0.1–10 T) can be produced. Naturally the cost of facilities and of electric power increase rapidly with the magnitude of the required magnetic field.

Magnetic fields can also be produced with *permanent magnets*. The advantages are first to avoid the use of electric power, and second to secure magnetic fields that do not vary with time. It is possible to produce fields of 50–100 kA m^{-1} (0.06–0.12 T) by using conventional permanent magnets. It is also possible to produce a field of about 0.8 MA/m (1 T) by using a large iron magnetic circuit with permanent magnets as a source of magnetomotive force. Such a facility is particularly useful for proton resonance experiments, because it gives an extremely uniform and time-independent magnetic field.[1] It is, however, a serious disadvantage in most cases that the field cannot be changed in magnitude or switched off.

Electromagnets are usually used to produce magnetic fields up to 0.8–1.6 MA m^{-1} (1–2 T). An electromagnet can be regarded as an apparatus to concentrate the magnetic flux produced by a coil into a small space, by using a magnetic circuit. If the total number of turns in the windings is N, the electric current i, the length of the magnetic circuit l_m, and the length of the air gap l_a (see Fig. 2.3), then we have

$$H_m l_m + H l_a = Ni, \tag{2.6}$$

where H_m is the field in the magnetic circuit. This field is small if the permeability of the circuit is sufficiently large. Then the first term can be neglected, so that (2.6) becomes

$$H = \frac{Ni}{l_a}. \tag{2.7}$$

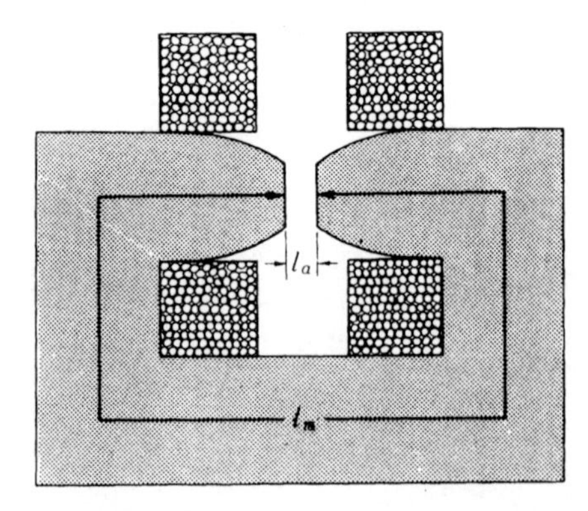

Fig. 2.3. Design of an electromagnet.

This simple linear relationship holds for any electromagnet up to some value of current or field, as shown in Fig. 2.4. The proportionality factor is given by $1/l_a$, independent of the properties and the shape of the magnetic circuit. As the core approaches magnetic saturation, the permeability decreases, so that the first term in (2.6) is no longer negligible. The magnetic field produced by the electromagnet then approaches a limiting value. In order to make the maximum field as high as possible, the pole pieces can be tapered as shown in Fig. 2.3. Theory shows that the maximum field is attained when the tapered surface is a cone whose half-angle is 54.7°, provided the magnetization is everywhere parallel to the cone axis. The most important factor in the design of an electromagnet is to avoid magnetic saturation in any portion of the magnetic circuit before the pole pieces begin to saturate.[2] It is easy to produce magnetic fields of 0.8–1.6 MA m^{-1} (1–2 T) using electromagnets, and possible to attain 2–2.4 MA m^{-1} (2.5–3 T) by careful design. However, the size, cost, and electric power consumption increase rapidly as the maximum field is increased beyond these values. Figure 2.5 shows a large electromagnet which produces a maximum field of 2.4 MA m^{-1} (3.0 T) in a 5 cm air gap.[2]

Fields ranging from 3.2 to 10 MA m^{-1} (4 to 12 T) can be produced by means of *superconducting magnets*. For example, a superconducting magnet provided with hybrid coils of multi-filament Nb–Ti alloy and Nb_3Sn compound can produce a maximum field of 8 MA m^{-1} (10 T) in a space of 4 cm diameter. The advantages of superconducting magnets are that:

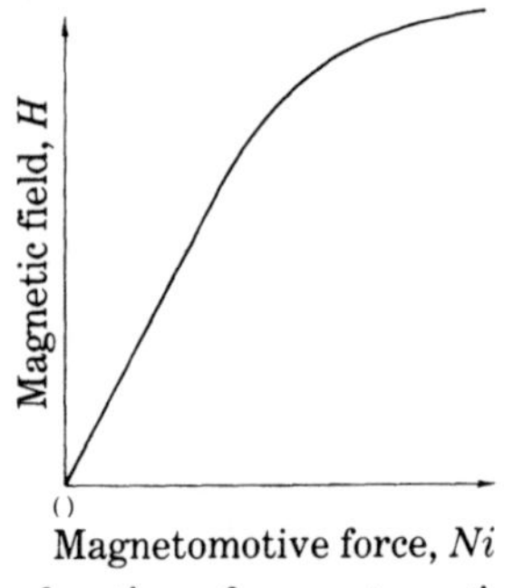

Fig. 2.4. Magnetic field as a function of magnetomotive force in an electromagnet.

Fig. 2.5. Bitter-type electromagnet (13 tons) which produces $2.5\,\mathrm{MA\,m^{-1}}$ (31 kOe) in a 5 cm air-gap.

Fig. 2.6. Superconducting magnet which produces $12\,\mathrm{MA\,m^{-1}}$ (150 kOe) in a 32 mm diameter bore. Magnet at right, cryostat at left.

(1) No electric power is needed to excite the magnet.
(2) No Joule heat is generated, so that no provision is necessary to remove heat from the coil, except for liquid helium which is used to keep the coil in a superconducting state.
(3) It is possible to keep the field completely time-independent by short-circuiting the coil with a superconducting shunt. This is known as *persistent current* or persistent-mode operation.

The disadvantages of superconducting coils are:

(1) They require cooling with liquid helium.
(2) If the maximum field is exceeded, the magnet 'quenches' from the superconducting state with the rapid generation of heat which very quickly evaporates the liquid helium.
(3) The uniformity of the field is perturbed by persistent eddy currents in the superconductors.
(4) When the field is changed, there can be sudden irregular small changes in field intensity caused by 'flux jumps'.

Figure 2.6 shows a superconducting coil and the cryostat in which the coil is cooled to liquid helium temperature. Recently a hybrid magnet in which an air-core copper magnet is installed inside a superconducting magnet has been successfully used to produce about $24\,\mathrm{MA\,m^{-1}}$ (30 T).[3]

Beyond this limit, steady (time-independent) fields are no longer practical, because they require enormous electric power and generate tremendous heat. Pulsed field methods are therefore usually used to produce fields higher than $8\,\mathrm{MA\,m^{-1}}$ (10 T).

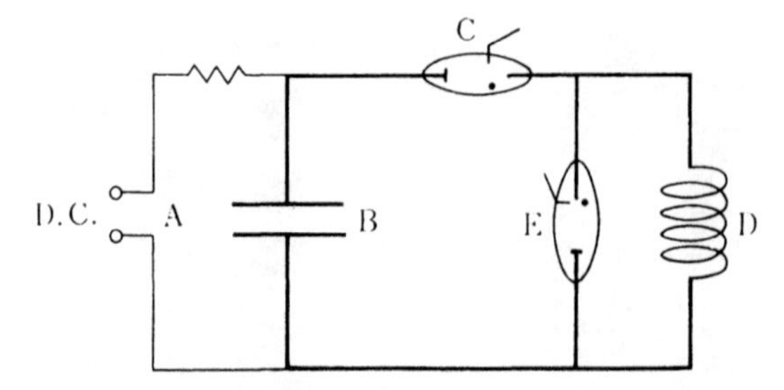

Fig. 2.7. Circuit for generating a pulsed high magnetic field.

The principle of this method is shown in Fig. 2.7. First a large condenser B is charged from a source A, then the switch C is closed and a large pulse current is sent to the coil D. When the current reaches its maximum value, the switch E is closed, in order to make the duration of the current pulse longer and also to prevent charging the capacitor in the opposite polarity. It is possible to produce magnetic fields of 20–32 MA m^{-1} (25–40 T) for times of about 1 ms by using a condenser bank of 3000 μF capacitance and a working voltage of 3000 V. This method is limited by the mechanical strength of the coil. For fields higher than 40 MA m^{-1} (50 T), the coil is damaged by the electromagnetic force acting on the lead wires. This force results from the Maxwell stresses given by (1.110). For instance, for $H = 56$ MA m^{-1} (70 T), the Maxwell stress is calculated to be

$$T = \tfrac{1}{2}\mu_0 H^2 = \tfrac{1}{2}(4\pi \times 10^{-7}) \times (5.6 \times 10^7)^2$$
$$= 2.0 \times 10^9 \text{ N m}^{-2} = 200 \text{ kgf mm}^{-2}. \tag{2.8}$$

Since the yield strength of a strong steel is at most 150 kg mm^{-2}, the coil is destroyed by the field even if it is made of high-strength materials.

Magnetic fields higher than this limit can be produced by *magnetic flux compression*, using either explosives or electromagnetic forces. The so-called *Cnare method*[4] utilizes the kinetic energy of the inward motion of a short metal cylinder called a liner. Figure 2.8 illustrates the experimental arrangement for the Cnare method. First a liner is placed in a one-turn coil which is connected to a condenser bank through a

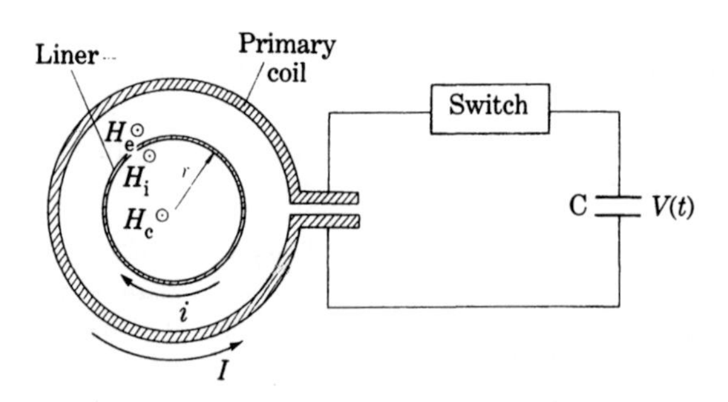

Fig. 2.8. Conceptual diagram of the Cnare method.

switch. When the switch is closed, the electric current flows in the one-turn coil, producing an increasing magnetic field inside the coil. Electromagnetic induction produces a current in the opposite sense in the liner, to prevent magnetic flux from penetrating inside the liner. Thus in a narrow space between the one-turn coil and the liner, there appears a high magnetic field which produces a Maxwell stress on the liner. This stress acts to compress the liner inward toward its axis. The liner accelerates inward, reaching a final speed as high as $1\,\mathrm{km\,s^{-1}}$ after about $10\,\mu$s. Meanwhile some of the magnetic flux has diffused into the liner. This trapped flux is squeezed into a narrow diameter by the inertial motion of the liner, thus resulting in an ultra high magnetic field. Fields as high as $400\,\mathrm{MA\,m^{-1}}$ (500 T) have been produced by this method.[5]

The major disadvantage of the Cnare method is that the specimen is destroyed by the collapse of the liner. An alternative approach is the so-called *one-turn coil method*, in which a fast current pulse from a low-impedance capacitor bank is sent through a small coil, about 1 cm in diameter. Fields up to $160\,\mathrm{MA\,m^{-1}}$ (2.0 MOe) can be produced before the coil fails. In this case the coil explodes outward (instead of collapsing inward or imploding as in the Cnare method), so that the specimen is not destroyed. The disadvantages are that the maximum field is somewhat lower and the rise time of the field is about ten times faster than in the Cnare method.[6]

2.2 MEASUREMENT OF MAGNETIC FIELDS

The methods for measuring magnetic fields can be classified into four categories:

(1) Measurement of the torque produced on a known magnetic moment by the magnetic field.
(2) Measurement of the electromotive force induced in a search coil be electromagnetic induction.
(3) Measurement of an electric signal induced in a galvanomagnetic probe.
(4) Measurement of a magnetic resonance such as electron spin resonance (ESR), ferromagnetic resonance (FMR, see Section 3.3) or nuclear magnetic resonance (NMR, see Section 4.1).

The first method, which was once common but is no longer popular, consists in measuring the torque given by (1.8) acting on a known magnetic moment M. Instead of measuring a torque directly, we can measure the period of oscillation, T, of a freely suspended permanent magnet, or

$$T = 2\pi\sqrt{\frac{I}{MH}}, \tag{2.9}$$

where I is the moment of inertia of the magnet.

The second category includes various methods utilizing the law of electromagnetic induction. Suppose that a search coil of n turns with cross-sectional area S ($\mathrm{m^2}$) is placed in a magnetic field H ($\mathrm{A\,m^{-1}}$) with the coil area perpendicular to the field. The magnetic flux which passes through the coil is given by

$$\Phi = \mu_0 nSH \qquad (\mathrm{Wb}). \tag{2.10}$$

Any change in the magnetic flux passing through the coil, caused by removing the search coil from the magnetic field, by rotating the coil by 180° about an axis perpendicular to the field, by reducing the intensity of the field to zero, or by reversing the sense of the field, results in the electromotive force $-\mathrm{d}\Phi/\mathrm{d}t$ in the search coil. The magnetic fluxmeter is an instrument that integrates this electromotive force with respect to time and produces a signal which is proportional to the change in the magnetic flux. If the electric circuit containing the search coil has resistance R (Ω) and inductance L (H), then from Kirchhoff's second law:

$$\frac{\mathrm{d}\Phi}{\mathrm{d}t} = Ri + L\frac{\mathrm{d}i}{\mathrm{d}t}. \tag{2.11}$$

If the flux has the value Φ (Wb) at $t = 0$ (so that $i = 0$), and becomes zero at $t = t_0$ (so that $i = 0$ again), integration of (2.11) results in

$$\Phi = R\int_0^{t_0} i\,\mathrm{d}t + L\,|i|_{t=0}^{t=t_0} = RQ. \tag{2.12}$$

Thus the total flux change is proportional to the electric charge Q which flows the circuit, irrespective of the value of the inductance L. An instrument that measures the electric charge Q by integrating the current i is called a *fluxmeter*.

One such instrument is the *ballistic galvanometer*. This instrument is a galvanometer having a moving coil with a large moment of inertia. When a pulse of current flows through this coil in a time interval much less than the natural period of oscillation of the coil, the maximum deflection of the galvanometer is proportional to the electric charge which passes through the coil. Therefore, when the search coil is connected to a ballistic galvanometer, the change in flux passing through the coil is measured by the deflection of the galvanometer. The proportionality factor can be determined using a standard mutual inductance M whose secondary coil is connected to the circuit (see Fig. 2.9). When an electric current in the primary coil i_0 is reversed, a flux change $2Mi_0$ occurs in the secondary coil. The resultant deflection of the galvanometer, Θ_0, is proportional to this flux change, or

$$\Theta_0 = 2kMi_0, \tag{2.13}$$

where k is the proportionality factor. Similarly when the flux change $\Delta\Phi$ occurs in the search coil, the deflection Θ is given by

$$\Theta = k\,\Delta\Phi. \tag{2.14}$$

Eliminating k from (2.13) and (2.14), we have

$$\Delta\Phi = 2Mi_0\frac{\Theta}{\Theta_0}. \tag{2.15}$$

If the flux given by (2.10) is reversed, putting $\Delta\Phi = 2\Phi$, we have

$$H = \frac{Mi_0}{\mu_0 nS}\frac{\Theta}{\Theta_0}, \tag{2.16}$$

from which we can determine the magnetic field H.

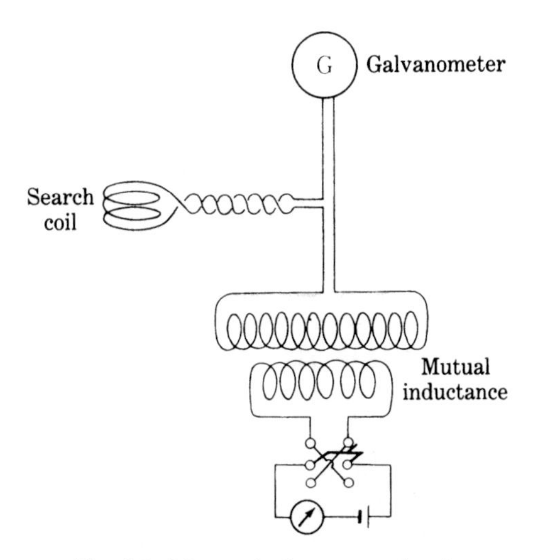

Fig. 2.9. Magnetic fluxmeter circuit.

A commercially available *fluxmeter* is provided with a moving coil suspended in the field of a permanent magnet. The suspension system provides a very small restoring torque on the coil. The flux change caused by a rotation of the moving coil is designed to be proportional to its deflection, or

$$\Phi_m = k\Theta. \tag{2.17}$$

When there is a change in the magnetic flux through a search coil connected with this fluxmeter, the electromotive force $d\Phi/dt$ causes a rotation of the moving coil and is balanced by the electromotive force caused by a change in Φ_m; that is

$$\frac{d\Phi}{dt} = k\frac{d\Theta}{dt}. \tag{2.18}$$

Integrating (2.18), we have

$$\Delta\Phi = k\Theta. \tag{2.19}$$

Thus the flux change is measured by the deflection of the moving coil. A fluxmeter is distinguished from a ballistic galvanometer by the fact that a fluxmeter ideally has no stable zero position; the pointer will remain at any position on the scale. A more automated recording fluxmeter will be described in Section 2.3.

The third category of instrument, based on the galvanomagnetic effect, is commonly found in portable devices known as *gaussmeters* or *field meters*. Two different physical phenomena, the Hall effect and the magnetoresistance effect, can be used. The principle of the Hall effect is illustrated in Fig. 2.10. When a magnetic field H ($\mathrm{A\,m^{-1}}$) is applied perpendicular to a semiconductor plate made from Ge, InSb, or InAs, the DC current J (A) which flows through the plate will produce a DC voltage V (v) across the plate given by

$$V = \frac{RJ}{d}H, \tag{2.20}$$

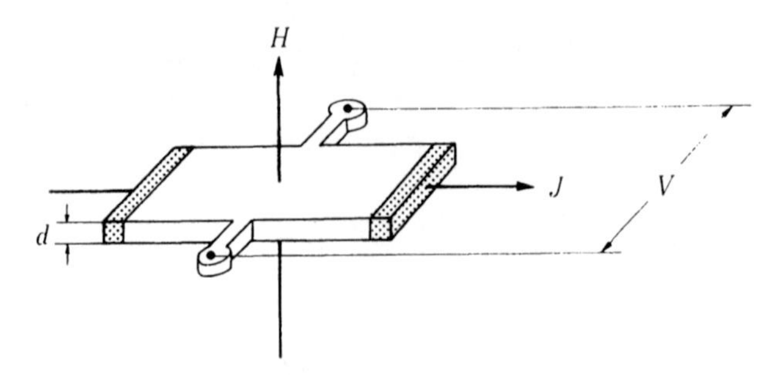

Fig. 2.10. Hall effect element.

where d is the thickness of the plate and R is the Hall coefficient. Direct-reading Hall gaussmeters covering the field range from $8\,\mathrm{A\,m^{-1}}$ (= 0.1 Oe) to $2.4\,\mathrm{MA\,m^{-1}}$ (= 3.0 T) or higher are commercially available.

The magnetoresistance effect refers to the change in electrical resistance of a conductor subjected to a magnetic field applied perpendicular to the current. The magnetoresistance effect is particularly large in Bi. This is because the current carriers consist of equal numbers of electrons and positive holes, and the two kinds of carriers are combined and annihilated after being driven in the same direction by the magnetic field. In this case the resistance change is proportional to H^2, which is inconvenient for a general-purpose instrument. Furthermore, galvanomagnetic effects are generally temperature dependent and also more or less subject to aging. For these reasons, magnetoresistance is rarely used for the direct measurement of magnetic fields. Magnetoresistive sensors have been used as read-out devices for magnetic bubble memories [see Chapter 22.3] and for magnetically recorded tapes and disks.

The fourth category, *magnetic resonance* phenomena, is used particularly for accurate calibration of other types of instruments. The principles will be explained in Chapter 3.3 for ESR and in Chapter 4.1 for NMR. In both cases, the resonance frequency is proportional to the intensity of the magnetic field. The standard material used for ESR is DPPH (α-diphenyl-β-picrylhydrazyl) which resonates at 2.804 ± 0.001 MHz in a field of $79.58\,\mathrm{A\,m^{-1}}$ (= 1.000 Oe). The standard isotopes used for NMR are H^1, Li^1, and D^2 contained in H_2O, LiCl, $LiSO_4$, and D_2O.

2.3 MEASUREMENT OF MAGNETIZATION

The methods for measuring the intensity of magnetization can be classified into three categories:

(1) measurement of the force acting on a magnetized body in a non-uniform field;
(2) measurement of the magnetic field produced by a magnetized body;
(3) measurement of the voltage produced in a search coil by electromagnetic induction caused by a change in the position or state of magnetization of a magnetized body.

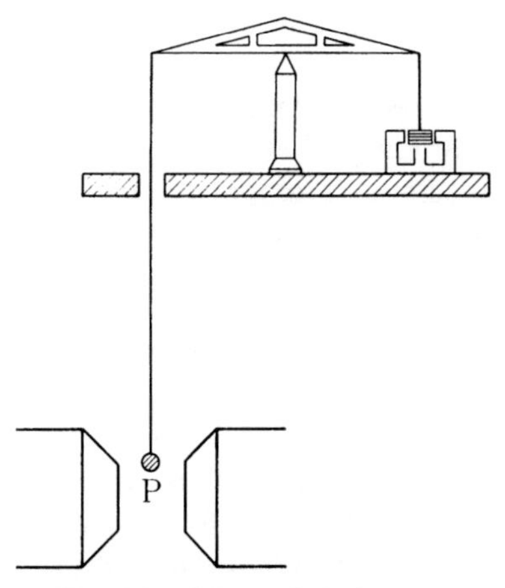

Fig. 2.11. Magnetic balance.

The *magnetic balance*[7] is the most common example of category 1. As illustrated in Fig. 2.11, a specimen suspended from one arm of a balance is attracted downwards by the inhomogeneous field produced by an electromagnet. This force is counterbalanced by an automatic device described later. The downward force is given by

$$F = I \frac{\partial H}{\partial z} v, \tag{2.21}$$

where I is the magnetization, v the volume of the specimen, and $(\partial H/\partial z)$ is the vertical gradient of the field. When I is proportional to H, we have from (2.21), using (1.33),

$$F = \tfrac{1}{2}\chi' \frac{\partial H^2}{\partial z} v, \tag{2.22}$$

from which we can determine the magnetic susceptibility χ'. If, however, $\bar{\chi}$ is not negligible as compared with 1 (say $\bar{\chi} > 10^{-2}$), the value of χ' in (2.22) is different from the true susceptibility, χ, because of the demagnetizing field. Since we have the relationship

$$\chi' = \frac{\chi}{1 + \bar{\chi}N}, \tag{2.23}$$

the true susceptibility is given by

$$\chi = \frac{\chi'}{1 - \bar{\chi}'N}. \tag{2.24}$$

The downward force on the sample is balanced by suspending a small coil from the opposite arm of the balance. The coil is placed in the radial magnetic field of a loudspeaker magnet, so that a current through the coil produces a vertical force on the coil. When the balance is in equilibrium, the current in the coil is proportional to the force acting on the specimen and accordingly to the susceptibility. The system is insensitive to magnetic disturbance from outside, because only a radial field exerts a vertical force on the coil. The system can be made automatic by providing a means to detect electrically the equilibrium position of the balance. The detector signal, which is proportional to the deflection, is amplified and sent to the hanging coil as a feedback signal so as to maintain the balance in equilibrium.

The Sucksmith ring balance[8] is a once-common apparatus utilizing the same principle. It is purely mechanical, without any automatic balancing feature.

A similar apparatus in which the magnetic force acts horizontally rather than vertically is known as the *magnetic pendulum*. It has the advantage that changes in the mass of the sample due to oxidation, absorption of water, etc., do not affect the reading. Figure 2.12 illustrates the construction of a magnetic pendulum with a wide range of sensitivity.[9] In this pendulum, the force acting on the specimen is compensated by two attracting coils so as to exert no lateral force on the support, which is a jewelled knife edge.

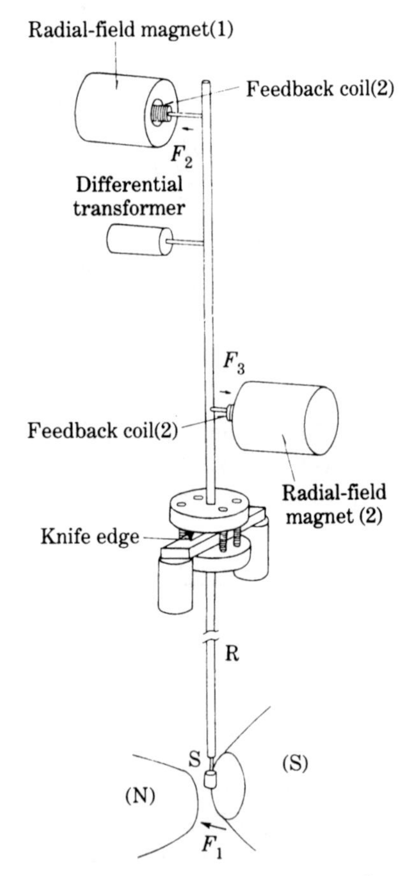

Fig. 2.12. Magnetic pendulum.[9]

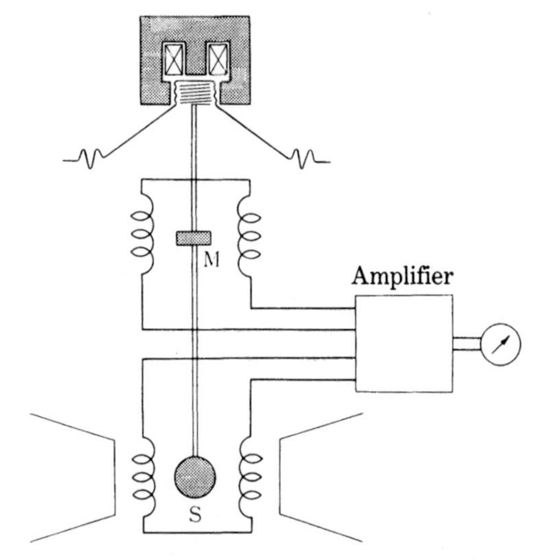

Fig. 2.13. Vibrating sample magnetometer (VSM).

The *astatic magnetometer* also falls into the second category. In this device, the magnetic field produced by the magnetization of the specimen is measured by the deflection of a pair of coupled magnetized needles suspended from a fine fiber. This apparatus is still being used for the measurement of weak residual magnetization of geological specimens.

The *vibrating sample magnetometer* (*VSM*)[10] is a modern version of this type of magnetometer. As shown in Fig. 2.13, the sample S is oscillated vertically in a region of uniform field. If the sample is driven by a loudspeaker mechanism, the frequency is usually near 80 Hz and the amplitude is 0.1–0.2 mm. A mechanical crank drive can also be used, in which case the frequency is normally somewhat lower and the amplitude larger, perhaps 1–2 mm. The AC signal induced in the pick-up coil by the magnetic field of the sample is compared with the signal from a standard magnet M and is converted to a number proportional to the magnetic moment. The advantages of this magnetometer are high sensitivity, ease of operation, and convenience for measurements above and below room temperature. Several manufacturers provide magnetometers of this type.

The principle of the third category (electromagnetic induction) is the same as that described in (2.2) above. When a rod specimen of cross-sectional area S' (m^2) is inserted into a search coil of n turns with cross-sectional area S (m^2) (Fig. 2.14), and is magnetized to I (T) by an external field H_{ex}, the magnetic flux which passes through this search coil is given by

$$\begin{aligned}\Phi &= n[IS' + \mu_0 H_{eff} S] \\ &= n[I(S' - NS) + \mu_0 H_{ex} S],\end{aligned} \tag{2.25}$$

where N is the demagnetizing factor of the specimen and H_{eff} is the effective field inside the specimen. Note that the magnetic field just outside the specimen is the

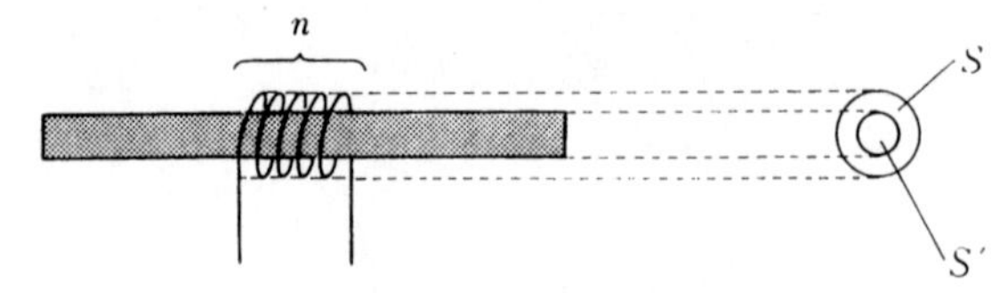

Fig. 2.14. Search coil.

same as the effective field H_{eff} inside the specimen, because the tangential component of the field is always continuous.

A long thin specimen is useful for the measurement of the magnetization curve of soft magnetic materials because of its small demagnetizing factor, but may be inconvenient to prepare, especially as a single crystal, and is also unsuitable for measurements at low and high temperatures or under high pressures. On the other hand, a spherical specimen has several advantages: it is relatively easy to prepare; requires only a small quantity of material (particularly advantageous for noble metals, rare earths and single crystals); it has a well-defined demagnetizing factor (refer to (1.46)), permitting measurement along many crystallographic directions in one single crystal sphere; and it is well suited for measurements at high and low temperatures or under pressure. The only disadvantages are a large demagnetizing factor and a certain difficulty in mounting and holding the sample during the measurement.

The following is an example of the measurement of a magnetization curve as a function of the effective field H_{eff} for a spherical specimen. The construction of the search coil is shown in Fig. 2.15. When a spherical specimen of radius r_s (m) magnetized to intensity I (T) in a magnetic field H (A m^{-1}) is placed at the center of a thin search coil of radius r (m) containing n turns, the flux which goes through the search coil is given by

$$\Phi = n\left[\frac{2\pi r_s^3}{3r} I + \pi\mu_0 r^2 H\right]. \tag{2.26}$$

Then the flux due to I is inversely proportional to r, while the flux due to H is proportional to r^2. A proper combination of two coils with radii r_1 and r_2 and number

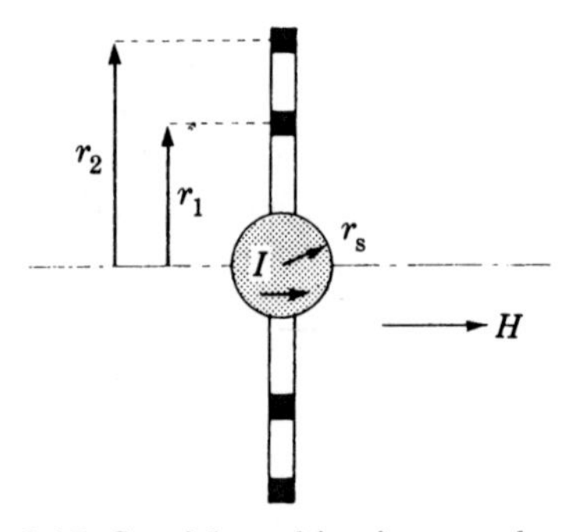

Fig. 2.15. Special combination search coils.

of turns n_1 and n_2 can mix the I and H terms in any desired ratio. For instance, if we make $n_1 : n_2 = 1/r_1^2 : 1/r_2^2$, the H term is cancelled, so that we have

$$\Phi = \frac{2\pi k}{3}\left[\frac{r_s^3}{r_1^3} - \frac{r_s^3}{r_2^3}\right]I, \tag{2.27}$$

where

$$k = n_1 r_1^2 = n_2 r_2^2. \tag{2.28}$$

Thus we have a search coil which picks up a signal proportional to I. If we select $n_1 : n_2 = r_1(r_2^3 + 2r_s^3) : r_2(r_1^3 + 2r_s^3)$, we have

$$\Phi = 2\pi k'(r_2^3 - r_1^3) r_s^3 \mu_0 H_{\text{eff}}, \tag{2.29}$$

where

$$H_{\text{eff}} = H - \frac{1}{3\mu_0} I, \tag{2.30}$$

and

$$k' = \frac{n_1}{r_1(r_2^3 + 2r_s^3)} = \frac{n_2}{r_2(r_1^3 + 2r_s^3)}. \tag{2.31}$$

Thus we have a search coil which picks up the effective field.[11]

By connecting these search coils to recording fluxmeters, we can draw a magnetization curve as a function of the effective magnetic field.

Figure 2.16 shows a circuit diagram of a *Cioffi-type recording fluxmeter*.[12] In this figure the specimen is shown as a ring, but it can be replaced by any combination of

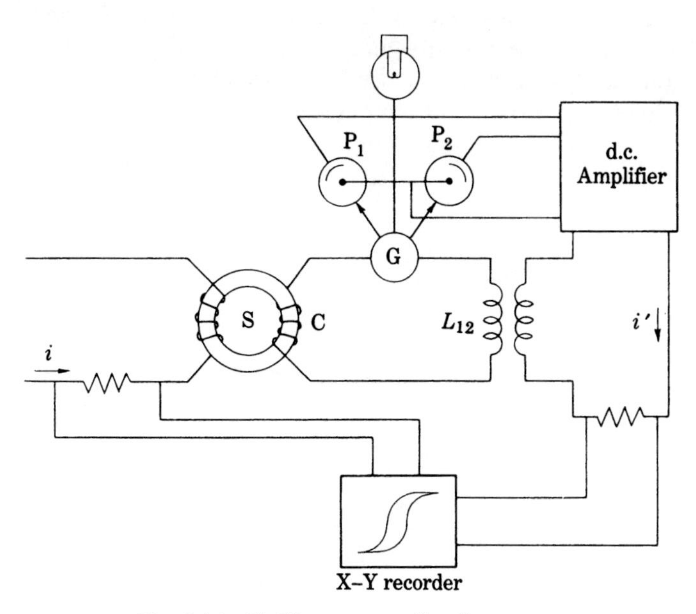

Fig. 2.16. Cioffi-type recording fluxmeter.

specimen and a search coil. First the current i is passed through the primary coil, so that the specimen S is magnetized. This gives rise to a voltage in the secondary coil

$$V = -\frac{\partial \Phi}{\partial t}. \tag{2.32}$$

This voltage deflects the galvanometer G. This deflection is sensitively detected by two photocells, P_1 and P_2, which are used to generate a feed-back current i' in the primary coil of the mutual inductance L_{12}, which induces a voltage in the secondary coil

$$V' = -L_{12}\frac{\partial i'}{\partial t}. \tag{2.33}$$

This voltage almost compensates the voltage given by (2.32), except for a small uncompensated voltage which keeps a current i' in the primary coil of L_{12}. Since this unbalanced voltage is negligibly small, we have

$$\frac{\partial \Phi}{\partial t} = L_{12}\frac{\partial i'}{\partial t}. \tag{2.34}$$

Integrating (2.34), we have the relationship

$$\Delta\Phi = L_{12}\,\Delta i', \tag{2.35}$$

which shows that the flux change is proportional to the current change $\Delta i'$. In the case of Fig. 2.16, the signals proportional to i and i' are fed to an x–y recorder so that the magnetization curve is drawn.

In the case of a spherical specimen which is placed in the air gap of an electromagnet together with a set of search coils as described by (2.31) and (2.28), the voltages induced in these coils are converted to signals proportional to H_{eff} and I by using two recording fluxmeters whose output signals are fed into the x- and y-axes of an x–y recorder to draw an I–H_{eff} curve. Figure 2.17 shows hysteresis loops obtained in this way for a Gd sphere at various low temperatures.[11]

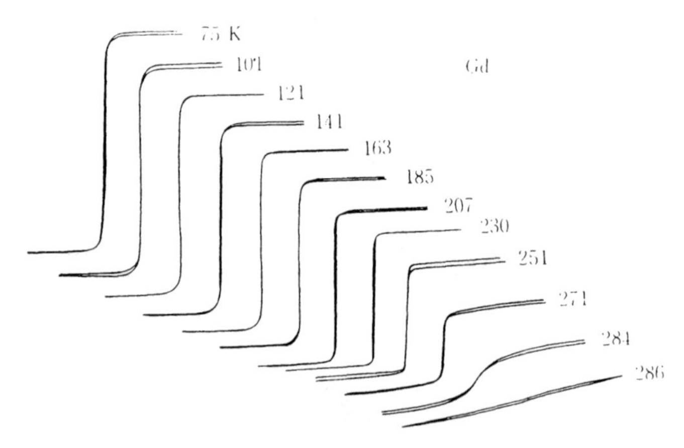

Fig. 2.17. Hysteresis loops measured for Gd at various temperatures using special combination search coils.

More accurate and stable measurement of magnetic flux is possible using a digital voltmeter. The principle of this method is to move a search coil with respect to a magnetized specimen by a fixed distance, measure the induced voltage with a digital voltmeter as a function of time, and then integrate these values with respect to time with an analog or digital integrator. In this case, the flux change is given by (2.25), provided that the search coil is removed completely from the sample and out of the field. If the search coil is removed from the specimen but remains in the magnetic field, or the specimen itself is removed from a fixed search coil, the flux change is given by the first term of the final line of (2.25) or the first term of (2.26). In favorable cases, four to five significant figures can be obtained by this method.[13-15]

The same principle has been successfully applied to measure the slowly changing magnetic fields produced by a hybrid air-core and superconducting coil.[16]

A very weak magnetic flux can be measured accurately by a *superconducting quantum interference device* (*SQUID*) magnetometer, which utilizes the Josephson effect. The Josephson effect is the name given to the fact that the flux change in a superconducting circuit interrupted by an insulating layer about 50 Å thick is quantized.[17] Counting these flux quanta gives a very sensitive measurement of the flux change and therefore of the magnetization of the sample. The fact that the measuring apparatus must be at liquid helium temperature is clearly a disadvantage in many cases, and a SQUID magnetometer requires a long time for each magnetization measurement, especially if the field is changed between readings.

PROBLEMS

2.1 Calculate the coil constant at the center of a one-layer solenoid of radius r (m), length $2\sqrt{3}\,r$ (m), and winding density n (turns m^{-1}).

2.2 A magnetomotive force Ni (A) is applied to an electromagnet made from high-permeability material. If the air gap is l (m) long, what is the magnetic field in the gap?

2.3 Describe typical methods for measuring magnetic moments and magnetic fields after classifying them according to the principle of measurements.

REFERENCES

1. M. Matsuoka and Y. Kakiuchi, *J. Phys. Soc. Japan*, **20**, (1965), 1174.
2. Y. Ishikawa and S. Chikazumi, *Japan. J. Appl. Phys.*, **1** (1962), 155.
3. Y. Nakagawa, K. Kido, A. Hoshi, S. Miura, K. Watanabe, and Y. Muto, *J. de Phys.*, **45** (1984), Cl-23.
4. E. C. Cnare, *J. Appl. Phys.*, **37** (1966), 3812.
5. H. Nojiri, T. Takamasu, S. Todo, K. Uchida, T. Haruyama, H. A. Katori, T. Goto, and N. Miura, *Physica*, B **201** (1994), 579.
6. K. Nakao, F. Herlach, T. Goto, S. Takeyama, T. Sakakibara, and N. Miura, *J. Phys. E, Sci. Instrum.*, **18** (1985), 1018.

7. Y. Nakagawa and A. Tasaki, *Lecture Series on Experimental Phys.*, **17** (12), (in Japanese) (Kyoritsu Publishing Co., Tokyo, 1968).
8. W. Sucksmith, *Proc. Roy. Soc.* (London), **170A** (1939), 551.
9. M. Matsui, H. Nishio, and S. Chikazumi, *Japan. J. Appl. Phys.*, **15** (1976), 299.
10. S. Foner, *Rev. Sci. Instr.*, **30** (1959), 548.
11. Y. Ishikawa and S. Chikazumi, *Lecture Series on Experimental Physics*, **17** (10) (in Japanese) (Kyoritsu Publishing Co., Tokyo, 1968).
12. P. P. Cioffi, *Rev. Sci. Instr.*, **21** (1950), 624.
13. T. R. McGuire, *J. Appl. Phys.*, **38** (1967), 1299.
14. K. Strnat and L. Bartimay, *J. Appl. Phys.*, **38** (1967), 1305.
15. J. P. Rebouillat, Thesis (Grenoble University, 1972); *IEEE Trans. Mag.*, **Mag-8** (1972), 630.
16. G. Kido and Y. Nakagawa, *Proc. 9th Int. Conf. Magnet. Tech.*, Zurich, (1985).
17. J. Clarke, *Proc. IEEE*, **61** (1973), 8.

Part II
MAGNETISM OF ATOMS

In most magnetic materials the carriers of magnetism are the magnetic moments of the atoms. In this Part, we discuss the origin of these atomic magnetic moments. Atomic nuclei also have feeble but non-zero magnetic moments. These have almost no influence on the magnetic properties of matter, but do provide useful information on the microscopic structure of matter through nuclear magnetic resonance (NMR) and Mössbauer spectroscopy. Neutron scattering and muon spin rotation are also useful tools for investigating microscopic magnetic structures.

3

ATOMIC MAGNETIC MOMENTS

3.1 STRUCTURE OF ATOMS

In the classical Bohr model of the atom, Z electrons are circulating about the atomic nucleus which carries an electric charge Ze (C), where Z is the atomic number, and e (C) is the elementary electric charge. One of the origins of the atomic magnetic moment is this orbital motion of electrons. Suppose that an electron moves in a circular orbit of radius r (m) at an angular velocity ω (s^{-1}) (Fig. 3.1). Since the electron makes $\omega/2\pi$ turns per second, its motion constitutes a current of $-e\omega/2\pi$ (A), where $-e$ is the electric charge of a single electron. The magnetic moment of a closed circuit of electric current i whose included area is S (m^2) is known from electromagnetic theory to be $\mu_0 iS$ (Wb m). Therefore the magnetic moment produced by the circular motion of the electron in its orbit is given by

$$M = -\mu_0\left(\frac{e\omega}{2\pi}\right)(\pi r^2)$$

$$= -\frac{\mu_0 e\omega r^2}{2}. \tag{3.1}$$

Since the angular momentum of this moving electron is given by

$$P = m\omega r^2, \tag{3.2}$$

where m (kg) is the mass of a single electron, (3.1) may be written as

$$M = -\frac{\mu_0 e}{2m}P. \tag{3.3}$$

Thus we find that the magnetic moment is proportional to the angular momentum, although their sense is opposite.

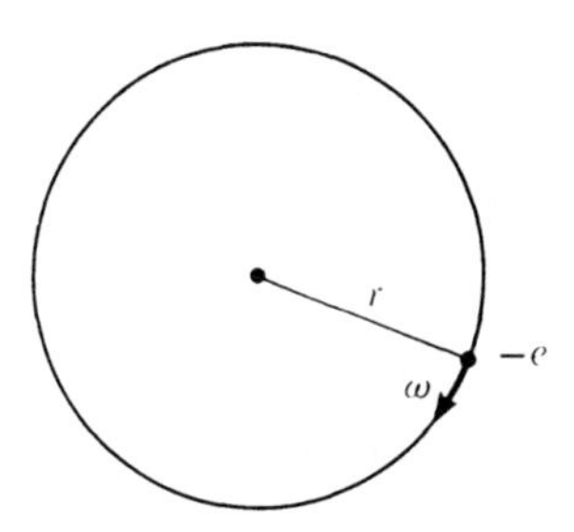

Fig. 3.1. Bohr's atomic model.

Now it is well known that the orbital motion of the electron is quantized, so that only discrete orbits can exist. In other words, the angular momentum is quantized, and is given by

$$P = l\hbar, \tag{3.4}$$

where $\hbar$ is Planck's constant h divided by 2π, or

$$\hbar = \frac{h}{2\pi} = 1.055 \times 10^{-34} \qquad (\mathrm{J\,s}) \tag{3.5}$$

and l is an integer called the *orbital angular momentum quantum number*. Substituting (3.4) for P in (3.3), we have

$$M = -\frac{\mu_0 e\hbar}{2m} l. \tag{3.6}$$

Thus the magnetic moment of an atom is given by an integer multiple of a unit

$$M_{\mathrm{B}} = \frac{\mu_0 e\hbar}{2m}$$

$$= 1.165 \times 10^{-29} \qquad (\mathrm{Wb\,m}), \tag{3.7}$$

which is called the *Bohr magneton.*

Why the angular momentum of an orbital electron is quantized by $\hbar$

In quantum mechanics, the s component of the angular momentum p_s is expressed by the operator*

$$p_s = -\mathrm{i}\hbar \frac{\partial}{\partial s}, \tag{3.8}$$

where s is the circular ordinate taken along the circular orbit, and is related to the azimuthal angle of the electron by

$$s = r\varphi. \tag{3.9}$$

The momentum along the orbit is given by

$$p_s = -\mathrm{i}\hbar \frac{1}{r} \frac{\partial}{\partial \varphi}. \tag{3.10}$$

Therefore the angular momentum along the z-axis, which is normal to the orbital plane, is given by

$$L_z = rp_s = -\mathrm{i}\hbar \frac{\partial}{\partial \varphi}. \tag{3.11}$$

The eigenvalue of the angular momentum, $m\hbar$, is obtained by solving the wave equation

$$L_z \psi = m\hbar \psi. \tag{3.12}$$

The general solution of this equation is given by

$$\psi = C\mathrm{e}^{\mathrm{i}m\varphi}. \tag{3.13}$$

* Although the angular momentum is a physical quantity related to time, it can be expressed in terms of only the ordinate s as long as we treat only stationary states.

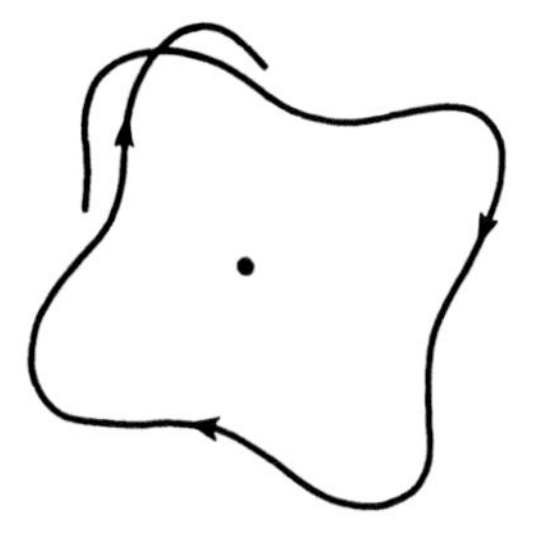

Fig. 3.2. Explanation of a stationary state.

The necessary condition for this solution to be a stationary state is that

$$\psi(\varphi + 2\pi) = \psi(\varphi). \tag{3.14}$$

Otherwise the wave function cannot be a unique function as shown in Fig. 3.2. Applying (3.14) to (3.13), we have

$$C\mathrm{e}^{\mathrm{i}m(\varphi+2\pi)} = C\mathrm{e}^{\mathrm{i}m\varphi}, \tag{3.15}$$

or

$$\mathrm{e}^{\mathrm{i}2\pi m} = \cos 2\pi m + \mathrm{i}\sin 2\pi m = 1. \tag{3.16}$$

In order to satisfy (3.16), we have

$$\cos 2\pi m = 1, \qquad \sin 2\pi m = 0, \tag{3.17}$$

or

$$m = 0, \pm 1, \pm 2, \ldots . \tag{3.18}$$

Thus the eigenvalue of L_z is an integer multiple of $\hbar$.

Besides the orbital angular momentum, the electron has a *spin angular momentum.* This concept was first introduced by Uhlenbeck and Goudsmit[1] for the purpose of interpreting the hyperfine structure of the atomic spectrum. In 1928, Dirac[2] provided a theoretical foundation for this concept by making a relativistic correction to the wave equation.

The magnitude of the angular momentum associated with spin is $\hbar/2$, so that its angular momentum is given by

$$P = s\hbar, \tag{3.19}$$

where s is the *spin angular momentum quantum number* and takes the values $\pm\frac{1}{2}$. The magnetic moment associated with spin angular momentum P is given by

$$M = -\frac{\pi_0 e}{m} P. \tag{3.20}$$

Comparing this equation with that of the orbital magnetic moment or (3.3), we find that a factor 2 is missing in the denominator. However, substituting P in (3.19) into (3.20), we find that the magnetic moment is again given by the Bohr magneton. These conditions were proved by Dirac[2] using the relativistic quantum theory.

Generally the relationship between M and P is given by

$$M = -g\frac{\mu_0 e}{2m}P, \tag{3.21}$$

where the *g-factor* is 2 for spin and 1 for orbital motion. The coefficient of P in (3.21) is calculated to be

$$\nu = g\frac{\mu_0 e}{2m} = 1.105 \times 10^5 g \qquad (\mathrm{m\,A^{-1}\,s^{-1}}), \tag{3.22}$$

which is called the *gyromagnetic constant*. Using (3.22), we can express (3.21) as

$$M = -\nu P. \tag{3.23}$$

Thus we find the magnetic moment of an atom is closely related to the angular momentum of the electron motion. A more exact definition of the g-factor is given by (3.39).

Now let us examine the relationship between the electronic structure of atoms and their angular momentum. As mentioned above, in a neutral atom Z electrons are circulating about a nucleus having an electric charge Ze (C). The size of the orbit of the electron is defined by the *principal quantum number* n, which takes the numerical values $1, 2, 3, 4 \ldots$. The groups of orbits corresponding to $n = 1, 2, 3, 4, \ldots$ are called the K, L, M, N, ... shells, respectively. The shape of the orbit is determined by the angular momentum, or in classical mechanics by the areal velocity. The angular momentum is defined by the *orbital angular momentum quantum number* in (3.5). The orbits which belong to the principal quantum number n can take n angular momenta corresponding to $l = 0, 1, 2, \ldots, n-1$. The electrons with $l = 0, 1, 2, 3, 4, \ldots$ are called the $s, p, d, f, g, \ldots$ electrons. The spin angular momentum is defined by (3.19) in which s can take the values $\pm\frac{1}{2}$.

According to the *Pauli exclusion principle*,[3] only two electrons, with $s = +\frac{1}{2}$ and $-\frac{1}{2}$, can occupy the orbit defined by n and l. The total angular momentum of one electron is defined by the sum of the orbital and spin angular momenta, so that

$$j = l + s, \tag{3.24}$$

where j is the *total angular momentum quantum number*.

When a magnetic field is applied to an atom, the angular momentum parallel to the magnetic field is also quantized and can take $2l + 1$ discrete states. This is called *spatial quantization*. Intuitively this corresponds to discrete tilts of the orbital planes relative to the axis of the magnetic field. The component of l parallel to the field, m, or the *magnetic quantum number* can take the values

$$m = l, l-1, l-2, \ldots, 0, \ldots, -(l-1), -l. \tag{3.25}$$

For instance, in the case of the d electron ($l = 2$), the orbital moment can take five possible orientations corresponding to $m = 2, 1, 0, -1, -2$ as shown in Fig. 3.3. In this case it is meaningless to discuss the azimuthal orientation of the orbit about the magnetic field, because the orbit precesses about the magnetic field. Figures 3.3(a) and 3.4 show such precessions for $m = 2$, 1, and 0, according to the classical model. It

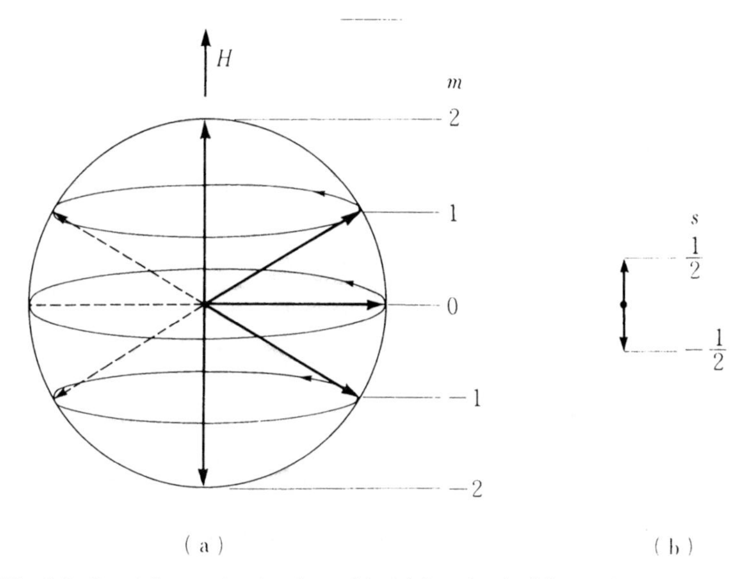

Fig. 3.3. Spatial quantization for orbital (a) and spin (b) angular momenta.

is interesting to note that these figures have some similarity with the atomic wave functions shown in Fig. 3.18. The spin can take up an orientation either parallel ($s = +\frac{1}{2}$) or antiparallel ($s = -\frac{1}{2}$) to the magnetic field (Fig. 3.3(b)).

When an atom contains many electrons, each electron can occupy one state defined by n, l, and s. Figure 3.5 shows possible states belonging to the M shell. Since the principal quantum number n of the M shell is 3, possible orbital angular momenta are $l = 0$, 1, and 2. In other words, there are s, p, and d orbital states. According to spatial quantization, each orbital states consists of $2l + 1$ orbits with different magnetic quantum numbers m. That is, the number of orbits is 1, 3, and 5 for s, p, and d, respectively. Therefore the total number of orbits belonging to one atom is given by

$$\sum_{l=0}^{n-1} (2l + 1) = n(n - 1) + n = n^2. \tag{3.26}$$

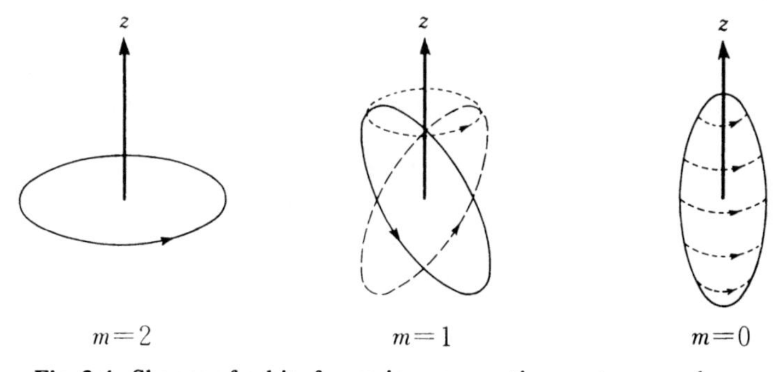

Fig. 3.4. Shapes of orbits for various magnetic quantum numbers.

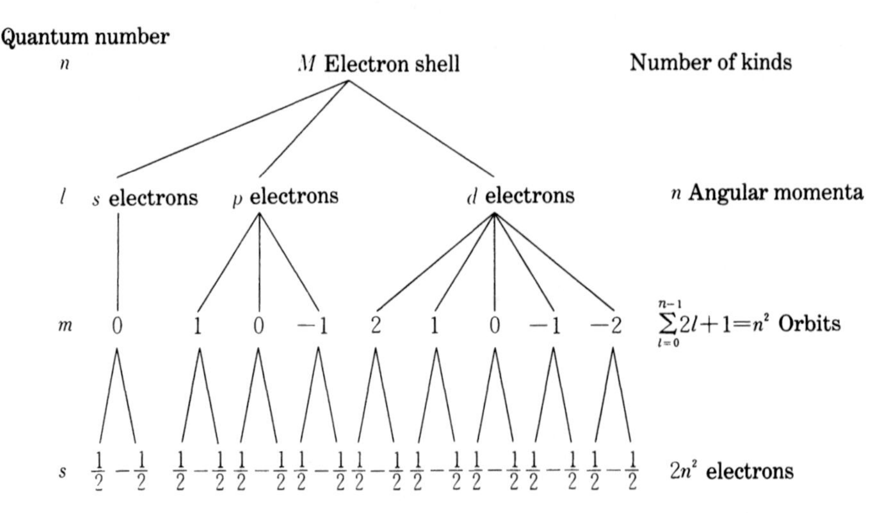

Fig. 3.5. Various electronic states belonging to the M electron shell.

Since two electrons with + and − spins can enter into one orbit, the total number of electrons belonging to one neutral atom is equal to $2n^2$. In the case of the M shell with $n = 3$, the total number is $2n^2 = 2 \times 3^2 = 18$.

In an actual atom with atomic number Z, Z electrons occupy the possible orbital states starting from the lowest energy state. Table 3.1 lists the electron configurations of the atoms which are most important in connection with magnetism. As seen in this table, the electrons occupy the states in the normal order, from the lower n states to the higher ones, up to argon ($Z = 18$). It must be noted that the energy defined by n is that of one isolated non-interacting electron. When a number of electrons are circulating around the same nucleus, we must take into consideration the interaction between these electrons. However, as long as the inner electrons are distributed with spherical symmetry about the nucleus, the ordering of the energy levels remains unchanged, because the effect of the inner electrons is simply to shield the electric field from the nucleus. Therefore, if the inner $Z - 1$ electrons form a spherical charge distribution about the nucleus, the outermost electron feels the difference in electric field between the nuclear charge $+Ze$ and the charge of the inner electron cloud $-(Z - 1)e$. This is the same as the electric field produced by a proton of charge $+e$. This is the reason why the size of the atom remains almost unchanged from that of hydrogen, irrespective of the number of electrons, up to heavy atoms with many electrons.

If, however, the charge distribution of the orbit deviates from spherical symmetry, the situation is changed. Figure 3.6 shows the shapes of various Bohr orbits. We see that the $3d$ orbit is circular, whereas the $3s$ orbit is elliptical, so that part of the orbit is close to the nucleus. In other words, the atomic wave function of s electrons is very large in the vicinity of the nucleus, as shown in Figure 3.21(c). Thus the energy of the $4s$ electron is lowered, because of a large Coulomb interaction with the unshielded nuclear charge. For this reason, the $4s$ orbits are occupied before the $3d$ orbits are

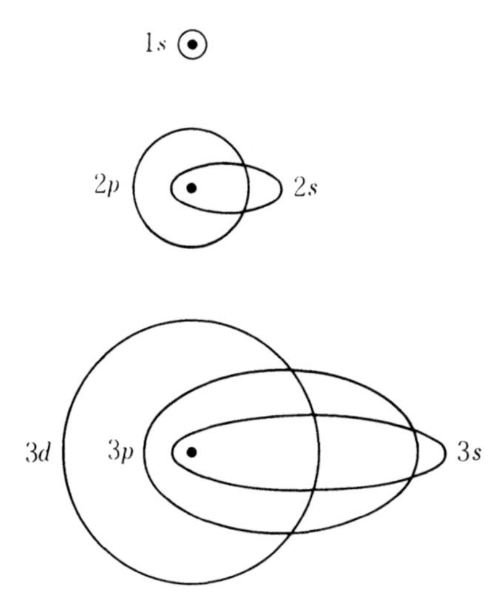

Fig. 3.6. Various Bohr orbits.

occupied for atoms heavier than potassium ($Z = 19$). As seen in Table 3.1, the $5s$ orbits are occupied before the $4d$ orbits for atoms heavier than rubidium ($Z = 37$). Similarly, the $5s$, $5p$, $5d$, and $6s$ orbits are occupied before the $4f$ orbits for atoms heavier than lanthanum ($Z = 57$). The same thing happens for atoms heavier than hafnium ($Z = 72$), in which the $6s$ orbits are filled before the $5d$ orbits are occupied. The elements which have incomplete electron shells exhibit abnormal chemical and magnetic properties, and are called *transition elements*. In the order mentioned above, they are called the $3d$, $4d$, rare-earth and $5d$ transition elements, respectively. The magnetic elements are found among these transition elements.

3.2 VECTOR MODEL

In this section we discuss how the orbital and spin magnetic moments of electrons in an incomplete electron shell form an atomic magnetic moment. Let the spin and orbital angular momentum vectors of the ith and jth electrons be $\boldsymbol{s}_i$, $\boldsymbol{l}_i$, $\boldsymbol{s}_j$ and $\boldsymbol{l}_j$, respectively. These vectors interact in a local scale. The most important of these interactions are those between spins ($\boldsymbol{s}_i$ and $\boldsymbol{s}_j$), and those between orbitals ($\boldsymbol{l}_i$ and $\boldsymbol{l}_j$). As a result, the spins of all the electrons are aligned by the strong spin–spin interactions, thus forming a *resultant atomic spin angular momentum*

$$\boldsymbol{S} = \sum_i \boldsymbol{s}_i. \tag{3.27}$$

Similarly, the orbitals of each electron are aligned by the strong orbit–orbit interactions, thus forming a *resultant atomic orbital angular momentum*

$$\boldsymbol{L} = \sum_i \boldsymbol{l}_i. \tag{3.28}$$

Table 3.1. Electronic configuration of elements.

	Elements		Levels and number of states																						
			K	L		M			N				O					P						Q	
			2	8		18			32				50					72						—	
			$1s$	$2s$	$2p$	$3s$	$3p$	$3d$	$4s$	$4p$	$4d$	$4f$	$5s$	$5p$	$5d$	$5f$	$5g$	$6s$	$6p$	$6d$	$6f$	$6g$	$6h$	$7s \cdots$	Ground
			2	2	6	2	6	10	2	6	10	14	2	6	10	14	18	2	6	10	14	18	22	2	terms
	1	H	1																						$^2S_{1/2}$
	2	He	2																						$^1S_0$
	3	Li	2	1																					$^2S_{1/2}$
	⋮		⋮	⋮																					⋮
	10	Ne	2	2	6																				$^1S_0$
	11	Na	2	2	6	1																			$^2S_{1/2}$
	⋮		⋮	⋮	⋮	⋮																			⋮
	18	Ar	2	2	6	2	6																		$^1S_0$
3d transition elements	19	K	2	2	6	2	6		1																$^2S_{1/2}$
	20	Ca	2	2	6	2	6		2																$^1S_0$
	21	Sc	2	2	6	2	6	1	2																$^2D_{3/2}$
	22	Ti	2	2	6	2	6	2	2																$^3F_2$
	23	V	2	2	6	2	6	3	2																$^4F_{3/2}$
	24	Cr	2	2	6	2	6	5	1																$^7S_3$
	25	Mn	2	2	6	2	6	5	2																$^6S_{5/2}$
	26	Fe	2	2	6	2	6	6	2																$^5D_4$
	27	Co	2	2	6	2	6	7	2																$^4F_{9/2}$
	28	Ni	2	2	6	2	6	8	2																$^3F_4$
	29	Cu	2	2	6	2	6	10	1																$^2S_{1/2}$
	30	Zn	2	2	6	2	6	10	2																$^1S_0$
	⋮		⋮	⋮	⋮	⋮	⋮	⋮	⋮																⋮
	36	Kr	2	2	6	2	6	10	2	6															$^1S_0$
4d transition elements	37	Rb	2	2	6	2	6	10	2	6			1												$^2S_{1/2}$
	38	Sr	2	2	6	2	6	10	2	6			2												$^1S_0$
	39	Y	2	2	6	2	6	10	2	6	1		2												$^2D_{3/2}$
	40	Zr	2	2	6	2	6	10	2	6	2		2												$^3F_2$
	41	Nb	2	2	6	2	6	10	2	6	4		1												$^6D_{1/2}$
	42	Mo	2	2	6	2	6	10	2	6	5		1												$^7S_3$
	43	Tc	2	2	6	2	6	10	2	6	5		2												$^6S_{5/2}$
	44	Ru	2	2	6	2	6	10	2	6	7		1												$^5F_5$
	45	Rh	2	2	6	2	6	10	2	6	8		1												$^4F_{9/2}$
	46	Pd	2	2	6	2	6	10	2	6	10		0												$^1S_0$
	47	Ag	2	2	6	2	6	10	2	6	10		1												$^2S_{1/2}$
	⋮		⋮	⋮	⋮	⋮	⋮	⋮	⋮	⋮	⋮		⋮												⋮
	54	Xe	2	2	6	2	6	10	2	6	10		2	6											$^1S_0$

Table 3.1. (*contd.*)

Elements			K 2	L 8		M 18			N 32				O 50					P 72						Q —	Ground terms
			$1s$ 2	$2s$ 2	$2p$ 6	$3s$ 2	$3p$ 6	$3d$ 10	$4s$ 2	$4p$ 6	$4d$ 10	$4f$ 14	$5s$ 2	$5p$ 6	$5d$ 10	$5f$ 14	$5g$ 18	$6s$ 2	$6p$ 6	$6d$ 10	$6f$ 14	$6g$ 18	$6h$ 22	$7s \cdots$ 2	
	55	Cs	2	2	6	2	6	10	2	6	10		2	6				1							$^2S_{1/2}$
	56	Ba	2	2	6	2	6	10	2	6	10		2	6				2							$^1S_0$
Rare earth elements	57	La	2	2	6	2	6	10	2	6	10		2	6	1			2							$^2D_{3/2}$
Rare earth elements	58	Ce	2	2	6	2	6	10	2	6	10	1	2	6	1			2							$^3H_4$
Rare earth elements	59	Pr	2	2	6	2	6	10	2	6	10	(3)	2	6	(0)			2							
Rare earth elements	60	Nd	2	2	6	2	6	10	2	6	10	4	2	6	0			2							$^5I_4$
Rare earth elements	61	Pm	2	2	6	2	6	10	2	6	10	(5)	2	6	(0)			2							
Rare earth elements	62	Sm	2	2	6	2	6	10	2	6	10	6	2	6	0			2							$^7F_4$
Rare earth elements	63	Eu	2	2	6	2	6	10	2	6	10	7	2	6	0			2							$^8S_{7/2}$
Rare earth elements	64	Gd	2	2	6	2	6	10	2	6	10	7	2	6	1			2							$^9D_2$
Rare earth elements	65	Tb	2	2	6	2	6	10	2	6	10	8	2	6	1			2							$^8H_{17/2}$
Rare earth elements	66	Dy	2	2	6	2	6	10	2	6	10	(9)	2	6	(1)			2							
Rare earth elements	67	Ho	2	2	6	2	6	10	2	6	10	(10)	2	6	(1)			2							
Rare earth elements	68	Er	2	2	6	2	6	10	2	6	10	(11)	2	6	(1)			2							
Rare earth elements	69	Tm	2	2	6	2	6	10	2	6	10	13	2	6	0			2							$^2F_{7/2}$
Rare earth elements	70	Yb	2	2	6	2	6	10	2	6	10	14	2	6	0			2							$^1S_0$
	71	Lu	2	2	6	2	6	10	2	6	10	14	2	6	1			2							$^2D_{3/2}$
5d transition elements	72	Hf	2	2	6	2	6	10	2	6	10	14	2	6	2			2							$^3F_2$
5d transition elements	73	Ta	2	2	6	2	6	10	2	6	10	14	2	6	3			2							$^4F_{3/2}$
5d transition elements	74	W	2	2	6	2	6	10	2	6	10	14	2	6	4			2							$^5D_0$
5d transition elements	75	Re	2	2	6	2	6	10	2	6	10	14	2	6	5			2							$^6S_{5/2}$
5d transition elements	76	Os	2	2	6	2	6	10	2	6	10	14	2	6	6			2							$^5D_4$
5d transition elements	77	Ir	2	2	6	2	6	10	2	6	10	14	2	6	7			2							$^4F_{9/2}$
5d transition elements	78	Pt	2	2	6	2	6	10	2	6	10	14	2	6	9			1							$^3D_3$
5d transition elements	79	Au	2	2	6	2	6	10	2	6	10	14	2	6	10			1							$^2S_{1/2}$
5d transition elements	80	Hg	2	2	6	2	6	10	2	6	10	14	2	6	10			2							$^1S_0$
	⋮		⋮	⋮	⋮	⋮	⋮	⋮	⋮	⋮	⋮	⋮	⋮	⋮	⋮			⋮							⋮
	86	Rn	2	2	6	2	6	10	2	6	10	14	2	6	10			2	6						$^1S_0$
	87	Fr	2	2	6	2	6	10	2	6	10	14	2	6	10			2	6					1	$^2S_{1/2}$
	⋮		⋮	⋮	⋮	⋮	⋮	⋮	⋮	⋮	⋮	⋮	⋮	⋮	⋮			⋮	⋮					⋮	
	102	No	2	2	6	2	6	10	2	6	10	14	2	6	10	(13)		2	6	(1)				2	

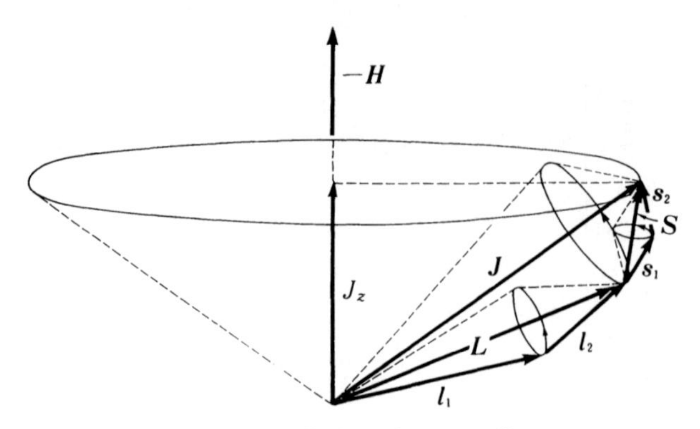

Fig. 3.7. Russell–Saunders coupling.

Now the senses of these resultant vectors $\boldsymbol{S}$ and $\boldsymbol{L}$ are governed by the so-called *spin–orbit interaction* energy

$$w = \lambda \boldsymbol{L} \cdot \boldsymbol{S}, \tag{3.29}$$

thus forming the total resultant angular momentum

$$\boldsymbol{J} = \boldsymbol{L} + \boldsymbol{S}. \tag{3.30}$$

This interaction is called the *Russell–Saunders interaction*[4] (see Fig. 3.7).

When a number of electrons exist in an atom, the arrangement of these vectors are governed by *Hund's rule*[5], which is described as follows:

(1) The spins $\boldsymbol{s}_i$ are arranged so as to form a resultant spin $\boldsymbol{S}$ as large as possible within the restriction of the Pauli exclusion principle. The reason is that the electrons tend to take different orbits owing to the Coulomb repulsion, and also the intra-atomic spin–spin interaction tends to align these spins parallel to each other. For instance, in the case of the $4f$ shell, which has the capacity of accepting 14 electrons, the electrons occupy states in the order of the numbers in Fig. 3.8. For example, when the $4f$ shell has 5 electrons, they occupy the states with a positive spin, thus forming $S = \frac{5}{2}$. When the number of electrons is 9, 7 electrons have positive spin, while the remaining 2 have negative spin, so that the resultant spin becomes $S = \frac{7}{2} - \frac{2}{2} = \frac{5}{2}$.

(2) The orbital vectors $\boldsymbol{l}_i$ of each electron are arranged so as to produce the maximum resultant orbital angular momentum $\boldsymbol{L}$ within the restriction of the Pauli exclusion principle and also of condition (1). The reason is that the electrons tend to circulate about the nucleus in the same direction so as to avoid approaching one another, which would increase the Coulomb energy. In the case of the $4f$ shell, the magnetic quantum number m can take the values $3, 2, 1, 0, -1, -2, -3$. In the case of 5 electrons, the maximum resultant orbit is $L = 3 + 2 + 1 + 0 - 1 = 5$. In the case of 9 electrons, the first 7 electrons occupy the half shell with positive spins, producing no orbital moment, while the remaining 2 electrons occupy the states with $m = 3$ and 2, thus resulting in $L = 3 + 2 = 5$ (Fig. 3.8).

(3) The third rule is concerned with the coupling between L and S. When the number of electrons in the $4f$ shell, n, is less than half the maximum number, or

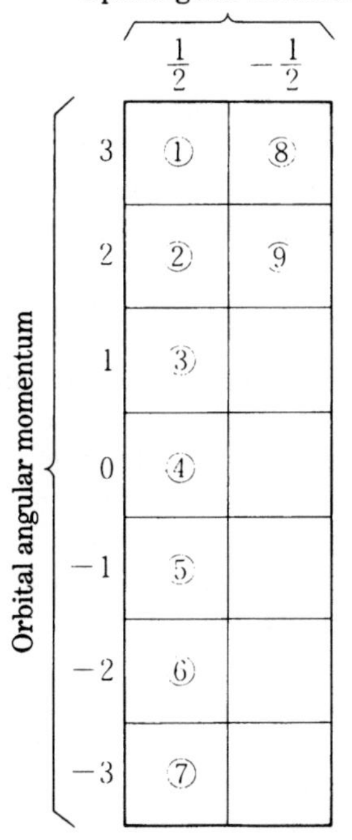

Fig. 3.8. Spin and orbital states of electrons in the $4f$ electron shell.

$n < 7$, $J = L - S$. When the shell is more than half filled, or $n > 7$, $J = L + S$. So when the number is 5, $J = 5 - \frac{5}{2} = \frac{5}{2}$, while when the number is 9, $J = 5 + \frac{5}{2} = \frac{15}{2}$. Such an interaction is based on the $s - l$ interaction of the same electron. When an electron is circulating about the nucleus (Fig. 3.9), this electron sees the nucleus circulating about itself on the orbit shown by the broken circle in the figure. As a result, the electron senses a magnetic field H pointing upwards produced by the circulating nucleus with positive electric charge. Then its spin points downwards, because the spin angular momentum is opposite to the spin magnetic moment (see (3.20)). Since the orbital angular momentum of this electron points upwards (see Fig. 3.9), it follows that l and s of a single electron are always opposite. Therefore when the number of electrons is less than half the maximum number, l and s are opposite for all the electrons, so that it follows that L and S are also opposite. However, when the

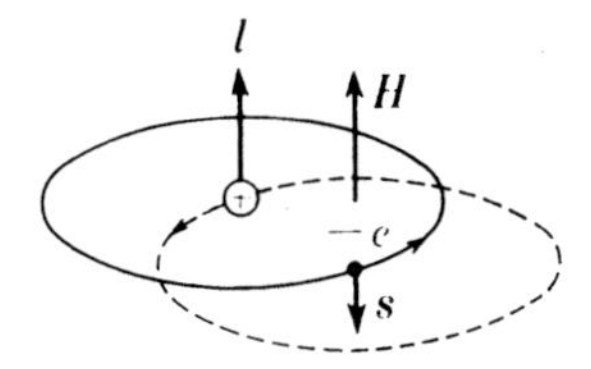

Fig. 3.9. Explanation of the l–s coupling.

number of electrons is more than half the maximum number, the orbital momentum for the 7 electrons with positive spin is zero, so that the only orbital momentum L comes from electrons with negative spin which point opposite to the resultant spin S, thus resulting in parallelism between L and S. In terms of w in (3.29), the sign of w is positive when the number of electrons $n < 7$, while it is negative when $n > 7$.

Now let us calculate the values of S, L and J for rare earth ions, which have incomplete $4f$ shells. The rare earth elements have an electronic structure expressed by

$$(4f)^n(5s)^2(5p)^6(5d)^1(6s)^2,$$

in which the incomplete $4f$ shell is well protected from outside disturbance by the outer $(5s)^2(5p)^6$ shell, so that its orbital magnetic moment is well preserved or 'unquenched' by the crystalline field. The outermost electrons $(5d)^1(6s)^2$ are easily removed from the neutral atom, thus producing trivalent ions in ionic crystals, and conduction electrons in metals or alloys. Therefore the atomic magnetic moments of rare earth elements are more or less the same in both compounds and in metals.

As mentioned above, as the number of $4f$ electrons, n, is increased, S increases linearly with n from La to Gd, which has $4f^7$, and then decreases linearly to Lu, which has $4f^{14}$. The value of L increases from 0 at La to values of 3, 5, 6; and then decreases towards 0 at Gd with $4f^7$, where a half-shell is just filled. This state and also a completely filled $4f^{14}$ state are called 'spherical'. Further increase in n results in a repetition of the same variation. The total angular momentum, J, changes as shown in Fig. 3.10, because $J = L - S$ for $n = 0$ to 7, while $J = L + S$ for $n = 7$ to 14. So J is relatively small for $n < 7$, while it is relatively large for $n > 7$.

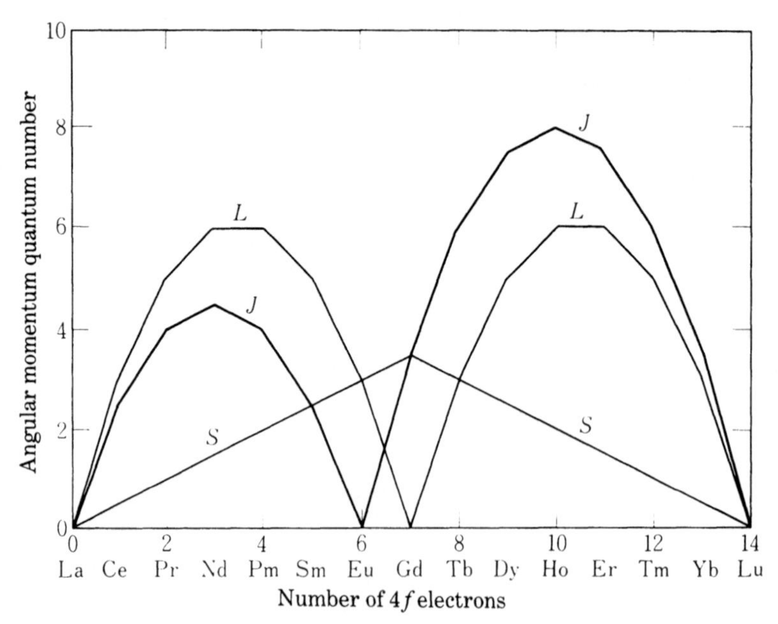

Fig. 3.10. Spin S, orbital L, and total angular momentum J as functions of the number of $4f$ electrons of trivalent rare earth ions.

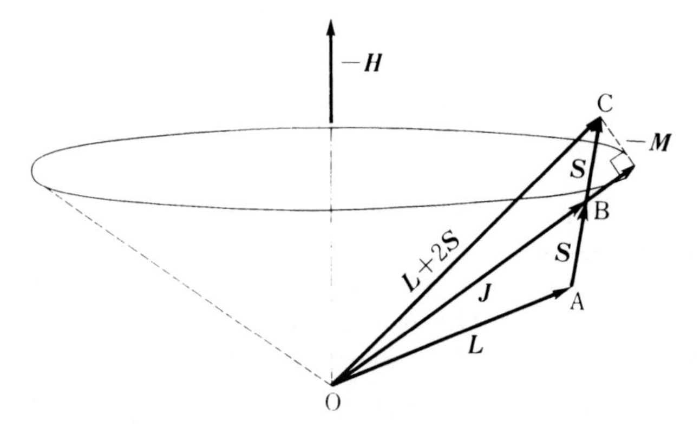

Fig. 3.11. Composition of atomic magnetic moment M.

Now let us discuss the magnetic moments associated with these angular momenta. Referring to (3.6) and (3.7), we have the orbital magnetic moment

$$\boldsymbol{M}_{\mathrm{L}} = -M_{\mathrm{B}}\boldsymbol{L}. \tag{3.31}$$

Referring to (3.19) and (3.20), we have the spin magnetic moment

$$\boldsymbol{M}_{\mathrm{S}} = -2M_{\mathrm{B}}\boldsymbol{S}. \tag{3.32}$$

Therefore, the total magnetic moment M_{R} is given by

$$\boldsymbol{M}_{\mathrm{R}} = \boldsymbol{M}_{\mathrm{L}} + \boldsymbol{M}_{\mathrm{S}} = -M_{\mathrm{B}}(\boldsymbol{L} + 2\boldsymbol{S}). \tag{3.33}$$

When vectors $\boldsymbol{L}$ and $\boldsymbol{S}$ take different orientations, the vector $\boldsymbol{L} + 2\boldsymbol{S}$ takes a direction different from $\boldsymbol{J}$ (Fig. 3.11). Since, however, $\boldsymbol{L}$ and $\boldsymbol{S}$ precess about $\boldsymbol{J}$, the vector $\boldsymbol{L} + 2\boldsymbol{S}$ also precesses about $\boldsymbol{J}$. Therefore the average magnetic moment becomes parallel to $\boldsymbol{J}$, and its magnitude is given by the projection on $\boldsymbol{J}$ or

$$M_{\mathrm{S}} = -gM_{\mathrm{B}}J. \tag{3.34}$$

The magnetic moment given by (3.34) is called the *saturation magnetic moment*. Comparing (3.34) and (3.33), and also referring to the geometrical relationship shown in Fig. 3.11, we have

$$gJ = |\boldsymbol{L} + 2\boldsymbol{S}| \cos \angle\mathrm{BOC} = J + S \cos \angle\mathrm{ABO}. \tag{3.35}$$

From the relationship between three sides and the angle $\angle$ABO in the triangle $\triangle$ABO, we have

$$L^2 = J^2 + S^2 - 2JS \cos \angle\mathrm{ABO}. \tag{3.36}$$

Eliminating $\cos \angle$ABO, we have

$$gJ = J + \frac{J^2 + S^2 - L^2}{2J}, \tag{3.37}$$

from which we obtain an expression for g:

$$g = 1 + \frac{J^2 + S^2 - L^2}{2J^2}. \tag{3.38}$$

In quantum mechanics, we must replace S^2, L^2 and J^2 by $S(S+1)$, $L(L+1)$ and $J(J+1)$, respectively.* Then we have

$$g = 1 + \frac{J(J+1) + S(S+1) - L(L+1)}{2J(J+1)}. \tag{3.39}$$

This relationship was first introduced by Landé empirically to explain the hyperfine structure of atomic spectra.[6] If $S = 0$, then $J = L$, so that it follows from (3.39) that $g = 1$, while if $L = 0$, then $J = S$, so that $g = 2$. This is exactly what we find in (3.1).

When a magnetic atom is placed in a magnetic field, J_Z can take the following discrete values as a result of spatial quantization of the vector $\boldsymbol{J}$:

$$J_z = J, J-1, J-2, \ldots, 0, \ldots, -J+2, -J+1, -J. \tag{3.40}$$

This fact affects the calculation of the statistical average of magnetization, as will be discussed in Part III. As a result, the magnitude of the atomic moment deduced from the thermal average of magnetization is given by

$$M_{\text{eff}} = gM_{\text{B}}\sqrt{J(J+1)}, \tag{3.41}$$

which is called the *effective magnetic moment*. The calculation will be shown in Chapter 5.

Figure 3.12 shows the effective magnetic moment calculated using (3.39) and (3.40) as a function of the number of $4f$ electrons. The solid curve represents the calculation based on Hund's rule. The shape of the curve is similar to that of L (see Fig. 3.10), except that the magnetic moment is much more enhanced by the g-factor for heavy rare earths than for light ones. Experimental values observed for trivalent ions

* Let us deduce this relationship for orbital angular momentum L. Let the x-, y- and z-components of L be L_x, L_y and L_z, respectively. Then

$$L^2 = L_x^2 + L_y^2 + L_z^2.$$

On the other hand, these components can be expressed in terms of the components of momentum p, or p_x, p_y and p_z, and the position coordinates x, y, and z, as

$$L_x = yp_z - zp_y, \ldots .$$

In quantum mechanics, p must be replaced by $-i\hbar(\partial/\partial q)$ (q is the positional variable), so that we have

$$L_x = -i\hbar\left(y\frac{\partial}{\partial z} - z\frac{\partial}{\partial y}\right), \ldots .$$

In polar coordinates, we can write

$$L^2 = -\hbar^2\left[\frac{1}{\sin\theta}\frac{\partial}{\partial\theta}\left(\sin\theta\frac{\partial}{\partial\theta}\right) + \frac{1}{\sin^2\theta}\frac{\partial^2}{\partial\varphi^2}\right].$$

Executing this operator to the atomic wave function given by (3.65), we have

$$L^2\psi = L(L+1)\psi.$$

(Examine this relationship for desired values of l and m in (3.65).) Thus it is concluded that the eigenvalue of L^2 is $L(L+1)$.

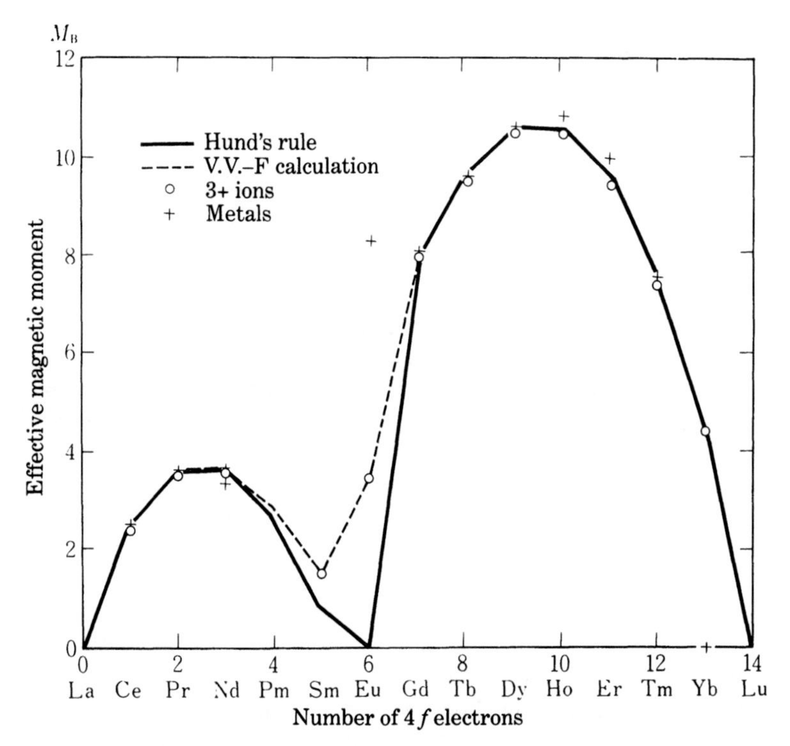

Fig. 3.12. Effective magnetic moment as a function of the number of $4f$ electrons measured for trivalent rare earth ions in compounds and rare earth metals, and comparison with the Hund and Van Vleck–Frank theories.

in the compounds are also shown as open circles in the figure. The agreement between theory and experiment is excellent, except for Sm and Eu. This discrepancy was explained by Van Vleck and Frank[7] in terms of multiplet terms of these elements. In these elements, S and L almost compensate one another in the ground state, while in the excited states, $\boldsymbol{S}$ and $\boldsymbol{L}$ make some small angle, thus producing a non-zero $\boldsymbol{J}$. The thermal excitation and the mixture of these excited states, therefore, result in an increase in the magnetic moment. The broken curve in the figure represents this correction, and reproduces the experiment well.

The experimental points for metals, shown as crosses in the figure, are also in excellent agreement with the theory, except for Eu and Yb. The reason is that these ions are divalent in metals, because there is a tendency for atoms or ions to become spherical. Therefore, Eu^{3+}, whose electronic structure is $4f^6$, tends to become Eu^{2+} or $4f^7$ by accepting an electron from the conduction band, whereas Yb^{3+}, which is $4f^{13}$, tends to become Yb^{2+} or $4f^{14}$ in the same way. These divalent ions have the same electronic structures as Gd^{3+} and Lu^{3+}, respectively, thus showing the same effective magnetic moments, as seen in the figure.

The electronic structures discussed in this section are often expressed in spectroscopic notation such as ${}^2S_{1/2}$, ${}^5D_0$, or ${}^4F_{9/2}$, in which $S, P, D, F, G, \ldots$ signify that

$L = 0, 1, 2, 3, 4$, respectively. A prefix to the capital letter represents $2S + 1$, and a suffix represents J. Electronic configuration and spectroscopic ground terms are given in Table 3.1.

3.3 GYROMAGNETIC EFFECT AND FERROMAGNETIC RESONANCE

As noted in the preceding section, there are two possible origins for magnetism in materials; spin and orbital magnetic moments. Many attempts* have been made to measure the g-factors of various magnetic materials in order to determine the contribution of these possible origins. The first such experiment was done by Maxwell.[8] This measurement was based on the simple idea: if a bar magnet supported horizontally at the center on pivots is rotated about the vertical axis, it is expected to tilt from the horizontal plane if it has an angular momentum associated with its magnetization. This experiment was, however, unsuccessful, because the effect is extremely small. In 1915, Barnett[9] first succeeded in determining the g-factor by comparing the magnetization of two identical magnetic rods, one rotating about its long axis and the other magnetized by an external magnetic field applied along its long axis. This experiment was based on the idea that if the magnetic atoms in the bar have an angular momentum, the rotation of the bar should drive this momentum towards the axis of rotation. This relationship between rotation and magnetization is called the *gyromagnetic effect*.

The most successful measurement technique is known as the *Einstein–de Haas effect*,[10] which was developed more precisely by Scott.[11] The principle is illustrated in Fig. 3.13. The specimen is suspended by a thin elastic fiber and is magnetized by a vertical field which reverses direction at a frequency corresponding to the natural frequency of mechanical oscillation of the system. Consider the moment when the field (and therefore the magnetization) change direction from upward to downward. Then the associated angular momentum must also change. Since this system is isolated mechanically, the total angular momentum must be conserved. Therefore the crystal lattice must rotate so as to compensate for the change in angular momentum of the magnetic atoms. The mechanism by which the magnetic atoms transfer their angular momentum to the crystal lattice will be discussed later in this section. In the actual experiment done by Scott, the natural period of oscillation was 26 seconds. The rate of decay of the amplitude of the oscillation was measured with and without an alternating magnetic field applied in synchronism with the mechanical oscillation. The difference gives the change in angular momentum. The magnetization was also measured, so that the gyromagnetic constant in (3.23), and accordingly the g-factor in (3.22), could be calculated. Since the values of the g-factor determined in this experiment are different from the values determined from a magnetic resonance

* The attempts are based on the gyroscopic effect: when a force is applied to change the axis of rotation of a spinning top, the axis always rotates perpendicular to the direction in which the force is applied (see Fig. 3.14).

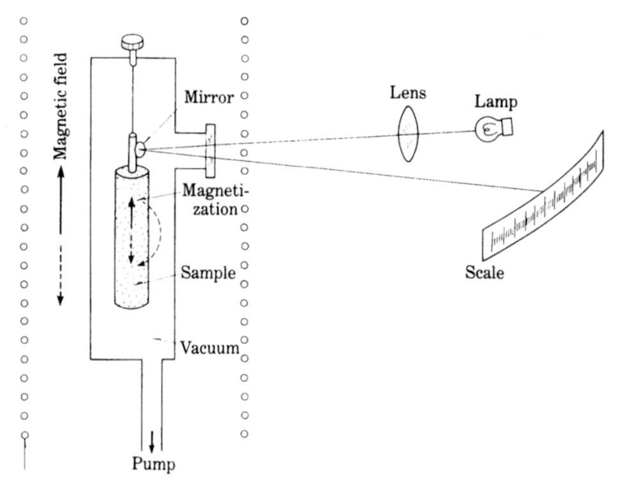

Fig. 3.13. Apparatus for observing the Einstein–de Haas effect.

experiment, as will be discussed below, the former is referred to as the *g'-factor*, while the latter is called simply the *g-factor*. Values of these two quantities are listed for various magnetic materials in Table 3.2.

The *g*-factor can also be determined by a *magnetic resonance* experiment. Suppose that the atomic magnetic moment, M, deviates from the direction of the applied magnetic field, H, by the angle θ. Then a torque

$$L = MH \sin\theta \tag{3.42}$$

will act on M. Since an angular momentum, $\boldsymbol{P}$, is associated with the magnetic moment, the direction of this vector must change by $\boldsymbol{L}$ per unit second or

$$\frac{\mathrm{d}\boldsymbol{P}}{\mathrm{d}t} = \boldsymbol{L}. \tag{3.43}$$

Table 3.2. Comparison between g- and g'-values.

Material	g	$g/(g-1)$	g'	ε (%) from g	ε (%) from g'
Fe	2.10	1.91	1.92	5	4
Co	2.21	1.83	1.85	10.5	7.5
Ni	2.21	1.83	1.84	10.5	8
FeNi	2.12	1.90	1.91	6	4.5
CoNi	2.18	1.85	1.84	9	8
Supermalloy	2.10	1.91	1.91	5	4.5
Cu_2MnAl	2.01	1.99	1.99	0.5	0.5
MnSb	2.10	1.91	1.98	5	1
$NiFe_2O_4$	2.19	1.84	1.85	9.5	7.5

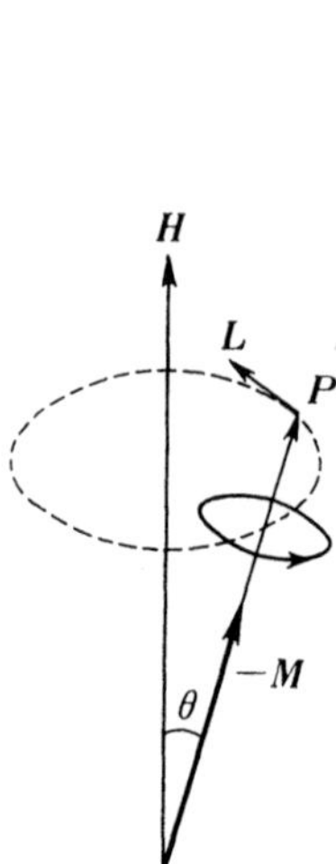
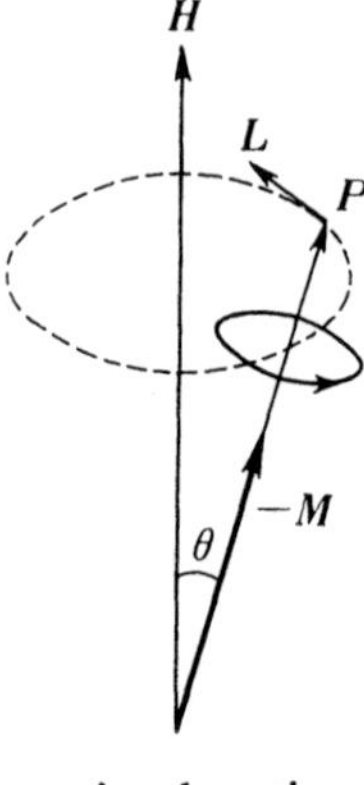
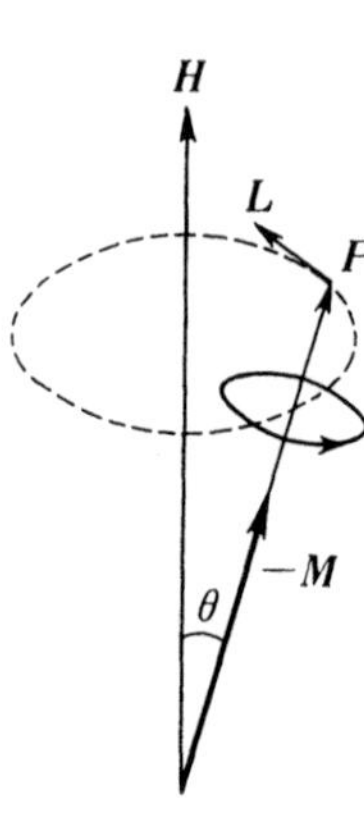

Fig. 3.14. Precessional motion of a magnetic moment.

Fig. 3.15. Ferromagnetic resonance experiment.

(This relationship is a modification of Newton's second law of motion saying that time derivative of momentum is a force. Applying this law to a rotational system, we have that the time derivative of the angular momentum must be given by a torque.) As shown in Fig. 3.14, the vector $\boldsymbol{L}$ is perpendicular to $\boldsymbol{P}$, and also to $\boldsymbol{H}$, so that $\boldsymbol{P}$ must rotate about $\boldsymbol{H}$ without changing its angle of tilt θ and its magnitude. The trace of the point of the vector $\boldsymbol{P}$ is a circle of radius $P \sin \theta$, so that the angular velocity of the precession is given by

$$\omega = \frac{L}{P \sin \theta} = \frac{MH \sin \theta}{P \sin \theta} = \frac{M}{P} H = \nu H. \tag{3.44}$$

It is interesting to note that the angular velocity given by (3.44) is independent of the angle θ. Accordingly, when a magnetic field is applied to a magnetic material, the magnetic moments of all the magnetic atoms in the material precess with the same angular frequency, ω, no matter how the magnetic moment tilts from the direction of the field. Therefore, if an alternating magnetic field of this frequency is applied perpendicular to the static magnetic field, precessional motion will be induced for all the magnetic atoms. In the actual experiment a specimen is attached to the wall of a microwave cavity, and the intensity of the magnetic field applied perpendicular to the $\boldsymbol{H}$ vector of the microwave is increased gradually (Fig. 3.15). When the intensity $\boldsymbol{H}$ of the magnetic field reaches the value which satisfies the condition (3.44), a precession is induced, so that the radio frequency (r.f.) permeability shows a sharp maximum.

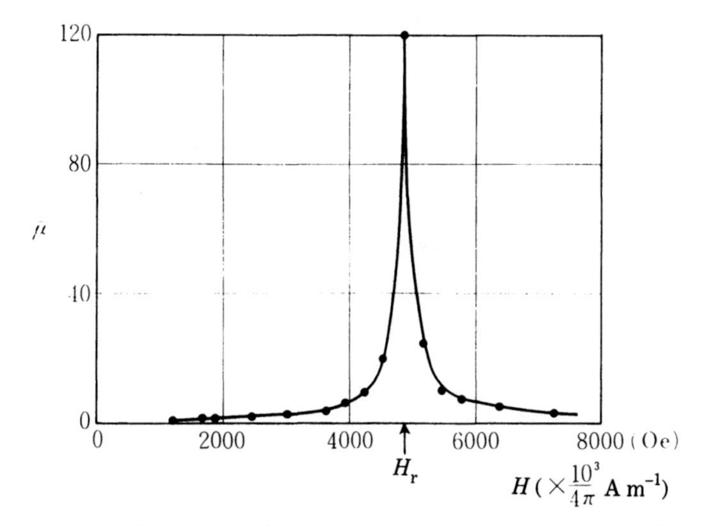

Fig. 3.16. Ferromagnetic resonance curve (H_r: resonance field).

Figure 3.16 shows experimental results for Permalloy. This phenomenon is known as *ferromagnetic resonance* (*FMR*). From the field at which resonance occurs, we can determine ν from (3.44), and accordingly the value of g from (3.22).

The precessional motion described above is somewhat idealized. The actual precession is associated with various relaxation processes by which the system loses energy. A more detailed discussion will be given in Chapter 20. In the preceding discussion of the Einstein–de Haas effect, we assumed that the magnetization reverses its sense upon the application of the magnetic field. This is, however, true only if some relaxation process absorbs energy from the precessional motion and allows the

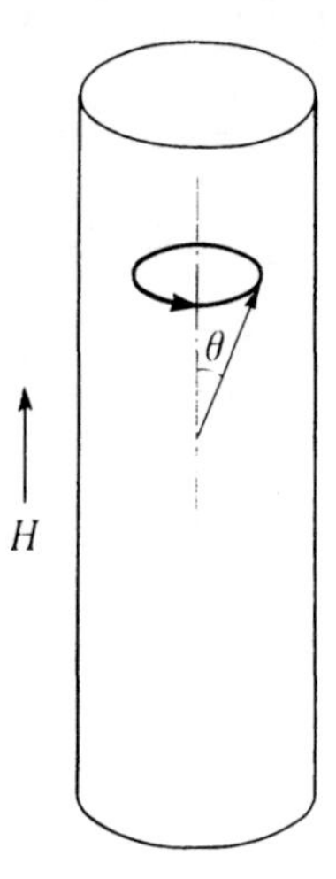

Fig. 3.17. Ferromagnetic resonance of a cylindrical ferromagnetic specimen.

magnetization to relax towards the direction of the field. The reaction of this relaxation mechanism causes the rotation of the crystal lattice of the specimen.

The ferromagnetic resonance experiment was first performed by Griffith[12] in 1946. The correct g-values were determined by Kittel[13] by taking into consideration the demagnetizing field caused by the precessional motion. For simplicity, let us consider a cylindrical specimen (Fig. 3.17). When the magnetization tilts from the cylinder axis by the angle θ, a demagnetizing field $-I_s(\sin\theta)/2\mu_0$ is produced perpendicular to the axis. The torque acting on the magnetization is, therefore, given by the sum of the torque from the external field and that from the demagnetizing field, or

$$L = -MH\sin\theta - \frac{MI_s}{2\mu_0}\cos\theta\sin\theta$$

$$\simeq -M\left(H + \frac{I_s}{2\mu_0}\right)\sin\theta \qquad (\text{because } \theta \ll 1). \tag{3.45}$$

Accordingly the resonance frequency (3.44) is modified to

$$\omega = \nu\left(H + \frac{I_s}{2\mu_0}\right). \tag{3.46}$$

If the specimen is a flat plate and the external field is applied parallel to its surface, a similar calculation shows that

$$\omega = \nu\left\{H\left(H + \frac{I_s}{\mu_0}\right)\right\}^{1/2}. \tag{3.47}$$

Equation (3.47) is obtained as follows: Set the x-axis normal to the sample surface, and the z-axis parallel to the external field. Then we have for the equations of motion

$$\begin{cases} \dfrac{dP_x}{dt} = L_x = +M_yH \\ \dfrac{dP_y}{dt} = L_y = -M_xH - M\dfrac{I_x}{\mu_0} = -M_x\left(H + \dfrac{I_s}{\mu_0}\right). \end{cases} \tag{3.48}$$

Eliminating P_y from the above equations by using (3.23), we have

$$\frac{d^2P_x}{dt^2} = \frac{dM_y}{dt}H = -\nu H\frac{dP_y}{dt}$$

$$= -\nu^2H\left(H + \frac{I_s}{\mu_0}\right)P_x. \tag{3.49}$$

Solving this equation, we have the angular frequency

$$\omega = \nu\left\{H\left(H + \frac{I_s}{\mu_0}\right)\right\}^{1/2}. \tag{3.50}$$

Using equations (3.46) and (3.47), we can calculate the gyromagnetic constant and accordingly the g-factor.

In addition to the demagnetizing field, the magnetic anisotropy also influences the resonance frequency, because it produces a restoring torque tending to hold the magnetization parallel to the easy axis (see Chapter 12).

The restoring force acting on the magnetization can be equated to a hypothetical magnetic field acting parallel to the easy axis. Let this field be H_1 when the magnetization deviates towards the x-axis, and H_2 when the magnetization deviates towards the y-axis. Then the resonance frequency is given by

$$\omega = \nu\sqrt{H_1 H_2}. \tag{3.51}$$

A high-frequency magnetic material called Ferroxplana takes advantage of this effect to increase the resonance frequency (see Section 20.3).

The g-values obtained from ferromagnetic resonance measurements are listed in Table 3.2 together with the g'-values determined from gyromagnetic experiments. Generally speaking, the values observed for $3d$ transition elements are close to 2, which tells us that the origin of the atomic magnetic moment is not orbital motion but mostly spin. We also note that the sign of the deviation from a value of 2 is different for g- and g'-values.

The deviation of the g'-factor from 2 is apparently caused by a small contribution from the orbital magnetic moment. Let the part of the saturation mangetization due to spin motion be $(I_s)_{spin}$, and that from orbital motion by $(I_s)_{orb}$. The corresponding angular momenta will be $(P_s)_{spin}$ and $(P_s)_{orb}$, respectively. Then we have

$$g' = \frac{2m}{\mu_0 e}\,\frac{(I_s)_{orb} + (I_s)_{spin}}{(P_s)_{orb} + (P_s)_{spin}}. \tag{3.52}$$

Considering that $g' = 1$ for orbital motion and $g' = 2$ for spin, (3.52) is modified to

$$\frac{1}{g'} = \frac{(I_s)_{orb} + (1/2)(I_s)_{spin}}{(I_s)_{orb} + (I_s)_{spin}}. \tag{3.53}$$

On the other hand, the g-values obtained from magnetic resonance are greater than 2, as seen in Table 3.2. If the magnetic atoms do not interact with the crystal lattice, the g-values determined from magnetic resonance should be equal to the g'-values determined from the gyromagnetic experiment for the same material. As pointed out by Kittel, Van Vleck and others,[13–16] when a magnetic atom is under the strong influence of the crystal lattice, it orbital angular momentum is not conserved, so that only spin angular momentum contributes to the expression for the g-factor:

$$\frac{1}{g} = \frac{(1/2)(I_s)_{spin}}{(I_s)_{orb} + (I_s)_{spin}}. \tag{3.54}$$

Adding both sides of (3.53) and (3.54), we have

$$\frac{1}{g} + \frac{1}{g'} = 1, \tag{3.55}$$

which can be written as

$$g' = \frac{g}{g-1}. \tag{3.56}$$

The g'-values calculated using (3.56) are listed in Table 3.2, together with the values directly determined from the gyromagnetic experiment. The values are in excellent agreement.

If we denote the ratio of the orbital contribution to the spin contribution by

$$\frac{(I_s)_{orb}}{(I_s)_{spin}} = \varepsilon, \tag{3.57}$$

we can write from (3.52)

$$g' \simeq 2(1-\varepsilon), \tag{3.58}$$

and from (3.53)

$$g \simeq 2(1+\varepsilon), \tag{3.59}$$

assuming $\varepsilon \ll 1$. The values calculated from (3.58) and (3.59) are also listed in Table 3.2. These values are all less than about 10%. The quantity g is often referred to as the *spectroscopic splitting factor*, while the g'-factor is known as the *magneto-mechanical factor*.

In addition to the ferromagnetic resonance discussed in this section, we can observe paramagnetic spin resonance, antiferromagnetic resonance, and ferromagnetic resonance (see Section 20.5). These resonances are all concerned with electron spins, so that they are called *electron spin resonance* (*ESR*).

3.4 CRYSTALLINE FIELD AND QUENCHING OF ORBITAL ANGULAR MOMENTUM

We have discussed the magnetism of atoms so far mainly in terms of Bohr's classical quantum theory. We have learned that the orbital magnetic moment is mostly quenched in materials composed of $3d$ magnetic atoms. The mechanism of quenching of the orbital moment must be known more precisely in order to understand the origin of magnetocrystalline anisotropy and magnetostriction, which will be discussed in Chapters 12–14.

Now let us consider the wave function of a hydrogen atom, in which a single electron is circulating about a proton. Since the electron is under the influence of the Coulomb field produced by a proton with electric charge $+e$ (C), the potential energy is given by

$$U(r) = -\frac{e^2}{4\pi\varepsilon_0 r} \quad \text{(J)}, \tag{3.60}$$

where r (m) is the distance between the electron and the proton, and ε_0 is the *permittivity of vacuum*

$$\varepsilon_0 = \frac{1}{c^2\mu_0} = 8.85 \times 10^{-12} \quad (\mathrm{F\,m^{-1}}). \tag{3.61}$$

The state of the orbital electron is described by the wave function $\psi(r)$, which must satisfy the Schrödinger equation

$$\mathscr{H}\psi = \varepsilon\psi, \tag{3.62}$$

where $\mathscr{H}$ is the Hamiltonian given by

$$\mathscr{H} = -\frac{\hbar^2}{2m}\Delta + U(r). \tag{3.63}$$

The first term is the operator corresponding to the kinetic energy of the electron. In classical dynamics, the kinetic energy is given by $\frac{1}{2}mv^2 = (1/2m)p^2$, where p is the momentum. In quantum mechanics, the momentum p is replaced by the operator $-\mathrm{i}\hbar(\partial/\partial s)$, so that the kinetic energy is given by

$$\frac{p^2}{2m} = \frac{1}{2m}\left(p_x^2 + p_y^2 + p_z^2\right) = \frac{-\hbar^2}{2m}\left(\frac{\partial^2}{\partial x^2} + \frac{\partial^2}{\partial y^2} + \frac{\partial^2}{\partial z^2}\right). \tag{3.64}$$

Since the operator Δ is expressed by

$$\Delta = \frac{\partial^2}{\partial x^2} + \frac{\partial^2}{\partial y^2} + \frac{\partial^2}{\partial z^2}, \tag{3.65}$$

we recognize that (3.64) is the same as the first term in (3.63).

The solution of the Schrödinger equation (3.62) is given in all textbooks on quantum mechanics. Here we simply refer to the final solution, which is

$$\psi = R_{nl}(r)\Theta_{lm}(\theta)\Phi_m(\varphi), \tag{3.66}$$

where $R_{nl}(r)$ is a function of the radial distance from the nucleus r, $\Theta_{lm}(\theta)$ is a function of the polar angle θ from the axis of quantization, and $\Phi_m(\varphi)$ is a function of the azimuthal angle φ about the axis of quantization. The suffixes n, l, and m are respectively the principal, orbital, and magnetic quantum numbers, as already explained in Section 3.1. The reason why these functions are characterized by these quantum numbers is that only functions characterized by these numbers express the stationary states, as shown for example by Fig. 3.2.

The specific forms of these functions involving the angles θ, φ are given by

$$\Theta_{lm}(\theta) = (-1)^{(m+|m|)/2}\sqrt{\frac{2l+1}{2}\frac{(l-|m|)!}{(l+|m|)!}}\,P_l^{|m|}(\cos\theta), \tag{3.67}$$

and

$$\Phi_m(\varphi) = \frac{1}{\sqrt{2\pi}}\mathrm{e}^{\mathrm{i}m\varphi}. \tag{3.68}$$

The term $P_l^{|m|}(\cos\theta)$ is the lth transposed Legendre polynomial and given for $l=0,1,2,3$ or s, p, d, f electrons

$$\left.\begin{array}{llll}
s, & l=0, & & \Theta_{00}=\dfrac{1}{\sqrt{2}} \\
p, & l=1, & m=0 & \Theta_{10}=\dfrac{\sqrt{3}}{\sqrt{2}}\cos\theta \\
 & & m=\pm1 & \Theta_{1\pm1}=\mp\dfrac{\sqrt{3}}{2}\sin\theta \\
d, & l=2, & m=0 & \Theta_{20}=\dfrac{\sqrt{5}}{2\sqrt{2}}(3\cos^2\theta-1) \\
 & & m=\pm1 & \Theta_{2\pm1}=\mp\dfrac{\sqrt{15}}{2}\sin\theta\cos\theta \\
 & & m=\pm2 & \Theta_{2\pm2}=\dfrac{\sqrt{15}}{4}\sin^2\theta \\
f, & l=3, & m=0 & \Theta_{30}=\dfrac{\sqrt{7}}{2\sqrt{2}}(2\cos^3\theta-3\sin^2\theta\cos\theta) \\
 & & m=\pm1 & \Theta_{3\pm1}=\mp\dfrac{\sqrt{21}}{4\sqrt{2}}\sin\theta(5\cos^2\theta-1) \\
 & & m=\pm2 & \Theta_{3\pm2}=\dfrac{\sqrt{105}}{4}\sin^2\theta\cos\theta \\
 & & m=\pm3 & \Theta_{3\pm3}=\mp\dfrac{\sqrt{35}}{4\sqrt{2}}\sin^3\theta.
\end{array}\right\} \quad (3.69)$$

The functions (3.69) express the eigenstates involving the azimuthal angle φ as discussed in Section 3.1. When the orbital magnetic moment remains unquenched, such wave functions are circulating about the axis of quantization, thus producing a circular current, either clockwise or counterclockwise. Accordingly, the azimuthal variations are smeared out, so that the directional distribution of the wave functions is expressed simply by rotating the functions (3.69) about the axis of quantization as shown in Fig. 3.18.

In this figure, we recognize that the wave functions for $m=0$ in d or f electrons stretch along the z-axis or the axis of quantization, while those for $m=1$ or the maximum values spread along the z-plane or the plane perpendicular to the axis of quantization. The reader may recognize a similarity between these pictures and those of the Bohr orbits as shown in Fig. 3.4. It is, however, meaningless to consider further details of this correspondence. Note that if the square of wave functions with different m are added, the sum becomes isotropic. This is also the case if the square of wave

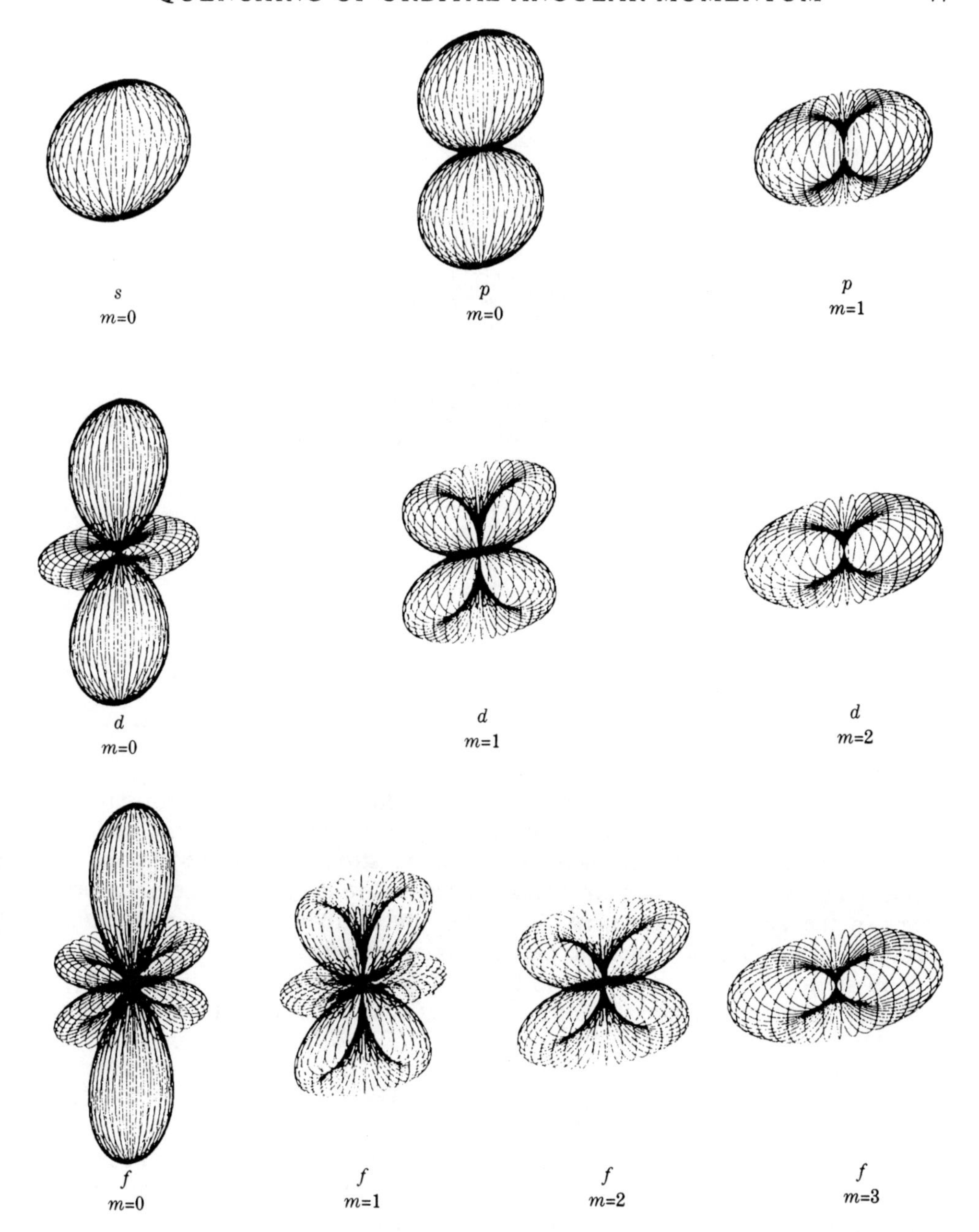

Fig. 3.18. Angular distribution of atomic wave functions with various orbital and magnetic quantum numbers.

functions with a finite positive m and a half with $m = 0$ are added. We call such an electron shell *spherical*, as already mentioned.

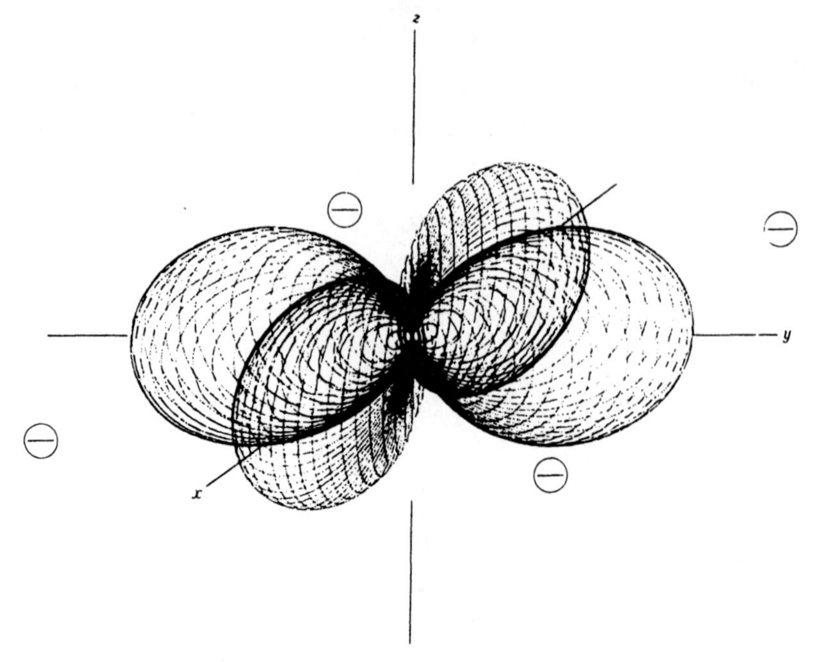

Fig. 3.19. The $3d$ atomic wave functions with $m = +2$ stabilized by neighboring anions or negatively charged ions.

When such magnetic atoms are assembled into crystals, their magnetic properties are influenced by these shapes. The magnetocrystalline anisotropy of most rare-earth ions can be interpreted in terms of the atom shapes (Chapter 12). In the case of $3d$ electron shells, the orbital magnetic moments are strongly influenced by the crystalline field, because the magnetic shells are exposed to the influence of neighboring atoms in the crystal environment. Sometimes their orbital angular momentum is totally quenched. We will elucidate the mechanism of quenching below.

Suppose that two wave functions with $m = +2$ and $m = -2$ are superposed, thus cancelling their angular momenta. Referring to (3.68), the resultant wave function is given by

$$\Phi'_2(\varphi) = \frac{1}{\sqrt{2}\sqrt{2\pi}}(e^{i2\varphi} + e^{-i2\varphi}) = \frac{1}{\sqrt{\pi}} \cos 2\varphi. \tag{3.70}$$

Figure 3.19 illustrates the angular distribution of this wave function, drawn by using Θ_{2+2} in (3.69) and (3.70). This figure represents a standing wave constructed by a superposition of the oppositely circulating wave functions.* Since this wave function spreads to avoid the anions (negatively charged ions) in the environment, as shown in the figure, the Coulomb energy will be lowered. Thus the quenched state of orbital moments is stabilized by the crystalline field.

* The fact that (3.70) is real means that the wave function is independent of time. Therefore neither angular momentum nor electric current is produced, and accordingly no orbital magnetic moment arises.

In general, when d electrons are placed in a crystalline field with cubic symmetry, the d wave function can be written in terms of the following orthogonal reai functions:

$$\left.\begin{aligned}\psi_\xi &= \sqrt{\frac{15}{16\pi}}\,\frac{2yz}{r^2}R_{32}(r)\\ \psi_\eta &= \sqrt{\frac{15}{16\pi}}\,\frac{2zx}{r^2}R_{32}(r)\\ \psi_\zeta &= \sqrt{\frac{15}{16\pi}}\,\frac{2xy}{r^2}R_{32}(r)\\ \psi_u &= \sqrt{\frac{15}{16\pi}}\,\frac{2z^2-x^2-y^2}{r^2}R_{32}(r)\\ \psi_v &= \sqrt{\frac{15}{16\pi}}\,\frac{x^2-y^2}{r^2}R_{32}(r).\end{aligned}\right\} \tag{3.71}$$

Figure 3.20 shows the angular distribution of these wave functions. As seen in the figure, the wave functions ψ_ξ, ψ_η, ψ_ζ extend along the $\langle 110\rangle$ directions, avoiding the principal axes with fourfold symmetry, while the functions ψ_u, ψ_v extend along the principal axes.* The former are called $d\varepsilon$, and the latter $d\gamma$. Since the $3d$ magnetic atoms or ions have different neighbors, or the same neighbors at different distances, along the $\langle 110\rangle$ and $\langle 100\rangle$ directions, the energy of the nearest-neighbor interaction is different for $d\varepsilon$ and $d\gamma$. The origin of this interaction is not only the Coulomb interaction, but also includes exchange interaction, covalent bonding, etc. A better name for this interaction is the *ligand field* rather than the crystalline field. If the state expressed by one of these wave functions is stabilized by a ligand field, the orbital magnetic moment will be quenched. We shall also use the ligand theory in Section 12.3 to discuss the mechanism of magnetocrystalline anisotropy.

Finally, we shall elucidate the radial part $R_{nl}(r)$ in (3.66), which is given by the general form

$$R_{nl}(x) = -\left\{\left(\frac{2Z}{na_0}\right)^3 \frac{(n-l-1)!}{2n[(n+l)!]^3}\right\}^{1/2} e^{-x/n}\left(\frac{2x}{n}\right)^l L_{n+1}^{2l+1}\left(\frac{2x}{n}\right) \tag{3.72}$$

where $x = Zr/a_0$, $a_0 = 4\pi\varepsilon_0\hbar^2/me^2$, and $L_{n+1}^{2l+1}(t)$ is the Laguerre polynomial given by

$$L_{n+l}^{2l+1}(t) = \frac{d^{2l+1}}{dt^{2l+1}}\left(e^t\frac{d^{n+l}}{dt^{n+l}}(t^{n+l}e^{-t})\right). \tag{3.73}$$

* This ψ_v is the same as that shown in Fig. 3.19, because

$$\cos 2\varphi = \cos^2\varphi - \sin^2\varphi = \left(\frac{x}{r}\right)^2 - \left(\frac{y}{r}\right)^2.$$

For a number or orbital electrons the functional form of (3.72) is given by

$$\left.\begin{aligned}
&1s; && n=1, \quad l=0; && R_{10}(x)=2\left(\frac{Z}{a_0}\right)^{3/2} e^{-x}\\
&2s; && n=2, \quad l=0; && R_{20}(x)=\frac{1}{2\sqrt{2}}\left(\frac{Z}{a_0}\right)^{3/2}(2-x)e^{-x/2}\\
&2p; && n=2, \quad l=1; && R_{21}(x)=\frac{1}{2\sqrt{6}}\left(\frac{Z}{a_0}\right)^{3/2} xe^{-x/2}\\
&3s; && n=3, \quad l=0; && R_{30}(x)=\frac{2}{81\sqrt{3}}\left(\frac{Z}{a_0}\right)^{3/2}\\
& && && \times(2x^2-18x+27)e^{-x/3}\\
&3p; && n=3, \quad l=1; && R_{31}(x)=\frac{4}{81\sqrt{6}}\left(\frac{Z}{a_0}\right)^{3/2} x(6-x)e^{-x/3}\\
&3d; && n=3, \quad l=2; && R_{32}(x)=\frac{4}{81\sqrt{30}}\left(\frac{Z}{a_0}\right)^{3/2} x^2e^{-x/3}\\
&4s; && n=4, \quad l=0; && R_{40}(x)=\frac{1}{768}\left(\frac{Z}{a_0}\right)^{3/2}(-x^3+24x^2\\
& && && -144x+192)e^{-x/4}\\
&4p; && n=4, \quad l=1; && R_{41}(x)=\frac{1}{256\sqrt{15}}\left(\frac{Z}{a_0}\right)^{3/2} x\\
& && && \times(x^2-20x+80)e^{-x/4}\\
&4d; && n=4, \quad l=2, && R_{42}(x)=\frac{1}{768\sqrt{5}}\left(\frac{Z}{a_0}\right)^{3/2} x^2\\
& && && \times(-x+24)e^{-x/4}\\
&4f; && n=4, \quad l=3; && R_{43}(x)=\frac{1}{768\sqrt{35}}\left(\frac{Z}{z_0}\right)^{3/2} x^3e^{-x/4}.
\end{aligned}\right\} \quad (3.74)$$

Figure 3.21 shows a computer-generated pattern representing the spatial distribution of the atomic wave function (3.66) calculated from (3.67), (3.68) and (3.72). In these patterns, the vertical axis is the z-axis or the axis of quantization. The magnitude of the wave function is expressed by the density of drawing, and its sign is distinguished by the direction of the lines: radial lines signify positive values and circumferential lines show negative values. The graphs shown below each pattern represent the radial variation of $R(r)$. It is seen that the s function is large at the nucleus, as already discussed with the Bohr models shown in Fig. 3.6.

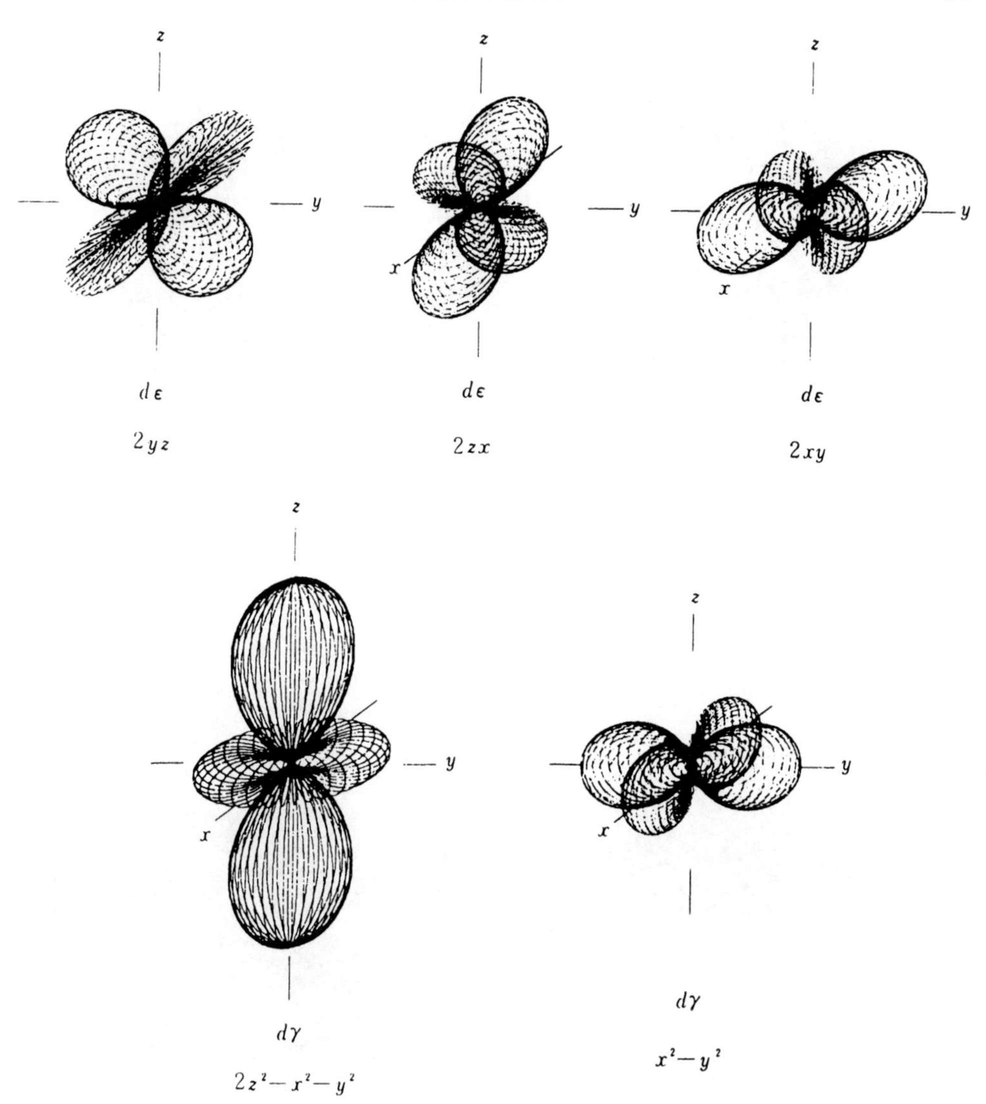

Fig. 3.20. The $d(\varepsilon)$ and $d(\gamma)$ wave functions of $3d$ electrons in a cubic ligand field.

PROBLEMS

3.1 Calculate the spectroscopic splitting factor (g-factor) for the neodymium (Nd) atom.

3.2 Assuming that a thin Permalloy rod can be magnetized to its saturation magnetization by applying a magnetic field of $100\,\mathrm{A\,m^{-1}}$ ($= 1.2\,\mathrm{Oe}$), calculate the angular velocity of the rotation of this rod about its long axis necessary to magnetize it to its saturation magnetization without applying any magnetic field. Assume that $g = 2$.

3.3 How strong a magnetic field must be applied parallel to the surface of an iron plate to

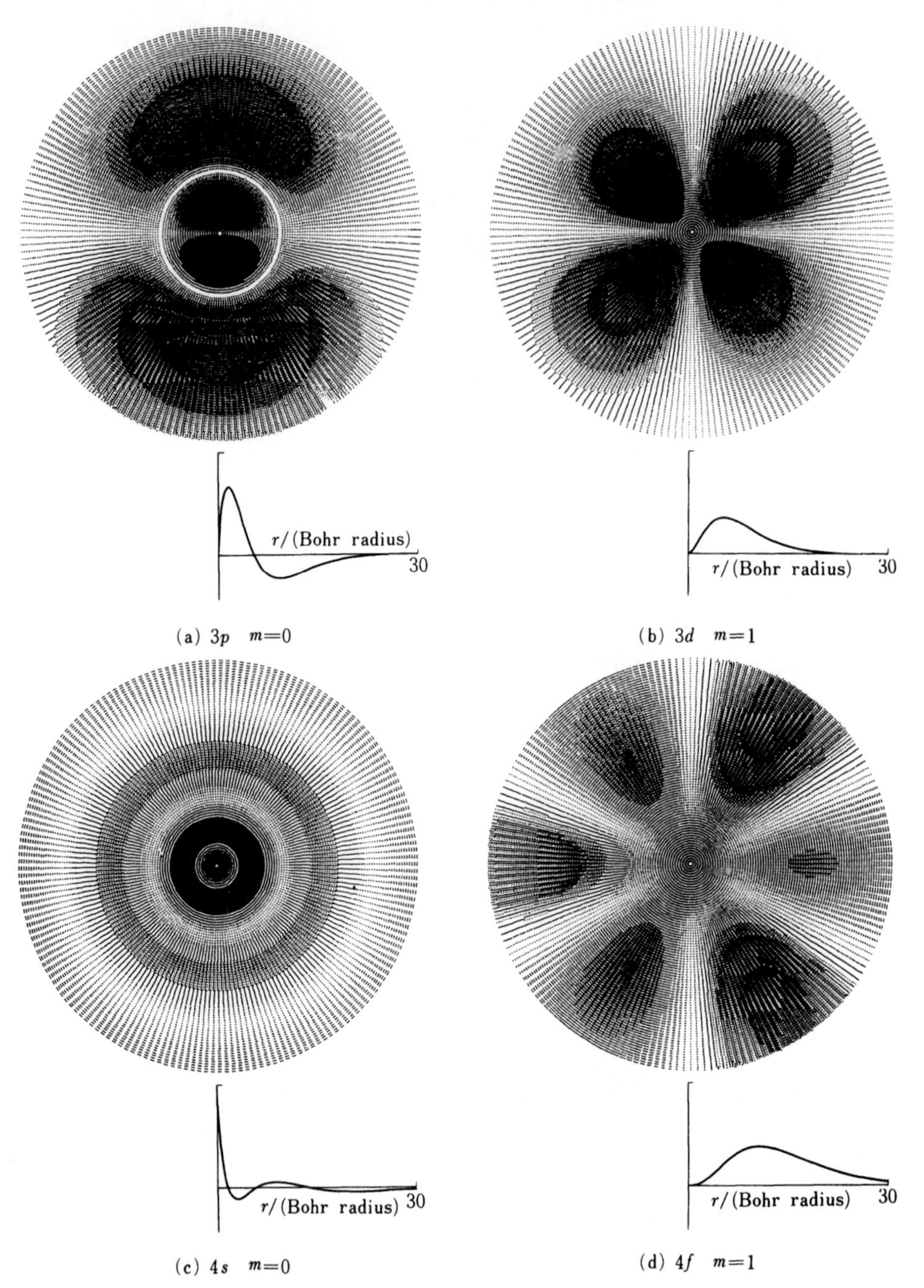

Fig. 3.21. Computer-drawn patterns for spacial distribution of atomic wave functions in a plane including the z-axis or the axis of quantization. (The magnitude of ψ is expressed by density of lines, while the sign of ψ is distinguished by the direction of drawing. Radial lines signify $\psi > 0$, while the circumferential line signifies $\psi < 0$.)

observe ferromagnetic resonance, using microwaves with a wavelength of 3.0 cm? Assume that $g = 2.10$ and $I_s = 2.12\,\mathrm{T}$, and ignore the effect of magnetocrystalline anisotropy.

3.4 Show that the $3d^5$ electron shell is spherical, by calculating the sum of the square of the wave functions given by (3.69) for $m = 2, 1, 0, -1$, and -2.

REFERENCES

1. G. E. Uhlenbeck and S. Goudsmit, *Die Naturwissenschaften*, **13** (1925), 953.
2. P. A. M. Dirac, *Proc. Roy. Soc.*, **A117** (1928), 610; **A118** (1928), 351.
3. W. Pauli, *Z. Physik*, **31** (1925), 765.
4. H. N. Russell and F. A. Saunders, *Astrophy. J.* **61** (1925), 38.
5. F. Hund, *Linien Spektren und periodisches System der Elemente* (Julius Springer, Berlin, 1927).
6. A. Landé, *Z. Physik*, **15** (1923), 189.
7. J. H. Van Vleck, *Theory of electric and magnetic susceptibilities* (Clarendon Press, Oxford, 1932), p. 245.
8. J. C. Maxwell, *Electricity and magnetism* (Dover, New York, 1954), p. 575.
9. S. J. Barnett, *Phys. Rev.*, **6**, 171 (1915), 239.
10. A. Einstein and W. J. de Haas, *Verhandl. Deut. Physik. Ges.*, **17** (1915), 152; A. Einstein, *Verhandl. Deut. Physik. Ges.*, **18** (1916), 173; W. J. de Haas, *Verhandl. Deut. Physik. Ges.*, **18** (1916), 423.
11. G. G. Scott, *Phys. Rev.*, **82** (1951), 542; *Proc. Int. Conf. Mag. and Cryst.*, Kyoto (*J. Phys. Soc. Japan*, **17**, Suppl. B-1) (1962), 372.
12. J. H. E. Griffith, *Nature*, **158** (1946), 670.
13. C. Kittel, *Phys. Rev.*, **71** (1947), 270; **73** (1948), 155; **76** (1949), 743.
14. D. Polder, *Phys. Rev.*, **73** (1948), 1116.
15. J. H. Van Vleck, *Phys. Rev.*, **78** (1950), 266.
16. C. Kittel and A. H. Mitchell, *Phys. Rev.*, **101** (1956), 1611.

4
MICROSCOPIC EXPERIMENTAL TECHNIQUES

4.1 NUCLEAR MAGNETIC MOMENTS AND RELATED EXPERIMENTAL TECHNIQUES

As discussed in Chapter 3, atomic magnetic moments originate from orbital or spin magnetic moments of electrons in unclosed electron shells. In addition to these moments, atomic nuclei possess small but non-zero magnetic moments. These moments are measured in *nuclear magnetons*. One nuclear magneton is given by

$$M_N = \frac{\mu_0 e\hbar}{2m_p} = 6.33 \times 10^{-33} \qquad (\text{Wb m}), \tag{4.1}$$

where m_p is the mass of the proton. Comparing (4.1) with the Bohr magneton given by (3.7), we see that the electron mass in the denominator is replaced by the proton mass, so that the nuclear magneton is smaller than the Bohr magneton by a factor which is the ratio of the electron mass to the proton mass, or $1/1836$. Therefore, nuclear magnetic moments make a negligibly small contribution to the magnetization of materials.

The spin angular momentum of a nucleus is measured in units of $\hbar$ as in the case of the electron spin, and is denoted by the symbol I. The nuclear spin and the nuclear magnetic moment for various isotopes are given in Table 4.1, from which we see that the magnetic moment is not necessarily proportional to the spin and sometimes even has opposite sign. Accordingly, the g-value defined by

$$g = \frac{M/M_N}{I/\hbar} \tag{4.2}$$

takes various values ranging from 0.1 to 5.6.

The nuclear magnetic moment is also accompanied by a quadrupole moment. The *quadrupole moment* of the nucleus is given by

$$eQ = \iiint \rho(3z^2 - r^2)\, \mathrm{d}v, \tag{4.3}$$

where ρ is the electric charge density, r the radial vector of the charge, and z the coordinate axis taken parallel to the nuclear spin. If the charge distribution has spherical symmetry it follows that $r^2 = x^2 + y^2 + z^2 = 3z^2$, so that $Q = 0$ as seen from (4.3). When $Q > 0$, the charge distribution of the nucleus stretches along the z-axis, so

Table 4.1. Various physical quantities of typical isotopes.

Element	Mass number A	Abundance[1] %	Spin, $I/\hbar$	Magnetic moment, M/M_N	g	Quadrupole moment, $Q/10^{-24}\,\mathrm{cm}^2$	Resonance frequency (MHz) for $H = 10\,\mathrm{kOe}$ ($= 0.7958\,\mathrm{MA\,m^{-1}}$)
n (neutron)	1	—	1/2	−1.91314	−3.82628	0	29.1658
H	1	99.985	1/2	+2.79277	+5.58554	0	42.5758
	2	0.015	1	+0.857406	+0.857406	+0.00282	6.5356
C	13	1.107	1/2	+0.702381	+1.404762	0	10.7078
N	14	99.273	1	+0.40361	+0.40361	+0.016	3.0765
O	17	0.0745	5/2	−1.89370	−0.757480	−0.0265	5.7739
F	19	100	1/2	+2.6287	+5.2574	0	40.0745
Al	27	100	5/2	+3.64140	+1.45656	0.146	11.1026
P	31	100	1/2	+1.13166	+2.26332	0	17.2522
Cl	35	75.529	3/2	+0.82183	+0.54789	−0.080	4.1763
V	51	99.76	7/2	+5.148	+1.4709	−0.052	11.212
Mn	55	100	5/2	+3.4678	+1.3871	+0.35	10.573
Fe	57	2.21	1/2	+0.0902	+0.1804	0	1.375
Fe* (14.4 ke V)	57	—	3/2	−0.1546	−0.1031	+0.300	0.7856
Co	59	100	7/2	+4.583	+1.3094	+0.404	9.981
Ni	61	1.25	3/2	−0.74868	−0.49912	0.134	3.8045
Ni* (67.4 ke V)	61	—	5/2	±0.3	±0.12		0.91
Cu	63	69.12	3/2	+2.2261	+1.4841	−0.24	11.312
	65	30.88	3/2	+2.3849	+1.5899	−0.22	12.119
Br	79	50.537	3/2	+2.1056	+1.4037	+0.33	10.670
	81	49.463	3/2	+2.2696	+1.5131	+0.28	11.534
Rh	103	100	1/2	−0.0883	−0.1766	0	1.346
Pd	105	22.6	5/2	−0.615	−0.246	+0.8	1.88
Ag	107	51.35	1/2	−0.113548	−0.227096	0	1.7310
	109	48.65	1/2	−0.130538	−0.261076	0	1.9953
Sb	121	57.25	5/2	+3.3590	+1.3436	−0.26	10.242
	123	42.75	7/2	+2.547	+0.7277	−0.68	5.547
I	127	100	5/2	+2.8091	+1.1236	−0.78	8.565
La	139	99.911	7/2	−2.7781	+0.7937	+0.21	6.050
Pr	141	100	5/2	+4.3	+1.72	−0.059	13.1
Nd	143	12.14	7/2	−1.064	−0.304	−0.482	2.317
	145	8.29	7/2	−0.653	−0.1866	−0.255	1.422
Sm	147	14.87	7/2	−0.80	−0.23	+1.9	1.75
	149	13.82	7/2	−0.65	−0.186	+0.060	1.42
Eu	151	47.86	5/2	+3.465	+1.386	+1.16	10.53
	153	52.14	5/2	+1.52	+0.608	+2.92	4.63
Gd	155	15.1	3/2	−0.242	−0.161	+1.1	1.84
	157	15.7	3/2	−0.323	−0.215	+1.0	1.64
Tb	159	100	3/2	+1.994	+1.329	+1.32	10.13
Dy	161	18.88	5/2	−0.47	−0.188	+2.36	1.43
	163	24.97	5/2	+0.65	+0.26	+2.46	1.98
Ho	165	100	7/2	+4.0	+1.14	+2.82	8.7
Er	167	22.94	7/2	−0.5647	−0.16134	+2.827	1.2300
Tm	169	100	1/2	−0.231	−0.462	0	3.52
Yb	171	14.4	1/2	+0.492	+0.984		7.50
	173	16.2	5/2	−0.678	−0.271	+3.1	2.06
Lu	175	97.412	7/2	+2.23	+0.637	+5.68	4.86
Ir	191	38.5	3/2	+0.16	+0.107	+1.5	0.82
	193	61.5	3/2	+0.17	+0.113	+1.5	0.86
Pt	195	33.8	1/2	+0.60602	+1.21204	0	9.239
Au	197	100	3/2	+1.4485	+0.09657	+0.58	0.7361
Bi	209	100	9/2	+4.0802	+0.90671	−0.34	6.9114

that the shape of the nucleus is a prolate ellipsoid. When $Q < 0$, the charge distribution of the nucleus spreads along the x–y plane, so that the shape of the nucleus is a flat oblate ellipsoid (see Fig. 4.1).

If the distribution of atoms surrounding the nucleus deviates from cubic symmetry,

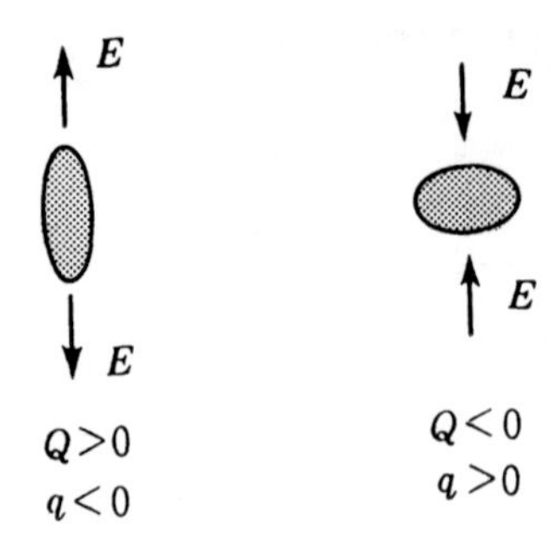

Fig. 4.1. Quadrupole moment of the nucleus and the field gradient at the nucleus.

the electric field changes from place to place; this produces a gradient of the electric field $\partial E/\partial z_0$ along some specific crystal axis z_0 at the nucleus. Since the electric field along the z_0-axis is given by $E = -\partial\phi/\partial z_0$, the negative field gradient is given by

$$-\frac{\partial E}{\partial z_0} = \frac{\partial^2\phi}{\partial z_0^2} = eq, \tag{4.4}$$

where q is the field gradient measured in units of $-e$.

If the nucleus is prolate, or $Q > 0$ and $q < 0$, the spin axis z tends to be parallel to the specific crystal axis z_0, so that the nuclear spin is forced to rotate towards the z_0-axis. This gives rise to a torque in addition to the torque produced by the magnetic field. On the contrary, if $q > 0$, the sign of the torque is opposite. When $Q < 0$, everything is reversed. The magnetude of this interaction is given by the energy e^2qQ. The value of Q is given in Table 4.1 for various isotopes. The sign and magnitude of q are determined by the atomic arrangement in the environment of the nucleus under consideration.

When a nucleus having a nuclear magnetic moment as well as a quadrupole moment is placed in a material, there are various interactions with the environment. The experimental techniques described in the following paragraphs can provide valuable information both about the identity of the nuclei existing in the materials and about the interactions between the nuclei and their environment.

4.1.1 Nuclear magnetic resonance (NMR)

When the nucleus is placed in a magnetic field H, the nuclear spin precesses about the axis of H, as in the case of the electron spin. The resonance frequency can be determined by a method analogous to that used in electron spin resonance (ESR). The resonance phenomenon and the experimental technique are known as *nuclear magnetic resonance* (*NMR*). Since in NMR the size of the magnetic moment is smaller by a factor of 1/1836, while the magnitude of spin angular momentum is more or less the same, the resonance frequency for NMR is much smaller than that for ESR. The resonance angular frequency for NMR is given by

$$\omega = g\frac{M_N}{\hbar}H. \tag{4.5}$$

Putting $H = 0.8\,\mathrm{MA\,m^{-1}}$ ($= 10\,000\,\mathrm{Oe}$), and $g = 1$, we have

$$f = \frac{(6.33 \times 10^{-33}) \times (0.8 \times 10^{6})}{2\pi \times (1.054 \times 10^{-34})} = 7.7\,\mathrm{MHz}, \tag{4.6}$$

which is in the radio-frequency range. The experimental procedure for NMR is, therefore, much simpler than that for ESR. Instead of using microwave guide tubes (wave guides), we can simply place a specimen in a static magnetic field, and surround it with a coil carrying an AC current at a frequency of a few megaherz.

The resonance frequency is quite different from isotope to isotope, because the g-factor varies with isotope. It is easy, therefore, to identify a specific nucleus just by observing the resonance frequency. If the nucleus is under the influence of an internal field, such as that produced by the polarization of magnetic shells, magnetic resonance occurs when the sum of the external field and the internal field satisfies condition (4.5). Therefore NMR provides information not only about the chemical species of the nucleus but also on the value of the internal field, from which we can sometimes deduce the value of the atomic magnetic moment (refer to Section 4.1.4).

Moreover, the resonance frequency is also modified by the field gradient, through the quadrupole moment. If nuclei of the same species occupy inequivalent lattice sites such that the direction of the field gradient is different in different sites, each site will have a slightly different resonance frequency. We say that the resonance is 'split' into several lines. From this *quadrupole splitting*, we can deduce which lattice sites are occupied by the resonating nuclei.

In the case of ferromagnetic materials, the nucleus experiences an extraordinarily large internal field through the polarization of the magnetic shell by the spontaneous magnetization. Therefore sometimes it is possible to observe the NMR signal with no external static field. Also the NMR signal is generally greatly enhanced for ferromagnetic materials. The reason is that the radio frequency field may oscillate the spontaneous magnetization, which in turn oscillates the nuclear moment through the strong internal fields. In particular, spins located in domain walls rotate through large angles, so the high frequency field which the nucleus inside the wall feels is greatly enhanced, say by a factor of 10^5.[4]

4.1.2 Spin echo

The *spin echo* technique[5] is one means for detecting NMR signals. The procedure is as follows: First we apply a high-frequency magnetic field pulse to the specimen for some time interval. After waiting for a time interval τ (s), we apply another high-frequency pulse for a time interval twice as long as the first pulse. Then we observe a sharp high-frequency signal after a further time interval τ (s). The mechanism of this phenomenon is illustrated in Fig. 4.2. Suppose that because of the first high-frequency pulse, the nuclear spins resonate, absorb energy, and then tilt towards the plane perpendicular to the static field. During the time interval τ (s), the spins precess in this plane. Because of the spin–spin interaction, some of the spins precess faster than others, so that the orientation of spins is spread over some angle.

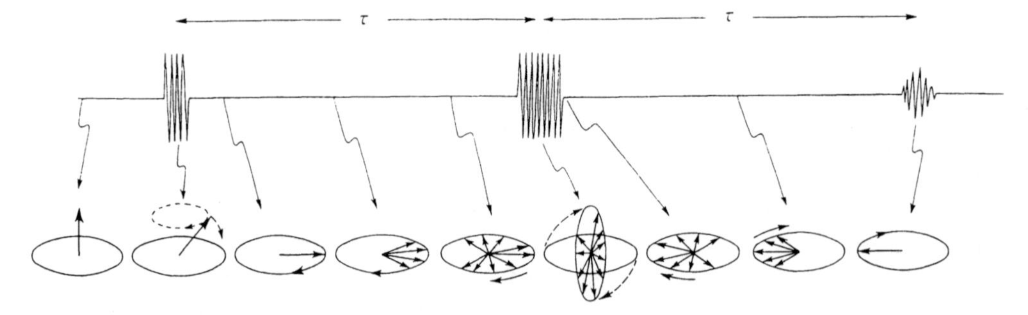

Fig. 4.2. Schematic illustration of spin echo technique.

By the application of the second high-frequency pulse with a double time-duration, the central spin tilts by 180° through the z-axis (the axis parallel to the static magnetic field). For the other spins, only the component parallel to the central spin is reversed, while the component perpendicular to the central spin remains unchanged, because this component has nothing to do with the absorption or the dissipation of the energy (see Section 20.5). Then the distribution of the spins is reversed from that before the application of the second pulse. The spins with relatively fast precessional speed lag behind the central spin, while those with relatively slow precessional speed precede the central spin, so that after a time τ (s), all the spins are precessing in phase with the central spin, thus giving rise to a sharp signal in the search coil.

The spin echo technique is useful for inhomogeneous materials, or in the case of a non-uniform static magnetic field. The normal NMR method requires homogeneity of the sample material as well as of the static magnetic field.

The spin echo method provides information not only on the internal fields, but also about the *spin–lattice relaxation* by which energy is transferred from precessional motion to the lattice, and about the *spin–spin relaxation* through which the energy is transferred between spins. The relaxation time involved in these processes is known as T_1 for the spin–lattice relaxation and as T_2 for the spin–spin relaxation. The value of T_1 can be obtained by observing the decay of the high-frequency signal associated with the rotation of the spins towards the z-axis. The value of T_2 can be obtained by observing the decay of the echo signal as a function of the time interval τ (s).

4.1.3 Mössbauer effect

The internal field for some isotopes can be measured by means of the *Mössbauer effect*. This effect utilizes the fact that a γ-ray can be emitted from a nucleus without recoil if the nucleus is bound into a solid. In this case the emitted γ-ray has an energy equal to the energy separation between the two states of the nucleus before and after the emission. This γ-ray can be absorbed by another nucleus of the same species, by selective absorption. This fact was first pointed out by Lamb[6] and was experimentally verified by Mössbauer.[7] To observe the Mössbauer effect, it is desirable to select an isotope with large mass which emits a low-energy γ-ray and to place the isotope in a

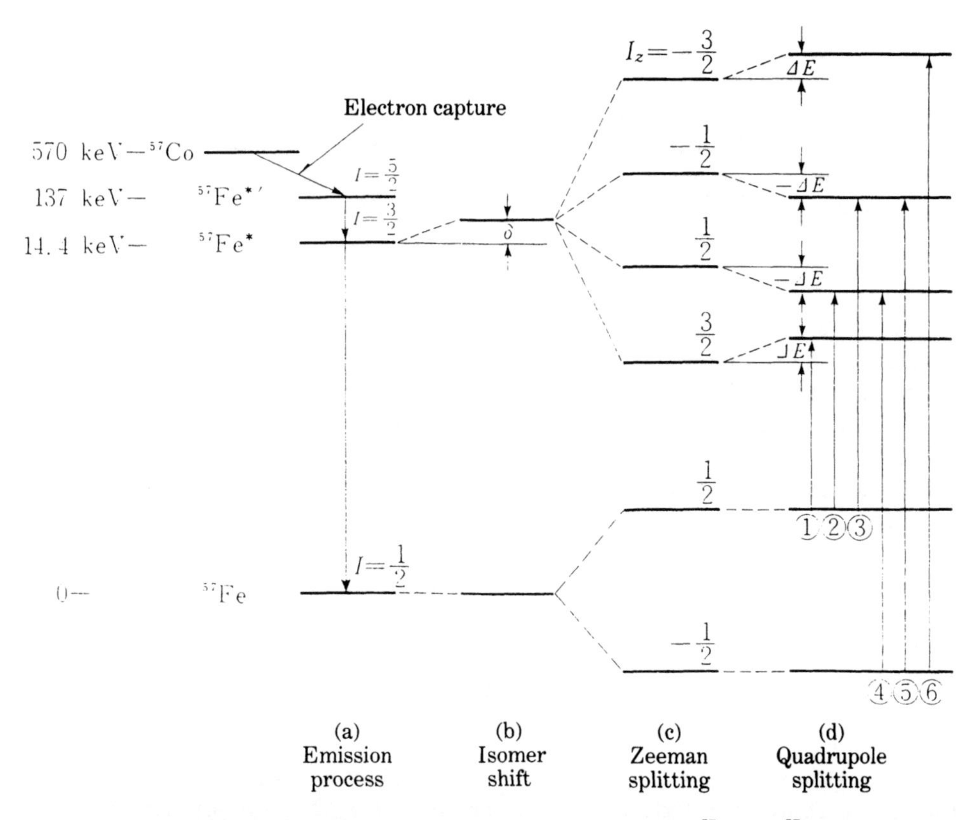

Fig. 4.3. Energe level scheme (a) of the nuclear transition of Co^{57} to Fe^{57} and energy levels (b, c, d) of the ground and excited states resulting from various interactions.

solid with a high Debye temperature so that the nucleus is firmly bound. Among various possibilities, Co^{57} (which decays to Fe^{57}) is particularly appropriate for the investigation of magnetism, because it allows detailed measurement of the internal field which acts on the nucleus of Fe^{57} in a magnetic material.

The energy scheme of the nuclear transition of Co^{57} to Fe^{57} is shown in Fig. 4.3. The isotope Co^{57}, which can be made by irradiation of Fe with 4 MeV deuterons, decays by electron capture with a half-life of 270 days to a second excited state of Fe^{57} with spin $I = \frac{5}{2}$. Then it makes a γ-transition to the first excited state with $I = \frac{3}{2}$ and finally transfers with a period of 10^{-7} s to the ground state with $I = \frac{1}{2}$. The energy separation between the last two levels is 14.4 keV.

If a nucleus of Fe^{57} exists in a magnetic material, the energy levels of the ground and the first excited are split by various effects into six levels, as shown in Fig. 4.3. One of the effects is the *isomer shift*, which is a shift of the excited level upwards by δ (Fig. 4.3). One of the reasons for this is the difference in the size of the nucleus in the ground and the excited states, which gives rise to a difference in Coulomb interaction between the nucleus and surrounding electrons. If for some reason the electron density around the nucleus is different between emitter and absorber, the difference in Coulomb interaction should result in a shift of energy levels. The isomer shift can also be caused by a temperature difference or a difference in Debye temperature between the emitter and the absorber, because the mass of the nucleus is changed during emission or absorption of the γ-ray. The valence of Fe^{57} can be estimated from the isomer shift.

The second effect is the *Zeeman splitting* by the internal magnetic field H_i which acts on the nucleus, as already mentioned in (a). In this case, the energy is changed by

$$E = -gM_N I_z H_i, \tag{4.7}$$

where I_z is the component of the nuclear spin parallel to H_i. Since $I = \frac{1}{2}$ in the ground state of Fe^{57}, the energy level is split into two levels with $I_z = \frac{1}{2}$ and $-\frac{1}{2}$. In the excited state with $I = \frac{3}{2}$, the level is split into four levels with $I_z = -\frac{3}{2}, -\frac{1}{2}, \frac{1}{2}$ and $\frac{3}{2}$ as shown in Fig. 4.3. The reason why $I_z = -\frac{1}{2}$ is lower than $I_z = \frac{1}{2}$ is that in most cases the sense of H_i is opposite to the atomic magnetic moment or the external magnetic field. On the other hand, the energy level of $I_z < 0$ is higher than $I_z > 0$ in the excited state, because the sense of I_z is reversed on excitation. The mechanism of the internal field is explained in Section 4.1.4.

The energy levels are split as shown in Fig. 4.3(a) by the quadrupole moment of Fe^{57} in the field gradient q. No quadrupole moment exists in the ground state, while the excited levels are split by

$$\Delta E = \tfrac{1}{4} e^2 qQ \frac{3\cos^2\theta - 1}{2}. \tag{4.8}$$

The sense of the splitting is positive for $I_z = \pm\frac{3}{2}$, while it is negative for $I_z = \pm\frac{1}{2}$. The sense of the splitting is independent of the sign of I_z because the shape of the nucleus is the same for $I_z > 0$ and for $I_z < 0$. From this shift we can determine the field gradient and deduce the identities of the lattice sites where Fe^{57} atoms are located. This is called the *quadrupole splitting*.

As a result of these three effects, the ground and the excited states are split into two- and fourfold levels (Fig. 4.3(d)). Between these levels there are six possible allowed transitions which satisfy the exclusion rule $\Delta I_z = 0$ or ± 1, or the rule of conservation of angular momentum.

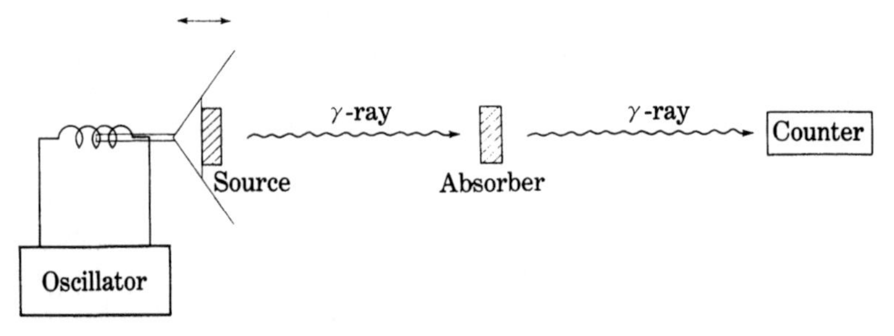

Fig. 4.4. Schematic illustration of the experimental arrangement for measurement of the Mössbauer effect.

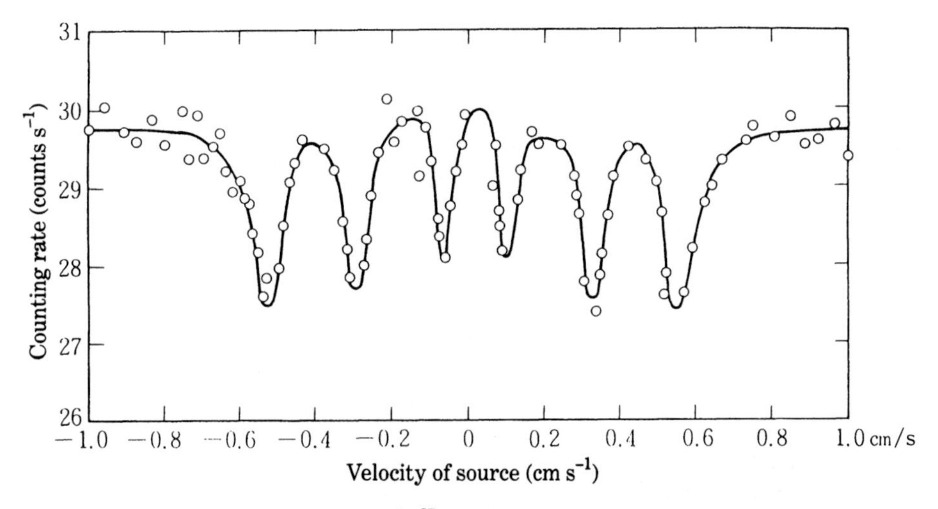

Fig. 4.5. The hyperfine spectrum of Fe^{57} in iron metal, produced by Mössbauer effect measurement using a stainless steel source and a natural iron absorber 0.025 mm thick. (After Wertheim[8])

The six transitions are observed as six absorption lines denoted by 1 to 6 in Fig. 4.3(d). The experimental procedure for observing the absorption lines is illustrated in Fig. 4.4. The γ-ray source, containing Co^{57}, is attached to a loudspeaker diaphragm. The oscillation of the speaker modulates the γ-ray wavelength through the Doppler effect. That is, when the source is moving with velocity v, the energy of the γ-ray is modified by $h\nu(v/c)$. It is therefore possible for the γ-rays emitted by Co^{57} to be absorbed by Fe^{57} in the magnetic sample material. The transmitted γ-rays are detected by a counter and recorded as a function of the velocity, which is of the order of $1\,\mathrm{cm\,s^{-1}}$. In this way we observe the six absorption lines shown in Fig. 4.5. By analyzing these absorption lines, we determine the various parameters characterizing the internal fields, the field gradient and the isomer shift. The most important of these is the internal field, which is discussed in the following section.

4.1.4 Internal fields

The values of internal fields measured by means of NMR and the Mössbauer effect are listed in Table 4.2 for various nuclei of magnetic atoms and ions. These values are generally tens of $\mathrm{MA\,m^{-1}}$ (hundreds of kOe), which is much larger than the Lorentz fields produced by these magnetic atoms or ions. Moreover, the sign of the internal field is usually negative, or opposite to the Lorentz field.

The internal field is thought to originate from the polarization of $1s$, $2s$, and $3s$ electrons.[10–11] In terms of classical physics, these s electrons can approach the nucleus (see Fig. 3.6), or in terms of quantum mechanics, the wave functions of these s electrons have high values at the nucleus (see Fig. 3.21(c)). These s electrons are

Table 4.2. The internal magnetic fields of several isotopes included in various magnetic materials (after Ishikawa[9]).

Nucleus	Host	Internal field		Temperature	Method
		(MA m^{-1})	(kOe)	(K)	
^{57}Fe	Fe	−27.3	−342	0	M
	Fe	|27.1|	|339|	0	NMR
	Co	−24.9 ± 0.4	−312 ± 5	0	M
	Ni	−22.3 ± 0.4	−280 ± 5	0	M
	Fe_3Al	−22.3	−280	78	M
		−17.6	−220	78	
	Fe_2Zr	−15.2 ± 0.8	−190 ± 10	room	M
	Fe_2Ti	< 0.8	< 10	room	M
	Fe_3N	−27.5	−345	room	M
		−17.2	−215	room	M
$^{57}Fe^{3+}$	YIG(tetra)	−36.7	−460	78	M
	YIG(tetra)	|37.3|	|468|	78	NMR
	YIG(octa)	−43.1	−540	78	M
	YIG(octa)	|43.9|	|550|	78	NMR
	α-Fe_2O_3	−41.1	−515	room	M
	γ-Fe_2O_3	−41.1 ± 1.6	−515 ± 20	85	M
	$NiFe_2O_4$	−40.7 ± 1.6	−510 ± 20	room	M
	MgO	−43.9	−550	1.3	ESR
	Fe_3O_4(tetra)	−40.7 ± 1.6	−510 ± 20	50	M
$^{57}Fe^{2+}$	Fe_3O_4(octa)	−37.1 ± 1.6	−465 ± 20	50	M
	CoO	−16.0 ± 0.8	−200 ± 10	169	M
	FeS	−25.5	−320	300	M
	$FeTiO_3$	−5.6	−70	0	M
^{59}Co	Co(fcc)	−17.36	−217.5	0	NMR
	Co(hcp)	−18.2	−228	0	NMR
	Fe	−23.1	−289	0	NMR
	Ni	−6.4	−80	0	C_v
^{61}Ni	Ni	−13.6	−170	room	M
^{119}Sn	Fe	−6.4 ± 0.3	−81 ± 4	100	M
	Co	−1.64 ± 0.12	−20.5 ± 1.5	100	M
	Ni	+1.48 ± 0.08	+18.5 ± 1.0	100	M
^{65}Ci	Fe	|17.0|	|212.7|	273	NMR
	Co	|12.6|	|157.5|	283	NMR
^{197}Au	Fe	|22.5|	|282|		M
^{161}Dy	DyIG	+279 ± 44	+3500 ± 550	85	M
^{159}Tb	Tb	+335 ± 80	+4200 ± 1000	0	NMR

polarized by the polarized *d*-electrons through the exchange interaction, and in turn they polarize the nucleus through a mechanism called the *Fermi contact*. Generally speaking, $1s$ and $2s$ electrons produce negative internal fields, while $3s$ electrons produce positive fields. For instance, in the case of Mn^{2+}, which has no orbital moment, the contribution of $1s$, $2s$, and $3s$ electrons are calculated to be −2.4, −112, and +59 MA m^{-1} (−30, −1400, and +740 kOe), respectively. These add up to −55 MA m^{-1} (−690 kOe),[11] in excellent agreement with the experimentally observed value −52 MA m^{-1} (−650 kOe).

The internal fields of Fe^{3+} in Table 4.2 range from −40.7 to −43.9 MA m^{-1} (−510 to −550 kOe). The calculated value on the basis of exchange polarization is −50.3 MA m^{-1} (−630 kOe), which is smaller than the calculated value for Mn^{2+}, in spite of the fact that the electronic structures are the same. The reason is that the radius of the $3d$ shell is smaller for Fe^{3+} than for Mn^{2+}. The internal field of Fe^{2+} is much smaller than Fe^{3+}, because the positive contribution from the partially unquenched orbital moment is added to the contribution of the s electrons. The orbital magnetic moment is caused by the orbital current, which produces a large internal field at the nucleus. It can be seen in Table 4.2 that ^{161}Dy or ^{159}Tb exhibit fairly large positive internal fields.[12]

In the case of the $3d$ transition elements, the orbital moment is almost quenched, so that the internal fields come mainly from the exchange polarization of s electrons. In magnetic insulators, the magnetic ions are always separated by anions, so that the internal field of an ion is caused by the polarization of its own d electrons. Therefore the internal field in this case is a good measure of the ionic magnetic moment. On the other hand, in magnetic metals and alloys, the polarization of the conducting s electrons is affected not only by their own ionic polarization but also by the polarization of their neighbors. It is seen in Table 4.2 that ^{119}Sn in the ferromagnetic metals Fe, Co and Ni exhibits fairly large internal fields in spite of its non-magnetic nature.

At temperatures above absolute zero, the internal field is proportional to the thermal average of the magnetic moment, because the frequency of thermal vibration is higher than that of the nuclear spins. Therefore the internal fields of paramagnets are generally very small. On the other hand, in antiferromagnets, the internal fields are as large as in ferromagnets. This is because the internal fields reflect the ionic magnetic moments, but not the overall magnetization.

4.2 NEUTRON DIFFRACTION

The neutron is an elementary particle which carries no electric charge, so that it penetrates matter without being influenced by the electric fields produced by electrons and ions. However, the neutron carries a nuclear magnetic moment of magnitude $-1.913M_N$, so that it interacts with and is scattered by the magnetic moments of magnetic atoms. By making use of this effect, the magnetic structure of a crystal or the magnitude of the magnetic moment of an atom can be determined.

Neutron beams can be obtained from a nuclear reactor, as illustrated schematically in Fig. 4.6. At the center of the reactor, there are a number of rods containing nuclear fuel such as uranium, which generate neutrons by chain reaction. A number of neutron-absorbing control rods limit the neutron density, to keep it below a critical value. The energy of neutrons produced by nuclear reaction is the order of 10 MeV, but this energy decays in heavy water at the center of the reactor to a value of the

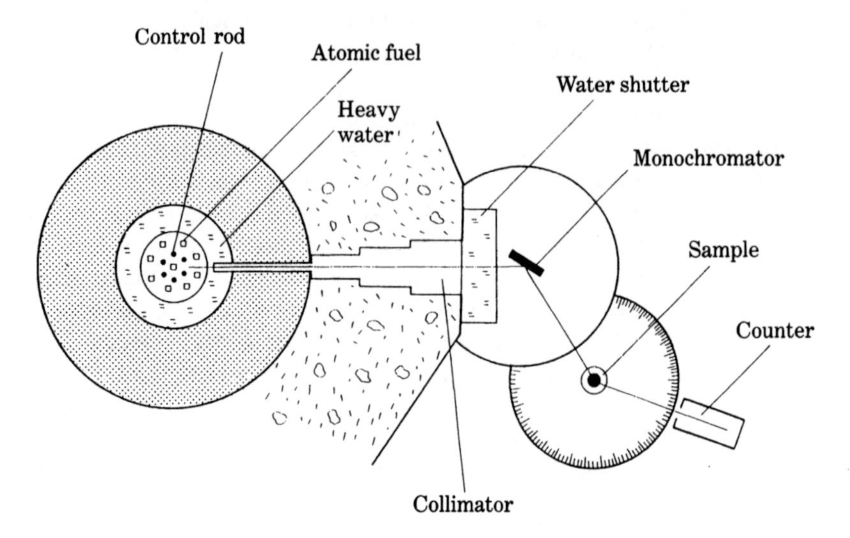

Fig. 4.6. Nuclear reactor and double axis goniometer for neutron diffraction.

order of kT, where k is the Boltzmann constant, and T is the absolute temperature of the heavy water (about 300 K). A neutron with energy near kT is called a *thermal neutron*. The density of thermal neutrons at the center of the reactor is in the order of $10^{14}\,\mathrm{cm}^{-2}\,\mathrm{s}^{-1}$. The wavelength of the thermal neutron is calculated by the de Broglie wave equation

$$\lambda = \frac{h}{mv}, \tag{4.9}$$

which gives a value of 1–2 Å, taking the neutron mass to be 1836 times the electron mass. The fact that the wavelength of thermal neutrons is close to the lattice constant of common crystals is quite fortunate for solid-state science. This fact was first pointed out by Elasser[13] and experimentally verified by Halban and Preiswerk.[14]

A neutron beam coming from the reactor through a collimator is reflected by Bragg diffraction from a monochromator crystal to produce a monochromatic beam containing neutrons of a single wavelength. This beam is diffracted by a sample crystal, and the angle of diffraction is detected by a counter. From this data the crystal structure and the magnetic structure can be analyzed, in a manner analogous to the interpretation of X-ray diffraction results. The first such experiment was done by Shull and Smart[15] to detect the antiferromagnetic spin arrangement in an MnO crystal (see Chapter 7). Generally speaking, the intensity of the thermal neutrons is 10^{12}–$10^{14}\,\mathrm{cm}^{-2}\,\mathrm{s}^{-1}$ in the beam as it comes from the reactor. This is reduced to 10^{6}–10^{8} after reflection from the monochromator crystal, and finally down to 10^{4}–10^{6} after diffraction from the specimen.

Neutrons are scattered by magnetic atoms in two ways: scattering by the atomic nucleus and scattering by the magnetic moment, either spin or orbital. The scattering amplitudes are different for different atoms, as shown in Table 4.3. The nuclear

Table 4.3. Differential cross sections for nuclear and magnetic scattering (after Bacon[16]).

Atom or ion	Nuclear scattering amplitude C, (10^{-12} cm)	Magnetic scattering amplitude D (10^{-12} cm)	
		For $\theta = 0$	For $\sin\theta/\lambda = 0.25$
Cr^{2+}	0.35	1.08	0.45
Mn^{2+}	−0.37	1.35	0.57
Fe (metal)	0.96	0.60	0.35
Fe^{2+}		1.08	0.45
Fe^{3+}		1.35	0.57
Co (metal)	0.28	0.47	0.27
Co^{2+}		1.21	0.51
Ni (metal)	1.03	0.16	0.10
Ni^{2+}		0.54	0.23

scattering amplitudes, C, are different not only in magnitude but also in sign. For example, C is negative for the Mn nucleus as a result of resonance scattering. The amplitude of the magnetic scattering, D, is given by a quantity called the *form factor*, f (see (4.11)), which varies with the scattering angle due to the size of the scattering body. Figure 4.7 shows the form factors for X-ray and neutron magnetic scattering as a function of $\sin\theta/\lambda$, where θ is the scattering angle. As seen in the figure, the form factor decreases with increasing scattering angle. The reason is as follows: When the diameter of the scattering atom is the same order of magnitude as the wavelength of the radiation, the beams going through the right and left edges of the atom interfere with each other after scattering as shown in Fig. 4.8, because a phase difference results. This phenomenon is quite similar to the diffraction of light after going through a small hole. The decay of the form factor is faster for neutron magnetic scattering than for X-ray scattering, as seen in Fig. 4.7, because neutrons are scattered magnetically principally by the outer electron shell, while X-rays are scattered by all the electrons. The effective size of the atom for X-ray scattering is therefore much less than for neutron magnetic scattering. On the contrary, the form factor for neutron scattering from the nucleus is independent of the scattering angle, because the nucleus can be regarded as an infinitesimally small point compared with the wavelength of thermal neutrons.

There are several methods for separating nuclear and magnetic scattering:

(1) The nuclear scattering intensity can be determined at high scattering angles where the magnetic scattering becomes negligibly small, and this value is subtracted from the total intensity to determine the magnetic intensity.
(2) The magnetic scattering can be suppressed by applying an external magnetic field parallel to the scattering vector, or perpendicular to the scattering plane.
(3) By using a polarized neutron beam and switching the direction of polarization, only the magnetic scattering contributes to the change in diffracted intensity.

First we discuss the nature of magnetic scattering in method (2). When spin

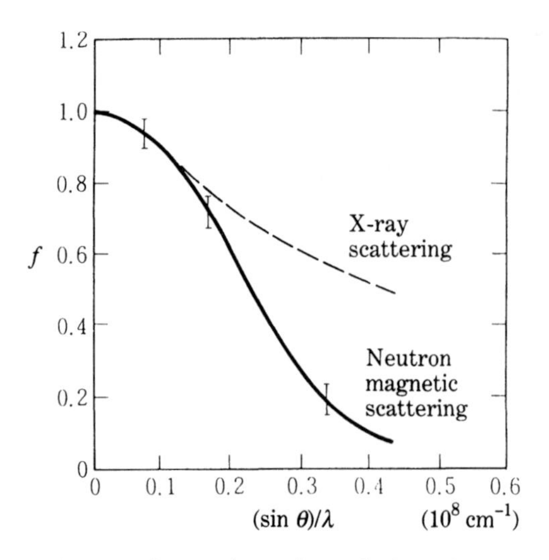

Fig. 4.7. Comparison of atomic form factors for X-ray and for neutron scattering. (After Bacon[16])

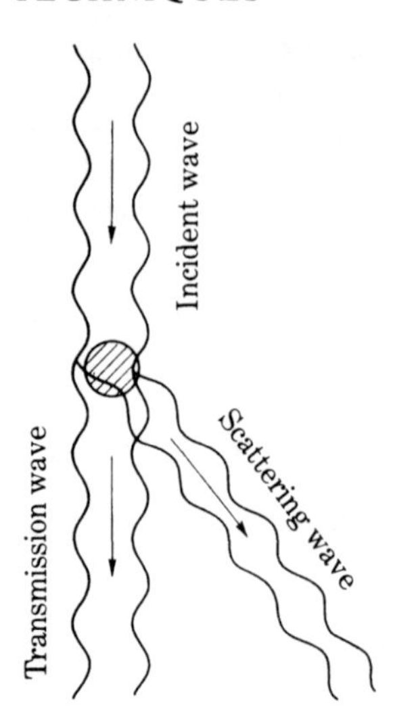

Fig. 4.8. Illustration of the reason for the decay in form factor with increasing scattering angle.

magnetic moments are arranged regularly, in either a ferromagnetic or antiferromagnetic structure, the differential scattering cross-section for neutrons is given by

$$\frac{\mathrm{d}\sigma}{\mathrm{d}\Omega} = D^2(\boldsymbol{q} \cdot \boldsymbol{\lambda})^2, \tag{4.10}$$

where D is the magnetic scattering amplitude and is given for spin S by

$$D = \frac{\mu_0 e^2 \gamma S}{4\pi m} f, \tag{4.11}$$

where γ is the magnetic moment of the neutron in units of nuclear magnetons, and f is the form factor amplitude. The scalar product in (4.10) is a factor which includes the unit vector parallel to the magnetic moment of the neutron, λ, and the reversed vector parallel to the projection of the unit vector of the atomic magnetic moment, $\boldsymbol{q}$, on the scattering plane. We define the scattering vector, $\boldsymbol{e}$, as

$$\boldsymbol{e} = \frac{\boldsymbol{k} - \boldsymbol{k}'}{|\boldsymbol{k} - \boldsymbol{k}'|}, \tag{4.12}$$

where $\boldsymbol{k}$ and $\boldsymbol{k}'$ are unit vectors parallel to the incident and scattering neutron beams, respectively. The scattering vector given by (4.12) is a unit vector perpendicular to the scattering plane pointing from the incident side. The vector $\boldsymbol{q}$ can be expressed as

$$\boldsymbol{q} = \boldsymbol{e}(\boldsymbol{e} \cdot \boldsymbol{\kappa}) - \boldsymbol{\kappa}, \tag{4.13}$$

as illustrated in Fig. 4.9. The magnetic scattering is proportional to the scalar product of $\boldsymbol{q}$ and $\boldsymbol{\lambda}$. For instance, if the atomic magnetic moment is perpendicular to the scattering plane, or $\boldsymbol{q} = \boldsymbol{0}$, no magnetic scattering occurs. Also if the magnetic moment of the neutron, $\boldsymbol{\lambda}$, is perpendicular to the vector $\boldsymbol{q}$, no magnetic scattering

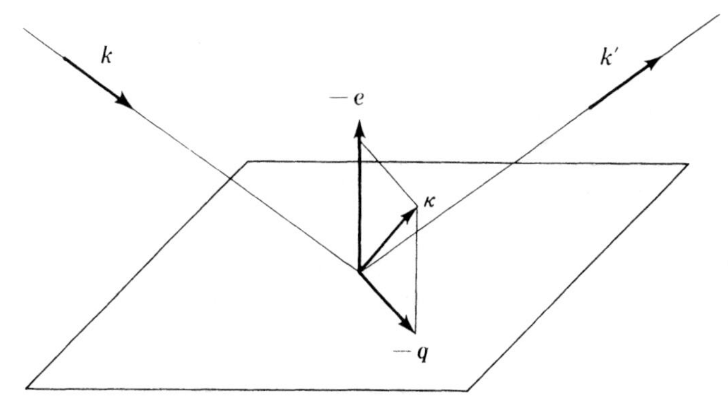

Fig. 4.9. Relationship between various vectors involved in neutron diffraction: incident and scattered wave vectors, $\boldsymbol{k}, \boldsymbol{k}'$, scattering vector, **e**, unit vector parallel to the magnetic moment, **κ** and the vector given by (4.13), $\boldsymbol{q}$.

occurs. This fact can be interpreted in terms of the magnetic dipolar interaction (1.14), which gives the maximum repulsion when the two dipoles are parallel to each other and perpendicular to the line joining them. In conclusion, magnetic scattering is effectively produced by neutrons polarized parallel to the projection of the atomic magnetic moments onto the scattering plane. (Note that if the atomic magnetic moments are perpendicular to some particular scattering plane, magnetic scattering will occur from other scattering planes.)

Now let us consider the method for detecting magnetic scattering in method (2). Suppose that the incident and scattering beams, and accordingly the scattering vector, $\boldsymbol{e}$, lie in the horizontal plane (see Fig. 4.10). When a magnetic field, H, is applied to magnetize the specimen parallel to the scattering vector (see (a)), the magnetic scattering vanishes because $\boldsymbol{q} = 0$. On the other hand, when the magnetic field is applied vertically (see (b)), the magnetic scattering is maximized because $\boldsymbol{q} = 1$. The difference between the two cases gives the intensity of magnetic scattering. If the

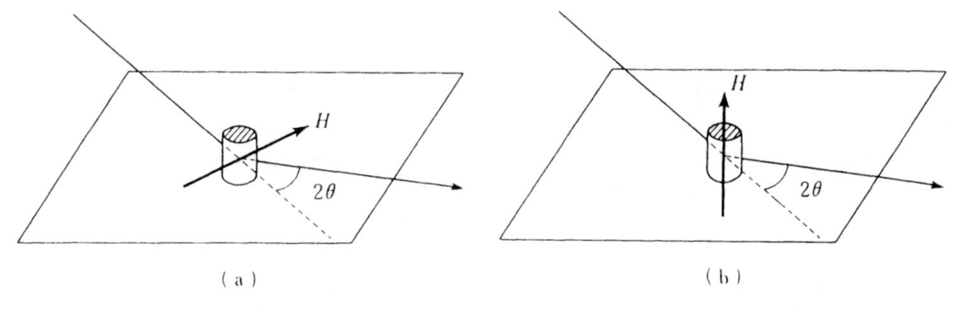

Fig. 4.10. One method for separating the magnetic scattering: (a) magnetic field $\boldsymbol{H}$ is applied perpendicular to the scattering plane, so the magnetic scattering is completely eliminated; (b) field $\boldsymbol{H}$ is applied parallel to the scattering plane, so both magnetic and nuclear scattering occur.

intensity of the magnetic field is adjusted to allow rotation of the magnetization towards the stable directions, we can get information on the magnetic anisotropy (see Chapter 12) or the canting of the spin system (see Section 7.3).

Including the nuclear scattering, we have for the total differential scattering

$$\frac{\partial \sigma}{\partial \Omega} = [C + D(\boldsymbol{q} \cdot \boldsymbol{\lambda})]^2 = C^2 + 2CD(\mathbf{q} \cdot \boldsymbol{\lambda}) + D^2 q^2, \tag{4.14}$$

where C is the nuclear scattering amplitude. When the incident neutrons are unpolarized, the second term vanishes after averaging over all possible directions of $\boldsymbol{\lambda}$. Then (4.14) leads to

$$\frac{\partial \sigma}{\partial \Omega} = C^2 + D^2 q^2. \tag{4.15}$$

Method (2) consists of altering q in (4.15). Therefore, unless D is much smaller than C, we can separate the magnetic from the nuclear scattering. Otherwise (as in the case of Ni metal, where D is about 10% of C), the second term in (4.15) becomes less than 1% of the first term, so that an accurate observation of magnetic scattering is impossible.

Alternatively, polarized neutrons can be used to observe magnetic scattering. This method was first proposed by Shull[17] and executed by Nathans *et al.*[18] They used Bragg scattering from a carefully selected monochromator crystal, in which D was nearly equal to C. (The form factor for magnetic scattering decreases with increasing scattering angle, so that D in (4.11) can become equal to C at some scattering angle, if there is an appropriate Bragg reflection.) When a vertical magnetic field is applied to the monochromator as in Fig. 4.10(b), $\boldsymbol{q} = -1$, so that

$$\frac{\mathrm{d}\sigma}{\mathrm{d}\Omega} = (C - D)^2, \tag{4.16}$$

for neutrons whose polarization is parallel to the magnetization of the monochromator crystal, while

$$\frac{\mathrm{d}\sigma}{\mathrm{d}\Omega} = (C + D)^2. \tag{4.17}$$

for oppositely polarized neutrons. Accordingly, if $C = D$, (4.16) vanishes, so that only the neutrons whose polarization is antiparallel to the magnetization of the monochromator can be diffracted. Possible monochromators for neutron polarization are the (220) reflection from an Fe_3O_4 crystal and the (111) and (200) reflections from an Fe–Co crystal. Figure 4.11 illustrates the experimental arrangement for generating the polarized neutron beam. The polarized neutron beam from the polarizing crystal retains its polarization direction in the collimating magnetic field, but the polarization can be reversed by the polarization inverter if necessary. The inverter provides a high-frequency magnetic field perpendicular to the static magnetic field or the precession field, and induces resonance precession of the neutron spins. The precession finally reverses the direction of neutron spins at the end of the apparatus. (The

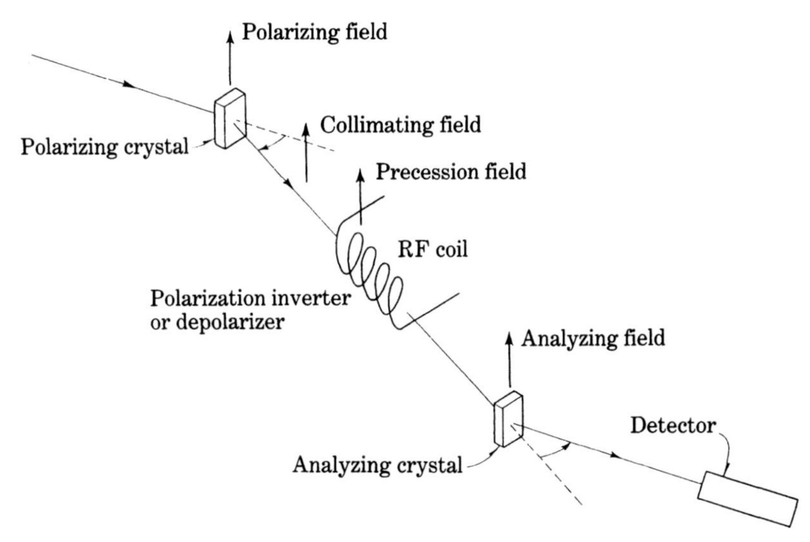

Fig. 4.11. Schematic diagram of a polarized neutron spectrometer. (After Nathans *et al.*[19])

intensity of the high-frequency field is adjusted so as to reverse the direction of neutron spins during the transit time of the high-frequency coil.)

By changing the direction of polarization of the neutron spins, $\boldsymbol{\lambda}$, from -1 to $+1$, the total differential cross-section given in (4.14) is changed by $4CDq$. The change amounts to 40% of C in the case of Ni metal, which is large enough to determine the intensity of magnetic scattering.

The wavelength of neutron beams remains unchanged by Bragg scattering. In other words, the velocity and therefore the kinetic energy of the neutrons is conserved. Such scattering is called *elastic scattering*. On the other hand, in *inelastic scattering* a part of the kinetic energy is lost by conversion to some other form of energy. For instance, the neutron can excite *spin waves* in magnetically ordered systems, such as ferromagnets, ferrimagnets, and antiferromagnets. By measuring the change in wavelength as a function of wave vector in inelastic scattering, we can elucidate the dispersion relationship of the *magnon* or the quantized spin wave.

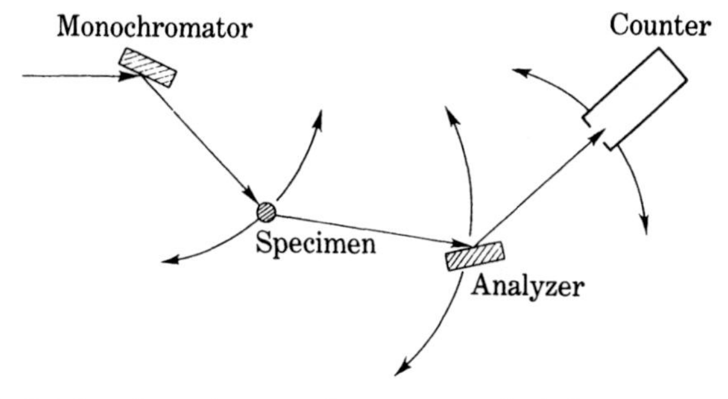

Fig. 4.12. Triple axis goniometer for energy analysis in neutron scattering.

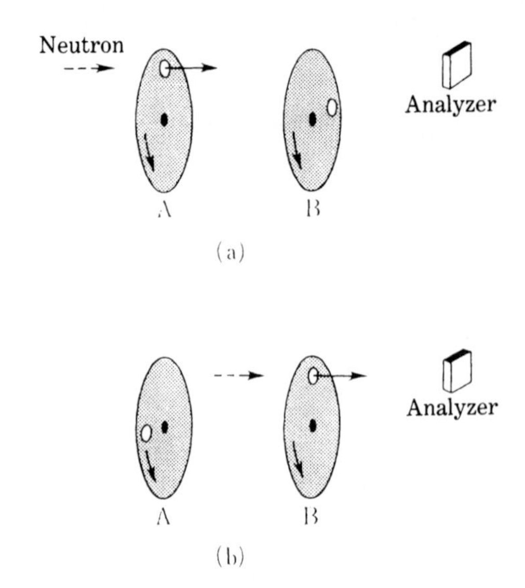

Fig. 4.13. Principle of the time-of-flight (TOF) method.

Figure 4.12 shows a schematic illustration of a triple-axis goniometer for the observation of inelastic scattering. The wavelength of the scattered neutrons is determined by changing the angle of incidence of the analyzer crystal for a known Bragg reflection. Another method to observe the change in wavelength is the time-of-flight (TOF) method, which is illustrated in Fig. 4.13. This method determines the velocity of the scattered neutrons, from which the wavelength is found by de Broglie's relationship (4.9). The velocity of the neutron is determined by adjusting the rotational speed of a pair of disks, each of which has a single hole, as shown in Fig. 4.13. When the rotational speed is correctly set, a neutron which passes through the hole in a disk A (see Fig. 4.13(a)) can pass through the hole in disk B (see Fig. 4.13(b)). Knowing this rotational speed, and the distance between the two disks, the velocity of the neutron is determined. Since a neutron of wavelength 1 Å has a velocity of 3.96 km s^{-1}, the time required to travel the distance, say 1 m, between the two disks is 252 μs, which can be measured accurately.

4.3 MUON SPIN ROTATION (μSR)

The muon is an artificially produced elementary particle, with mass about 207 times larger than the electron mass. It has a magnetic moment whose magnitude is between that of the Bohr magneton and the nuclear magneton. This particle can be used as a microscopic probe for detecting local magnetic fields inside magnetic materials.

Muons are produced as follows: First high-energy protons collide with nuclei to produce mesons, which disintegrate into neutrinos and muons. The muons thus produced have high energy (about 100 MeV), and have their spin axes antiparallel to their velocity. After these muons are introduced into a specimen, they lose energy and finally come to rest at some point in the material.

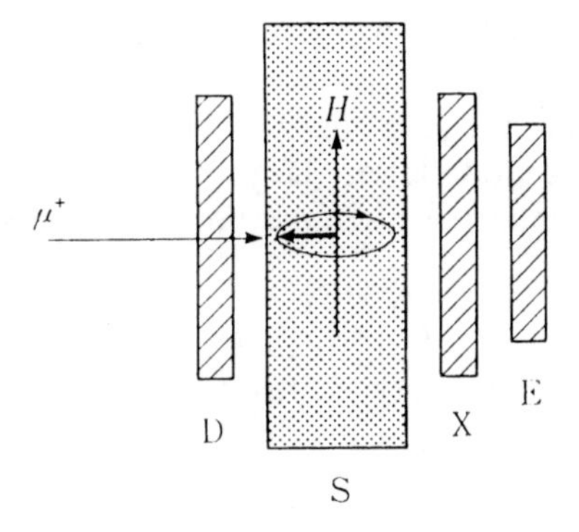

Fig. 4.14. Conceptual diagram of the μSR apparatus.

There are two species of muons: the positive muon, μ^+, with positive electric charge $+e$ (C); and the negative muon, μ^-, with negative charge $-e$ (C). The behavior of these muons in the specimen is quite different: The positive muons come to rest at an interstitial site in the crystal, being repelled by the positively charged nuclei. The negative muons are attracted to nuclei and take up orbital motion around a nucleus. Finally they stabilize to the $1s$ ground state, whose radius is very small because of the large muon mass. This orbital muon adds a large magnetic moment to the nuclear magnetic moment of the host nucleus, and also reduces the nuclear charge by e as a result of electrostatic shielding, thus producing a completely new artificial nucleus.

The positive muon which stops at an interstitial site has its spin antiparallel to the incident direction, and then makes a precession motion about the magnetic field at this point. The precession frequency is proportional to the intensity of the magnetic field. This phenomenon is called the μSR.

Figure 4.14 is a schematic illustration of the μSR apparatus. A positive muon travelling from left to right passes through the counter D and stays in the specimen S.

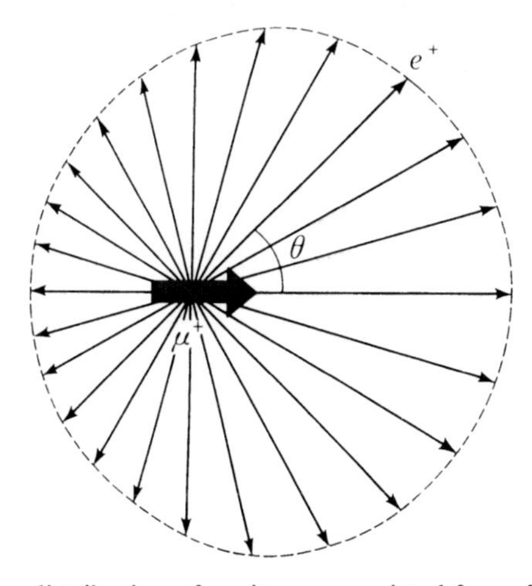

Fig. 4.15. Angular distribution of positrons emmitted from disintegrated μ^+.

A muon which passes through the specimen is detected by the counter X and is ignored. The spin of a muon which stays at an interstitial site begins its precession motion at almost the same time as its passage is detected by the counter D, because the velocity of the incident muon is very large.

During the precession motion, the muon eventually decays with a lifetime of 2.2 μs, and by the reaction

$$\mu^+ \rightarrow e^+ + \bar{\nu}_\mu + \nu_e \qquad (4.18)$$

turns into a positron e^+, an antimuon neutrino $\bar{\nu}_\mu$, and an electron neutrino ν_e. The

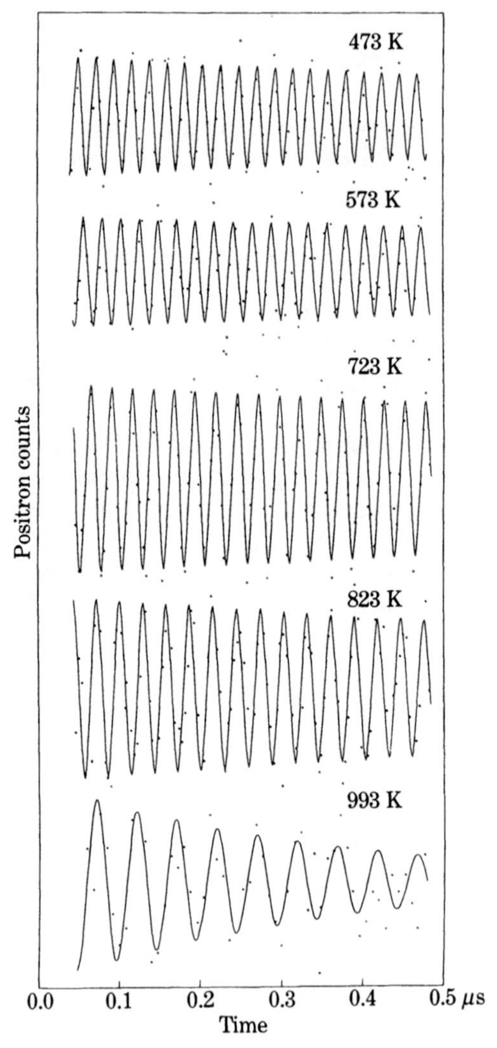

Fig. 4.16. μSR signal for iron observed at various temperatures. (After Yamazaki *et al.*[20] and Nishida[21])

positron tends to be emitted parallel to the direction of the spin of the muon. If $W(\theta)$ is the probability that the positron is emitted along the direction making an angle θ with the direction of the spin, then

$$W(\theta) = 1 + A \cos \theta. \tag{4.19}$$

where A is a coefficient with a value of about $\frac{1}{3}$. Figure 4.15 shows a polar diagram of $W(\theta)$. The number of positrons is counted by the counter E and plotted as a function of the time since the detection at the counter D. Then we can observe a precession oscillation as shown in Fig. 4.16. In this case the rate of arrival of muons was low enough so that the decay of each muon was complete before the next muon arrived. The data shown in Fig. 4.16 were obtained for metallic iron at various temperatures. The experimental points scatter a great deal, but the frequencies obtained after Fourier analysis of the data are quite sharp, as shown in the Fourier spectra of Fig.

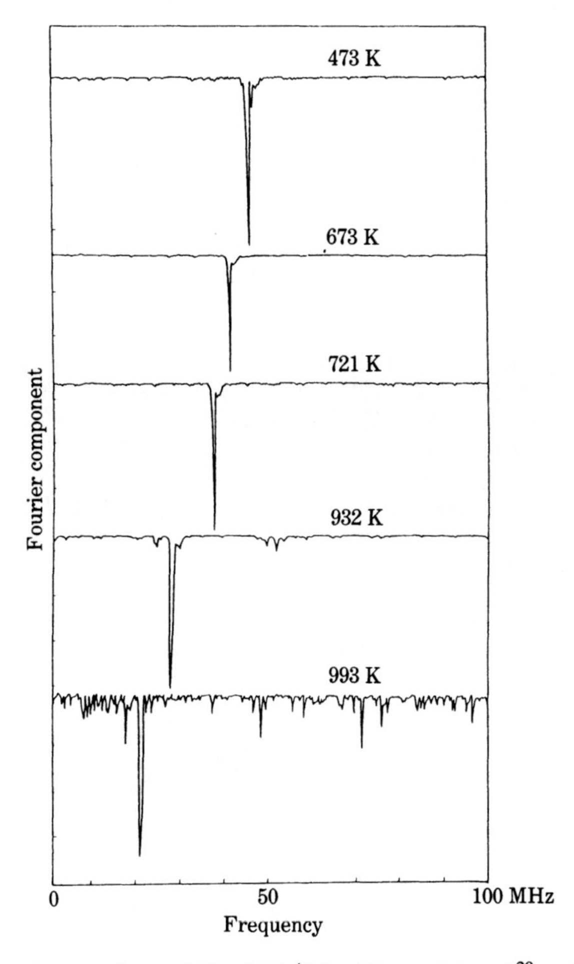

Fig. 4.17. Fourier transform of Fig. 4.16. (After Yamazaki *et al.*[20] and Nishida[21])

4.17. The relaxation mechanism of μSR can also be investigated by observing the decay of the precession, as seen in Fig. 4.16.

For further details, the reader may refer to a reference book on this subject.[22]

PROBLEMS

4.1 Knowing that the nucleus ^{57}Fe has nuclear spin $\frac{1}{2}$, and nuclear magnetic moment $0.090M_N$, calculate the NMR resonance frequency for ^{57}Fe in an internal field of $-27\,\mathrm{MA\,m^{-1}}$.

4.2 Using the Mössbauer spectrum shown in Fig. 4.5, calculate the internal field ($\mathrm{A\,m^{-1}}$), the quadrupole shift (J), and the isomer shift ($\mathrm{mm\,s^{-1}}$).

4.3 In order to produce a neutron beam which contains neutrons with an average wave length of 4 Å, what temperature is required for the neutron bath?

REFERENCES

1. *American Inst. Phys. handbook*, 2nd edn. (McGraw-Hill, New York, 1963), 8-6–13.
2. E. Mattias and D. A. Shirley, *Hyperfine structure and nuclear radiations* (North-Holland, Amsterdam, 1968), pp. 988–1011.
3. Lederer, Hollander and Perlman, *Table of isotopes* (6th edn., Wiley, New York, 1967).
4. A. C. Gossard and A. M. Portis, *Phys. Rev. Letters*, **3** (1959), 164; A. M. Portis and A. C. Gossard, *J. Appl. Phys.*, **31** (1960), 205S.
5. E. L. Hahn, *Phys. Rev.*, **80** (1950), 580.
6. W. E. Lamb, *Phys. Rev.*, **55** (1939), 190.
7. R. L. Mössbauer, *Z. Physik*, **151** (1958), 124; *Z. Naturforsch.*, **149** (1959), 211.
8. G. K. Wertheim, *J. Appl. Phys.*, **32** (1961), 110S.
9. Y. Ishikawa, *Metal Physics*, **8** (1962), 65.
10. W. Marshall, *Phys. Rev.*, **110** (1958), 1280.
11. R. E. Watson and A. J. Freeman, *Phys. Rev.*, **123** (1961), 1091; *J. Appl. Phys.*, **32** (1961), 118S.
12. J. Kondo, *J. Phys. Soc. Japan*, **16** (1961), 1690.
13. W. M. Elsasser, *Compt. Rend.*, **202** (1936), 1029.
14. H. Halban and P. Preiswerk, *Compt. Rend.*, **203** (1936), 73.
15. C. G. Shull and J. S. Smart, *Phys. Rev.*, **76** (1949), 1256.
16. G. E. Bacon, *Neutron diffraction* (Oxford, Clarendon Press, 1955).
17. C. G. Shull, *Phys. Rev.*, **81** (1951), 626.
18. R. Nathans, M. T. Pigott, and G. G. Shull, *Phys. Chem. Solids*, **6** (1958), 38.
19. R. Nathans, C. G. Shull, G. Shirane, and A. Anderson, *Phys. Chem. Solids*, **10** (1959), 138.
20. T. Yamazaki, S. Nagamiya, O. Hashimoto, K. Nagamine, K. Nakai, K. Sugimoto, and K. M. Crowe, *Phys. Letters*, **53B** (1974), 117.
21. N. Nishida, Thesis (Tokyo University, 1977).
22. A. Schenck, *Muon spin rotation spectroscopy* (Adam Hilger, Bristol & Boston, 1985).

Part III

MAGNETIC ORDERING

In this Part, we shall learn how the atomic magnetic moments, which we studied in a previous Part, are arranged in magnetic materials to produce three-dimensional structures. In Chapter 5, we treat magnetic disorder, in which the atomic magnetic moments are absent or oriented at random; in Chapter 6 we treat Ferromagnetism, in which the atomic magnetic moments are set parallel, thus creating spontaneous magnetization; and in Chapter 7 we treat Antiferromagnetism and ferrimagnetism, in which the atomic magnetic moments are arranged antiparallel to one another.

5

MAGNETIC DISORDER

5.1 DIAMAGNETISM

Diamagnetism means a feeble magnetism,[1] which occurs in a material containing no atomic magnetic moments. The relative susceptibility of such a material is negative and small, typically $\bar{\chi} \simeq 10^{-5}$.

The mechanism by which the magnetization is induced opposite to the magnetic field is the acceleration of the orbital electrons by electromagnetic induction caused by the penetration of the external magnetic field to the orbit (Fig. 5.1). According to Lenz's law, the magnetic flux produced by this acceleration of an orbital electron is always opposite to the change in the external magnetic field, so that the susceptibility is negative.

For simplicity, let us assume that the orbit is a circle of radius r (m), and that a magnetic field H ($\mathrm{A\,m^{-1}}$) is applied perpendicular to the orbital plane. According to the law of electromagnetic induction, an electric field E ($\mathrm{V\,m^{-1}}$) is produced in such a way that

$$\oint E_s \, \mathrm{d}s = 2\pi r E_s = -\pi r^2 \mu_0 \frac{\mathrm{d}H}{\mathrm{d}t}, \tag{5.1}$$

so that

$$E_s = -\frac{\mu_0}{2} r \frac{\mathrm{d}H}{\mathrm{d}t}. \tag{5.2}$$

The electron is accelerated by this field, and the velocity change Δv during the time interval Δt is given by

$$\Delta v = -\frac{e}{m} E_s \, \Delta t = \frac{e\mu_0}{2m} r \Delta H, \tag{5.3}$$

where ΔH is the change in the magnetic field during Δt. We assume that the radius r remains unchanged, which is in fact the case. According to (5.3), the change in the centrifugal force acting on the electron is given by

$$\begin{aligned} \Delta\left(m\frac{v^2}{r}\right) &= 2m\frac{v}{r}\Delta v \\ &= ev\mu_0 \, \Delta H, \end{aligned} \tag{5.4}$$

which is just balanced by an increase in the Lorentz force, $ev \, \Delta B$. This is always true, even when the orbital plane is inclined. In other words, the orbit precesses about the applied field without changing its shape, with angular velocity

$$\omega_L = \frac{v}{r} = \frac{\mu_0 e}{2m} H. \tag{5.5}$$

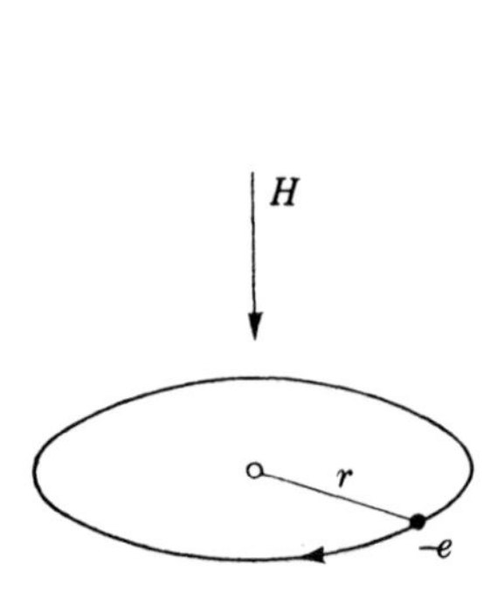

Fig. 5.1. Mechanism of atomic diamagnetism.

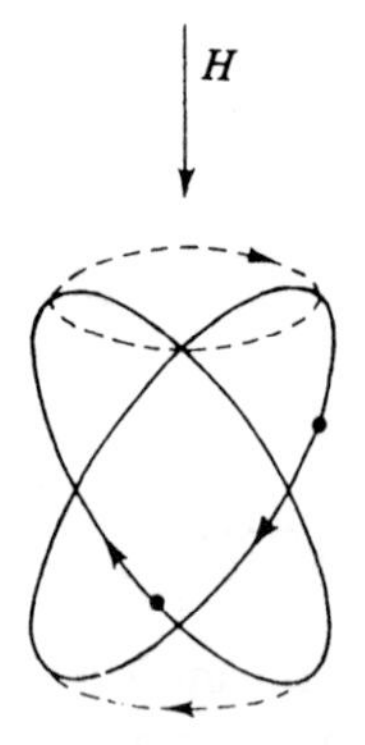

Fig. 5.2. Larmor precession of a tilted orbit.

This motion is called the *Larmor precession* (Fig. 5.2). Referring to (1.12) and (5.3), the magnetic moment produced by the motion shown in Fig. 5.1 is given by

$$M = -\mu_0\left(\frac{e\,\Delta v}{2\pi r}\right)(\pi r^2)$$
$$= -\frac{\mu_0^2 e^2}{4m} r^2 H. \tag{5.6}$$

In the case of a closed shell, electrons are distributed on a spherical surface with radius a (m), so that r^2 in (5.6) is replaced by $\overline{x^2+y^2}$, where the z-axis is parallel to the magnetic field (Fig. 5.3). Considering spherical symmetry, we have $\overline{x^2}=\overline{y^2}=\overline{z^2}=a^2/3$, so that $\overline{r^2}=\overline{x^2}+\overline{y^2}=\frac{2}{3}a^2$. Therefore (5.6) becomes

$$M = -\frac{\mu_0^2 e^2}{6m} a^2 H. \tag{5.7}$$

When a unit volume of the material contains N atoms, each of which has Z orbital electrons, the magnetic susceptibility is given by

$$\chi = -\frac{N\mu_0^2 e^2 Z}{6m}\overline{a^2}. \tag{5.8}$$

where $\overline{a^2}$ is the average a^2 for all the orbital electrons. This relationship holds fairly well for materials containing atoms or ions with closed shells.[2]

Diamagnetism is also exhibited by inorganic compounds. For instance, benzene rings, in which π electrons are circulating just like orbital electrons, act as closed shells. They exhibit fairly strong diamagnetism when the field is applied perpendicular to the rings, but not when the field is parallel to the plane of the ring.[3]

In the case of diamagnetic metals, conduction electrons play an important role in producing diamagnetism, which is influenced by the band structure. Table 5.1 lists the

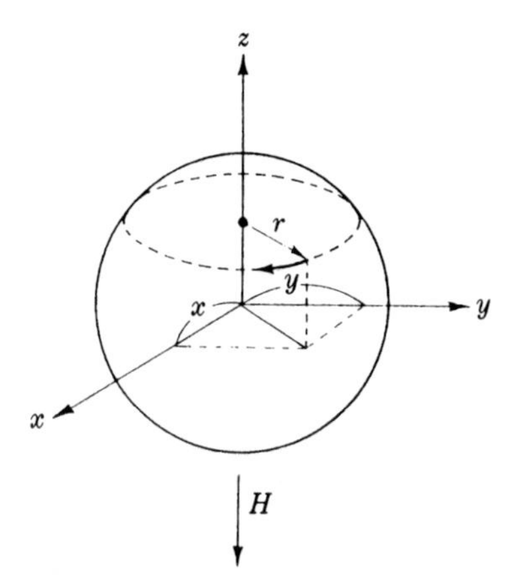

Fig. 5.3. Diamagnetism of spherically distributed electrons.

values of diamagnetic susceptibility at room temperature for various materials. Generally speaking, diamagnetic susceptibility is not strongly temperature-dependent.

Superconductivity is the disappearance of the electrical resistivity at some (usually low) temperature called the critical temperature, T_c. Superconductivity is also characterized by perfect diamagnetism: that is, the magnetic flux density, $\boldsymbol{B}$, is always zero in the superconducting state, even in the presence of an external field $\boldsymbol{H}$ (*Meissner effect*).

Since in this case

$$B = I + \mu_0 H = 0, \tag{5.9}$$

the magnetization, I, is given by

$$I = -\mu_0 H. \tag{5.10}$$

In other words,

$$\bar{\chi} = -1 \quad \text{or} \quad \bar{\mu} = 0. \tag{5.11}$$

Table 5.1. Relative diamagnetic susceptibility for various materials at room temperature.

Materials	$\bar{\chi}$ (= $4\pi\chi$ in cgs)	Reference
Cu	-9.7×10^{-6}	4
Ag	-25	
Au	-35	
Pb	-16	
C (graphite)	-14	
Al_2O_3	-18	5
H_2O	-9.05	
SiO_2 (quartz)	-16.4	
Benzene	-7.68	
Ethyl Alcohol	-7.23	

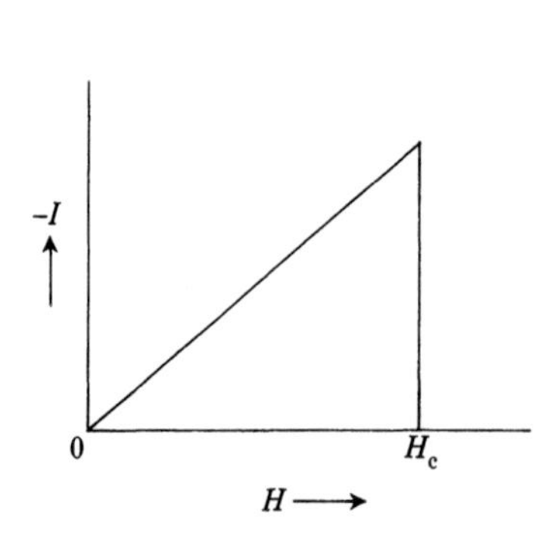

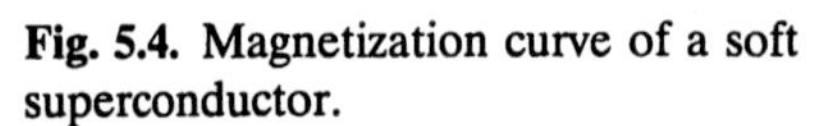

Fig. 5.4. Magnetization curve of a soft superconductor.

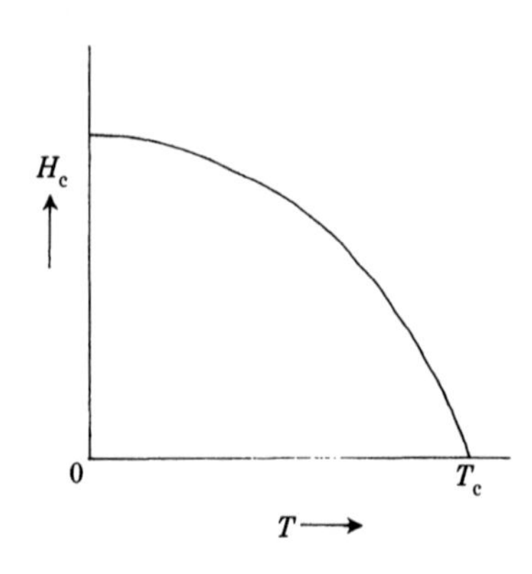

Fig. 5.5. Temperature dependence of the critical field, H_c, of a soft superconductor.

When the external field is increased to a critical field, H_c, the superconducting state is destroyed, so that magnetization, I, vanishes. The magnetization curve of a superconductor is shown in Fig. 5.4. This type of superconductor is classified as a Type I or soft superconductor. The critical field, H_c, increases with decreasing temperature, as shown in Fig. 5.5.

There is another kind of superconductor, called a Type II or *hard superconductor*. In this case, the magnetization does not drop to zero at the first critical field, H_{c1}, but decreases gradually with increasing field as shown in Fig. 5.6, and finally vanishes at the second critical field, H_{c2}. These critical fields increase with decreasing temperature as shown in Fig. 5.7. Table 5.2 lists typical superconductors, with their critical temperatures and critical fields. This table includes oxide superconductors with relatively high critical temperatures as first discovered by Bednorz and Müller.[6]

5.2 PARAMAGNETISM

Paramagnetism describes a feeble magnetism which exhibits positive susceptibility of the order of $\bar{\chi} = 10^{-5}$–10^{-2}. This magnetic behavior is found in materials that contain magnetic atoms or ions that are widely separated so that they exhibit no appreciable interaction with one another.

Let us assume a paramagnetic system containing N magnetic atoms each with magnetic moment M (Wb m), which we will refer to simply as spin. At a temperature above absolute zero, the position of each atom undergoes thermal vibration, as does the direction of spin. At absolute temperature T (K), the thermal energy shared for one degree of freedom is given by $kT/2$, where k is the Boltzmann constant, 1.38×10^{-23} J K^{-1}. At room temperature,

$$\tfrac{1}{2}kT = \tfrac{1}{2} \times 1.38 \times 10^{-23} \times 300 = 2.1 \times 10^{-21}\ \text{J}. \tag{5.12}$$

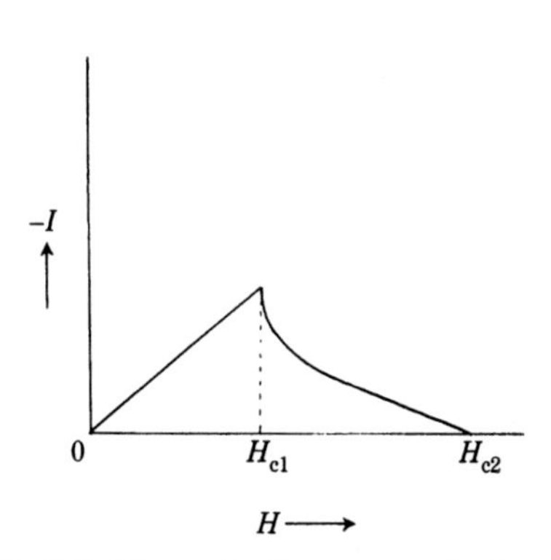

Fig. 5.6. Magnetization curve of a hard superconductor.

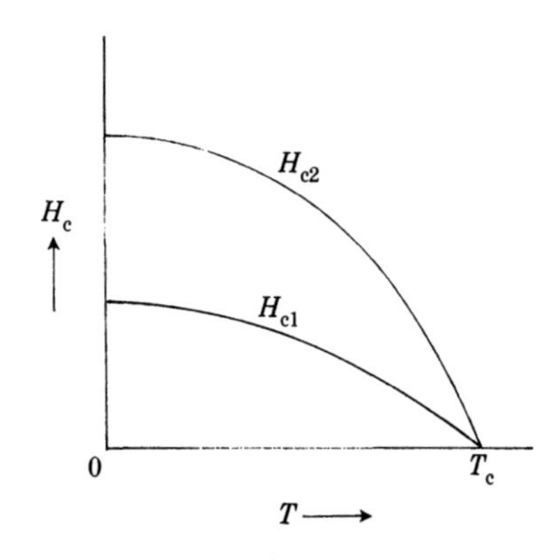

Fig. 5.7. Temperature dependence of the critical fields H_{c1} and H_{c2} of a hard superconductor.

To what extent can the spin system be influenced when a fairly high magnetic field is applied? If a magnetic moment is placed in a magnetic field, H ($\mathrm{A\,m^{-1}}$), the potential energy is given by (1.9). Assuming that $M = 1M_B$ and $H = 1\,\mathrm{MA\,m^{-1}}$, which is a moderately high field such as that produced by an electromagnet, the potential energy is given by

$$MH \approx 1.2 \times 10^{-29} \times 10^{6} = 1.2 \times 10^{-23}\,\mathrm{J}, \tag{5.13}$$

which is about $\frac{1}{200}$ of the thermal energy at room temperature given by (5.12). Such a magnetic field barely influences a thermally agitated spin system at ordinary temperatures.

We can calculate the magnetization of such a system more quantitatively using the *Langevin theory*. The angular distribution of the ensemble of spins can be expressed by unit vectors drawn from the center of a sphere with a unit radius (Fig. 5.8). In the absence of a magnetic field, the spins are distributed uniformly over all possible

Table 5.2. Typical superconductors.

Superconductors	Type	T_c (K)	H_c or H_{c2} (0 K)	
V	I	5.40	0.113 $\mathrm{MA\,m^{-1}}$	(= 1420 Oe)
Nb	I	9.25	0.164	(= 2060 Oe)
Pb	I	7.20	0.064	(= 803 Oe)
Nb_3Sn	II	18.1	20.7	(= 260 kOe)
Nb–Ti	II	10	8.8	(= 110 kOe)
La–Ba–Cu–O	II	40	—	
Y–Ba–Cu–O	II	100	—	
Bi–Sr–Ca–Cu–O	II	110	—	

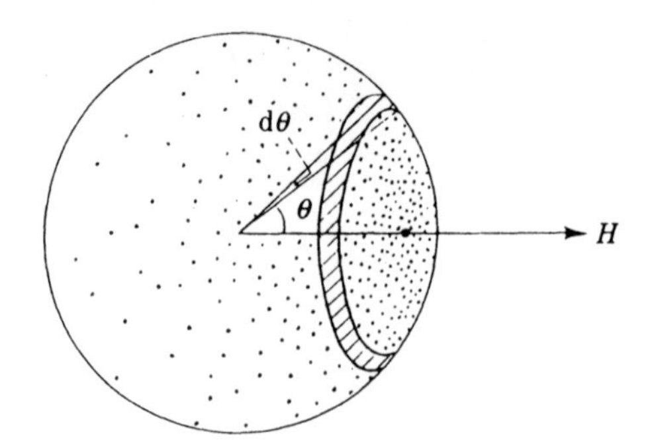

Fig. 5.8. Angular distribution of paramagnetic spins in a magnetic field.

orientations, so that the points of the unit vectors cover the sphere uniformly. When a magnetic field, H, is applied, the points are gathered slightly towards H. For a spin which makes an angle θ with H, the potential energy is given by (1.9), so that the probability for a spin to take this direction is proportional to the *Boltzmann factor*

$$\exp\left(-\frac{U}{kT}\right)=\exp\left(\frac{MH}{kT}\cos\theta\right). \tag{5.14}$$

On the other hand, the *a priori probability* for a spin to make an angle between θ and $\theta+\mathrm{d}\theta$ with the magnetic field is proportional to the shaded area in Fig. 5.8, or $2\pi\sin\theta\,\mathrm{d}\theta$. The physical probability for a spin to make an angle between θ and $\theta+\mathrm{d}\theta$ is, therefore, given by

$$p(\theta)\,\mathrm{d}\theta=\frac{\exp\left(\dfrac{MH}{kT}\cos\theta\right)\sin\theta\,\mathrm{d}\theta}{\displaystyle\int_0^\pi \exp\left(\frac{MH}{kT}\cos\theta\right)\sin\theta\,\mathrm{d}\theta}. \tag{5.15}$$

Since such a spin contributes an amount $M\cos\theta$ to the magnetization parallel to the magnetic field, the magnetization due to the whole spin system is given by

$$\begin{aligned} I&=NM\overline{\cos\theta}=NM\int_0^\pi \cos\theta\,p(\theta)\,\mathrm{d}\theta\\ &=NM\frac{\displaystyle\int_0^\pi \exp\left(\frac{MH}{kT}\cos\theta\right)\cos\theta\sin\theta\,\mathrm{d}\theta}{\displaystyle\int_0^\pi \exp\left(\frac{MH}{kT}\cos\theta\right)\sin\theta\,\mathrm{d}\theta}. \end{aligned} \tag{5.16}$$

If we put $MH/kT=\alpha$ and $\cos\theta=\mathrm{x}$, we get $-\sin\theta=\mathrm{d}x$, so that (5.16) becomes

$$I=NM\frac{\displaystyle\int_{-1}^{1}\mathrm{e}^{\alpha x}x\,\mathrm{d}x}{\displaystyle\int_{-1}^{1}\mathrm{e}^{\alpha x}\,\mathrm{d}x}. \tag{5.17}$$

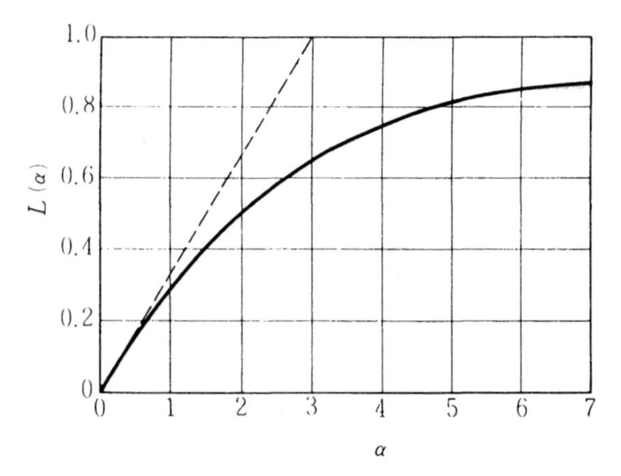

Fig. 5.9. Langevin function.

The integral in the denominator is calculated to be

$$\int_{-1}^{1} e^{\alpha x}\,dx = \frac{1}{\alpha}|e^{\alpha x}|_{-1}^{1} = \frac{1}{\alpha}(e^{\alpha} - e^{-\alpha}). \tag{5.18}$$

By differentiating both sides of (5.18) with respect to α, we have the numerator of (5.17), or

$$\int_{-1}^{1} e^{\alpha x} x\,dx = \frac{1}{\alpha}(e^{\alpha} + e^{-\alpha}) - \frac{1}{\alpha^2}(e^{\alpha} - e^{-\alpha}). \tag{5.19}$$

Therefore, (5.17) is reduced to

$$\begin{aligned} I &= NM\left(\frac{e^{\alpha} + e^{-\alpha}}{e^{\alpha} - e^{-\alpha}} - \frac{1}{\alpha}\right) \\ &= NM\left(\coth\alpha - \frac{1}{\alpha}\right), \end{aligned} \tag{5.20}$$

where the function in parentheses is called the *Langevin function* and denoted by $L(\alpha)$. Figure 5.9 is a plot of $L(\alpha)$ as a function of α. As α is increased, $L(\alpha)$ approaches 1. This means that as H, and accordingly α, is increased, the magnetization, I, given by (5.20) approaches NM, which corresponds to the complete alignment of all the spins. This saturation of the paramagnetic system, however, cannot be realized in practice unless we use extra-high fields or cool the sample to extremely low temperatures. For a field of the order of $1\,\mathrm{MA\,m^{-1}}$, referring to (5.13) and (5.12), we have

$$\alpha = \frac{MN}{\frac{1}{2}kT} = \frac{1.2 \times 10^{-23}}{2.1 \times 10^{-21}} = 0.005. \tag{5.21}$$

Therefore, only the linear part of $L(\alpha)$ near the origin is observed. For $\alpha \ll 1$, the Langevin function can be expanded as

$$L(\alpha) = \frac{\alpha}{3} - \frac{\alpha^3}{45} - \cdots , \tag{5.22}$$

so that retaining only the first term, we have from (5.20)

$$I = \frac{NM}{3}\alpha = \frac{NM^2}{3kT}H. \tag{5.23}$$

This means that the magnetization is proportional to the magnetic field. The susceptibility is, therefore, given by

$$\chi = \frac{I}{H} = \frac{NM^2}{3kT}. \tag{5.24}$$

Thus we find that the susceptibility is inversely proportional to the absolute temperature. This relationship is known as the *Curie law*.

In the above calculation, we assumed that the spin can take all possible orientations. In reality, a spin can have only discrete orientations because of spatial quantization, as shown in (3.25). In particular, if we set the z-axis parallel to the magnetic field, the z-component of M is given by

$$M_z = gM_B J_z, \tag{5.25}$$

where J_z can take only $2J+1$ values

$$J_z = J, J-1, \ldots, -(J-1), -J. \tag{5.26}$$

Therefore, the average magnetization in a magnetic field, H, is given by

$$I = NgM_B \frac{\sum_{J_z=-J}^{J} J_z \exp\left(\frac{gM_B}{kT}HJ_z\right)}{\sum_{J_z=-J}^{J} \exp\left(\frac{gM_B}{kT}HJ_z\right)}$$

$$= NgJM_B\left(\frac{2J+1}{2J}\coth\frac{2J+1}{2J}\alpha - \frac{1}{2J}\coth\frac{\alpha}{2J}\right), \tag{5.27}$$

where the function in parentheses is called the *Brillouin function* and is denoted by $B_J(\alpha)$. The functional form of $B_J(\alpha)$ is similar to that of the Langevin function and in the limit of $J \to \infty$, they coincide. Figure 5.10 shows the magnetization curves of three paramagnetic salts containing Cr^{3+}, Fe^{3+}, and Gd^{3+}. The solid curves represent $JB_J(\alpha)$ for $J = \frac{3}{2}$, $\frac{5}{2}$, and $\frac{7}{2}$ (note that in these ions $L = 0$, so $J = S$). All of the theoretical curves reproduce the experimental points quite well. For $\alpha \ll 1$, $B_J(\alpha)$ can be expanded as

$$B_J(\alpha) = \frac{J+1}{3J}\alpha - \frac{[(J+1)^2 + J^2](J+1)}{90J^3}\alpha^3 + \cdots . \tag{5.28}$$

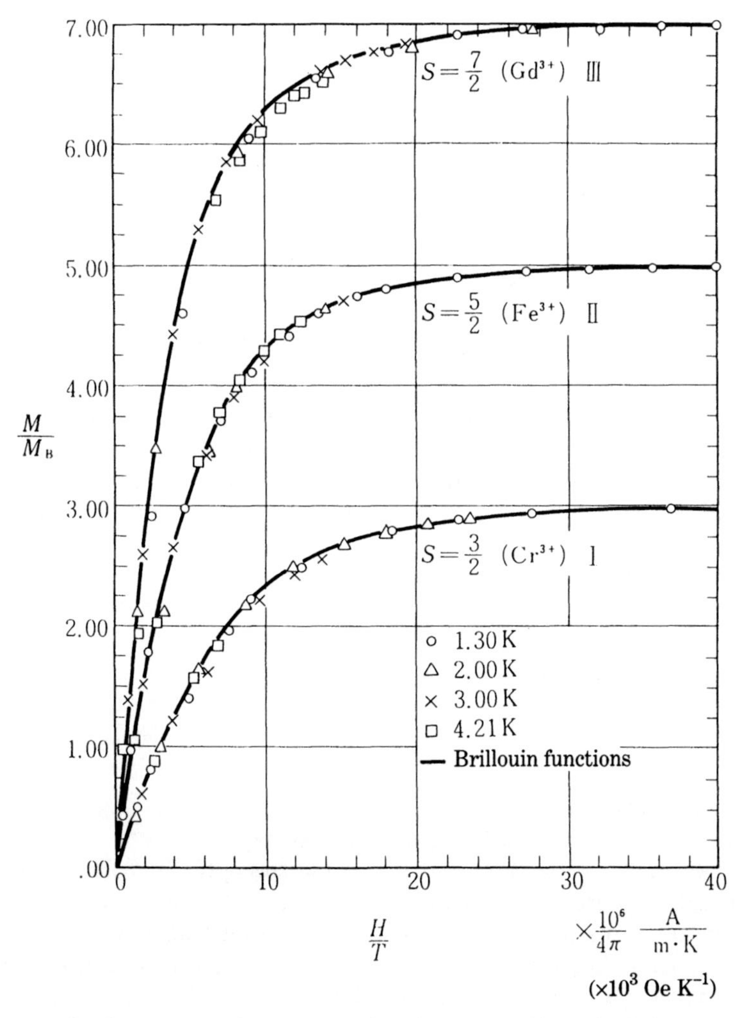

Fig. 5.10. Magnetization curves of paramagnetic salts: I potassium chromium alum, II ferric ammonium alum, and III gadolinium sulfate octahydrate. (After W. E. Henry[7])

If we put $J = \infty$ in (5.28), we find that (5.28) becomes equal to (5.22). Thus the Brillouin function includes the Langevin function as a special case ($J = \infty$). Considering that $\alpha = JM_B H/kT$, if we adopt only the first term in (5.28) for the Brillouin function in (5.27), we have

$$\chi = \frac{Ng^2 J(J+1) M_B^2}{3kT}. \tag{5.29}$$

Comparing this equation with (5.24), we find that if we replace M in (5.24) by

$$M_{\text{eff}} = g\sqrt{J(J+1)}\, M_B, \tag{5.30}$$

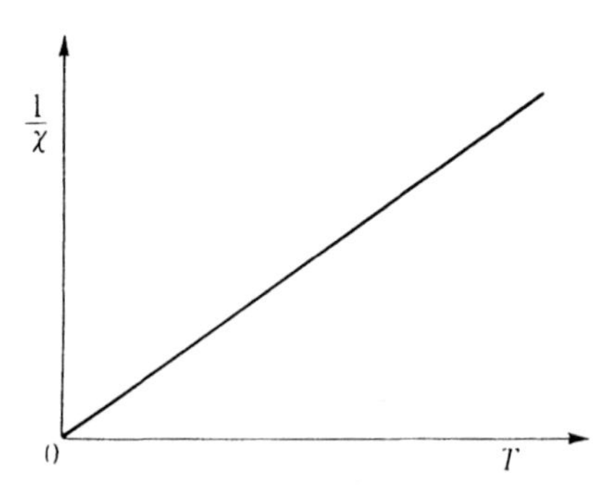

Fig. 5.11. $1/\chi$ versus T curve for Langevin paramagnet.

(5.24) holds. The quantity M_{eff} is the *effective magnetic moment* as already given by (3.41). Equation (5.29), which gives the linear relationship between $1/\chi$ and T, as shown in Fig. 5.11, is known as the *Curie law*. From the slope of this curve, we can calculate the effective magnetic moment. On the other hand, from the saturation magnetization given in (5.27), or

$$I_s = NgJM_B, \tag{5.31}$$

we can get the *saturation magnetic moment*

$$M_s = gJM_B, \tag{5.32}$$

as already given by (3.34).

In addition to Langevin paramagnetism, paramagnetic metals and alloys exhibit nearly temperature-independent paramagnetic susceptibility. This is called *Pauli paramagnetism.*[8] The susceptibility in this case is proportional to the density of states at the Fermi level, which corresponds to several thousand K (refer to Chapter 8). Therefore, a change in temperature of a few hundred kelvin has only a small effect on the Fermi level. *Orbital paramagnetism* produced by orbital magnetic moments[9] is also temperature-independent. When the effective magnetic moment is deduced from the temperature dependence of the paramagnetic susceptibility, this temperature-independent term must be subtracted before applying the Curie law. For this purpose, we must measure the temperature dependence over a wide range of temperature, so as to separate the Langevin term from the temperature independent one.

PROBLEMS

5.1 Knowing that copper has atomic number 29, atomic mass 63.54, density $8.94\,\text{g}\,\text{cm}^{-3}$, and average orbital radius 0.5 Å, calculate its relative diamagnetic susceptibility. Use the following values: Avogadro's number $N = 6.02 \times 10^{23}\,\text{mol}^{-1}$, electric charge of electron $e = 1.60 \times 10^{-19}\,\text{C}$, and electron mass $m = 9.11 \times 10^{-31}\,\text{kg}$.

5.2 Calculate the relative paramagnetic susceptibility for an ideal gas, in which each molecule has a magnetic moment with $J = 1$, $g = 2$, at 1 atmosphere pressure and 0 °C. The 1 mol ideal gas takes a volume of 22.4 l at the above-mentioned temperature and pressure. Avogadro's number is $6.02 \times 10^{23}\ \text{mol}^{-1}$.

REFERENCES

1. J. H. Van Vleck, *The theory of electric and magnetic susceptibilities* (Clarendon Press, Oxford, 1932).
2. W. R. Myers, *Rev. Mod. Phys.*, **24** (1952), 15.
3. L. Pauling, *J. Chem. Phys.*, **4** (1936), 673.
4. E. Vogt, *Magnetism and metallurgy* I (ed. Berkowitz and Kneller, Academic Press, New York, 1969) p. 252.
5. *Handbook of physics & chemistry* (53rd edn.) (Chem. Rubber Co., Cleveland, 1972).
6. J. G. Bednorz and K. A. Müller, *Z. Phys.*, **B64** (1986), 189.
7. W. E. Henry, *Phys. Rev.*, **88** (1952), 559.
8. W. Pauli, *Z. Physik*, **41** (1926), 81.
9. R. Kubo and Y. Obata, *J. Phys. Soc. Japan*, **11** (1956), 547.

6

FERROMAGNETISM

The term *ferromagnetism* is used to characterize strongly magnetic behavior, such as the strong attraction of a material to a permanent magnet. The origin of this strong magnetism is the presence of a spontaneous magnetization produced by a parallel alignment of spins. Instead of a parallel alignment of all the spins, there can be an anti-parallel alignment of unequal spins. This also results in a spontaneous magnetization, which we call *ferrimagnetism*. In this book we describe either of these cases, both of which lead to the presence of spontaneous magnetization, as ferromagnetism in a wide sense; this term then includes both ferromagnetism and ferrimagnetism. In this section we discuss the origin and nature of ferromagnetism in a narrow sense.

6.1 WEISS THEORY OF FERROMAGNETISM

The mechanism for the appearance of spontaneous magnetization was first clarified by P. Weiss[1] in 1907. He assumed that in a ferromagnetic material there exists an effective field which he called the *molecular field*. This field was considered to align the neighboring spins parallel to one another. As discussed in Section 5.2, the ensemble of non-interacting spins is subject to thermal agitation and can be magnetized only if an extremely high magnetic field is applied. Weiss considered that a molecular field could be produced at the site of one spin by the interaction of the neighboring spins (Fig. 6.1). He assumed that the intensity of the molecular field is proportional to the magnetization, or

$$H_m = wI. \tag{6.1}$$

Note that this is just like the Lorentz field given by (1.52) acting in a spherical hole in a magnetized body. It was found (as you will see later) that the molecular field coefficient w in (6.1) is much larger than the Lorentz field coefficient $1/(3\mu_0)$, so that the molecular field cannot be attributed to a classical magnetostatic interaction. The physical origin of this interaction will be discussed in (6.3).

The average magnetization under the action of an external field, H, and a molecular field, wI, is given by

$$I = NM \frac{\displaystyle\int_0^\pi \exp\left(\frac{M(H+wI)}{kT}\right)\cos\theta\sin\theta\,d\theta}{\displaystyle\int_0^\pi \exp\left(\frac{M(H+wI)}{kT}\right)\sin\theta\,d\theta}$$

$$= NML(\alpha), \tag{6.2}$$

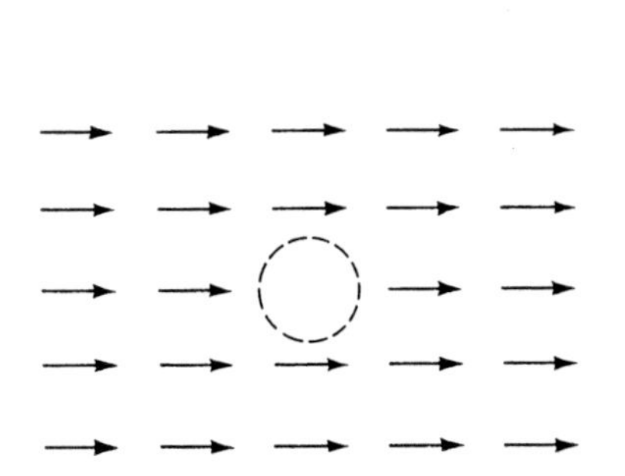

Fig. 6.1. The molecular field acting at the site from which a spin is removed.

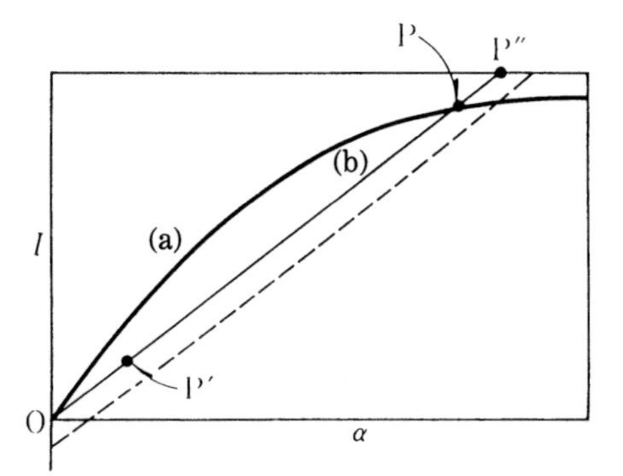

Fig. 6.2. Graphical solution for spontaneous magnetization: (a) Langevin function (6.2); (b) function (6.4).

similar to (5.16). The parameter in this case is given by

$$\alpha = \frac{M(H + wI)}{kT}. \tag{6.3}$$

Since (6.3) includes I, we can solve (6.3) with respect to I, and obtain

$$I = \frac{kT}{Mw}\alpha - \frac{H}{w}. \tag{6.4}$$

The solution of (6.2) must satisfy (6.2) and (6.4) simultaneously with the same value of α.

Such a solution can be obtained graphically. Figure 6.2 shows plots of (6.2) and (6.4) as a function of α. Equation (6.2) represents the Langevin function as shown by curve (a) in the figure. Let us first assume that $H = 0$, because an external field, H, is not necessary to produce a spontaneous magnetization. In this case, the relationship (6.4) becomes a straight line (b) through the origin. The intersection points O and P of curves (a) and (b) represent solutions. However, the point O represents an unstable solution, because if the magnetization happens to take some non-zero value, say P′, near the origin, the state P′ must rise along the curve b (this follows from the definition of α, which requires that the state P′ must always stay on the line (b)). The state of thermal equilibrium represented by curve (a) is above point P′ until P′ reaches P. Physically, point O represents a completely random orientation of the spins. Once there is some alignment of spins by random processes, the alignment produces a non-zero molecular field which leads to more complete alignment. On the other hand, point P represents a stable solution, because if the state changes from P to P″, the equilibrium state is always closer to P.

Now let us consider the temperature dependence of spontaneous magnetization on the basis of this graphical solution. At $T = 0$, the slope of line (b) is zero, so that the point P goes far to the right, where $L(\alpha) = 1$ and $I = NM$. This state is called *absolute saturation magnetization* and denoted by I_{s0}, or

$$I_{s0} = NM. \tag{6.5}$$

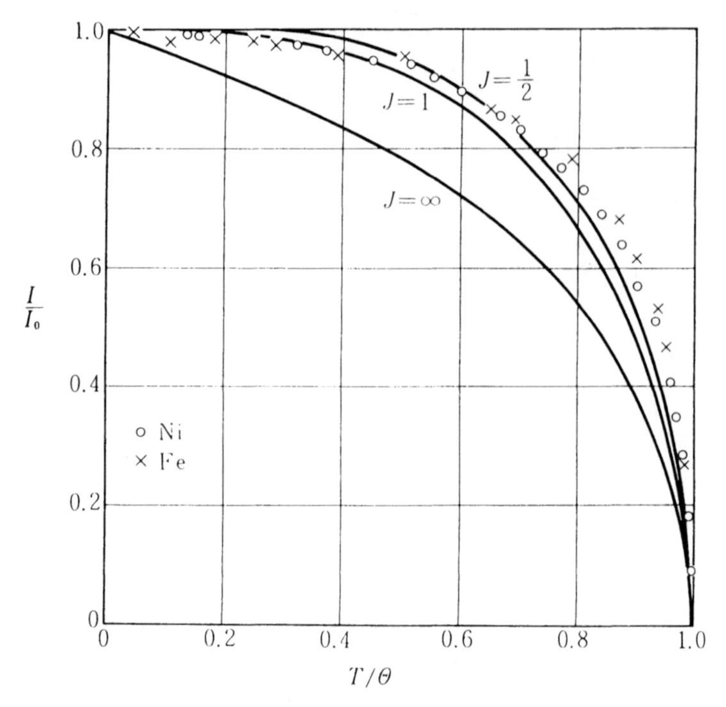

Fig. 6.3. Temperature dependence of spontaneous magnetization. The points are experimental values for Ni and Fe, compared with theoretical curves calculated from the Weiss theory using Brillouin functions with $J = \frac{1}{2}$, 1, and ∞. (After Becker and Döring[2])

As the temperature increases, the slope of line (b) increases. The point P moves down along curve (a) and finally approaches O as the gradient of line (b) approaches the initial gradient of curve (a). This final temperature is called the *Curie point*, and denoted by Θ_f. At any temperature higher than the Curie point, the solution is always $I = 0$, meaning that the spontaneous magnetization vanishes.

The temperature dependence of spontaneous magnetization obtained by this graphical method is plotted as a function of temperature in Fig. 6.3 (see the curve denoted by $J = \infty$.) The curves denoted by $J = 1$ and $J = \frac{1}{2}$ are the solutions using the Brillouin functions $B_1(\alpha)$ and $B_{1/2}(\alpha)$, respectively, instead of the Langevin function $L(\alpha)$. The experimental points for Ni and Fe are in better agreement with the curves $J = \frac{1}{2}$ and $J = 1$ than with the curve $J = \infty$. This means that the spatial quantization of spins is a more realistic picture than the classical view of unlimited spin orientations.

In order to determine the Curie point, Θ_f, the gradient of the initial slope of curve (a) in Fig. 6.2, or

$$\frac{\partial I}{\partial \alpha} = NM\left(\frac{\partial L(\alpha)}{\partial \alpha}\right)_{\alpha=0} = \frac{NM}{3} \tag{6.6}$$

(see (5.22)), must be equal to the gradient of line (b) at $T = \Theta_f$, or

$$\left(\frac{\partial I}{\partial \alpha}\right)_{\alpha=0} = \frac{k\Theta}{Mw}. \tag{6.7}$$

Equating (6.6) with (6.7), we have

$$\Theta_f = \frac{NM^2 w}{3k}. \tag{6.8}$$

If we use the Brillouin function (5.27) for the Langevin function, we have

$$\Theta_f = \frac{(J+1)NM^2 w}{3Jk} = \frac{NM_{\text{eff}}^2 w}{3k} \tag{6.9}$$

(see (5.28)). Thus the Curie point is a good measure of the molecular field coefficient w. For iron, using the values $\Theta_f = 1063\,\text{K}$, $M = 2.2M_B$, $N = 8.54 \times 10^{28}\,\text{m}^{-3}$, $J = 1$, we have

$$w = \frac{3Jk\Theta_f}{(J+1)NM^2} = \frac{3(1.38 \times 10^{-23})(1063)}{2(8.54 \times 10^{28})(2.2)^2(1.17 \times 10^{-29})^2} = 3.9 \times 10^8,$$

from which the molecular field is estimated to be

$$\begin{aligned} H_m &= wI \simeq wNM = (3.9 \times 10^8)(8.54 \times 10^{28})(2.2)(1.17 \times 10^{-29}) \\ &= 0.85 \times 10^9\,\text{A}\,\text{m}^{-1} (\simeq 1.1 \times 10^7\,\text{Oe}). \end{aligned} \tag{6.10}$$

This value is very much larger than the Lorentz field

$$H_m = \frac{1}{3\mu_0} I = 5.8 \times 10^5\,\text{A}\,\text{m}^{-1} (\simeq 7400\,\text{Oe}), \tag{6.11}$$

which results from the magnetostatic interaction. The physical origin of this enormously large molecular field is the exchange interaction, which will be discussed later.

So far we have ignored the effect of the external field. Usually the external field is so weak compared with the exchange field that is has very little effect on the magnitude of the spontaneous magnetization. The effect of the external field is given by the second term of (6.4), which causes a shift of line (b) in Fig. 6.2 downwards and accordingly a shift of point P upwards. Thus the magnetization I increases slightly. The susceptibility in this case is given by

$$\chi = \frac{dI}{dH} = NM \frac{\partial L(\alpha)}{\partial \alpha} \frac{d\alpha}{dH}. \tag{6.12}$$

On the other hand, from the definition of α in (6.3), we have

$$\frac{d\alpha}{dH} = \frac{M}{kT} + \frac{Mw}{kT} \frac{dI}{dH}. \tag{6.13}$$

Eliminating $d\alpha/dH$ from (6.12) and (6.13), we have

$$\chi = \frac{NM^2 L'(\alpha)}{k(T - 3\Theta_f L'(\alpha))}, \tag{6.14}$$

where $L'(\alpha)$ is the derivative of $L(\alpha)$ with respect to α.

At a temperature $T \ll \Theta_f$, $L'(\alpha)$ is small enough to give a very small value for the susceptibility (6.14). In order to measure this susceptibility accurately we must use a high magnetic field. In this sense, we call it the *high field susceptibility*. When,

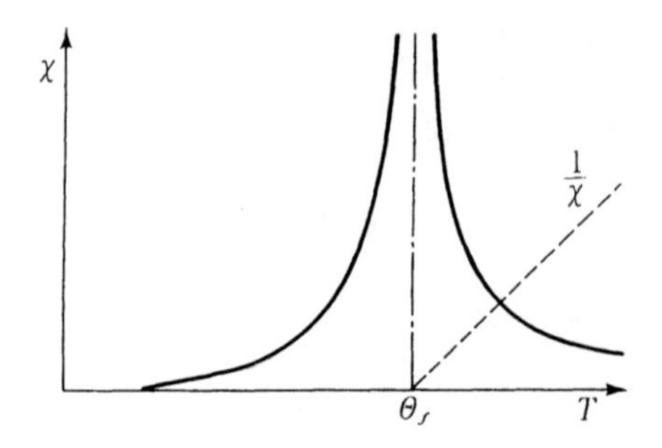

Fig. 6.4. Divergence of magnetic susceptibility in the vicinity of the Curie point.

however, T approaches Θ_f, $L'(\alpha) = \frac{1}{3}$, so that the denominator of (6.14) tends to vanish and the susceptibility approaches infinity. At $T > \Theta_f$, the point P stays near the origin in Fig. 6.2, where $L'(\alpha) = \frac{1}{3}$, so that (6.14) becomes

$$\chi = \frac{NM^2}{3k(T - \Theta_f)}. \tag{6.15}$$

That is, the susceptibility is inversely proportional to the deviation of T from Θ_f. This is called the *Curie–Weiss law*. Taking into account spatial quantization, the quantity M in (6.15) must be replaced by the effective magnetic moment given by (5.30). The temperature dependence of the susceptibility as predicted from these arguments is shown over the complete temperature range in Fig. 6.4, which shows a divergence of the susceptibility at the Curie point. At $T < \Theta_f$, some effects of technical magnetization may overlap with the high-field susceptibility. Separation of the two phenomena is difficult, and may require the use of extremely high fields.

Let us discuss in more detail the shape of the magnetization curve near the Curie point, where the magnetization is relatively small in comparison with the absolute spontaneous magnetization. According to (5.22) or (5.28), the magnetization in this range can be approximated as

$$I = A\alpha - B\alpha^3, \tag{6.16}$$

where the coefficients A and B are given by

$$A = \frac{NM}{3}, \qquad B = \frac{NM}{45}, \tag{6.17}$$

for the Langevin function and

$$A = \frac{NgM_B(J+1)}{3}, \qquad B = \frac{NgM_B[(J+1)^2 + J^2](J+1)}{90J^2}, \tag{6.18}$$

for the Brillouin function. Substituting (6.3) in (6.16), and noting that $wI \gg H$, we have

$$I = \frac{AM}{kT}H + \frac{AMw}{kT}I - \frac{BM^3w^3}{k^3T^3}I^3;$$

therefore

$$\frac{BM^3w^3}{k^3T^3}I^3 - \left(\frac{\Theta_f}{T} - 1\right)I = \frac{AM}{kT}H. \tag{6.19}$$

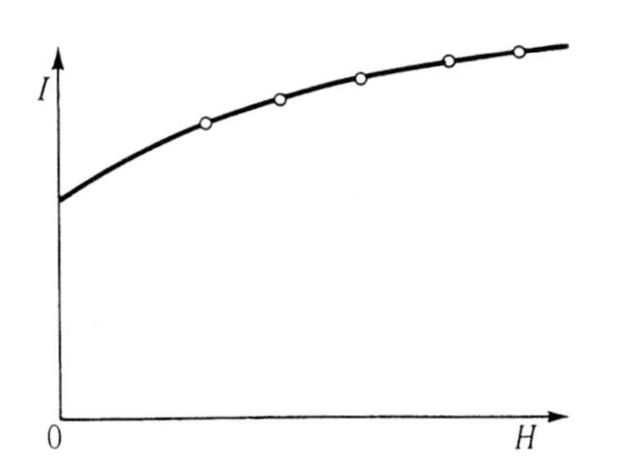

Fig. 6.5. Magnetization curve at a temperature near the Curie point.

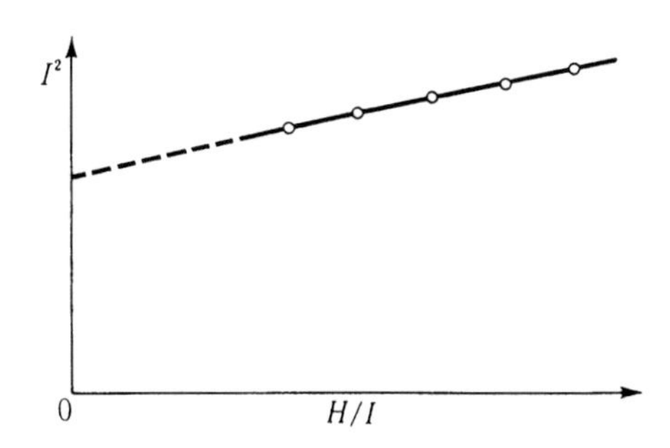

Fig. 6.6. Arrott plot.

This relationship between I and H is shown graphically in Fig. 6.5. In practice, the magnetization is also affected by rotation against the magnetocrystalline anisotropy (Chapter 12), so that the spontaneous magnetization must be determined by extrapolation from the high-field region. For this purpose, we divide both sides of (6.19) by I to give

$$I^2 = \frac{k^3T^2}{BM^3w^3}(\Theta_f - T) + \frac{A}{B}\frac{k^2T^2}{M^2w^3}\frac{H}{I}. \tag{6.20}$$

Therefore if I^2 is plotted against H/I, we get the linear relationship shown in Fig. 6.6. Extrapolating to $H/I = 0$, we can determine the value of I_s. This plot is called the *Arrott plot*.[3] From the first term of (6.20), we know that the temperature dependence of spontaneous magnetization is given by

$$I_s = \sqrt{\frac{k^3}{BM^3w^3}}\, T(\Theta_f - T)^{1/2}. \tag{6.21}$$

Therefore the Curie point can be determined by extrapolating the line representing I_s^2 versus T towards $I_s^2 \to 0$. However, such a relationship does not necessarily hold in every case. For more details the reader may refer to Kouvel and Fisher,[4] Kouvel and Rodbell,[5] and Arrott and Noakes.[6]

The Curie point can also be determined by observing the anomalous specific heat. The work necessary to increase the spontaneous magnetization from I to $I + \delta I$ in the presence of molecular field (no external field) is given by

$$\delta E = -wI\,\delta I, \tag{6.22}$$

so that the internal energy of the state associated with spontaneous magnetization I is given by

$$E = \int \mathrm{d}E = -\int_0^I wI\,\mathrm{d}I = -\frac{w}{2}I^2. \tag{6.23}$$

This internal energy is plotted as a function of temperature in Fig. 6.7 (solid curve). By differentiating (6.23) with respect to temperature, we obtain the specific heat

$$C_v = \frac{\mathrm{d}E}{\mathrm{d}T} = -wI\frac{\mathrm{d}I}{\mathrm{d}T}, \tag{6.24}$$

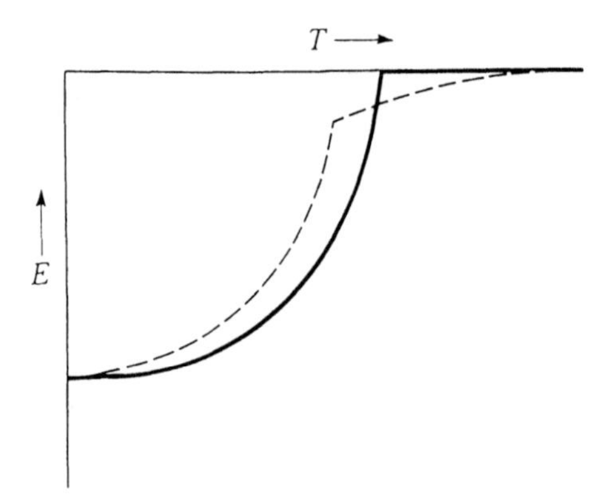

Fig. 6.7. Temperature dependence of the magnetic energy. The solid curve was calculated by the Weiss theory, while the dashed curve represents experiment.

which has a sharp peak as shown by the solid curve in Fig. 6.8. From the relationship (6.24), we can also estimate the molecular field coefficient w.

6.2 VARIOUS STATISTICAL THEORIES

Experiment shows that the anomalous specific heat is accompanied by a tail above the Curie point, as shown by the dashed curve in Fig. 6.8. By integrating the specific heat with respect to temperature, we know that the internal energy is increasing even above the Curie point. This is due to the presence of clusters of parallel spins above the Curie point. These clusters exist because the molecular field is due to short-range forces caused by the exchange interaction.

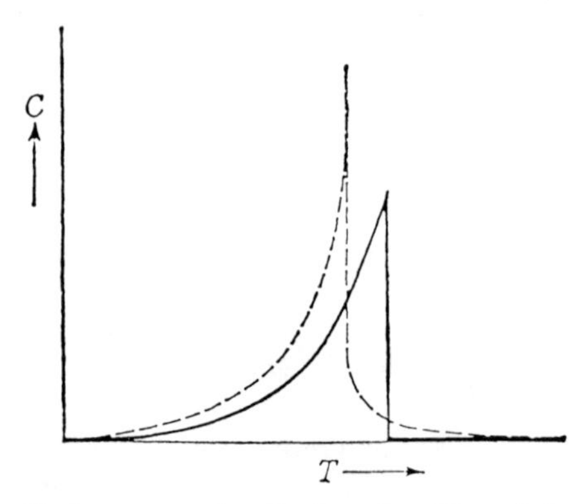

Fig. 6.8. Anomalous specific heat near the Curie point. The solid curve was calculated by the Weiss theory, while the dashed curve represents experiment.

The exchange interaction was first proposed by Heisenberg[7] in 1928. He suggested that between atoms with spins S_i and S_j there is an interaction with energy

$$w_{ij} = -2J\boldsymbol{S}_i \cdot \boldsymbol{S}_j, \tag{6.25}$$

where J is the exchange integral. We will discuss the physical meaning of this interaction in Section 6.3. The magnitude of J is in the order of $10^3\,\text{cm}^{-1}$ ($\sim 10^3$ K), which is 10^3 times larger than the dipole interaction given by (1.17). If $J > 0$, the energy is lowest when $\boldsymbol{S}_i \parallel \boldsymbol{S}_j$, so that a ferromagnetic alignment is stable, while if $J < 0$, an antiferromagnetic alignment of $\boldsymbol{S}_i$ and $\boldsymbol{S}_j$ is stable and antiferromagnetism results. Since the exchange interaction is short-range, the value of J is largest for nearest-neighbor spins. This tendency to align the nearest-neighbor spins parallel (for positive J) causes complete parallel alignment of the entire spin system, which results in ferromagnetism. In the Weiss theory of ferromagnetism discussed in Section 6.1, the molecular field is assumed to be proportional to the average magnetization. This is equivalent to the assumption that the value of J is the same for all spin pairs, not just nearest-neighbor pairs. In reality, however, the exchange interaction is short-range, so that when the temperature is near the Curie point, where the parallelism of the spins is considerably disturbed, near-neighbor spins still tend to remain parallel, thus forming spin clusters.

A similar phenomenon appears in a superlattice, in which two kinds of atoms, A and B, are arranged in a regular alternating pattern. This problem was treated by Bragg and Williams,[8] using statistical mechanics in a similar way to the Weiss theory. Considering that the tendency to form a superlattice is equivalent to a short-range force acting to make A atoms tend to attract B atoms as their nearest neighbors, Bethe[9] succeeded in calculating the degree of short-range order. Applying this method to a spin system, Peierls[10] treated spin clusters in ferromagnetic theory. Therefore this approximation for treating short-range order is referred to as the *Bethe–Peierls method*.

Let us assume the *Ising model*[11]* in which $S = \frac{1}{2}$, so that the z-component of S, S_z, can take values only of $\frac{1}{2}$ or $-\frac{1}{2}$, and show using the Bethe–Peierls approximation how the spin clusters are created.

We assume that a particular spin S_i can have the value of $+\frac{1}{2}$ or $-\frac{1}{2}$ under the influence of the exchange interaction with the nearest neighbor spins, S_j. The situation is also the same for S_j, except that the exchange interaction from the other spins is replaced by the molecular field, which is determined by the average value of the spin S (see Fig. 6.9). This model is called *Bethe's first approximation*.

Then the exchange energy associated with the spins S_i and all the spins S_j is given by

$$U = -2JS_i \sum_{j=1}^{z} S_j - 2M_B H_m \sum_{j=1}^{z} S_j. \tag{6.26}$$

* The case in which the spins can be oriented in all directions, as treated in Section 6.1, is called the *Heisenberg Model*.

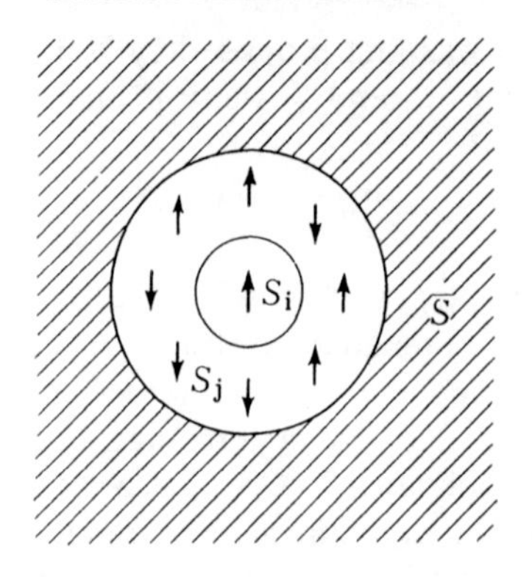

Fig. 6.9. Model for Bethe's first approximation.

If p spins of the total of the z values of S_j take the value $\frac{1}{2}$, while q spins take the value $-\frac{1}{2}$, then (6.26) becomes

$$U = -(JS_i + M_B H_m)(p-q). \tag{6.27}$$

Let us denote U in (6.27) as U_p^+ if $S_i = \frac{1}{2}$, and as U_p^- if $S_i = -\frac{1}{2}$. Then the probability that S_i has the value $\frac{1}{2}$ is given by

$$p_{i+} = \sum_{p=0}^{z} \frac{z!}{p!q!} \exp\left(-\frac{U_p^+}{kT}\right) = 2^z \cosh^z \frac{J + 2M_B H_m}{2kT},^* \tag{6.28}$$

and the probability that S_i has the value $-\frac{1}{2}$ is given by

$$p_{i-} = \sum_{p=0}^{z} \frac{z!}{p!q!} \exp\left(-\frac{U_p^-}{kT}\right) = 2^z \cosh^z \frac{-J + 2M_B H_m}{2kT}. \tag{6.29}$$

Therefore the average value of S_i is given by

$$\langle S_i \rangle = \frac{p_{i+}(1/2) + p_{i-}(-1/2)}{p_{i+} + p_{i-}}, \tag{6.30}$$

where the undetermined parameter, H_m, remains in p_{i+} and p_{i-} as shown in (6.28) and (6.29). Similarly the average value of S_j is given by

$$\langle S_j \rangle = \frac{p_{j+}(1/2) + p_{j-}(-1/2)}{p_{j+} + p_{j-}}, \tag{6.31}$$

which also includes the undetermined parameter H_m.

$$^*p_{i+} = \exp\left(-\frac{J + 2M_B H_m}{2kT} z\right) + \frac{z!}{1!(z-1)!} \exp\left(-\frac{J + 2M_B H_m}{2kT}(z-1)\right) + \cdots + \exp\left(\frac{J + 2M_B H_m}{kT} z\right)$$

$$= \left\{\exp\left(-\frac{J + 2M_B H_m}{2kT}\right) + \exp\left(\frac{J + 2M_B H_m}{2kT}\right)\right\}^z = 2^z \cosh^z \frac{J + 2M_B H_m}{2kT}.$$

Since S_i and S_j must be equivalent, $\langle S_i \rangle = \langle S_j \rangle$, and finally we get

$$\left(\frac{\cosh(J + 2M_B H_m/2kT)}{\cosh(-J + 2M_B H_m/2kT)} \right)^{z-1} = \exp\left(\frac{2M_B H_m}{kT} \right). \tag{6.32}$$

From this relationship we obtain H_m as a function of temperature T, and using (6.30), we can calculate the spontaneous magnetization

$$I_s = 2NM_B \langle S_i \rangle. \tag{6.33}$$

At a temperature near the Curie point, H_m becomes small, so that $M_B H_m \ll kT$, and we have

$$\frac{k\Theta_f}{J} = \frac{1}{\log[z/(z-2)]}. \tag{6.34}$$

In the case of a two dimensional lattice, $z = 4$, so that

$$\frac{k\Theta_f}{J} = \frac{1}{\log 2} = 1.443. \tag{6.35}$$

In the case of a body-centered cubic lattice, $z = 6$, so that

$$\frac{k\Theta_f}{J} = \frac{1}{\log 1.5} = 2.446. \tag{6.36}$$

In order to compare this result with the Weiss theory, the energy per spin for $S = \frac{1}{2}$ is calculated as

$$wNM_B^2 = 2Jz(\tfrac{1}{2})^2, \tag{6.37}$$

so that using (6.9) and taking $M_{eff} = 3M_B$, we get

$$\frac{k\Theta_f}{J} = \tfrac{1}{2}z. \tag{6.38}$$

This gives the values 2 for $z = 4$, and 3 for $z = 6$, both of which are larger than the values given by (6.35) and (6.36). In Table 6.1, values of $k\Theta_f/J$ are listed for various approximations.

Bethe's second approximation rigorously takes into consideration the influence of the second-nearest neighbors. The exact solution is that given by Onsager[12] for a

Table 6.1. The values of $k\Theta_f/J$ for $S = \frac{1}{2}$ calculated by various approximations (after Oguchi[15]).

Approximations	$z = 4$	$z = 6$
Weiss theory	2	3
Bethe's 1st	1.443	2.466
Bethe's 2nd	1.312	2.372
Exact solution	1.135	

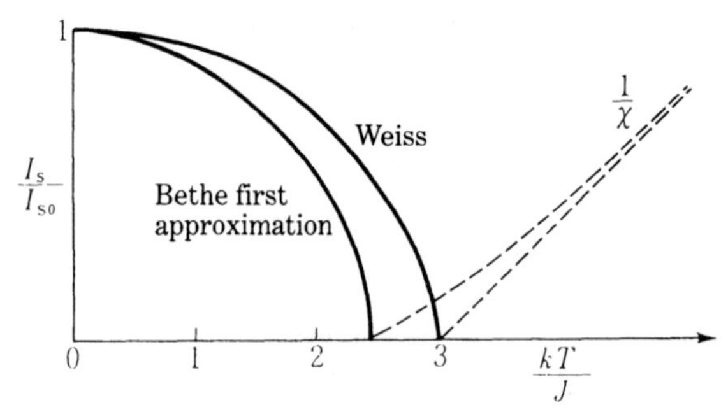

Fig. 6.10. Comparison of I_s versus T and $1/\chi$ versus T curves for the Weiss theory and for Bethe's first approximations.

two-dimensional lattice. It can be seen that the value of $k\Theta_f/J$ gets smaller as the order of approximation increases. This means that the Curie point is reduced for a given J. For instance, Fig. 6.10 shows the temperature dependence of I_s for Weiss and Bethe's first approximations, and we can clearly see a difference in the Curie points between the two approximations. This fact must be taken into consideration when we estimate the value of J or the value of the molecular field from the value of the Curie point.

The $1/\chi$ versus T curve calculated by the Bethe approximation shows a deviation from the linear Curie–Weiss law, which is clearly due to cluster formation above the Curie point. Experiments also show such a deviation. Thus the linear extrapolation of the high-temperature part of the $1/\chi$ versus T curve to the abscissa differs from the real Curie point. The extrapolated intersection is called the *asymptotic Curie point* (symbol Θ_a) to distinguish it from the real Curie point where the spontaneous magnetization vanishes. The Bethe approximation also explains the influence of cluster formation on the temperature dependence of energy and specific heat, as shown by the dashed curves in Figs 6.7 and 6.8.

The temperature dependences of various physical quantities are altered by the appearance of parallel spin clusters. Let us assume that the temperature dependence of the specific heat, C, spontaneous magnetization, I_s, and magnetic susceptibility, χ, near the Curie point are given by

$$\left.\begin{array}{ll} C(T) \propto (\Theta_f - T)^{-\alpha'} & T \lesssim \Theta_f \\ C(T) \propto (T - \Theta_f)^{-\alpha} & T \gtrsim \Theta_f \\ I_s(T) \propto (\Theta_f - T)^{\beta} & T \lesssim \Theta_f \\ \chi(T) \propto (\Theta_f - T)^{-\gamma'} & T \lesssim \Theta_f \\ \chi(T) \propto (T - \Theta_f)^{-\gamma} & T \gtrsim \Theta_f \\ \text{and also at the Curie point} & \\ I \propto H^{1/\delta} & \end{array}\right\} \quad (6.39)$$

where the exponents take the values given in Table 6.2 for Fe, Ni, and Gd. These exponents are called the *critical indices*

According to the Weiss approximation, $\beta = \frac{1}{2}$, as shown by (6.21), and also $\gamma =$

Table 6.2. Critical indices (after Oguchi[15]).

Materials	θ_f (K)	α	α'	β	γ	γ'	δ
Fe	1043.0	$\lesssim 0.17$	$\lesssim 0.13$	0.34 ± 0.02	1.33	—	—
Ni	629.6	0.104 ± 0.050	-0.262 ± 0.060	0.51 ± 0.04	1.32 ± 0.02	—	4.2 ± 0.1
Gd	292.5	—	—	—	1.33	—	4.0 ± 0.1
Ising model ($S = \frac{1}{2}$, 3 dimens.)		0.125	0.063	0.313	1.25	1.31	5.2
Weiss theory		—	—	0.5	1	1	3

$\gamma' = 1$ as shown by (6.14) and (6.15). These values are not in very good agreement with the experimental results. However, more accurate approximations based on the three-dimensional Ising model with $S = \frac{1}{2}$ give better values, as listed also in Table 6.2.

According to the *scaling law*, there exist the following relationships between these critical indices:

$$\left.\begin{aligned} \alpha' + 2\beta + \gamma' &= 2 \\ \alpha' + \beta(\delta + 1) &= 2 \\ \alpha = \alpha', \qquad \gamma &= \gamma' \end{aligned}\right\}. \tag{6.40}$$

The first two formulas hold well for various approximations, as verified by the numbers for the Weiss and Ising model given in Table 6.2. This law was deduced by Kadanoff *et al.*[13] by assuming that the magnitude of the unit moment and the effective temperature depend on the size of the cluster.

If we assume that $z = 2$ in (6.34), we find that the right-hand side vanishes. In other words, in a one-dimensional ferromagnet, in which the spins are connected in a chain, no ferromagnetism appears. Bloch[14] treated the thermal disturbance of the spin arrangement at low temperatures using a *spin wave model*, in which he assumed that the reversed spins propagate just like waves. He concluded that no ferromagnetism appears for one- and two-dimensional spin lattices. Other strict statistical treatments also prove that no ferromagnetism appears in a two-dimensional lattice in the Heisenberg model, but that it does appear in the Ising model or in the presence of a uniaxial anisotropy, both of which tend to develop a parallel alignment of the spin system. Bloch showed that at low temperature the spontaneous magnetization depends on temperature as

$$I_s = I_{s0}\left[1 - C\left(\frac{kT}{J}\right)^{3/2}\right], \tag{6.41}$$

where J is the exchange integral and C is a constant which depends on the crystal structure: 0.1174 for simple cubic, 0.0587 for body-centered cubic, and 0.0294 for face-centered cubic lattices. This $T^{3/2}$ law holds fairly well experimentally, except for the values of the proportionality constant C.

6.3 EXCHANGE INTERACTION

The *exchange interaction* was first treated by Heisenberg[7] in 1928 to interpret the origin of the enormously large molecular fields acting in ferromagnetic materials. This interaction is due to a quantum mechanical effect, so that it is rather difficult to explain it in terms of classical physics. If, however, one accepts the Pauli exclusion principle, the exchange interaction may be understood as follows: Suppose that two atoms with unpaired electrons approach each other. If the spins of these two electrons are antiparallel to each other, the electrons will share a common orbit, thus increasing the electrostatic Coulomb energy. If, however, the spins of these two electrons are parallel, they cannot share a common orbit because of the Pauli exclusion principle, so that they form separate orbits, thus reducing the Coulomb interaction. The order of magnitude of the Coulomb energy involved in this case is estimated as

$$U_e = \frac{e^2}{4\pi\varepsilon_0 r} = \frac{(9\times 10^9)(1.6\times 10^{-19})^2}{(1\times 10^{-10})} = 2.1\times 10^{-18}\ \mathrm{J}^*$$

$$= 1.0\times 10^5\ \mathrm{cm}^{-1} = 1.4\times 10^5\ \mathrm{K}, \tag{6.42}$$

where r is the average distance between the two electrons, assumed to be 1 Å. The value estimated from (6.42) is 10^5 times larger that the magnetic dipolar interaction calculated from (1.17). Therefore if this Coulomb energy is disturbed by the Pauli exclusion principle by a small factor, say 1%, the change in Coulomb energy, ΔU_e, is the order of 1400 K, which can explain the magnitude of the molecular field.

The Pauli exclusion principle was first introduced to explain the multiplicity of atomic spectra.[15] Let us elucidate how this principle is expressed in terms of the atomic wave function.

The atomic wave function of an atom with a single electron is a function of a spatial coordinate $\boldsymbol{r}(x, y, z)$ and a spin variable σ $(= \pm\frac{1}{2})$, both of which can be expressed by a general coordinate q. Now let us consider the wave function of an atom with two electrons, $\psi(q_1, q_2)$, where q_1 and q_2 are the general coordinates of electrons 1 and 2, respectively. It is difficult to express this function exactly, but it can be approximated by the product of two wave functions $\psi_1(q_1)$ of electron 1, and $\psi_2(q_2)$ of electron 2, or

$$\psi(q_1, q_2) = \psi_1(q_1)\psi_2(q_2). \tag{6.43}$$

Since the two electrons are indistinguishable, the probability of realizing this state, $\psi^2(q_1, q_2)$, must be invariant for the exchange of the two electrons. In other words,

$$\psi(q_2, q_1) = \psi(q_1, q_2), \tag{6.44}$$

or

$$\psi(q_2, q_1) = -\psi(q_1, q_2). \tag{6.45}$$

The Pauli principle asserts that condition (6.45) is correct, rather than condition (6.44). The reasoning is as follows: suppose that electrons 1 and 2 occupy the same state, ψ_1, which can be written as $\psi_1(q_1)\psi_1(q_2)$. By exchange of the two electrons, this wave function becomes $\psi_1(q_2)\psi_1(q_1)$, which is the same as before. On the other hand,

* Since the velocity of light is given by $c = \dfrac{1}{\sqrt{\varepsilon_0 \mu_0}} = 3\times 10^8\ \mathrm{m\,s^{-1}}$, it follows that

$$\frac{1}{4\pi\varepsilon_0} = \frac{c^2\mu_0}{4\pi} = \frac{9\times 10^{16}\times(4\pi\times 10^{-7})}{4\pi} = 9\times 10^9.$$

according to condition (6.45), the sign of these wave functions must be different. In order to satisfy these conditions, it follows that

$$\psi_1(q_1)\psi_1(q_2) = 0. \tag{6.46}$$

This is another expression of the Pauli exclusion principle.[16]

The relationship (6.45) asserts that the wave function of the two-electron system, $\psi(q_1, q_2)$, must be antisymmetric upon the exchange of the two electrons. The wave function $\psi(q_1, q_2)$ can be expressed as a product of two parts, one containing the spatial coordinate, $\varphi(\boldsymbol{r}_1, \boldsymbol{r}_2)$, and one containing the spin variables, $\chi(\sigma_1, \sigma_2)$; that is

$$\psi(q_1, q_2) = \varphi(\boldsymbol{r}_1, \boldsymbol{r}_2)\chi(\sigma_1, \sigma_2). \tag{6.47}$$

According to (6.45),

$$\psi(q_2, q_1) = \varphi(\boldsymbol{r}_2, \boldsymbol{r}_1)\chi(\sigma_2, \sigma_1) = -\varphi(\boldsymbol{r}_1, \boldsymbol{r}_2)\chi(\sigma_1, \sigma_2). \tag{6.48}$$

In order to realize this condition, it must be that

$$\left.\begin{aligned} \varphi(\boldsymbol{r}_2, \boldsymbol{r}_1) &= \varphi(\boldsymbol{r}_1, \boldsymbol{r}_2), \\ \chi(\sigma_2, \sigma_1) &= -\chi(\sigma_1, \sigma_2); \end{aligned}\right\} \tag{6.49}$$

or

$$\left.\begin{aligned} \varphi(\boldsymbol{r}_2, \boldsymbol{r}_1) &= -\varphi(\boldsymbol{r}_1, \boldsymbol{r}_2), \\ \chi(\sigma_2, \sigma_1) &= \chi(\sigma_1, \sigma_2). \end{aligned}\right\} \tag{6.50}$$

These conditions mean that the state which is antisymmetric for the exchange of spin variables is symmetric for the exchange of spatial coordinates, while the state which is symmetric for the exchange of spin variables is antisymmetric for the exchange of spatial coordinates. Thus the symmetry of the spin variable can affect the symmetry of the part of the wave function containing the spatial coordinates, resulting in a change in electrostatic interaction between the two electrons.

Example

For example, let us consider the case in which two electrons occupy the $1s$ state of an He atom. The Hamiltonian in this case is given by

$$\mathcal{H} = \mathcal{H}_1 + \mathcal{H}_2 + \mathcal{H}_{12}, \tag{6.51}$$

where the first term concerns only electron 1, the second term concerns only electron 2, and the third term concerns the interaction between the two electrons. The actual forms of these three terms are given by

$$\left.\begin{aligned} \mathcal{H}_1 &= -\frac{\hbar^2}{2m}\Delta_1 - \frac{Ze^2}{4\pi\varepsilon_0 r_1} \\ \mathcal{H}_2 &= -\frac{\hbar^2}{2m}\Delta_2 - \frac{Ze^2}{4\pi\varepsilon_0 r_2} \\ \mathcal{H}_{12} &= \frac{e^2}{4\pi\varepsilon_0 r_{12}} \end{aligned}\right\}, \tag{6.52}$$

where

$$\Delta_j = \frac{\partial^2}{\partial x_j^2} + \frac{\partial^2}{\partial y_j^2} + \frac{\partial^2}{\partial z_j^2} \qquad (j = 1, 2), \tag{6.53}$$

Z is the atomic number ($Z = 2$ for He), and r_{12} is the distance between the two electrons.

Suppose that the spins of two electrons are antiparallel. Since the wave function is antisymmetric for the exchange of spin variables, the wave function must be symmetric for the exchange of spatial coordinates. A wave function symmetric for the exchange of spatial coordinates is

$$\varphi_s(\boldsymbol{r}_1, \boldsymbol{r}_2) = \frac{1}{\sqrt{2}}[\varphi_1(\boldsymbol{r}_1)\varphi_2(\boldsymbol{r}_2) + \varphi_2(\boldsymbol{r}_1)\varphi_1(\boldsymbol{r}_2)], \tag{6.54}$$

where $\boldsymbol{r}_1$ and $\boldsymbol{r}_2$ are the radial vectors of the electron 1 and 2, respectively.

On the other hand, if the spins of the two electrons are parallel, the spatial wave function must be antisymmetric, so that

$$\varphi_a(\boldsymbol{r}_1, \boldsymbol{r}_2) = \frac{1}{\sqrt{2}}[\varphi_1(\boldsymbol{r}_1)\varphi_2(\boldsymbol{r}_2) - \varphi_2(\boldsymbol{r}_1)\varphi_1(\boldsymbol{r}_2)]. \tag{6.55}$$

The suffixes s and a signify that the function is symmetric and antisymmetric, respectively.

The energy of the state can be calculated by

$$U = \int \varphi^* \mathscr{H} \varphi \, \mathrm{d}V. \tag{6.56}$$

Assuming that φ_1 and φ_2 are normalized and also orthogonal, the energies of the states given by (6.54) and (6.55) are calculated as

$$\begin{aligned}
U &= \frac{1}{2}\iint [\varphi_1^*(\boldsymbol{r}_1)\varphi_2^*(\boldsymbol{r}_2) \pm \varphi_2^*(\boldsymbol{r}_1)\varphi_1^*(\boldsymbol{r}_2)](\mathscr{H}_1 + \mathscr{H}_2 + \mathscr{H}_{12}) \\
&\quad \times [\varphi_1(\boldsymbol{r}_1)\varphi_2(\boldsymbol{r}_2) \pm \varphi_2(\boldsymbol{r}_1)\varphi_1(\boldsymbol{r}_2)] \, \mathrm{d}V_1 \, \mathrm{d}V_2 \\
&= \frac{1}{2}\left\{\int \varphi_1^*(\boldsymbol{r}_1)\mathscr{H}_1\varphi_1(\boldsymbol{r}_1)\,\mathrm{d}V_1 + \int \varphi_2^*(\boldsymbol{r}_1)\mathscr{H}_1\varphi_2(\boldsymbol{r}_1)\,\mathrm{d}V_1\right. \\
&\quad + \int \varphi_1^*(\boldsymbol{r}_2)\mathscr{H}_2\varphi_1(\boldsymbol{r}_2)\,\mathrm{d}V_2 + \int \varphi_2^*(\boldsymbol{r}_2)\mathscr{H}_2\varphi_2(\boldsymbol{r}_2)\,\mathrm{d}V_2 \\
&\quad + \iint \varphi_1^*(\boldsymbol{r}_1)\varphi_2^*(\boldsymbol{r}_2)\mathscr{H}_{12}\varphi_1(\boldsymbol{r}_1)\varphi_2(\boldsymbol{r}_2)\,\mathrm{d}V_1\,\mathrm{d}V_2 \\
&\quad + \iint \varphi_2^*(\boldsymbol{r}_1)\varphi_1^*(\boldsymbol{r}_2)\mathscr{H}_{12}\varphi_2(\boldsymbol{r}_1)\varphi_1(\boldsymbol{r}_2)\,\mathrm{d}V_1\,\mathrm{d}V_2 \\
&\quad \pm \iint \varphi_1^*(\boldsymbol{r}_1)\varphi_2^*(\boldsymbol{r}_2)\mathscr{H}_{12}\varphi_2(\boldsymbol{r}_1)\varphi_1(\boldsymbol{r}_2)\,\mathrm{d}V_1\,\mathrm{d}V_2 \\
&\quad \left.\pm \iint \varphi_2^*(\boldsymbol{r}_1)\varphi_1^*(\boldsymbol{r}_2)\mathscr{H}_{12}\varphi_1(\boldsymbol{r}_1)\varphi_2(\boldsymbol{r}_2)\,\mathrm{d}V_1\,\mathrm{d}V_2\right\} \\
&= I_1 + I_2 + K_{12} \pm J_{12},
\end{aligned} \tag{6.57}$$

where

$$\left.\begin{aligned}
I_1 &= \int \varphi_1^*(\boldsymbol{r}_i)\mathscr{H}_i\varphi_1(\boldsymbol{r}_i)\,\mathrm{d}V_i \\
I_2 &= \int \varphi_2^*(\boldsymbol{r}_i)\mathscr{H}_i\varphi_2(\boldsymbol{r}_i)\,\mathrm{d}V_i \\
K_{12} &= \iint \varphi_1^*(\boldsymbol{r}_i)\varphi_2^*(\boldsymbol{r}_2)\mathscr{H}_{12}\varphi_1(\boldsymbol{r}_1)\varphi_2(\boldsymbol{r}_2)\,\mathrm{d}V_1\,\mathrm{d}V_2 \\
J_{12} &= \iint \varphi_1^*(\boldsymbol{r}_1)\varphi_2^*(\boldsymbol{r}_2)\mathscr{H}_{12}\varphi_2(\boldsymbol{r}_1)\varphi_1(\boldsymbol{r}_2)\,\mathrm{d}V_1\,\mathrm{d}V_2
\end{aligned}\right\}. \tag{6.58}$$

In equation (6.57), the sign is + for the symmetric function (6.54) and − for the antisymmetric function (6.55).

In (6.58), I_1 and I_2 are the energies of electrons 1 and 2, respectively, and K_{12} is the Coulomb interaction between these two electrons. However, J_{12} has no corresponding concept in classical physics. This term signifies the energy produced by the exchange of two electrons between two orbits, and is called the *exchange integral*. In (6.57), we find that the energies of the parallel and antiparallel spin pairs differ by $2J_{12}$. The same situation was expressed in terms of the spins of two electrons, S_1 and S_2, in (6.25).]

This treatment, which was first given by Heisenberg,[7] clarified not only the nature of the molecular field in ferromagnetic materials, but also the difference between ortho- and para-helium, which correspond to antiparallel and parallel spins, respectively. Moreover, a similar treatment was applied to the hydrogen molecule, H_2, and showed that $J_{12} < 0$, so that an antiparallel spin pair is stable and forms a bonding orbit.

PROBLEMS

6.1 Knowing that nickel has the Curie point $\Theta_f = 628.3\,K$, $J = \frac{1}{2}$, and saturation moment $M_s = 0.6 M_B$, calculate the molecular field at 0 K according to the Weiss theory. Would the accuracy of the value of the molecular field be increased or reduced when more exact statistical treatments were applied?

6.2 Calculate a magnetization curve at the Curie point, according to the Weiss theory. Use the first and second terms in the power series of the Langevin function with respect to α.

REFERENCES

1. P. Weiss, *J. Phys.* **6** (1907), 661.
2. R. Becker and W. Döring, *Ferromagnetismus* (Springer, Berlin, 1939), p. 32.
3. A. Arrott, *Phys. Rev.*, **108** (1957), 1394.
4. J. S. Kouvel and M. E. Fisher, *Phys. Rev.*, **136** (6A) (1964), A1626.
5. J. S. Kouvel and D. S. Rodbell, *Phys. Rev. Letters*, **18** (1967), 215.
6. A. Arrott and J. E. Noakes, *Phys. Rev. Letters*, **19** (1967), 786.
7. W. Heisenberg, *Z. Physik*, **49** (1928), 619.
8. W. L. Bragg and E. J. Williams, *Proc. Roy. Soc.*, **145** (1934), 699; **151** (1935), 540.
9. H. A. Bethe, *Proc. Roy. Soc.*, **A150** (1935), 552.
10. R. Peierls, *Proc. Camb. Phil. Soc.*, **32** (1936), 477.
11. E. Ising, *Z. Physik*, **31** (1925), 253.
12. L. Onsager, *Phys. Rev.*, **65** (1944), 117.
13. L. P. Kadanoff, W. Gotze, D. Hamblen, R. Hecht, E. Lewis, V. Palciaukas, W. Rayl, J. Swift, D. Aspness, and J. Kane, *Rev. Mod. Phys.*, **39** (1967), 395.
14. F. Bloch, *Z. Physik*, **61** (1630), 206.
15. T. Oguchi, *Statistical theory of magnetism* (in Japanese) (Syokabo Publishing Co., Tokyo, 1970).
16. W. Pauli, *Z. Physik*, **31** (1925), 765.

7

ANTIFERROMAGNETISM AND FERRIMAGNETISM

7.1 ANTIFERROMAGNETISM

In antiferromagnetism, neighboring spins are aligned antiparallel to one another so that their magnetic moments cancel. Therefore an antiferromagnet produces no spontaneous magnetization and shows only a feeble magnetism. The relative magnetic susceptibility of antiferromagnetic materials, $\bar{\chi}$, ranges from 10^{-5} to 10^{-2}, the same as for paramagnets. The only difference is the presence of an ordered spin structure, as shown in Fig. 7.1. When an external magnetic field is applied parallel to the spin axis, the spins which are parallel and antiparallel to the field experience almost no torque and so keep their ordered spin arrangement. Therefore the susceptibility in this case is smaller than for a normal paramagnet. As the temperature increases, the ordered spin structure tends to be destroyed, and the susceptibility increases, contrary to the case of the normal paramagnet. However, above some critical temperature, such spin ordering disappears completely, so that the temperature dependence of the susceptibility becomes similar to that of an ordinary paramagnet. Therefore, the susceptibility shows a sharp maximum at the critical temperature, as shown in Fig. 7.2. This is the characteristic feature of an antiferromagnet. This critical temperature is called the *Néel point*, denoted by Θ_N.

Antiferromagnetic spin ordering was first verified experimentally in MnO by Shull and Smart,[1] using neutron diffraction. The crystal structure of MnO is such that Mn ions form a face-centered cubic lattice, and O ions are located between each Mn–Mn pair (Fig. 7.3). The spin magnetic moments of the Mn ions are arranged antiferromagnetically, as shown in the figure. Since the magnetic scattering of a neutron differs for + spins and − spins, extra diffraction lines appear if magnetic ordering is present.

Figure 7.4 shows the neutron diffraction lines observed for MnO below and above the Néel point. As expected, a number of extra lines are seen below the Néel point, where the antiferromagnetic structure appears. Such extra lines are called superlattice lines, because they reflect the presence of a structure larger than the basic

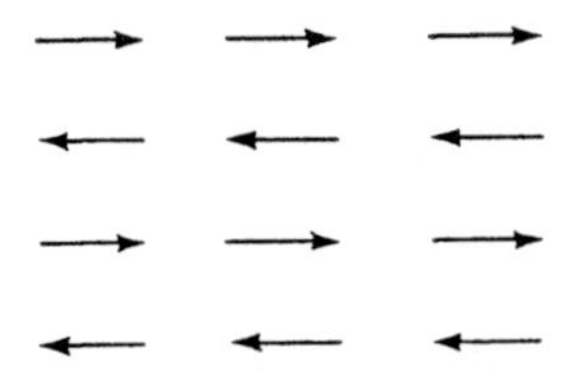

Fig. 7.1. Antiferromagnetic spin structure.

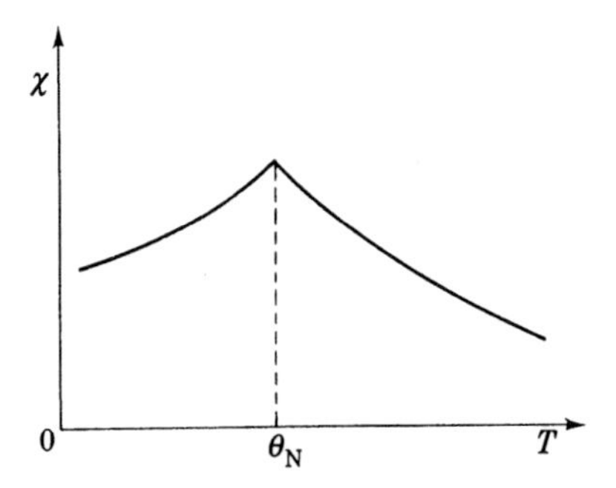

Fig. 7.2. Temperature dependence of magnetic susceptibility of antiferromagnetic materials.

crystallographic unit cell. The antiferromagnetic structure as shown in Fig. 7.3 was deduced by analyzing these superlattice lines.

The direct exchange interaction between Mn ions is very weak, because it is interrupted by the interstitial O^{2-} ions. However, a superexchange interaction acts between Mn ions through the O^{2-} ion, an idea first introduced by Kramers[2] and theoretically interpreted by Anderson.[3] The essential point of this mechanism is as follows: The O^{2-} ion has electronic structure expressed by $(1s)^2(2s)^2(2p)^6$. The p-orbit stretches towards the neighboring Mn ions, M_1 and M_2, as shown in Fig. 7.5.

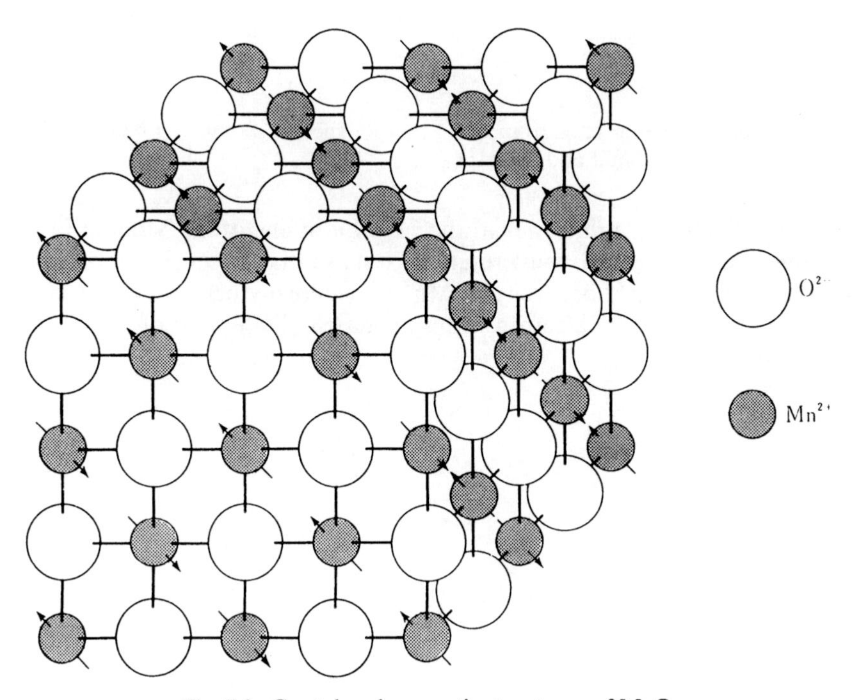

Fig. 7.3. Crystal and magnetic structures of MnO.

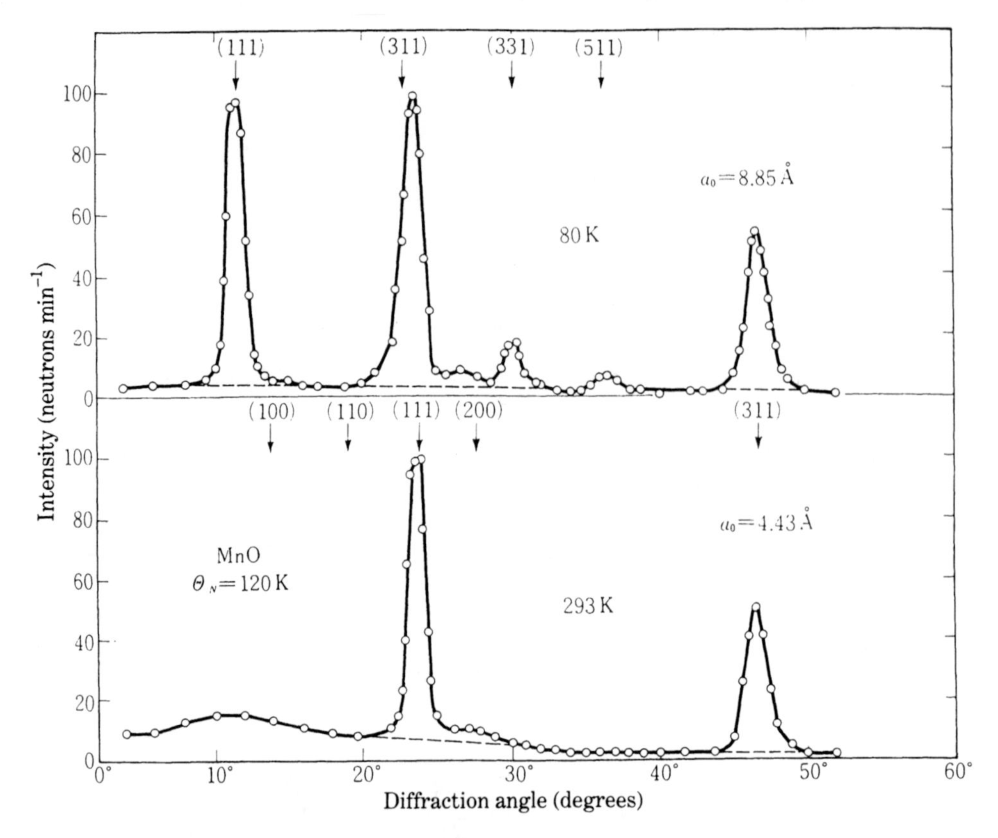

Fig. 7.4. Neutron diffraction lines from MnO powder specimen observed below and above its Néel point, 120 K. (After Shull and Smart[1]).

One of the p-electrons can transfer to the $3d$ orbit of one of the Mn ions (say the M_1 ion). In this case the electron must retain its spin, so that its sense will be antiparallel to the total spin of Mn^{2+}, because the Mn^{2+} has already had five electrons and the vacant orbit must accept an electron with spin antiparallel to that of the five electrons

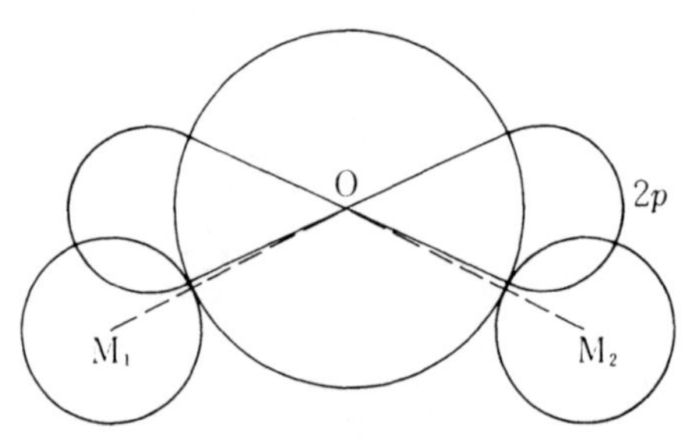

Fig. 7.5. The p-orbit of the O^{2-} ion through which exchange interaction acts between the spins on magnetic ions M_1 and M_2.

(Hund's rule, Section 3.2). On the other hand, the remaining electron in the *p*-orbit must have spin antiparallel to that of the transferred electron because of the Pauli exclusion principle. The exchange interaction with the other Mn ion (M_2) is therefore negative. As a result, the total spin of M_1 becomes antiparallel to that of M_2. The superexchange interaction is strongest when the angle M_1–O–M_2 is 180°, and becomes weaker as the angle becomes smaller. When the angle is 90°, the interaction tends to become positive. The superexchange interaction acts also through S^{2-}, Se^{2-}, Cl^{1-}, and Br^{1-} ions. It is experimentally verified[4] that positive superexchange interaction at 90° increases in the order O^{2-}, S^{2-}, and Se^{2-}. For more details the reader may refer to the review by Nagamiya *et al.*[5]

The first theoretical treatment of antiferromagnetism was made by Van Vleck[6] with later advances by Néel[7] and Anderson.[8]

Let us designate as A sites the lattice sites on which + spins are located in a completely ordered arrangement, and as B sites the locations of the − spins. It is expected that the spins on the A sites are subject to the superexchange interaction with the spins on the B sites as well as with the other A sites. These interactions can be expressed as molecular fields, as in the Weiss theory of ferromagnetism. The molecular field acting on the A site spins is given by

$$H_{mA} = w_{AA} I_A + w_{AB} I_B, \tag{7.1}$$

where the inter-site interaction coefficient, w_{AB}, must be negative, and I_A and I_B are the magnetizations due to all the spins on the A sites and B sites, respectively. Such a magnetization is called a *sublattice magnetization.* Similarly the molecular field acting on the B site is given by

$$H_{mB} = w_{BA} I_A + w_{BB} I_B. \tag{7.2}$$

Since the antiferromagnetic spin structure in the absence of an external field is symmetric with respect to the A and B sites, as shown in Fig. 7.1, we can assume that $w_{AA} = w_{BB}$, and $w_{AB} = w_{BA}$. For simplicity, we put

$$\left.\begin{aligned} w_{AA} &= w_{BB} = w_1 \\ w_{AB} &= w_{BA} = w_2 \end{aligned}\right\}. \tag{7.3}$$

Moreover, the sublattice magnetizations of the A- and B-sublattices are equal in magnitude and opposite in sign, so that

$$I_B = -I_A. \tag{7.4}$$

Therefore, (7.1) and (7.2) can be written as

$$H_{mA} = (w_1 - w_2) I_A, \tag{7.5}$$

and

$$H_{mB} = (w_1 - w_2) I_B. \tag{7.6}$$

The formulas (7.5) and (7.6) are of the same form as (6.1) for ferromagnetism.

Therefore, as in (6.2), the thermal equilibrium values of sublattice magnetizations are given by

$$I_A = \frac{NM}{2} L\left(\frac{M(w_1 - w_2)I_A}{kT}\right) \tag{7.7}$$

and

$$I_B = \frac{NM}{2} L\left(\frac{M(w_1 - w_2)I_B}{kT}\right), \tag{7.8}$$

where N is the number of magnetic atoms per unit volume, M is the atomic magnetic moment, and $L(\alpha)$ is the Langevin function. The sublattice magnetizations given by (7.7) and (7.8) decrease with increasing temperature (see Fig. 6.3), similar to the spontaneous magnetization in ferromagnetism, and vanish at a critical temperature, Θ_N, which is the Néel point. The Néel point is given by

$$\Theta_N = \frac{NM^2(w_1 - w_2)}{6k}, \tag{7.9}$$

which is similar to equation (6.8).

When an external field is applied parallel to the spin axis, I_A and I_B are no longer symmetrical, so that equation (7.4) does not hold. In this case, the sublattice magnetizations are given by

$$I_A = \frac{NM}{2} L\left(\frac{M(H + w_1 I_A + w_2 I_B)}{kT}\right), \tag{7.10}$$

$$I_B = \frac{NM}{2} L\left(\frac{M(H + w_2 I_A + w_1 I_B)}{kT}\right). \tag{7.11}$$

Differentiating both sides with respect to H, we have

$$\frac{\partial I_A}{\partial H} = \frac{NM^2}{2kT} L'(\alpha)\left(1 + w_1 \frac{\partial I_A}{\partial H} + w_2 \frac{\partial I_B}{\partial H}\right), \tag{7.12}$$

$$\frac{\partial I_B}{\partial H} = \frac{NM^2}{2kT} L'(\alpha)\left(1 + w_2 \frac{\partial I_A}{\partial H} + w_1 \frac{\partial I_B}{\partial H}\right). \tag{7.13}$$

Adding the two equations side by side, and solving with respect to $(\partial I_A/\partial H) + (\partial I_B/\partial H)$, we have the magnetic susceptibility

$$\chi = \frac{\partial I}{\partial H} = \frac{\partial I_A}{\partial H} + \frac{\partial I_B}{\partial H}$$
$$= \frac{NM^2}{kT} L'(\alpha) \bigg/ \left\{1 - \frac{NM^2}{2kT} L'(\alpha)(w_1 + w_2)\right\}. \tag{7.14}$$

If we put

$$\Theta_a = \frac{NM^2(w_1 + w_2)}{6k}, \tag{7.15}$$

and

$$C = \frac{NM^2}{3k}, \tag{7.16}$$

(7.14) can be expressed in a simple form

$$\chi = \frac{3CL'(\alpha)}{T - 3L'(\alpha)\Theta_a}. \tag{7.17}$$

If the temperature is well above the Néel point, Θ_N, the spin arrangement becomes random, so that from (5.22) we have

$$L'(\alpha) = \tfrac{1}{3}. \tag{7.18}$$

Then (7.17) becomes

$$\chi = \frac{C}{T - \Theta_a}, \tag{7.19}$$

which has the same form as the Curie–Weiss law (6.15).

If we plot $1/\chi$ as a function of T, (7.19) gives a straight line which intersects the abscissa at $T = \Theta_a$ (see Fig. 7.6). Unlike the case of ferromagnetism, the asymptotic Curie point, Θ_a, is entirely different from the Néel point Θ_N. The reason is that the intersite interaction coefficient, w_2, is always negative, so that (7.15) gives a negative value provided that the intrasite interaction coefficient, w_1, is relatively small. If

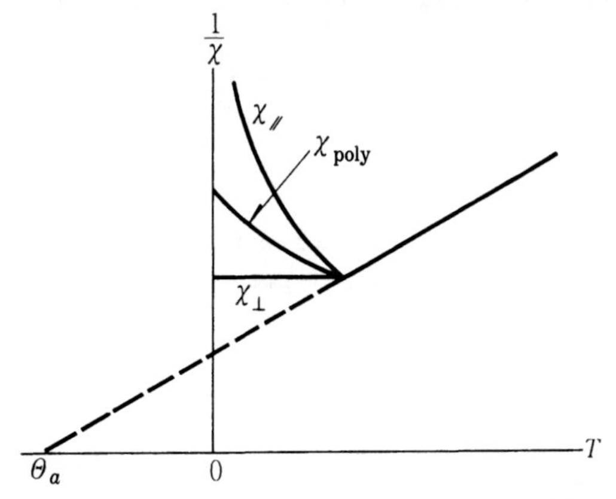

Fig. 7.6. Temperature dependence of reciprocal magnetic susceptibility for antiferromagnetic materials. ($\chi_{\parallel}$ and $\chi_{\perp}$ are the magnetic susceptibilities measured by applying magnetic field parallel and perpendicular to the spin axis. χ_{poly} is the susceptibility for polycrystalline material calculated by averaging $\chi_{\parallel}$ and $\chi_{\perp}$.)

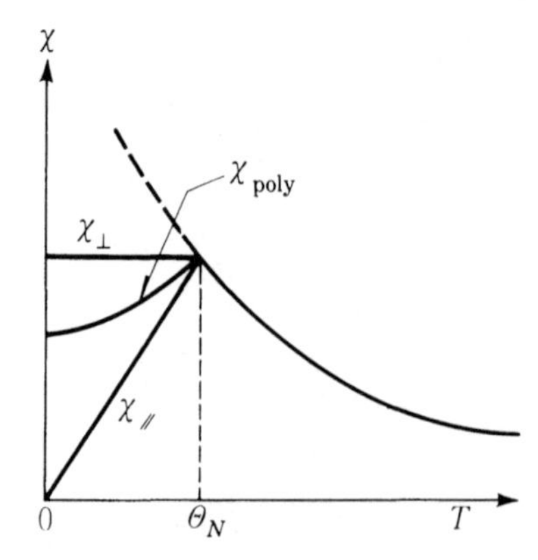

Fig. 7.7. Temperature dependence of magnetic susceptibility of antiferromagnetic materials. (Symbols are the same as Fig. 7.6.)

$w_1 = 0$, Θ_a is negative and its magnitude is equal to the Néel point, as easily found from (7.9) and (7.15). The χ–T curve corresponding to Fig. 7.6 is shown in Fig. 7.7. Typical antiferromagnetic materials and their Néel and asymptotic Curie points are listed in Table 7.1.

The susceptibility at the Néel point reaches a maximum, as is given by (7.19) with T replaced by Θ_N, or

$$\chi_{\max} = \frac{C}{\Theta_N - \Theta_a} = -\frac{1}{w_2}. \tag{7.20}$$

As the temperature decreases from the Néel point, the sublattice magnetization increases, so that the point P in Fig. 6.2 rises along the curve $L(\alpha)$ and the value of $L'(\alpha)$ decreases from $\frac{1}{3}$. As deduced from (7.17), the susceptibility decreases with decreasing temperature, and finally in the limit $T \to 0$, we have

$$\lim_{T\to 0} \chi = \frac{3CL''(\alpha)\dfrac{\partial \alpha}{\partial T}}{1 - 3L''(\alpha)\Theta_a \dfrac{\partial \alpha}{\partial T}} = 0. \tag{7.21}$$

This behavior matches well the χ–T curve in Fig. 7.7. If we plot equation (7.21) as $1/\chi$–T, we find that the curve for $\chi_{\parallel}$ diverges as T tends to zero (Fig. 7.6).

Table 7.1. Various constants for typical antiferromagnetic materials.[9]

Material	Crystal type	Θ_N (K)	Θ_a (K)	M_{eff}/M_B
MnO	NaCl	122	−610	5.95*
NiO	NaCl	520	−2000	4.6
Cr_2O_3	Corundum	307	−1070	3.86
FeS	NiAs	593	−917	5.22

* Landolt-Börnstein, *Magnetische Eigenshaften* Vol II, Part 9, 3–148, (1962).

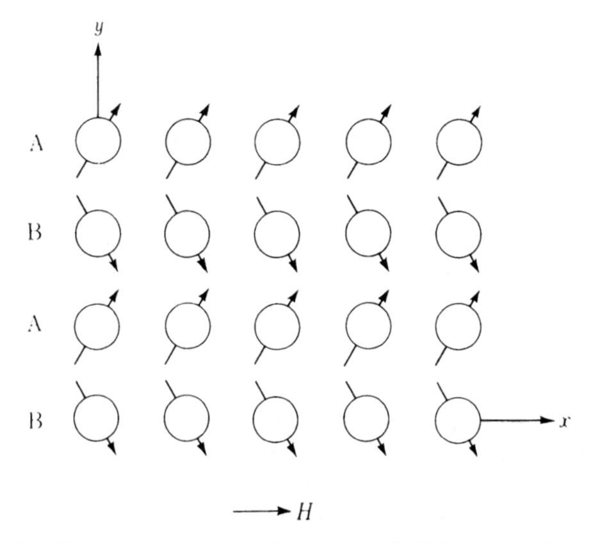

Fig. 7.8. Rotation of antiferromagnetic spins in a magnetic field perpendicular to the spin axis.

When a magnetic field is applied perpendicular to the spin axis, the susceptibility may not vanish even at $T=0$, because the rotation magnetization of the sublattice magnetizations occurs even at 0 K (see Fig. 7.8). Suppose an external magnetic field, H_{ex}, is applied parallel to the x-axis. The sublattice magnetizations, which are originally parallel to the y-axis, will rotate toward the x-axis as shown in Fig. 7.8. Then the x- and y-components of the molecular field acting on the A-sublattice are given by

$$\left.\begin{aligned}(H_{mA})_x &= w_1 I_{Ax} + w_2 I_{Bx}\\ (H_{mA})_y &= w_1 I_{Ay} + w_2 I_{By}\end{aligned}\right\}. \tag{7.22}$$

Since the A- and B-sublattices are symmetrical, it follows that

$$I_{Bx} = I_{Ax}, \qquad I_{By} = -I_{Ay}. \tag{7.23}$$

Using this relationship, (7.22) becomes

$$\left.\begin{aligned}(H_{mA})_x &= (w_1 + w_2) I_{Ax}\\ (H_{mA})_y &= (w_1 - w_2) I_{Ay}\end{aligned}\right\}. \tag{7.24}$$

The direction of the sublattice magnetization must coincide with the resultant magnetic field, composed of the external field and the molecular field. Therefore

$$\frac{H_x}{H_y} = \frac{H + (H_{mA})_x}{(H_{mA})_y} = \frac{H + (w_1 + w_2) I_{Ax}}{(w_1 - w_2) I_{Ay}} = \frac{I_{Ax}}{I_{Ay}}. \tag{7.25}$$

Eliminating I_{Ay} from (7.25), we have

$$H + (w_1 + w_2) I_{Ax} = (w_1 - w_2) I_{Ax}$$

$$\therefore \quad I_{Ax} = -\frac{H}{2w_2}. \tag{7.26}$$

Therefore the susceptibility in this case is given by

$$\chi = \frac{I_x}{H} = \frac{I_{Ax} + I_{Bx}}{H} = -\frac{1}{w_2}, \tag{7.27}$$

which is equal to the maximum susceptibility given by (7.20). Actually the susceptibility in this case remains constant from the Néel point to 0 K as shown in Figs 7.6 and 7.7. In the case of a polycrystalline material, the measured susceptibility is the value averaged over the crystallites of all possible orientations, as shown by the curve labelled 'poly' in Figs 7.6 and 7.7.

7.2 FERRIMAGNETISM

In ferrimagnets, the A- and B-sublattices are occupied by different magnetic atoms and sometimes by different numbers of atoms, so that the antiferromagnetc spin arrangement results in an uncompensated spontaneous magnetization (see Fig. 7.9). Such magnetism is called *ferrimagnetism* and was treated by Néel.[7]

For simplicity, let us assume that only one kind of magnetic atom, with magnetic moment M, contributes to ferrimagnetism. The total number, $2N$, of magnetic atoms is distributed on the A- and B-sites in the ratio $\lambda : \mu$, where

$$\lambda + \mu = 1. \tag{7.28}$$

If all the magnetic moments are completely ordered, the sublattice magnetizations are given by

$$\left.\begin{aligned} I_A &= 2N\lambda M \\ I_B &= -2N\mu M \end{aligned}\right\}, \tag{7.29}$$

so that the spontaneous magnetization is

$$I = I_A + I_B = 2NM(\lambda - \mu). \tag{7.30}$$

At any temperature above absolute zero, the sublattice magnetizations are thermally disturbed; and even at 0 K, sometimes complete order may not be realized. To describe this disordering, we introduce the normalized sublattice magnetizations, I_a and I_b, or

$$\left.\begin{aligned} I_A &= \lambda I_a \\ I_B &= \mu I_b \end{aligned}\right\}. \tag{7.31}$$

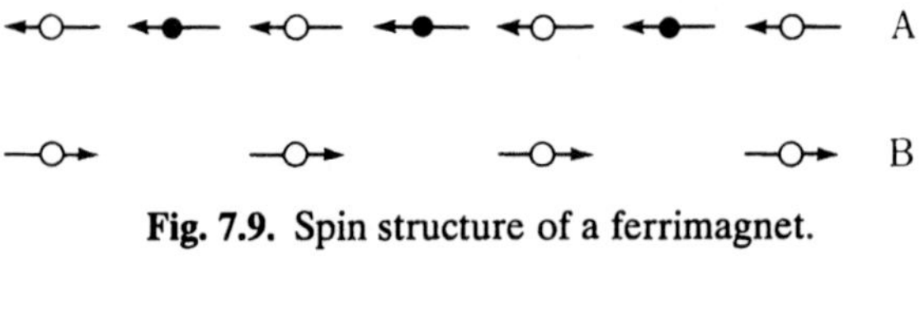

Fig. 7.9. Spin structure of a ferrimagnet.

The quantities I_a and I_b are interpreted as the average magnetic moment at each sublattice site, multiplied by $2N$. The molecular field acting on the magnetic atoms is caused by the atoms on the same sublattice as well as the atoms on the other sublattice, or

$$\left.\begin{aligned} H_{mA} &= w(\alpha\lambda I_a - \mu I_b) \\ H_{mB} &= w(\beta\mu I_b - \lambda I_a) \end{aligned}\right\}, \tag{7.32}$$

where w is the absolute value of the molecular field coefficient between the A- and B-sublattices, and α and β are the intra-site coefficients normalized by w. Under the action of these molecular fields, the sublattice magnetizations are calculated using common statistical procedures as

$$\left.\begin{aligned} I_a &= 2NML\left\{\frac{Mw(\alpha\lambda I_a - \mu I_b)}{kT}\right\} \\ I_b &= 2NML\left\{\frac{Mw(\beta\mu I_b - \lambda I_a)}{kT}\right\} \end{aligned}\right\}. \tag{7.33}$$

In the presence of an external magnetic field H, the term MH/kT must be added inside the braces { }.

At temperatures above the Curie point, using the approximation $L'(\alpha) = \frac{1}{3}$, we have

$$\left.\begin{aligned} \frac{\partial I_a}{\partial H} &= \frac{2}{3}\frac{NM^2}{kT}\left\{1 + w\left(\alpha\lambda\frac{\partial I_a}{\partial H} - \mu\frac{\partial I_b}{\partial H}\right)\right\} \\ \frac{\partial I_b}{\partial H} &= \frac{2}{3}\frac{NM^2}{kT}\left\{1 + w\left(\beta\mu\frac{\partial I_b}{\partial H} - \lambda\frac{\partial I_a}{\partial H}\right)\right\} \end{aligned}\right\}. \tag{7.34}$$

Solving (7.34) with respect to $\partial I_a/\partial H$, $\partial I_b/\partial H$,

$$\begin{aligned} \frac{\partial I}{\partial H} &= \lambda\frac{\partial I_a}{\partial H} + \mu\frac{\partial I_b}{\partial H} \\ &= C\frac{T - Cw\lambda\mu(\alpha + \beta + 2)}{T^2 - Cw(\alpha\lambda + \beta\mu)T + C^2w^2\lambda\mu(\alpha\beta - 1)}, \end{aligned} \tag{7.35}$$

where C is given by

$$C = \frac{2NM^2}{3k}. \tag{7.36}$$

The equation (7.35) can be simplified to

$$\frac{1}{\chi} = \frac{T}{C} + \frac{1}{\chi_0} - \frac{\sigma}{T - \Theta}, \tag{7.37}$$

where

$$\left.\begin{aligned} \frac{1}{\chi_0} &= w(2\lambda\mu - \lambda^2\alpha - \mu^2\beta) \\ \sigma &= Cw^2\lambda\mu\{\lambda(\alpha + 1) - \mu(\beta + 1)\}^2 \\ \Theta &= Cw\lambda\mu(\alpha + \beta + 2) \end{aligned}\right\}. \tag{7.38}$$

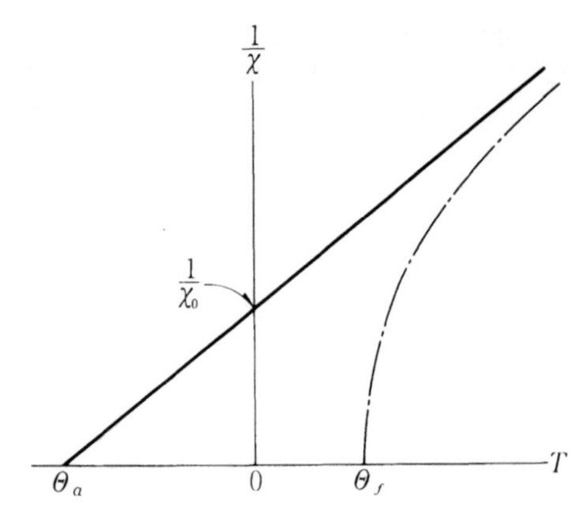

Fig. 7.10. Temperature dependence of reciprocal susceptibility of a ferrimagnet.

The relationship between $1/\chi$ and T given in (7.37) is graphically shown in Fig. 7.10. At high temperatures, the third term on the right-hand side of (7.37) becomes negligibly small, so that the remaining terms are approximated by a straight line, which has slope $1/C$ and intersects the ordinate at $1/\chi_0$. When the temperature is lowered from well above the Curie point, the magnitude of the third term in (7.37) increases as the temperature approaches Θ, so that the $1/\chi$–T curve departs from the asymptotic line and finally drops to zero at the Curie point, Θ_f. The *asymptotic Curie point* in this case is given by

$$\Theta_a = -\frac{C}{\chi_0}, \tag{7.39}$$

while at the ferrimagnetic Curie point, $1/\chi = 0$, so that from (7.37) we have

$$\Theta_f^2 - Cw(\alpha\lambda + \beta\mu)\Theta_f + C^2w^2\lambda\mu(\alpha\beta - 1) = 0. \tag{7.40}$$

Solving this equation with respect to Θ_f, we have

$$\Theta_f = \frac{Cw}{2}\left\{\alpha\lambda + \beta\mu \pm \sqrt{(\alpha\lambda - \beta\mu)^2 + 4\lambda\mu}\right\}. \tag{7.41}$$

If $\Theta_f < 0$, the paramagnetic state persists down to 0 K, while if $\Theta_f > 0$, the susceptibility becomes infinite at Θ_f, and ferrimagnetism appears. The spontaneous magnetization in this case is given by

$$I_s = \lambda I_a + \mu I_b, \tag{7.42}$$

where I_a and I_b can be found by solving (7.33) under the assumption that $T < \Theta_f$.

The condition for producing this ferrimagnetism is $\Theta_f > 0$, so that the critical condition $\Theta_f = 0$ follows from (7.41)

$$\alpha\lambda + \beta\mu \pm \sqrt{(\alpha\lambda - \beta\mu)^2 + 4\lambda\mu} = 0;$$

therefore

$$\alpha\beta = 1. \tag{7.43}$$

For $\alpha > 0$ and $\beta > 0$, it is evident that ferrimagnetism appears, while for $\alpha < 0$ and $\beta < 0$, (7.43) gives the limiting conditions for realizing ferrimagnetism or paramagnetism: $\alpha\beta < 1$. If the absolute values of α and β become larger than the values that satisfy (7.43), the spin arrangement on the A and B sublattices becomes antiferromagnetic or non-magnetic. If the absolute values of α and β are smaller than the values that satisfy (7.43), negative intrasite interactions may be suppressed, giving rise to spontaneous magnetization.

The most conspicuous macroscopic feature of ferrimagnetism is the appearance of various forms for the curve of the temperature dependence of spontaneous magnetization. The shape of the I_s–T curve depends on the combination of α, β, λ, and μ. To see how this works, let us investigate how the values of I_a and I_b vary with the values of these variables. If there is no external field, the internal energy is given by

$$E = -\tfrac{1}{2}\lambda I_a H_{mA} - \tfrac{1}{2}\mu I_b H_{mB}, \tag{7.44}$$

which by using (7.32) is converted to

$$E = -\tfrac{1}{2}w(\alpha\lambda^2 I_a^2 - 2\lambda\mu I_a I_b + \beta\mu^2 I_b^2). \tag{7.45}$$

The stable values of I_a and I_b are calculated by minimizing (7.45) with respect to I_a and I_b in the following four cases:

(I) *Paramagnetism.* In this case we must have

$$I_a = I_b = 0. \tag{7.46}$$

The internal energy is obtained by inserting (7.46) in (7.45), or

$$E = 0. \tag{7.47}$$

(II) *Saturation in both A and B sites.* In this case, I_a and I_b take their maximum values, so that

$$\left.\begin{aligned} I_a &= 2NM \\ I_b &= -2NM \end{aligned}\right\}. \tag{7.48}$$

The internal energy is then given by

$$E = -2wN^2M^2(\alpha\lambda^2 + 2\lambda\mu + \beta\mu^2). \tag{7.49}$$

(III) *Saturation in A sites only.* By putting

$$I_a = 2NM, \tag{7.50}$$

and minimizing (7.45) with respect to I_b, we have

$$I_b = \frac{\lambda}{\beta\mu} 2NM. \tag{7.51}$$

The internal energy in this case is given by

$$E = -2wN^2M^2\lambda^2\left(\alpha - \frac{1}{\beta}\right). \tag{7.52}$$

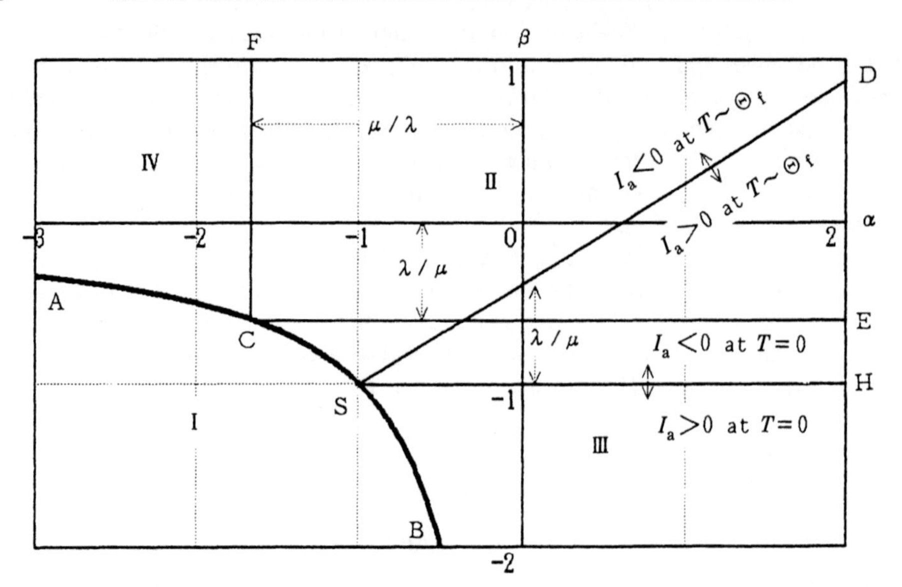

Fig. 7.11. Equilibrium regions for four stable spin configurations I, II, III, and IV in the α–β plane (see text).

(IV) *Saturation in B sites only.* The B site magnetization in this case is fully saturated, and given by

$$I_b = -2NM. \tag{7.53}$$

Then the A site magnetization takes the equilibrium value

$$I_a = -\frac{\mu}{\alpha\lambda} 2NM, \tag{7.54}$$

so that the internal energy is given by

$$E = -2wN^2M^2\mu^2\left(\beta - \frac{1}{\alpha}\right). \tag{7.55}$$

By comparing internal energies in the four cases, we can determine which case is realized for a given set of the parameters α, β, λ, and μ. Let us discuss the problem on the α–β plane shown in Fig. 7.11. First we assume that $\lambda < \mu$. In order to determine the regions in which the cases (I)–(IV) are realized, we draw the line FC for which $\alpha = -\mu/\lambda$, and the line EC for which $\beta = -\lambda/\mu$. The paramagnetic case I is realized below the curve ACB, or $\alpha\beta > 1$, $\alpha < 0$ and $\beta < 0$, because in this case the energies in the cases II, III, and IV all give positive values. (For case II, consider that the expression in parenthesis in (7.49) is equal to $\{(\alpha\lambda + \mu)\lambda + (\beta\mu + \lambda)\}\mu$, which is negative unless $\alpha > -\mu/\lambda$ and $\beta > -\lambda/\mu$.) This is consistent with the condition for the appearance of ferrimagnetism as mentioned just below (7.43). Ferrimagnetism appears in the region above and to the right of curve ACB.

Case (IV) is realized in the region FCA, because $I_a < 2NM$, so that from (7.54) we know that $\alpha < -\mu/\lambda$. Similarly, case (III) is realized in the region ECB. Case (II) appears in region ECF, where $\alpha > -\mu/\lambda$ and $\beta > -\lambda/\mu$.

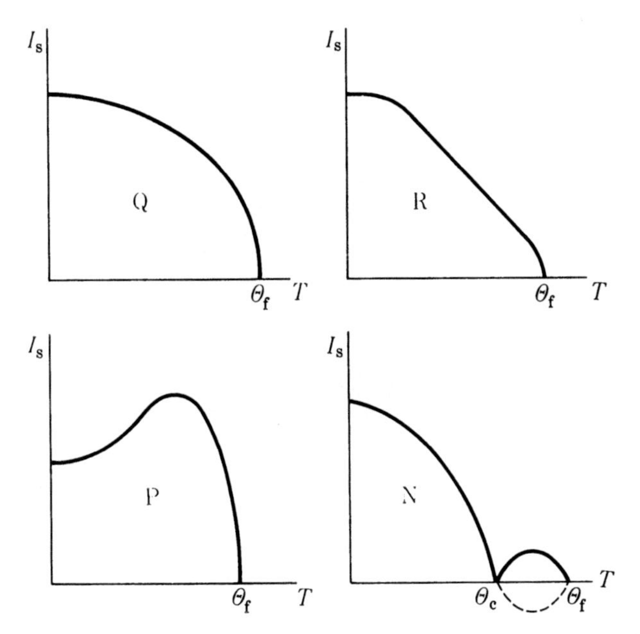

Fig. 7.12. Four types of temperature dependence of spontaneous magnetization in ferrimagnets.

In case (II), both A and B sublattices are saturated, and the spontaneous magnetization decreases as in the case of ferromagnetism, resulting in Q-type temperature dependence (see Fig. 7.12). In cases (III) and (IV), one sublattice is unsaturated and is easily thermally disturbed, producing various complicated temperature dependences of the spontaneous magnetization shown as R-, P- and N-types in Fig. 7.12. The R-type appears when the sublattice with the larger moment is thermally disturbed more easily. This occurs when $\beta > -1$ in case (III). The P-type appears when the sublattice with smaller moment is thermally disturbed more easily; this occurs in all of region IV or the area FCA in Fig. 7.11. The most characteristic temperature dependence is the N-type, in which the sign of the spontaneous magnetization reverses at the temperature of the *compensation point*, Θ_c. The possibility of the occurrence of N-type behavior can be examined by investigating the sign of $I_A - I_B$ near the Curie point.

At $T \approx \Theta_f$ the Langevin function in (7.33) can be approximated by the first term in (5.22), so that we have

$$\left.\begin{aligned} I_a &\simeq \frac{Cw}{T}(\alpha\lambda I_a - \mu I_b) \\ I_b &\simeq \frac{Cw}{T}(\beta\mu I_b - \lambda I_a) \end{aligned}\right\}, \tag{7.56}$$

where C is given by (7.36). At the compensation point,

$$\lambda I_a + \mu I_b = 0, \tag{7.57}$$

Fig. 7.13. Demonstration showing a reversal of spontaneous magnetization of a ferrimagnet with a change in temperature.

so that using (7.56) we have

$$I_s = \lambda I_a + \mu I_b \simeq \frac{Cw}{T}\{(\alpha\lambda^2 - \lambda\mu)I_a + (\beta\mu^2 - \lambda\mu)I_b\}$$

$$= \frac{Cw}{T}\{(\alpha\lambda - \mu) - (\beta\mu - \lambda)\}\lambda I_a = 0, \qquad (7.58)$$

or

$$\lambda(\alpha + 1) - \mu(\beta + 1) = 0. \qquad (7.59)$$

If the left side of (7.59) is positive, $I_s > 0$. The boundary condition (7.59) gives the line SD which goes through the point S, or $\alpha = \beta = -1$, and has tangent λ/μ. Above the line SH, $\beta > -1$, so that $I_s < 0$, as seen from (7.51), not only in region (III) but also in region (II). Therefore in the region between the lines SD and SH, $I_s < 0$ at $T = 0$, while $I_s > 0$ near Θ_f. In other words, N-type temperature dependence results.

In the experimental determination of the temperature dependence, the saturation magnetization is measured in a fairly strong magnetic field. The magnetization always is aligned parallel to the field, and is therefore always measured as positive, as shown by the solid N-type curve in Fig. 7.12. If, however, an N-type ferrimagnet is suspended freely by a thin thread in a weak magnetic field, as shown in Fig. 7.13, and heated through the compensation point, the sample will rotate through 180° as the spontaneous magnetization changes sign.

In Fig. 7.14, the various forms of I_s vs T curves are illustrated by computer simulation[10] for Ga-doped Yttrium–Iron–Garnet (YIG) (see Section 9.3).

7.3 HELIMAGNETISM

In a *helimagnetic structure*, all spins in a layered crystal are aligned parallel within each c-plane, but the spin direction varies from plane to plane such that the tips of

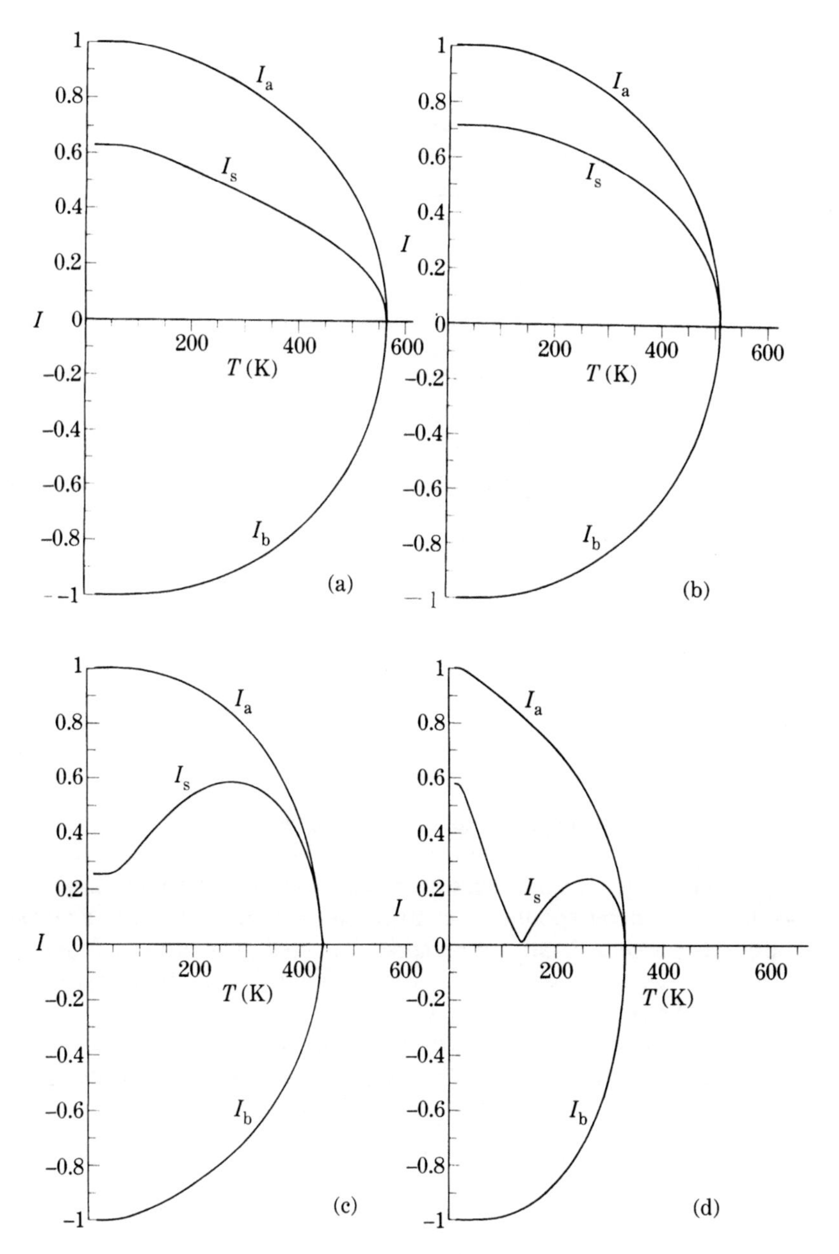

Fig. 7.14. Computer simulation of temperature dependence of spontaneous and sublattice magnetizations in Ga-doped YIG: (a) $Y_3Fe_5O_{12}$ (R-type); (b) $Y_3Ga_{.5}Fe_{4.5}O_{12}$ (Q-type); (c) $Y_3Ga_{1.1}Fe_{3.9}O_{12}$ (P-type); (d) $Y_3Ga_2Fe_3O_{12}$ (N-type). $J_{AA} = 8.45\,cm^{-1}$, $J_{BB} = 11.86\,cm^{-}1$, $J_{AB} = 25.36\,cm^{-1}$ were assumed.[10]

the spin vectors of the spins along any line parallel to the c-axis describe a spiral or helix (see Fig. 7.15). This spin structure is also called a *screw structure*. The resultant

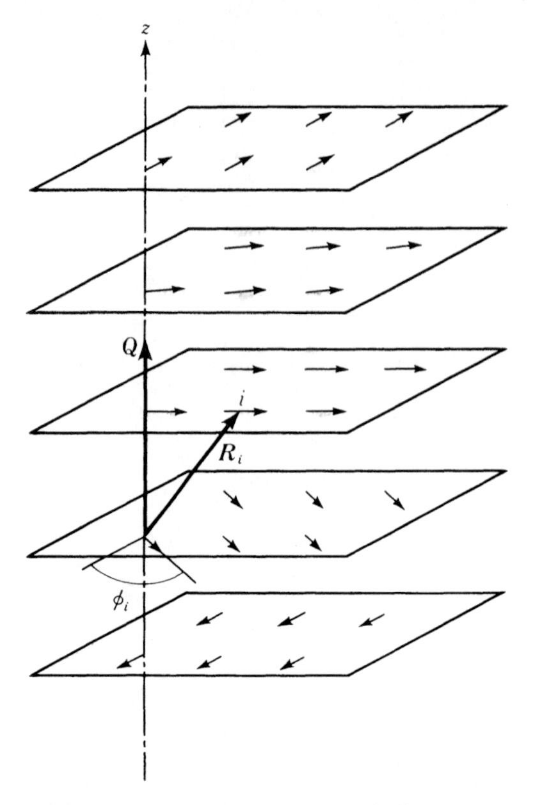

Fig. 7.15. Screw spin structure in helimagnet.

magnetic moment of such a spin structure is zero over a sufficiently large volume of sample, so that there is no spontaneous magnetization. In this sense, helimagnetism can be considered a kind of antiferromagnetism. Helimagnetism is found in many magnetic materials. A screw spin structure was first found in MnO by Yoshimori,[11] and in $MnAu_2$ by Villan[12] and later by Kaplan.[13]

Let us suppose that all the spins are aligned parallel to one another in an x–y plane, and that the spins in the ith x–y plane make an angle ϕ_i with the x-axis. Then the x-, y- and z-components of the spin S are give by

$$\left.\begin{aligned} S_{ix} &= S\cos\phi_i \\ S_{iy} &= S\sin\phi_i \\ S_{iz} &= 0 \end{aligned}\right\}. \tag{7.60}$$

The rotation of the spins is expressed by

$$\phi_i = \phi_0 + \boldsymbol{Q}\cdot\boldsymbol{R}_i, \tag{7.61}$$

where $\boldsymbol{R}_i$ is the vector drawn from the origin to a particular spin in the ith x–y plane, and $\boldsymbol{Q}$ is a vector parallel to the z-axis with magnitude

$$Q = \frac{2\pi}{nC}, \tag{7.62}$$

where n is the number of x–y planes required for the spins to rotate by 2π or 360°, and C is the separation between neighboring x–y planes. If $n = 1$, or ∞, the spin arrangement is ferromagnetic, while if $n = 2$, it is antiferromagnetic. In the general case, n (which is not necessarily an integer) expresses the pitch of the screw spin structure. Note that the second term of (7.61) is constant for all spins in the same x–y plane, so that the angle ϕ_i is constant with each x–y plane.

The origin of such an unusual spin structure can be deduced by assuming that the exchange interaction between nearest-neighbor planes as well as between spins in a single plane is ferromagnetic, while that between second nearest-neighbor planes is antiferromagnetic. Applying the general expression for exchange interaction given by (6.25) to the helimagnetic case, the exchange interaction energy stored per unit volume is given by

$$E_{ex} = -2NS^2(J_1 \cos QC + J_2 \cos 2QC), \tag{7.63}$$

where N is the number of magnetic atoms and S is the spin.

The pitch of a stable helical spin arrangement is given by minimizing (7.63) with respect to Q, or

$$\frac{\partial E_{ex}}{\partial Q} = +2NS^2C(J_1 \sin QC + 2J_2 \sin 2QC) = 0, \tag{7.64}$$

so that we have

$$\cos QC = -\frac{J_1}{4J_2}. \tag{7.65}$$

Therefore the condition for the existence of a helical spin structure is

$$|J_2| > \frac{J_1}{4}. \tag{7.66}$$

In other words, the second nearest-neighbor interaction must be strong enough to satisfy (7.66).

The helical spin structure appears also in rare earth metals. The mechanism in this case is the oscillatory variation of polarization of the spins of the conduction electrons brought about by the RKKY interaction (see Section 8.3).

7.4 PARASITIC FERROMAGNETISM

Parasitic ferromagnetism is a term used by Néel[14] to describe a weak ferromagnetism associated with antiferromagnetism in α-Fe_2O_3. The magnetization curve of this material is composed of a ferromagnetic part, I_s, which saturates in a sufficiently large field, and a paramagnetic part, χH, which is proportional to the field (see Fig. 7.16). The susceptibility, χ, of the latter part shows a maximum at some temperature, Θ, which demonstrates that this material is antiferromagnetic. The saturation magnetization, I_s, vanishes at the same temperature, Θ (see Fig. 7.17). Observing this phenomenon, it appears that ferromagnetism is parasitic on antiferromagnetism.

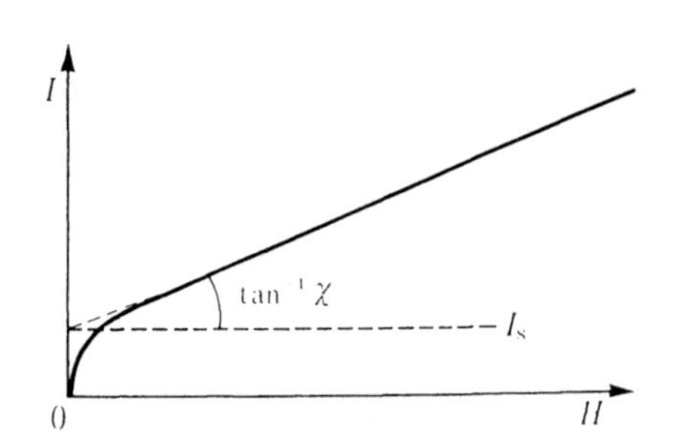

Fig. 7.16. Magnetization curve of a parasitic ferromagnet.

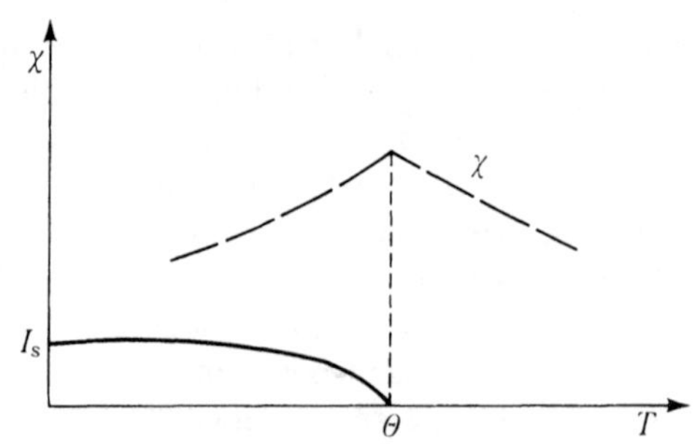

Fig. 7.17. Magnetic susceptibility in a parasitic ferromagnet exhibits a maximum at the critical temperature at which saturation magnetization disappears.

Néel suggested that the ferromagnetism in this material is caused by a small amount of Fe_3O_4, precipitated along its common crystallographic plane with the antiferromagnetic lattice of α-Fe_2O_3. However, the actual reason was found to be the spontaneous canting of + and − antiferromagnetic spins. This idea was deduced phenomenologically by Dzyaloshinsky,[15] and theoretically by Moriya.[16] The *antisymmetric interaction* which causes this spin-canting is expressed by

$$w_D = \boldsymbol{D} \cdot (\boldsymbol{S}_1 \times \boldsymbol{S}_2), \tag{7.67}$$

where $\boldsymbol{S}_1$ and $\boldsymbol{S}_2$ are the canting spins and $\boldsymbol{D}$ is a vector coefficient. If $D < 0$, the angle between $\boldsymbol{S}_1$ and $\boldsymbol{S}_2$ is not 0 or 180° (see Fig. 7.18). A non-zero value of D arises from an excited state and its perturbation mixed through the intra- and inter-atomic l–s interaction between the two spins. The order of magnitude of D is given by

$$D \approx \frac{\Delta g}{g} J_{\text{super}}, \tag{7.68}$$

where Δg is the deviation of the g-factor from 2, and J_{super} is the superexchange interaction between the two spins. The order of magnitude of $\Delta g/g$ for $3d$ ions is 10^{-2}–10^{-3}, so that D is such a weak interaction that it causes only a small deviation (say a few degrees at most) of the spin angle from 180°. The $\boldsymbol{D}$-vector lies in a direction which is determined by crystal symmetry. For instance, in α-Fe_2O_3, the $\boldsymbol{D}$-vector lies parallel to the c-axis, so that the spin-canting occurs in the c-plane. This spin arrangement is almost antiferromagnetic, but the spin-canting also produces a small spontaneous magnetization in the c-plane. This is the origin of parasitic ferromagnetism. It is interesting that the material α-Fe_2O_3 exhibits a phenomenon which demonstrates this mechanism: on cooling through a temperature of 250 K, which is called the *Morin point*, the antiferromagnetic spin axis rotates from the c-plane to the c-axis and the spontaneous magnetization simultaneously disappears.[17] This proves the validity of the interaction given by (7.67), because any tilting of spins from the c-axis makes the vector $\boldsymbol{S}_1 \times \boldsymbol{S}_2$ parallel to the c-plane, so that w_D vanishes. Table 7.2 lists several materials which exhibit parasitic ferromagnetism and compares the values of $\Delta S/S$ and $\Delta g/g$, where ΔS is the tilt component of spin which contributes the parasitic ferromagnetism. The agreement between the two quantities proves the validity of this mechanism.

Fig. 7.18. Diagram showing the relationship between the coefficient of antisymmetric interaction, D, and spins S_1 and S_2.

Table 7.2. Typical helimagnetic materials and comparison between tilt of spin, $\Delta S/S$ and $\Delta g/g$. (After Moriya[18])

Material	$\Delta S/S$	$\Delta g/g$
α-Fe_2O_3	1.4×10^{-3}	$\sim 1 \times 10^{-3}$
$MnCo_3$	$2-6 \times 10^{-3}$	$\sim 1 \times 10^{-3}$
$CoCO_3$	$2-6 \times 10^{-2}$	—
CrF_3	1×10^{-2}	$\sim 1 \times 10^{-2}$
FeF_3	2×10^{-3}*	$\sim 1 \times 10^{-3}$

* Livinson.[19]

Parasitic ferromagnetism appears also in NiF_2, for which we expect that $D=0$ from the crystal symmetry. Moriya suggested that a strong crystalline field may be responsible for causing the spin canting.[20]

For further details of this topic, see the review article.[20] Sometimes parasitic ferromagnetism is called weak ferromagnetism, but this invites confusion with the weak ferromagnetism of metals. A better term for parasitic ferromagnetism would be *spin-canted magnetism*, which suggests the mechanism.

7.5 MICTOMAGNETISM AND SPIN GLASSES

'Mictomagnetism' is a term coined by Beck (Beck[21,22]; Waber and Beck[23]; Beck and Chakrabarti[24]) to describe a spin system in which various exchange interactions are mixed. The prefix *micto* is a Greek prefix signifying 'mixing'. When the temperature is lowered, the spin system is frozen with no ordered structure. Such magnetism is observed in Cu–Mn, Fe–Al and Ni–Mn alloys.

The distinguishing experimental feature of mictomagnetism is that the magnetization drops abruptly when the material is cooled in the absence of magnetic field, as shown in Figs 7.19 and 7.20 for Cu–Mn[24] and Ni–Mn[25,26] alloys. A similar phenomenon was observed for Fe–Al.[27–29] The reason for this decrease is the reversal of magnetization clusters caused by antiferromagnetic interactions. If the material is cooled in a magnetic field, the decrease in magnetization disappears (see Figs 7.19 and 7.20); instead there is a shift of the hysteresis loop along the H-axis[30] as shown in Fig. 7.21. The torque curve in this case contains a component with a period of 360°, showing the presence of unidirectional magnetic anisotropy (see Section 13.4). This phenomenon is interpreted to mean that the ferromagnetic spins are interacting with the antiferromagnetic spins, which are fixed with respect to the lattice.

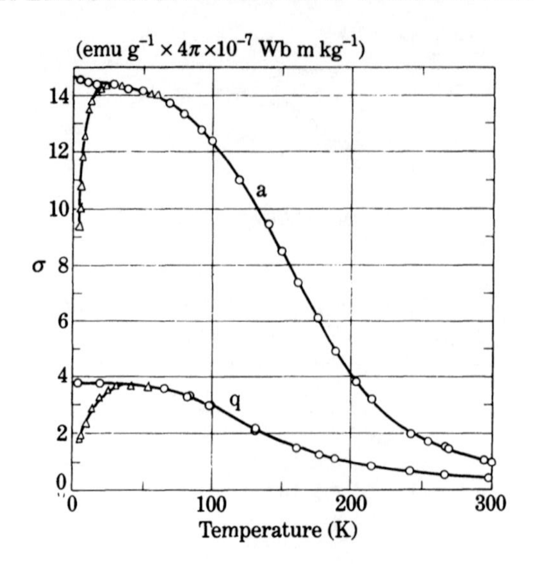

Fig. 7.19. Temperature dependence of magnetization for 16.7% Mn–Cu alloys: (a) heat-treated at 100°C; (q) quenched; (circles) cooled in a magnetic field; (triangles) cooled in absence of magnetic field.[23,24]

It should be remarked here that the antiferromagnetic arrangement is not strongly fixed in a face-centered cubic lattice. Suppose a face-centered cubic lattice contains spins $\boldsymbol{S}_A$, $\boldsymbol{S}_B$, $\boldsymbol{S}_C$ and $\boldsymbol{S}_D$ in the A-, B-, C- and D-sublattices, as shown in Fig. 7.22. If

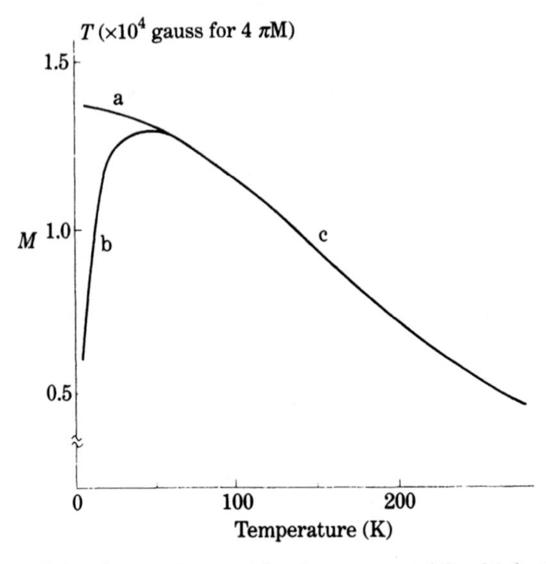

Fig. 7.20. Temperature dependence of magnetization measured for 24.6 at% Mn–Ni alloys in a magnetic field of 0.64 MA m^{-1} (8 kOe): curve ac: cooled in a magnetic field; curve bc: cooled without magnetic field.[26]

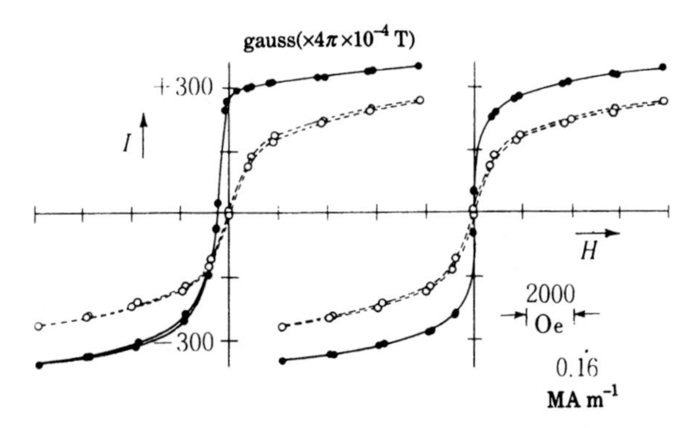

Fig. 7.21. Hysteresis loops measured for disordered 26.5 at% Mn–Ni alloys at 1.8 K. Solid curves: cooled in a magnetic field of $0.4\,\mathrm{MA\,m^{-1}}$ (5 kOe) from 300 K to 1.8 K. Broken curves: similarly cooled without magnetic field. Left: measured along the axis parallel to the field during cooling. Right: measured perpendicular to the field during cooling.[30]

the antiferromagnetic exchange interactions between all neighboring spins are the same in magnitude, then any spin configuration satisfying

$$\boldsymbol{S}_\mathrm{A} + \boldsymbol{S}_\mathrm{B} + \boldsymbol{S}_\mathrm{C} + \boldsymbol{S}_\mathrm{D} = 0 \tag{7.69}$$

keeps the total exchange energy constant, as will be shown below. Let us assume that spins $\boldsymbol{S}_\mathrm{A}$, $\boldsymbol{S}_\mathrm{B}$, $\boldsymbol{S}_\mathrm{C}$ and $\boldsymbol{S}_\mathrm{D}$ all have magnitude S, and their direction cosines are given by $(\alpha_\mathrm{A}, \beta_\mathrm{A}, \gamma_\mathrm{A})$, $(\alpha_\mathrm{B}, \beta_\mathrm{B}, \gamma_\mathrm{B})$, $(\alpha_\mathrm{C}, \beta_\mathrm{C}, \gamma_\mathrm{C})$ and $(\alpha_\mathrm{D}, \beta_\mathrm{D}, \gamma_\mathrm{D})$. Then the x-, y- and z-components of (7.69) are reduced to

$$\left.\begin{aligned} \alpha_\mathrm{A} + \alpha_\mathrm{B} + \alpha_\mathrm{C} + \alpha_\mathrm{D} &= 0 \\ \beta_\mathrm{A} + \beta_\mathrm{B} + \beta_\mathrm{C} + \beta_\mathrm{D} &= 0 \\ \gamma_\mathrm{A} + \gamma_\mathrm{B} + \gamma_\mathrm{C} + \gamma_\mathrm{D} &= 0 \end{aligned}\right\}. \tag{7.70}$$

Squaring the left-hand side of each equation in (7.70), and summing up by considering the relationship $\alpha_i^2 + \beta_i^2 + \gamma_i^2$, we have

$$\sum_{i \neq j} \alpha_i \alpha_j + \sum_{i \neq j} \beta_i \beta_j + \sum_{i \neq j} \gamma_i \gamma_j = -2. \tag{7.71}$$

The total exchange energy is then given by

$$\begin{aligned} E_\mathrm{ex} &= -4NJ(\boldsymbol{S}_\mathrm{A} \cdot \boldsymbol{S}_\mathrm{B} + \boldsymbol{S}_\mathrm{B} \cdot \boldsymbol{S}_\mathrm{C} + \boldsymbol{S}_\mathrm{C} \cdot \boldsymbol{S}_\mathrm{D} + \boldsymbol{S}_\mathrm{D} \cdot \boldsymbol{S}_\mathrm{A} + \boldsymbol{S}_\mathrm{A} \cdot \boldsymbol{S}_\mathrm{C} + \boldsymbol{S}_\mathrm{B} \cdot \boldsymbol{S}_\mathrm{D}) \\ &= -4NJS^2\{(\alpha_\mathrm{A}\alpha_\mathrm{B} + \beta_\mathrm{A}\beta_\mathrm{B} + \gamma_\mathrm{A}\gamma_\mathrm{B}) + (\alpha_\mathrm{B}\alpha_\mathrm{C} + \beta_\mathrm{B}\beta_\mathrm{C} + \gamma_\mathrm{B}\gamma_\mathrm{C}) \\ &\quad + (\alpha_\mathrm{C}\alpha_\mathrm{D} + \beta_\mathrm{C}\beta_\mathrm{D} + \gamma_\mathrm{C}\gamma_\mathrm{D}) + (\alpha_\mathrm{D}\alpha_\mathrm{A} + \beta_\mathrm{D}\beta_\mathrm{A} + \gamma_\mathrm{D}\gamma_\mathrm{A}) \\ &\quad + (\alpha_\mathrm{A}\alpha_\mathrm{C} + \beta_\mathrm{A}\beta_\mathrm{C} + \gamma_\mathrm{A}\gamma_\mathrm{C}) + (\alpha_\mathrm{B}\alpha_\mathrm{D} + \beta_\mathrm{B}\beta_\mathrm{D} + \gamma_\mathrm{B}\gamma_\mathrm{D})\} \\ &= -4NJS^2\left\{\sum_{i \neq j} \alpha_i \alpha_j + \sum_{i \neq j} \beta_i \beta_j + \sum_{i \neq j} \gamma_i \gamma_j\right\} \\ &= 8NJS^2 \end{aligned} \tag{7.72}$$

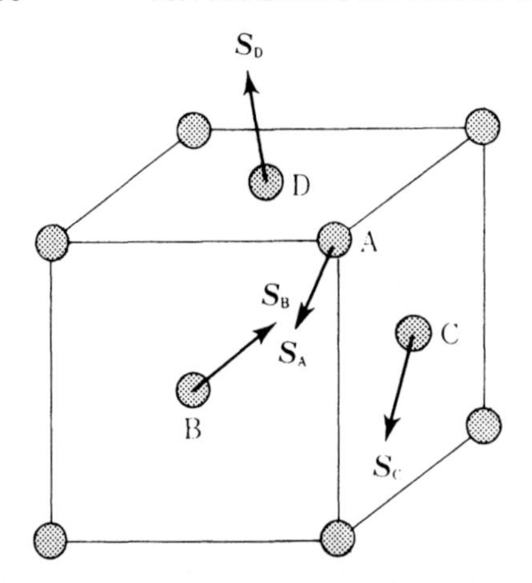

Fig. 7.22. Illustration of spin vectors in four sublattices in face-centered cubic antiferromagnet.

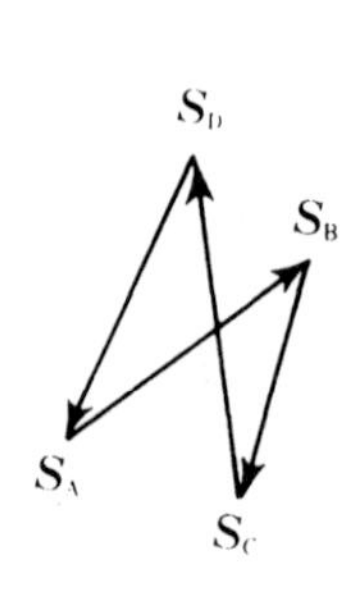

Fig. 7.23. The relationship between four sublattice spin-vectors under the condition of equation (7.69).

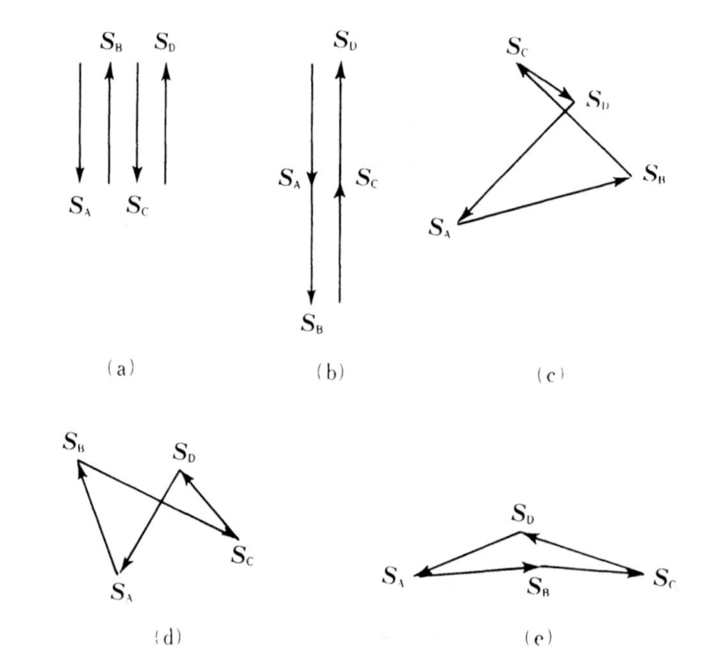

Fig. 7.24. Special cases in Fig. 7.23: (a) and (b) spin axis is parallel to [001]; (c) spin axes are parallel to different ⟨110⟩ crystal axes; (d) spin axes are parallel to different ⟨111⟩ axes; (e) general case.

(see (7.71)), where N is the number of unit cells in the system. Thus as long as condition (7.69) is satisfied, the exchange energy remains constant. The condition (7.69) is shown as a vector diagram in Fig. 7.23. Figure 7.24 shows special cases of Fig.

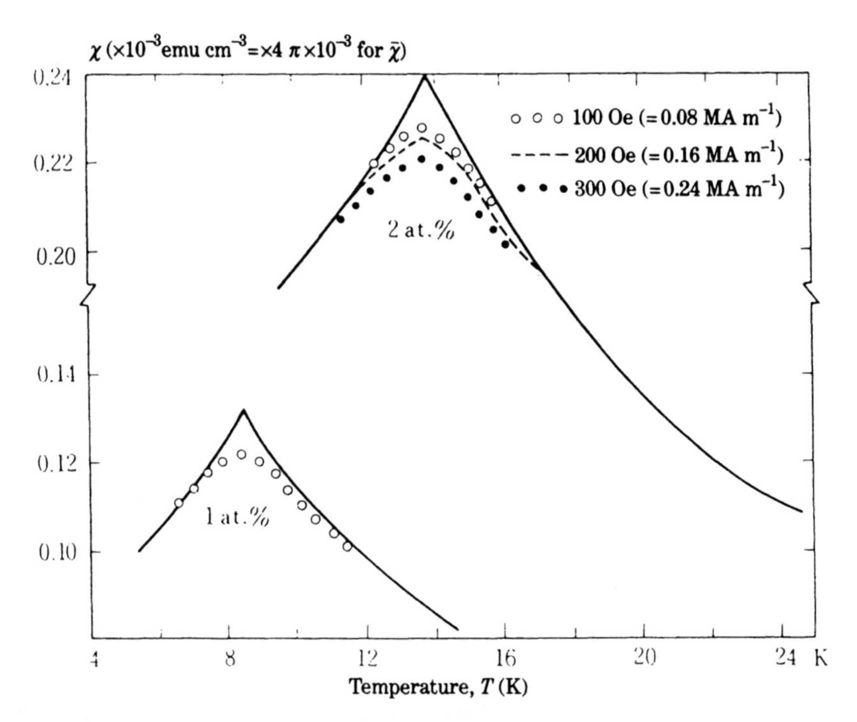

Fig. 7.25. Temperature dependence of magnetization measured for 1 and 2 at% Fe–Au alloys in a weak magnetic field. Solid curves: extrapolated to zero magnetic field.[32]

7.23: (a) and (b) show all spins aligned parallel to the z-axis or [001], while (c) and (d) show all spins parallel to one of the ⟨110⟩ or ⟨111⟩. [These spin configurations were already suggested by Kouvel and Kasper[31] for Fe–Mn alloys.] Figure 7.24(e) shows the case in which none of the spins are parallel to the principal crystal axes. Thus the antiferromagnetic spin arrangement in a face-centered cubic crystal has freedom to change its spin orientation without changing the exchange energy. Therefore, if the magnetic atoms in the A-, B-, C- and D-sites are different chemical species, the value and sign of the exchange integral J_{ij} or the dipole interaction l_{ij} (see Section 12.3) depends on the chemical identity of the atomic pairs, and the local spin configuration is easily disturbed, thus resulting in mictomagnetism.

A *spin glass* state occurs in dilute alloys in which spins of magnetic atoms are frozen randomly by the oscillatory RKKY exchange interaction (see (8.38)). Experimentally, a sharp maximum is observed in the temperature dependence of susceptibility measured in a weak field. Figure 7.25 shows an example observed for dilute Fe–Au alloys.[32] Below the temperature of this maximum, T_c, the random spin arrangement is considered to be fixed. A similar phenomenon is also observed in mictomagnetism. The difference between a spin glass and mictomagnetism is clarified if the critical temperature, T_c, is plotted as a function of the magnetic impurity content. Figure 7.26 shows an example for Fe–Au alloys. In the spin glass region below 12 at% Fe, the

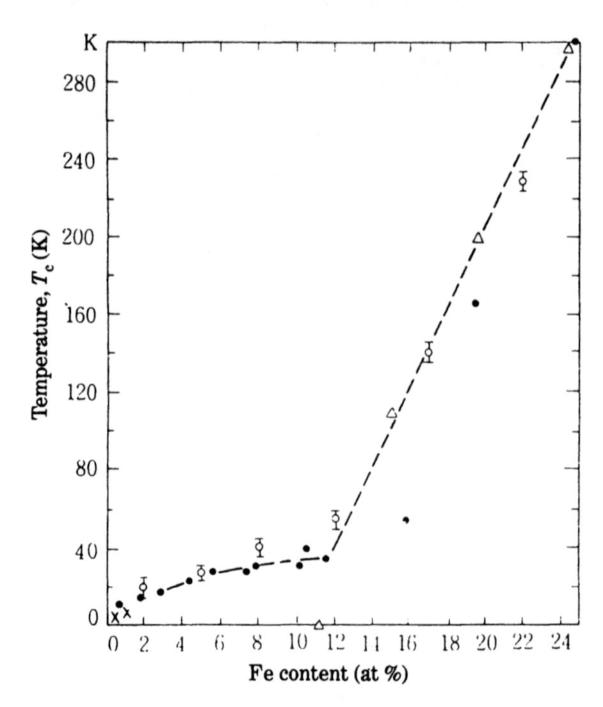

Fig. 7.26. Compositional dependence of the transition temperature, T_c, for Fe–Au dilute alloys.[32]

critical temperature, T_c, remains low because individual spins are randomly fixed. But T_c increases rapidly with Fe content above 12 at%, where mictomagnetism exists. The reason is that ferromagnetic clusters are coupled antiferromagnetically, so that the interactions are strong. However, it is not certain whether or not the spin glass state is actually realized in real dilute magnetic alloys. For more details the reference book[33] should be consulted.

PROBLEMS

7.1 Consider an antiferromagnetic material which has a susceptibility, χ_0, at its Néel point Θ_N. Assuming that the exchange interactions within the A and B sites are negligibly small compared to the exchange interaction between A and B sites, find the values of the susceptibility which would be measured when a magnetic field is applied perpendicular to the spin axis at $T = 0$, $\Theta_N/2$, and $2\Theta_N$.

7.2 Consider a ferrimagnetic material which has the same kind of magnetic ions in A and B sites, in the ratio 3:2. Assuming that the exchange interactions within A and B sites are negligibly small compared to those between A and B sites, find the nature of the spin arrangement at 0 K and the ratio of the ferrimagnetic Curie point to the asymptotic Curie point, and the possibility of having a compensation point.

7.3 In a helimagnetic material, in which J_2 is negative and has magnitude $\sqrt{3}/6J_1$, what will be the pitch of the screw structure, measured in units of the interplanar spacing?

REFERENCES

1. C. G. Shull and J. S. Smart, *Phys. Rev.*, **76** (1949), 1256.
2. H. A. Kramers, *Physica*, **1** (1934), 182.
3. P. W. Anderson, *Phys. Rev.*, **79** (1950), 350.
4. P. K. Baltzer, *Solid State Phys.* (*Kotai Butsuri*, in Japanese, Sci. Tech. Center, Tokyo), **2** (1967), 19.
5. T. Nagamiya, K. Yosida, and R. Kubo, *Adv. Phys.*, **4** (1955), 1.
6. J. H. Van Vleck, *J. Chem. Phys.*, **9** (1941), 85; *J. de Phys. Rad.*, **12** (1951), 262.
7. L. Néel, *Ann. de Physiq.* [12] **3** (1948), 137.
8. P. W. Anderson, *Phys. Rev.*, **79** (1950), 705.
9. S. Chikazumi *et al.* (eds): *Handbook on magnetic substances* (Asakura Publishing Co., Tokyo, 1975).
10. N. Miura, I. Oguro, S. Chikazumi, *J. Phys. Soc. Japan*, **45** (1978), 1534.
11. A. Yoshimori, *J. Phys. Soc. Japan*, **14** (1959), 807.
12. J. Villain, *Chem. Phys. Solids*, **11** (1959), 303.
13. T. A. Kaplan, *Phys. Rev.*, **116** (1959), 888.
14. L. Néel: *Ann. Physiq.*, **4** (1949), 249.
15. I. Dzyaloshinsky, *J. Phys. Chem. Solids*, **4** (1958), 241.
16. T. Moriya, *Phys. Rev. Letters*, **4** (1960), 228.
17. F. J. Morin, *Phys. Rev.*, **78** (1950), 819.
18. T. Moriya, *Magnetism I* (Academic Press, 1963, ed. by Rado & Suhl), p. 86.
19. L. M. Livinson, *J. Phys. Chem. Solids*, **29** (1968), 1331.
20. T. Moriya, *Phys. Rev.*, **117** (1960), 635.
21. P. A. Beck, *Met. Trans.*, **2** (1971), 2015.
22. P. A. Beck, *J. Less Common Metals*, **28** (1972), 193.
23. J. T. Waber and P. A. Beck, *Magnetism in alloys* (TMS, AIME, 1972).
24. P. A. Beck and D. J. Chakrabarti, *Amorphous magnetism* (ed. H. O. Hooper and A. M. de Graaf), Plenum Press, New York, 1973), p. 273.
25. J. S. Kouvel, C. D. Graham, Jr., and J. J. Becker, *J. Appl. Phys.*, **29** (1958), 518.
26. T. Satoh and C. E. Patton, *AIP Conf. Proc.*, **34** (1976), 361.
27. H. Sato and A. Arrott, *J. Appl. Phys.*, **29** (1958), 515.
28. A. Arrott and H. Sato, *Phys. Rev.*, **114** (1959), 1420.
29. H. Sato and A. Arrott, *Phys. Rev.*, **114** (1959), 1427.
30. J. S. Kouvel and C. D. Graham, Jr., *J. Appl. Phys.*, **30** (1959), 312S.
31. J. S. Kouvel and J. S. Kasper, *J. Phys. Chem. Solids*, **24** (1963), 539.
32. V. Cannella and J. A. Mydosh, *Phys. Rev.*, **B6** (1972), 4420.
33. R. A. Levy and R. Hasegawa (eds), *Amorphous magnetism* II (Plenum Press, New York, 1977).

Part IV

MAGNETIC BEHAVIOR AND STRUCTURE OF MATERIALS

Since magnetism is closely related to the electronic structure of materials, the magnetic behavior of metals, oxides and compounds is not the same. In this Part, we discuss these phenomena: the band structure of metals and their magnetic behavior is treated in Chapter 8, crystal structures of oxides and their magnetic behavior in Chapter 9, bonding of various compounds and their magnetic behavior in Chapter 10, and the structure and magnetic behavior of amorphous materials in Chapter 11.

8

MAGNETISM OF METALS AND ALLOYS

8.1 BAND STRUCTURE OF METALS AND THEIR MAGNETIC BEHAVIOR

Of all the metallic elements, ferromagnetism occurs only in three of the $3d$ transition metals (Fe, Co, and Ni), and in heavy rare-earth metals such as Gd, Tb, Dy, etc. The $3d$ transition metals have high Curie points and exhibit ferromagnetism with large spontaneous magnetizations at room temperature, so that alloys containing these metals are used as magnetic materials in a wide range of practical applications. The carriers of the magnetism, the $3d$ electrons, are located relatively far from the atomic core, and are considered to be moving among the atoms (or itinerant), rather than localized at individual atoms. In other words, they form a band structure. On the other hand, the carriers of magnetism in rare-earth metals are $4f$ electrons, which are located deep inside the atoms (see Section 3.2), so that their magnetic moments are well localized at individual atoms. We shall discuss such cases in Sections 8.2 and 8.3.

Metals are characterized by free electrons moving or itinerating in the crystal lattice. The most naive model of free electrons is to regard them as randomly moving particles, like the molecules of an ideal gas. Using this model, we can explain Ohm's law (electric current proportional to the electric field), and Wiedemann–Franz law (thermal conductivity proportional to electrical conductivity). However this model is inadequate to interpret the electronic heat capacity, which is commonly observed in metals at low temperatures, and Pauli paramagnetism, which is observed only for metals and alloys (see Section 5.2). These phenomena were explained only after the quantum theory of metals was developed.

In wave mechanics, a particle moving with momentum p is replaced by a plane wave with wavelength

$$\lambda = \frac{h}{p}, \tag{8.1}$$

where h is Planck's constant. The wave function is expressed as

$$\psi \sim e^{i\boldsymbol{k}\cdot\boldsymbol{r}}, \tag{8.2}$$

where $\boldsymbol{r}$ is the positional vector, and $\boldsymbol{k}$ is the wave number vector given by

$$k = \frac{2\pi}{\lambda}. \tag{8.3}$$

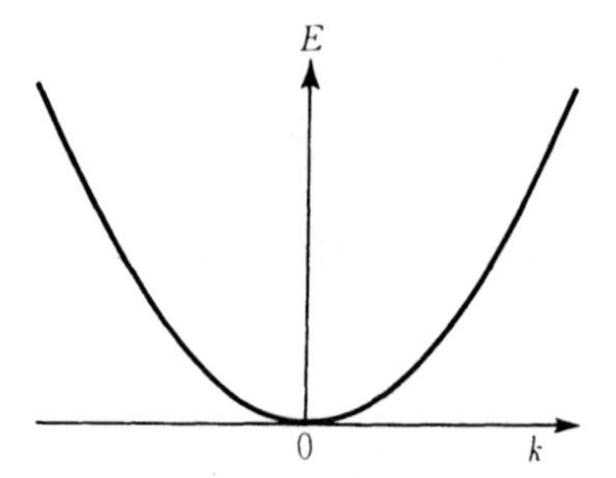

Fig. 8.1. Energy (E) vs wave vector (k) curve for free electrons.

The kinetic energy of this particle is given by

$$E = \frac{1}{2m} p^2, \tag{8.4}$$

which is modified, using (8.1) and (8.3), to

$$E = \frac{1}{2m}\left(\frac{h}{\lambda}\right)^2 = \frac{\hbar^2}{2m} k^2, \tag{8.5}$$

where $\hbar$ is equal to $h/2\pi$. The dependence of E on k is graphically shown in Fig. 8.1.

Now suppose an electron is moving in a cubic box with edge length L. The condition for which the wave function forms a stationary wave is given by

$$\boldsymbol{k} = \frac{\pi}{L}\boldsymbol{n}, \tag{8.6}$$

where $\boldsymbol{n}$ is a vector with components (n_x, n_y, n_z), where n_x, n_y, and n_z are integers such as $0, \pm 1, \pm 2, \ldots$. Thus the $\boldsymbol{k}$-vectors of free electrons are quantized. Each stationary state can be occupied by two electrons with $+$ and $-$ spins, owing to the Pauli exclusion principle. Using (8.6), (8.5) becomes

$$E = \frac{h^2}{8mL^2} n^2. \tag{8.7}$$

Thus the kinetic energy of an electron increases with an increase of $\boldsymbol{n}$. Therefore, when N electrons exist in a unit volume, pairs of electrons occupy the states successively from $n = 0$ (the lowest energy state) up to some non-zero maximum n with finite energy. Thus in metals the electrons with non-zero kinetic energy are moving even at absolute zero, which cannot be expected in a classical picture. The energy of the electron which occupies the highest energy state is called the *Fermi level* and is denoted by E_f. The Fermi level can be calculated by equating the total number of electrons, NL^3, to twice the number of states with energy less than E_f. Since the states can be identified with lattice sites in $\boldsymbol{n}$-space with positive values of $\boldsymbol{n}$, we can write

$$\frac{\pi}{3} n_f^3 = NL^3, \tag{8.8}$$

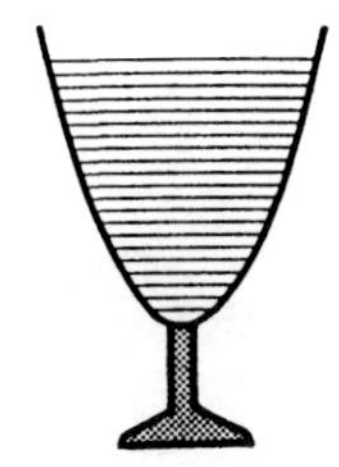

Fig. 8.2. Similarity of water in a glass to free electrons in metals.

where n_f is the value corresponding to the Fermi level. Using n_f and referring to (8.7), the Fermi level is given by

$$E_f = \frac{h^2}{8mL^2} n_f^2. \tag{8.9}$$

Using the relationship (8.8), (8.9) becomes

$$E_f = \frac{h^2}{8m}\left(\frac{3N}{\pi}\right)^{2/3} = \frac{\hbar^2}{2m}(3\pi^2 N)^{2/3}. \tag{8.10}$$

The value of E_f is estimated from (8.10) to be 20 000–50 000 K, much larger than the thermal energy kT at room temperature.

The situation is analogous to a glass of water (see Fig. 8.2): the water level corresponds to the Fermi level, and the total volume of water corresponds to the total number of electrons. In order to calculate the total volume of water, we need to know the cross-sectional area of the glass at each level. Corresponding to this quantity, we define the *density of states*, $g(E)$. The number of states between energy E and $E + \mathrm{d}E$ is given by $g(E)\mathrm{d}E$. Referring to (8.10), the total number of electrons is given by

$$\int_0^{E_f} g(E)\,\mathrm{d}E = N = \frac{1}{3\pi^2}\left(\frac{2mE_f}{\hbar^2}\right)^{3/2}. \tag{8.11}$$

This relationship holds for any value of E. Differentiating (8.11) with respect to E, we have

$$g(E) = \frac{1}{2\pi^2}\left(\frac{2m}{\hbar^2}\right)^{3/2} E^{1/2}. \tag{8.12}$$

This relationship $g(E)$ is graphically shown in Fig. 8.3.

In real metals we have, in addition to the free electrons, positively charged metallic atoms forming a crystal lattice. Slow electrons with long wave length or small k can move without being disturbed by such a crystal lattice. If, however, the wavelength becomes nearly equal to the lattice constant a, the electron wave tends to be reflected in a Bragg reflection. Consider a one-dimensional crystal lattice with lattice constant a along the x-axis. The Bragg reflection occurs when

$$k = \pm\frac{\pi}{a}. \tag{8.13}$$

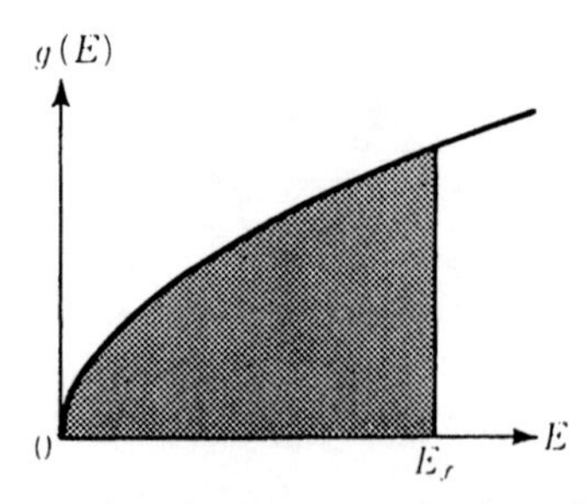

Fig. 8.3. Density of states curve for free electrons.

If a wave e^{ikx} propagates in the $+x$-direction, then a wave e^{-ikx} is produced by Bragg reflection. As a result, the stationary waves

$$\left.\begin{aligned}\psi_1 &\sim \sin kx \sim (e^{ikx} - e^{-ikx}) \\ \psi_2 &\sim \cos kx \sim (e^{ikx} + e^{-ikx})\end{aligned}\right\} \tag{8.14}$$

are formed. As shown in Fig. 8.4, the first wave has its maximum amplitude between the lattice points, while the second wave has its maximum amplitude at the lattice points. Since ψ^2 signifies the probability of the existence of electrons, the former wave has a higher Coulomb energy than the latter. Let this energy difference be ΔE.

Let us consider how such an energy difference modifies the E–k curve in Fig. 8.1. The modification is negligible for small k, because there is no influence of the Bragg reflection. As k approaches π/a, the effect of the Coulomb interaction with the lattice lowers the energy and so increases the probability of free electrons existing at metal ions. At $k = \pi/a$, the electron wave resonates with the lattice and the energy E increases discontinuously by ΔE (Fig. 8.5). The reason for this discontinuity is that the reflected wave changes its phase angle by 180°, as is usual in resonance phenomena, so that the standing wave changes its mode from $\cos kx$ to $\sin kx$ (see (8.14) and Fig. 8.4).

As a result of the appearance of the energy gap, the density of states curve, $g(E)$, is changed from Fig. 8.3 to Fig. 8.6, where we find two energy regions are separated by ΔE. We call such a division into limited energy regions the *band structure*.

Next we consider how the situation changes when a magnetic field is applied. Figure 8.7 illustrates a band structure in which the + spin and − spin bands are shown separately. In this case, the ordinate represents energy, while the abscissa

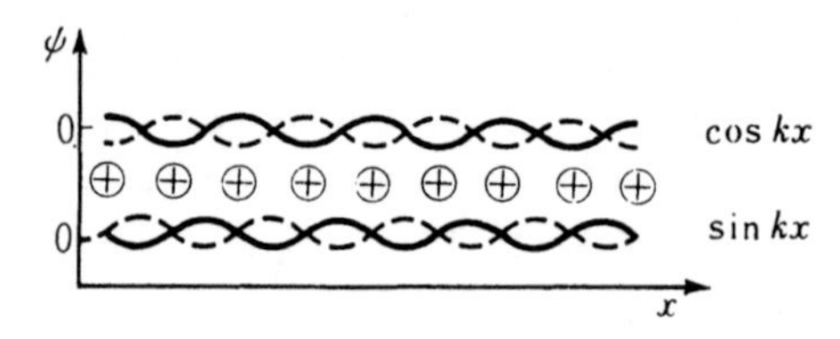

Fig. 8.4. Two modes of standing waves of wave functions in a crystal lattice.

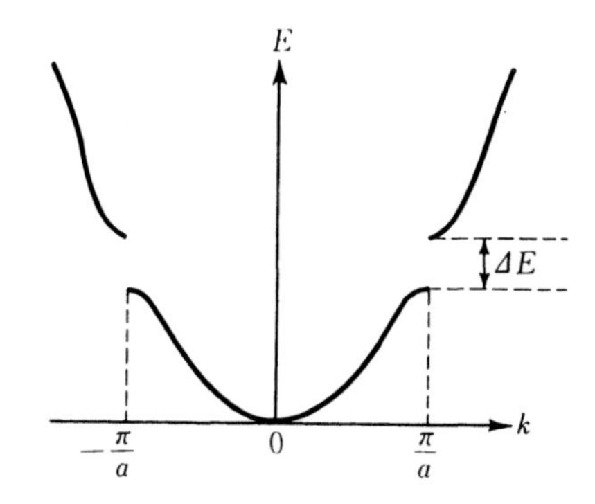

Fig. 8.5. The E vs k curve for electrons in a metallic lattice.

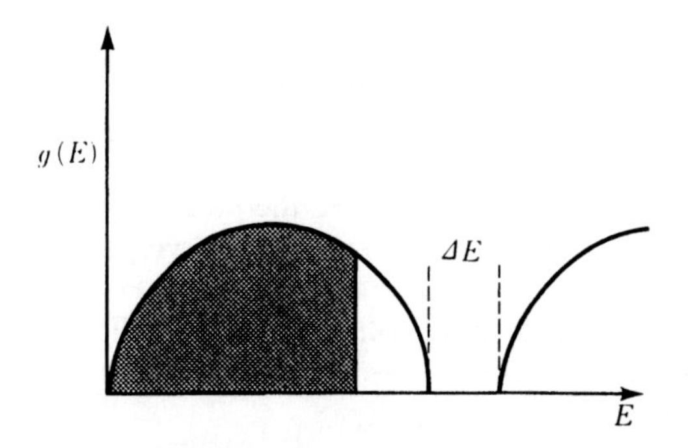

Fig. 8.6. Density of states curve for electrons in a metallic lattice.

represents the density of states of the + spin and − spin bands. When a magnetic field H is applied parallel to the + spin (here we use the word spin for spin magnetic moment), the + spin band is lowered by an amount

$$E_H = 2M_B H. \tag{8.15}$$

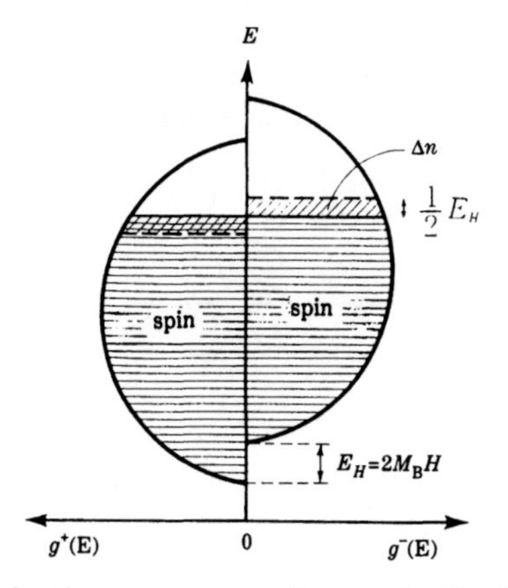

Fig. 8.7. Mechanism for the appearance of magnetization by hand polarization.

Therefore the electrons at the top of the − spin band will be transferred to the + spin band by reversing their spins, bringing their Fermi levels to a common value. The number of electrons, Δn, which will be transferred is given by the area between the old and new Fermi levels, so that

$$\Delta n = g(E_\mathrm{f}) \cdot \frac{E_H}{2} = g(E_\mathrm{f}) M_\mathrm{B} H. \tag{8.16}$$

The resulting increase in magnetization is given by

$$\Delta I = 2M_\mathrm{B}\,\Delta n = 2g(E_\mathrm{f}) M_\mathrm{B}^2 H. \tag{8.17}$$

Therefore the susceptibility is given by

$$\chi_p = 2g(E_\mathrm{f}) \cdot M_\mathrm{B}^2. \tag{8.18}$$

As will be mentioned later, the Fermi level is not strongly temperature-dependent, so the susceptibility given by (8.18) is approximately independent of temperature. This is the origin of the *Pauli paramagnetism.*[1]

When the temperature is raised from 0 K, the electrons are excited to levels higher than the Fermi level. As mentioned previously, however, the Fermi level corresponds to a temperature of order ten-thousand K, so the thermal excitation caused by a temperature of several hundred K will have a very slight effect on the electron distribution. The probability that an electron will occupy a state with energy E which is higher than the Fermi level is given by

$$f = \frac{1}{\exp\left(\dfrac{E - E_\mathrm{f}}{kT}\right) + 1}. \tag{8.19}$$

This $f(E)$ is called the *Fermi–Dirac distribution function*, and is shown graphically Fig. 8.8 for $T = 0$ (solid line) and for $T > 0$ (broken curve).

Such a modification of the electron distribution will be reflected in various physical properties. For instance, it produces a special low-temperature heat capacity which is characteristic of metals. When the temperature is increased from 0 K, the electrons in the vicinity of the Fermi level are excited across the Fermi level, as seen in Fig. 8.8, so

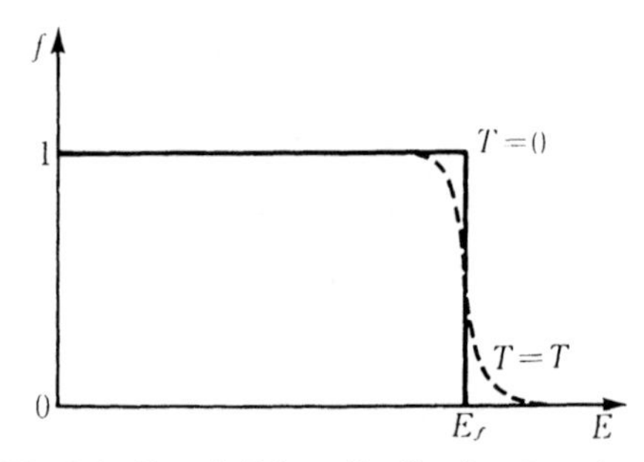

Fig. 8.8. Fermi–Dirac distribution function.

that the electronic heat capacity should be a good measure of the density of states at the Fermi level. Let us next calculate this heat capacity.

The energy of free electrons at a non-zero temperature is given by

$$U = \int_0^\infty f(E,T)g(E)E\,dE. \tag{8.20}$$

The heat capacity per unit volume can be obtained by differentiating (8.20) with respect to T. Since the derived function $df(E,T)/dT$ in the integrand of (8.20) is non-zero only in the vicinity of E_f, we can regard $g(E)$ as constant, so that

$$C_v = \int_0^\infty \frac{df(E,T)}{dT} g(E)E\,dE$$

$$= g(E_f)\int_0^\infty \frac{d}{dT}\left(\frac{1}{\exp\left(\frac{E-E_f}{kT}\right)+1}\right)E\,dE. \tag{8.21}$$

Putting $(E-E_f)/kT = x$, we have the relationships $E = E_f + kTx$, $dE = kT\,dx$, and $dT/T = -dx/x$, so that

$$-\int_{-x_0}^\infty \frac{d}{dx}\left(\frac{1}{e^x+1}\right)(E_f + kTx)kx\,dx = \int_{-x_0}^\infty \frac{e^x}{(e^x+1)^2}(kE_f x + k^2Tx^2)\,dx$$

$$= k^2T\int_{-x_0}^\infty \frac{e^x x^2}{(e^x+1)^2}\,dx = \frac{\pi^2k^2}{3}T, \tag{8.22}$$

where $x_0 = E_f/kT$. Then, from (8.21), we have

$$C_v = \frac{\pi^2k^2}{3}g(E_f)T. \tag{8.23}$$

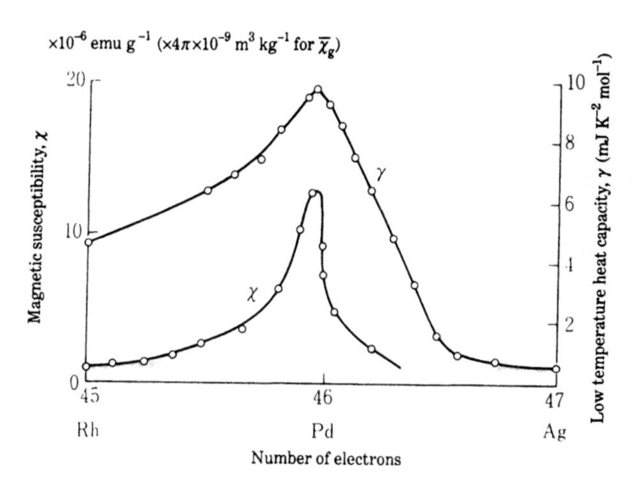

Fig. 8.9. Comparison between magnetic susceptibility and the coefficient of low temperature specific heat for Rh–Pd and Pd–Ag alloys.[2]

This heat capacity increases linearly with increasing T, and is called the *electronic heat capacity* or the *electronic specific heat*. At low temperatures it exceeds the heat capacity due to lattice vibration because the latter is zero at 0 K and increases only slowly with T (proportional to T^3). The proportionality factor γ in (8.23),

$$\gamma = \frac{\pi^2 k^2}{3} g(E_\mathrm{f}) \tag{8.24}$$

is a good measure of the density of states at the Fermi level, $g(E_\mathrm{f})$. The susceptibility χ given by (8.18) is equally a measure of the density of states at the Fermi level. Figure 8.9 shows a comparison between γ and χ observed for Rh–Pd and Pd–Ag alloys.[2] We see a similarity between the two quantities.

As a result of modification of the electron distribution near the Fermi level, the Fermi level will shift slightly, provided the density of states curve, $g(E)$, has non-zero gradient at E_f.

The Fermi level of completely free electrons at $T=0$ is given by (8.10). When the density of states is given by $g(E)$, the Fermi level E_f at $T=0$ is defined by

$$N = \int_0^{E_\mathrm{f}(0)} g(E)\,\mathrm{d}E. \tag{8.25}$$

At non-zero temperature, this condition becomes

$$N = \int_0^\infty f(E) g(E)\,\mathrm{d}E, \tag{8.26}$$

which is modified by partial integration to

$$\begin{aligned}
N &= \left| f(E) \int_0^E g(E)\,\mathrm{d}E \right|_0^\infty - \int_0^\infty f'(E) \left(\int_0^E g(E)\,\mathrm{d}E \right) \mathrm{d}E \\
&= -\int_0^\infty f'(E) \left[\int_0^{E_\mathrm{f}(T)} g(E)\,\mathrm{d}E + g(E_\mathrm{f})(E-E_\mathrm{f}) + \tfrac{1}{2}(g'(E))_{E_\mathrm{f}(T)}(E-E_\mathrm{f})^2 + \cdots \right] \mathrm{d}E \\
&= \int_0^{E_\mathrm{f}(T)} g(E)\,\mathrm{d}E - \tfrac{1}{2}(g'(E))_{E_\mathrm{f}(T)} \int_0^\infty f'(E)(E-E_\mathrm{f})^2\,\mathrm{d}E \\
&= \int_0^{E_\mathrm{f}(T)} g(E)\,\mathrm{d}E + \frac{\pi^2}{6}(kT)^2 (g'(E))_{E_\mathrm{f}(T)}.
\end{aligned} \tag{8.27}$$

(See (8.22)). Subtracting (8.25) from (8.27), we have

$$\int_{E_\mathrm{f}(0)}^{E_\mathrm{f}(T)} g(E)\,\mathrm{d}E + \frac{\pi^2}{6}(kT)^2 (g'(E))_{E_\mathrm{f}(T)} = 0. \tag{8.28}$$

Since $g(E)$ is almost constant between $E_f(0)$ and $E_f(T)$, the first term in (8.28) can be approximated as $(E_f(T) - E_f(0))g(E_f)$, so that

$$E_f(T) = E_f(0)\left[1 - \frac{\pi^2}{6}\frac{(kT)^2}{E_f(0)}\left(\frac{g'(E_f)}{g(E_f)}\right)\right]. \tag{8.29}$$

Such a small shift of the Fermi level will cause only a weak temperature dependence of the Pauli paramagnetic susceptibility.

In the case of ferromagnetic metals, the exchange field H_m is stronger than ordinary external fields by a factor of 10^2 to 10^3, so that the splitting of the bands is much larger than in the case of paramagnetic metals. In general, the number of electrons in the + spin and − spin bands is

$$\left.\begin{aligned} N_+ &= \int_{-\infty}^{+\infty} g(E) f(E_f + M_B H_m)\,dE \\ N_- &= \int_{-\infty}^{+\infty} g(E) f(E_f - M_B H_m)\,dE \end{aligned}\right\}. \tag{8.30}$$

The magnetization induced by this polarization of bands is

$$I = M_B(N_+ - N_-). \tag{8.31}$$

Since the molecular field H_m is

$$H_m = wI, \tag{8.32}$$

the spontaneous magnetization can be found from a solution which satisfies (8.30), (8.31), and (8.32). For this purpose, we must know the functional form of $g(E)$. However, whether or not ferromagnetism appears can be determined only by $g(E_f)$. The increase of the band energy by the transfer of Δn electrons is given by

$$\tfrac{1}{2}\Delta n E_H = \frac{I^2}{2\chi_p}, \tag{8.33}$$

so that, referring to (8.32) and (8.33), the energy of the system is expressed by

$$\begin{aligned} E &= \frac{I^2}{2\chi_p} - \tfrac{1}{2} w I^2 \\ &= \frac{I^2}{2\chi_p}(1 - \chi_p w). \end{aligned} \tag{8.34}$$

Therefore the criterion for the appearance or non-appearance of ferromagnetism is given by

$$\chi_p \gtrless \frac{1}{w}. \tag{8.35}$$

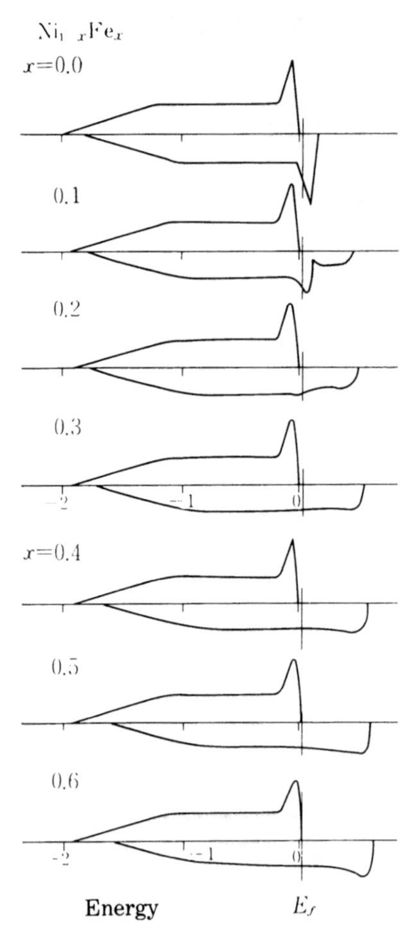

Fig. 8.10. Density of states curves for + and − spins for various Fe–Ni alloys, as calculated by the coherent potential approximation.[8]

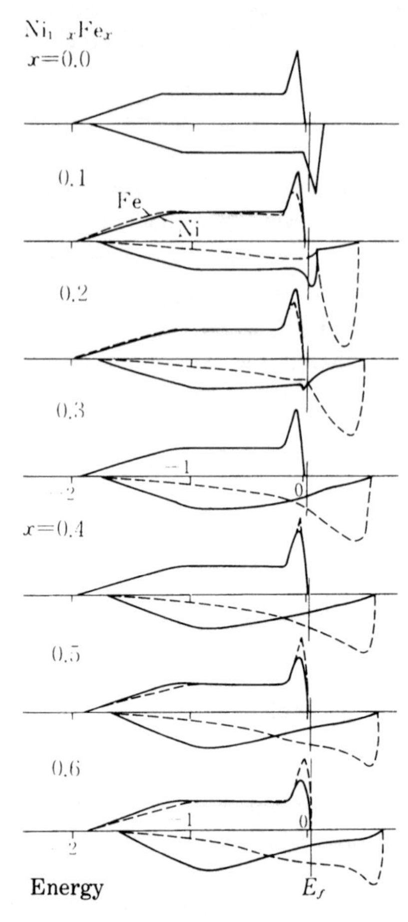

Fig. 8.11. Density of states curves for Ni (solid curve) and for Fe (broken curves) as calculated for Ni–Fe alloys by the coherent potential approximation.[8]

If the > sign is satisfied, the energy (8.34) is lowered as the magnetization I is increased, while if the < sign is satisfied, the energy increases as I appears and ferromagnetism is unstable. This is the criterion for the appearance of ferromagnetism proposed by Stoner,[3,4] who treated ferromagnetism in metals in this way.[5,6] In this case, he assumed that the density of states curve remains unchanged even when ferromagnetism appears. Such a treatment is called the *rigid band model*.

On the other hand, by means of the *coherent potential approximation*,[7] we can treat the band structure by taking into consideration the electronic states of individual atoms. According to this calculation, we can get information on the electronic states and magnetic moments in the vicinity of different kinds of atoms in ferromagnetic alloys.[8,9] In A–B alloys, first the electronic structure of A atoms is calculated by assuming an average electronic structure in the neighborhood of an A atom. The electronic structure of B atoms is calculated in the same way. Then the average electronic structure is calculated by taking the weighted average of the A and B atom structures. Equating this with the structures assumed in the calculation of the electronic structure of the individual atoms, we have a consistent (coherent) solution. This method is somewhat similar to the Bethe approximation in the statistical treatment of ferromagnetism (see Chapter 6). Figure 8.10 shows the density of states curves calculated by this method for Fe_xNi_{1-x} alloys.[8] The important feature of this result is that the shapes of the density of states curves are different for + and − spins, and also for different alloys constituents. The vertical lines signify the Fermi levels. Figure 8.11 shows the band structure for Ni and Fe separately. Looking at this graph, we see that the + spin bands of Ni and Fe are not very different, whereas there are big differences in the − spin bands. Comparison with experiment will be made in the next section.

8.2 MAGNETISM OF 3*d* TRANSITION METALS AND ALLOYS

Only the three transition elements Fe, Co, and Ni exhibit ferromagnetism at room temperature. Most magnetic alloys made for engineering uses contains one or more of these elements. The carrier of magnetism in this case is the 3*d* electrons, which form band structures together with the 4*s* electrons.

In Chapter 2 we discussed the electronic structure of the elements and learned that the 3*d* elements ($Z = 21{-}30$) have an incomplete 3*d* electron shell which produces magnetic moments in accordance with Hund's rule. The three ferromagnetic elements Fe, Co, and Ni are atomic numbers 26, 27, and 28, and have respectively 4, 3, and 2 vacancies in the 3*d* shell (see Table 3.1). According to Hund's rule, we expect spin magnetic moments of 4, 3, and 2 Bohr magnetons, respectively (in addition to the orbital magnetic moment). Actually, however, these elements exhibit saturation magnetic moments of only 2.2, 1.7, and 0.6 Bohr magnetons, respectively, at 0 K.

The situation is well described by plotting the saturation magnetic moment at 0 K as a function of number of electrons per atom. Figure 8.12 shows such curves plotted for various 3*d* transition-metal alloys. This is a famous graph, known as the *Slater–Pauling curve*.[10] The non-integral Bohr magneton numbers of 2.2, 1.7, and 0.6 for Fe, Co, and Ni are smoothly connected by two straight lines.

The experimental points for Ni–Co alloys fall on the straight line connecting the points (27, 1.7) for Co and (28, 0.6) for Ni. It is possible to interpret this behavior to mean that Co atoms with $1.7M_B$ and Ni atoms with $0.6M_B$ are simply mixed in the alloys, with each atom keeping its individual moment. However, in the case of Ni–Cu alloys, the experimental points lie not on the straight line connecting the points (28, 0.6) for Ni and (29, 0) for Cu, but on the straight line connecting (28, 0.6) and (28.6, 0).

If we assume that the number of 4*s* electrons is 0.6 per atom, then we can subtract 0.6 and 18 (corresponding to the electrons in the filled shells of argon) from 28.6 to obtain 10 electrons, which means that the 3*d* electron shell is just filled. This situation

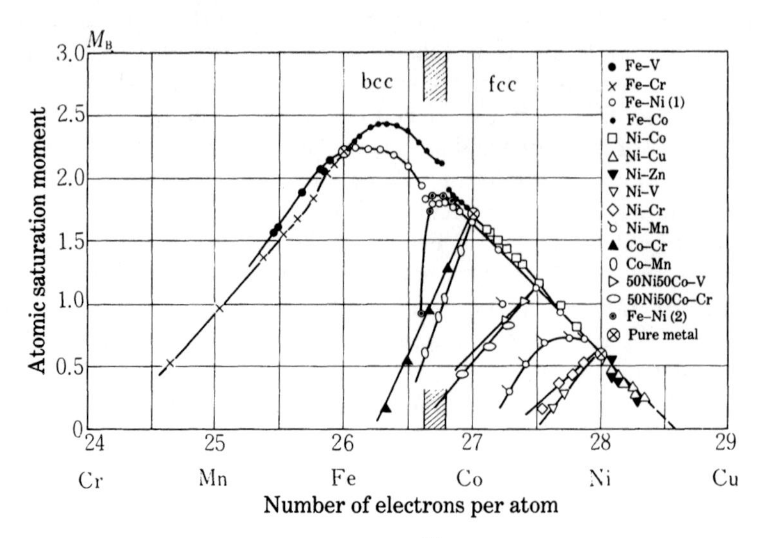

Fig. 8.12. Slater–Pauling curve. (After Bozorth[10], except for NiCo–V, NiCo–Cr[11] and Fe–Ni(2).)[12]

cannot be interpreted in terms of the localized model in which we assume that the 3*d* electrons contributing to the magnetic moment are well localized on individual atoms. The situation can be explained by assuming that the 3*d* electrons are itinerating in the 3*d* band which is common to all the atoms.

As the electron number decreases from 28.6, where the + spin band and − spin band are both filled, electron vacancies appear only in the − spin band. Therefore the magnetic moment in Bohr magnetons, which is given by the difference in the number of electrons in the + spin and the − spin band, increases with a decreasing number of electrons at a rate of one Bohr magneton per one electron. This means that the magnetic moment versus number of electrons is given by a straight line with a slope of −45° in the Slater–Pauling curve. This is actually the case for Co–Ni, Ni–Cu, Ni–Zn and Ni-rich Fe–Ni alloys, as seen in Fig. 8.12. The electron vacancies appear only in the − spin band because the density of states has a sharp peak at the top of the 3*d* band as shown by Fig. 8.13.[13] The shift of the − spin band relative to the + spin band is very small, because the electron vacancies can be accommodated in a narrow space at the top of the − spin band. Therefore the increase in total kinetic energy of the electrons (band energy) is relatively small. Since, however, the total space in the peak at the top of 3*d* band is about 1.5 electrons per atom, any decrease in the number of electrons above 1.5 must produce vacancies also in the + spin band. In this situation, nature prefers a change to a crystal structure with a different density of states curve to an unnaturally large polarization of bands. Figure 8.14 shows the density of states curve calculated by Wakoh and Yamashita[14] for body-centered cubic iron. In this case the density of states versus energy curve exhibits double peaks and the Fermi level in the − spin band is located in the valley between the two peaks. Since the density of states at the Fermi level in the + spin band is fairly high, any

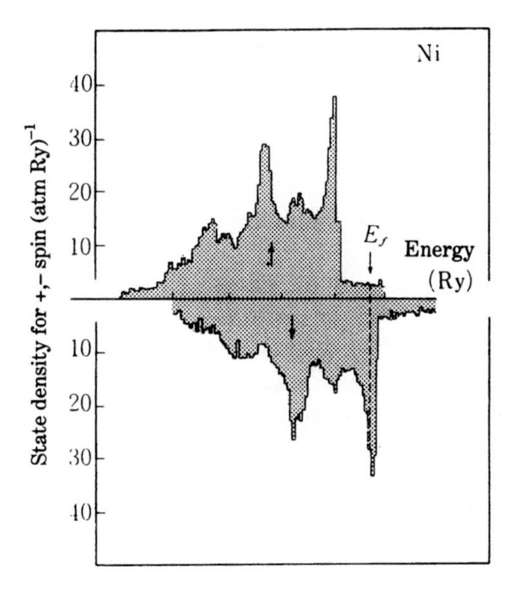

Fig. 8.13. Density of states curve of Ni. (After Connoly[13].)

further decrease in the electron number occurs mainly in the + spin band, resulting in a decrease in magnetic moment. This is the reason why the Slater–Pauling curve has a slope of +45° at the point representing Fe.

It is very interesting that the crystal structure changes from face-centered cubic to body-centered cubic at an electron concentration of 26.7, irrespective of the chemical

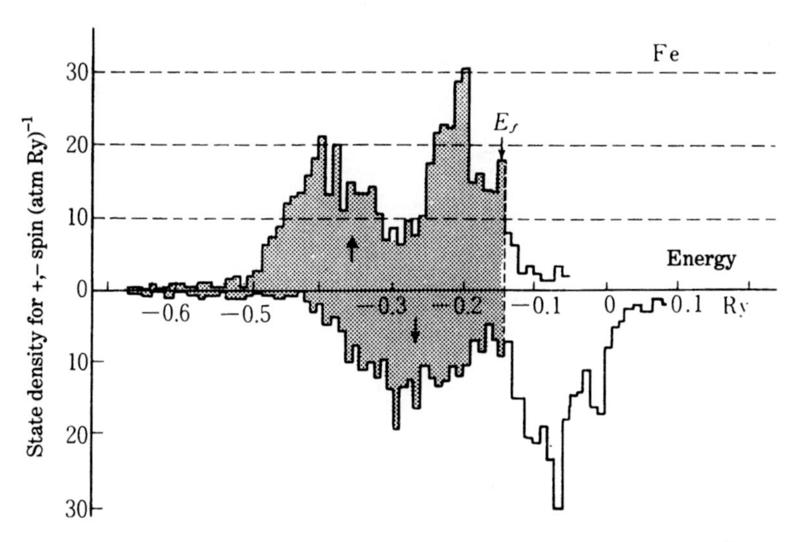

Fig. 8.14. Density of states curve of Fe. (After Wakoh and Yamashita[14].)

species in the alloy (see Fig. 8.12). In the case of Fe–Ni, the saturation moment drops sharply as the electron concentration is decreased and approaches the phase boundary [the points labelled Fe–Ni(2) in Fig. 8.12 are more recent than those labelled Fe–Ni(1)]. At the peak of this curve, located at 35 at% Ni in Fe, a very low thermal expansion coefficient is measured at room temperature. This alloy is called Invar, and its unusual thermal behavior is known as the *Invar effect*. The thermal expansion of Invar is apparently related to the instability of its ferromagnetism. This problem is further discussed later.

The magnetic moments localized on atoms of different chemical species can be determined by means of small-angle scattering of neutrons from disordered alloys. If the spins S_A and S_B associated with A and B atoms are different in magnitude, the intensity of neutron small-angle magnetic scattering should be proportional to $(S_A - S_B)^2$. On the other hand, the saturation magnetization should be proportional to $C_A S_A + C_B S_B$, where C_A and C_B are the concentration of A and B atoms. From this information, we can solve for S_A and S_B for each alloy composition. Figure 8.15 shows the magnetic moments of Fe and Ni atoms determined in this way, as a function of composition in Ni–Fe alloys.[15] It is found that the magnetic moment of Fe is about $2.8M_B$, while that of Ni is about $0.6M_B$, and also that these values change gradually with alloy composition. This behavior can be explained neither by a localized electron model nor by a rigid band model.

It was shown by Hasegawa and Kanamori[8] that this behavior can be explained using the *coherent potential approximation (CPA)* which was discussed in Section 8.1.

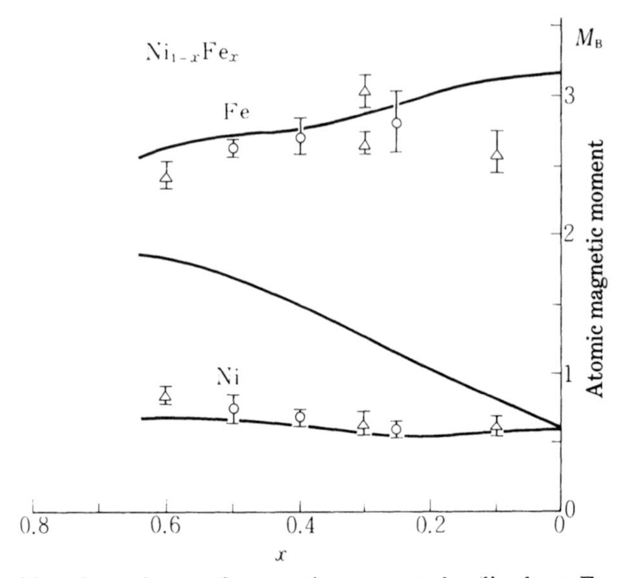

Fig. 8.15. Composition dependence of magnetic moments localized on Fe and Ni atoms in Ni–Fe alloys. Experimental points are obtained by means of neutron small-angle scattering.[16] The solid curves represent the CPA calculation for magnetic moments of Fe and Ni atoms and their average.[8]

The solid curves in Fig. 8.15 represent the CPA calculation for the individual atoms and also the average magnetic moment. We see that not only the individual moments but also the behavior of the average magnetic moment are well reproduced by the CPA calculation. The distribution of electrons in the $3d$ band is not the same in different atom species because the $3d$ electrons in the $-$ spin band tend to shield the excess nuclear charge, so that the difference in nuclear charge $2e$ between Ni and Fe results in a difference in magnetic moment of $2M_B$. (Note that there is a vacancy accommodating two electron vacancies at the top of the $-$ spin band (broken curve in Fig. 8.11)). The gradual change in magnetic moment with alloy composition is due to the fact that the density of states changes gradually as a function of the number of electrons.

There are many branches from the right-hand straight line of the Slater–Pauling curve (see Fig. 8.12). Each branch describes a sharp decrease in saturation magnetic moment produced by the addition of impurity atoms with fewer positive nuclear charges, i.e. Mn, Cr, V, or Ti. This behavior was first treated theoretically by Friedel,[16] and later explained more quantitatively by Akai *et al.*[17] using a CPA calculation. They showed that the magnetic moments of these impurity atoms are coupled antiferromagnetically with the ferromagnetic matrix moment, which results in a sharp decrease in the average magnetic moment. Manganese impurities are an exception, since they sometimes couple ferromagnetically.

The local magnetic disturbance caused by impurity atoms was investigated experimentally by Low and Collins[18-20] by means of magnetic scattering of cold (long-wavelength) neutrons (see Problem 4.3). As explained in Section 4.2, the magnetic form factor for the scattering of neutrons depends strongly on the size of the $3d$ shell (see Figs 4.7 and 4.8). Therefore by measuring the form factor we can analyze not only the size of magnetic atoms but also the extent of the magnetic disturbance around the atoms. The advantage of using long-wavelength neutrons is that we can determine the shape of the form-factor curve over a relatively large range of small scattering angles without interference from elastic scattering (*small-angle diffuse scattering*). Figure 8.16 summarizes the difference in the magnetic moment, ΔM, between impurity and matrix atoms as a function of the difference in electron number, Δn, between them. The experimental points scattered along the straight line in the first and fourth quadrants are all consistent with the right-hand and left-hand straight lines of the main Slater–Pauling curve (Fig. 8.12). The points for Fe-base alloys in the third quadrant, such as $\underline{\text{Fe}}$–Co, $\underline{\text{Fe}}$–Ni, $\underline{\text{Fe}}$–Pd, and $\underline{\text{Fe}}$–Pt (underline means the main alloy constituent), are all inconsistent with the Slater–Pauling curve, because the addition of these impurities increases the average magnetic moment to some extent. It was found by neutron scattering experiments that this increase in average magnetic moment in Fe-base alloys occurs not at the impurity atoms but in the Fe atoms surrounding the impurities.[19] A similar situation occurs in Ni-base alloys with non-transition elements such as Al, Ga, Sb, Si, Ge, or Sn. The addition of these impurities reduces the average magnetic moment as if the valence electrons of the impurity atoms filled up the vacancies in the $3d$ band. Actually, however, a neutron scattering experiment revealed that the reduction of magnetic moment occurs not at the impurity atom but at Ni atoms surrounding the impurity atom.[21] The reason is

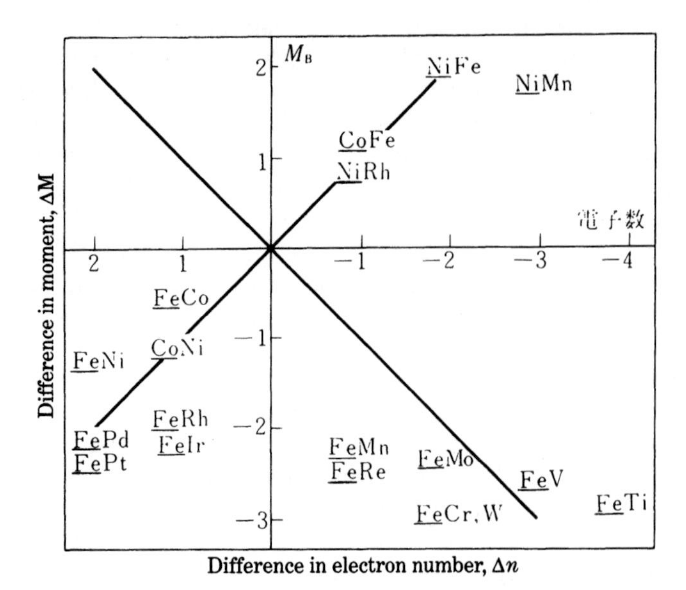

Fig. 8.16. Difference in magnetic moment of impurity atoms and of matrix metals (underlined) as determined by neutron small-angle scattering as a function of difference in number of electrons.[19]

that the shielding of the nuclear charges of the impurity atoms must be done by the $3d$ band of the matrix atoms, which has a high density of states, because the Fermi level of the non-transition impurity is located in the s-band with a very low density of states.

Generally speaking, the crystal structure of transition metals depends on the Fermi level in the d-band. In $4d$ and $5d$ transition metals, the crystal structure changes in the order hexagonal closed packed (hcp), body-centered cubic (bcc), hcp and finally face-centered cubic (fcc) structures, as the Fermi level moves higher. On the other hand, in $3d$ transition metals, this sequence is modified and the crystal structure changes in the order hcp–bcc–αMn–bcc–hcp–fcc; the cycle is doubled compared with the case of $4d$ and $5d$ metals. The reason is that when ferromagnetism appears, the Fermi level splits into different positions in the + spin and − spin bands, resulting in a repetition of the sequence. Thus the crystal structure and magnetism of $3d$ transition metals are mutually correlated.

One example of this correlation is Invar behavior, which appears at a composition near a change in crystal structure. As mentioned earlier, the saturation moment of Fe–Ni alloys disappears at the phase boundary between bcc and fcc (see Fig. 8.12). The Invar characteristic (the thermal expansion coefficient becomes negligibly small at room temperature) is one manifestation of a correlation between magnetism and crystal structure. Phenomenologically, the temperature dependence of the thermal expansion coefficient of Invar can be well explained by assuming the thermal excitation of the low spin state of the Fe atom.[22,23]. More physical explanations based on the band theory have been developed by many theoreticians.[24-26,17]

A complicated phase transition in pure Fe is also caused by a correlation between magnetism and crystal structure. With increasing temperature, pure Fe transforms from bcc to fcc at 910°C, and then once again from fcc to bcc at 1390°C. Zener[27] pointed out that this phenomenon may be explained by assuming an additional 'magnetic' free energy. Since, however, 910°C is above the Curie point, this has nothing to do with magnetic ordering. This phenomenon was also explained theoretically in terms of a disordered local moment theory.[28]

Manganese metal (Mn) is located next to Fe in the periodic table, and undergoes complicated crystal and magnetic phase transitions. Below 705°C, it has the α-Mn structure with 29 atoms in the unit cell. Below 95 K, Mn develops an antiferromagnetic structure with four different atomic moments, 1.90, 1.78, 0.60, and $0.25M_B$, each of which takes a different direction.[29] With increasing temperature, it transforms at 705°C to the β-phase with no magnetic ordering, and then to the fcc γ-phase at 1100°C. The γ-phase is paramagnetic in this temperature range. However, this phase can be retained to lower temperatures by alloying with Cu, and is found to exhibit antiferromagnetism with a magnetic moment of $2.25M_B$ per atom below the Néel point of 207°C.[30]

Chromium (Cr) is located next to Mn in the periodic table. Its crystal structure is bcc at all temperatures, and its magnetism is characterized by a *spin density wave*, in which the magnitude of the spin forms a spatial wave. This is a kind of antiferromagnetism, with Néel point 312 K. The maximum amplitude of the wave is $0.50M_B$ at 0 K and the spatial propagation of the wave is parallel to $\langle 100 \rangle$ with the spin axis perpendicular to the propagation above 122 K and parallel below this temperature.

The spin density wave was treated theoretically by Overhauser[31] and later by Lomer[32] who took an interaction between the Fermi surface of 3*d* electrons and holes into consideration. Kanamori and Teraoka[33] proposed an entirely different interpretation of this phenomenon.

In Chapter 6, we learned that the Curie–Weiss law holds in many ferromagnetic metals and alloys. It seems that this fact supports the localized electron model, because this law was deduced from the Weiss theory which assumes a fixed magnetic moment per atom. It is known, however, that there are many ferromagnetic metals and alloys for which the effective magnetic moment, M_{eff}, and saturation moment, M_s, are not consistently described by the same J (see (5.30) and (5.32)). Figure 8.17 shows the variation of P_c/P_s as a function of Curie point for various magnetic materials, where P_s is gJ, deduced from M_s, and P_c is gJ, calculated by using J deduced from M_{eff}.[34,35] It is seen that for ferromagnetic metals with high Curie points, P_c/P_s is nearly unity, as expected from the Weiss theory, while for metals with low Curie points, this ratio becomes very large. Typical examples of the latter category are $ZrZn_2$ with $M_s = 0.12M_B$ and $\Theta_f = 22$ K, and Sc_3In with $M_s = 0.057M_B$ and $\Theta_f = 7.5$ K. These are called *weak ferromagnets*. Even in these weak ferromagnets, the Curie–Weiss law holds.

Such behavior has been explained by Moriya[35] in terms of *spin-fluctuation theory*,

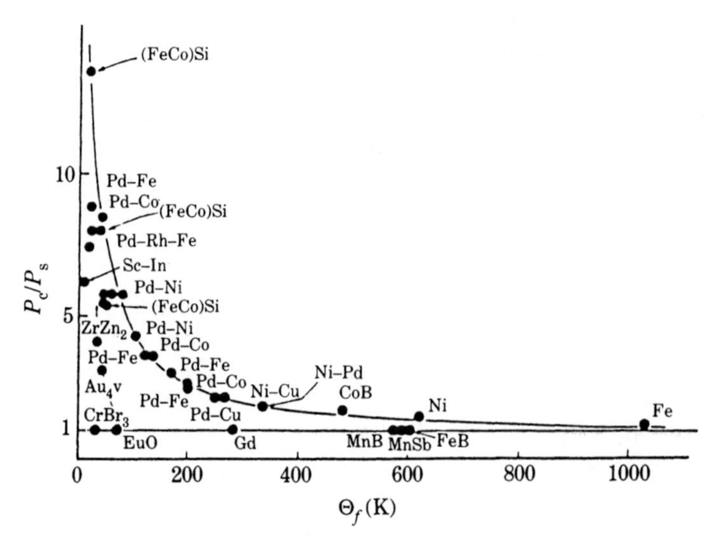

Fig. 8.17. The ratio P_c/P_s (see text) of ferromagnetic metals and alloys as a function of their Curie points. (After Rhodes and Wohlfarth[34])

which treats the band calculation of metal magnetism at elevated temperatures. The thermally disturbed spin system can be described as an assembly of spin density waves with different wave vectors $\boldsymbol{q}$. The feature of this theory is to take proper account of the interaction between spin density waves with different modes. In weak ferromagnetism, the spin density waves with large q have much higher energies, so that they

Table 8.1. Classification of magnetic materials by the spin fluctuation concept (after Moriya[36]).

Wave vector q Magnetic moment	Localized Non-localized		... ...	Non-localized Localized	
Amplitude					
small	(a)				
⋮	(b)	(c)			
⋮	MnSi	α-Mn	$CeFe_2$		
⋮	Cr	CrB_2	Fe_3Pt		
⋮	γ-Mn	MnP	CoS_2		
⋮	Ni	Co	Fe		
⋮			$MnPt_3$	$FePd_3$	
saturated					(d)

(a) Nearly ferromagnetic metals and alloys such as Pd, $HfZn_2$, $TiBe_2$, YRh_6B_4, $CeSn_3$, Ni–Pt, etc.
(b) Weakly ferromagnetic metals and alloys such as Sc_3In, $ZrZn_2$, Ni_3Al, $Fe_{0.5}Co_{0.5}Si$, $LaRh_6B_4$, $CrRh_3B_2$, Ni–Pt, etc.
(c) Antiferromagnetic metals and alloys such as β-Mn, V_3Se_4, V_3S_4, V_5Se_8, etc.
(d) Localized magnetic moment system such as insulating magnetic compounds, $4f$-metals, and Heusler alloys (Pd_2MnSn etc.).

are hardly created by thermal excitation. In other words, spin density waves in weak ferromagnets mostly have small q or long wavelengths. Moriya classified all magnetic materials from the point of view of spin fluctuation as shown by Table 8.1. The columns are arranged from left to right in order of increasing non-localization of q or increasing spatial localization of magnetic moment, while the rows are arranged in order of increasing amplitude of the spin density wave. In the limit of a localized moment system, the spin fluctuation is localized in real space and so is of a short-range nature and its amplitude is large and saturated. On the other hand, in the limit of weak ferromagnetism, the spin fluctuation is localized in q-space and limited to small q, and its amplitude is small and variable. All magnetic materials can be classified between the two limiting cases as shown in Table 8.1.[36] For further details, the review article[35,36] should be consulted.

8.3 MAGNETISM OF RARE EARTH METALS

As mentioned in Section 3.2, the carriers of magnetism in the rare earth elements are the $4f$ electrons, which are located deep inside the atoms. When these rare earth atoms condense to form metals, three electrons in the outer shells $[(5d)^1(6s)^2]$ are shared by many atoms and contribute to electrical conductivity and metallic bonding. At the same time, these electrons serve as a medium of exchange interaction which produces ordered arrangements of the $4f$ spins. Pure rare earth metals with less than half the maximum number (fourteen) of $4f$ electrons, such as La, Ce, Pr, Nd, Sm, and Eu, are known as the light rare earths and exhibit only weak magnetism. The elements with more than half, such as Gd, Tb, Dy, Ho, Er, and Tm, are called the heavy rare earths, and are ferromagnetic at low temperatures. In contrast to the $3d$ transition metals, most of these heavy rare earth metals exhibit helimagnetism (see Section 7.3) over some temperature range above the Curie point before becoming paramagnetic. Gadolinium has the highest saturation magnetization, with a value at 0 K almost the same as that of Fe. However, the Curie points of the pure rare earth metals are all lower than room temperature, so they cannot be used as practical magnetic materials.

Figure 8.18 shows the asymptotic Curie point, Néel point, and ferromagnetic Curie point of the rare earth metals as a function of the number of $4f$ electrons. As seen in this graph, the asymptotic Curie point is mostly small and negative for light rare earth metals, indicating that the exchange interaction is weak and negative. In fact, these metals exhibit antiferromagnetic structures with very low Néel points. On the other hand, the heavy rare earth metals have relatively high Curie points, Néel points, and asymptotic Curie points, all of which decrease as the number of $4f$ electrons increases. This behavior is quite different from that of the effective magnetic moment shown in Fig. 3.12, but is rather similar to the variation of spin S shown in Fig. 3.10. This is reasonable, because the exchange interaction is related to the spin S and has nothing to do with the orbital momentum L. However, J rather than S is a good quantum number, so that all the vectors precess about J. Therefore the effective spin contributing to the exchange interaction is the spin component parallel to J. As seen

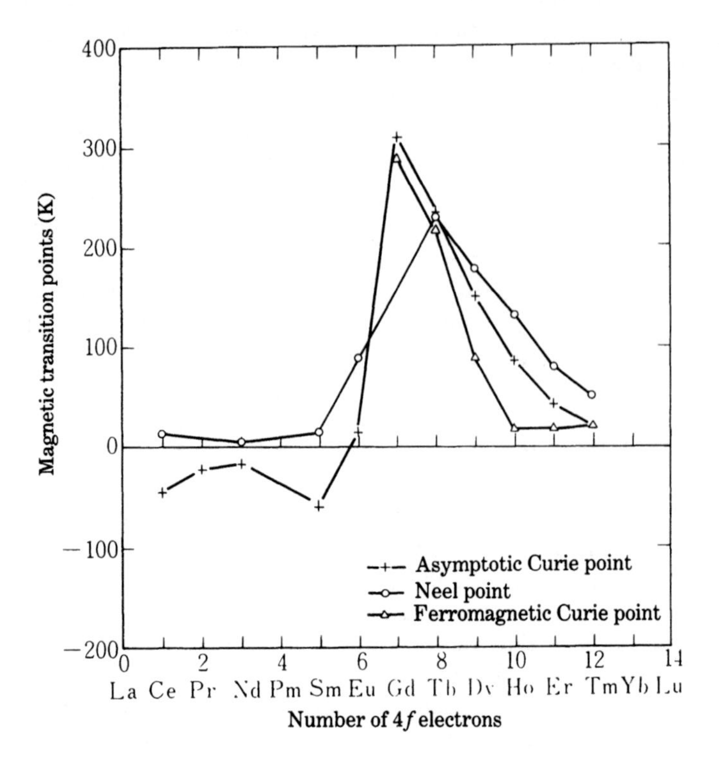

Fig. 8.18. Variation of magnetic transition points of rare earth metals with number of $4f$ electrons.

in Fig. 3.11, $-M$ is given by gJ, which is equal to the component of $J + S$ parallel to J. Therefore the spin component parallel to J is given by $gJ - J$ or $(g - 1)J$. Since the Curie point is proportional to $J(J + 1)$, it is appropriate to describe the various transition temperatures in terms of

$$\xi = c(g - 1)^2 J(J + 1), \tag{8.36}$$

which is called the *de Gennes factor*. The Néel points of many rare earth metals and alloys are plotted as a function of ξ in Fig. 8.19, which shows that all the Néel points lie on curves of $\xi^{2/3}$, indicating the validity of the de Gennes factor. However, no theoretical explanation has been given for this $\xi^{2/3}$ law. The Curie points of Gd alloys (except for the Gd–La system) deviate from the Néel points for ξ less than 11.5 $(\omega = 0)$,* indicating the appearance of helimagnetism.

Various magnetic constants and the magnetic structures of rare earth metals are summarized in Table 8.2 and Fig. 8.20. The magnetic properties of the individual metals are summarized next, starting with the heavy rare earth metals.

* This means no spin rotation, i.e. ferromagnetic.

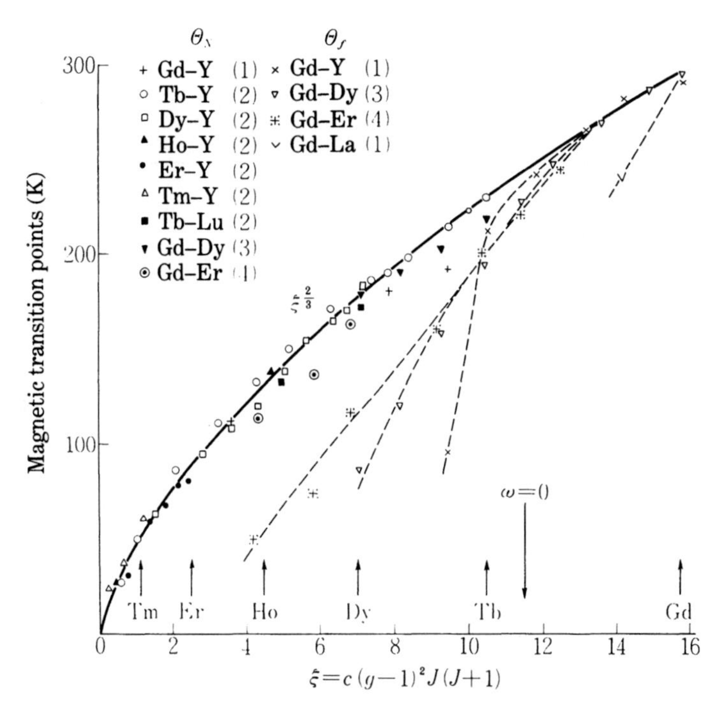

Fig. 8.19. Magnetic transition of rare earth metals and alloys as a function of de Gennes factor (see equation (8.38)).

Gadolinium (Gd) exhibits ferromagnetism below the Curie point $\Theta_f = 293\,K$. It is different from other heavy rare earth metals in that it has no orbital momentum L, so the magnetocrystalline anisotropy is relatively small. It is therefore relatively easy to magnetize. According to older data, the saturation magnetization at 0 K was $7.12M_B$,[51] which is very close to the theoretical value of $7M_B$. However, recent measurements on a purified sample (99.99%) gave $7.55M_B$,[52] which is much larger than the theoretical value. The saturation moment of a Gd atom in Gd–Y and Gd–La alloys (calculated from the measured saturation magnetization of the alloys by assuming that only Gd is magnetic), is also found to be much larger than $7M_B$.[52,53] This may be due to some contribution to the saturation magnetization from the $5d$ electrons.

Terbium (Tb) has moments $S = 3$, $L = 3$ and $J = 6$, and is predicted to exhibit a saturation moment $gJ = 9.0$. The measured saturation magnetic moment at 0 K is $9.34M_B$,[55] which is larger than the theoretical value. Ferromagnetism disappears above the Curie point $\Theta_f = 215\,K$, and helimagnetism is observed above Θ_f up to $\Theta_N = 230\,K$. The crystal structure is hexagonal closed-packed and the spin rotates about the c-axis with a period of about 18 atomic layers. Note that the distance required for a complete rotation of the spins is not required to be an integral number of atomic layers, and may change with temperature.

Dysprosium (Dy) has moments $S = \frac{5}{2}$, $L = 5$ and $J = \frac{15}{2}$, and is predicted to have $gJ = 10$. The actual saturation magnetization at 0 K is $10.7M_B$, slightly larger than the

Table 8.2. Various properties of rare earth metals.

R	Density	Crys. type	Tr. pt.	Me. pt.	Θ_f	Θ_N	Θ_a	Electronic structure of free ions: No. of 4f el.	S	L	J	$(g-1)J$	de Genne factor
	($g\,cm^{-3}$)	(r.t.)	(°C)	(°C)	(K)	(K)	(K)						
Sc	2.992	hcp	1335	1539									
Y	4.478	hcp	1459	1509									
La	6.174	hcp	310	920				0	0	0	0	0	0
	6.186	fcc	868										
Ce	6.771	fcc	725	795		12.5	−46	1	1/2	3	$2\frac{1}{2}$	−0.36	0.182
Pr	6.782	hex	798	935			−21	2	1	5	4	−0.80	0.80
Nd	7.004	hex	862	1024		7.5	−16	3	3/2	6	$4\frac{1}{2}$	−1.23	1.84
Pm	—	—		1035				4	2	6	4	−1.60	3.20
Sm	7.536	rhomb	917	1072		14.8		5	5/2	5	$2\frac{1}{2}$	−1.78	4.44
Eu	5.259	bcc		826		(90)	15	6	3	3	0	0	—
Gd	7.895	hcp	1264	1312	289		310	7	7/2	0	$3\frac{1}{2}$	3.5	15.75
Tb	8.272	hcp	1317	1356	218	230	236	8	3	3	6	3.0	10.50
Dy	8.536	hcp		1407	90	179	151	9	5/2	5	$7\frac{1}{2}$	2.5	7.08
Ho	8.803	hcp		1461	20	133	87	10	2	6	8	2.0	4.50
Er	9.051	hcp		1497	20	80(53)	41.6	11	3/2	6	$7\frac{1}{2}$	1.5	2.55
Tm	9.332	hcp		1545	22	53	20	12	1	5	6	1.0	1.17
Yb	6.977	fcc	798	824				13	1/2	3	$3\frac{1}{2}$	0.5	0.32
Lu	9.842	hcp		1652				14	0	0	0	0	0
Ref.	37		37	37									

theoretical value.[55] Ferromagnetism disappears at $\Theta_f = 85\,K$ and helimagnetism exists up to $\Theta_N = 178.5\,K$. The pitch of the helimagnetic structure (the rotational angle per atomic layer) changes from 25°/layer to 43°/layer as the temperature increases.

Holmium (Ho) has moments $S = 2$, $L = 6$ and $J = 8$, so that $gJ = 10$. The actual saturation magnetization observed at 0 K is $10.34M_B$. In the temperature range from 20 to 133 K, helimagnetism is observed, pitch varying from 35° to 50°/layer with increasing temperature. Below 20 K, the spins deviate from the c-plane and at the same time develop a screw structure, so that the spin structure is conical, with a net spontaneous magnetization along the c-axis. It was observed by neutron diffraction that the component of spontaneous magnetization is $1.7M_B$, while the screw component in the c-plane is $9.5M_B$.

Erbium (Er) has a more complicated magnetic structure: between 52 and 80 K the spin oscillates parallel to the c-axis with a half period of seven layers; between 20 and 52 K, a c-plane component is added; and below 20 K a conical screw structure appears. The component parallel to the c-axis is $7.9M_B$, while the screw component in the c-plane is $4.3M_B$. The saturation magnetization observed by applying a strong magnetic field is $8.8M_B$, which is in good agreement with $gJ = 9.0$.

Thulium (Tm) shows spin oscillation parallel to the c-axis with a half-period of seven layers between 40 and 50 K, while below 40 K, four spins point in the $+c$-direction and three spins point in the $-c$-direction, thus forming an usual kind of ferrimagnetic structure.

Table 8.2. (*cont.*) Magnetic moments of rare earth metals.

	Magnetic moment (M_B)						
	M_{eff}				M_s		
	Theory		Experiment		Theory		
R	Hund	V.V.-F.	3+ Ion	Metals	(gJ)	Experiment	Ref.
La	0	0	0	0			
Ce	2.54	2.56	2.52	2.51	2.14		38
Pr	3.58	3.62	3.60	2.56	3.20		39
Nd	3.62	3.68	3.50	3.3–3.71	3.27		40, 41
Pm	2.68	2.83	—	—	2.40		
Sm	0.85	1.55	—	1.74	0.72		38
Eu	0.00	3.40	—	8.3	0.0		42–4
Gd	7.94	7.94	7.80	7.98	7.0	7.55	45
Tb	9.72	9.70	9.74	9.77	9.0	9.34	46
Dy	10.64	10.6	10.5	10.65	10.0	10.20	47
Ho	10.60	10.6	10.6	11.2	10.0	10.34	48
Er	9.58	9.6	9.6	9.9	9.0	8.0	49
Tm	7.56	7.6	7.1	7.6	7.0	3.4	50
Yb	4.53	4.5	4.4	0.0	4.0		
Lu	0	0	0	0	0		

Ytterbium (Yb) has the electronic structure $4f$,[14] so that it is non-magnetic like lutetium (Lu), as already explained in Chapter 3.

The origin of the screw spin structures in heavy rare earth metals is in oscillatory polarization of conduction electrons caused by exchange interaction with the $4f$ spins. The polarization of conduction electrons $\sigma(r)$ at a distance r from the localized spin S is given by

$$\sigma(r) \propto SF(2k_F r), \tag{8.37}$$

where k_F is the wave vector at the Fermi surface or the radius of the Fermi sphere in k-space. The function $F(x)$ is given by

$$F(x) = \frac{1}{x^4}(x \cos x - \sin x), \tag{8.38}$$

which describes a damped oscillation as shown in Fig. 8.21. The exchange interaction through such oscillatory polarization is called the *RKKY interaction*, a name originating from initials of the authors Ruderman and Kittel,[56] Kasuya,[57] and Yosida.[58] As discussed in Section 7.3, a screw structure is caused by the coexistence of a positive exchange interaction between nearest neighbor planes and a negative interaction between second nearest neighbor planes. The oscillatory nature of the RKKY interaction is equivalent to such a coexistence of positive and negative interactions. The magnetocrystalline anisotropy and magnetostriction also affect the pitch of the screw structure and may modify the spin structure. Theories on helimagnetism in heavy rare earth metals are discussed in various papers.[59-62]

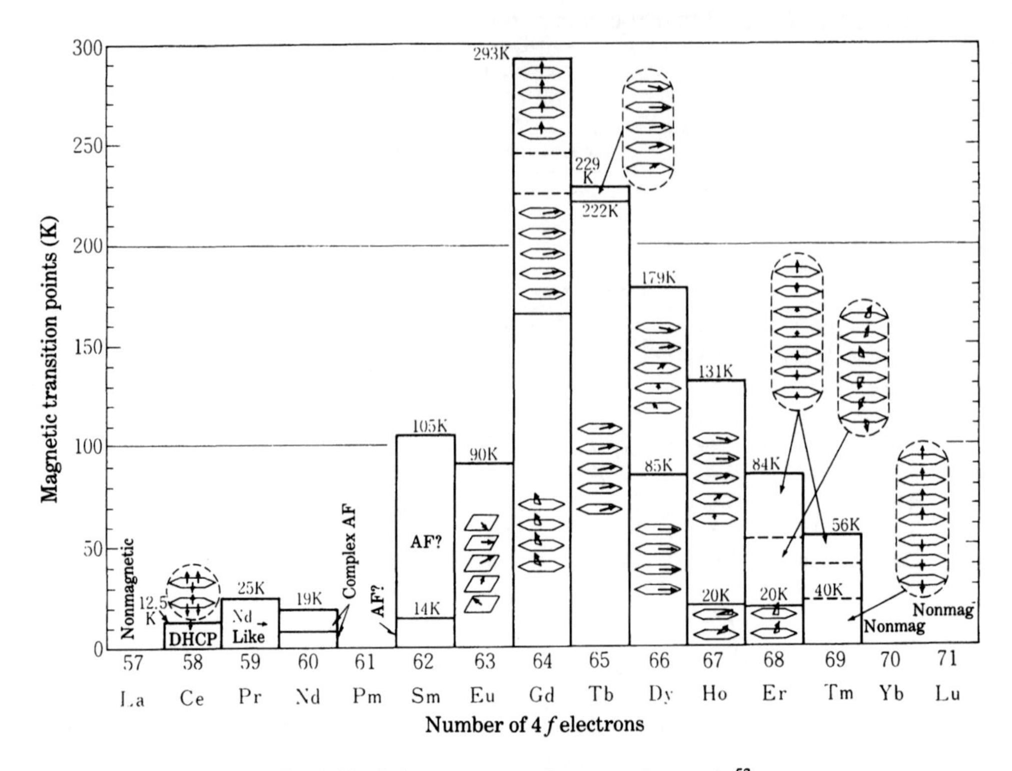

Fig. 8.20. Spin structures of rare earth metals.[53]

In contrast to the heavy rare earth metals, light rare earth metals with less than half-filled $4f$ shells exhibit only weak magnetism, and the variation of magnetic structure with the number of electrons is not so regular as in the heavy rare earth metals.

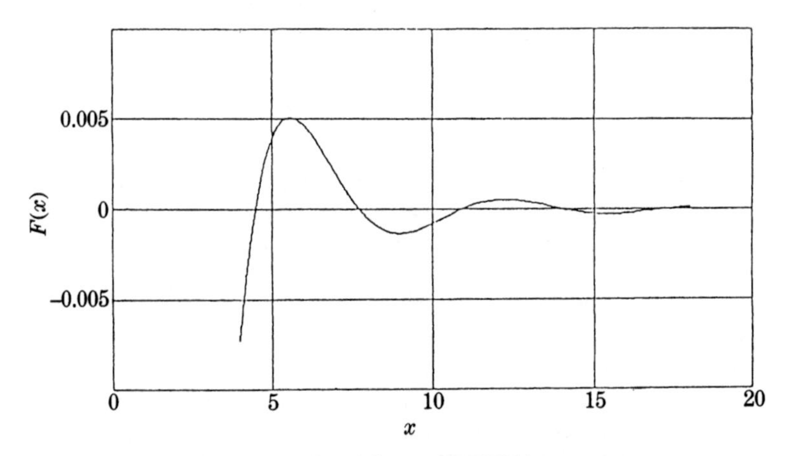

Fig. 8.21. Functional form of RKKY interaction.

Lanthanum (La) has no $4f$ electrons and accordingly is non-magnetic. Its outer electron arrangement ($5d^16s^2$) is similar to those of scandium (Sc) ($3d^14s^2$) and yttrium (Y) ($4d^15s^2$), so that Sc and Y are usually classified as rare earths.

Cerium (Ce) has a magnetic moment of about $0.6M_B$, which is much smaller than $gJ = 2.14$. The spins are arranged ferrimagnetically in the c-plane and antiferromagnetically along the c-axis below 12.5 K, producing zero spontaneous magnetization.

Praseodymium (Pr) has no ordered spin structure. When a magnetic field is applied parallel to [110], the measured magnetization corresponds to a value of $1.6M_B$ per atom, which is much smaller than the theoretical value $gJ = 3.20$.

Neodymium (Nd) has an antiferromagnetic structure below 19 K, with the spins arranged ferromagnetically in the c-planes but antiferromagnetically along the c-axis. This spin structure is changed below 7.5 K. The magnetization in a magnetic field of $4.8\,\mathrm{MA\,m^{-1}}$ (= 60 kOe) is $1.6M_B$ per atom, which is substantially less than the theoretical value of $3.27M_B$.

Promethium (Pm) is an unstable element, and no information is available on its metallic state.

Samarium (Sm) develops a complicated antiferromagnetic structure below 106 K, which is similar to that of Nd, and undergoes a further change in structure below 14 K.[63]

Europium (Eu) has a body centered cubic lattice. Below its Néel point of 91 K, it has a screw spin structure with the spins lying in the (100) plane[64] with a pitch which varies with temperature from 51.4° to 50.0°/layer.

Many investigations have been made on rare earth alloys. When non-magnetic rare earth metals such as La, Y and Sc are added to ferromagnetic heavy rare earth metals, the Néel and Curie points are generally lowered. However, La often stabilizes

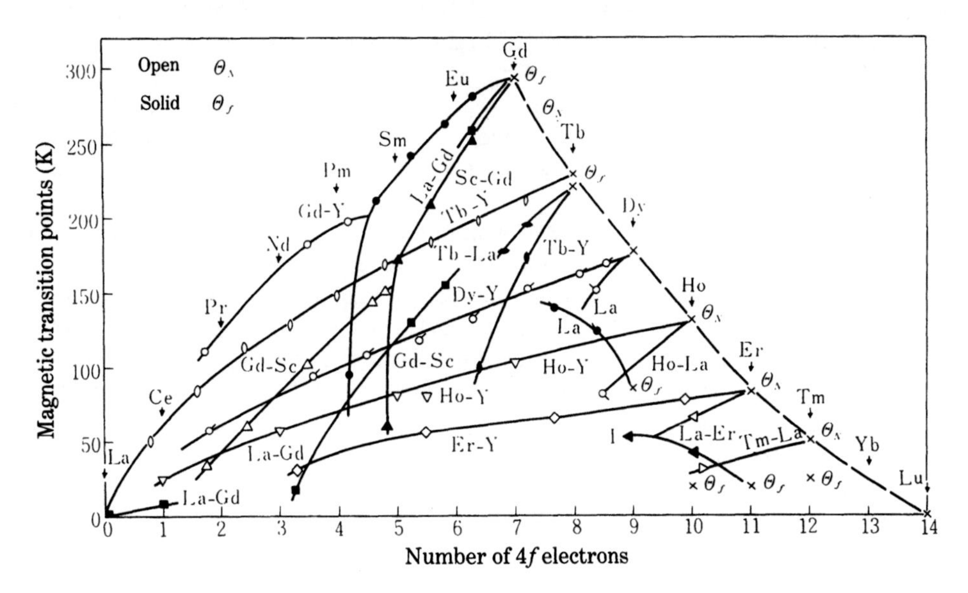

Fig. 8.22. Magnetic transition points of rare earth alloys R–La, R–Y and R–Sc.[53]

ferromagnetism. For instance, the Curie points of Dy–La and Er–La alloys increase with La content, as shown in Fig. 8.22. As already shown in Fig. 8.19, the Néel points of rare earth alloys vary as $\xi^{2/3}$. On the other hand, the Curie point decreases more rapidly with a decrease in the de Gennes factor, thus increasing the temperature range in which helimagnetism is observed. (The de Gennes factors of alloys are calculated as a weighted average of the factors for the constituent elements.)

Considering alloys between heavy rare earth metals, we find that the Néel point of Gd–Dy and Gd–Er alloys changes as $\xi^{2/3}$, as shown in Fig. 8.19. This is, however, not a general rule. For instance, the Curie point of Dy–Er alloys undergoes a discontinuous change with composition. The Néel point of Ho–Er alloys changes monotonically, while the Curie point shows a maximum as a function of composition. It was observed that in Ho–Er alloys, the spins of both constituent atoms rotate in the same way about the *c*-axis.

Further topics are considered in several reviews.[65–7]

8.4 MAGNETISM OF INTERMETALLIC COMPOUNDS

Intermetallic compounds are composed of metallic elements in fixed integer ratios, such as Ni_3Al or MnBi, which usually exhibit metallic properties such as high electrical conductivity and metallic luster. Generally speaking, they have complicated crystal and magnetic structures. It is rather difficult to explain intermetallic compounds by any unified theory. In this section, we discuss only the intermetallic compounds which show interesting or useful magnetic properties, and we consider only elements which belong to the 3*d* transition metals, the rare earth metals, and the actinide metals, plus Be. Compounds which contain the elements belonging to the IIIb group, such as B, Al, Ga, In, Tl; the IVb group, such as C, Si, Ge, Sn, Pb; and the Vb group such as N, P, As, Sb, Bi, will be treated in Chapter 10.

First we discuss the intermetallic compounds which contain only 3*d*, 4*d*, and 5*d* transition metals. They form σ-, χ-, Laves-, and CsCl-phases, and occur at an electron concentration near that of Mn. As discussed in Section 8.2, this fact is related to the irregularity in the relationship between crystal structure and electron concentration caused by band polarization.

The crystal structure of the σ-phase is complex: the unit cell contains 30 atoms, each of which has a high coordination number (number of nearest neighbors), such as 12, 14 and 15[68] (see Fig. 8.23). The only ferromagnetic σ-phase compounds are V_xFe_{1-x} ($x = 0.39$–0.545) and Cr_xFe_{1-x} ($x = 0.435$–0.50). Other σ-phase compounds such as V–Co, V–Ni and Cr–Co exhibit only Pauli paramagnetism. The maximum saturation magnetic moment in the Fe–V system is $0.5M_B$ per atom, and the maximum Curie point is 240 K.

The χ-phase has a complicated crystal structure with a unit cell containing 29 atoms, which is the same as that of α-Mn described in Section 8.2. The magnetic structure of this phase has not been investigated in detail except for α-Mn itself.

The *Laves phase* is an AB_2-type compound, described by Laves.[69] The atomic radius of the A atom is 1.225 times larger than that of the B atom. There are three

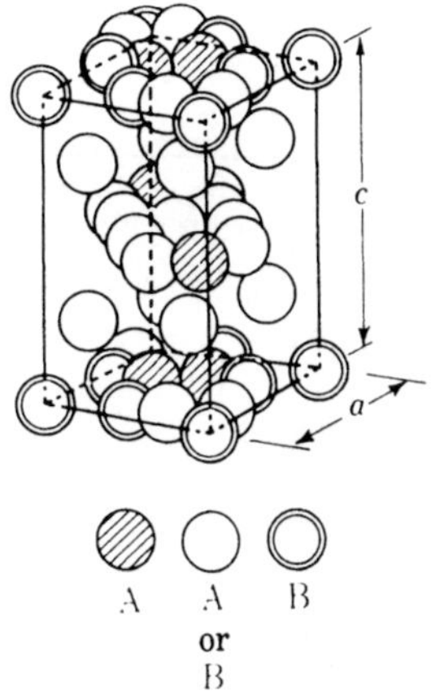

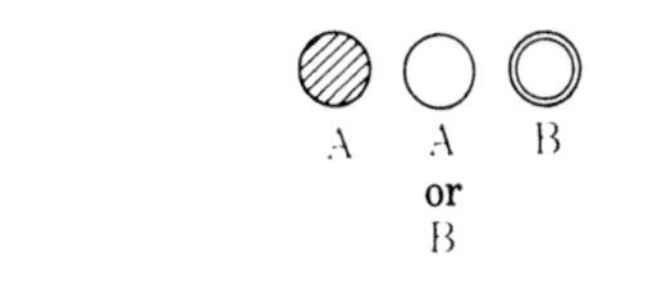

Fig. 8.23. Crystal structure of σ-phase.[54]

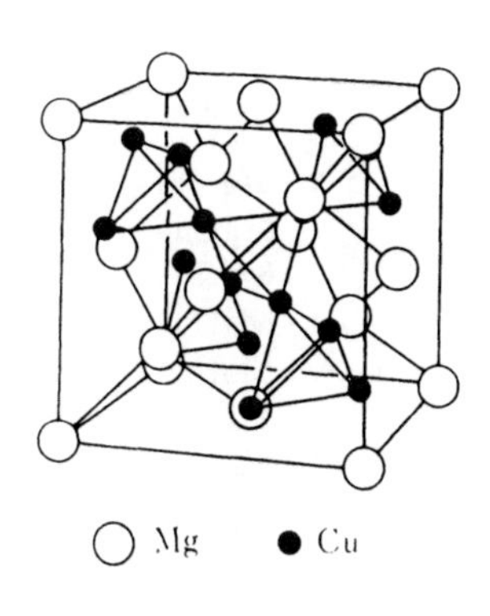

Fig. 8.24. Crystal structure of Laves-phase $MgCu_2$-type compounds.[55]

kinds of complicated crystal structures: $MgCu_2$-type cubic (Fig. 8.24), $MgZn_2$-type hexagonal, and $MgNi_2$-type hexagonal. The common feature of these crystals is that A and A atoms, and B and B atoms, are in contact with each other, while A and B atoms are not. All of the ferromagnetic Laves compounds are listed in Table 8.3, from which we see that the B atoms are mostly Fe or Co. The paramagnetic Laves phases have high values of magnetic susceptibility when B is Fe or Co, and they tend to be ferromagnetic for B contents higher than stoichiometric.[70]

$ZrZn_2$ is a weak ferromagnet containing no ferromagnetic elements, and is a typical itinerant-electron ferromagnet. Even at 0 K, the magnetization of this compound increases by the application of high magnetic fields, and does not saturate even in a field as strong as $5.6\,\mathrm{MA\,m^{-1}}$ (70 kOe).[91]

It has been found that the saturation moment of the Fe atom in various AFe_2 compounds changes linearly with the distance between A atom neighbors.[92] There are many ABe_2-type compounds, but only $FeBe_2$ is ferromagnetic.

The ferromagnetic CsCl-type intermetallic compounds containing only transition metals are listed in Table 8.4. Interesting magnetic behavior is observed in FeRh, as shown in Fig. 8.25: as the temperature is increased through room temperature, a saturation magnetization suddenly appears as the result of a transition from antiferromagnetism to ferromagnetism. Taking into consideration the fact that the electronic configuration of Rh is similar to that of Co, Kanamori and Teraoka[100] suggested that Rh atoms become magnetic by transferring some electrons to the Fe atoms, which simultaneously results in ferromagnetic coupling between Fe–Fe pairs.

MnZn is known to exhibit a large spin canting.[97,98] MnRh seems to be antiferromagnetic below room temperature, but it becomes ferrimagnetic with excess Mn content.

In rare earth–transition metal compounds, a negative exchange interaction exists through the conduction electrons. Therefore the light rare earths, where J is opposite

Table 8.3. Magnetic properties of Laves phase intermetallic compounds.

Compounds	Crystal-type	Curie point (K)	Mag. moment per formula (M_B)	Ref.
$ZrFe_2$	$MgCu_2$	588, 633	(Fe) 1.55	70–74
$HfFe_2$	$MgCu_2$	591	(Fe) 1.46	75
$ZrZn_2$	$MgCu_2$	35	($ZrZn_2$) 0.13	76
$FeBe_2$	$MgCu_2$	823	(Fe) 1.95	77, 78
$ScFe_2$	$MgNi_2$	—	—	79
YFe_2	$MgCu_2$	550	(Fe) 1.45_5	80
$CeFe_2$	$MgCu_2$	878	($CeFe_2$) 6.97	80
$SmFe_2$	$MgCu_2$	674		80
$GdFe_2$	$MgCu_2$	813		80
$DyFe_2$	$MgCu_2$	663	($DyFe_2$) 5.44	80
$HoFe_2$	$MgCu_2$	608	($HoFe_2$) 6.02	80
$ErFe_2$	$MgCu_2$	473	($ErFe_2$) 5.02	80
$TmFe_2$	$MgCu_2$	613	($TmFe_2$) 2.94	80
$PrCo_2$	$MgCu_2$	44	($PrCo_2$) 2.9	80
$NdCo_2$	$MgCu_2$	116, 109	($NdCo_2$) 3.83, 3.6	81, 82
$SmCo_2$	$MgCu_2$	203	($SmCo_2$) 1.7	83
$GdCo_2$	$MgCu_2$	412	($GdCo_2$) 4.8	83
$TbCo_2$	$MgCu_2$	256, 230	($TbCo_2$) 6.72, 6.0	82, 83
$DyCo_2$	$MgCu_2$	146	($DyCo_2$) 7.1	82
$HoCo_2$	$MgCu_2$	95, 90	($HoCo_2$) 7.81, 7.7	82, 83
$ErCo_2$	$MgCu_2$	36, 37	($ErCo_2$) 7.00, 6.6	82, 83
UFe_2	$MgCu_2$	172	(Fe) 0.51	84–88
$NpFe_2$	$MgCu_2$	600?		89

to S, tend to align their magnetic moment parallel to that of the transition metal. On the other hand, the heavy rare earths, where J is parallel to S, tend to align their magnetic moment opposite to that of transition metals. Using this pattern, the saturation magnetic moments of the RFe_2 and RCo_2 compounds (R: rare earth) can be well accounted for, assuming M_s to be $2.2M_B$ for Fe and $1.0M_B$ for Co.

Table 8.4. Magnetic properties of ferromagnetic CsCl-type intermetallic compounds.

Compounds	Composition range	Curie point Θ_f (K)	Saturation mag. moment per formula (M_B)	Ref.
FeRh	20 ~ 53% Rh	673 ~ 950	Fe: 3.2, Rh: 0.6 (in the case of 48% Rh)	83–95
MnZn	50 ~ 56.5% Zn	> 550	Mn: 1.7 (ferromagnetic component) Mn: 2.9 (antiferromagnetic component)	97, 98
SmZn		125	(SmZn) 0.07	99
GdZn		270	(GdZn) 6.7	99
TbZn		206	(TbZn) 6.0	99
DyZn		144	(DyZn) 4.9	99
HoZn		80	(HoZn) 4.7	99
ErZn		50	(ErZn) 2.3	99

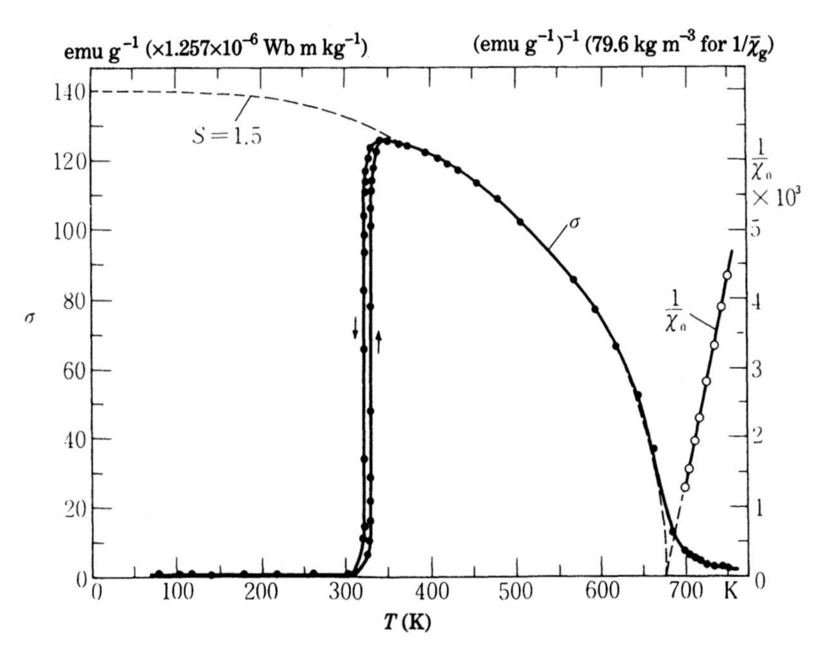

Fig. 8.25. Temperature dependence of specific saturation magnetization and reciprocal susceptibility of FeRh.[82]

Compounds with the CsCl-type structure and the formula RM are found for M = Cu, Ag and Au in the Ib group; Mg (IIa), Zn, Cd, and Hg in the IIb group; and Al, Ga, In and Tl in the IIIb group. There is a general tendency that RM(Ib) are antiferromagnetic, RM(IIb) are ferromagnetic, and RM(IIIb) are again antiferromagnetic. An attempted explanation of this change in the exchange interaction according to the number of conduction electrons in terms of the RKKY interaction was not successful.[101] Ferromagnetic RM(IIb) compounds are listed in Table 8.4.

The RCo_5-type compounds have a hexagonal crystal structure, as shown by Fig. 8.26. As in the case of RCo_2, the light rare earth moments couple ferromagnetically with those of Co, while the heavy rare earth moments couple ferrimagnetically with Co.[102] In these crystals the orbital moment L of the rare earth atoms remains

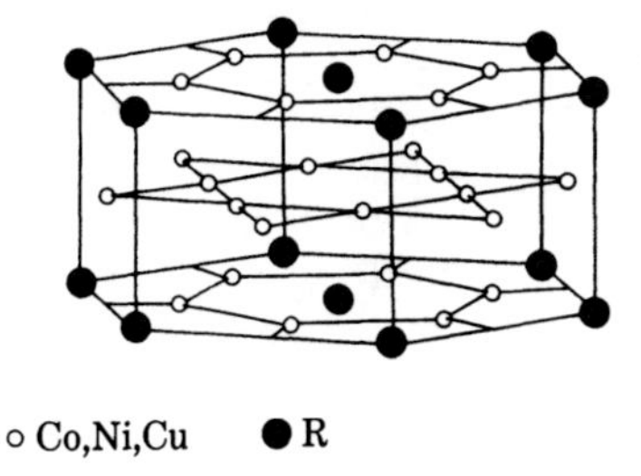

Fig. 8.26. Crystal structure of RCo_5-type compounds.[92]

Table 8.5. Magnetic properties of RCo_5 and similar compounds.

Compounds	Crystal type	Curie point (K)	I_s (T)	Sat. moment in M_B at 0 K	Ref.
YCo_5	$CaCu_2$	921	1.06		104
$LaCo_5$	$CaCu_2$	840	0.909		104
$CeCo_5$	$CaCu_2$	647	0.77		104
$PrCo_5$	$CaCu_2$	885	1.2		104
$SmCo_5$	$CaCu_2$	997	0.965		104
Sm_2Co_{17}	Th_2Zn_{17}	920	1.2		104
Gd_2Co_{17}	Th_2Zn_{17}	930	0.73		104
Th_2Fe_{17}	Th_2Zn_{17}	295		(Fe) 1.76	105
Th_2Co_{17}	Th_2Zn_{17}	1053		(Co) 1.42	105
$ThCo_5$	$CaCu_2$	415		(Co) 0.94	105
Th_2Fe_7	Ce_2Ni_7	570		(Fe) 1.37	105
$ThFe_3$	$PuNi_3$	425		(Fe) 1.37	105

unquenched, so that the magnetocrystalline anisotropy (see Chapter 12) is quite large. For this reason some of these compounds show excellent permanent magnet characteristics. The R_2Co_{17}-type compounds have even better permanent magnet properties in some ways (see Table 8.5).

The actinide elements U, Np, and Pu, which require careful handling because of their radioactivity, form intermetallic compounds with transition metals. Some Laves

Table 8.6. Magnetic and crystal properties of ferromagnetic superlattice alloys.[93]

Super-lattice	Crystal type	Order–disorder Curie point (°C)	Θ_f in ordered state (K)	I_s (0 K) emu g^{-1} ($\times 4\pi \times 10^{-7}$ Wb m kg^{-1})	M_s per formula in M_B
FeV	CuZn			51	0.98 (Fe 0.73, V 0.03)
Ni_3Mn	Cu_3Au	510	750	105	4.4 (Mn 3.83, Ni 0.47)
FeCo	CuZn	730	1390	230	4.70 (Fe 3.0, Co 1.9)
FeNi	CuAu	~320			
Ni_3Fe	Cu_3Au	500	983	113	4.8 (Fe 2.99, Ni 0.62)
UAu_4	Ni_4Mo	~565	53	2.8	0.41
$MnAu_4$	Ni_4Mo	~420	371	27.5	4.15
$CrPt_3$	Cu_3Au		687	19.1	2.18 (Cr 2.33, Pt 0.27)
$MnPt_3$	Cu_3Au	~1000	370	35.2	4.04 (Mn 3.64, Pt 0.26)
FePd	CuAu	~700	749	105	3.06 (Fe 2.9, Pd 0.30)
$FePd_3$	Cu_3Au	~800	529	65.2	4.38 (Fe 2.72, Pd 0.51)
Fe_3Pt	Cu_3Au	835	430	138	8.95 (Fe 3.5, Pt 1.5)
FePt	CuAu	~1300	750	33.8	1.52
CoPt	CuAu	825		44.2	2.01
$CoPt_3$	Cu_3Au	~750	~290		
Ni_3Pt	Cu_3Au	580	370	23.4	1.66

compounds listed in Table 8.3 show fairly high Curie points, but have a rather small Fe moment, as seen in UFe_2. The ferromagnetic ThM_5- and Th_2M_{17}-type compounds (M = Fe and Co) are also listed in Table 8.5.

Finally we will describe the ferromagnetic superlattice alloys containing only transition metals, as listed in Table 8.6. Ni_3Mn if ferromagnetic when it forms a superlattice, because the exchange interaction between Ni–Mn pairs is ferromagnetic. The spin ordering is disturbed when the alloy is quenched from high temperature to attain the atomically disordered state, because the Mn–Mn interaction is antiferromagnetic.

The saturation magnetization of ordered FeCo is 4% larger than that of the disordered state. As for Ni_3Fe, detailed investigations have been made on the induced magnetic anisotropy (see Chapter 13). Fe_3Pt exhibits Invar characteristic when it is atomically disordered. CoPt makes an excellent (although expensive) machinable permanent magnet. Many investigations have been made on magnetic superlattice alloys, because they allow an examination of the relationship between magnetism and atomic pairs.

PROBLEMS

8.1 Assuming that the density of states of a non-magnetic metal is given by $g(E) = g(E_f)(1 + \alpha(E - E_f))$ near the Fermi level, E_f, calculate the temperature dependence of the Pauli-paramagnetic susceptibility of this metal.

8.2 Referring to the Slater–Pauling curve in Fig. 8.12, find the alloy compositions of Ni–Co, Co–Cr, and Fe–Cr alloys whose average saturation moment is equal to $1M_B$.

8.3 Discuss the reason why the magnetic transition points of rare earth metals and alloys are well described in terms of the de Gennes factor.

8.4 Explain why the magnetic moment of light rare earth atoms is parallel to that of transition metal atoms in intermetallic compounds, whereas that of heavy rare earth atoms is antiparallel.

REFERENCES

1. W. Pauli, *Z. Physik*, **41** (1927), 81.
2. D. W. Budworth, F. E. Hoare, and J. Preston, *Proc. Roy. Soc.*, **A257** (1960), 250.
3. E. C. Stoner, *Proc. Roy. Soc.*, **A154** (1936), 656.
4. E. C. Stoner, *Proc. Roy. Soc.*, **A165** (1938), 372; **A169** (1939), 339.
5. E. C. Stoner, *Rept. Progr. Phys.*, **11** (1948), 43.
6. E. C. Stoner, *Acta Metal.*, **2** (1954), 259.
7. R. J. Elliot, J. A. Krumhansel, P. L. Leath, *Rev. Mod. Phys.*, **46** (1974), 465.
8. H. Hasegawa and J. Kanamori, *J. Phys. Soc. Japan*, **31** (1971), 382; **33** (1972), 1599, 1607.
9. T. Jo, H. Hasegawa, and J. Kanamori, *J. Phys. Soc. Japan*, **35** (1973), 57.
10. R. M. Bozorth, *Ferromagnetism* (Van Nostrand, New York, 1951) p.441.

11. Y. Kono and S. Chikazumi, *Kobayashi Riken Rept.* (in Japanese) **9** (1959), 12.
12. J. S. Kouvel and R. H. Wilson, *J. Appl. Phys.*, **32** (1961), 435.
13. J. W. D. Connoly, *Phys. Rev.*, **150** (1967), 415.
14. S. Wakoh and J. Yamashita, *J. Phys. Soc. Japan*, **21** (1966), 1712.
15. C. G. Shull and M. K. Wilkinson, *Phys. Rev.*, **97** (1955), 304.
16. J. Friedel, Nuovo Cim. (Suppl.), **7** (1958); *Theory of magnetism in transition metals* (ed. by W. Marshall, Academic Press, New York and London, 1957) p.283.
17. A. Akai, P. H. Dedrichs, and J. Kanamori, *J. de Phys.*, **49** (1988), C8-23.
18. G. G. Low and M. F. Collins, *J. Appl. Phys.*, **34** (1963), 1195.
19. G. G. Low, *Adv. Phys.*, **XVIII** (1968), 371.
20. M. F. Collins and G. G. Low, *Proc. Phys. Soc.*, **86** (1965), 535.
21. J. B. Comly, T. M. Holden, and G. G. Low, *J. Phys. C. (Proc. Phys. Soc.)*, [2] **1** (1968), 458.
22. R. J. Weiss, *Proc. Phys. Soc.* (London), **82** (1963), 281.
23. M. Matsui and S. Chikazumi, *J. Phys. Soc.*, **45** (1978), 458.
24. T. Moriya and K. Usami, *Solid State Comm.*, **34** (1980), 95.
25. H. Hasegawa, *J. Phys.*, **C14** (1981), 2793; *J. Phys. Soc. Japan*, **51** (1982), 767; *Physica* **B + C 119** (1983), 15.
26. Y. Kakehashi, *J. Phys. Soc. Japan*, **50** (1981), 2236; **51** (1982), 3183.
27. C. Zener, *Trans. AIME*, **203** (1955), 619.
28. H. Hasegawa and D. G. Pettifor, *Phys. Rev. Lett.*, **50** (1983), 130, *Proc. 3d-Met Mag.* ILL Grenoble p. 203.
29. T. Yamada, N. Kunitomi, and Y. Nakai, *J. Phys. Soc. Japan*, **28** (1970), 615.
30. D. Meneghetti and S. S. Sidhu, *Phys. Rev.*, **105** (1957), 130; G. E. Bacon, I. E. Dunmur, J. H. Smith and R. Street, *Proc. Roy. Soc.* (London), **241** (1957), 223.
31. A. W. Overhauser, *Phys. Rev.*, **126** (1962), 517; **128** (1962), 1437.
32. W. M. Lomer, *Proc. Phys. Soc.* (London), **80** (1962), 489.
33. J. Kanamori and Y. Teraoka, *Physica*, **91B** (1977), 199.
34. R. R. Rhodes and E. P. Wohlfarth, *Proc. Roy. Soc.*, **273** (1963), 247; E. P. Wohlfarth: *J. Mag. Mag. Mat.*, **7** (1978), 113.
35. T. Moriya, *J. Mag. Mag. Mat.*, **14** (1979), 1.
36. T. Moriya, *J. Mag. Mag. Mat.*, **31–34** (1983), 10.
37. F. H. Spedding and A. H. Daane, *The Rare Earths* (John Wiley & Sons, 1961).
38. J. M. Lock, *Proc. Phys. Soc.* (London), **B70** (1957), 566.
39. C. H. La Blanchetais, *Compt. Rend.*, **234** (1952), 1353.
40. J. F. Elliot, S. Legvold, and F. H. Spedding, *Phys. Rev.*, **94** (1954), 50.
41. D. R. Behrendt, S. Legvold, and F. H. Spedding, *Phys. Rev.*, **106** (1957), 723.
42. W. Klemm and H. Bonmer, *Z. Anorg. u. Allegem. Chem.*, **231** (1937), 138; **241** (1939), 264.
43. R. N. Bozorth and J. H. Van Vleck, *Phys. Rev.*, **118** (1960), 1493.
44. M. K. Wilkinson, W. C. Koehler, E. O. Wollan, and J. W. Cable, *J. Appl. Phys.*, **32** (1961), S48.
45. H. E. Nigh, S. Legvold, and F. H. Spedding, *Phys. Rev.*, **132** (1963), 1092.
46. D. E. Hegland, S. Legvold, and F. H. Spedding, *Phys. Rev.*, **131** (1963), 158.
47. D. R. Behrendt, S. Ledvold, and F. H. Spedding, *Phys. Rev.*, **109** (1958), 1544.
48. D. L. Strandburg, S. Legvold, and F. H. Spedding, *Phys. Rev.*, **127** (1962), 2046.
49. R. W. Green, S. Legvold, and F. H. Spedding, *Phys. Rev.*, **122** (1961), 827.
50. B. L. Rhodes, S. Legvold, and F. H. Spedding, *Phys. Rev.*, **109** (1958), 1547.
51. W. C. Thoburn, S. Legvold, and F. H. Spedding, *Phys. Rev.*, **110** (1958), 1298.
52. H. E. Nigh, S. Legvold, and F. H. Spedding, *Phys. Rev.*, **132** (1963), 1092.
53. W. C. Thoburn, S. Legvold, and F. H. Spedding, *Phys. Rev.*, **110** (1958), 1298.
54. K. Toyama and S. Chikazumi, *J. Phys. Soc. Japan*, **35** (1973), 47.

55. J. J. Rhyne, S. Foner, E. J. McNiff, and R. Doclo, *J. Appl. Phys.*, **39** (1968), 807.
56. M. Ruderman and C. Kittel, *Phys. Rev.*, **96** (1954), 99.
57. T. Kasuya, *Prog. Theor. Phys.*, **16** (1956), 45.
58. K. Yosida, *Phys. Rev.*, **106** (1957), 893.
59. H. Miwa and K. Yosida, *Prog. Theor. Phys.*, **26** (1961), 693.
60. H. Miwa, *Prog. Theor. Phys.*, **28** (1962), 208.
61. R. J. Elliot, *Phys. Rev.*, **124** (1961), 340.
62. R. J. Elliot and F. A. Wedgwood, *Proc. Phys. Soc.* (London), **81** (1963), 846; **84** (1964), 63.
63. W. C. Koehler, R. M. Moon, J. W. Cable, and H. R. Child, *Acta Cryst.*, **A28** (1972), S1197.
64. N. G. Nereson, C. E. Olsen, and G. P. Arnold, *Phys. Rev.*, **135** (1964), A176.
65. K. Yosida, *Prog. Low Temp. Phys.*, **IV** (1964), 265.
66. W. C. Koehler, *J. Appl. Phys.*, **36** (1965), 1078.
67. R. M. Bozorth and C. D. Graham, Jr., *G.E. Tech. Inf. Ser.*, **66-C-225** (1966).
68. G. Bergman and D. P. Shoemaker, *Acta Cryst.*, **7** (1954), 857.
69. F. Laves, *Naturwiss.*, **27** (1939), 65.
70. C. W. Kocher and P. J. Brown, *J. Appl. Phys.*, **33** (1962), S1091.
71. E. Piegger and R. S. Craig, *J. Chem. Phys.*, **39** (1963), 137.
72. K. Kanematsu, *J. Appl. Phys.*, **39** (1968), 465.
73. W. Bruckner, R. Perthel, K. Kleinstuck, and G. E. R. Schulze, *Phys. Stat. Sol.*, **29** (1968), 211.
74. K. Kanematsu, *J. Phys. Soc. Japan*, **27** (1969), 849; K. Kanematsu and Y. Fujita, *J. Phys. Soc. Japan*, **29** (1970), 864.
75. T. Nakamichi, K. Kai, Y. Aoki, K. Ikeda, and Y. Yamamoto, *J. Phys. Soc. Japan*, **29** (1970), 794.
76. B. T. Matthias and R. M. Bozorth, *Phys. Rev.*, **109** (1958), 604.
77. L. Misch, *Z. Phys. Chem.*, **29** (1935), 42.
78. K. Ohta and Y. Kobayashi, *Rep. Kobayashi Riken.*, **11** (1961), 61 (in Japanese).
79. M. V. Nevitt, C. W. Kimball, and R. S. Preston, *Proc. Int. Conf. on Magnetism*, Nottingham (1964), 137.
80. W. E. Wallace and E. A. Skrabek, *Proc. 3rd R. E. Conf.* (Gordon & Breech, N.Y., 1964), 431.
81. G. K. Wertheim and J. H. Wernick, *Phys. Rev.*, **125** (1962), 1937.
82. R. M. Moon, W. C. Koehler, and J. Farrel, *J. Appl. Phys.*, **36** (1965), 978.
83. J. Crangle and J. W. Ross, *Proc. Int. Conf. on Magnetism*, Nottingham (1964), 240.
84. P. Gordon, *Atom. Energy Comm. U.S.A. Nr.* (1952), 1833.
85. S. Komura, N. Kunitomi, Y. Hamaguchi, and M. Sakamoto, *J. Phys. Soc. Japan*, **16** (1961), 1486.
86. B. I. Chechernikov, V. A. Pletynshkim, T. M. Shavishvili, and V. K. Slovyanshikh, *Sov. Phys. JETP.*, **31** (1970), 44.
87. S. T. Lin and R. E. Ogilvic, *J. Appl. Phys.*, **34** (1963), 1372.
88. S. Komura and N. Shikazono, *J. Phys. Soc. Japan*, **18** (1963), 323.
89. S. Blow, *J. Phys. C. Sol. Stat. Phys.*, **3** (1970), 835.
90. T. Nakamichi, *Nihon Kinzoku Gakkaiho*, **7** (1968), 63 (in Japanese).
91. S. Ogawa, *Nihon Buturi Gakkaishi*, **24** (1969), 450 (in Japanese).
92. K. Kai and T. Nakamichi, *J. Phys. Soc. Japan*, **30** (1971), 1755.
93. J. S. Kouvel and C. C. Hartelius, *J. Appl. Phys.*, **33** (1962), S1343.
94. G. Shirane, C. W. Chen, and P. A. Flinn, *Phys. Rev.*, **131** (1963), 183.
95. E. F. Bertaut, A. Delapalme, F. Fowat, G. Roult, F. de Bergevin, and R. Pauthenet, *J. Appl. Phys.*, **33** (1962), S1123.
96. J. S. Kouvel, *G. E. Report No. 64-RL-3740M* (1964).

97. T. Hori and Y. Nakagawa, *J. Phys. Soc. Japan*, **19** (1964), 1255.
98. Y. Nakagawa and T. Hori, *J. Phys. Soc. Japan*, **19** (1964), 2082.
99. K. Kanematsu, G. T. Alfieri, and E. Banks, *J. Phys. Soc. Japan*, **26** (1969), 244.
100. J. Kanamori and Y. Teraoka, *Physica*, **91B** (1977), 199.
101. D. Mattice, A. Anthony, and L. Horvitz, *IBM Res. Rept. RC-945* (1963).
102. E. A. Nesbitt, H. J. Williams, J. H. Wernick, and R. C. Sherwood, *J. Appl. Phys.*, **33** (1962), 1674.
103. G. Hoffer and K. J. Strnat, *IEEE Trans. Mag.*, **MAG-2** (1966), 487.
104. K. J. Strnat, *IEEE Trans. Mag.*, **MAG-8** (1972), 511, 516.
105. K. H. J. Bushow, *J. Appl. Phys.*, **42** (1971), 3433.
106. E. A. Nesbitt and J. H. Wernick, *Rare earth magnets* (Academic Press, N.Y., 1973) p. 44.
107. S. Chikazumi *et al.*, ed, *Handbook on magnetic substances* (in Japanese) (Asakura, Tokyo, 1975) #7.5 Y. Nakamura and M. Mekata, Table 7.30, p. 357.

9
MAGNETISM OF FERRIMAGNETIC OXIDES

Most oxides are electrically insulating. Therefore magnetic oxides such as ferrites exhibit very low losses when magnetized in high-frequency fields. This characteristic has been very useful in magnetic materials for communication and information devices. In Section 9.1 we note the general features of oxide crystals, then discuss magnetic and crystalline properties of spinel ferrites in Section 9.2, garnet ferrites in Section 9.3, hexagonal ferrites in Section 9.4, and finally other oxides in Section 9.5.

9.1 CRYSTAL AND MAGNETIC STRUCTURE OF OXIDES

When metallic ions with valence $+n$, or M^{n+}, are chemically combined with divalent oxygen ions O^{2-}, the oxide MO_x is produced, where $x = n/2$. In most cases, $n = 2$ or 3, but some oxides contain additional metal ions with $n = 4$ or 1. Therefore oxides are expressed by the chemical formula MO_x, where $x = 1$–2. When the oxides form solids, the unit of structure is not the MO_x molecule, but individual ions M^{n+} and O^{2-}, which are arranged in an ionic crystal. The cohesive force in this crystal is supplied by an electrostatic Coulomb interaction between the charge $+ne$ of M^{n+} ions and the charge $-2e$ of O^{2-} ions.

The metal and oxygen ions are quite different in size. The radius of O^{2-} is 1.32 Å, while the radius of an M^{n+} ion may be small as 0.6–0.8 Å. In oxide crystals, large O^{2-} ions are in contact with each other, thus forming close-packed crystals. Small M^{n+} ions occupy interstitial sites between them.*

The close-packed structure is illustrated in Fig. 9.1, where (a) shows the two-dimensional dense packing of spheres in the first layer, while (b) shows a possible placement of the dense packed second layer on the first layer. When the third layer is placed on the second layer, one possibility is shown in (c), where the position of atoms in the third layer is different from that in either the first layer or the second layer. Such a close-packed structure is face-centered cubic. If the atomic position in the third layer is the same as the first layer, the structure becomes hexagonal closed packed.

In order to construct oxides, we must insert M^{n+} ions in such a close-packed oxygen lattice. There is more than one kind of interstitial site, as shown in Fig. 9.2.

* This does not mean that the size of the O atom is much larger than that of M atoms. The size of O^{2-} is large because it has two extra electrons outside the neutral atom, while the M^{n+} ion is small because it loses n electrons from the neutral atom.

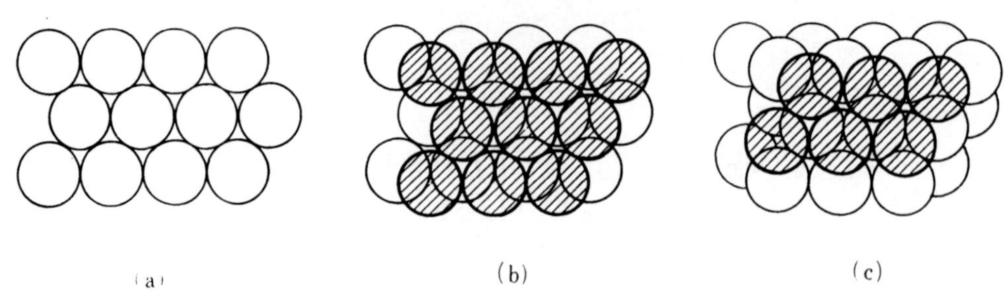

Fig. 9.1. Stacking of atomic layers in close-packed structures.

Site A is surrounding by four oxygen ions and is called the tetrahedral site, because the surrounding four oxygen ions form a tetrahedron. Site B is surrounded by six oxygen ions and is called the octahedral site, because the surrounding six oxygen ions form an octahedron. In addition to these two, there is a dodecahedral site. There are many kinds of oxides, as will be discussed later, but in all cases the metallic ions occupy one of these interstitial sites.

The size of these interstitial sites is generally fairly small, so that the oxygen ions tend to be pushed outward by the interstitial metal ions. This displacement of oxygen ions is expressed in terms of the u-parameter, as described in Section 9.2. It should be noted, however, that even when the interstitial metal ion is absent, oxygen ions are still pushed outward because of their Coulomb repulsion.

Table 9.1 lists many magnetic oxides MO_x in order of increasing x, with their crystal structures and some additional information. As mentioned in Section 7.1, magnetic moments are aligned by the superexchange interaction through the oxygen ions. When the angle M–O–M is nearly 180° the magnetic moments on the metal ions tend to be antiparallel, thus giving a ferrimagnetic structure. When the angle is nearly 90°, the moments tend to be parallel, thus forming a ferromagnetic structure. In some cases, the moments are neither parallel nor antiparallel, but aligned at some arbitrary angle, thus producing a canted magnetic structure (see Section 7.3).

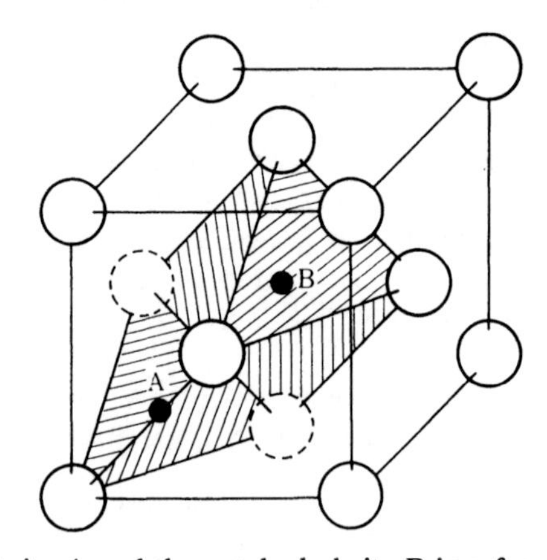

Fig. 9.2. The tetrahedral site A and the octahedral site B in a face-centered cubic lattice.

Table 9.1. Crystal and magnetic structure of magnetic oxides MO_x.

x Formula	Crystal-type	Name	Examples (F: Ferro-, FR: Ferri- CA: Cant-Magnetism)
1.0 MO	NaCl		M = Eu(F)
1.33 MFe_2O_4	spinel	ferrite	M = Mn, Fe, Co, Ni, Cu, Mg, $Li_{.5}Fe_{.5}$(FR)
1.33 MCo_2O_4	spinel	cobaltite	M = Mn, Fe, Co, Ni (FR)
1.33 MCr_2O_4	spinel	chromite	M = Mn, Fe, Co, Ni, Cu (FR)
1.33 MMn_2O_4	spinel	manganite	M = Cr, Mn, Fe, Co, Ni, Cd (FR)
1.33 MV_2O_4	spinel		M = Mn, Fe, Co (FR)
1.33 M_2VO_4			M = Mn, Fe, Co (FR)
1.33 M_2TiO_4			M = Mn, Fe, Co (FR)
1.40 $CaFe_4O_7$	hexagonal	Ca-diferrite	(FR)
1.46 $MFe_{12}O_{19}$	magnetoplumbite (M)	hexagonal ferrite	M = Ba, Pb, Sr, Ca, $Ni_{.5}La_{.5}$ $Ag_{.5}La_{.5}$ (FR)
1.42 $M_2BaFe_{16}O_{27}$	magnetoplumbite (W)	hexagonal ferrite	M = Mn, Fe, Ni, $Fe_{.5}Zn_{.5}$, $Mn_{.5}Zn_{.5}$ (FR)
1.41 $M_2Ba_3Fe_{24}O_{41}$	magnetoplumbite (Z)	hexagonal ferrite	M = Co, Ni, Cu, Mg, $Co_{.75}Fe_{.25}$ (FR)
1.38 $M_2Ba_2Fe_{12}O_{22}$	magnetoplumbite (Y)	hexagonal ferrite	M = Mn, Co, Ni, Mg, Zn, $Fe_{.25}Zn_{.75}$ (FR)
1.50 $M_{.8}Ti_{.8}Fe_{.4}O_3$	ilmenite	mixed ilumenite	M = Mn, Fe, Co, Ni (FR)
1.50 $MMnO_3$	ilmenite		M = Co, Ni (FR)
1.50 $R_3Fe_5O_{12}$	garnet	RIG	R = Y, Sm, Eu, Gd, Tb, Dy, Ho, Er, Tm, Yb, Lu (FR)
1.50 $RFeO_3$	perovskite	orthoferrite	R = Y, La, Nd, Sm, Eu, Gd, Tb, Dy, Ho, Er, Tm, Yb, Lu (CA)
1.50 $MMnO_3$	perovskite		M = Bi, $La_{.7}Ca_{.3}$, $La_{.7}Sr_{.3}$, $La_{.7}Ba_{.3}$ $La_{.6}Pb_{.4}$, $La_{.7}Cd_{.3}$ (F)
1.50 M_3MnO_6	perovskite		M_3 = Gd_2Co, Ba_2Fe, Ca_2Fe (FR)
2.0 MO_2	rutile		M = Cr(F)

The following sections describe spinel ferrites (section 9.2), garnet ferrites (Section 9.3), hexagonal ferrites (Section 9.4), and other oxides (Section 9.5).

9.2 MAGNETISM OF SPINEL-TYPE OXIDES

Spinel ferrites have the general formula $MO \cdot Fe_2O_3$, where M represents one or more divalent metal ions such as Mn, Fe, Co, Ni, Cu, Zn, Mg, etc. These ferrites are typical spinel-type oxides, with the crystal structure shown in Fig. 9.3. The unit cell contains 32 O^{2-} ions, 8 metal ions on A-sites, and 16 metal ions on B-sites, for a total of 56 ions. The open circles in the figure represent O^{2-} ions, which form a face-centered cubic lattice. There are two kinds of interstitial sites in the oxygen lattice: A or $8a$ sites which are surrounded by four O^{2-} and B or $16d$ sites which are surrounded by six O^{2-} ions. Since the numbers of nearest-neighbor O^{2-} ions for A and B sites are in the ratio of 2:3, the occupation of A sites by M^{2+} ions and B sites by Fe^{3+} will give a net electrical charge of zero, thus minimizing the electrostatic energy. Such a configuration is called a *normal spinel*. However, many magnetic ferrites are *inverse*

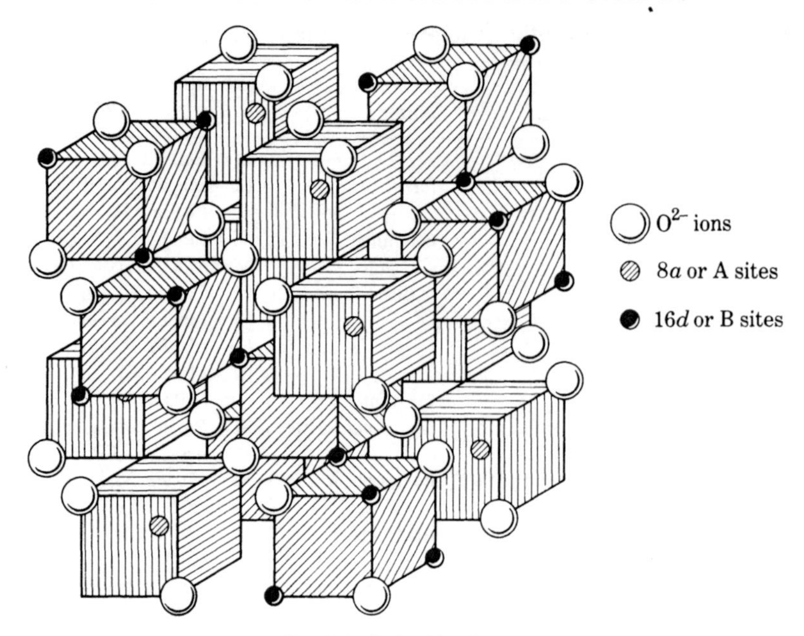

Fig. 9.3. Spinel lattice.

spinels, in which half of the Fe^{3+} ions occupy the A-sites and the remaining Fe^{3+} plus the M^{2+} ions occupy the B-sites.

Zinc ferrite, with M = Zn, is a normal spinel. The Mn ferrites are 80% normal; this means that 80% of the Mn ions occupy A sites, while the other 20% occupy B sites. It is interesting to note that ions occupying the A sites mostly have a spherical configuration: the electronic structure of Zn^{2+} is $3d^{10}$, and that of both Mn^{2+} and Fe^{3+} is $3d^{5+}$.

The radii of the spheres which fit in the A and B sites are calculated to be

$$\left.\begin{aligned} r_A &= \frac{\sqrt{3}}{8}a - R_0 \\ r_B &= \tfrac{1}{4}a - R_0 \end{aligned}\right\}, \tag{9.1}$$

where a is the lattice constant of the spinel lattice, and R_0 is the radius of the O^{2-} ions. Since $R_0 = 1.32\,\text{Å}$, and $a = 8.50\,\text{Å}$ in most ferrites, (9.1) gives $r_A = 0.52\,\text{Å}$, and $r_B = 0.81\,\text{Å}$. Thus the A sites are much smaller than most metallic ions, whose radii range from 0.6 to 0.8 Å. Therefore when a metallic ion occupies an A site, it pushes the surrounding O^{2-} ions outward. Such a deformation of the lattice is expressed by the *u-parameter*, which is defined by the coordinate of the O^{2-} ion as shown in Fig. 9.4. In an undistorted lattice $u = \frac{3}{8} = 0.375$, while in real ferrites $u = 0.380{-}0.385$. Using this u-parameter, (9.1) can be rewritten as

$$\left.\begin{aligned} r_A &= (u - \tfrac{1}{4})a\sqrt{3} - R_0 \\ r_B &= (\tfrac{5}{8} - u)a - R_0 \end{aligned}\right\}. \tag{9.2}$$

If we assume that $u = 0.385$, (9.2) gives $r_A = 0.67\,\text{Å}$, $r_B = 0.72\,\text{Å}$, which are suitable

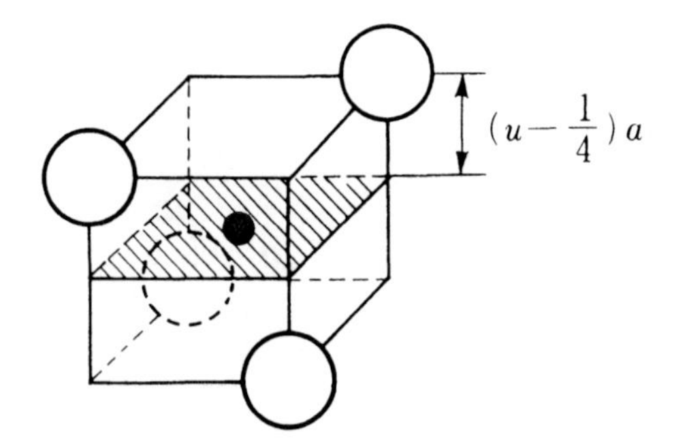

Fig. 9.4. Definition of the u-parameter.

sizes to accept metal ions. Values of u are listed in Table 9.2 for various spinel ferrites.

Before discussing the magnetism of ferrites, let us consider what kind of superexchange interactions are expected in this crystal structure. As mentioned in Section 7.1, the sign and magnitude of the superexchange interaction between M_1 and M_2 ions depend upon the angle of the path M_1–O–M_2. Figure 9.5 shows the angle and the distance between ions for AB, BB, and AA pairs. It is seen that in general the angle A–O–B is closer to 180° than the angles B–O–B or A–O–A, so that we expect the AB pair to have a stronger negative interaction than the AA or BB pairs. Based on this idea, let us discuss a possible magnetic structure of the inverse spinel. The following explanation was given by Néel[2], who first developed the theory of ferrimagnetism. In the inverse spinel ferrite, the main negative interaction acts between A and B sites, thus resulting in the magnetic arrangement given by

$$\left(\overrightarrow{\mathrm{Fe}}^{3+}\right)\mathrm{O}\cdot\left(\overleftarrow{\mathrm{Fe}}^{3+}\cdot\overleftarrow{\mathrm{M}}^{2+}\right)\mathrm{O}_3, \tag{9.3}$$

Table 9.2. Magnetic and physical properties of spinel ferrites MFe_2O_4.[8]

M	Density ($\mathrm{g\,cm^{-3}}$)	Lattice const. a (Å)	Resistivity (Ω cm)	M_{mol} at 0 K (M_B)	I_s r.t. (T)	Θ_f (K)	u-parameter
Zn	5.33	8.44	10^2	—	—	—	0.385
Mn	5.00	8.51	10^4	4.55	0.50	300	0.385
Fe	5.24	8.39	4×10^{-3}	4.1	0.60	585	0.379
Co	5.29	8.38	10^7	3.94	0.53	520	0.381
Ni	5.38	8.34	10^9	2.3	0.34	585	—
Cu (quenched)	5.42	8.37	10^5	2.3	—	455	0.380
Cu (slow cool)	5.35	c: 8.70 a: 8.22	—	1.3	0.17	—	—
Mg	4.52	8.36	10^7	1.1	0.15	440	0.381
Li	4.75	8.33	10^2	2.6	0.39	670	0.382
γ-Fe_2O_3	—	8.34	—	2.3	—	575	—

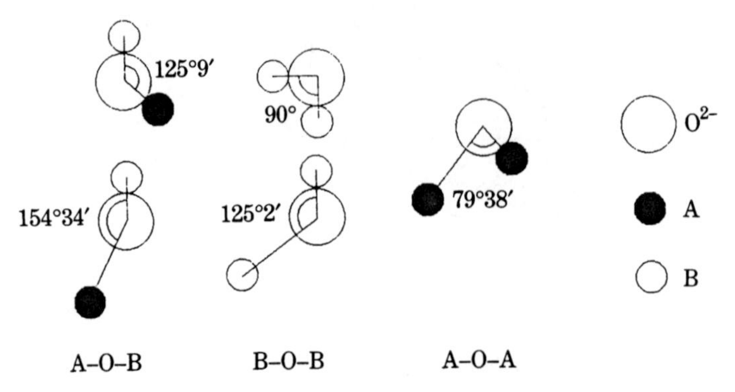

Fig. 9.5. Various paths for superexchange interactions in the spinel lattice. (Modified from Fig. 32.1 in Smit and Wijn.[1])

in which Fe^{3+} is $3d^5$ with a spin magnetic moment of $5M_B$ and M^{2+} has a magnetic moment of nM_B. Therefore the saturation magnetic moment per formula unit (the molar magnetic moment) at 0 K is given by

$$M = \{(5+n) - 5\}M_B$$
$$= nM_B. \tag{9.4}$$

This results in a moment of nM_B because the spins of the two Fe^{3+} ions in A- and B-sites cancel one another. If the identity of the M ion changes in the order M = Mn, Fe, Co, Ni, Cu, and Zn, the number of $3d$ electrons changes in order from 5 to 10, and the magnetic moment n of the M ion changes from 5 to 0. Then we expect that the molar magnetic moment changes linearly from $5M_B$ to zero as shown by the thick solid line in Fig. 9.6. Experimental points are very close to this theoretical line. The tendency for the experimental points to deviate upward from the theoretical line is due to the contribution of some orbital magnetic moment remaining unquenched by a crystalline field. The molar magnetic moment of Mn ferrite is $4.6M_B$, smaller than the theoretical value $5M_B$. This can be explained as follows: since M^{2+} and Fe^{3+} ions are both $3d^5$ and have $5M_B$, a partly normal spinel, or a partial mixture of Mn^{2+} ions, does not result in a change in molar magnetic moment. The real reason is that one electron is transferred from Mn^{2+} to Fe^{3+}, thus resulting in Mn^{3+} and Fe^{2+}, both of which have $4M_B$. If the ionic arrangement is a normal spinel, such an electron transfer results in a molar moment of $4+5-4=5M_B$, while if the arrangement is an inverse spinel, the molar moment is given by $4+4-5=3M_B$. Since Mn-ferrite is an 80% normal spinel, the molar moment should be given by

$$M = (0.8 \times 5 + 0.2 \times 3)M_B = 4.6M_B, \tag{9.5}$$

which agrees with experiment.

An interesting feature of ferrimagnetic oxides is that the addition of a non-magnetic oxide will sometimes result in an increase in molar magnetic moment. This occurs in the mixed MZn-ferrites: if x moles of Zn-ferrite (a normal spinel) is

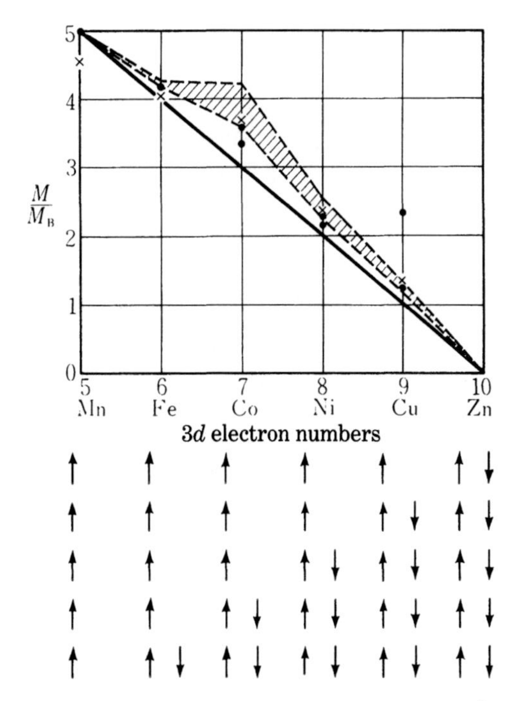

Fig. 9.6. Molar saturation moments of various simple spinel ferrites.[3] (Mark × is from Smit and Wijn.[1])

added to $1-x$ moles of M-ferrite (an inverse spinel), then the ionic arrangement is given by

$$\begin{array}{rl} x & (Zn^{2+})O\cdot(Fe^{3+}\cdot Fe^{3+})O_3 \\ 1-x & (Fe^{3+})O\cdot(Fe^{3+}\cdot M^{2+})O_3 \quad (+ \\ \hline \multicolumn{2}{c}{(Zn_x^{2+}\cdot Fe_{1-x}^{3+})O\cdot(Fe^{3+}\cdot Fe_x^{3+}\cdot M_{1-x}^{2+})O_3} \end{array} \tag{9.6}$$

Therefore the molar magnetic moment of the mixed ferrite is given by

$$\begin{aligned} M &= \{(1+x)5+(1-x)n\}M_B-(1-x)5M_B \\ &= \{n+(10-n)x\}M_B. \end{aligned} \tag{9.7}$$

We see that M increases with increasing x, heading toward a value of $10M_B$ at $x=1$. Figure 9.7 shows the variation of M with x for various kinds of mixed Zn ferrites. For small x, M increases along the line given by (9.7). For large x, however, M falls below the line (9.7). Yafet and Kittel[5] explained this deviation in terms of a triangular arrangement of spins. On the other hand, Ishikawa[6] interpreted this phenomenon in terms of superparamagnetic spin clusters produced by the breaking of exchange paths by non-magnetic Zn ions (see Section 20.1).

The R-type temperature dependence is commonly observed in spinel-type magnetic oxides. Sometimes the N-type temperature dependence appears, as in Li–Cr ferrites

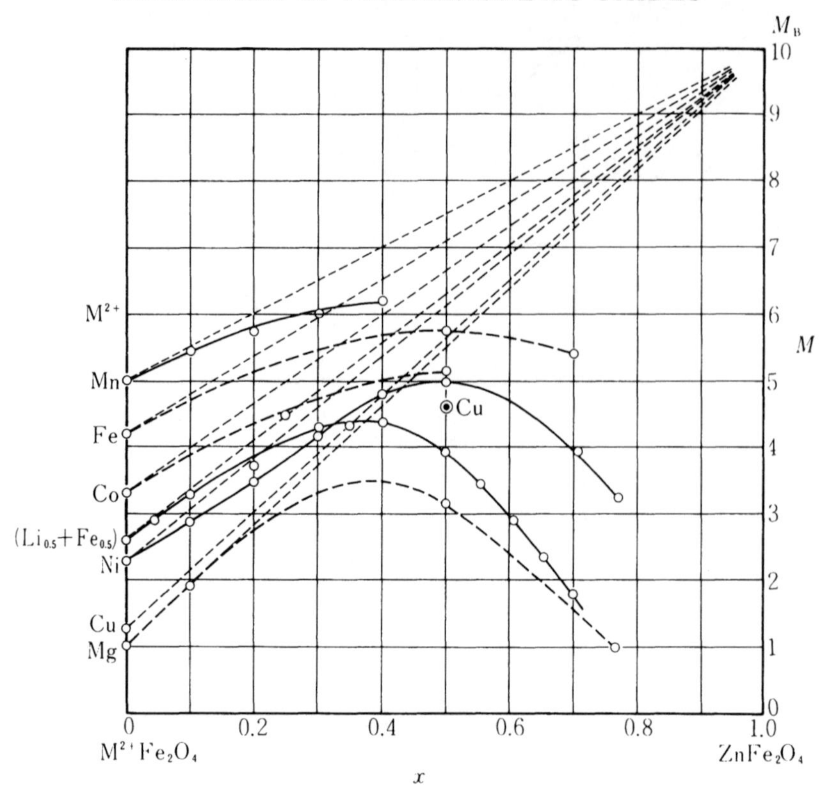

Fig. 9.7. Variation of molar saturation moment of various spinel ferrites with addition of Zn-ferrite.[4]

(for $x > 1$ in $Li_{0.5}Fe_{2.5-x}Cr_xO_4$; see Fig. 9.8). Similar behavior was found in Ni–Al ferrites.[7]

Table 9.2 summarizes the physical and magnetic properties of various spinel ferrites. To end this section we discuss features of some particular ferrites.

Zn-ferrite is an antiferromagnet with a Néel point near 10 K. When cooled from high temperatures, it exhibits weak ferrimagnetism. Mn–Zn and Ni–Zn mixed ferrites are widely used as soft magnetic materials for high-frequency applications because of their high electrical resistivity. The Zn is added partly to increase saturation magnetization at 0 K, as shown in Fig. 9.7. However, since the Curie point is decreased by the addition of Zn, the saturation magnetization at room temperature remains almost unchanged. A more significant reason for adding Zn is to increase the magnetic permeability by lowering the high permeability temperature range to room temperature (Hopkinson effect, see Fig. 18.20).

Mn-ferrite has the highest saturation magnetization at 0 K of the simple ferrites. Since the resistivity is relatively low, Mn–Zn ferrites are used in relatively low-frequency applications.

Fe-ferrite or *magnetite* (Fe_3O_4) is the oldest magnetic material known by man.

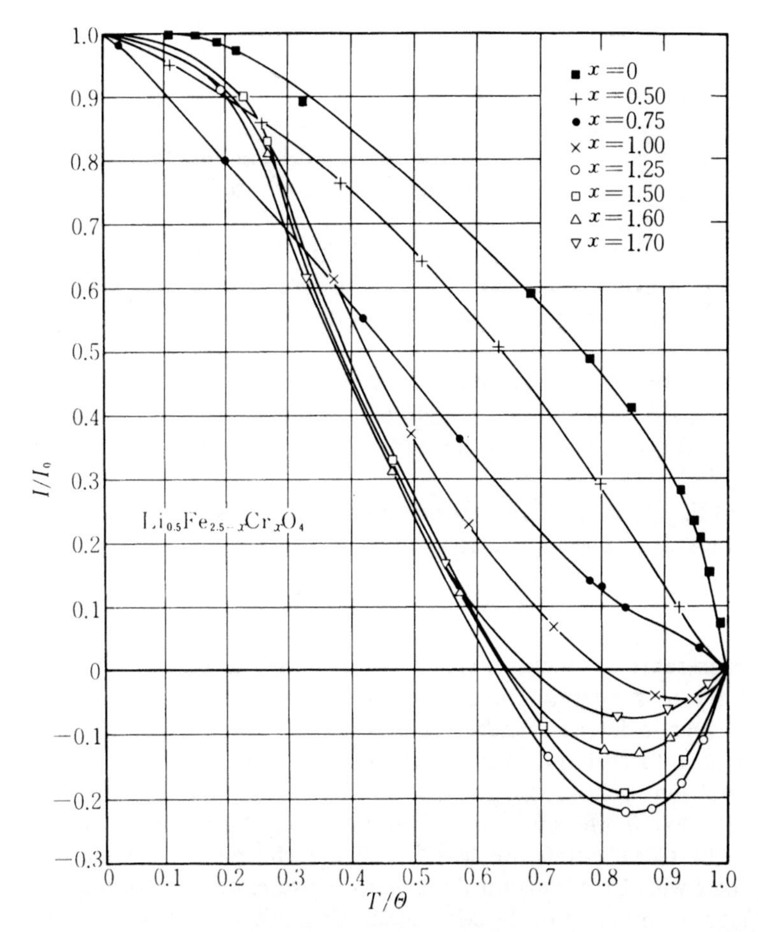

Fig. 9.8. Temperature dependence of saturation magnetization of various Li–Cr-ferrites.[4]

Since Fe^{3+} and Fe^{2+} coexist on B-sites, electrons 'hop' between these ions, so that the resistivity is extraordinarily low for an insulating material. The temperature coefficient of resistivity even becomes positive (like a metal) at high temperatures. On cooling, this electron hopping ceases at 125 K, below which the resistivity increases abruptly and the crystal structure transforms to a lower symmetry. This temperature is called the *Verwey point*, honoring the name of the discoverer of this transition. Verwey proposed an ordered arrangement of Fe^{3+} and Fe^{2+} ions in the low temperature phase,[9] but this has been disproved by recent experiments.[10] The ionic arrangement and crystal symmetry at low temperatures are not yet certain.

As seen in Table 9.2 and Fig. 9.6, Co-ferrite has a saturation moment of $3.94M_B$, which is much higher than the theoretical value of $3M_B$. This is due to a contribution from the orbital magnetic moment remaining unquenched by the crystalline field. Because of this contribution, Co-ferrite exhibits an extraordinary large induced anisotropy (see Section 13.1).

In Ni-ferrite, no electron hopping occurs because Ni exists only in the divalent state; the resistivity is therefore very high. For this reason, Ni–Zn ferrites are used as magnetic core materials at high frequencies. The frequency limit depends on the Zn content (see Section 20.3, Fig. 20.17).

Cu-ferrite has a cubic structure when quenched from high temperatures, but if it is slowly cooled, the lattice becomes tetragonal with a 5–6% distortion from cubic. This is the *Jahn–Teller distortion*, as will be discussed in Section 14.7.

Mg-ferrite is an incomplete inverse spinel, in which some Mg^{2+} ions occupy the A-sites, so that it exhibits ferrimagnetism. The ionic distribution and accordingly the saturation magnetization and the Curie point depend on the temperature from which it is quenched.

Li-ferrite is given by the formula $Li_{0.5}Fe_{2.5}O_4$ and has an inverse spinel lattice, whose B-sites are occupied by Li^{1+} and Fe^{3+} in the ratio of 1:3. It is known that two kinds of ions in the B-sites form an ordered arrangement when it is cooled below about 735°C.[11]

γ-(gamma)-Fe_2O_3 is called *maghemite* and its crystal structure is an inverse spinel containing vacancies. The ionic arrangement is given by

$$(Fe^{3+})O\cdot\left(Fe^{3+}_{5/3}V_{1/3}\right)O_3, \tag{9.8}$$

where V represents a vacancy. This ferrite, in the form of very small elongated particles, is the most common magnetic recording material.

The Fe^{3+} ions in ferrites can be replaced by Cr^{3+}; the resulting oxides are called *chromites*, whose formula is given by $MO\cdot Cr_2O_3$ or MCr_2O_4, where M represents a divalent metal ion such as Mn, Fe, Co, Ni, Cu, or Zn. The crystal structures are all cubic spinel. Table 9.3 lists various magnetic and crystal constants for several simple chromites. The Curie points are low for all the chromites. In most cases, they are normal spinels with Cr^{3+} ions on the B sites. If we assume according to Néel theory that the magnetic moment of the M^{2+} ions, nM_B, aligns antiparallel to that of $2Cr^{3+}$ on the B-site, we expect that the molar saturation moment should be given by

$$M = (6-n)M_B. \tag{9.9}$$

Table 9.3. Magnetic and crystalline constants of various chromites.[12]

Material	Θ_a (K)	Θ_f (K)	M_{mol} (M_B)	Lattice const. (Å)	u-parameter
$MnCr_2O_4$	−310	55, 43	1.2	8.437	0.3892 ± 0.00005
$FeCr_2O_4$	-400 ± 30	90 ± 5	0.8 ± 0.2	8.377	0.386
$CoCr_2O_4$	−650	100	0.15	8.332	0.387 ± 0.00005
$NiCr_2O_4$	−570	$80 \pm 10, 60$	0.14, 0.3	8.248	
$CuCr_2O_4$	−600	135	0.51	$a = 8.532$, $c = 7.788$	
$ZnCr_2O_4$	−380	$\Theta_N = 10$	—	8.327	

Table 9.4. Magnetic and crystalline constants of various manganites.[14]

Material	Θ_f (K)	Θ_a (K)	Θ_N (K)	M_{mol} (M_B)	M_{eff} (M_B)	c/a
$CrMn_2O_4$	65	−300		1.08		1.05
$MnMn_2O_4$	30 ~ 43			1.4 ~ 1.85		1.14 ~ 1.16
	42	−564			5.27	
$FeMn_2O_4$	390 ~ 395			1.55		1.06
$CoMn_2O_4$	85 ~ 105	−550 ~ −590		0.1 ~ 0.14	4.8 ~ 5.0	1.06 ~ 1.15
$NiMn_2O_4$	115 ~ 165	−392				
$Cu_{1.5}Mn_{1.5}O_4$	80				3.35	
$ZnMn_2O_4$		−450	≈ 200		4.74	1.14
$CdMn_2O_4$	< 4					1.20
$MgMn_2O_4$		−450			4.9	1.13 ~ 1.15
$LiMn_2O_4$		−164				

The experimental values, however, are in poor agreement with (9.9). The reason is that the chromites develop conical spiral spin structures at low temperatures. For instance, manganese chromite, $MnCr_2O_4$, shows a spiral spin structure with the spontaneous magnetization parallel to ⟨110⟩ from which the Mn moment tilts by 68°, and the Cr moments tilt by 94° and 47°.[13]

Manganites are spinel oxides containing Mn^{3+} with the general formula $MO \cdot Mn_2O_3$ or MMn_2O_4, where M represents a divalent ion such as Cr, Mn, Fe, Co, Ni, Cu, Zn, Cd, Mg, etc. Various magnetic and crystal data are listed in Table 9.4. The molar saturation moment is generally small, probably owing to the formation of canting spin configurations.

Cobaltites are spinel oxides continuing Co^{3+} with the general formula $MO \cdot Co_2O_3$ or MCo_2O_4, where M represents a divalent ion. For M = Mn, Fe, Ni, the Curie points are as high as 170, 450 and 350 K, respectively, while the molar saturation moments are as low as 0.1, 1.0 and $1.5M_B$, respectively.[15]

In these spinel-type oxides, not only divalent M^{2+} ions but also trivalent ions can be combined to make mixed spinel oxides.[16]

9.3 MAGNETISM OF RARE EARTH IRON GARNETS

Ferrimagnetic *rare earth iron garnets* (*RIG*) are oxides with the formula $3R_2O_3 \cdot 5Fe_2O_3$, or $R_3Fe_5O_{12}$, where R stands for one or more of the rare earth ions such as Y, Sm, Eu, Gd, Tb, Dy, Ho, Er, Tm, Yb, Lu, etc. The crystal structure is the garnet-type cubic oxide with a unit cell of 160 atoms[17] based on a framework consisting of 96 O^{2-} ions. The trivalent rare earth ions, R^{3+}, occupy the $24c$ sites (or dodecahedral sites) which are surrounded by O^{2-} ions forming a dodecahedron. Since the radius of the rare earth ions is 1.3 Å, which is much larger than that of the transition metal ions, the rare earths push the surrounding O^{2-} ions outward, thus distorting the framework of the close-packed oxygen lattice. The Fe^{3+} ions occupy $24d$, or octahedral, sites, and also $16a$, or tetrahedral, sites.

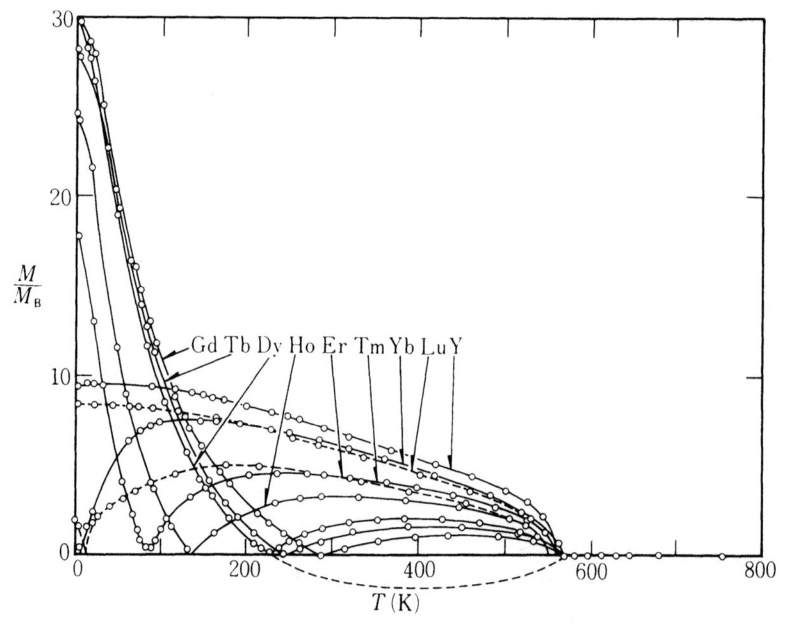

Fig. 9.9. Temperature dependence of molar saturation moment of various rare earth iron garnets.[19]

As explained by Néel,[18] the magnetic structure of rare earth iron garnets consists of strongly coupled Fe^{3+} on $24d$ sites and Fe^{3+} on $16a$ sites, and loosely coupled R^{3+} on $24c$ sites which point antiparallel to to the ferrimagnetic component of Fe^{3+}. Generally speaking, the magnetic moments of the R^{3+} ions are large, so that the resultant saturation magnetization is parallel to the magnetic moment of R^{3+} at low temperatures. Since the intrasite interaction between R^{3+} ions is very weak, the R^{3+} moments behave paramagnetically under the action of exchange fields produced by the Fe^{3+} spins. Therefore, as the temperature increases, the sublattice moment of the R^{3+} ions decreases sharply, resulting in an N-type temperature dependence as shown in Fig. 9.9. As seen in this graph, the Curie point is almost the same for all RIGs, since only the exchange interaction between Fe^{3+} survives at high temperatures.

Néel explained this situation by the formulation given below: the sublattice magnetization of the rare earth site under the action of the external field, H, is given by

$$I_R = \chi(H + wI_{Fe}), \tag{9.10}$$

where I_{Fe} is the total sublattice magnetization of Fe^{3+} ions on $24d$ and $16a$ sites, w is the molecular field coefficient, and χ is the paramagnetic susceptibility of rare earth ions. Then the total magnetization is given by

$$I = I_{Fe} + I_R = I_{Fe} + \chi(H + wI_{Fe}) = I_{Fe}(1 + \chi w) + \chi H. \tag{9.11}$$

The first term on the right-hand side is the saturation magnetization, while the second

Table 9.5. Magnetic and crystalline properties of rare earth iron garnets.[19,20]

R^{3+}	M_{mol} (0 K, M_B)	I_S (300 K, T)	$M(R^{3+})$ (0 K, M_B)	Θ_f (K)	Θ_c (K)	Θ_a (K)	Lattice const. (Å)	Density (g cm^{-3})
Y	5.0	0.170	0	560	—	—	12.38	5.169
Sm	5.43	0.160	0.14	578	—	—	12.52	6.235
Eu	2.78	0.110	−0.74	566	—	—	12.52	6.276
Gd	16.0	0.005	−7.0	564	286	−24	12.48	6.436
Tb	18.2	0.019	−7.7	568	246	−8	12.45	6.533
Dy	16.9	0.040	−7.3	563	226	−32	12.41	6.653
Ho	15.2	0.078	−6.7	567	137	−6	12.38	6.760
Er	10.2	0.110	−5.1	556	83	−8	12.35	6.859
Tm	1.2	0.110	−1.3	549	—	—	12.33	6.946
Yb	0	0.150	−1.7	548	—	—	12.29	7.082
Lu	5.07	0.150	0	539	—	—	12.28	7.128

term is the incremental magnetization induced by external fields when the material is nominally saturated. Looking at the first term, we find that the saturation magnetization vanishes either when

$$I_{Fe} = 0, \tag{9.12}$$

which is realized above the Curie point, or when

$$1 + \chi w = 0, \qquad \text{therefore} \qquad \chi = -\frac{1}{w}. \tag{9.13}$$

Since χ is the paramagnetic susceptibility, it decreases inversely with the absolute temperature and it will satisfy the condition at some temperature (note that $w < 0$, so that $-1/w > 0$). This temperature is the compensation point, Θ_c. As seen in Fig. 9.9 and Table 9.5, the value of Θ_c decreases monotonically with the increase in the number of $4f$ electrons in R, in the order Gd, Tb, Dy, etc. This is because the spin S decreases, so that the value of w decreases and the condition given by (9.13) is satisfied at a lower temperature. As seen in Table 9.5, the magnetic moment of R^{3+}, as deduced from the molar saturation moment at 0 K, is rather smaller than the value expected from Hund's rule for most rare earths. This fact tells us that most of the orbital magnetic moment is quenched in this type of crystal.

The garnet-type oxides contain only trivalent ions, with no divalent ions. Therefore no electron hopping occurs and the resistivity is very high, giving low magnetic losses even at high frequencies. Thin garnet crystals are, therefore, transparent, so that ferromagnetic domains can be observed by means of the Faraday effect.

It is possible to mix several kinds of rare earth ions on $24c$ sites, or to replace Fe^{3+} on $24d$ or $16a$ sites by Al^{3+} or Ga^{3+}. It is also possible to introduce Si^{4+} or Ge^{4+} ions together with the same number of Ca^{2+} or Mg^{2+} ions.[21] By such an introduction of non-magnetic ions, we can control the compensation point, saturation magnetization, magnetocrystalline anisotropy, g-factor, lattice constant, etc. This rather sophisticated materials technology has been utilized in developing bubble domain devices (see Section 17.3).

9.4 MAGNETISM OF HEXAGONAL MAGNETOPLUMBITE-TYPE OXIDES

Magnetoplumbite-type oxides contain 2+ ions such as Ba^{2+}, Sr^{2+} or Pb^{2+}, in addition to Fe^{3+} and divalent metal ions, M^{2+}, where M represents Mn, Fe, Co, Ni, Cu, Zn, Mg, etc. The ionic radii of Ba^{2+}, Sr^{2+} and Pb^{2+} are 1.43, 1.27 and 1.32 Å, respectively, and are therefore comparable to the radius of O^{2-} (1.32 Å). Accordingly, these large metal ions occupy substitutional sites rather than the usual interstitial sites of the close-packed oxygen lattice. The crystal layers containing these large metal ions alternate with spinel layers containing M^{2+} and Fe^{2+} ions with boundaries parallel to {111} planes of the cubic spinel structure. The result is a hexagonal lattice for the layered structure. The hexagonal oxides thus produced are classified as M-, W-, Y- and Z-types, according to the structures and concentrations of the layers containing the large metal ions.

Figure 9.10 shows a ternary phase diagram of these hexagonal oxides as a function of composition expressed as a combination of Fe_2O_3, BaO and MO. The 1:1 mixture of Fe_2O_3 and MO, denoted by S, is a common cubic spinel. The crystal structure of S shown in Fig. 9.11 is simply a rotation of Fig. 9.3 with the vertical axis perpendicular to (111) or an oxygen layer. It is seen that an inter-oxygen layer contains either B-site ions exclusively or A- and B-site ions half and half.

The hexagonal oxides are composed of spinel layers (S-type) and hexagonal R- and/or T-type layers, which contain large 3+ metal ions such as Ba^{3+} on the substitutional sites in the oxygen layers (Figs. 9.12 and 9.13).

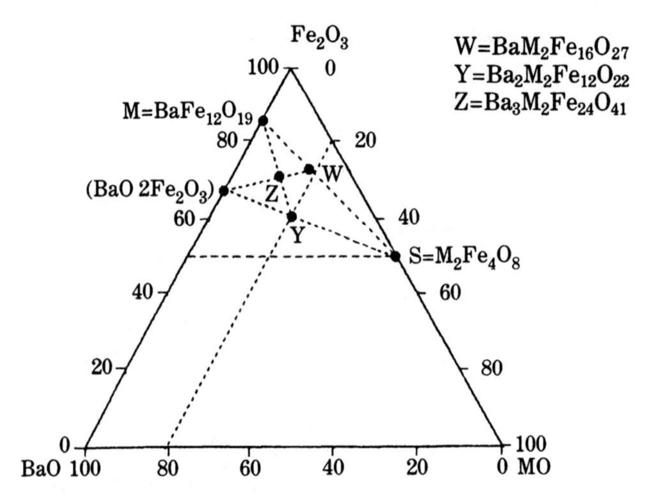

Fig. 9.10. Ternary phase diagram of hexagonal ferrites.[22]

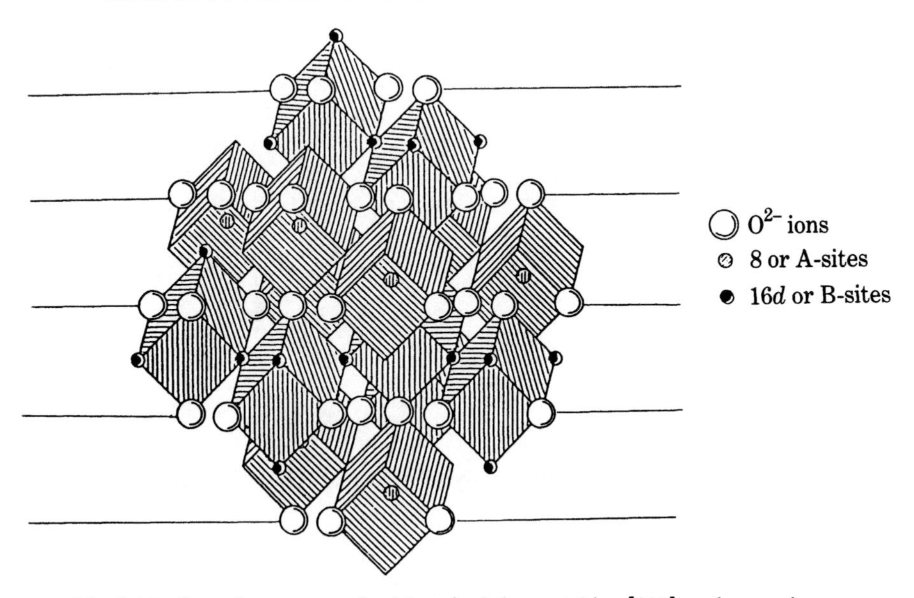

Fig. 9.11. Crystal structure of cubic spinel drawn taking [111] as the *c*-axis.

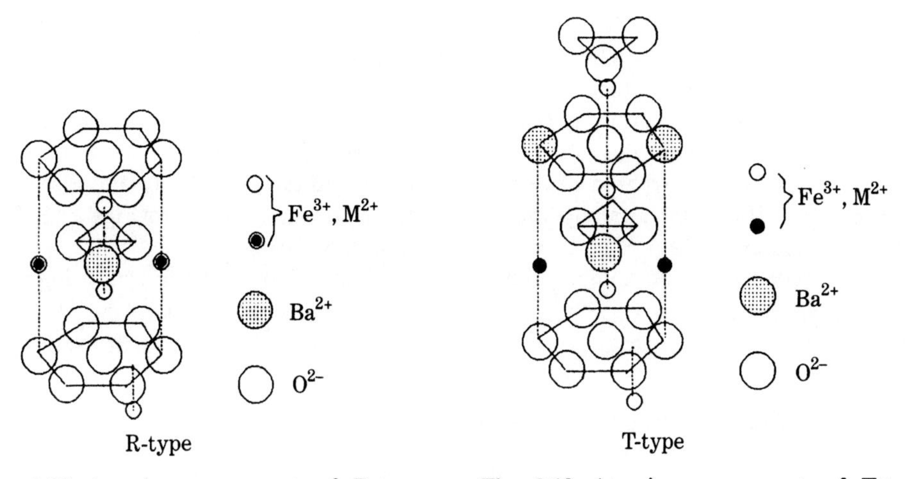

Fig. 9.12. Atomic arrangement of R-type layers.

Fig. 9.13. Atomic arrangement of T-type layers.

The M-type oxide has the formula $BaFe_{12}O_{19}$ or $BaO \cdot 6Fe_2O_3$, which is indicated as M on the left-hand edge of the phase diagram in Fig. 9.10. This oxide is a simple mixture of BaO and Fe_2O_3 and contains no MO. The crystal structure of the M-type oxide is shown in Fig. 9.14 as alternate layers of spinel blocks S, containing only Fe^{3+} metal ions, and R blocks containing Ba^{2+} and Fe^{3+}. The layers denoted S* and R* have the atomic arrangement obtained by rotating the layers S and R by 180° about the *c*-axis. The small circles with arrows indicate magnetic ions and the directions of

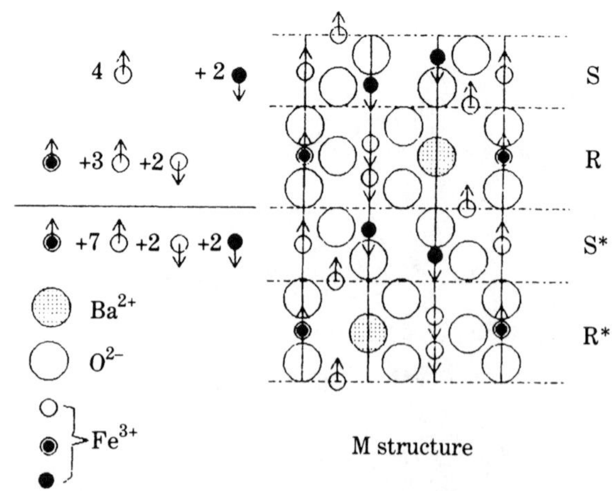

Fig. 9.14. Atomic arrangement of M-type crystal structure. (Modified from Fig. 37.4 in Smit and Wijn[22].)

their spin magnetic moments. The numbers of circles are those per two chemical formulas.

The magnetic properties of the M-type oxides are listed in Table 9.6. The values of saturation moment per mol formula are very close to the theoretical value, $20M_B$, for all the M-type oxides. The Curie points are about 450°C, and the saturation magnetizations at room temperature are nearly the same as those of spinel ferrites. The magnetocrystalline anisotropy is fairly large, so that some M-type oxides are used as permanent magnet materials (see Sections 12.4.1(d) and 22.2.2).

The W-, Y- and Z-type oxides contain MO in addition to Fe_2O_3 and BaO (see Figure 9.10). The atomic arrangements of these oxides are shown in Figs 9.15–9.17. These structures contain R and/or T layers in addition to S layers.

The W-type oxides are composed of alternate S and R layers as shown in Fig. 9.15. Their magnetic properties are listed in Table 9.7. By replacing some of the M^{2+} ions by Zn^{2+}, which occupy A sites as in cubic spinels, we can increase the saturation

Table 9.6. Magnetic properties of M-type oxides.

	Saturation magnetization			
	0 K		Room temperature	
Materials	(emu g^{-1}) ($\times 4\pi \times 10^{-7}$ Wb m kg^{-1})	(M_B mol^{-1})	$4\pi I_s$ (G) $I_s \times 10^4$ (T)	Curie point (°C)
BaM	100	20	4780	450
PbM	80	18.6	4020	452
SrM	108	20.6	4650	460
CaM	—	—	—	445
$Na_{.5}La_{.5}M$	—	21.5	—	440 ± 10
$Ag_{.5}La_{.5}M$	—	—	—	435

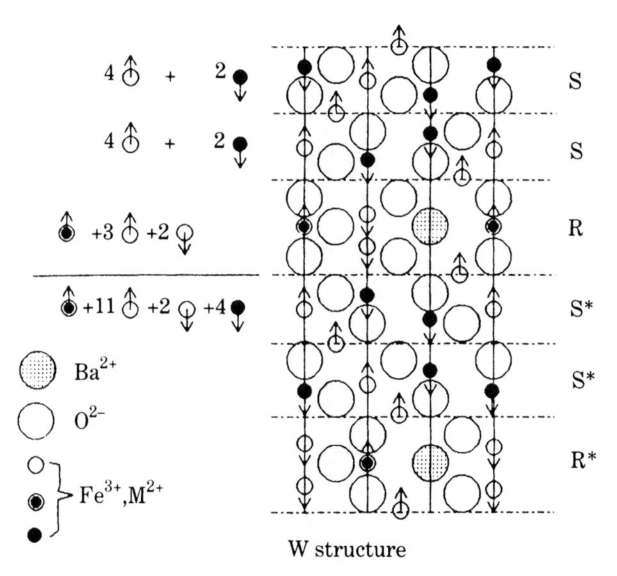

Fig. 9.15. Atomic arrangement of W-type crystal structure. (Modified from Fig. 37.7 in Smit and Wijn[22].)

magnetization. The saturation magnetic moments per mol formula at 0 K are in fair agreement with the theoretical value calculated by adding the moments for the two S formulas with that of the M-type. The slight differences may be due to the mixing of M^{2+} in the R-layers.

The Y-type oxides are composed of S and T layers, the latter of which contains equal numbers of divalent ions with opposite spins (Fig. 9.16). It these ions were exclusively Fe^{2+}, the spins would cancel, resulting in a molecular saturation moment equal to that of the two spinel formulas. The magnetic properties of various Y-type oxides are listed in Table 9.8, from which we see that the molecular saturation moments are mostly larger than the expected values. This shows that some M^{2+} ions intrude into the T-layers. In this type of oxide, the introduction of Zn^{2+} ions, which

Table 9.7. Magnetic properties of W-type oxides.

Materials	Saturation magnetization 0 K (emu g^{-1}) ($\times 4\pi \times 10^{-7}$ Wb m kg^{-1})	0 K (M_B mol^{-1})	Room temperature $4\pi I_s$ (G) $I_s \times 10^4$ (T)	Curie point (°C)
Mn_2W	97	27.4	3900	415
Fe_2W	98	27.4	5220	455
NiFeW	79	22.3	3450	520
ZnFeW	108	30.7	4800	430
$Ni_{.5}Zn_{.5}FeW$	104	29.5	4550	450

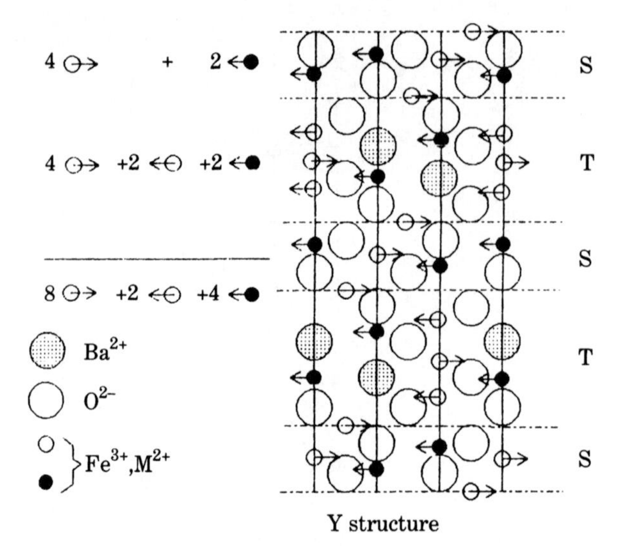

Fig. 9.16. Atomic arrangement of Y-type crystal structure. (Modified from Fig. 37.6 in Smit and Wijn[22])

occupy the A-sites, causes an increase in saturation magnetization as is the case for other ferrimagnetic oxides.

The Z-type oxides are composed of S-, R-, and T-layers, as shown in Fig. 9.17. The structure can be regarded as a superposition of M- and Y-types. As a matter of fact, the saturation moment per mol formula is about $30M_B$, which is approximately equal to the sum of those of the M- and Y-types (Table 9.9). The small difference may be due to the intrusion of M^{2+} ions into R-layers, as in the case of W-type oxides.

The magnetoplumbite-type oxides discussed in this section exhibit fairly large magnetocrystalline anisotropies, because of their low crystal symmetry (see Section 12.4.1(d)). Because of this characteristic, these oxides are used as permanent magnet materials, and are regarded as possible core materials for extremely high-frequency use (see Section 20.3).

Table 9.8. Magnetic properties of Y-type oxides.

Materials	Saturation magnetization 0 K (emu g^{-1}) ($\times 4\pi \times 10^{-7}$ Wb m kg^{-1})	0 K (M_B mol^{-1})	Room temperature $4\pi I_s$ (G) $I_s \times 10^4$ (T)	Curie point (°C)
Mn_2Y	42	10.6	2100	290
Co_2Y	39	9.8	2300	340
Ni_2Y	25	6.3	1600	390
Cu_2Y	28	7.1	—	—
Mg_2Y	29	6.9	1500	280
Zn_2Y	72	18.4	2850	130

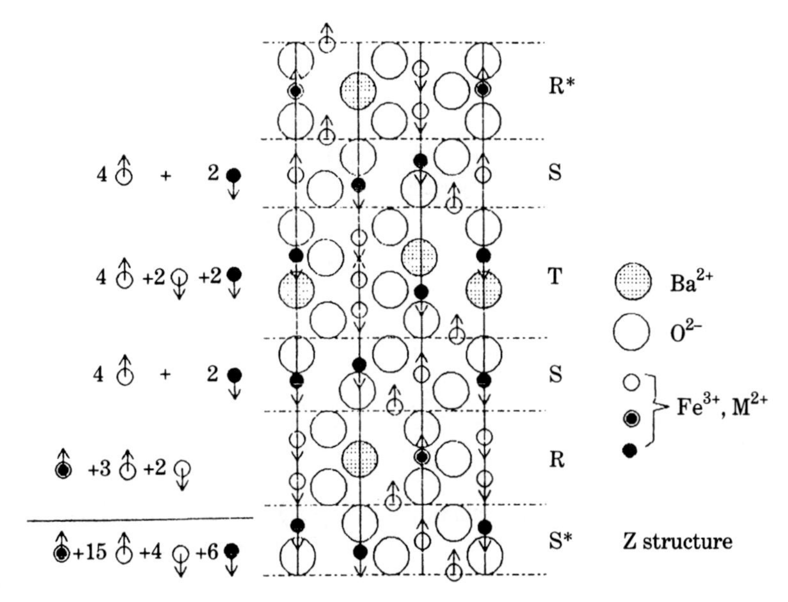

Fig. 9.17. Atomic arrangement of Z-type crystal structure. (Modified from Fig. 37.8 in Smit and Wijn[22])

Table 9.9. Magnetic properties of Z-type oxides.

	Saturation magnetization			
	0 K		Room temperature	
Materials	(emu g^{-1}) ($\times 4\pi \times 10^{-7}$ Wb m kg^{-1})	(M_B mol^{-1})	$4\pi I_s$ (G) $I_s \times 10^4$ (T)	Curie point (°C)
Co_2Z	69	31.2	3350	410
Ni_2Z	54	24.6	—	—
Cu_2Z	60	27.2	3100	440
Zn_2Z	—	—	3900	360
Mg_2Z	55	24	—	—

The descriptions in this section are largely based on Smit and Wijn.[22] For further details, this work should be consulted.

9.5 MAGNETISM OF OTHER MAGNETIC OXIDES

In this section, we discuss the crystalline and magnetic properties of various oxides not previously described.

9.5.1 Corundum-type magnetic oxides

In addition to magnetite (Fe_3O_4) and maghemite (γ-Fe_2O_3), (see Section 9.2), there

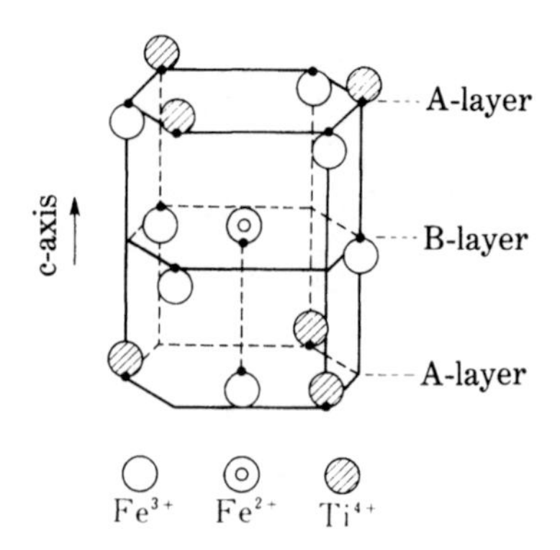

Fig. 9.18. Atomic arrangement of metal ions in a unit cell of 50:50 ilmenite–hematite solid solution.

are two well-known iron oxides, *hematite* (α-Fe_2O_3) and *ilmenite* ($FeTiO_3$), both of which have the corundum-type crystal structure. Corundum is the name of the oxide Al_2O_3. In this crystal, the oxygen ions form a closed-packed hexagonal lattice with two-thirds of the octahedral sites occupied by metal ions. Figure 9.18 shows the positions of the metal ions in a unit cell of a 50–50 mol% mixture of ilmenite and hematite. Each of the metal ions is displaced by about 0.3 Å either upward or downward along the c-axis. These small displacements are caused by the presence of vacancies on one-third of the octahedral sites. A combination of electrostatic and mechanical forces causes the metal ions to move towards the vacancies.

The atomic arrangement in α-Fe_2O_3 is given by replacing all the metal ions in Fig. 9.18 by Fe^{3+}. It has an antiferromagnetic spin arrangement, in which all spins on the same c-plane are aligned parallel, while the spins of the alternate planes, the A and B layers are aligned antiparallel to each other. The spin axis is parallel to the c-axis below 250 K, which is called the *Morin point*,[23] while it rotates to the c-plane above this temperature, and at the same time produces a canted spin structure with weak parasitic magnetism through an antisymmetric interaction (Section 7.4).

The atomic arrangement of ilmenite is obtained by replacing Fe^{3+} on the A layers in Fig. 9.18 by Ti^{4+}, and Fe^{3+} on the B layers by Fe^{2+}, so that all the A layers are occupied by non-magnetic Ti^{4+} ions ($3d^0$), while all the B layers are occupied by magnetic Fe^{2+} ions, each with a moment of $4M_B$. This crystal has also an antiferromagnetic structure, in which spins on alternate B layers are aligned antiparallel.

It is interesting to note that ferrimagnetic saturation magnetization appears in ilmenite–hematite solid solutions. The introduction of magnetic Fe^{3+} ions onto the B layers by mixing α-Fe_2O_3 into $FeTiO_3$ may induce an antiferromagnetic interaction between A and B layers. At the same time, as seen from Fig. 9.18, the A and B layers are occupied by different numbers and different species of magnetic ions, which gives rise to a ferrimagnetic spin arrangement.[24-26] Figure 9.19 shows the variation of saturation magnetic moment per mol formula as a function of the ratio of α-Fe_2O_3 in $M^{2+}Ti^{4+}O_3$. The straight line represents the theory described above. The experimental values for the oxides with M = Fe are very close to this theoretical line.

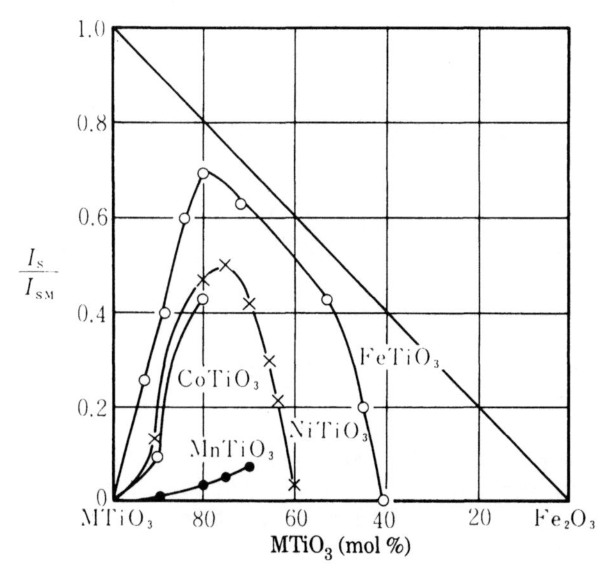

Fig. 9.19. Saturation magnetization as a function of $MTiO_3$ content for various $MTiO_3$–Fe_2O_3 solid solutions.[30]

Natural rocks which contain this series of oxides exhibit the so-called self-reversed thermal remanence. When rocks of this kind are cooled in the Earth's magnetic field, the resultant remanent magnetization at room temperature is opposite to the Earth's field.[28-29] The mechanism of this phenomenon has not yet been clarified.

For more details the reader may refer to the review articles.[27-30]

9.5.2 Perovskite-type magnetic oxide

Perovskite is the name of a mineral with the composition $CaTiO_3$. Replacing Ti by Fe^{3+}, we have the ferrimagnetic perovskite-type oxide $MFeO_3$, where M represents a large ion such as La^{3+}, Ca^{2+}, Ba^{2+} or Sr^{2+}. The basis of the crystal structure is an NaCl-type lattice composed of O^{2-} and M^{3+}, and a small Fe^{3+} ion goes into an octahedral site surrounded by six O^{2-} ions (Fig. 9.20). Replacing Fe^{3+} by Mn^{3+} or Co^{3+}, we also have other *perovskite-type oxides* $MMnO_3$ and $MCoO_3$. These oxides are antiferromagnetic, but in solid solutions of $La^{3+}Mn^{3+}O_3$ and $Ca^{2+}Mn^{4+}O_3$; $La^{3+}Mn^{3+}O_3$ and $Sr^{2+}Mn^{4+}O_3$; $La^{3+}Mn^{3+}O_3$ and $Ba^{2+}Mn^{4+}O_3$; and $La^{3+}Co^{3+}O_3$ and $Sr^{2+}Co^{4+}O_3$; ferromagnetism appears.[31] Two examples are shown in Figs. 9.21 and 9.22, in which the broken lines are calculated by assuming that the magnetic moments of Mn^{3+} ($4M_B$) and Mn^{4+} ($3M_B$) are aligned ferromagnetically. In the composition range 20–40 Mn^{4+}, the experimental points are very close to the theoretical lines in both cases. In this range the electrical conductivity becomes very large. Zener[32] explained this phenomenon in terms of *double exchange interaction*: in this composition range, Mn^{3+} and Mn^{4+} coexist, so that conduction electrons carry

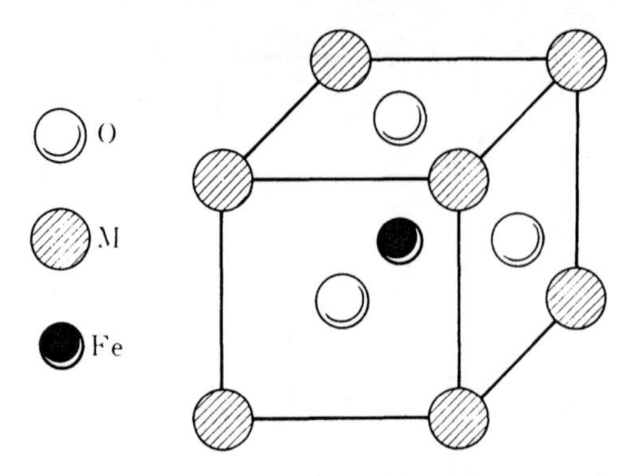

Fig. 9.20. Crystal structure of Perovskite-type magnetic oxides.

electric charges in the common d band, thus resulting in high conductivity. Since conduction electrons keep their spins during itinerant motion, Mn ions align their spins parallel to those of conduction electrons through the s–d interaction, thus resulting in ferromagnetic alignment. This idea was formulated by Anderson and Hasegawa[33] later. Different from the usual exchange or superexchange interaction, this double exchange interaction is proportional to $\cos(\theta/2)$, where θ is the angle between the spins of two Mn ions. Therefore, if antiferromagnetic superexchange interaction acts in addition to double exchange interaction, spin canting is expected to result.[34]

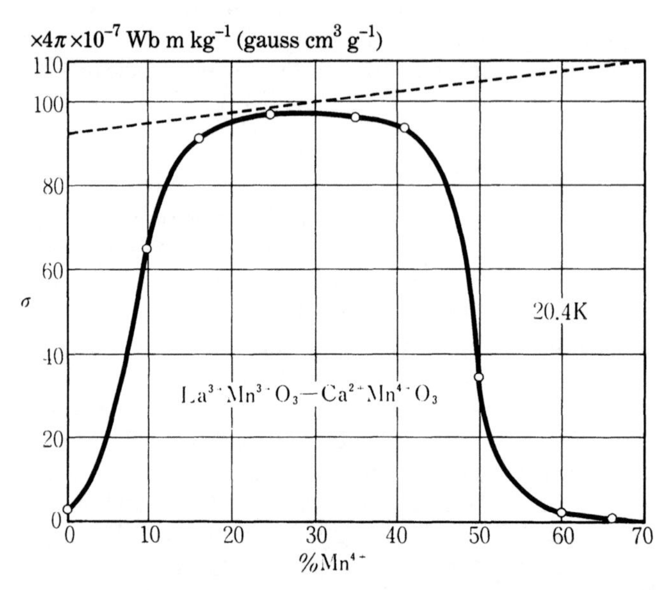

Fig. 9.21. Variation of specific saturation magnetization of an $LaMnO_3$–$CaMnO_3$ series with change in Mn^{4+} content.[31] The broken line is a theoretical curve assuming ferromagnetic alignment of magnetic moments of Mn^{3+} and Mn^{4+}.

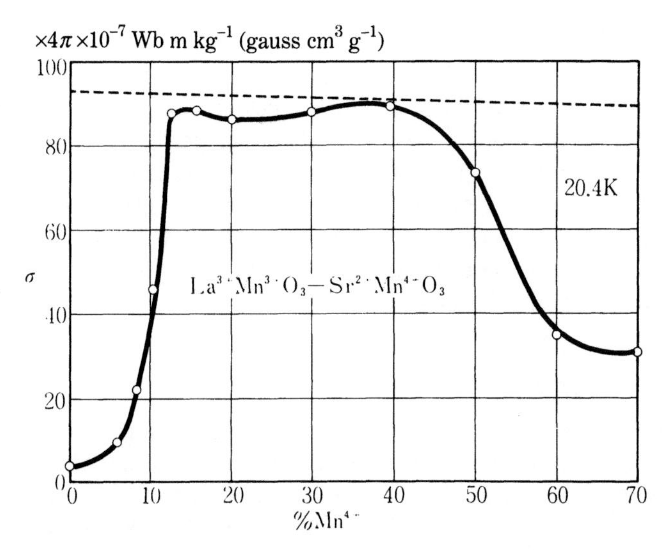

Fig. 9.22. Variation of specific saturation magnetization of an $LaMnO_3$–$SrMnO_3$ series with change in Mn^{4+} content. The broken line is a theoretical curve assuming ferromagnetic alignment of magnetic moments of Mn^{3+} and Mn^{4+}.[31]

Orthoferrite of the formula $RFeO_3$ (R: rare earths) is also a kind of Perovskite-type oxide. The crystal structure is deformed slightly from cubic to orthorhombic. The spins of Fe^{3+} ions are aligned antiferromagnetically through a strong superexchange interaction, because the angle Fe^{3+}–O–Fe^{3+} is 180°. The Néel point of this oxide is 700 K, which is quite high. Because of the orthorhombic deformation, an antisymmetric exchange interaction (see Section 7.4) acts between Fe^{3+} pairs, thus resulting in a spin canted magnetism with a feeble saturation moment of $0.05M_B$ per mol formula. This material was used initially for bubble domain devices (see Section 17.3).

As a result of the fact that the superexchange interaction through Mn^{3+}–O–Mn^{3+} or Fe^{4+}–O–Fe^{4+} is positive, the oxides $BiMnO_3$,[35] $SrFeO_3$[36] and $BaFeO_3$[37] are all ferromagnetic with Curie points of 103, 160, and 180 K, respectively.

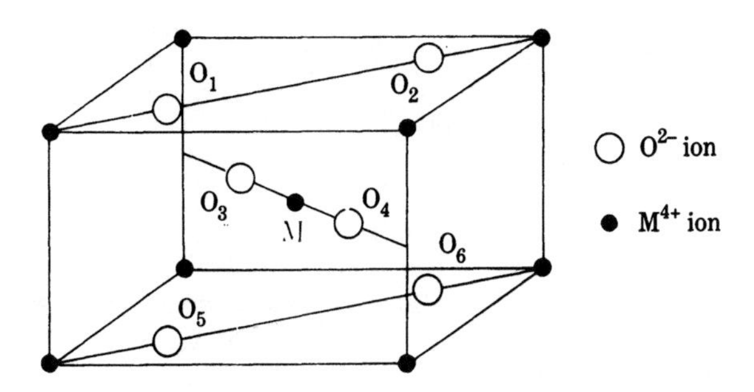

Fig. 9.23. Crystal structure of Rutile-type oxides.

9.5.3 Rutile-type magnetic oxides

Rutile is the name of the oxide TiO_2. Its crystal structure is shown in Fig. 9.23. The crystal symmetry is tetragonal, with $c < a$. Replacing Ti by $3d$ transition metal ions, we have *rutile-type magnetic oxides*. A typical example is CrO_2, which is ferromagnetic[38,39] with Curie point 390–400 K and specific saturation magnetization of 1.23×10^{-4} Wb m kg^{-1} (98 emu g^{-1}) at 300 K. It can be prepared either by heating Cr_2O_3 in a high-pressure oxygen atmosphere or by hydrothermal synthesis. This oxide is used as a magnetic tape material (see Table 22.4). MnO_2 is another example of this type of oxide, but it is antiferromagnetic.

9.5.4 NaCl-type magnetic oxides

A typical example of this type of oxide is MnO, whose crystal and spin structure was shown in Fig. 7.3 as a representative antiferromagnetic oxide. Other similar oxides FeO, CoO, and NiO are all antiferromagnets. The only exception is EuO, which is a ferromagnet[40] with a Curie point of 77 K. In this case the electronic structure of Eu is $4f^7$, which has no orbital moment, so that this material is often regarded as an ideal ferromagnet.

PROBLEMS

9.1 Describe crystal and magnetic structures of spinel ferrites and discuss the variation of molar saturation moment by the addition of Zn^{2+} ions.

9.2 Describe the mechanism by which the compensation point appears in rare earth iron garnets.

9.3 Explain why ferrimagnetism appears in ilmenite–hematite solid solutions.

9.4 Explain why ferromagnetism appears in $LaMnO_3$–$CaMnO_3$ solid solutions.

REFERENCES

1. J. Smit and H. P. J. Wijn, *Ferrites* (Wiley, New York, 1959), p. 149.
2. L. Néel, *Ann. Physique* [12] **3** (1948), 137.
3. L. Néel, *Proc. Phys. Soc.* (London), **65A** (1952), 869.
4. E. W. Gorter, *Philips Res. Rept.*, **9** (1954), 295, 321, 403.
5. Y. Yafet and C. Kittel, *Phys. Rev.*, **87** (1952), 290.
6. Y. Ishikawa, *J. Phys. Soc. Japan*, **17** (1962), 1877.
7. J. Smit and H. P. J. Wijn, *Ferrites* (Wiley, New York, 1959), p. 159.
8. T. Tsushima, T. Teranishi, and K. Ohta, *Handbook on magnetic substances* (ed. by S. Chikazumi *et al.*, Asakura Publishing Co., Tokyo, 1975), 9.2, p. 612, Table 9.3.
9. E. J. W. Verwey, P. W. Haayman, and F. C. Romeijn, *J. Chem. Phys.*, **15** (1947), 181.
10. S. Chikazumi, *AIP Proceedings*, No. 29 (1975), 382.

11. P. B. Braun, *Nature*, **170** (1952), 1123.
12. T. Tsushima, T. Teranishi, and K. Ohta, *Handbook on magnetic substances* (ed. by S. Chikazumi *et al.*, Asakura Publishing Co., Tokyo, 1975), 9.2, p. 629, Table 9.16.
13. T. Tsushima, Y. Kino, and S. Funahashi, *J. Appl. Phys.*, **39** (1968), 626.
14. T. Tsushima, T. Teranishi, and K. Ohta, *Handbook on magnetic substances* (ed. by S. Chikazumi *et al.* Asakura Publishing Co., Tokyo, 1975) 9.2, p. 630, Table 9.20.
15. G. Blasse, *Philips Res. Rept.*, **18** (1963), 383.
16. T. Tsushima, T. Teranishi, and K. Ohta, *Handbook on magnetic substances* (ed. by S. Chikazumi *et al.*, Asakura Publishing Co., Tokyo, 1975), 9.2, pp. 607–33.
17. S. Geller and M. A. Gilleo, *Acta Cryst.*, **10** (1957), 787.
18. L. Néel, *Comp. Rend.*, **239** (1954), 8.
19. F. Bertaut and R. Pauthenet, *Proc. IEEE Suppl.*, **B104** (1957), 261; R. Pauthenet, *J. Appl. Phys.*, **29** (1958), 253.
20. S. Miyahara and T. Miyadai, *Handbook on magnetic substances* (ed. by S. Chikazumi *et al.*, Asakura Publishing Co., Tokyo, 1975) 9.4.2., p. 667, Table 9.41.
21. S. Geller, *Z. f. Kristgr.*, **125** (1967), 1.
22. J. Smit and H. P. J. Wijn, *Ferrites* (Wiley, New York, 1959), pp. 177–211.
23. J. Morin, *Phys. Rev.*, **78** (1950), 819.
24. Y. Ishikawa and S. Akimoto, *J. Phys. Soc. Japan*, **13** (1958), 1298.
25. R. M. Bozorth, D. E. Walsh, and A. J. Williams, *Phys. Rev.*, **108** (1957), 157.
26. Y. Ishikawa, *J. Phys. Soc. Japan*, **17** (1962), 1835.
27. Y. Ishikawa and Y. Syono, *Handbook on magnetic substances* (ed. by S. Chikazumi *et al.*, Asakura Publishing Co., Tokyo, 1975) 9.4.1, pp. 645–56.
28. T. Nagata, S. Ueda, and S. Akimoto, *J. Geomag. Geoelect.*, **4** (1952), 22.
29. Y. Ishikawa and Y. Syono, *J. Phys. Chem. Solids*, **24** (1963), 517.
30. Y. Ishikawa, *Metal Phys.* (in Japanese), **6** (1960), 19; *Progress on Phys. of Mag.* (in Japanese) (Chikazumi ed. Agne Pub. Co, Tokyo, 1964), 329.
31. G. H. Jonker and J. H. van Santan, *Physica*, **16** (1950), 337; **19** (1953), 120; G. H. Jonker, *Physica*, **22** (1956), 707.
32. C. Zener, *Phys. Rev.*, **82** (1951), 403.
33. P. W. Anderson and H. Hasegawa, *Phys. Rev.*, **100** (1955), 675.
34. P. G. de Gennes, *Phys. Rev.*, **118** (1960), 141.
35. F. Sugawara, S. Iida, Y. Syono, and S. Akimoto, *J. Phys. Soc. Japan*, **25** (1968), 1553.
36. J. B. MacChesney, R. C. Sherwood, and J. P. Potter, *J. Chem. Phys.*, **43** (1965), 1907.
37. S. Mori, *J. Phys. Soc. Japan*, **28** (1970), 44.
38. C. Guillard, A. Michel, J. Bernard, and M. Fallot, *Comp. Rend.*, **219** (1944), 58.
39. T. J. Swoboda, *J. Appl. Phys.*, **32** (1961), 374S.
40. B. T. Matthias, R. M. Bozorth, and J. H. Van Vleck, *Phys. Rev. Lett.*, **7** (1961), 160.

10
MAGNETISM OF COMPOUNDS

In this chapter, we discuss the magnetism of compounds composed of electropositive $3d$ transition elements or $4f$ rare earth elements combined with electronegative elements which belong to the IIIb, IVb, Vb, VIb and VIIb groups. The electronegative elements which belong to these groups are listed in Table 10.1. Some elements such as Al, Ga, In, Tl, Sn and Pb, which belong to IIIb and IVb groups, have metallic character, so that they could have been treated as intermetallic compounds in Section 8.4. However, since it is very hard to draw a sharp line between metallic and nonmetallic elements, it is reasonable to treat them in this chapter. We have already treated oxides in Chapter 9, so that oxides are omitted from this chapter. The compounds considered in this chapter include borides, Heusler alloys, MnBi, MnAl and various chalcogenides, all of which have useful engineering applications. We have limited the treatment to compounds with Curie points above room temperature, following a general rule applied throughout this book.

In this chapter we discuss compounds in six categories:

(1) $3d$–IIIb compounds;
(2) $3d$–IVb compounds;
(3) $3d$–Vb compounds;
(4) $3d$–VIb compounds;
(5) $3d$–VII (halogen) compounds;
(6) rare earth compounds.

As a general rule, with an increase in the group number, the electrical properties of the compounds tend to change from metallic to semimetallic, then to semiconducting and finally to insulating; while with an increase in the period number, the atomic size increases so that the crystal tends to change from interstitial to substitutional. From the point of view of magnetism, the present classification based on the electronegativity of the elements may not be proper, but it has the merit of clarity. When a

Table 10.1. IIIb–VIIb Group elements with atomic number.

Period	IIIb Group		IVb Group		Vb Group		VIb Group		VIIb Group	
2	5	B	6	C	7	N	8	O	9	F
3	13	Al	14	Si	15	P	16	S	17	Cl
4	31	Ga	32	Ge	33	As	34	Se	35	Br
5	49	In	50	Sn	51	Sb	52	Te	53	I
6	81	Tl	82	Pb	83	Bi	84	Pb	85	At

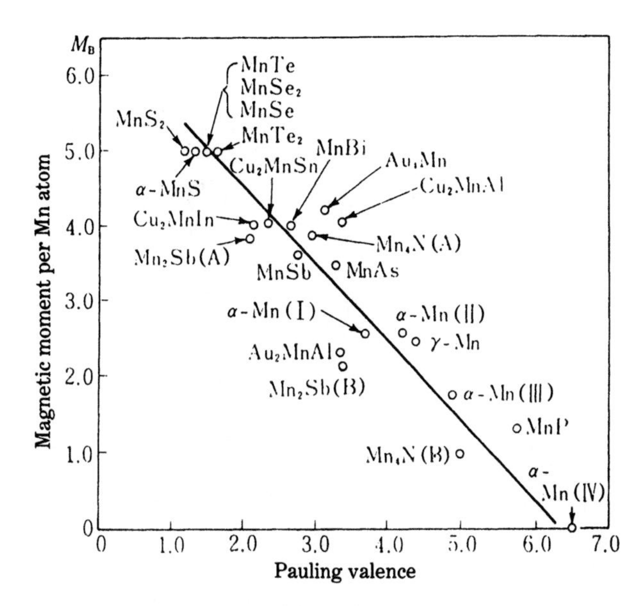

Fig. 10.1. Relationship between the atomic magnetic moment on an Mn atom in various magnetic compounds and the Pauling valence.[1]

compound includes more than one electronegative element, it is classified by the most electronegative element.

There is no general guiding principle to interpret the variety of magnetic behavior in these compounds. The only clear feature is that the magnetic behavior is influenced by the covalency. Mori and Mitui[1] showed that the magnetic moment of an Mn atom in various Mn compounds is well described as a function of Pauling valence, which is calculated from the atomic distance and coordination numbers, as shown in Fig. 10.1. They interpreted this result to show that electrons which contribute to the covalency have compensated spin pairs and make no contribution to the magnetic moments. This is the reason why the magnetic moment of an atom of one element can be different in different compounds, as seen in the following sections.

10.1 3*d* TRANSITION VERSUS IIIb GROUP MAGNETIC COMPOUNDS

10.1.1 Borides

Boron combines with transition elements M to form compounds of formulae M_3B, M_2B, M_3B_2, MB, M_3B_4, MB_2, etc. All these compounds have metallic luster and conductivity, and are stable at high temperatures. The magnetic borides with Curie points above room temperature are listed in Table 10.2. If the saturation moment of

Table 10.2. Crystal structure and magnetic properties of 3*d*-B compounds.[4]

Material	Crystal structure	Curie point Θ_f (K)	Magnetic moments per M atom M_s (M_B)	M_{eff} (M_B)
Co_3B	Fe_3C	747	1.12	—
Co_2B	Fe_2B	433	0.806	—
Fe_2B	C16	1015	1.9	—
MnB	FeB	578	1.92	2.71
FeB	B27	598	1.12	1.84
$Co_{20}Al_3B_6$	$Cr_{23}C_6$	406	0.6	2.0
$Co_{21}Ge_2B_6$	$Cr_{23}C_6$	511	0.56	—
Mn_5SiB_2	Cr_5B_3	398	1.5	2.6

the transition element M in these magnetic borides is plotted as a function of the number of electrons in M, we have the curve as shown in Fig. 10.2,[2-4] which is similar to the Slater–Pauling curve (see Fig. 8.12). The difference from the curve for compositions containing only metallic elements is that the curve for M_2B is shifted to the left by one electron, and that for MB is shifted by two electrons, from the curve for the metallic alloys. This means that boron acts as electron donor which contributes two electrons per atom to the lattice. Other than this, boron has little influence on the magnetic properties of magnetic alloys. This is also the case for amorphous alloys which contain B or P, as will be discussed in Chapter 11.

Atomic-scale investigations on borides have been exclusively made by means of NMR or Mössbauer experiments; neutron diffraction experiments are almost impossible because boron strongly absorbs neutrons. It was observed that the internal field of

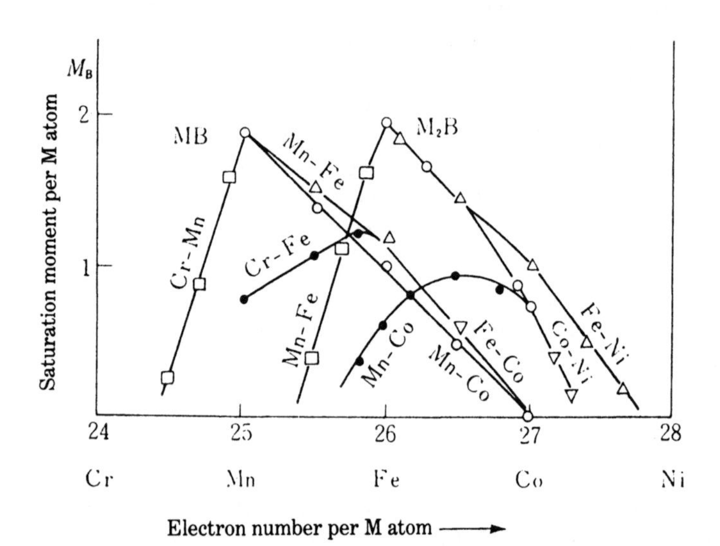

Fig. 10.2. Saturation atomic magnetic moment of M atom in 3*d*-B compounds with the formulas M_2B and MB at 20 K, as a function of the number of electrons in the M atom.[3,4]

B in Fe_2B has the large value of 2.24 $MA\,m^{-1}$ (28.2 kOe).[5] This was interpreted[6] to mean that the outer electrons of B are polarized through covalent bonds by the magnetized Fe atoms.

10.1.2 Al compounds

Elemental aluminum is purely metallic, but its chemical properties are sometimes electropositive and sometimes electronegative, so that it forms various compounds.

Iron–aluminum alloys exhibit ferromagnetism in the Fe-rich composition range. The superlattice Fe_3Al has the unit cell shown in Fig. 10.3., and is ferromagnetic below its Curie point of 750 K. Neutron diffraction shows that FeI, which is surrounded by four Fe and four Al nearest neighbors, has a moment of $1.46M_B$, while FeII, which is surrounded by eight Fe nearest neighbors, has $2.14M_B$.[7] Another superlattice, FeAl, is paramagnetic. It was observed that Fe–Al alloys with compositions between the two superlattices have an antiferromagnetic exchange interaction.[8-10] The superlattice Ni_3Al is ferromagnetic below its Curie point of 75 K, while NiAl and also CoAl are paramagnetic. On the other hand, MnAl is a metastable compound and is ferromagnetic below its Curie point of 650 K,[11] with an effective moment of $2.31M_B$ per Mn atom. This material is used as a permanent magnet material (see Section 22.2).

The alloy of composition Cu_2MnAl is known as the Heusler alloy; it is notable because it contains no ferromagnetic elements but nevertheless exhibits ferromagnetism[12] in the ordered state, as shown in Fig. 10.3. The Curie point of this alloy is 610 K, and the saturation moment per Mn atom is $3.20M_B$ at room temperature. According to neutron diffraction measurements, the magnetic moment is concentrated on the Mn atoms, and the localized moment at a Cu atom is no more than $0.1M_B$.[13] The κ-phase alloys with the general formula $M_{0.8}Mn_{1.2}Al_2$ (M = Cu, Ni, Co, Fe) are also ferromagnetic[14] with Curie points 300–400 K. The saturation moment per Mn atom is about $1.5M_B$, while the M atoms are nonmagnetic in all cases.[15]

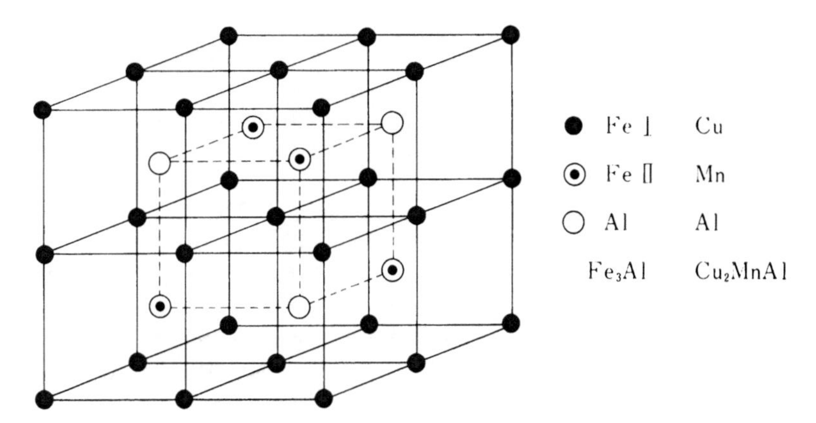

Fig. 10.3. Atomic arrangement of the superlattice Fe_3Al and the Heusler alloy Cu_2MnAl.

10.1.3 Ga compounds

Many ferro- or ferrimagnetic phases have been found in the Mn–Ga series. The ordered hexagonal ε-phase exists at high temperature in the composition range 72.5–70 at% Mn. It is ferromagnetic below its Curie point of 470 K. The $1/\chi$–T curve deviates downward from a linear relationship as the temperature is lowered from high temperatures to the Curie point, so that it is rather ferrimagnetic. The saturation magnetic moment extrapolated to 0 K is $0.02 M_B$ per Mn, while the effective moment is calculated from the linear part of the $1/\chi$–T curve is $3.1 M_B$.[16] There is also an ε'-phase having the CuAu-type face-centered tetragonal lattice at low temperature in the composition range 72.0–68.0 at% Mn. It is ferrimagnetic below its Curie point of 690 K. The saturation moment extrapolated to 0 K is $0.78 M_B$ per Mn, but the absolute value of the moment was estimated to be $1.5 M_B$ to account for its ferrimagnetic arrangement.[17] The ζ-phase has a γ-brass type cubic lattice at the composition 62 at% Mn. It is ferromagnetic below its Curie point of 210 K and has the moments $M_s = 1.0 M_B$ and $M_{eff} = 2.05 M_B$.[18] The face-centered tetragonal η-phase with the CuAu-type ordered structure exists in the composition range 60–54.5 at% Mn. The excess Mn atoms align their magnetic moments antiparallel to those of the other Mn atoms, thus exhibiting ferrimagnetism. Therefore the saturation magnetization increases from 7.04 to 7.48×10^{-5} Wb m kg^{-1} (56.0 to 59.5 emu g^{-1}) and the Curie point decreases from 640 to 605 K as the excess Mn content decreases. This corresponds to a change in saturation moment of Mn from 1.58 to $1.76 M_B$.[19]

10.1.4 In compounds

The compound Mn_3In has the β-brass type crystal structure. It is weakly ferrimagnetic below its Curie point at 583 K.[20] There are Heusler-type ferromagnetic compounds Cu_2MnIn[21] ($\Theta_f = 500$ K) and Ni_2MnIn[22] ($\Theta_f = 323$ K).

10.2 3*d*-IVb GROUP MAGNETIC COMPOUNDS

10.2.1 Magnetic carbides

All of these carbides are mechanically brittle, and have metallic luster, good electrical conductivity, and high melting points. This suggests that the binding force is partly metallic and partly covalent.

Cementite, Fe_3C, is a typical ferromagnetic carbide. There are many other ferromagnetic carbides, which are listed in Table 10.3 together with their magnetic properties. As seen from this table, some of them have ferromagnetic Curie points, Θ_f, fairly different from their asymptotic Curie points, Θ_a. This fact indicates that their spin arrangement is ferrimagnetic. According to a Mössbauer experiment by Moriya *et al.*,[23] the C atoms tend to go into small octahedral sites rather than large tetrahedral sites in the body-centered cubic iron. It was found that the internal field of the Fe atoms that are nearest neighbors of the octahedral C atoms is reduced from

Table 10.3. Crystal structures and magnetic properties of $3d$–C compounds.[4]

Materials	Crystal structure	Curie point Θ_f (K)	Asymptotic Curie point Θ_a (K)	Magnetic moments per magnetic atom M_s (M_B)	M_{eff} (M_B)	Note
Fe_3C	DO_{11}	483	233	1.78	3.89	
Fe_2C	hex.	653		1.72		
Fe_2C	monoc.	520	246	1.75	5.55	
Mn_3ZnC	perov.	368		1.0		ferro–ferri trans. 353 K
Co_2Mn_2C	perov.	733	723	4.0, 0.4	1.70	

the value in pure iron, and that the amount of reduction is proportional to the number of nearest-neighbor C atoms. Moreover, this reduction rate is almost the same for all compounds in the Fe–C system. This result agrees with the prediction by Bernas *et al.*[24] for the compounds $Fe_3C_{1-x}B_x$ $(0 \le x \le 0.54)$.

10.2.2 Magnetic silicides

As mentioned in Section 8.2, the Fe–Si system forms a solid solution in the Si-poor composition range, beyond which it forms a number of silicides. The ferromagnetic silicides are listed in Table 10.4. In these compounds, magnetic atoms occupy crystallographically different lattice sites and their magnetic moments are different from each other. The MSi-type compounds retain the same cubic crystal structure (B 20 type) when M is Cr, Mn, Fe, Co, or Ni, and even when these metallic elements are mixed as solid solutions. As the electron concentration of M increases from Cr to Ni, the magnetic properties change from a special diamagnetism, to a weak helimagnetism, to a peculiar Pauli paramagnetism, to a weak ferromagnetism, and finally to diamagnetism. At the same time, the electrical properties change from metallic to semiconducting, and finally to semimetallic. For further details, a recent review[25] (mainly on NMR investigations) should be consulted.

Table 10.4. Crystal structure and magnetic properties of ferromagnetic $3d$-Si compounds.[4]

Materials	Crystal structure	Curie point (K)	Magnetic moments per magnetic atom M_s (M_B)	M_{eff} (M_B)
Fe_3Si	Mn_3Si	823	2.40, 1.20	—
Fe_5Si_3	Mn_5Si_3 (hex.)	373	1.05, 1.55	2.4
Co_2MnSi	Cu_2MnAl (Heusler)	985	0.75(Co), 3.57(Mn)	

10.2.3 Ge magnetic compounds

In the Fe–Ge system, there are a number of phases around the composition Fe_3Ge: the ε phase (hexagonal) compound has a Curie point of 640 K and a saturation moment of $2.2M_B$ per Fe atom; the ε_1 phase (cubic) has a Curie point of about 760 K. In the vicinity of Fe_2Ge, there are the β phase (hexagonal) and the η phase, both of which are ferromagnetic, but the saturation moment decreases abruptly with increasing Ge content. The compounds FeNiGe, MnNiGe, and $Mn_{3.4}Ge$ are ferromagnetic below their Curie points, 770 K, 360 K, and 870 K, respectively.[26–29]

10.2.4 Sn magnetic compounds

Sn forms a number of compounds with 3*d* transition elements, similar to the Ge compounds. Thus the hexagonal compounds near Fe_3Sn are ferromagnetic below the Curie point of 743 K, and have magnetic moments $M_s = 1.90M_B$ and $M_{eff} = 2.27M_B$ per Fe atom. $Fe_{1.67}Sn$ is hexagonal (B 8_2 type) and ferromagnetic. Its Curie point and the saturation moment are reported by different authors to be 553 or 583 K and 2.10 or $1.81M_B$, respectively. Fe_3Sn_2 is a monoclinic compound, which is ferromagnetic below its Curie point of 612 K. $Mn_{1.45-2.0}Sn$ is ferromagnetic below a Curie point of about 260 K with saturation moment per formula unit of about $2M_B$. NiCoSn is ferromagnetic below a Curie point higher than 830 K.[28]

By replacing Al by Sn in the Heusler alloy Cu_2MnAl (see Section 10.1), we have ferromagnetic Heusler-like alloys. For example, Cu_2MnSn has a Curie point higher than 500 K and $M_s = 4.11M_B$ per Mn atom. Pd_2MnSn has a Curie point of 189 K and $M_s = 4.23M_B$ per Mn. Ni_2MnSn has a Curie point of 342 K and $M_s = 3.69M_B$ per Mn. Co_2MnSn has a Curie point of 811 K and $M_s = 4.79M_B$ per Mn atom.[30]

10.3 3*d*–Vb GROUP MAGNETIC COMPOUNDS

10.3.1 Magnetic nitrides

Ferromagnetic nitrides are listed in Table 10.5 with their magnetic properties. The nitrogen atom is small, so that it tends to go into the interstitial sites. Materials containing even fairly large amounts of nitrogen retain a metallic luster and have magnetic properties qualitatively the same as the nitrogen-free material. Most of the nitrides listed in Table 10.5 have the face-centered cubic structure of the Fe_4N type, as shown in Fig. 10.4. We see that one N atom occupies the body-centered site in face-centered cubic iron. As is well-known, fcc iron is antiferromagnetic, composed of low-spin-state iron atoms. It is interesting to note that the introduction of only one N atom in a unit cell changes the weak antiferromagnetism to ferromagnetism, with a Curie point as high at 761 K. A possible interpretation is that the N atom acts as an acceptor of electrons from Fe, thus creating electron vacancies in the Fe band.[31] Another possibility is that the expansion of the lattice caused by the introduction of N atoms may reduce the overlapping of 3*d* wave functions between Fe atoms, thus

Table 10.5. Crystal and magnetic properties of magnetic 3*d* nitrides.[31]

Compound	Lattice type	Magnetism	Curie point (K)	Average M_s (M_B)	Sublattice M_s (M_B)
Mn_4N	fcc	ferri	738	1.14 (mol)	3.5, −0.7 (0 K)
Fe_4N	fcc	ferro	761	9.02 (mol)	2.98, 201 (300 K)
Fe_8N	bct	ferro	ca. 573	2.6–2.8	
$Mn_4N_{.75}C_{.25}$	fcc	ferri	850	0.88 (mol)	3.52, −0.98
$Mn_4N_{.5}C_{.5}$	fcc	ferri	899	0.479	
$Fe_4N_{1-x}C_x$	fcc	ferro	743–865		
Fe_3NiN	fcc	ferro	1033	7.15	
Fe_3PtN	fcc	ferro	640	7.90	
$Fe_2N_{.78}$	hcp	ferro	398	2.15	1.5 (Fe)

suppressing antiferromagnetic interaction. In fact, the lattice constants of fcc Fe–Ni–Cr, Fe–Mn–C and Fe–Mn alloys, which exhibit low-spin antiferromagnetism, are about 3.6 Å, while that of Fe_4N is much larger, 3.8 Å. It was discovered by Kim and Takahashi[32] that metallic Fe films evaporated in a nitrogen atmosphere have much higher saturation magnetic moment than pure Fe: 2.6–2.8M_B as compared to 2.2M_B for pure iron. They ascribed this result to the appearance of the compound Fe_8N, but still there is a possibility that this effect is caused by a lattice distortion or expansion. For more details on magnetic nitrides a number of reviews[31–34] may be consulted.

10.3.2 Magnetic phosphides

There are three types of phosphides: MP, M_2P and M_3P. MnP has a screw spin structure below 50 K which has been the subject of many investigations.[35] Some of the M_2P type solid solutions containing different metals are ferromagnetic above room temperature. Thus $Fe_{1.2}Co_{0.8}P$ has the Curie point $\Theta_f = 450$ K and saturation moment $M_s = 2.0 M_B$ per formula unit; $Fe_{1.8}Ni_{0.2}P$ has $\Theta_f = 376$ K; CoMnP has $\Theta_f = 583$ K and $M_s \simeq 3.0 M_B$ per formula; FeMnP is ferrimagnetic below $\Theta_f = 320$ K. Many M_3P

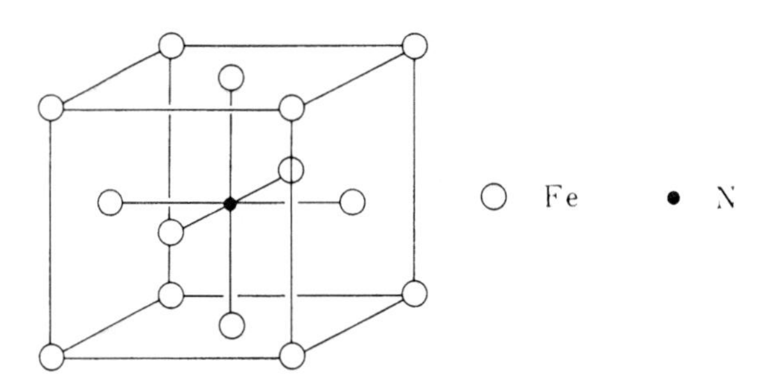

Fig. 10.4. Crystal structure of Fe_4N-type compounds.

Table 10.6. Magnetic properties of M_3P-type compounds.[35]

Compound	Curie point Θ_f (K)	σ (emu g^{-1}) $\times 4\pi \times 10^{-7}$ Wb m kg^{-1}	M_s (M_B)
$Fe_{2.4}Mn_{.6}P$	680	34	—
Fe_3P	716	155.4	1.84
$Fe_{2.25}Ni_{.75}P$	525	120	—
$Fe_{1.5}Ni_{1.5}P$	320	79	—
$Fe_{2.4}Co_{.6}P$	670	145	—

type compounds are ferromagnetic, as listed in Table 10.6. The saturation moment of M in M_3P is plotted in Fig. 10.5 as a function of the number of electrons in the M atom. This curve is quite similar to the Slater–Pauling curve for $3d$ alloys (see Fig. 8.12), but slightly shifted to the left. This may be due to a transfer of some electrons from the P atoms to the matrix. This situation is similar to the case of borides (see Fig. 10.2) and also to the amorphous alloys containing B and P (see Chapter 11).

For further details on magnetic phosphides, the reviews by Watanabe and Shinohara[34] and Hirahara[35] should be consulted.

10.3.3 Magnetic arsenides

There are many magnetic compounds which have the NiAs-type crystal structure shown in Fig. 10.6. MnAs exhibits an interesting magnetic behavior: on cooling, it

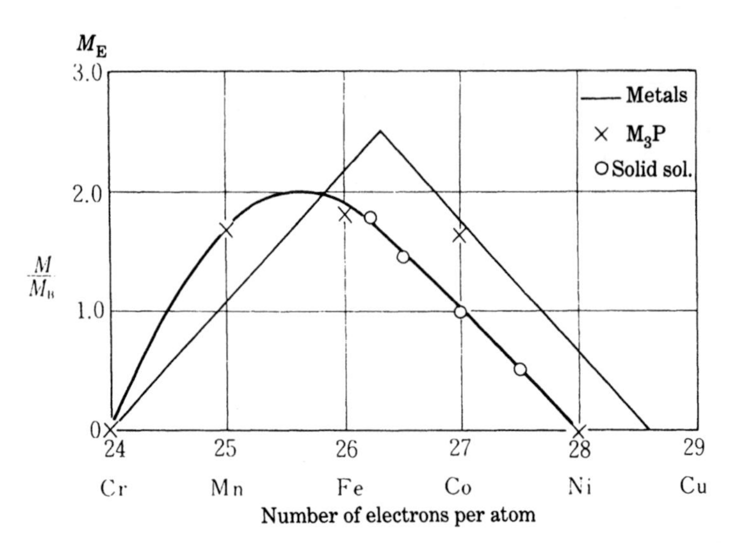

Fig. 10.5. Saturation moment of M in M_3P as a function of the number of electrons in the metal atom M, and comparison with the Slater–Pauling curve.[35]

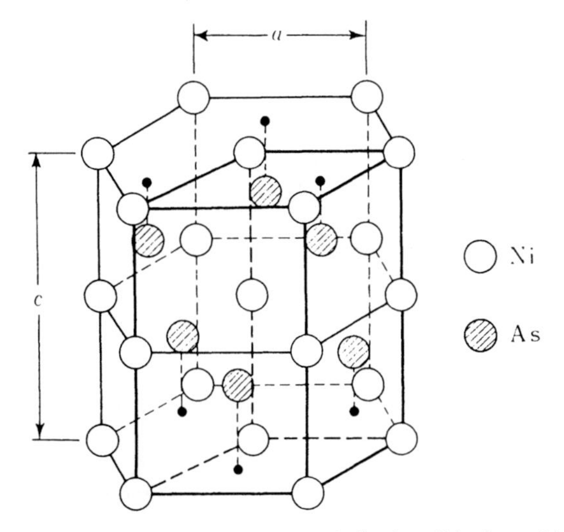

Fig. 10.6. Crystal structure of NiAs-type compounds (unit cell indicated by heavy lines).

undergoes a first-order magnetic transition at 313 K from paramagnetism to ferromagnetism, with saturation magnetization 1.76×10^{-4} Wb m kg^{-1} (= 140 emu g^{-1}) at 0 K. This phenomenon was explained in terms of the *exchange inversion*, which is caused by a combination of a distance-dependent exchange interaction and a spontaneous lattice deformation.[37–8]

10.3.4 Magnetic antimonides

Magnetic antimonides are summarized in Table 10.7. In Cr-rich MnSb–CrSb solid solutions, ferrimagnetism is realized at low temperatures and antiferromagnetism appears at high temperatures (Fig. 10.7).

10.3.5 Magnetic bismuthides

MnBi has the NiAs-type crystal structure, which is stable below 633 K and exhibits

Table 10.7. Crystal and magnetic properties of magnetic antimonides.[38–41]

Compound	Crystal type	Curie point Θ_f (K)	Magnetic moments (M_B)	
			M_s at 0 K	M_{eff}
MnSb	NiAs	586	3.5	3.23
$MnAs_{0.5}Sb_{0.5}$	NiAs	313		2.0
CoMnSb	CaF_2	490	2.0	4.0/formula
NiMnSb	CaF_2	750	1.9	
PdMnSb	CaF_2	500	1.9	
$Ni_{1.6}MnSb$	CaF_2	470	1.6	
$Ni_{2.0}MnSb$	Cu_2MnAl	410	1.2	
NiCoSb	Ni_2In	830	0.6	

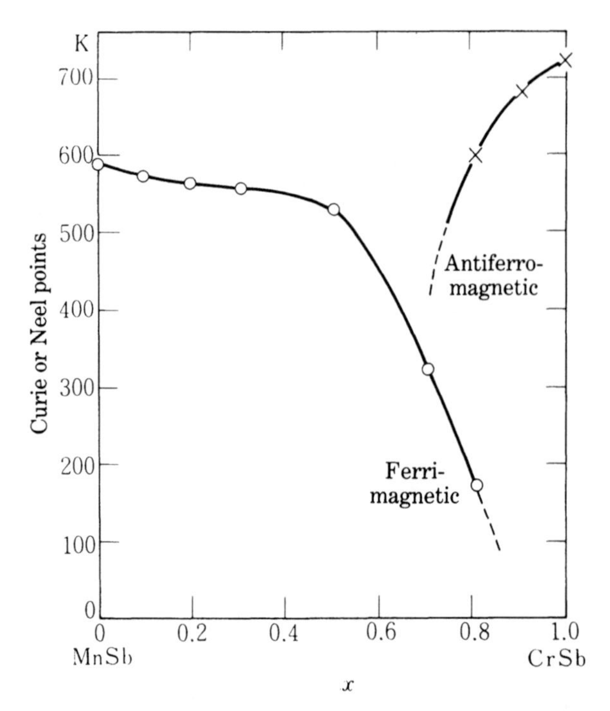

Fig. 10.7. Magnetic transition points as a function of composition x in $(MnSb)_{1-x}(CrSb)_x$.[42]

ferromagnetism with a saturation moment of $3.95M_B$ per Mn atom at 0 K. The Curie point determined by extrapolation is 750 K.[43] Magnetic writing of patterns was first achieved on evaporated thin films of MnBi.[44]

10.4 3*d*–VIb GROUP MAGNETIC COMPOUNDS

10.4.1 Magnetic sulfides

The iron sulfide FeS has the NiAs-type crystal structure (see Fig. 10.6) and its stoichiometric compound has antiferromagnetic spin ordering. However, pyrrhotite, which is a natural mineral, has the composition $Fe_{1-x}S$ ($x = 0$–0.2) with the vacancies located in alternate *c*-planes, thus producing a ferrimagnetic spin arrangement. The compound with $x = 0.125$, or Fe_7S_8, has a Curie point $\Theta_f = 578$ K, an effective moment $M_{eff} = 5.93M_B$, and an average saturation moment per Fe atom of $0.28M_B$. The compound with $x \simeq 0.1$ is also ferrimagnetic, but its saturation magnetization disappears once again on cooling to 483 K. For further information on this topic the review article[45] should be consulted.

Similar magnetic behavior is observed also for the magnetic sulfides $Cr_{1-x}S$. Figure 10.8 shows the temperature dependence of saturation magnetization for $CrS_{1.17}$, from which we see that the saturation magnetization disappears abruptly on cooling to 150 K.[46] A detailed description is given in the review.[47]

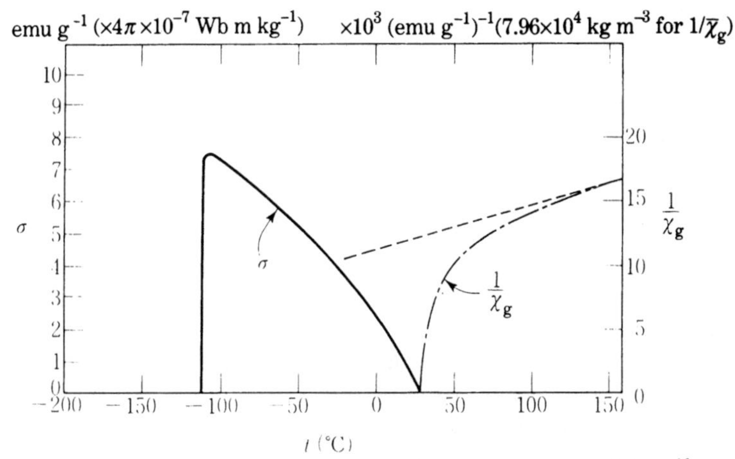

Fig. 10.8. Temperature dependence of magnetization of $CrS_{1.17}$.[46]

The MS_2 compounds are known as pyrites, and their magnetic properties have been extensively investigated. The crystal structure of this compound is shown in Fig. 10.9. As M changes from Fe to Co, Ni and Cu in increasing order of electron number, the electrical properties change: FeS_2 is metallic but the resistivity is fairly high; CoS_2 is more metallic and its resistivity is as low as that of Nichrome, which is used for heating elements; NiS_2 is semiconducting and highly resistive; and CuS_2 is again metallic. The magnetic properties also change: FeS_2 is paramagnetic but its solid solution with CoS_2 becomes ferromagnetic. The solid solution with NiS_2 exhibits antiferromagnetism. This situation tells us that spin correlations in these compounds affect the electronic structure pronouncedly. Several reviews[48–51] are available on this topic.

Similar to O, S forms the spinel-type compounds $A^{2+}B_2^{3+}S_4$. These compounds belong to the *chalcogenide spinels*, together with the compounds with Se or Te which will be treated later. Some of the sulfide spinels are ferrimagnetic above room

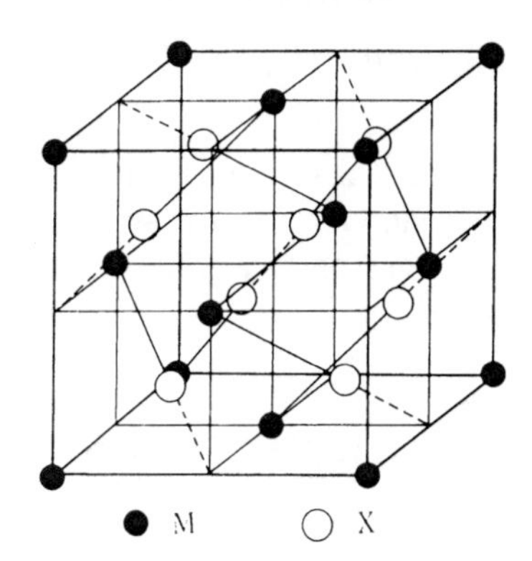

Fig. 10.9. Crystal structure of pyrite-type compounds MX_2.[48]

temperature. $CuCr_2S_4$ ($\Theta_f = 420\,K$)[52,53] is a normal spinel in which Cu^{2-} is on the A-site while Cr^{3+} is on the B site.[54,55] The saturation moment per formula unit is $5M_B$, which may be explained as a difference between the moment of $2Cr^{3+}$ or $3M_B \times 2 = 6M_B$, and that of Cu^{2+} or $1M_B$. Neutron diffraction, however, provided no evidence for such localized magnetic moments. Since this material has a high electrical conductivity, its magnetic properties may be interpreted on the basis of band magnetism. The mineral Fe_3S_4 is called *greigite*, which corresponds to magnetite Fe_3O_4. In contrast to Fe_3O_4, Fe_3S_4 remains cubic down to low temperatures without any phase transition, so that it retains metallic conductivity at all temperatures.[56] One of the reasons is that the ionic radius of S^{2-} is 1.74 Å, which is significantly larger than that of O^{2-}, 1.32 Å. Therefore the interstitial sites are large enough to accept metallic ions without any distortion of the lattice. In fact, the u-parameter of Fe_3S_4 is 0.375, which is the ideal value.[57] Greigite is ferrimagnetic below the Curie point $\Theta_f = 606\,K$, and its saturation magnetization is $2.2M_B$ per mol formula unit.[58] This material is also piezoelectric.[59]

The bonding in chalcogenide spinels has more covalent nature than in oxide spinels, so that most of the chalcogenide spinels are semiconducting. They have a fairly large magnetoresistance effect (see Section 21.2). For more details, a number of reviews[60-62] should be consulted.

10.4.2 Magnetic selenides

There are many magnetic selenides, which are similar to the magnetic sulfides: Fe_7Se_8 is ferrimagnetic below the Curie point $\Theta_f = 449\,K$ and has average saturation moment per Fe atom of $0.32M_B$; Fe_3Se_4 is ferrimagnetic below 314 K; $CuCr_2Se_4$ ($\Theta_f = 460\,K$), $CdCr_2Se_4$ ($\Theta_f = 129.5\,K$) and $HgCr_2Se_4$ ($\Theta_f = 120\,K$) are chalcogenide spinels. The latter two are ferromagnets with parallel alignment of Cr^{3+} spins ($3M_B$), so that they have saturation moment $6M_B$ per mol. This parallel alignment is caused by a positive exchange interaction through the path Cr–Se–Cr, making an angle of 90° (see Fig. 9.5), as explained in Section 7.1. They are representative magnetic semiconductors and their resistivity changes several orders of magnitude at their Curie points. The reason for this behavior has not yet been clarified.[60-62]

10.4.3 Magnetic tellurides

CrTe is ferromagnetic below the Curie point 333 K, with $M_{eff} = 5.58M_B$. The tellurides Cr_7Te_8, Cr_5Te_6, Cr_3Te_4, and Cr_2Te_3 are all ferromagnetic. The chalcogenide spinels $TiCr_2Te_4$ and $CuCr_2Te_4$ are both ferromagnetic below their Curie points of 214 K and 365 K, respectively.[60-62]

10.5 3d–VIIb (HALOGEN) GROUP MAGNETIC COMPOUNDS

Most of the halides with 3d transition elements are antiferromagnetic. There are ferromagnetic halides such as $RbNiF_3$, K_2CuF_4, $CrBr_3$, CrI_2 and $TiNiF_3$, but their

Curie points are all below room temperature. Therefore they have no engineering applications. However, many magnetic halides exhibit one- or two-dimensional spin ordering, which has attracted much academic interest. Concerning this topic, Hirakawa[63] and Urya and Hirakawa[64] should be consulted.

A part of the chalcogen X in a chalcogenide spinel $CuCr_2X_4$ (X: S, Se) can be replaced by a halogen Y. The resultant $CuCr_2X_{4-x}Y_x$ is called a *chalcogenide halide spinel*, in which part of the Cu^{2+} is changed to Cu^{1+}, thus reducing vacancies in the conduction band and also decreasing resistivity. In the limit of $x \rightarrow 1$, resistivity increases abruptly, and at the same time the Curie point decreases abruptly. This fact indicates that vacancies in the $3d$ band, which contribute to conductivity, contribute also to the exchange interaction.[61]

10.6 RARE EARTH COMPOUNDS

All Laves compounds RAl_2 (R: rare earths) are ferromagnetic, except for those containing only nonmagnetic R = La, Yb, and Lu. The maximum Curie point 175 K is realized at R = Gd. The saturation moments are slightly smaller than Hund's values.[65]

Generally speaking, many ferromagnetic rare earth compounds have Curie points below room temperature. Exceptional cases are the Th_3P-type compounds Gd_4Bi_3 and Gd_4Sb_3, which have Curie points of 340 K and 260 K, respectively.[66]

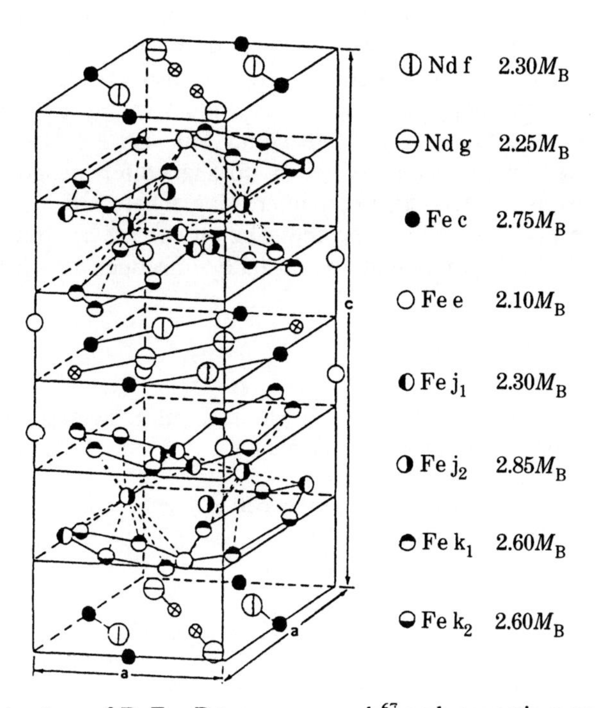

Fig. 10.10. Crystal structure of $R_2Fe_{14}B$-type compounds[67] and magnetic moments of individual atoms as determined by neutron diffraction.[69]

Table 10.8. Crystal and magnetic properties of $R_2Fe_{14}B$ compounds.[68]

Compound	Lattice constant a (Å)	Lattice constant c (Å)	Density (g cm^{-3})	I_s (T)	M_s (M_B mol)	H_a (MA m^{-1})	Θ_f (K)
$Ce_2Fe_{14}B$	8.77	12.11	7.81	1.16	22.7	3.7	424
$Pr_2Fe_{14}B$	8.82	12.25	7.47	1.43	29.3	10	564
$Nd_2Fe_{14}B$	8.82	12.24	7.55	1.57	32.1	12	585
$Sm_2Fe_{14}B$	8.80	12.15	7.73	1.33	26.7	basal	612
$Gd_2Fe_{14}B$	8.79	12.09	7.85	0.86	17.3	6.1	661
$Tb_2Fe_{14}B$	8.77	12.05	7.93	0.64	12.7	28	639
$Dy_2Fe_{14}B$	8.75	12.00	8.02	0.65	12.8	25	602
$Ho_2Fe_{14}B$	8.75	11.99	8.05	0.86	17.0	20	576
$Er_2Fe_{14}B$	8.74	11.96	8.24	0.93	18.1	basal	554
$Tm_2Fe_{14}B$	8.74	11.95	8.13	1.09	21.6	basal	541
$Y_2Fe_{14}B$	8.77	12.04	6.98	1.28	25.3	3.1	565

Other exceptional cases are the $R_2Fe_{14}B$-type compounds. The compound with R = Nd is the main constituent of the Nd–Fe–B permanent magnet (see Section 22.2.2). The crystal structure is tetragonal ($P4_2/mnm$ space group)[67] as shown in Fig. 10.10. The crystal and magnetic properties of $R_2Fe_{14}B$-type compounds are listed in Table 10.8.[68] From this table we see that the Curie points of these compounds are all fairly high and the saturation magnetizations are much higher than ferrites. Some of the anisotropy fields, H_a (see Section 12.2), are also extremely large. Figure 10.10 shows also the saturation moments of individual atoms determined by neutron diffraction.[69] We see that not only Fe atoms but also Nd atoms have fairly large saturation magnetic moments, all of which are aligned ferromagnetically. The atomic arrangement in this lattice is also very interesting. There are σ-phase-like Fe groups composed of distorted hexagons, and these Fe groups are separated by B–Nd layers. The Curie point of this compound was raised by introducing such B atoms.[68]

PROBLEMS

10.1 Describe the relationship between the electronegativity of electronegative ions and the electromagnetic properties of magnetic compounds.

10.2 Describe the relationship between the period number and the crystal and magnetic structures of magnetic compounds.

REFERENCES

1. N. Mori and T. Mitsui, *J. Phys. Soc. Japan*, **25** (1968), 82.
2. N. Lundqvist and A. J. P. Myers, *Ark. Fys.*, **20** (1961), 463.
3. M. C. Cadeville and A. J. P. Myers, *Compt. Rend.* **255** (1962), 3391.

4. H. Watanabe, *Handb. Mag. Subst.* (ed. by S. Chikazumi *et al.*, Asakura Publishing Co., Tokyo, 1975) p. 520, Tables 8.31, 8.33, Fig. 8.28.
5. H. Abe, H. Yasuoka, and A. Hirai, *J. Phys. Soc. Japan*, **21** (1966), 77.
6. T. Shinohara and H. Watanabe, *Sci. Rept. RITU*, **A18**, Suppl. (1966), 385.
7. R. Nathans, M. T. Pigott, and C. G. Shull, *J. Phys. Chem. Solids*, **6** (1958), 38.
8. H. Sato and A. Arrott, *J. Appl. Phys.*, **29** (1958), 515.
9. A. Arrott and H. Sato, *Phys. Rev.*, **114** (1959), 1420.
10. H. Sato and A. Arrott, *Phys. Rev.*, **114** (1959), 1427.
11. H. Kono, *J. Phys. Soc. Japan*, **13** (1958), 1444; **17** (1962), 1092.
12. O. Heusler, *Ann. Phys.*, **19** (1934), 155.
13. G. P. Felcher, G. W. Cable, and M. K. Wilkinson, *J. Phys. Chem. Solids*, **24** (1963), 1663.
14. I. Tsuboya and M. Sugihara, *J. Phys. Soc. Japan*, **16** (1961), 571; **16** (1961), 1875; **15** (1960), 1534; **16** (1961), 1257; **17S B-I** (1962), 172; **17** (1962), 410.
15. H. Katsuraki, H. Takada, and K. Suzuki, *J. Phys. Soc. Japan*, **18** (1963), 93.
16. I. Tsuboya and M. Sugihara, *J. Phys. Soc. Japan*, **18** (1963), 143.
17. I. Tsuboya and M. Sugihara, *J. Phys. Soc. Japan*, **20** (1965), 170.
18. I. Tsuboya and M. Sugihara, *J. Phys. Soc. Japan*. **18** (1963), 1096.
19. H. Hasegawa and I. Tsuboya, *Rev. El. Chem. Lab.*, **16** (1968), 605.
20. K. Aoyagi and M. Sugihara, *J. Phys. Soc. Japan*, **17** (1962), 1072.
21. B. R. Coles, W. Hume-Rothery, and H. P. Myers, *Proc. Roy. Soc.*, **A196** (1949), 125.
22. R. S. Tebble and D. J. Craik, *Mag. Materials* (Wiley, New York, 1969), p. 152.
23. T. Moriya, H. Ito, F. E. Fujita, and Y. Maeda, *J. Phys. Soc. Japan*, **24** (1968), 60.
24. H. Bernas, I. A. Campbell, and R. Fruchart, *J. Phys. Chem. Solids*, **28** (1967), 17.
25. H. Yasuoka, *Kotaibutsuri* (Solid State Phys.) (in Japanese), **12** (1977), 664.
26. K. Kanematsu and T. Ohoyama, *J. Phys. Soc. Japan*, **20** (1965), 236.
27. T. Ohoyama and K. Kanematsu, *Handbook on magnetic substances* (ed. by S. Chikazumi *et al.*, Asakura Publishing Co., Tokyo, 1975) 8.4.4, p. 543.
28. M. Asanuma, *Kinzokubutsuri* (Metal Phys.) (in Japanese) **7** (1961), 3.
29. T. Ohoyama, K. Yasukochi, and K. Kanematsu, *J. Phys. Soc. Japan*, **16** (1961), 352.
30. Y . Nakagawa, *Handbook on magnetic substances* (ed. by S. Chikazumi *et al.*, Asakura Publishing Co., Tokyo, 1975) 7.6.1, p. 371.
31. M. Mekata, *Handbook on magnetic substances* (ed. by S. Chikazumi *et al.*, Asakura Publishing Co., Tokyo, 1975) 8.4.3, p. 540, Table 8.40.
32. T. K. Kim and M. Takahashi, *Appl. Phys. Lett.*, **20** (1972), 492; M. Takahashi, Kotaibutsuri (Solid State Phys.) (in Japanese) **7** (1972), 483.
33. G. W. Wiener and J. A. Berger, *J. Metals*, **Feb.** (1955), 1.
34. H. Watanabe and T. Shinohara, *Nihon Kinzoku Gakkaishi* (J. Japan Metal Ass.) (in Japanese) **7** (1968), 433.
35. E. Hirahara, *Handbook on magnetic substances* (ed. by S. Chikazumi *et al.*, Asakura Publishing Co., Tokyo, 1975) 8.4.2, p. 525.
36. S. Nagase, H. Watanabe, and T. Shinohara, *J. Phys. Soc. Japan*, **34** (1973), 908.
37. C. Kittel, *Phys. Rev.*, **120** (1960), 335.
38. C. P. Bean and D. S. Rodbell, *Phys. Rev.*, **126** (1962), 104.
39. K. Sato and K. Adachi, *Handbook on magnetic substances* (ed. by S. Chikazumi *et al.*, Asakura Publishing Co., Tokyo, 1975) 8.4.5, p. 549.
40. K. Sato and K. Adachi, Tokyo, *Nihon Kinzoku Gakkaiho* (*Rept. Japan Metal Ass.*), **11** (1972), 447.
41. K. Endo, Y. Fujita, R. Kimura, T. Ohoyama, and M. Terada, *J. de Phys., Colloq. CI Suppl.*, **32** (1971), 1.

42. T. Hirone, S. Maeda, I. Tsubokawa, and N. Tsuya, *J. Phys. Soc. Japan*, **11** (1956), 1083.
43. B. W. Roberts, *Phys. Rev.*, **104** (1956), 607.
44. H. J. Williams, R. C. Sherwood, F. G. Foster, and E. M. Kelley, *J. Appl. Phys.*, **28** (1957), 1181.
45. N. Tsuya, I. Tsubokawa, and M. Yuzuri, *Metal Phys.* (in Japanese) **4** (1958), 140.
46. M. Yuzuri, T. Hirone, H. Watanabe, S. Nagasaki, and S. Maeda, *J. Phys. Soc. Japan*, **12** (1957), 385.
47. M. Yuzuri, *Solid State Phys.* (in Japanese), **11** (1976), 539.
48. K. Adachi, *Buturi* (in Japanese), **24** (1969), 518.
49. K. Adachi, *Solid State Phys.* (in Japanese), **10** (1975), 3, 101.
50. S. Ogawa, *Buturi* (in Japanese), **29** (1974), 688.
51. S. Ogawa, *Solid State Phys.*, **12** (1977), 657.
52. H. Hahn, C. de Lorent, and B. Harder, *Z. anorg. Chem.*, **283** (1956), 138.
53. F. K. Lotgering, *Proc. Int. Conf. Mag.* (Nottingham, 1964), 533.
54. C. Colominus, *Phys. Rev.*, **153** (1967), 558.
55. M. Robbins, H. W. Lehmann, and J. G. White, *J. Phys. Chem. Solids*, **28** (1967), 897.
56. H. Nozaki, *J. Appl. Phys.*, **51** (1980), 486.
57. M. Uda, *Sci. Papers. Inst. Phys. Chem. Res.*, **62** (1968), 14.
58. M. R. Spender, J. M. D. Coey, and A. H. Morrish, *Can. J. Phys.*, **50** (1972), 2313.
59. S. Yamagushi, *Buturi* (in Japanese), **28** (1973), 42.
60. P. J. Wojtowicz, *IEEE Trans. Mag.*, **MAG-5** (1969), 840.
61. K. Miyatani, *Solid State Phys.* (in Japanese), **5** (1970), 11, 251.
62. K. Miyatani, *Handbook on magnetic substances* (ed. by S. Chikazumi *et al.*, Asakura Publishing Co., Tokyo, 1975) 8.4.6, p. 557.
63. K. Hirakawa, *Handbook on magnetic substances* (in Japanese) (ed. by S. Chikazumi *et al.*, Asakura Publishing Co., Tokyo, 1975) 10, p. 707.
64. N. Urya and K. Hirakawa, *Buturi* (in Japanese), **25** (1970), 441.
65. K. Sekizawa, *Handbook on magnetic substances* (in Japanese) (ed. by S. Chikazumi *et al.*, Asakura Publishing Co., Tokyo, 1975) 8.2.2, p. 513.
66. F. Holtzberg, T. R. McGuire, S. Methfessel, and J. C. Suits, *J. Appl. Phys.*, **35** (1964), 1033.
67. J. F. Herbst, J. J. Croat, F. E. Pinkerton, and W. B. Yelon, *Phys. Rev.*, **B29** (1984), 4176.
68. M. Sagawa, S. Fujimura, H. Yamamoto, Y. Matsuura, and K. Hiraga, *IEEE Trans. Mag.*, **MAG-20** (1984), 1584.
69. D. Givord, H. S. Li, and F. Tasset, *J. Appl. Phys.*, **57** (1984) 4100.

11

MAGNETISM OF AMORPHOUS MATERIALS

In Chapters 8–10, we treated metals, oxides and compounds, all in crystalline form with atoms arranged on regular lattices. There is another group of materials, *amorphous materials*, in which the atoms are distributed in an irregular manner. Ordinary glass is a representative example of an amorphous material. When a beam of monochromatic X-rays is scattered by such an amorphous material, there is no well-defined diffraction pattern, as in the case for crystalline materials. Instead there are only diffuse halos, from which we can deduce the statistical distribution of atoms.

The magnetism of amorphous materials is interesting at least concerning the following two points: First, how such a random arrangement of atoms affects the magnetic properties. As different crystal structures affect magnetic behavior differently, an amorphous form of matter might produce a special type of magnetism. Second, amorphous materials are mesoscopically homogeneous. The magnetic properties of materials, especially technical magnetization processes, are quite structure-sensitive. For instance, the presence of grain boundaries in polycrystalline materials sometimes interferes with domain wall motion. One possible way to avoid this is to eliminate grain boundaries by using a single crystal. However, the preparation of a perfect single crystal without any imperfections requires extremely high technology. It is very interesting to note that completely perfect single crystals and completely random amorphous materials both provide us with mesoscopically homogeneous magnetic media. This fact is very important for engineering applications of amorphous materials. Some amorphous materials have various other useful features: they are mechanically strong, isotropic (no directional properties), and may be produced by relatively simple manufacturing processes.

There are several ways for preparing amorphous materials: evaporation onto a cold substrate, electroplating, electroless plating, rapid quenching, sputtering, etc. However, almost all results on transition-metal amorphous alloys have been obtained from samples made by rapid solidification. In order to stabilize $3d$ metal-base amorphous materials, it is usually necessary to add 10–20% so-called metalloid elements, such as B, C, N, Si, and P. Generally speaking, the amorphous state thus obtained is changed to a crystalline state by heating above a temperature called the *crystallization temperature*. Even below this temperature there are annealing effects, by which various properties are changed substantially even though the amorphous structure is retained. These changes are caused by two related but separable processes: diffusion and structural relaxation.[1,2] A similar phenomenon[3] was also reported in rare earth-base sputtered amorphous materials.

We shall discuss magnetism of 3*d* metal-base amorphous materials in (11.1), and that of rare earth-base amorphous materials in (11.2).

11.1 MAGNETISM OF 3*d* TRANSITION METAL-BASE AMORPHOUS MATERIALS

Pioneering work on this subject was carried out by Mizoguchi *et al.*[4] for a series $(Fe_{1-x}M_x)_{0.8}B_{0.1}P_{0.1}$, where M = Ni, Co, Mn, Cr, and V. These raw materials were mixed, melted in a plasma-jet furnace, and then quenched from the melt by rapid compression between cold copper plates.[5] Figure 11.1 shows the average saturation magnetic moment of the magnetic atoms of these amorphous alloys as a function of average number of electrons of transition elements. As seen from this graph, the Fe–Ni and Fe–Co systems follow an almost straight line parallel to the right-hand line of the Slater–Pauling curve (see Fig. 8.12), shifted to the left by 0.4 electrons per atom. This shift can be accounted for by considering that B and P act as electron donors. This is analogous to the case of borides (see Fig. 10.2) and Mn_3P (see Fig. 10.5). The steep decreases of the saturation moment caused by the addition of Mn, Cr, and V to Fe are interpreted by these authors as resulting from the fact that Mn, Cr, and V atoms have moments of 3, 4, and $5M_B$, respectively, and their magnetic moments are coupled antiferromagnetically with the Fe moment.

Figure 11.2 shows the Curie point of $Fe_{1-x}M_x$ amorphous alloys as a function of average number of electrons. The behavior of the Curie point is quite different from

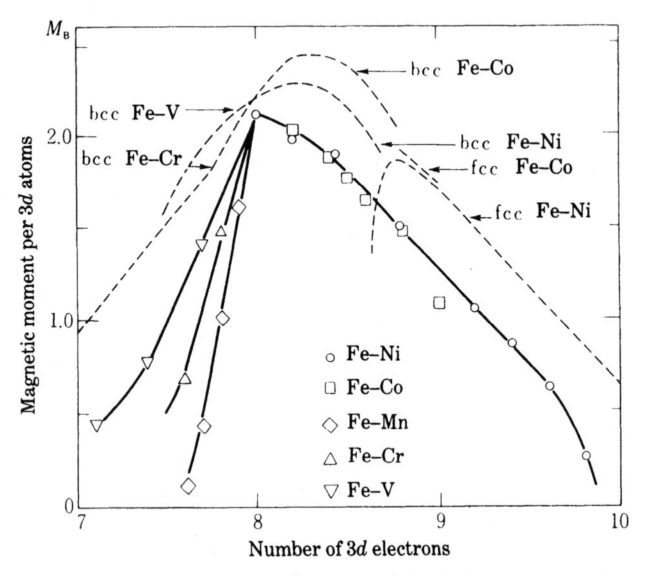

Fig. 11.1. Saturation magnetic moment per $(Fe_{1-x}M_x)$ in Bohr magneton for amorphous alloys $(Fe_{1-x}M_x)_{0.8}B_{0.1}P_{0.1}$ as a function of average number of electrons in $(Fe_{1-x}M_x)$.[4]

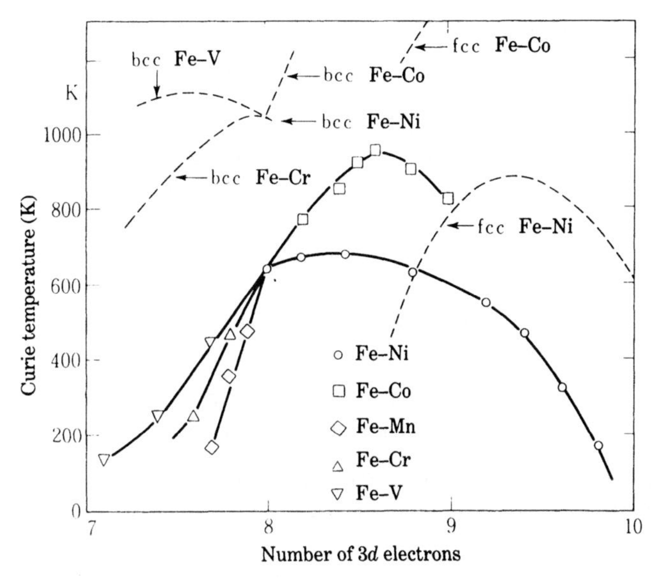

Fig. 11.2. Curie point, Θ_f, of amorphous alloys of $(Fe_{1-x}M_x)_{0.8}B_{0.1}P_{0.1}$ as a function of average numbers of electrons in $(Fe_{1-x}M_x)$.[4]

the crystalline alloys, which are plotted by dashed curves in the same figure with respect to the following points:

(1) There are no discontinuities as found in crystalline alloys at phase boundaries.
(2) The changes in Curie point by the addition of various elements are rather smooth functions of electron concentration for amorphous alloys, while these are quite different for different crystalline alloys.

It should be noted that the Curie points of amorphous alloys are subject to change by annealing below the crystallization temperature. Figure 11.3 shows the experimental results for the $Fe_{27}Ni_{53}P_{14}B_6$ system.[1] As seen in this graph, the Curie temperature is as low as 85°C in an as-quenched sample, while it increases as high as 112°C after annealing for 10 000 min at 200°C. There is a tendency that the effect of annealing is more pronounced by increasing the annealing time at low temperatures. This was suggested to be due to a compositional short-range ordering of different elements.

As mentioned above, the saturation moment of amorphous alloys is influenced by metalloid elements which act as electron donors to the matrix. It is interesting to see how the magnetic moment changes when the metalloid element content is reduced. Figure 11.4 shows the saturation magnetic moment of an Fe atom in Fe–Si amorphous alloys which were evaporated onto a substrate cooled to liquid H_2 temperature (20 K), as a function of Si content.[6,7] It is seen in this graph that the Fe moment drops sharply when the Si content is reduced below 0.6 at%. This behavior is quite similar to that of Invar alloys (see Section 8.2). The reason may be due to the appearance of

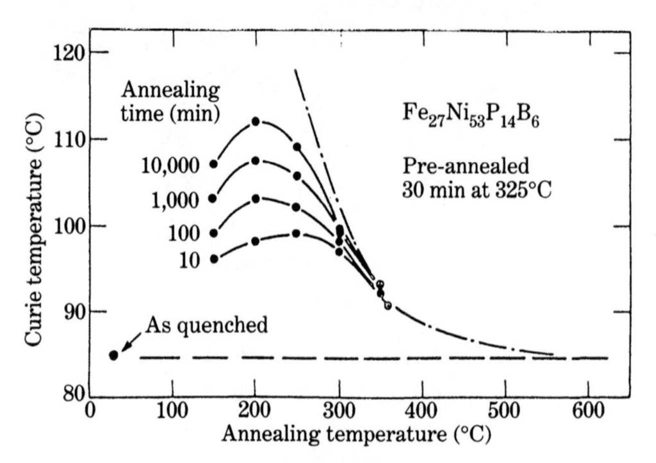

Fig. 11.3. Change in Curie temperature of amorphous $Fe_{27}Ni_{53}P_{14}B_6$ as a function of annealing temperature. The annealing times are indicated in the figure.[1]

the low-spin state of Fe atoms caused by an increase in the number of Fe–Fe pairs. The similarity between fcc Invar and amorphous alloys seems reasonable, because the number of nearest neighbors in amorphous alloys is 11–13,[8] which is close to the value of 12 in the fcc structure. Then a question arises why the magnetic moment of

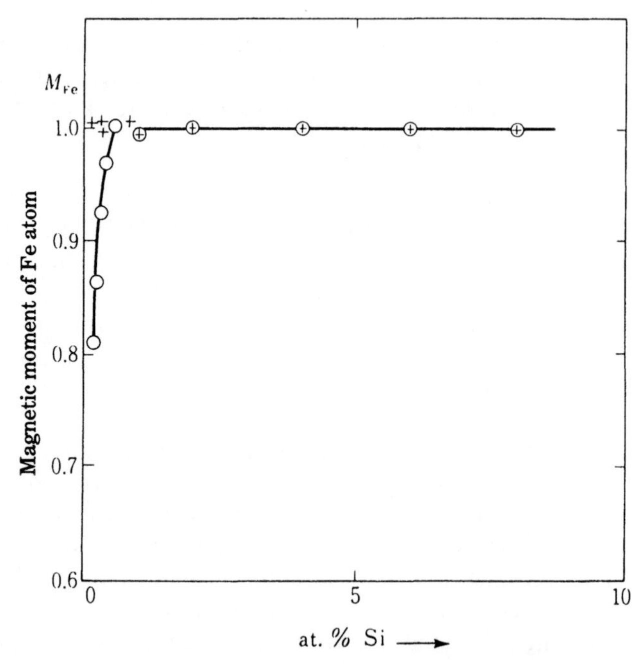

Fig. 11.4. Saturation magnetic moment, M, on Fe atoms in Fe–Si amorphous alloys evaporated on cold substrates, as a function of Si content. The ordinate is normalized to M_{Fe} $(=2.2M_B)$.[6]

Fe in amorphous alloys containing more than 0.6 at% Si is $2.2M_B$ (see Fig. 11.4), which is the same as that of bcc Fe where the number of nearest neighbors is 8. However, when Si is replaced by Au, the magnetic moment of Fe becomes $2.8M_B$, which is the same as that in fcc alloys.

If the appearance of low spin states in Invar alloys is the origin of their low thermal expansion coefficients, as discussed in Section 8.2, we would expect that the amorphous iron-base alloys would also exhibit some Invar anomaly. In fact, Fukamichi *et al.*[9] discovered that $Fe_{1-x}B_x$ ($x = 0.09$–0.21) amorphous alloys exhibit anomalous thermal expansion, and an expansion coefficient as low as that of Invar is found at $x = 0.17$. The temperature dependence of spontaneous magnetization of these amorphous alloys was found to be almost linear, similar to that of Invar.

Generally speaking, ferromagnetism of amorphous materials is not very different from that of crystalline materials. For instance, the magnetic critical phenomena were investigated in $Co_{0.7}B_{0.2}P_{0.1}$,[10] $Fe_{0.8}P_{0.13}C_{0.07}$,[11] etc., and found to be completely the same as those of the crystalline materials: The Curie point can be determined uniquely and the critical indices are quite normal (Section 6.2). The spin–wave dispersion relation determined by means of neutron diffraction[12] for amorphous materials of composition $(Fe_{0.93}Mo_{0.07})_{0.8}B_{0.1}P_{0.1}$ was also found to be quite normal. From these facts, we know that the exchange interaction acts over long distances even between atomic moments arranged in an irregular manner, thus forming the usual ferromagnetic spin arrangement. However, the internal fields of $Fe_{0.75}P_{0.15}C_{0.1}$ determined by means of the Mössbauer effect were found to be distributed over a fairly wide range. Judging from this and other magnetic data, it was concluded that the atomic magnetic moments and exchange interactions in these amorphous materials have fairly large fluctuations.[13]

For more details, refer to various reviews.[14,15]

11.2 MAGNETISM OF 3*d* TRANSITION PLUS RARE EARTH AMORPHOUS ALLOYS

The important feature of these amorphous systems is the appearance of antiferromagnetic or ferrimagnetic spin arrangements, in spite of the irregular atomic arrangement. Since the first report[16] of ferrimagnetic Gd–Co and Gd–Fe films sputtered onto glass plates, many investigations have been reported on this system. Figure 11.5 shows the temperature dependence of magnetization measured for $(Gd_{0.15}Co_{0.85})_{0.86}Mo_{0.14}$ in a magnetic field of $0.8\,MA\,m^{-1}$ (10 kOe). This is apparently the temperature dependence of magnetization for an N-type ferrimagnet (see Section 7.2). It is interesting that two kinds of atomic magnetic moments are aligned antiferromagnetically in spite of the irregular spatial arrangement. The dashed curves in the same figure represent the temperature dependence of the sub-magnetizations of Gd and Co deduced theoretically.

The situation is, however, not so simple. In R–Fe amorphous alloys, the atomic magnetic moment of Fe deduced from the internal field, measured by Mössbauer effect, depends considerably on the composition.[18] The exchange interaction in this system has been interpreted in terms of a 'mean field model'.

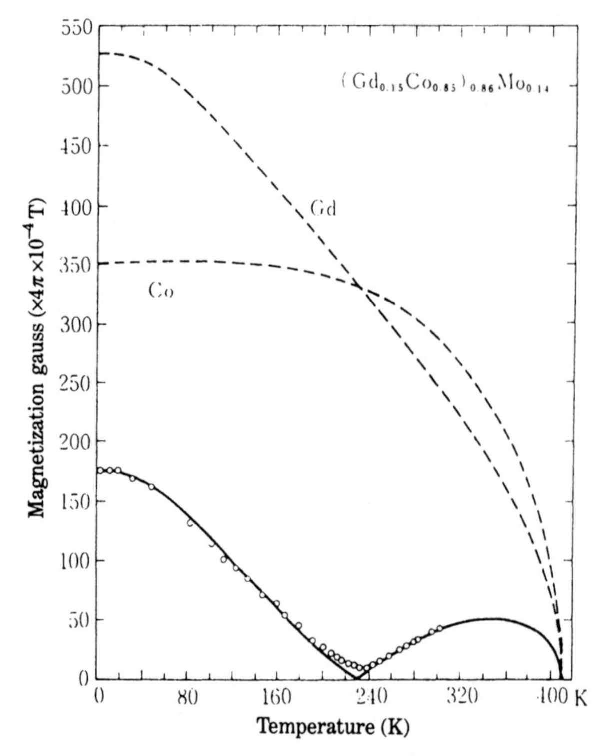

Fig. 11.5. Temperature dependence of saturation magnetization of the amorphous alloy $(Gd_{0.15}Co_{0.85})_{0.86}Mo_{0.14}$ (circles) and the submagnetizations of Gd and Co (calculated).[17]

It is interesting to compare the amorphous RM_2 alloy prepared by sputtering and the crystalline Laves RM_2 phase. For instance, in amorphous $GdCo_2$, the Co moment is $1.4 \pm 0.15 M_B$, which is much higher than the value of $1.02 M_B$ in Laves $GdCo_2$, but still lower than the value $1.7 M_B$ in Co metal. This situation has been explained in terms of a transfer of electrons from Gd to Co.[19] The number of electrons transferred is smaller in amorphous alloys than in the crystalline state, because the separation between Gd–Gd is larger in the former than the latter.[14] However, $GdFe_2$ has an Fe moment of $1.55 M_B$ in both the amorphous and crystalline states.[20] Such a decrease in local moment is suspected to be partly due to a disturbance of the parallel alignment of spins by local anisotropy.[21] It was observed by inelastic neutron diffraction that the density of magnetic states in amorphous $TbFe_2$ is considerably shifted towards lower energies compared to the crystalline state.[22]

PROBLEM

11.1 State the major differences in magnetism between amorphous and crystalline materials and discuss possible reasons.

REFERENCES

1. C. D. Graham, Jr. and T. Egami, *Ann. Rev. Mater. Sci.*, **8** (1978), 423.
2. T. Egami, *Rep. Prog. Phys.*, **47** (1984), 1601.
3. F. E. Luborsky, J. T. Furey, R. B. Skoda, and B. C. Wagner, *IEEE Trans. Mag.*, **MAG-21** (1985), 1618.
4. T. Mizoguchi, K. Yamauchi, and H. Miyajima, *Amorphous magnetism* (ed. by H. O. Hooper and A. M. de Graaf, Plenum Press, New York, 1973) p. 325.
5. K. Yamauchi and Y. Nakagawa, *Japanese J. Appl. Phys.*, **10** (1971), 1730.
6. W. Felsch, *Z. Phys.*, **219** (1969), 280.
7. W. Felsch, *Z. f. angew. Phys.*, **29** (1970), 218.
8. G. S. Cargill, *Solid State Phys.*, **30** (1975), 227.
9. K. Fukamichi, M. Kikuchi, S. Arakawa, and T. Masumoto, *Solid State Comm.*, **23** (1977), 955.
10. T. Mizoguchi, N. Ueda, K. Yamauchi, and H. Miyajima, *J. Phys. Soc. Japan*, **34** (1973), 1691.
11. K. Yamada, Y. Ishikawa, and Y. Endo, *Solid State Comm.*, **16** (1975), 1335.
12. J. D. Axe, G. Shirane, T. Mizoguchi, and Y. Yamauchi, *Phys. Rev.*, **B15** (1977), 2763.
13. C. C. Tsuei and H. Lilienthal, *Phys. Rev.*, **13B** (1976), 4899.
14. G. S. Cargill, *AIP Conf. Proc.*, **24** (1975), 138.
15. T. Mizoguchi, *AIP Conf. Proc.*, **34** (1976), 286.
16. P. Chaudhari, R. J. Gambino, and J. J. Cuomo, *Appl. Phys. Lett.*, **22** (1973), 337.
17. R. Hasegawa, B. E. Argyle, and L. T. Tao, *AIP Conf. Proc.*, **24** (1974), 110.
18. N. Heiman, K. Lee, and R. I. Potter, *AIP Conf. Proc.*, **29** (1975), 130.
19. L. J. Tao, R. J. Gambino, S. Kirkpatrick, J. J. Cuomo, and H. Lilienthal, *AIP Conf. Proc.*, **18** (1974), 641; *Solid State Comm.*, **13** (1973), 1491.
20. J. J. Rhyne, J. Schelling, and N. Koon, *Phys. Rev.*, **B10** (1974), 4672.
21. D. W. Forester, R. Abbundi, R. Segnan, and R. Sweger, *AIP Conf. Proc.*, **24** (1974), 115.
22. J. J. Rhyne, S. J. Pickart, and H. A. Alperin, *AIP Conf. Proc.*, **18** (1974), 563.

Part V

MAGNETIC ANISOTROPY AND MAGNETOSTRICTION

The exchange interaction between spins in ferro- or ferrimagnetic materials is the main origin of spontaneous magnetization. This interaction is essentially isotropic, so that the spontaneous magnetization can point in any direction in the crystal without changing the internal energy, if no additional interaction exists. However, in actual ferro- or ferrimagnetic materials, the spontaneous magnetization has an easy axis, or several easy axes, along which the magnetization prefers to lie. Rotation of the magnetization away from the easy axis is possible only by applying an external magnetic field. This phenomenon is called magnetic anisotropy.

Furthermore, the size or shape of a ferromagnet is more or less changed by magnetization; this phenomenon is called magnetostriction.

In this Part we discuss the physical origins of these phenomena, and consider some representative data.

12

MAGNETOCRYSTALLINE ANISOTROPY

12.1 PHENOMENOLOGY OF MAGNETOCRYSTALLINE ANISOTROPY

The term *magnetic anisotropy* is used to describe the dependence of the internal energy on the direction of spontaneous magnetization. We call an energy term of this kind a *magnetic anisotropy energy*. Generally the magnetic anisotropy energy term has the same symmetry as the crystal structure of the material, and we call it a *magnetocrystalline anisotropy*.

The simplest case is *uniaxial magnetic anisotropy*. For example, hexagonal cobalt exhibits uniaxial anisotropy with the stable direction of spontaneous magnetization, or easy axis, parallel to the c-axis of the crystal at room temperature. As the magnetization rotates away from the c-axis, the anisotropy energy initially increases with θ, the angle between the c-axis and the magnetization vector, then reaches a maximum value at $\theta = 90°$, and decreases to its original value at $\theta = 180°$. In other words, the anisotropy energy is minimum when the magnetization points in either the + or − direction along the c-axis. We can express this energy by expanding it in a series of powers of $\sin^2\theta$:

$$E_a = K_{u1}\sin^2\theta + K_{u2}\sin^4\theta + K_{u3}\sin^6\theta + K_{u4}\sin^6\theta\cos 6\varphi + \cdots, \qquad (12.1)$$

where φ is the azimuthal angle of the magnetization in the plane perpendicular to the c-axis. Using the relationships

$$\sin^2\theta = \tfrac{1}{2}(1-\cos 2\theta), \qquad \cos^2\theta = \tfrac{1}{2}(1+\cos 2\theta),$$

(12.1) is converted to a series in $\cos n\theta$ ($n = 2, 4, 6, \ldots$) as

$$\begin{aligned} E_a = {} & \tfrac{1}{2}K_{u1}(1-\cos 2\theta) + \tfrac{1}{8}K_{u2}(3 - 4\cos 2\theta + \cos 4\theta) \\ & + \tfrac{1}{32}K_{u3}(10 - 15\cos 2\theta + 6\cos 4\theta - \cos 6\theta) \\ & + \tfrac{1}{32}K_{u4}(10 - 15\cos 2\theta + 6\cos 4\theta - \cos 6\theta)\cos 6\varphi + \cdots. \end{aligned} \qquad (12.2)$$

The coefficients K_{un} ($n = 1, 2, \ldots$) in these equations are called *anisotropy constants*.

The values of uniaxial anisotropy constants of cobalt at 15°C[1] are

$$\left.\begin{aligned} K_{u1} &= 4.53 \times 10^5\ \mathrm{J\,m^{-3}}\ (= 4.53 \times 10^6\ \mathrm{erg\,cm^{-3}}) \\ K_{u2} &= 1.44 \times 10^5\ \mathrm{J\,m^{-3}}\ (= 1.44 \times 10^6\ \mathrm{erg\,cm^{-3}}) \end{aligned}\right\}. \qquad (12.3)$$

The higher-order terms are small and their values are not reliably known. If these

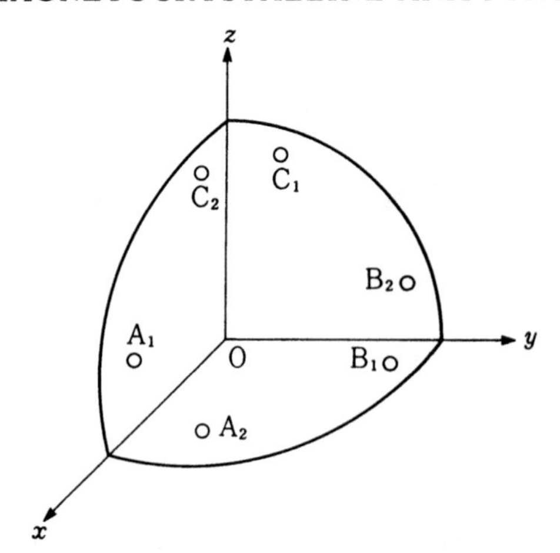

Fig. 12.1. Equivalent directions in a cubic crystal.

constants are positive as in the case of cobalt, the anisotropy energy, E_a, increases with increasing angle θ, so that E is minimum at $\theta = 0$. In other words, the spontaneous magnetization is stable when it is parallel to the c-axis. Such an axis is called an *axis of easy magnetization* or simply an *easy axis*. If these constants are negative, the anisotropy energy is maximum at the c-axis, so that it becomes unstable. Such an axis is called an *axis of hard magnetization* or simply a *hard axis*.

In the latter case, the magnetization is stable when it lies in any direction in the c-plane ($\theta = 90°$). Such a plane is called a *plane of easy magnetization* or an *easy plane*. If $K_{u1} > 0$ and $K_{u2} < 0$, the stable direction of magnetization forms a cone, which is called a *cone of easy magnetization* or an *easy cone*.

For cubic crystals such as iron and nickel, the anisotropy energy can be expressed in terms of the direction cosines ($\alpha_1, \alpha_2, \alpha_3$) of the magnetization vector with respect to the three cube edges. There are many equivalent directions in which the anisotropy energy has the same value, as shown by the points A_1, A_2, B_1, B_2, C_1, and C_2 on an octant of the unit sphere in Fig. 12.1. Because of the high symmetry of the cubic crystal, the anisotropy energy can be expressed in a fairly simple way: We expand the anisotropy energy in a polynomial series in α_1, α_2, and α_3. Those terms which include the odd powers of α_i must vanish, because a change in sign of any of the α_i should bring the magnetization vector to a direction which is equivalent to the original direction. The expression must also be invariant to the interchange of any two α_is so that the terms of the form $\alpha_i^{2l}\alpha_j^{2m}\alpha_k^{2n}$ must have, for any combination of l, m, n, the same coefficient for any interchange of i, j, k. The first term, therefore, should have the form $\alpha_1^2 + \alpha_2^2 + \alpha_3^2$, which is always equal to 1. Next is the fourth-order term which can be reduced to the form $\Sigma_{i>j}\, \alpha_i^2\alpha_j^2$ by the relationship

$$\alpha_1^4 + \alpha_2^4 + \alpha_3^4 = 1 - 2(\alpha_1^2\alpha_2^2 + \alpha_2^2\alpha_3^2 + \alpha_3^2\alpha_1^2). \tag{12.4}$$

Thus we have the expression

$$E_a = K_1(\alpha_1^2\alpha_2^2 + \alpha_2^2\alpha_3^2 + \alpha_3^2\alpha_1^2) + K_2\,\alpha_1^2\alpha_2^2\alpha_3^2 + K_3(\alpha_1^2\alpha_2^2 + \alpha_2^2\alpha_3^2 + \alpha_3^2\alpha_1^2)^2 + \cdots, \tag{12.5}$$

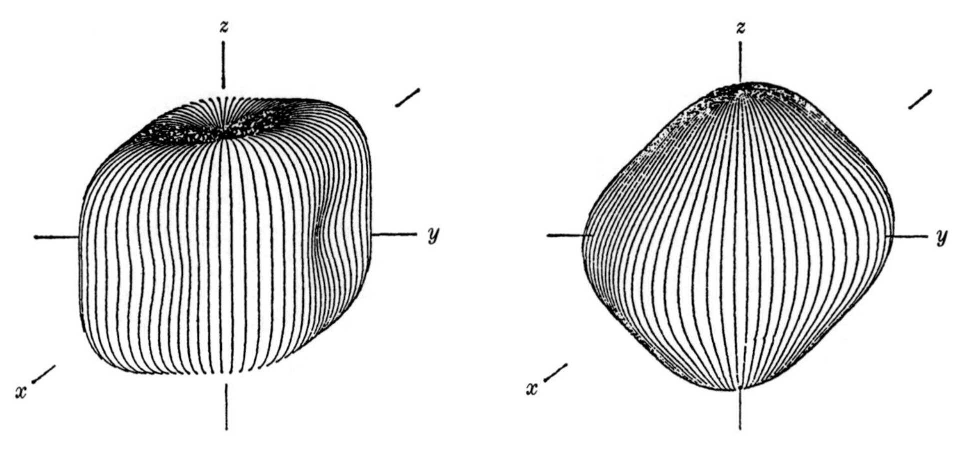

Fig. 12.2. Polar diagram of the cubic anisotropy energy for $K_1 > 0$ and $K_2 = 0$. (Radial vector is equal to $E_a + \frac{2}{3}K_1$.)

Fig. 12.3. Polar diagram of the cubic anisotropy energy for $K_1 < 0$ and $K_2 = 0$. (Radial vector is equal to $E_a + 2|K_1|$.)

where K_1, K_2, and K_3 are the cubic anisotropy constants. For iron at 20°C,[2]

$$\left.\begin{aligned} K_1 &= 4.72 \times 10^4\,\mathrm{J\,m^{-3}}\ (= 4.72 \times 10^5\,\mathrm{erg\,cm^{-3}}) \\ K_2 &= -0.075 \times 10^4\,\mathrm{J\,m^{-3}}\ (= -0.075 \times 10^5\,\mathrm{erg\,cm^{-3}}) \end{aligned}\right\}. \tag{12.6}$$

and, for nickel at 23°C,[3]

$$\left.\begin{aligned} K_1 &= -5.7 \times 10^3\,\mathrm{J\,m^{-3}}\ (= -5.7 \times 10^4\,\mathrm{erg\,cm^{-3}}) \\ K_2 &= -2.3 \times 10^3\,\mathrm{J\,m^{-3}}\ (= -2.3 \times 10^4\,\mathrm{erg\,cm^{-3}}) \end{aligned}\right\}. \tag{12.7}$$

For [100], $\alpha_1 = 1$, $\alpha_2 = \alpha_3 = 0$, so that the value of E_a given by (12.5) is

$$E_0 = 0, \tag{12.8}$$

and, for [111], $\alpha_1 = \alpha_2 = \alpha_3 = 1/\sqrt{3}$, so that

$$E_a = \tfrac{1}{3}K_1 + \tfrac{1}{27}K_2 + \tfrac{1}{9}K_3 + \cdots. \tag{12.9}$$

If $K_1 > 0$ as in the case of iron, and ignoring the K_2 and K_3 terms, E_a for [111] is higher than that for [100], so that [100] becomes the easy axis. Considering the cubic symmetry, [010] and [001] are also easy axes. Figure 12.2 shows a polar diagram of the anisotropy energy in this case. This diagram is a locus of the vector drawn from the origin in the direction of the spontaneous magnetization with length equal to the anisotropy energy given by (12.5) plus a constant term equal to $\frac{2}{3}K_1$. It is seen that the surface is concave, or in other words the energy is minimum, in the *x*-, *y*-, and *z*- (or [100], [010], and [001]) directions. If $K_1 < 0$ as in the case of nickel, $E_a < 0$ for [111] as we see in (12.9) (ignoring K_2 and K_3 terms), so that E_a for [111] is lower than that for [100] and [111] and its equivalent [$\bar{1}$11], [1$\bar{1}$1], and [11$\bar{1}$] directions are easy axes. Figure 12.3 shows a polar diagram of the anisotropy energy, plus a constant term $2|K_1|$, for $K_1 < 0$ and $K_2 = K_3 = 0$. It is seen that the surface is convex for cube axes

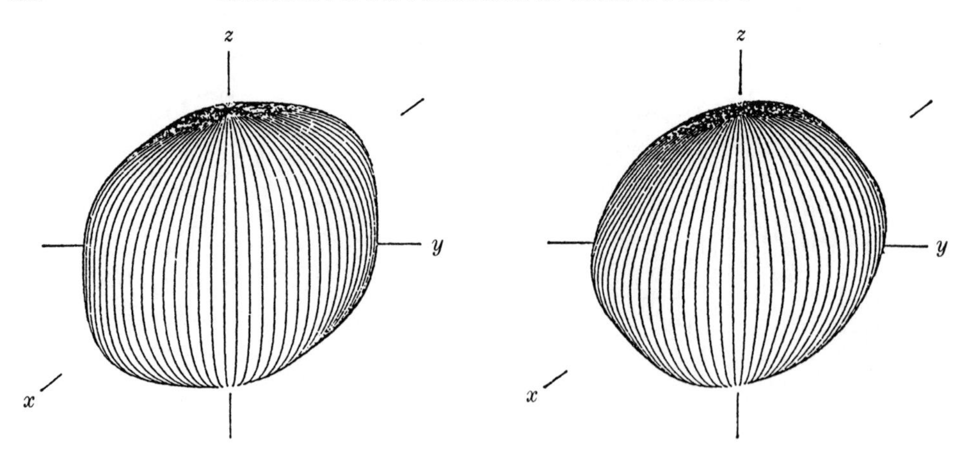

Fig. 12.4. Polar diagram of the cubic anisotropy energy for $K_1 = 0$ and $K_2 > 0$. (Radial vector is equal to $E_a + \frac{1}{2}K_2$.)

Fig. 12.5. Polar diagram of the cubic anisotropy energy for $K_1 = 0$ and $K_2 < 0$. (Radial vector is equal to $E_a + \frac{9}{16}K_2$.)

and concave for the $\langle 111 \rangle$ axes. Figures 12.4 and 12.5 show the anisotropy energy surfaces for $K_1 = K_3 = 0$ and $K_2 > 0$ and for $K_1 = K_3 = 0$, and $K_2 < 0$, respectively. In these cases, the energy is equal to zero for all directions in the x–y, y–z, and z–x planes, but E_a for $\langle 111 \rangle$ is highest in the case of positive K_2 and lowest in the case of negative K_2. However, in many ferro- or ferrimagnetic materials $K_1 > K_2$, and even if $K_1 = K_2$, the change in a formula for the K_2 term is only $\frac{1}{9}$ of that from the K_1 term, so that the contribution of the K_2 term can be ignored.

When, however, the magnetization rotates in some particular crystallographic plane, the K_2 term is not necessarily negligible. This is the case if the magnetization is confined to the {111} plane.

Before examining the {111} plane, let us consider the case in which the magnetization rotates in the x–y or (001) plane. If θ is the angle between the magnetization and the x-axis (see Fig. 12.6), then we have

$$\left.\begin{aligned} \alpha_1 &= \cos\theta \\ \alpha_2 &= \sin\theta \\ \alpha_3 &= 0 \end{aligned}\right\}. \tag{12.10}$$

Using this relationship, we have from (12.5)

$$\begin{aligned} E_a &= K_1 \cos^2\theta \sin^2\theta + K_3 \cos^4\theta \sin^4\theta + \cdots = \frac{K_1}{4}\sin^2 2\theta + \frac{K_3}{16}\sin^4 2\theta + \cdots \\ &= \frac{K_1}{8}(1 - \cos 4\theta) + \frac{K_3}{128}(3 - 4\cos 4\theta + \cos 8\theta) + \cdots. \end{aligned} \tag{12.11}$$

Note that there is no contribution from the K_2 term.

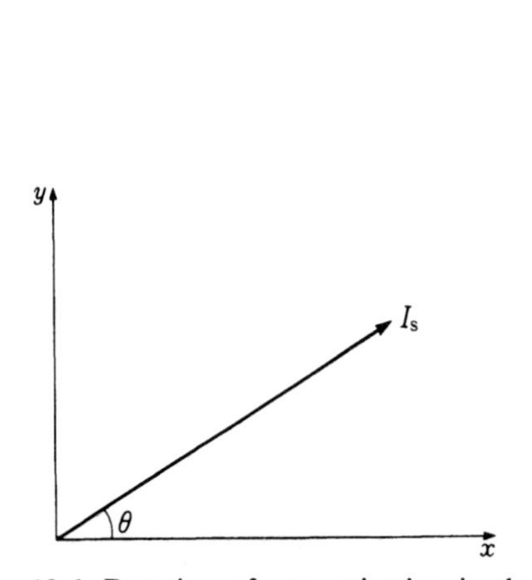

Fig. 12.6. Rotation of magnetization in the (001) plane out of the x-direction.

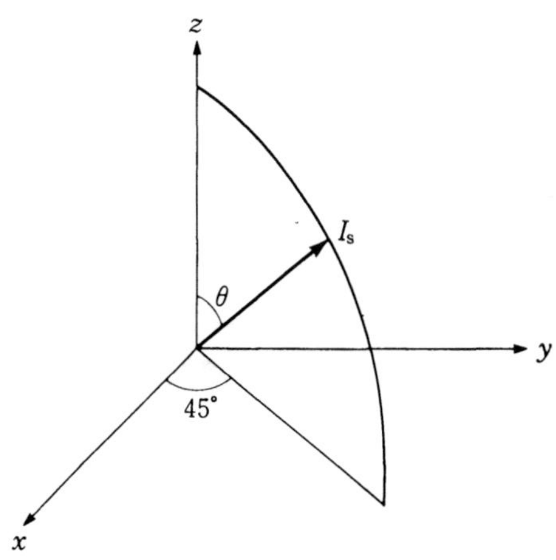

Fig. 12.7. Rotation of magnetization in the $(1\bar{1}0)$ plane out of the z-direction.

Next we consider the $(1\bar{1}0)$ plane. If θ is the angle of rotation of the magnetization from the z-axis in the $(1\bar{1}0)$ plane (see Fig. 12.7), then we have the relationship

$$\left.\begin{aligned} \alpha_1 = \alpha_2 &= \frac{1}{\sqrt{2}} \sin\theta \\ \alpha_3 &= \cos\theta \end{aligned}\right\}. \tag{12.12}$$

Using this relationship, we have from (12.5) the anisotropy energy

$$\begin{aligned} E_a &= K_1\left(\tfrac{1}{4}\sin^4\theta + \sin^2\theta\cos^2\theta\right) + \tfrac{1}{4}K_2\sin^4\theta\cos^2\theta \\ &\quad + K_3\left(\tfrac{1}{4}\sin^4\theta + \sin^2\theta\cos^2\theta\right)^2 + \cdots \\ &= \tfrac{1}{32}K_1(7 - 4\cos 2\theta - 3\cos 4\theta O) + \tfrac{1}{128}K_2(2 - \cos 2\theta - 2\cos 4\theta \\ &\quad + \cos 6\theta) + \tfrac{1}{2048}K_3(123 - 88\cos 2\theta - 68\cos 4\theta + 24\cos 6\theta \\ &\quad + 9\cos 8\theta) + \cdots, \end{aligned} \tag{12.13}$$

which includes contributions from all three (K_1, K_2, and K_3) terms.,

Next let us calculate the anisotropy energy in the (111) plane. For this purpose, we set up a new coordinate system (x', y', z'), in which the z'-axis is parallel to the [111] axis, the y'-axis is in the $(1\bar{1}0)$ plane, and the x'-axis parallel to the $[1\bar{1}0]$ axis (see Fig. 12.8). The direction cosines between coordinate axes of the old (x, y, z) and the new (x', y', z') systems are listed in Table 12.1*. The (111) plane in the old coordi-

* To construct such a table, first we determine two axes, say $x'\left(\frac{1}{\sqrt{2}}, -\frac{1}{\sqrt{2}}, 0\right)$ and $z'\left(\frac{1}{\sqrt{3}}, \frac{1}{\sqrt{3}}, \frac{1}{\sqrt{3}}\right)$, and then by using the relationships $\Sigma_i\,\alpha_i^2 = 1$, $\Sigma_i\,\beta_i^2 = 1$, $\Sigma_i\,\gamma_i^2 = 1$, $\Sigma_i\,\alpha_i\beta_i = 0$, $\Sigma_i\,\beta_i\gamma_i = 0$, $\Sigma_i\,\gamma_i\alpha_i = 0$, we can determine other direction cosines.

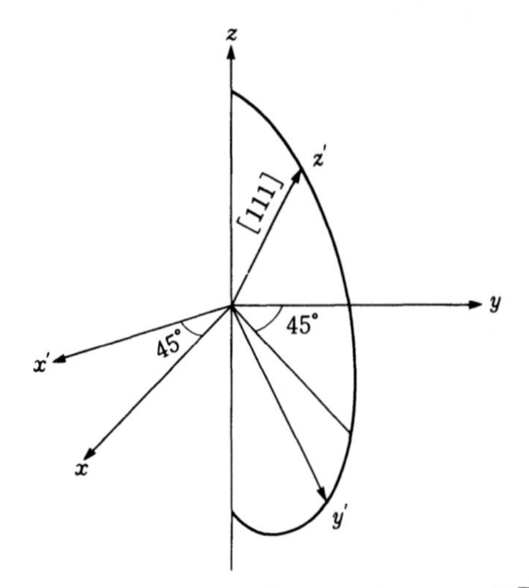

Fig. 12.8. New cubic coordinate system (x', y', z') with $x' \parallel [1\bar{1}0]$ and $z' \parallel [111]$.

nates is the $x'y'$ plane in the new one. Let θ be the angle between the magnetization and the x'-axis in the x'–y' plane; then we have

$$\left.\begin{aligned} \alpha_1' &= \cos\theta \\ \alpha_2' &= \sin\theta \\ \alpha_3' &= 0 \end{aligned}\right\}. \tag{12.14}$$

Referring to Table 12.1, we have the direction cosines of magnetization in the old coordinates

$$\left.\begin{aligned} \alpha_1 &= \frac{1}{\sqrt{2}}\alpha_1' + \frac{1}{\sqrt{6}}\alpha_2' + \frac{1}{\sqrt{3}}\alpha_3' = \frac{1}{\sqrt{2}}\cos\theta + \frac{1}{\sqrt{6}}\sin\theta \\ \alpha_2 &= -\frac{1}{\sqrt{2}}\alpha_1' + \frac{1}{\sqrt{6}}\alpha_2' + \frac{1}{\sqrt{3}}\alpha_3' = -\frac{1}{\sqrt{2}}\cos\theta + \frac{1}{\sqrt{6}}\sin\theta \\ \alpha_3 &= -\frac{\sqrt{2}}{\sqrt{3}}\alpha_2' + \frac{1}{\sqrt{3}}\alpha_3' = -\frac{\sqrt{2}}{\sqrt{3}}\sin\theta \end{aligned}\right\}. \tag{12.15}$$

Using (12.15) in (12.5), we have the anisotropy energy in the (111) plane

$$\begin{aligned} E_a &= \frac{K_1}{4} + \frac{K_2}{54}(9\sin^2\theta - 24\sin^4\theta + 16\sin^6\theta) + \frac{K_3}{16} \\ &= \frac{K_1}{4} + \frac{K_2}{108}(1 - \cos 6\theta) + \frac{K_3}{16} + \cdots, \end{aligned} \tag{12.16}$$

in which only the K_2 term has angular dependence and contributes to the anisotropy energy.

Table 12.1.

	x	y	z
x'	$\frac{1}{\sqrt{2}}$	$-\frac{1}{\sqrt{2}}$	0
y'	$\frac{1}{\sqrt{6}}$	$\frac{1}{\sqrt{6}}$	$-\frac{\sqrt{2}}{\sqrt{3}}$
z'	$\frac{1}{\sqrt{3}}$	$\frac{1}{\sqrt{3}}$	$\frac{1}{\sqrt{3}}$

In summary, the anisotropy energy can be expressed in the general form

$$E_a = A_2 \cos 2\theta + A_4 \cos 4\theta + A_6 \cos 6\theta + A_8 \cos 8\theta + \cdots. \quad (12.17)$$

In the case of uniaxial anisotropy, by comparing (12.17) with (12.2), we have

$$\left.\begin{aligned} A_2 &= -\tfrac{1}{2}K_{u1} - \tfrac{1}{2}K_{u2} - \tfrac{15}{32}K_{u3}(1+\cos 6\varphi) + \cdots \\ A_4 &= \tfrac{1}{8}K_{u2} + \tfrac{3}{16}K_{u3}(1+\cos 6\varphi) + \cdots \\ A_6 &= -\tfrac{1}{32}K_{u3}(1+\cos 6\varphi) + \cdots \end{aligned}\right\}. \quad (12.18)$$

In the case of the (001) plane in a cubic crystal, by comparison with (12.11), we have

$$\left.\begin{aligned} A_2 &= 0 \\ A_4 &= -\tfrac{1}{8}K_1 - \tfrac{1}{32}K_3 + \cdots \\ A_6 &= 0 \\ A_8 &= \tfrac{1}{128}K_3 + \cdots \end{aligned}\right\}. \quad (12.19)$$

In the case of the $(1\bar{1}0)$ plane, we have

$$\left.\begin{aligned} A_2 &= -\tfrac{1}{8}K_1 - \tfrac{1}{128}K_2 - \tfrac{11}{256}K_3 - \cdots \\ A_4 &= -\tfrac{3}{32}K_1 - \tfrac{1}{64}K_2 - \tfrac{17}{512}K_3 - \cdots \\ A_6 &= \tfrac{1}{128}K_2 + \tfrac{3}{256}K_3 + \cdots \\ A_8 &= \tfrac{9}{2048}K_3 + \cdots \end{aligned}\right\}. \quad (12.20)$$

In the case of the (111) plane, we have

$$\left.\begin{aligned} A_2 &= 0 \\ A_4 &= 0 \\ A_6 &= -\tfrac{1}{108}K_2 \\ A_8 &= 0 \end{aligned}\right\}. \quad (12.21)$$

Anisotropy energy is also produced by magnetostatic energy due to magnetic free poles appearing on the outside surface or internal surfaces of an inhomogeneous magnetic material. Referring to (1.99), the magnetostatic energy due to free poles on the outside surface of a magnetic body of volume v magnetized to intensity I along an axis with demagnetizing factor N is given by

$$U = \frac{1}{2\mu_0} N I^2 v. \tag{12.22}$$

If the specimen is in the form of an ellipsoid of revolution with the long axis parallel to the z-axis, the demagnetizing factor parallel to the x- and y-axes is given by $N_x = N_y = \frac{1}{2}(1 - N_z)$, where N_z is the demagnetizing factor along the z-axis. Let θ be the angle between the magnetization and the z-axis, and φ be the angle between the x-axis and the projection of the magnetization onto the x–y plane. Then applying (12.22) to the x-, y- and z-components, we have the magnetostatic energy

$$\begin{aligned} U &= \frac{1}{2\mu_0} I_s^2 v \left(N_x \sin^2\theta \cos^2\varphi + N_y \sin^2\theta \sin^2\varphi + N_z \cos^2\theta \right) \\ &= \frac{1}{4\mu_0} I_s^2 v (3N_z - 1) \cos^2\theta + \text{const.} \end{aligned} \tag{12.23}$$

which depends on the direction of magnetization. This kind of anisotropy is called *shape magnetic anisotropy.*

If precipitate particles of magnetization I_s', different from that of the matrix I_s ($I_s' \parallel I_s$), have demagnetizing factor N_z ($N_z < \frac{1}{3}$), the magnetostatic energy is given by

$$U = \frac{1}{4\mu_0} (I_s - I_s')^2 v (3N_z - 1) \cos^2\theta. \tag{12.24}$$

The easy axis of this anisotropy is the z-axis, irrespective of the magnitude of I_s' relative to I_s. One example of this kind of shape anisotropy is found in Alnico 5, in which elongated precipitates produce the uniaxial anisotropy (see Section 13.3.1).

12.2 METHODS FOR MEASURING MAGNETIC ANISOTROPY

The most accurate means for measuring magnetic anisotropy is the torque magnetometer. The principle of this method is as follows: the ferromagnetic specimen, in the form of a disk or a sphere, is placed in a reasonably strong magnetic field which magnetizes the specimen to saturation. If the easy axis is near the direction of magnetization, the magnetic anisotropy tends to rotate the specimen to bring the easy axis parallel to the magnetization, thus producing a torque on the specimen. If the torque is measured as a function of the angle of rotation of the magnetic field about the vertical axis, we can obtain the *torque curve*, from which we can deduce the anisotropy constants.

There are various types of torque magnetometer.[4] Figure 12.9 shows a typical automatic torque magnetometer. The specimen S is suspended by a thin metal wire in

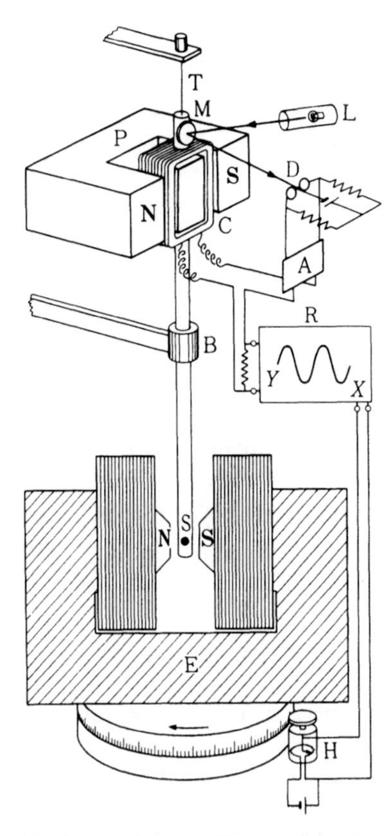

Fig. 12.9. Automatic recording torque magnetometer.

the magnetic field between the pole pieces of the electromagnet E, which can be rotated about the vertical axis. The lateral displacement of the specimen holder is prevented by a frictionless bearing B. Various kinds of frictionless bearings are used, including air-bearings,[5] taut-wire suspensions,[6] multiple elastic plates,[7] etc. The specimen holder is attached to a mirror M and a moving coil C, which is placed in the field of a permanent magnet P. The deflection of the specimen holder is detected by the displacement of the light beam from L reflected by the mirror M onto two phototransistors D. The signal from the phototransistors is amplified by the amplifier A, and fed back to the moving coil, so as to produce a torque to counterbalance the torque exerted on the specimen by the field. Therefore the current in the coil C is always proportional to the torque exerted on the specimen. The torque is recorded by the x–y recorder R as a function of the rotational angle of the magnet, which is measured by a helical multi-turn potentiometer resistor H. The amplifier A should have high gain to keep the deflection of the specimen holder as small as possible, and also should have good dynamic properties to prevent unstable oscillations of the specimen.[8]

Suppose that the anisotropy energy is increased by $\delta E(\theta)$, when the magnetization is rotated by the angle $\delta\theta$. Then the torque $L(\theta)$ must act on a unit volume of the specimen in a sense to decrease θ. The work done by the torque must equal the decrease in the anisotropy energy, so that

$$-L(\theta)\,\delta\theta = \delta E(\theta),$$

or

$$L(\theta) = -\frac{\partial E(\theta)}{\partial \theta}.^{*} \qquad (12.25)$$

If the anisotropy energy is given by (12.17), the torque can be calculated from (12.25) as

$$L(\theta) = -2A_2 \sin 2\theta - 4A_4 \sin 4\theta - 6A_6 \sin 6\theta - 8A_8 \sin 8\theta - \cdots. \qquad (12.26)$$

If the magnetization is rotated in a plane which includes the c-axis in a uniaxial crystal, the torque is given by (12.26), where the coefficients $A_2, A_4, \ldots$ are given by (12.18). In the case of cubic anisotropy, the torque is given by (12.26), where the coefficients are given by (12.19), (12.20), and (12.21) for (001), $(1\bar{1}0)$, and (111) planes, respectively. For example, ignoring the K_2 term, the torque in the (001) plane is given by

$$L(\theta) = \tfrac{1}{2}K_1 \sin 4\theta. \qquad (12.27)$$

Figure 12.10 shows a torque curve (a plot of the torque vs. angle) measured for a single crystal disk of 4% Si–Fe cut parallel to (001), as a function of the angle of rotation, θ, of the field measured from the [001] axis, for $\theta = 0$–180°. This angle almost coincides with that of the magnetization as long as the field is strong enough. The experimental points are well fitted by (12.27), which oscillates twice in the range

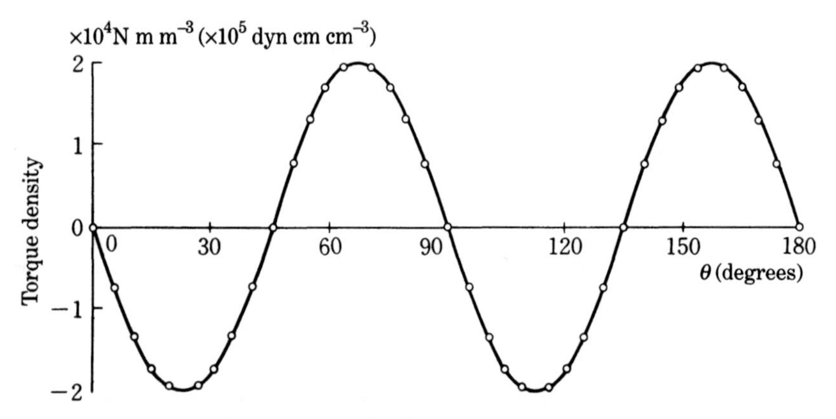

Fig. 12.10. Torque curve measured in (001) plane of 4% Si–Fe single crystal at room temperature (Chikazumi and Iwata).

* This relationship is analogous to the force $-F(x)$ which acts on a body placed on a slope. If the potential energy is increased by $\delta U(x)$ when the body is displaced by δx along the slope, we have the relationship $-F(x)\,\delta x = \delta U(x)$, so that we have $F(x) = -\partial U(x)/\partial x$.

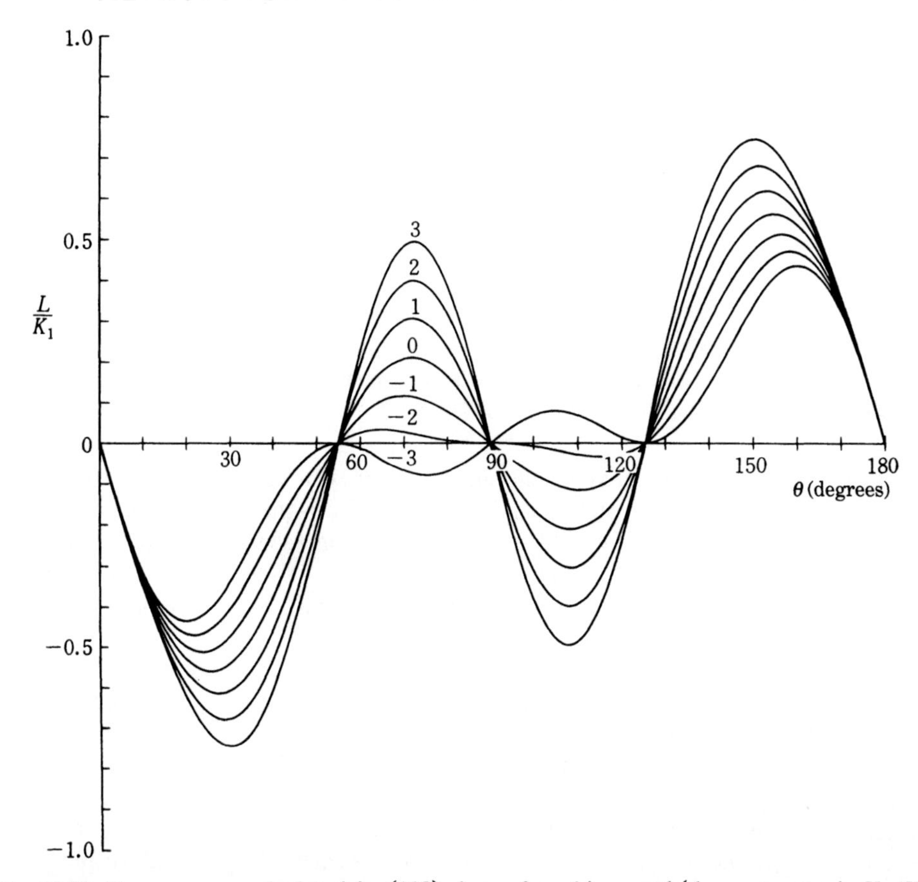

Fig. 12.11. Torque curve calculated for (110) plane of a cubic crystal (the parameter is K_2/K_1).

$\theta = 0-180°$. Since the amplitude is equal to $K_1/2$, the anisotropy constant K_1 can be determined from the torque curve.

In the case of the $(1\bar{1}0)$ plane, ignoring the K_3 term, the torque curve is given by

$$L(\theta) = (\tfrac{1}{4}K_1 + \tfrac{1}{64}K_2)\sin 2\theta + (\tfrac{3}{8}K_1 + \tfrac{1}{16}K_2)\sin 4\theta + \tfrac{3}{64}K_2 \sin 6\theta. \quad (12.28)$$

Figure 12.11 shows the torque curve, or L/K_1 vs. θ curve, calculated from (12.28), where the parameter is the value of K_2/K_1. When $K_2 = 0$, the ratio of the first peak to the second peak is 2.67, and these peaks appear at $\theta = 25.5°$ and $71.3°$. As the value of K_2 increases, the ratio of the first to the second peak, and also the value of θ, the angle at which peaks appear, are both changed. It is, however, dangerous to estimate the value of K_2 from these changes, because such changes also occur if the plane of the disk is tilted from a {110} plane, or if some induced magnetic anisotropy, as discussed in Chapter 13, is present.

The most reliable method to determine the value of K_2 is to measure the torque curve for a (111) plane specimen. Using the value of (12.21) in (12.26), we have

$$L(\theta) = \tfrac{1}{18}K_2 \sin 6\theta, \quad (12.29)$$

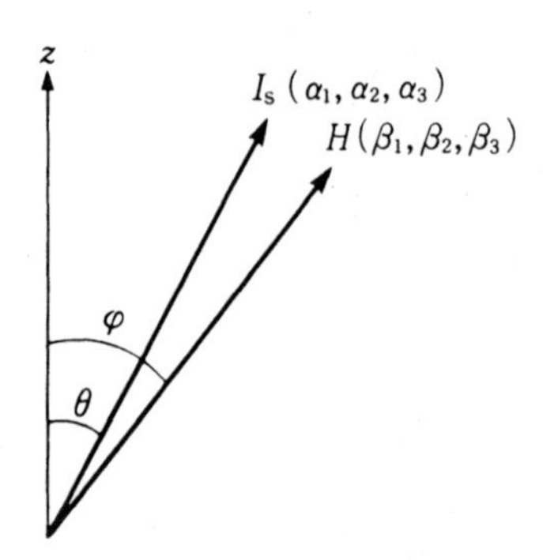

Fig. 12.12. Definition of the angle θ between spontaneous magnetization, I_s, and the z-axis (easy axis); and the angle φ between the magnetic field, $\boldsymbol{H}$, and the z-axis.

which represents a torque curve oscillating three times in the range $\theta = 0$ to 180°. In this case, even if the plane of the sample is tilted from the (111) plane, or even if some induced anisotropy is present, the value of K_2 can accurately be determined by Fourier analysis of the torque curve to give the term in (12.29). The Fourier analysis can be performed by the calculation given by Chikazumi,[9] or by a simple computer program.[10]

It must be noted that when the magnetic field is not strong enough to align the internal magnetization in the field direction, the uniaxial anisotropy may give rise to an apparent fourth-order anisotropy. As shown in Fig. 12.12, if the magnetic field H is applied in a direction φ from the easy axis z, the magnetization rotates towards the field by the angle θ, which is slightly smaller than φ. We assume that the uniaxial anisotropy has the anisotropy constant K_u, and the easy axis is parallel to the z-axis. Let $(\alpha_1, \alpha_2, \alpha_3)$, and $(\beta_1, \beta_2, \beta_3)$ be the direction cosines of I_s and H, respectively. Then the anisotropy energy is given by

$$E_a = K_u \cos^2 \theta = K_u \alpha_3^2. \tag{12.30}$$

In addition, we must consider the field energy (see equation (1.9)), because the magnetization makes a nonzero angle $\varphi - \theta$ with the field, so that we have

$$E_H = -I_s H \cos(\varphi - \theta) = -I_s H(\alpha_1 \beta_1 + \alpha_2 \beta_2 + \alpha_3 \beta_3). \tag{12.31}$$

Therefore, the total energy of this system is given by

$$E = E_a + E_H = -K_u \alpha_3^2 - I_s H(\alpha_1 \beta_1 + \alpha_2 \beta_2 + \alpha_3 \beta_3). \tag{12.32}$$

The direction of magnetization can be found by minimizing (12.32) with respect to the angle θ.

The procedure for minimizing the energy (12.32) is as follows: If the direction cosines of the magnetization, I_s are changed virtually by $\delta\alpha_1, \delta\alpha_2, \delta\alpha_3$, the associated change in energy (12.32) should be zero when the direction of I_s is in an equilibrium state. Therefore, we have

$$\delta E = -I_s H \beta_1 \, \delta\alpha_1 - I_s H \beta_2 \, \delta\alpha_2 - (2K_u \alpha_3 + I_s H \beta_3) \, \delta\alpha_3 = 0. \tag{12.33}$$

Since $\alpha_1^2+\alpha_2^2+\alpha_3^2=1$, there must be a relationship between $\delta\alpha_1$, $\delta\alpha_2$, and $\delta\alpha_3$, given by

$$\alpha_1\,\delta\alpha_1+\alpha_2\,\delta\alpha_2+\alpha_3\,\delta\alpha_3=0. \tag{12.34}$$

Multiplying (12.34) by an undetermined parameter λ and adding (12.33), we have

$$(-I_sH\beta_1+\lambda\alpha_1)\,\delta\alpha_1+(-I_sH\beta_2+\lambda\alpha_2)\,\delta\alpha_2+(-2K_u\alpha_3-I_sH\beta_3+\lambda\alpha_3)\,\delta\alpha_3=0. \tag{12.35}$$

In order that (12.35) be satisfied for arbitrary changes in $\delta\alpha_1$, $\delta\alpha_2$, and $\delta\alpha_3$, the coefficients of $\delta\alpha_1$, $\delta\alpha_2$, and $\delta\alpha_3$ must be zero. Therefore

$$\lambda=\frac{\beta_1}{\alpha_1}I_sH=\frac{\beta_2}{\alpha_2}I_sH=\frac{\beta_3}{\alpha_3}I_sH+2K_u. \tag{12.36}$$

If the field is strong enough to satisfy $I_sH\gg 2K_u$, we put $2K_u/I_sH=p$ $(p\ll 1)$, and after dividing each side of (12.36) by I_sH, we have

$$\frac{\beta_1}{\alpha_1}=\frac{\beta_2}{\alpha_2}=\frac{\beta_3}{\alpha_3}+p. \tag{12.37}$$

Putting each side of (12.37) equal to $1+\mu$ $(\mu\ll 1)$, we have

$$\left.\begin{aligned}\alpha_1&=\frac{\beta_1}{1+\mu}=\beta_1(1-\mu+\mu^2)\\ \alpha_2&=\frac{\beta_2}{1+\mu}=\beta_2(1-\mu+\mu^2)\\ \alpha_3&=\frac{\beta_3}{1+\mu-p}=\beta_3\{1-\mu+p+(\mu-p)^2\}\end{aligned}\right\}. \tag{12.38}$$

Using (12.38), we have

$$\begin{aligned}\alpha_1^2+\alpha_2^2+\alpha_3^2&=(\beta_1^2+\beta_2^2+\beta_3^2)(1-2\mu+3\mu^2)+p(2+3p-6\mu)\beta_3^2=1\\ \therefore\quad &3\mu^2-2(1+3p\beta_3^2)\mu+(2+3p)p\beta_3^2=0\\ \therefore\quad &\mu=p\beta_3^2(1+\tfrac{3}{2}p-\tfrac{3}{2}p\beta_3^2).\end{aligned} \tag{12.39}$$

Substituting (12.39) for (12.38), we have

$$\left.\begin{aligned}\alpha_1&=\beta_1-p\beta_2\beta_3^2-\tfrac{3}{2}p^2\beta_1\beta_3^2+\tfrac{5}{2}p^2\beta_1\beta_3^4+\cdots\\ \alpha_2&=\beta_2-p\beta_2\beta_3^2-\tfrac{3}{2}p^2\beta_2\beta_3^2+\tfrac{5}{2}p^2\beta_2\beta_3^4+\cdots\\ \alpha_3&=(1+p+p^2)\beta_3-p\beta_3^3-\tfrac{7}{2}p^2\beta_3^3+\tfrac{5}{2}p^2\beta_3^5+\cdots\end{aligned}\right\}, \tag{12.40}$$

where $p=2K_u/I_sH$.

Substituting $(\alpha_1, \alpha_2, \alpha_3)$, which minimize the total energy (12.32), into (12.32), we have

$$\begin{aligned}E&=-K_u(1+2p)\beta_3^2+2K_up\beta_3^4-I_sH\left(1-\tfrac{1}{2}p^2\beta_3^2+\tfrac{1}{2}p^2\beta_3^4\right)\\ &=-K_u(1+p)\beta_3^2+K_up\beta_3^4+\text{const}.\end{aligned} \tag{12.41}$$

This energy is expressed as a function of the direction of the external magnetic field, not the direction of magnetization. This form is proper to express the torque curve, because as seen in Fig. 12.12, the actual rotation of the specimen from the direction of the field is given by φ, not θ. Thus the torque is deduced from (12.41) as

$$\begin{aligned} L &= -\frac{\partial E}{\partial \varphi} \\ &= -2K_u(1+p)\cos\varphi\sin\varphi + 4K_u p\cos^3\varphi\sin\varphi \\ &= -K_u\sin 2\varphi + \tfrac{1}{2}K_u p\sin 4\varphi. \end{aligned} \tag{12.42}$$

As seen in (12.42), the term with twofold symmetry is not affected by the intensity of the applied magnetic field. The term with fourfold symmetry does include p, which is a function of H. In order to correct for such a higher-order term, the rotation of the magnetization vector from the field direction, $\varphi - \theta$, is determined by the relationship

$$L = I_s H \sin(\varphi - \theta), \tag{12.43}$$

and then the torque curve is corrected by shifting the angle by $\varphi - \theta$.[8] Figure 12.13 shows an example of such a correction carried out on the torque curve measured for a Gd alloy containing a small amount of Tb.[11] In the case of a disk-shaped specimen, which will have some domains remaining at the edge even in high applied fields, the torque curve shows some field dependence.[12]

When the magnetic field is so weak that the magnetization cannot rotate reversibly with the magnetic field, the torque curve exhibits hysteresis, which is called *rotational hysteresis*. If the rotation of magnetization is associated with some loss of energy,

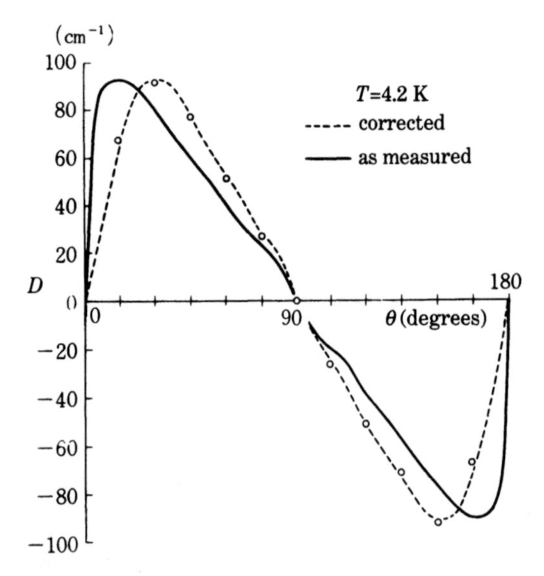

Fig. 12.13. Torque curve normalized per one Tb atom in a dilute Tb–Gd alloy. The dashed curve represents the corrected torque curve as a function of the angle of the Tb moment.[11]

whether internal or external, rotational hysteresis appears even when the field is strong enough to rotate the magnetization reversibly. This phenomenon will be treated again in Section 13.1.

The second term in (12.42) shows the appearance of a higher-order term when the applied magnetic field is not strong enough. The reason for this phenomenon is that the angle of deviation of the magnetization from the field direction changes sign every time the field passes through the easy or hard axes. If the easy axis of the uniaxial anisotropy is unique, the second term is simply a small correction as in the case of (12.42). But if the easy axes are distributed on the x-, y-, or z-axes in different places, the uniaxial terms must cancel one other, resulting in the dominant contribution to the anisotropy from the second term. The fourth-order anisotropy thus produced is field-dependent, so that value of K_1 changes its magnitude or sometimes changes its sign with a change in the intensity of applied field. This phenomenon is called the *torque reversal.*

Let us formulate the torque reversal by using (12.41). If the easy axis is parallel to the x- or y-axis, β_3 in (12.41) should be replaced by β_1, or β_2, so that the average anisotropy is given by

$$\begin{aligned} E &= \tfrac{1}{3}K_u(1+p)(\beta_1^2+\beta_2^2+\beta_3^2) + \tfrac{1}{3}K_u p(\beta_1^4+\beta_2^4+\beta_3^4) \\ &= -\frac{4}{3}\frac{K_u^2}{I_s H}(\beta_1^2\beta_2^2+\beta_2^2\beta_3^2+\beta_3^2\beta_1^2) + \text{const.} \end{aligned} \tag{12.44}$$

We see that the fourth-order term* in (12.44) has the same form as the cubic magnetocrystalline anisotropy (12.5). Therefore if the material has its own inherent K_{10}, the resultant field-dependent K_1 will be given by

$$K_1 = K_{10} - \frac{4}{3}\frac{K_u^2}{I_s H}. \tag{12.45}$$

If K_{10} is small and positive, K_1 changes its sign from positive to negative as the field increases. Torque reversals have been observed for Fe_2NiAl,[14] Alnico 5[15] and Co ferrite.[16]

Magnetic anisotropy can also be measured by ferromagnetic resonance. In ferromagnetic resonance, the spontaneous magnetization precesses about the applied magnetic field with a frequency proportional to the magnetic field. If the frequency of the microwave radiation field applied to the specimen coincides with the resonance frequency, a part of the microwave power is absorbed to excite the precession (see Section 3.3).

If the applied magnetic field is parallel to the easy axis, the magnetic anisotropy influences the resonance field, because the anisotropy gives an additional torque to the spontaneous magnetization to rotate it towards the easy axis. This effect is equivalent to a presence of the so-called *anisotropy field.*

* In the original paper,[13] the coefficient is different, because higher order terms were ignored.

Let us calculate the anisotropy field for a uniaxial anisotropy. Ignoring the higher-order terms in (12.1), we can simplify the anisotropy energy for small θ:

$$E_a = K_{u1} \sin^2 \theta \simeq K_{u1}\theta^2 + \cdots . \tag{12.46}$$

On the other hand, if an anisotropy field exists, the energy is expressed for small θ by

$$E_a = -I_s H_a \cos \theta = -I_s H_a\left(1 - \tfrac{1}{2}\theta^2 - \cdots\right). \tag{12.47}$$

Comparing the terms in θ^2 in (12.46) and (12.47), we have

$$H_a = \frac{2K_{u1}}{I_s}. \tag{12.48}$$

Then the resonance field should be given by

$$\omega = \nu(H + H_a), \tag{12.49}$$

where the coefficient ν is the gyromagnetic constant (see (3.22)) which is given by

$$\nu = 1.105 \times 10^5 g \quad (\mathrm{m\,A^{-1}\,s^{-1}}). \tag{12.50}$$

The g-factor was defined in (3.21) and (3.39). Therefore we conclude that the resonance is expected to occur at a field less than the usual value by an amount H_a, when magnetic anisotropy is present.

In the case of a cubic crystal, when the magnetization is in a direction near [001], the direction cosines of magnetization are given by

$$\left.\begin{aligned} \alpha_1 &= \sin\theta\cos\varphi \simeq \theta\cos\varphi \\ \alpha_2 &= \sin\theta\sin\varphi \simeq \theta\sin\varphi \\ \alpha_3 &= \cos\theta \simeq 1 - \tfrac{1}{2}\theta^2 \end{aligned}\right\}, \tag{12.51}$$

where θ is the polar angle from the z-axis and φ is the azimuthal angle about the z-axis (see Fig. 12.14). Substituting α_i in (12.5) by (12.51), we have

$$E_a = K_1\left\{\theta^4 \sin^2\varphi\cos^2\varphi + \left(1 - \tfrac{1}{2}\theta^2\right)^2\theta^2\right\} \simeq K_1\theta^2, \tag{12.52}$$

which has the same form as (12.47), so that the anisotropy field is given by

$$H_a = \frac{2K_1}{I_s}. \tag{12.53}$$

When the magnetization is in a direction near [111], we first change the coordinate

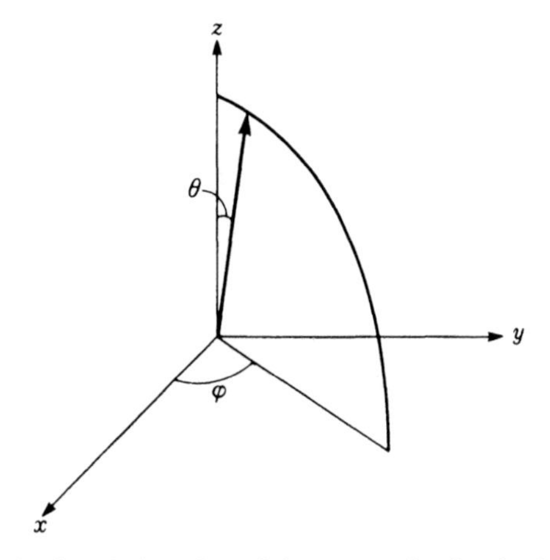

Fig. 12.14. The polar and azimuthal angles of the magnetization in the vicinity of the z-axis.

system as shown in Fig. 12.8, then use polar coordinates, as in (12.51), and finally substitute using (12.15). Then we have

$$\left.\begin{aligned}\alpha_1 &= \frac{1}{\sqrt{2}}\,\theta\cos\varphi + \frac{1}{\sqrt{6}}\,\theta\sin\varphi + \frac{1}{\sqrt{3}}\left(1-\frac{1}{2}\,\theta^2\right)\\ \alpha_2 &= -\frac{1}{\sqrt{2}}\,\theta\cos\varphi + \frac{1}{\sqrt{6}}\,\theta\sin\varphi + \frac{1}{\sqrt{3}}\left(1-\frac{1}{2}\,\theta^2\right)\\ \alpha_3 &= -\frac{\sqrt{2}}{\sqrt{3}}\,\theta\sin\varphi + \frac{1}{\sqrt{3}}\left(1-\frac{1}{2}\,\theta^2\right)\end{aligned}\right\}. \tag{12.54}$$

Using this expression in the first term in (12.5), we finally have

$$E_a = \frac{K_1}{3} - \frac{2}{3}K_1\theta^2. \tag{12.55}$$

Comparing this result with (12.47), we have the anisotropy field

$$H_a = -\frac{4K_1}{3I_s}. \tag{12.56}$$

Therefore when the magnetic field is rotated from [001] to [111], the resonance field should shift by

$$\Delta H = \frac{2K_1}{I_s} - \left(-\frac{4K_1}{3I_s}\right) = \frac{10K_1}{3I_s}. \tag{12.57}$$

Figure 12.15 shows the shift of resonance field observed for a single crystal of Fe_3O_4, when the magnetic field is rotated in the (001) or $(1\bar{1}0)$ planes.[17] From such experiments, we can deduce the anisotropy constant.

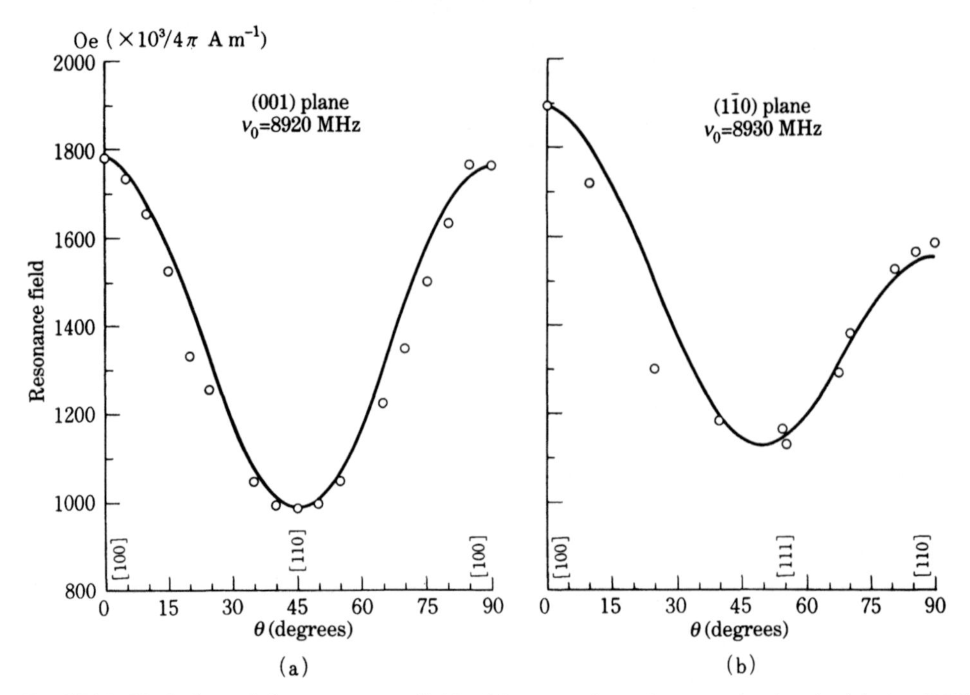

Fig. 12.15. Variation of the resonance field with a rotation of magnetization in (a) the (001) plane and (b) the $(1\bar{1}0)$ plane, observed for a single crystal of Fe_3O_4. (After Bickford[17])

It should be noted that a torque measurement measures the first derivative of the anisotropy energy with respect to the angle of rotation, whereas the resonance field measures the second derivative of the anisotropy energy. Moreover, if the specimen is inhomogeneous, so that the anisotropy is different in different places, the torque measures only the average of the local anisotropy, whereas the magnetic resonance exhibits different absorption peaks corresponding to the local anisotropies.

The magnetic anisotropy can also be determined by measuring magnetization curves. We shall discuss this method in Section 18.3.

12.3 MECHANISM OF MAGNETIC ANISOTROPY

Magnetic anisotropy is a change in the internal energy of a magnetic material with a change in the direction of magnetization. Figure 12.16 shows the rotation of ferromagnetic spontaneous magnetization, which is composed of parallel spins, from one direction (a) to another (b). The reason that neighboring spins remain parallel is that a strong exchange interaction acts between spins (see Sections 6.2, 6.3). According to Heisenberg's expression,[18] the exchange interaction between spins S_i and S_j is given by

$$w_{ij} = -2JS_i \cdot S_j = -2JS^2 \cos \varphi, \tag{12.58}$$

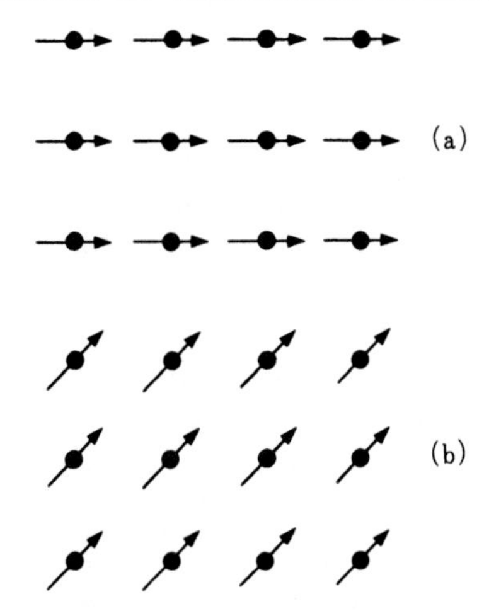

Fig. 12.16. Rotation of spins in ferromagnetic spontaneous magnetization.

where S is the magnitude of the spin and φ is the angle between $\boldsymbol{S}_i$ and $\boldsymbol{S}_j$. When the magnetization rotates as shown in Fig. 12.16 from (a) to (b), all the spins remain parallel, so that $\varphi = 0$ in (12.58) and the exchange energy does not change. Thus the exchange energy is isotropic.

Therefore to explain magnetic anisotropy, we need an additional energy term which includes the crystal axes. Suppose that neighboring spins make the angle φ with the bond axis (see Fig. 12.17). Then the energy of the spin pair is expressed, after expanding in Legendre polynomials, as

$$w(\cos\varphi) = g + l\left(\cos^2\varphi - \tfrac{1}{3}\right) + q\left(\cos^4\varphi - \tfrac{6}{7}\cos^2\varphi + \tfrac{3}{35}\right) + \cdots . \tag{12.59}$$

The first term is independent of the angle φ, so that it corresponds to the exchange interaction. The second term is called the *dipole–dipole interaction* term, because it has the same form as the magnetic dipole–dipole interaction (see (1.15)) if the coefficient is given by

$$l = -\frac{3M^2}{4\pi\mu_0 r^3}. \tag{12.60}$$

However, we shall find later that the value of l corresponding to the actual measured magnetic anisotropy is 100 to 1000 times larger than the magnetic interaction given by (12.60). The real mechanism is believed to be as follows: a partially unquenched magnetic orbital moment coupled with the spins lead to a variation in the exchange or electrostatic energy with a rotation of the magnetization, through a change in the

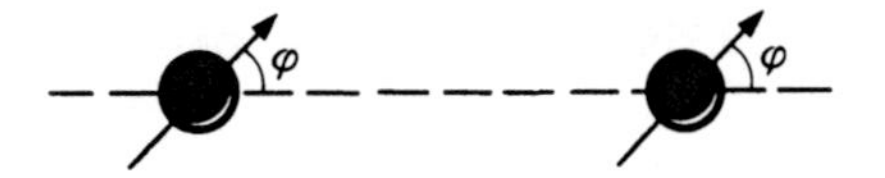

Fig. 12.17. A pair of spins.

overlap of wave functions. Such an interaction is called a *pseudodipolar interaction.* The third term in (12.59) is a higher-order term of the same origin, and is called the *quadrupolar interaction.* The magnetic anisotropy can be calculated by summing up the pair energy given by (12.59) for all the spin-pairs in the crystal. Such a model is called a *spin-pair model.* This model is useful to discuss the effect of crystal symmetry on the magnetic anisotropy and to calculate the induced magnetic anisotropy (see Chapter 13).

Before going into details of the spin-pair model, let us consider how magnetic anisotropy is produced by the magnetic dipole–dipole interaction. The sum of dipole–dipole interactions between all the spin pairs in a ferromagnetic crystal is called the *dipole sum.* In order to calculate the dipole sum, we first calculate the magnetic field produced at one atom by all the surrounding magnetic dipoles, and then calculate the magnetostatic energy by using (1.94). Since the dipole field is inversely proportional to the third power of distance, it is tempting to assume that the main contribution comes only from the neighboring dipoles. In reality, however, the convergence of the dipole sum with distance is not very rapid, because the number of dipoles in a spherical shell with inner and outer radii of r and $r + \mathrm{d}r$ increases as $r^2\,\mathrm{d}r$ with increasing r, so that the dipole sum increases only as $(1/r^3)(r^2\,\mathrm{d}r) = \mathrm{d}r/r$. Therefore the contribution of all the dipoles in the sphere of the radius r increases as $\ln r$ with increasing r. Such a logarithmic function increases only slowly, so we must calculate the contribution of all the dipoles in a large sphere. If we extend the range of sum to the whole volume of the specimen, the dipole sum depends on the shape of the specimen. This corresponds to the shape anisotropy given by (12.23). Therefore in order to calculate the dipole sum free from shape anisotropy, we must calculate the contribution from a spherical shell with radius r, and then increase r to infinity.

Although special techniques have been used for such a calculation of the dipole sum, the convergence is still unsatisfactory. A real calculation of magnetic anisotropy has been made for hexagonal cobalt, but the result can explain only a part of the experimental value.[19] Some published papers show the calculation of the dipole sum only in the vicinity of an atomic site. We cannot trust even the sign of such calculated results, let along the magnitude.

In the case of pseudodipolar interaction, however, the range of interaction is truly short range, so that it is not necessary to worry about the range of the sum.

Let us elucidate how magnetic anisotropy is constructed from the spin-pair energy given by (12.59). For simplicity, first we consider a simple cubic lattice (Fig. 12.18). The magnetic anisotropy is calculated by summing up all the spin pairs included in a unit volume of this crystal. This is simply

$$E_a = \sum_i w_i, \tag{12.61}$$

where i identifies the spin pair. Since the pair energy for distant pairs is small, we consider only the interactions between the first or at most the second nearest neighbors. In the present case, we consider only the first nearest neighbors. Let $(\alpha_1, \alpha_2, \alpha_3)$ be the direction cosines of parallel spins. For spin pairs with bonding

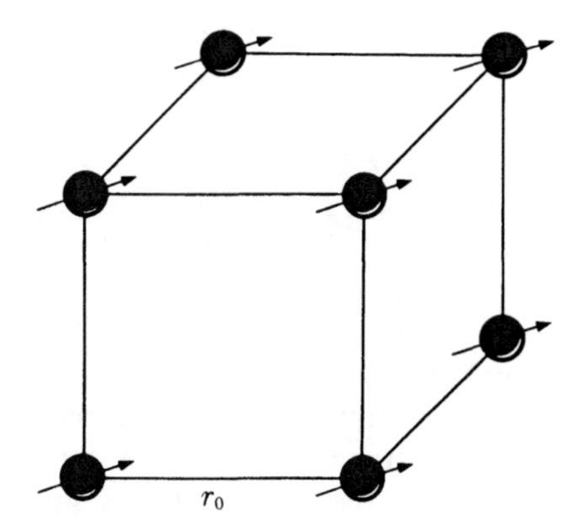

Fig. 12.18. Ferromagnetic spins on a simple cubic lattice.

parallel to the x-axis, $\cos\varphi$ is (12.59) is simply replaced by α_1. Similarly for y- or z-pairs it is replaced by α_1 and α_3. Thus we have

$$\begin{aligned} E_a &= N\{l(\alpha_1^2 - \tfrac{1}{3}) + q(\alpha_1^4 - \tfrac{6}{7}\alpha_1^2 + \tfrac{3}{35}) + \cdots \\ &\quad + l(\alpha_2^2 - \tfrac{1}{3}) + q(\alpha_2^4 - \tfrac{6}{7}\alpha_2^2 + \tfrac{3}{35}) + \cdots \\ &\quad + l(\alpha_3^2 - \tfrac{1}{3}) + q(\alpha_3^4 - \tfrac{6}{7}\alpha_3^2 + \tfrac{3}{35}) + \cdots\} \\ &= Nq(\alpha_1^4 + \alpha_2^4 + \alpha_3^4) + \text{const.} \\ &= -2Nq(\alpha_1^2\alpha_2^2 + \alpha_2^2\alpha_3^2 + \alpha_3^2\alpha_1^2) + \text{const.}, \end{aligned} \tag{12.62}$$

where N is the total number of atoms in a unit volume. Comparing this with (12.5), we have

$$K_1 = -2Nq. \tag{12.63}$$

Similar calculations for the body-centered cubic lattice lead to

$$K_1 = \tfrac{16}{9}Nq, \tag{12.64}$$

and for the face-centered cubic lattice

$$K_1 = Nq. \tag{12.65}$$

It should be noted here that the second term or the dipole term in (12.59) makes no contribution to the anisotropy, since $\sum_i \alpha_i^2 = 1$. This is always the case for any cubic crystal. On the other hand, in structures with uniaxial symmetry, such as hexagonal crystals, the dipole term does contribute to the anisotropy. The K_{u1} term in (12.1) is caused by such an interaction. Generally speaking, l is one or two orders of magnitude larger than q, so that crystals with low symmetry exhibit relatively large magnetocrystalline anisotropy. For instance, the magnitude of K_{u1} for hexagonal cobalt is of the order of $10^5\,\mathrm{J\,m^{-3}}$, which is much larger than K_1 ($=10^3$–$10^4\,\mathrm{J\,m^{-3}}$) for cubic iron or nickel.

This spin-pair model is applicable to metals, in which atomic spins are close to one another. This is not, however, the case for oxides and compounds in which the magnetic atoms are separated by large negative ions. In such a case, the magnetic anisotropy results from the behavior of non-spherical magnetic atoms in the crystalline field produced by the surrounding ions. This model is called the *one-ion model.* In general, the theory that treats the behavior of transition metal ions in a crystalline field is called *ligand field theory.*[20] We shall discuss in terms of the one-ion model how the magnetic anisotropy is produced in a cubic crystal.

We have seen in Chapter 3 that the wave functions of d-electrons of transition metal atoms which are placed in a cubic lattice are separated into three $d\varepsilon$ and two $d\gamma$ functions. The $d\varepsilon$ functions stretch along $\langle 110 \rangle$ axes, while the $d\gamma$ functions stretch along the cubic axes $\langle 100 \rangle$ (see Fig. 3.20). We shall discuss how such wave functions result in a magnetic anisotropy in an oxide crystal.

In an oxide crystal, divalent oxygen ions (O^{2-}) are in direct contact to one another, thus forming a close-packed structure. Since the divalent or trivalent metal ions (M^{2+} or M^{3+}) are much smaller than the O^{2-} ions (radius 0.6–0.8 Å as compared to 1.32 Å for O^{2-}), they are squeezed into interstitial sites of the oxygen lattice. There are two kinds of interstitial sites, as shown in Fig. 12.19: one is a *tetrahedral site*, surrounded by four nearest neighbor O^{2-} ions, and the other is an *octahedral site*, surrounded by six O^{2-} ions. In all the oxides, metal ions occupy these small interstitial sites. For example, ferrites have the spinel lattice (see Fig. 9.3) consisting of a close-packed oxygen framework with face-centered cubic structure, plus metal ions which occupy the tetrahedral sites and octahedral sites in a ratio of 1:2. Cobalt-ferrite has the composition $Co^{2+}Fe_2^{3+}O_4$, in which one Fe^{3+} occupies a tetrahedral site, while the other Fe^{3+} and the Co^{2+} occupy two octahedral sites. We shall discuss the behavior of the Co^{2+} ions in the octahedral site, which are expected to produce a large anisotropy.

Generally speaking, the energy levels of d-electrons which are degenerate in the free ion state ((a) in Fig. 12.20) are split into doubly degenerate $d\gamma$ levels and triply degenerate $d\varepsilon$ levels when they are placed in an octahedral site (see (b) in Fig. 12.20).

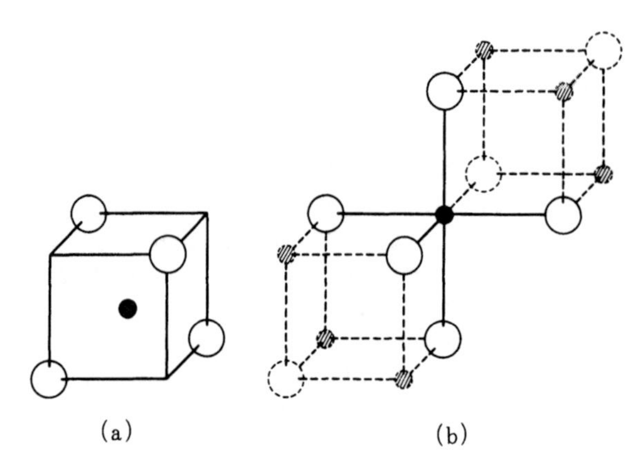

Fig. 12.19. Interstitial sites in an oxide lattice: (a) tetrahedral site; (b) octahedral sites.

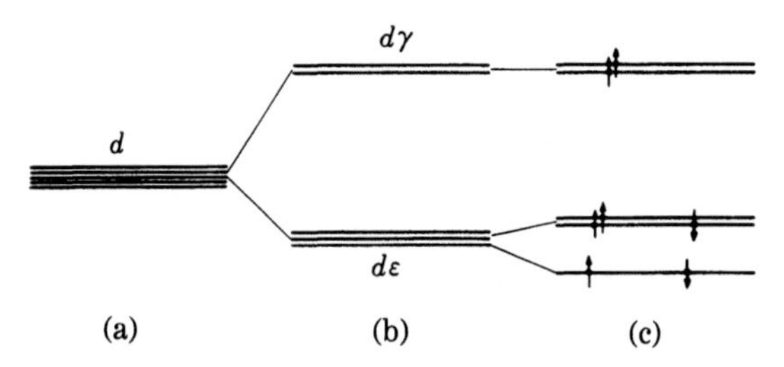

Fig. 12.20. Splitting of energy levels of $3d$ electrons by crystalline fields: (a) free ions; (b) cubic fields; (c) trigonal field (arrows represent spins in a Co^{2+} ion).

The reason for this behavior is that the $d\gamma$ wave function stretches along a cubic axis on which the nearest neighbor O^{2-} ion is located, so that because of the Coulomb interaction between the negatively charged electron and the O^{2-} ion, the energy level of $d\gamma$ is increased; while the $d\varepsilon$ wave function stretches between two cubic axes and avoids the O^{2-} ions, so that the Coulomb energy is relatively small and the $d\varepsilon$ level is lowered. In addition to this, the second nearest neighbor metal ions surrounding an octahedral site (hatched circles in Fig. 12.19(b)) are arranged symmetrically about the trigonal axis (the [111] axis in the same figure), so that they produce a trigonal field which tends to attract the electron along the trigonal axis. In consequence, the triply degenerate $d\varepsilon$ levels are split into an isolated lower single level, which corresponds to the wave function stretching along the trigonal axis and the doubly degenerate higher levels which correspond to the wave functions stretching perpendicular to the trigonal axis.

According to Hund's rule, five electrons out of seven in the Co^{2+} ion in an octahedral site fill up the + spin levels, while the remaining two electrons occupy the − spin levels (see Fig. 12.20(c)). The last electron which occupies the doubly degenerate levels can alternate between the two possible wave functions, thus realizing a circulating orbit. This orbit produces an orbital magnetic moment $\pm \boldsymbol{L}$ parallel to the trigonal axis. This orbital magnetic moment $\boldsymbol{L}$ interacts with the total spin $\boldsymbol{S}$ of the Co^{2+}.

Such spin orbit coupling (see (3.29)) is expressed as

$$w = \lambda \boldsymbol{L} \cdot \boldsymbol{S}. \tag{12.66}$$

Since the number of electrons in a Co^{2+} ion is more than half the number required for a filled shell, $\boldsymbol{L}$ is parallel to $\boldsymbol{S}$ (Hund rule (iii)), and $\lambda < 0$. When, therefore, $\boldsymbol{S}$ has a positive component parallel to the trigonal axis, $\boldsymbol{L}$ points in the + direction of this axis. When $\boldsymbol{S}$ is rotated so that it has a negative component, $\boldsymbol{L}$ is reversed. Therefore the interaction energy in this case is given by[21,22]

$$w = \lambda LS|\cos\theta|. \tag{12.67}$$

In cubic crystal there are four $\langle 111 \rangle$ axes. If Co^{2+} ions are distributed equally on octahedral sites with different $\langle 111 \rangle$ axes, the anisotropy energy produced by (12.67) becomes

$$E_a = \tfrac{1}{4} N\lambda LS(|\cos\theta_1| + |\cos\theta_2| + |\cos\theta_3| + |\cos\theta_4|), \tag{12.68}$$

where θ_1, θ_2, θ_3, and θ_4 are the angles between S and the four $\langle 111 \rangle$ axes. By Fourier expansion, $|\cos\theta|$ is reduced to

$$
\begin{aligned}
|\cos\theta| &= \frac{\pi}{3}\cos 2\theta - \frac{\pi}{15}\cos 4\theta + \cdots \\
&= \frac{\pi}{3}(2\cos^2\theta - 1) - \frac{\pi}{15}(8\cos^4\theta - 8\cos^2\theta + 1) + \cdots \\
&= \frac{18\pi}{15}\cos^2\theta - \frac{8\pi}{15}\cos^4\theta + \cdots .
\end{aligned}
\tag{12.69}
$$

Therefore (12.68) becomes

$$
\begin{aligned}
E_{\mathrm{a}} = \tfrac{1}{4}N\lambda LS\Big[&\frac{18\pi}{45}\{(\alpha_1+\alpha_2+\alpha_3)^2 + (\alpha_1+\alpha_2-\alpha_3)^2 \\
&+(\alpha_1-\alpha_2+\alpha_3)^2 + (-\alpha_1+\alpha_2+\alpha_3)^2\} \\
&-\frac{18\pi}{135}\{(\alpha_1+\alpha_2+\alpha_3)^4 + (\alpha_1+\alpha_2-\alpha_3)^4 \\
&+(\alpha_1-\alpha_2+\alpha_3)^4 + (-\alpha_1+\alpha_2+\alpha_3)^4\}\Big] \\
= -\frac{32\pi}{135} &N\lambda LS(\alpha_1^2\alpha_2^2 + \alpha_2^2\alpha_3^2 + \alpha_3^2\alpha_1^2).
\end{aligned}
\tag{12.70}
$$

Since $\lambda < 0$, the anisotropy constant in (12.70) is positive. This explains the fact that many ferrites have negative K_1, but the addition of Co tends to make K_1 positive.

In the case of metal ions other than Co^{2+}, the orbital moment L is induced through the LS coupling, and this induced L gives rise to the magnetic anisotropy through the LS coupling. For further discussion of various mechanisms of magnetic anisotropy, refer to Yosida and Tachiki,[23] Slonczewski,[24] and Kanamori.[25]

Before closing this section, let us discuss the temperature dependence of magnetic anisotropy. Generally speaking, magnetic anisotropy is produced through the interaction between spontaneous magnetization and the crystal lattice, so that the temperature dependence of spontaneous magnetization should give rise to a change in magnetic anisotropy. In fact, in any ferro- or ferrimagnetic material, the magnetic anisotropy vanishes when the spontaneous magnetization disappears at the Curie point. The temperature dependence of the anisotropy is stronger than that of the spontaneous magnetization. We shall discuss this problem along the line proposed by Zener[26] and developed by Carr.[27]

Consider a ferromagnetic material with spontaneous magnetization produced by parallel spin alignment. We assume that parallel spin clusters survive up to high temperatures because of a strong exchange interaction, so that the pair energy given by (12.59) is applicable over a wide range of temperature. At temperatures above absolute zero, the direction cosines of a local spin cluster, $(\beta_1, \beta_2, \beta_3)$, are not the same as those of the total spontaneous magnetization, $(\alpha_1, \alpha_2, \alpha_3)$, and the deviations increase with increasing temperature. Therefore, even if all the coefficients in (12.59) are independent of temperature, the magnetic anisotropy averaged over all the spin clusters decreases with increasing temperature. The cubic anisotropy at temperature T is given by

$$E_a(T) = K_1(0)\langle \beta_1^2\beta_2^2 + \beta_2^2\beta_3^2 + \beta_3^2\beta_1^2 \rangle, \tag{12.71}$$

where $K(0)$ is the anisotropy constant at 0 K and $\langle\ \rangle$ is the average of the angular function for all the spin clusters. The larger the power of the angular function in $\langle\ \rangle$, the more rapidly the function $\langle\ \rangle$ decreases with increasing temperature. According to an accurate calculation for the nth power function, we have the relationship

$$\langle K^{(n)} \rangle \propto I_s^{n(n+1)/2}, \tag{12.72}$$

where $K^{(n)}$ is the anisotropy constant for the nth power angular function. Therefore we have for $n = 2$ (uniaxial anisotropy)

$$\frac{K_u}{K_{u0}} = \left(\frac{I_s}{I_{s0}}\right)^3, \tag{12.73}$$

and for $n = 4$ (cubic anisotropy)

$$\frac{K_1}{K_{10}} = \left(\frac{I_s}{I_{s0}}\right)^{10}. \tag{12.74}$$

Figure 12.21 shows a comparison between the observed temperature dependence of K_1 for iron and the function (12.74). The agreement is satisfactory.

The magnetic anisotropy is also influenced by other factors, such as thermal expansion of the lattice,[27] thermal excitation of the electronic states of magnetic atoms, temperature dependence of the valence states, etc. It should be noted, therefore, that the mechanism described above gives the general features of the temperature dependence and is not always applicable to particular materials.

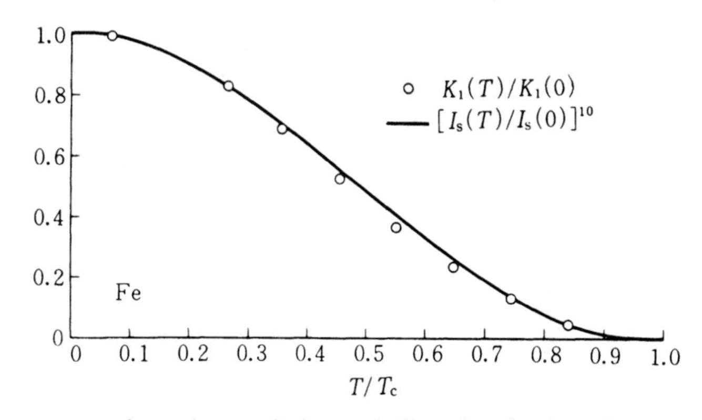

Fig. 12.21. Temperature dependence of observed K_1 value for iron in comparison with the 10th power rule. (After Carr[27])

12.4 EXPERIMENTAL DATA

Magnetocrystalline anisotropy constants have been measured for many ferro- and ferrimagnetic materials. In this section we describe some of these results. First we consider the case of materials with uniaxial anisotropy, because generally the anisotropy constants are large, and the origin of the anisotropy is comparatively clear.

12.4.1 Uniaxial anisotropy

The magnetocrystalline anisotropies of rare earth metals and alloys are extraordinarily large, because the orbital magnetic moments remain unquenched in these materials (see Section 8.3). After discussing this topic, we proceed to consider uniaxial $3d$ transition metals, oxides, and compounds.

(a) Magnetocrystalline anisotropy of rare-earth metals and alloys

In rare-earth metals, the orbital magnetic moments remain unquenched by crystalline fields, because the magnetic moments are produced by $4f$ electrons which are located relatively far inside the atoms, and are well protected from the influence of the surrounding atoms. The magnetic moments in this case originate from the total angular momentum, J, which is composed of the spin angular momentum S and the orbital angular momentum L according to Hund's rules (see Section 3.2). Table 12.2 lists the angular momenta and magnetic moments for fifteen rare earth ions and

Table 12.2. Angular momentum and magnetic moment of rare earth ions and metals.

R	No. of $4f$ el.	3+ Ions					Metals		Ref. on M_s
		S	L	J	$g\sqrt{J(J+1)}$	M_{eff}/M_B	gJ	M_s/M_B	
Sc		0	0	0	0	0	0	0	
Y		0	0	0	0	0	0	0	
La	0	0	0	0	0	0	0	0	
Ce	1	1/2	3	5/2	2.54	2.52	2.14	—	
Pr	2	1	5	4	3.58	3.60	3.20	—	
Nd	3	3/2	6	9/2	3.62	3.50	3.27	—	
Pr	4	2	6	4	2.68	—	2.40	—	
Sm	5	5/2	5	5/2	0.85	—	0.72	—	
Eu	6	3	3	0	0.00	—	0.00	—	
Gd	7	7/2	0	7/2	7.94	7.80	7.0	7.55	28
Tb	8	3	3	6	9.72	9.74	9.0	9.34	29
Dy	9	5/2	5	15/2	10.64	10.5	10.0	10.20	30
Ho	10	2	6	8	10.60	10.6	10.0	10.34	31
Er	11	3/2	6	15/2	9.58	9.6	9.0	8.0	32
Tm	12	1	5	6	7.56	7.1	7.0	3.4	33
Tb	13	1/2	3	7/2	4.53	4.4	4.0		
Lu	14	0	0	0	0	0	0		

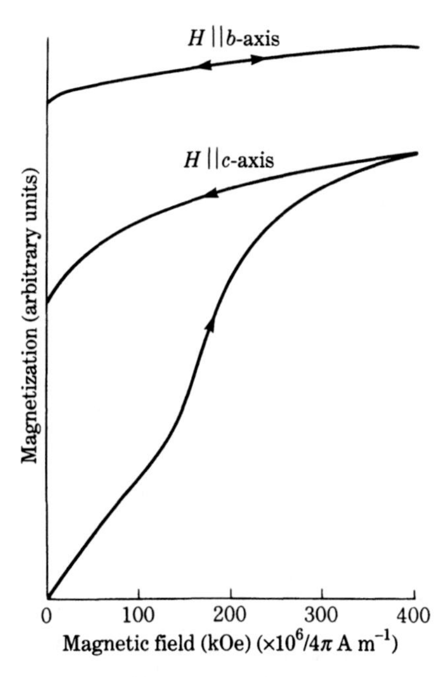

Fig. 12.22. Magnetization curve measured for a Tb metal single crystal at 77 K.[34]

metals (see also Table 8.2). In Table 12.2, we see that the observed values of the effective magnetic moments of 3+ ions and the saturation moments of ferromagnetic metals are in good agreement with the theoretical values of $g\sqrt{J(J+1)}$ and gJ, respectively. This proves that the orbital magnetic moments survive almost unchanged in the crystalline solids.

In such materials, where the orbital magnetic moment remains unquenched, rotation of the spontaneous magnetization tends to rotate the orbits in the crystal lattice, thus giving rise to a change in electrostatic interaction between orbits and the lattice. For example, let us consider how Tb metals develops its enormous magnetocrystalline anisotropy. Terbium has hexagonal crystal structure, as do many other ferromagnetic rare earth metals. When a magnetic field is applied parallel to the c-plane, the magnetization is increased rather easily, whereas when the field is applied parallel to the c-axis, the sample is magnetized only with great difficulty: only 80% of the saturation magnetization is attained in a field as high as $32\,\mathrm{MA\,m^{-1}}$ (400 kOe)[34] (see Fig. 12.22). We can say that the spontaneous magnetization is confined in the c-plane by a strong magnetocrystalline anisotropy. The anisotropy constant in this case is approximately $K_u = 6 \times 10^7\,\mathrm{J\,m^{-3}}$ ($6 \times 10^8\,\mathrm{erg\,cm^{-3}}$). This large anisotropy can be explained in terms of the shape of the $4f$ orbit and the crystal symmetry. The orbital moment of Tb is $L = 3$, which is the maximum value, so that the orbital plane spreads perpendicular to J forming pancake-like electron clouds (see Fig 12.23). On the other hand, the c/a value of the hexagonal lattice of Tb in 1.59, which is much smaller than the ideal value of 1.633 for a closed-packed hexagonal lattice. In other words, the lattice is compressed along the c-axis. If the c/a ratio were the ideal value

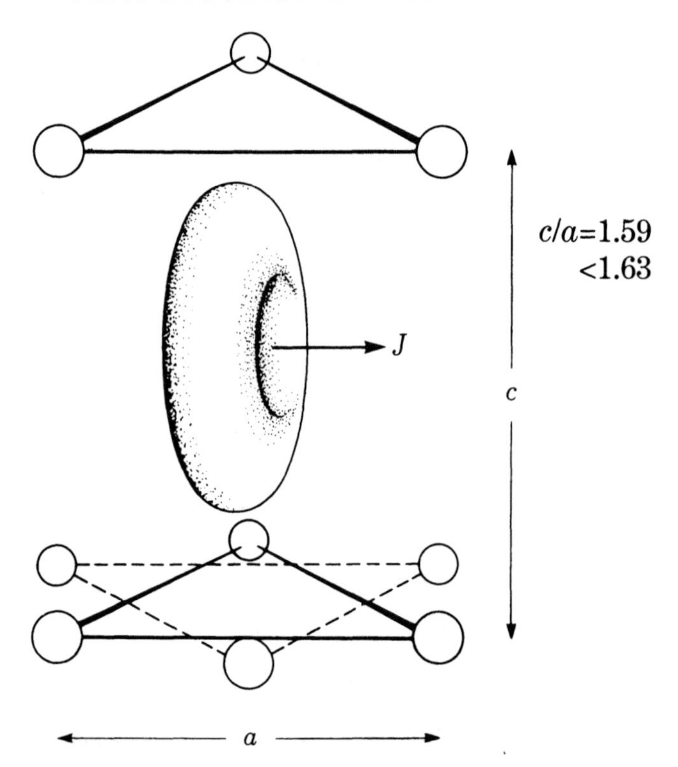

Fig. 12.23. Electron cloud of the $4f$ electron of a Tb atom in a hexagonal lattice.

and also if three nearest neighbor ions underneath a reference ion were rotated 60° about the c-axis (from the solid triangle to the broken one in Fig. 12.23), the reference ion and its neighbors would have face-centered cubic symmetry, so that no uniaxial anisotropy should be produced. The situation remains unchanged with respect to the uniaxial crystalline field about the c-axis even if the triangular atom group is rotated back to its original position. However, if the lattice is compressed along the c-axis, the neighboring $3+$ ions approach the electron clouds of the reference ion from above and below, thus attracting the clouds as shown in Fig. 12.23. Therefore J is forced to point parallel to the c-plane.

It is interesting to note that the magnetization does not retrace back along the initial magnetization curve after the maximum magnetic field is applied parallel to the c-axis (see Fig. 12.22), in spite of the fact that the magnetization process is expected to occur exclusively by reversible rotation magnetization. At the same time, the lattice is plastically deformed: elongated by 7% along the c-axis and compressed by 6% parallel to the c-plane. The same phenomenon was observed for Dy metal, and was attributed to the formation of mechanical twins which have the easy axis parallel to the applied field.[35]

Since Gd has seven $4f$ electrons, which just half fill the $4f$ shell, it has no orbital moment ($L = 0$), and accordingly exhibits no large magnetocrystalline anisotropy.

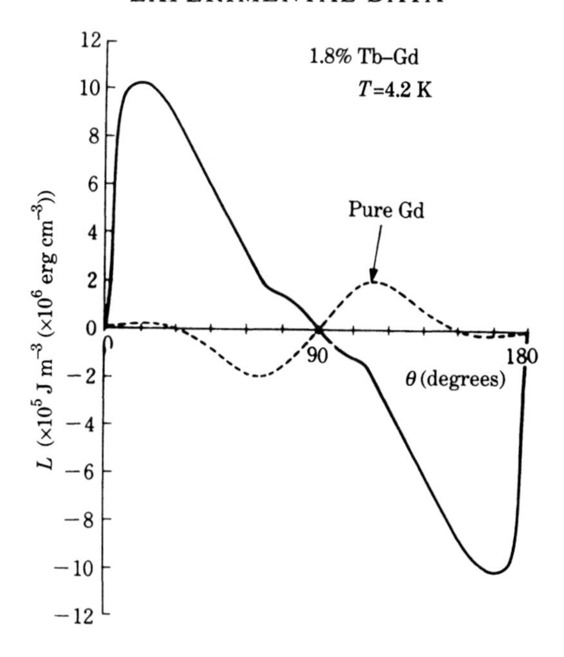

Fig. 12.24. Torque curve measured at 4.2 K for a 1.8% Tb–Gd alloy single crystal in a plane containing the c-axis. The broken curve is for pure Gd.[11]

Moreover, Gd has the maximum spin ($S = \frac{7}{2}$), so that its exchange interaction is quite large. Therefore, when a small amount of another rare earth element, R, is alloyed with Gd, the magnetic moment of R is strongly coupled by the exchange interaction with the moment of Gd, so that the two rotate together when a field is applied. The result is that Gd tends to exhibit an anisotropy characteristic of R.[11]

Figure 12.24 shows the torque curve measured for an alloy of 1.8% Tb in Gd. The dashed curve is that for pure Gd. By adding only 1.8% Tb, the amplitude of the torque curve is increased by a factor of five, showing that a large anisotropy is produced by Tb. From the uniaxial anisotropy measured in this case, the anisotropy constant per one R atom, D, is deduced and plotted as a function of the number of $4f$ electrons of R in Fig. 12.25. The solid curve is from a theory based on a rigorous calculation of $4f$ orbits and their interaction with the crystalline field of Gd.[36] The general form of this curve can be interpreted intuitively in terms of the electron clouds of $4f$ electrons, as shown in Fig. 12.26. According to Hund's rule (Section 3.2), the $4f$ electron successively takes states with magnetic quantum numbers $m = 3, 2, 1, 0, -1, -2, -3$ as the number of $4f$ electrons per atom increases (see Fig. 12.26). The shapes of the electron clouds are independent of the sign of m, as shown in Fig. 12.26. Since the c/a value of Gd metal is less than the ideal value of 1.633, a pancake-like electron cloud for $m = 3$ results in an easy plane parallel to the c-plane as in the case of Tb ($D > 0$). On the other hand, the electron cloud for $m = 0$ extends along the c-axis, thus making the c-axis an easy axis ($D < 0$). When all the states less than half are occupied with seven electrons, the resultant electron clouds become

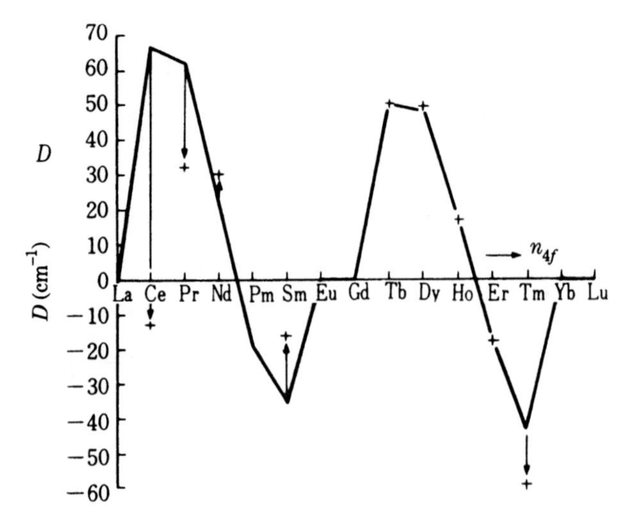

Fig. 12.25. Uniaxial anisotropy constant D per one rare earth ion dissolved in Gd metal, as a function of the number of $4f$ electrons in the dissolved rare earth.[11]

spherical, because $L = 0$. Since the shape of the electron clouds is independent of the sign of m, a hypothetical group composed of one electron in each of the states $m = 3$, 2, and 1 and a half in $m = 0$ will also result in a spherical electron cloud. In fact, the

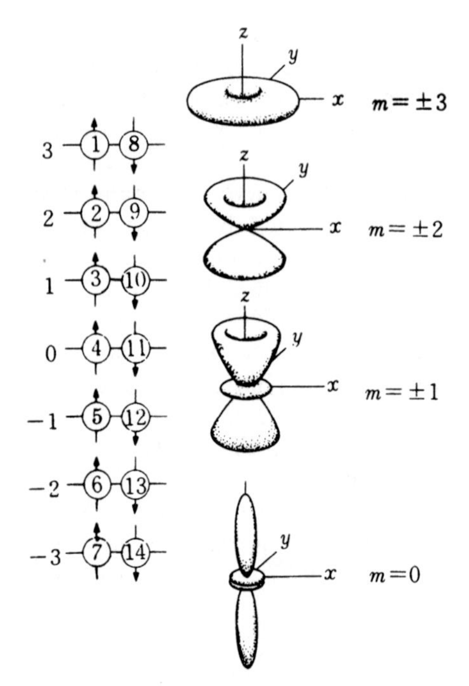

Fig. 12.26. Configurations of $4f$ electron clouds for various magnetic quantum numbers.[11]

theoretical curve in Fig. 12.25 intersects the abscissa between Nd and Pm ($n_{4f} = 3.5$) and between Ho and Er ($n_{4f} = 10.5 = 7 + 3.5$). The values of D are generally larger for $n_{4f} < 7$ than for $n_{4f} > 7$. This is because the nuclear charge is relatively smaller for $n_{4f} < 7$, so that the average radius of the $4f$ orbits is relatively larger and the interaction with the crystalline field is larger. The D-value is zero for Eu and Yb, because they are divalent in the metallic state, and accordingly are identical to Gd and La, respectively (see Fig. 3.12). The experimental points shown by crosses are in good agreement with the theory, except for Ce, Pm, and Sm. The reason for this disagreement may be that the effect of crystalline fields is more influential than the exchange interaction on the orbital states in these elements.

(b) Magnetocrystalline anisotropy of Co-alloys

Metallic cobalt is a $3d$ transition metal with hexagonal closed-packed structure which exhibits a fairly large magnetocrystalline anisotropy. One of the reasons for this large anisotropy may be the low crystal symmetry. In this connection, it is of interest to investigate how large an anisotropy would be found in hypothetical hexagonal iron or nickel. Although it is difficult or impossible to produce hexagonal ferromagnetic iron or nickel, we can see the anisotropy of iron or nickel in a hexagonal crystal by alloying these elements in hexagonal cobalt and measuring the change in the anisotropy constant.[37]

Figure 12.27 shows the temperature dependences of the uniaxial anisotropy constants, K_{u1}, of cobalt alloys containing a small amount of Cr, Mn, Fe, Ni, or Cu. Extrapolating these curves to zero absolute temperature, the values of K_{u1} at $T = 0$ were determined, and are plotted against the atomic percentage of added element M in Fig. 12.28. The dot-dash line shows the calculated effect of simple dilution, that is, the replacement of Co atoms by nonmagnetic atoms. The line for M = Cu is very close to that for simple dilution; this is expected, because Cu is nonmagnetic. The lines for all the other added elements except Fe have slopes steeper than that for

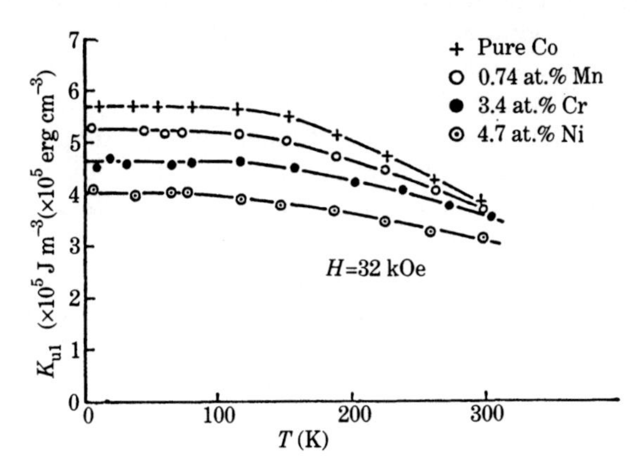

Fig. 12.27. Temperature dependences of the uniaxial anisotropy constants, K_{u1}, measured for pure Co and for Co-alloys containing small amounts of Mn, Cr and Ni.[37]

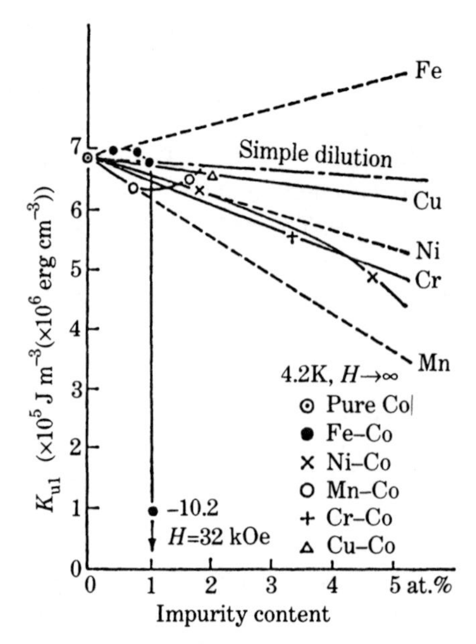

Fig. 12.28. Uniaxial anisotropy constants, K_{u1}, at 0 K for various Co-alloys as a function of the $3d$ metal alloys content.[37]

simple dilution. This means that these elements make negative contributions to K_{u1}. In the case of the Co–Fe system, the value of K_{u1} first increases and then decreases slowly with the addition of Fe up to 1.05 at% Fe, at which point K_{u1} decreases discontinuously, changes sign to negative, and finally reaches a large negative value of $-10.2 \times 10^5\,\mathrm{J\,m^{-3}}$. At the same time the crystal structure changes from closed-packed hexagonal to double hexagonal. The difference between the two is that the stacking sequence of close-packed planes along the c-axis is in the sequence ABAB... in close-packed hexagonal, and in the sequence ABACABAC... in double hexagonal. The atoms in an A-layer between B- and C-layers have the same nearest neighbors as in the face-centered cubic lattice (see Fig. 9.1). However, since K_1 for face-centered cubic Co is negative,[38] which contributes a positive K_{u1}, this change in crystal structure does not explain the dramatic change in K_{u1}. The origin of this phenomenon must be due to atomic interactions of longer range than the second-nearest neighbors. The actual origin, however, has not yet been clarified.

Figure 12.29 shows a plot of the anisotropy constant per M atom, deduced from the slope in Fig. 12.28, as a function of the number of $3d$ holes in M. It is interesting to note that Fe and Mn exhibit an anisotropy larger than Co. This behavior has been interpreted qualititatively in terms of the pseudo-localized model, in which the crystalline field effect results in the following level scheme: $m = \pm 1$ are the lowest, $m = \pm 2$ the next, and $m = 0$ is the highest.[39]

(c) Magnetocrystalline anisotropy of rare earth–3d metal compounds

The rare earth–cobalt compounds of the compositions RCo_5 and R_2Co_{17} have

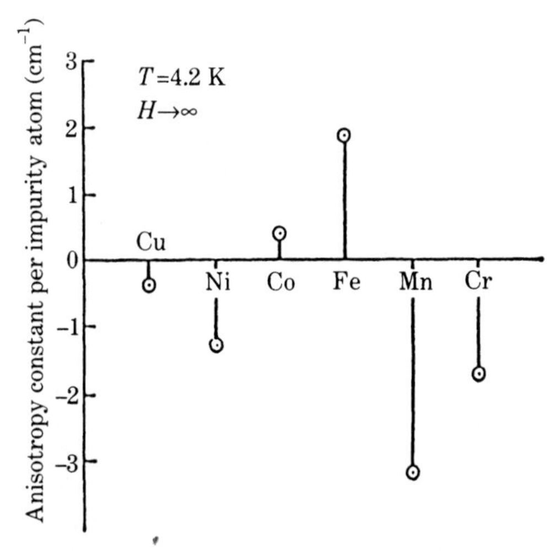

Fig. 12.29. Uniaxial anisotropy constant per added $3d$ alloy atom in Co as a function of vacancies in the $3d$ band.[37]

hexagonal crystal structure, so that they exhibit strong magnetocrystalline anisotropies. Table 12.3 list various magnetic data[40] for the compounds of the RCo_5-type. As seen in this table, the anisotropy constants amount to 10^6–$10^7\,J\,m^{-3}$ (10^7–10^8 erg cm^{-3}). This is due to a very strong crystalline field acting in a crystal of a very uniaxial nature (see Fig. 8.26). The compounds of the R_2Co_{17}-type exhibit comparable values of magnetocrystalline anisotropy.[41]

The rare earth compound $Nd_2Fe_{14}B$, which is the main constituent of NdFeB magnets, has the tetragonal structure (see Section 10.6), and exhibits a very strong magnetocrystalline anisotropy. The value of K_{u1} is $-6.5\times10^6\,J\,m^{-3}$ at 0 K, but it changes sign at 126 K, and reaches $+3.7\times10^6\,J\,m^{-3}$ at 275 K.[42]

(d) Magnetocrystalline anisotropy of hexagonal ferrites

The crystal structures and magnetic properties of hexagonal ferrites were described in detail in Section 9.4. The crystal structure is the magnetoplumbite-type hexagonal, which contains Fe^{3+}, M^{2+} and Ba^{2+}, Sr^{2+}, or Pb^{2+}. The ionic radii of Ba^{2+}, Sr^{2+}, and Pb^{2+} are 1.43, 1.27, and 1.32 Å, respectively, which are very large as compared with

Table 12.3. Saturation moment and anisotropy constants of RCo_5.[40]

Compounds	Curie point (K)	M_s/M_B at 0 K per RCo_5	M_s/M_B at 0 K per Co	K_{u1} at 0 K ($MJ\,m^{-3}$)	K_{u2} at 0 K ($MJ\,m^{-3}$)
YCo_5	978	7.9	1.57	6.5	~0
$CeCo_5$	673	6.6	1.32	5.5	~0
$PrCo_5$	921	10.4	1.44	−7	18
$NdCo_5$	913	10.4	1.43	−40	19
$SmCo_5$	984	7.7	1.40	10.5	~0

Table 12.4. Magnetocrystalline anisotropy of hexagonal ferrites at 20°C.

Symbols	Composition	K_{u1} ($10^5\,J\,m^{-3}$)	$K_{u1}+2K_{u2}$ ($10^5\,J\,m^{-3}$)	I_s (T)	Θ_f (°C) (Ref. 44)	Ref.
M	$BaFe_{12}O_{19}$	3.3		0.478	450	43
Fe_2W	$BaFe_{18}O_{27}$	3.0		0.395	455	43
FeCoW	$BaCoFe_{17}O_{27}$	0.21	$K_{u2}=0.56$			44
Co_2W	$BaCo_2Fe_{16}O_{27}$	−1.86	$K_{u2}=0.75$			44
FeZnW	$BaZnFe_{17}O_{27}$	2.4		0.478	430	43
Ni_2W	$BaNi_2Fe_{16}O_{27}$	2.1		0.415		43
Mg_2Y	$Ba_2Mg_2Fe_{12}O_{22}$		−0.6	0.150	280	43
Ni_2Y	$Ba_2Ni_2Fe_{12}O_{22}$		−0.9	0.160	390	43
Zn_2Y	$Ba_2Zn_2Fe_{12}O_{22}$		−1.0	0.285	130	43
Co_2Y	$Ba_2Co_2Fe_{12}O_{22}$		−2.6	0.232	340	43
Co_2Z	$Ba_3Co_2Fe_{24}O_{41}$		−1.8	0.339	410	43

the radii of Fe^{3+} and M^{2+}, (0.6–0.8 Å), and are comparable with that of O^{2+} (1.32 Å). Therefore these large metal ions cannot enter the interstitial sites in the closed-packed oxygen lattice, but occupy substitutional sites in the layer parallel to (111) of the spinel phase, thus forming hexagonal crystal structures. Depending on the structures of the layers containing Ba^{2+}, etc., and their ratios to the spinel phase, the compounds are classified into M-, W-, Y-, and Z-types. These compounds are all ferrimagnetic with Curie points about 400°C and with saturation magnetizations at room temperature from 0.2 to 0.5 T.

The magnetocrystalline anisotropy of these hexagonal ferrites is expressed by (12.1), because in spite of the presence of a cubic spinel phase the overall symmetry of the crystal is uniaxial. Table 12.4 lists magnetocrystalline anisotropy constants measured at room temperature for various hexagonal ferrites, together with other magnetic properties. In this case, the anisotropy constants were determined by means of a magnetic torsion pendulum[45] from the frequency of the torsional vibration of a specimen with the *c*-axis perpendicular to the torsional axis. In the case of positive K_{u1}, the *c*-axis is the easy axis, while for negative K_{u1} the *c*-plane is the easy plane. The frequency is related to the second derivative of the anisotropy energy (12.1) with respect to the polar angle, θ, so that it is proportional to K_{u1} for a positive K_{u1} and to $K_{u1}+2K_{u2}$ for a negative K_{u1}.

As seen in Table 12.4, the anisotropy constants for these hexagonal ferrites are all of the order of $10^5\,J\,m^{-3}$ ($10^6\,erg\,cm^{-3}$). It is interesting that such a large anisotropy is observed in spite of the fact that the main magnetic constituent of these ferrites is Fe^{3+} with no orbital magnetic moment. From the calculation of a dipole sum, the anisotropy constant for a Y-type compound was expected to be $K_{u1} = -5$ to $-7 \times 10^5\,J\,m^{-3}$ ($\times 10^6\,erg\,cm^{-3}$) at 0 K.[46] However, a similar calculation for the M-type compound led to $K_{u1} = -1.5 \times 10^5\,J\,m^{-3}$, which has the opposite sign of the observed value. The actual origin of this anisotropy is considered to be an Fe^{3+} ion surrounded by five O^{2-} ions.[46] Kondo[47] showed that an Mn^{2+} ion, which has the same electronic configuration as an Fe^{3+} ion, can produce the expected anisotropy; he considered a distortion of the ion, overlapping of the electron clouds with the surrounding O^{2+}, and mixing of the excited states.

When the anisotropy constants K_{u1} and K_{u2} have opposite signs, and also the relationship $2|K_{u1}| > |K_{u2}|$ holds, the anisotropy results in an easy cone of magnetization with polar half-angle $\sin^{-1}\sqrt{-K_{u1}/2K_{u2}}$.[46] The mechanism of the easy cone in Co_2Y and Co_2Z has been explained in terms of the Co^{2+} anisotropy[48] discussed in Section 12.3.

Barium and strontium ferrites, which have the easy axis parallel to the *c*-axis, are used as permanent magnet materials (see Section 22.2). Ferroxplana, which has an easy plane parallel to the *c*-plane, has excellent high-frequency characteristics (see Section 20.3). These useful properties are based on the strong magnetocrystalline anisotropy of these materials.

The uniaxial anisotropy observed in the low-temperature phase of magnetite (Fe_3O_4) will be discussed in relation to the cubic anisotropy of this material (see Section 12.4.2(b)).

(e) Uniaxial anisotropy of magnetic compounds

Magnetism of ferro- and ferrimagnetic compounds other than oxides was described in Chapter 10. The magnetocrystalline anisotropy of typical uniaxial compounds will be described here.

Pyrrhotite ($Fe_{1-x}S$) has the NiAs-type crystal structure and exhibits ferrimagnetism as a result of the ordering of vacancies in the stoichiometric compound FeS, which has antiferromagnetic spin structure (see Section 10.4). The compound Fe_7Se_8 has a similar spin structure and also exhibits ferrimagnetism.

The feature of the magnetocrystalline anisotropy of these compounds is that the easy axis is parallel to or near the *c*-axis at low temperatures, and gradually approaches the *c*-plane with increasing temperature. In the case of FeS, which is antiferromagnetic, the spin axis is parallel to the *c*-axis in the temperature range from 0 to 400 K, and then rotates discontinuously to the *c*-plane at 400 K. The spin axis of the ferrimagnetic Fe_7S_8 tilts by 20° at 0 K, rotates gradually towards the *c*-plane with increasing temperature, and reaches the *c*-plane at about 80 K. It was observed for Fe_7Se_8 that the temperature range for the rotation of the easy axis, and the shape of the curve of easy axis angle vs temperature, depend on the ordering scheme of the lattice vacancies. These phenomena were interpreted by the one-ion model of the Fe^{2+} ion surrounded by lattice vacancies.[49]

The compound MnBi, which is a NiAs-type ferromagnetic compound (Section 10.3), exhibits a very strong uniaxial anisotropy with the easy axis parallel to the *c*-axis. The anisotropy constants measured for a polycrystalline material with the *c*-axes aligned parallel are $K_{u1} = 9.1 \times 10^5\,\mathrm{J\,m^{-3}}$ $(9.1 \times 10^6\,\mathrm{erg\,cm^{-3}})$ and $K_{u2} = 2.6 \times 10^5\,\mathrm{J\,m^{-3}}$ $(2.6 \times 10^6\,\mathrm{erg\,cm^{-3}})$.[50] When this compound is evaporated onto a glass substrate and annealed, the *c*-axis is oriented perpendicular to the substrate surface. It has been demonstrated that 'magnetic writing' can be performed by using a magnetic needle to locally magnetize such a grain-oriented MnBi film.[51]

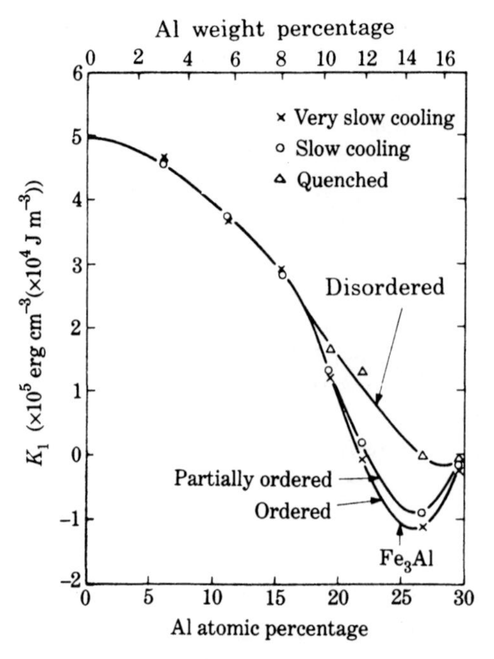

Fig. 12.30. Variation of K_1 with composition in Fe–Al alloys (room temperature). (After Hall[52])

12.4.2 Cubic anisotropy

Generally speaking, cubic anisotropy is smaller than uniaxial anisotropy, because of its higher crystal symmetry. Therefore most soft magnetic materials, which exhibit high permeability and low coercive force, have cubic crystal structures with low cubic anisotropy (see Section 22.1). The origin of the cubic anisotropy is not as clear as the uniaxial anisotropy, because many higher-order terms caused by various mechanisms can contribute to this anisotropy. Some of the experimental results will be summarized below.

(a) Magnetocrystalline anisotropy of 3d transition metals and alloys

First we will discuss the anisotropy of iron alloys with body-centered cubic structure.

The anisotropy constant K_1 of Fe–Al alloys decreases monotonically with the addition of nonmagnetic Al as shown in Fig. 12.30.[52] The exact nature of the variation depends on the annealing of the alloy, which produces a superlattice as already explained in Chapter 10 (see Fig. 10.3). The value of K_1 becomes more negative as the degree of order increases.

The K_1 value of Fe–Co alloys also decreases with an increase of Co content. The exact form of the decrease also depends on the annealing treatment, suggesting the formation of the superlattices Fe_3Co and FeCo (Fig. 12.31).

Figure 12.32 shows the composition dependence of K_1 of Fe alloys with the

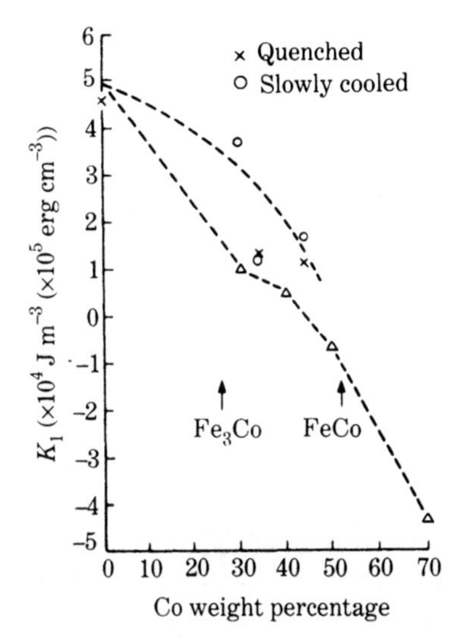

Fig. 12.31. Variation of K_1 with composition in Fe–Co alloys (room temperature). (After Hall[52])

nonmagnetic elements Ti[53] and Si.[54] The saturation magnetization of Fe–Si alloys decreases with the addition of Si as if Si simply dilutes the magnetization with the atomic magnetic moment of Fe remaining constant, as already discussed in Chapter 8. The variation of K_1, however, is much more steep. It is interesting to note that in the case of Fe–Ti alloys, the value of K_1 increases with the addition of nonmagnetic Ti.

The Ni–Fe alloys with face-centered cubic structures form the Cu_3Au-type superlattice Ni_3Fe, as shown in Fig. 12.33. The ordering temperature of this superlattice is about 600°C, below which the ordering develops.[55] When the alloy with composition Ni_3Fe is quenched from about 600°C, it retains the disordered state, in

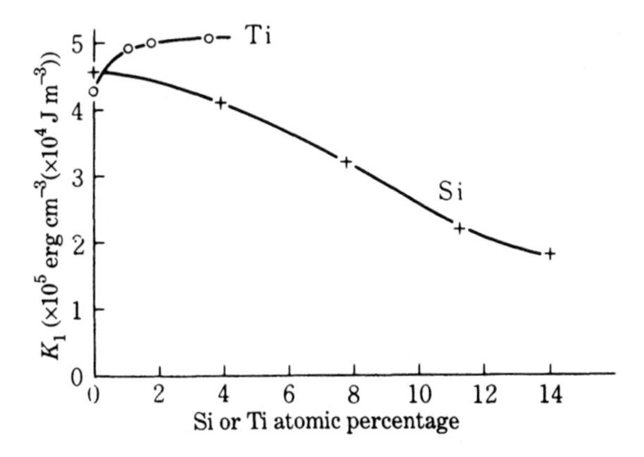

Fig. 12.32. Variation of K_1 with composition in Fe–Ti and Fe–Si alloys (room temperature).

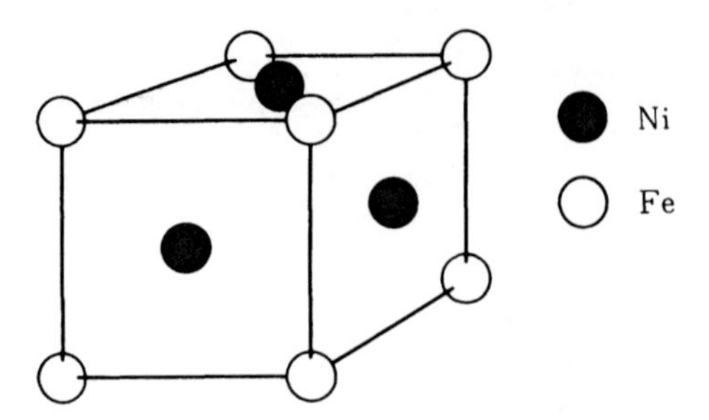

Fig. 12.33. The superlattice Ni_3Fe.

which the lattice sites are occupied by Ni and Fe at random, with the probability of 3:1. The anisotropy constant K_1 in this state is very small or almost zero, but it increases with increased ordering. Figure 12.34 shows the variation of K_1 as a function of logarithm of the cooling rate in the temperature range from 600°C to 300°C, during which the superlattice is formed.[56] Figure 12.35 shows the composition dependence of K_1 for the ordered and disordered states.[56] The alloy corresponding to $K_1 = 0$ is called Permalloy, which exhibits extraordinarily large permeability (see Fig. 18.27 and Section 22.1). The K_1 values of the Fe-rich alloys containing more than 65% Fe decrease once again with increasing Fe content, reflecting the instability of ferromagnetism due to the Invar characteristic[57] (see Section 8.2).

In Ni–Co alloys, K_1 changes sign from negative to positive at 4 at% Co, and then once again from positive to negative at 20 at% Co[52,58,59] (Fig. 12.36). This complicated compositional dependence of K_1 is quite different from the monotonic change in the saturation magnetization as seen in the Slater–Pauling curve (see Fig. 8.12). Similar behavior can be seen for Ni–Cr and Ni–V[60] in Fig. 12.36.

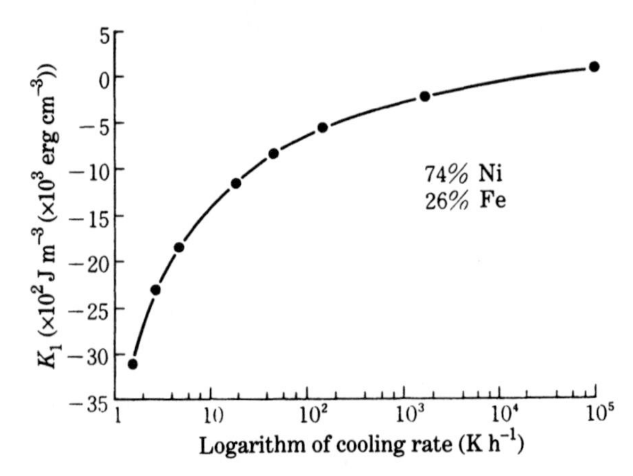

Fig. 12.34. Variation of K_1 in Ni_3Fe as a function of logarithm of cooling rate in the range 600–300°C (room temperature). (After Bozorth[56])

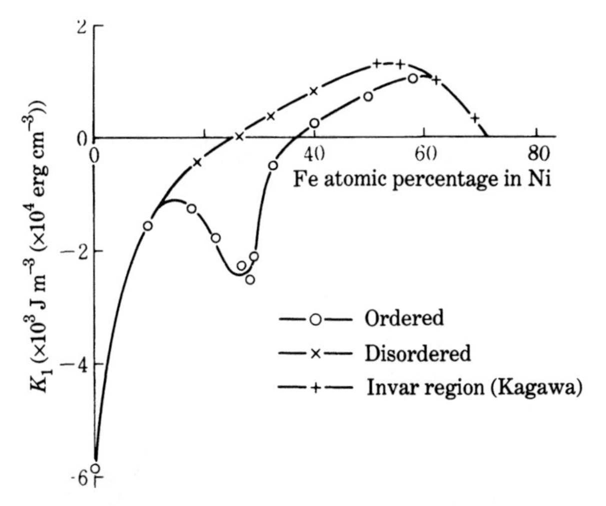

Fig. 12.35. Composition dependence of K_1 at room temperature for ordered and disordered Ni–Fe alloys. (After Bozorth[56] and Kagawa and Chikazumi[57])

(b) Magnetocrystalline anisotropy of spinel-type ferrites

The crystal and magnetic structures of spinel-type ferrites were discussed in detail in Section 9.2. The essential features may be summarized as follows.

The composition of these oxides is represented by the chemical formula $M^{2+}O \cdot Fe_2^{3+}O_3$ (M = Mn, Fe, Co, Ni, Cu, Zn, Cd and Mg). The crystal structure is the spinel-type, whose unit cell is composed of 32 close-packed O^{2-} ions and 24 interstitial metallic ions, such as M^{2+} and Fe^{3+}, located on two sites called the A and B sites (see Fig. 9.3). In a normal spinel, 8 M^{2+} ions occupy A sites, while the other 16 Fe^{3+} ions occupy B sites. In an inverse spinel, 8 Fe^{3+} ions occupy A sites, while the

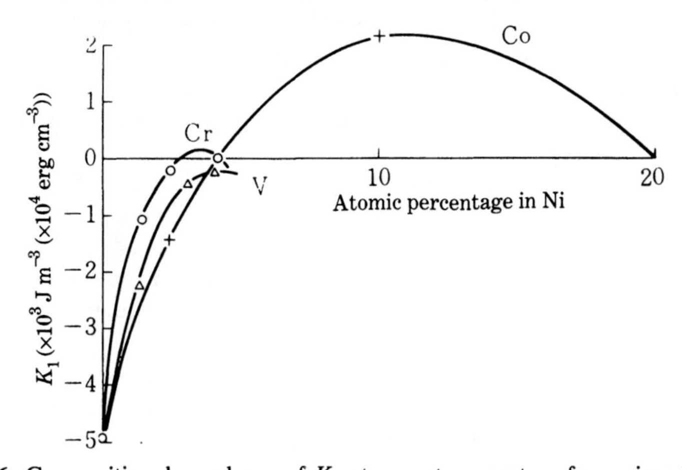

Fig. 12.36. Composition dependence of K_1 at room temperature for various Ni alloys.

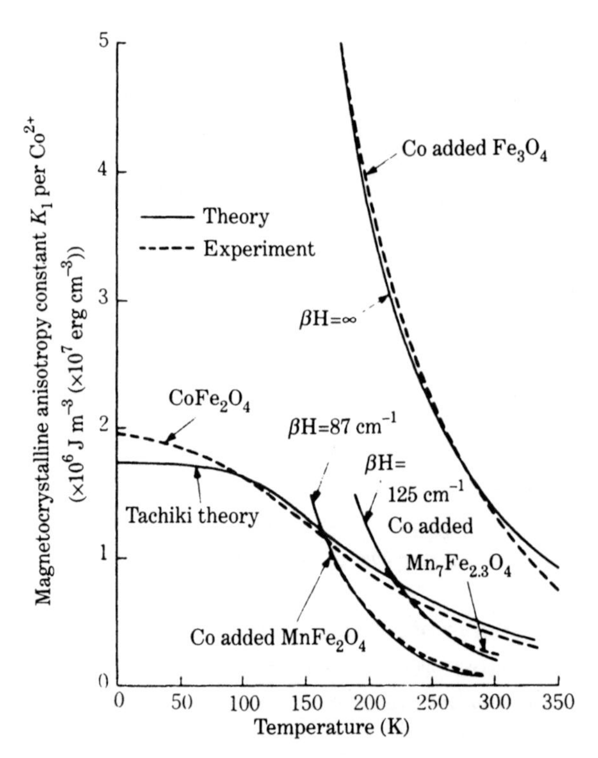

Fig. 12.37. Temperature dependence of K_1 for various spinel ferrites containing Co^{2+}. (After Slonczewski[24])

remaining 8 Fe^{3+} ions and 8 M^{2+} ions occupy B sites. Most ferrimagnetic ferrites have the inverse spinel structure, in which magnetic moments of Fe^{3+} ions on A sites couple antiparallel to those of Fe^{3+} and M^{2+} on B sites, so that only the magnetic moments of M^{2+} ions contribute to the net spontaneous magnetization. Generally speaking, the shape of the ions which occupy A sites, such as Zn^{2+}, Mn^{2+} and Fe^{3+}, are spherical (see Section 3.2), so that these ions do not make a large contribution to the magnetic anisotropy.

The magnetocrystalline anisotropy of ferrites is interpreted by the one-ion model as explained in Section 12.3. As explained there, the Co^{2+} ions on B sites produce a positive K_1. Most ferrites have negative K_1, so that the addition of Co tends to decrease the magnitude of K_1. Figure 12.37 shows the contributions of Co^{2+} ions to K_1 for Fe_3O_4, $Mn_{0.7}Fe_{2.3}O_4$, $MnFe_2O_4$, etc. when Co is added. It is seen that the contribution of Co^{2+} ions to K_1 is positive for all the ferrites.

The Fe^{2+} ions also make a large contribution to the magnetic anisotropy. One example is the anomalous behavior of K_1 of magnetite or Fe_3O_4, as as shown in Fig. 12.38. The value of K_1 is negative at room temperature, decreases with decreasing temperature, begins to increase at 230 K and changes sign to positive at 130 K. Below about 125 K, or the Verwey point, T_v, the crystal transforms to a structure of lower

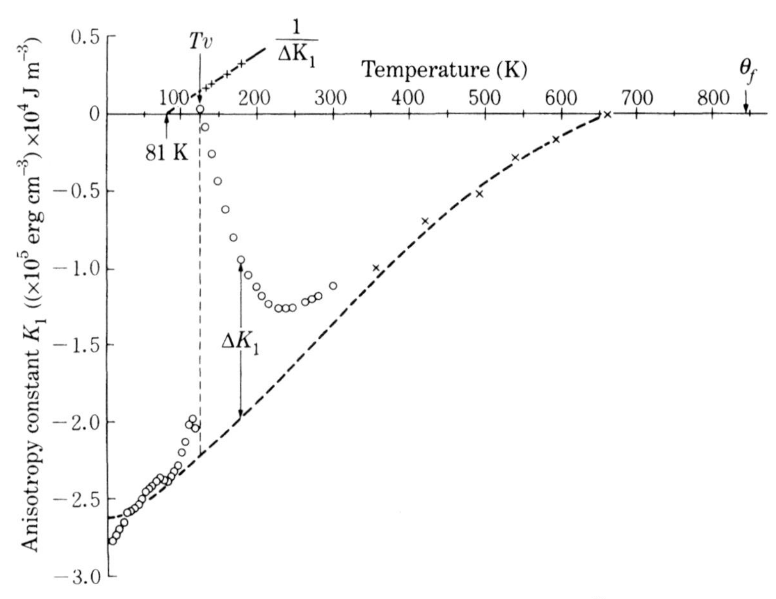

Fig. 12.38. Temperature dependence of K_1 for magnetite (Fe_3O_4).[61] The K_1 values below T_v were calculated by (12.77).

symmetry (see Section 9.2), so that the magnetocrystalline anisotropy becomes uniaxial. The a-, b-, and c-axes of the low temperature phase are almost parallel to the [1$\bar{1}$0], [110] and [001] axes of the original cubic structure. Let α_a, α_b, and α_c be the direction cosines of spontaneous magnetization referred to the a-, b-, and c-axes. Then the magnetocrystalline anisotropy energy is expressed as

$$E_a = K_0 + K_a\alpha_a^2 + K_b\alpha_b^2 - K_u\alpha_{111}^2 + K_{aa}\alpha_a^4 + K_{bb}\alpha_b^4 + K_{ab}\alpha_a^2\alpha_b^2, \quad (12.75)$$

where α_{111} is the direction cosine referred to the longest body-diagonal or [$\bar{1}$11]. The temperature dependence of these anisotropy constants,[61] as shown in Fig. 12.39, was found to be expressed by the Arrhenius equation,

$$\Delta K_i = K_{0i}e^{-Q/kT}. \quad (12.76)$$

The activation energy Q was found to be 0.20–0.23 eV for K_a, K_b, K_u. The electrical conductivity has a similar temperature dependence, with the same activation energy. It seems clear that this phenomenon is caused by a hopping motion of electrons between Fe^{2+} and Fe^{3+} on the B-sites. Therefore the origin of the magnetic anisotropy of the low-temperature phase is considered to also be related to Fe^{2+} ions on B-sites.

The higher order constants K_{aa}, K_{bb}, and K_{ab} correspond to K_1 of the cubic phase. If the K_1 term is converted to the coordinate system for the low temperature phase, the magnitude of these constants should be in the ratio 3:3:10. The actual

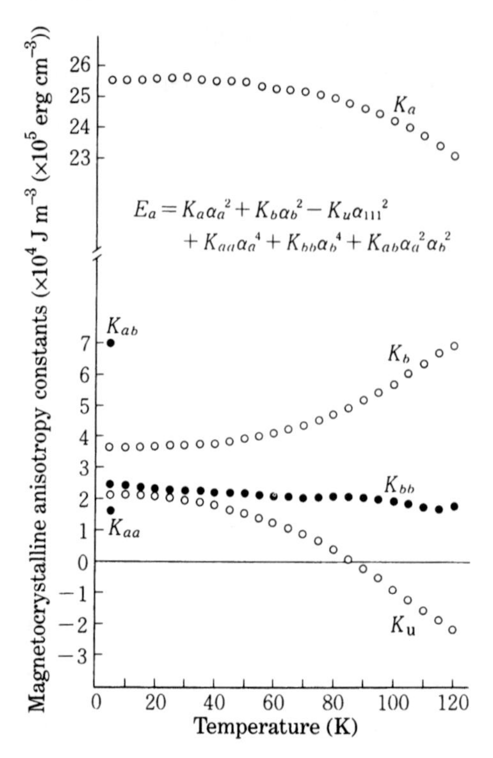

Fig. 12.39. Temperature dependence of magnetic anisotropy constants for low-temperature phase of magnetite (Fe_3O_4).[61]

ratio observed at 4.2 K was 2.6 : 3.4 : 10, which is very close to the prediction. Then the K_1 value can be calculated by the relationship

$$K_1 = \tfrac{1}{3}\left(-\tfrac{4}{3}K_{aa} - \tfrac{4}{3}K_{bb} - \tfrac{2}{5}K_{ab}\right), \tag{12.77}$$

which is plotted in Fig. 12.38 in the temperature range $T < T_v$.[61] It is to be noted that these points fit well with the extrapolation of the K_1 vs. T curve above 350 K. We can regard the deviation of K_1, ΔK_1, from the broken curve connecting the points at high and low temperatures as caused by some clustering related to the Verwey transition. In the same figure, the reciprocal of ΔK_1 is plotted against temperature as a solid line, which extrapolates to the abscissa at 81 K.[62]

The low-temperature transition of magnetite was first found by Verwey,[63] who assumed an ordered arrangement of Fe^{2+} and Fe^{3+} ions on B-sites as a result of a cessation of the electron hopping between these ions. This model was, however, disproved by later experiments.[64] This transition may be a sort of structural transition, in which some phonons accompanying a charge ordering are softened with decreasing temperature, and finally condensed to a low-temperature phase. As a matter of fact, various physical quantities are condensed towards 81 K.[62] The variation of K_1 is regarded as one of the critical phenomena associated with this transition.[65]

Miyata[66] measured the temperature dependence of K_1 for various mixed ferrites.

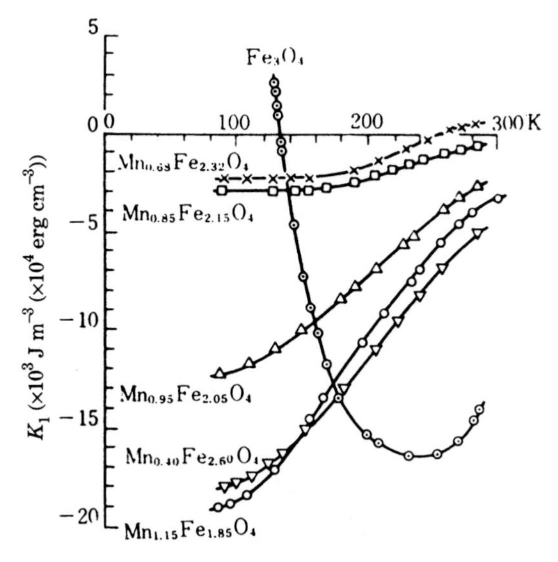

Fig. 12.40. Temperature dependence of K_1 for $MnFe_2O_4$–Fe_3O_4 mixed ferrites. (After Miyata[66])

Figures 12.40–12.42 show the results for Fe_3O_4 with added Mn, Ni, and Zn. Comparing these graphs, we see that the anomalous increase of ΔK_1 observed in Fe_3O_4 is

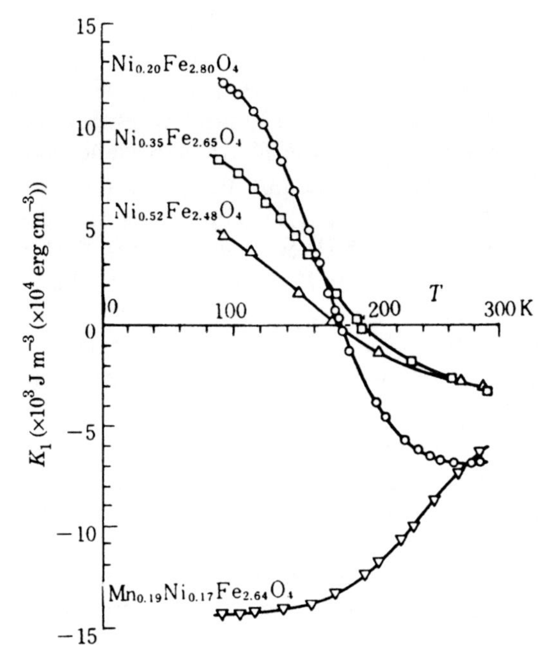

Fig. 12.41. Temperature dependence of K_1 for $NiFe_2O_4$–Fe_3O_4 mixed ferrites. (After Miyata[66])

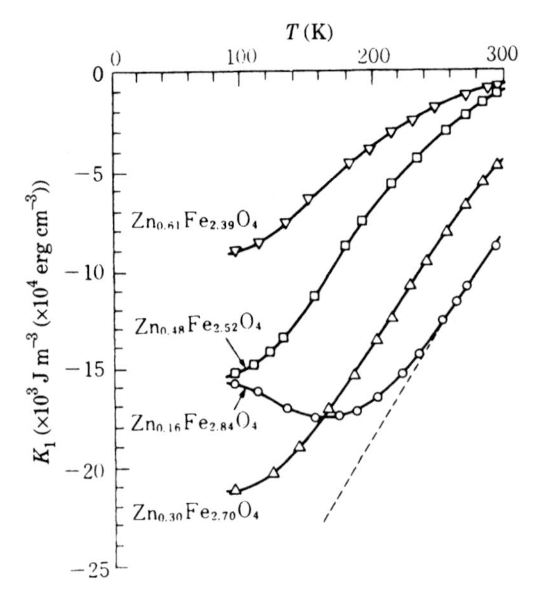

Fig. 12.42. Temperature dependence of K_1 for $ZnFe_2O_4$–Fe_3O_4 mixed ferrites. (After Miyata[66])

not very much influenced by the addition of Ni, while this anomaly is completely suppressed by the addition of Mn or Zn. The Ni^{2+} ions go into B-sites and do not disturb Fe^{3+} ions on A-sites, whereas the Mn^{2+} or Zn^{2+} ions go into narrow A-sites and change the instability of the lattice, because they have ionic sizes larger than Fe^{+3}. This fact suggests that the ΔK_1 anomaly may possibly be caused by Fe^{2+} ions coupled with the lattice instability related to the low temperature transition.

In the mixed ferrites $Mn_xFe_{3-x}O_4$, the K_1 value becomes more positive as soon as x is reduced from 1, as shown in Fig. 12.43.[67] For $x = 1$, the composition is $MnFe_2O_4$, in which no Fe^{2+} ion exists, while Fe^{2+} ions appear as soon as x is reduced from 1. When Ti^{4+} ions are added to Fe_3O_4, they go into B-sites, thus increasing the number of Fe^{2+} ions. A large positive shift of K_1 was observed by this addition.[68] These facts tell us that the Fe^{2+} ions on B-sites contribute to a positive K_1.

Magnetic ions other than Co^{2+} and Fe^{2+} do not contribute strongly to the magnetic anisotropy. Yosida and Tachiki[23] summarized the origin of magnetic anisotropy as follows:

(1) The contribution from magnetic dipole interaction is small.
(2) Deformation of the $3d$ shells of magnetic ions produces a magnetic anisotropy through a dipole interaction in the shell (a kind of shape anisotropy (see Section 12.1)).
(3) An excited state with an orbital magnetic moment is mixed through the spin–orbit interaction and this orbital magnetic moment produces magnetic anisotropy through the interaction with spins.

If should be noted that mechanisms (2) and (3) are not realized for magnetic ions

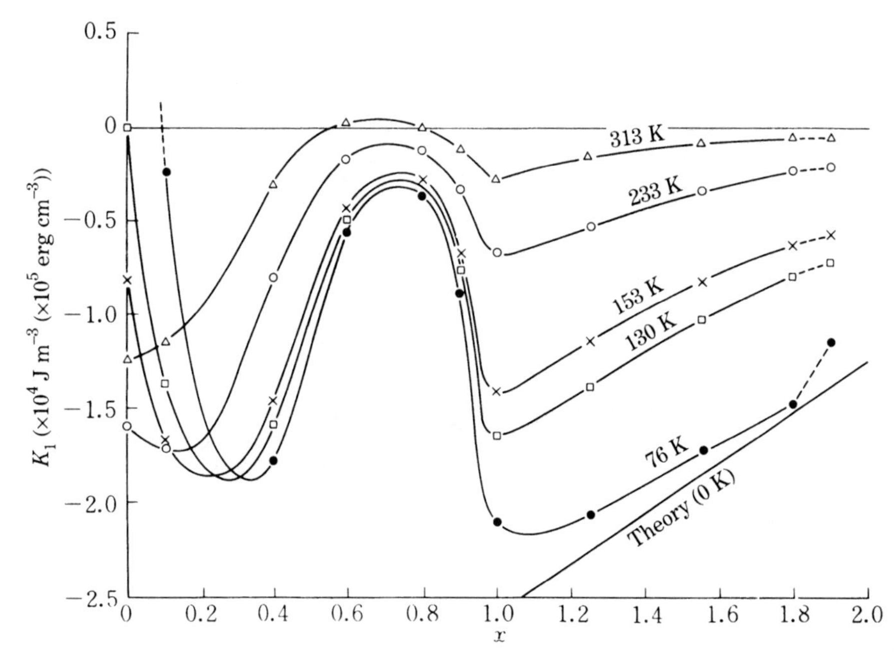

Fig. 12.43. Variation of K_1 for $Mn_xFe_{3-x}O_4$ ($0 \leq x \leq 2$) with x measured at various temperature. (After Penoyer and Shafer[67])

with $S = \frac{1}{2}$, 1 and $\frac{3}{2}$, because these magnetic shells do not deform. For instance, Ni^{2+} with $3d^8$ ($S = 1$) and Co^{2+} with $3d^7$ ($S = \frac{3}{2}$) do not develop magnetic anisotropy from these mechanisms. In the case of Mn^{2+} and Fe^{3+} with $3d^5$ ($S = \frac{5}{2}$), the $3d$ shell is deformable. Figure 12.44 shows the temperature dependence of K_1/I_s observed for Mn-ferrite or $MnFe_2O_4$ and the comparison with the theoretical curve calculated on this basis.[23] We see that the agreement is excellent.

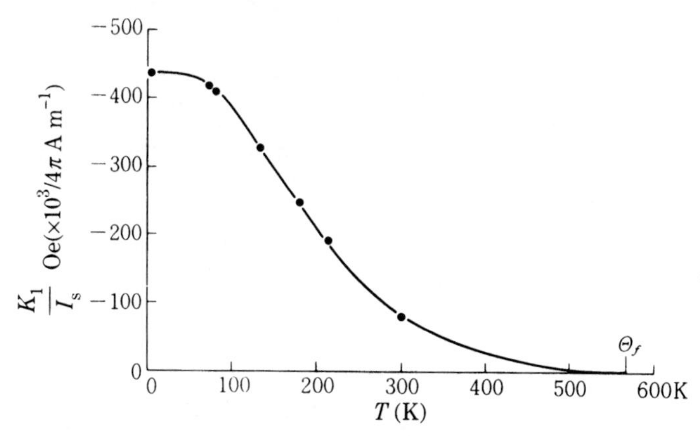

Fig. 12.44. Temperature dependence of K_1/I_s for $MnFe_2O_4$ and the comparison with theory (solid curve). (After Yosida and Tachiki[23])

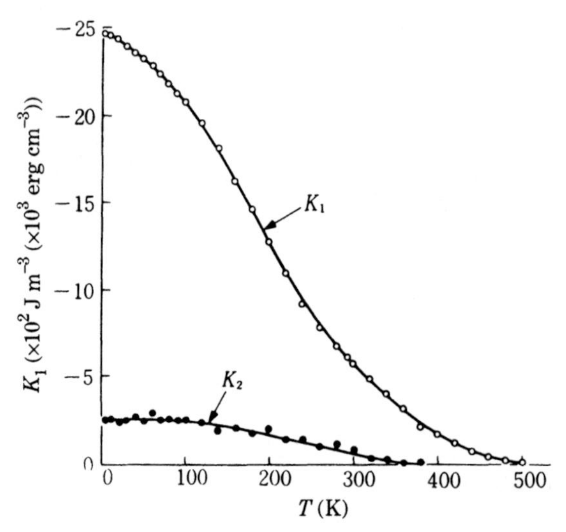

Fig. 12.45. Temperature dependence of K_1 and K_2 for YIG ($Y_3Fe_5O_{12}$). (After Hansen[69])

(c) Magnetocrystalline anisotropy of garnet-type ferrites

The details of this class of oxide have been described in Chapter 9. The essential points are as follows:

The garnet-type ferrites are oxides containing rare earth (R) and iron (Fe) ions with a complicated crystal structure represented by a chemical formula $3R_2O_3 \cdot Fe_2O_3$ or $R_3Fe_5O_{12}$. All the metal ions are trivalent and no divalent ions exist, so that there is no electron hopping. Five Fe^{3+} ions in a formula unit of $R_3Fe_5O_{12}$ occupy $24d$ and $16a$ lattice sites in the ratio 3:2, with their spins antiparallel, thus exhibiting ferrimagnetism. The Curie point is 540–580 K. Since the Fe^{3+} ion has $5M_B$ (Bohr

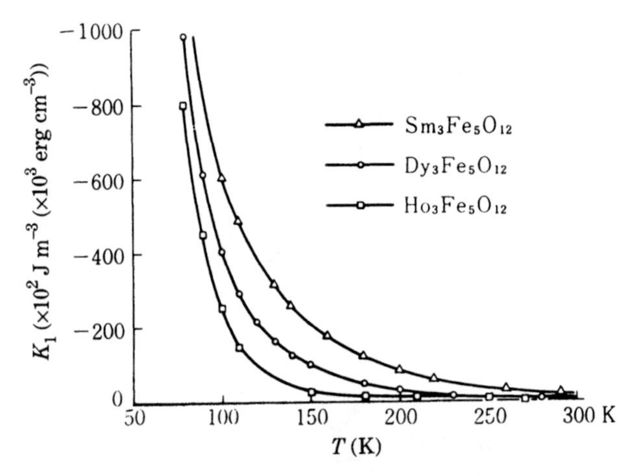

Fig. 12.46. Temperature dependence of K_1 for $Sm_3Fe_5O_{12}$, $Dy_3Fe_5O_{12}$ and $Ho_3Fe_5O_{12}$.[70]

magnetons), these five Fe^{3+} ions contribute $5M_B$ to the resultant magnetic moment per formula unit at 0 K. In addition, the magnetic moment of three R^{3+} ions on the 24c sites couple with this resultant Fe moment ferrimagnetically, thus producing a large spontaneous magnetization at 0 K. However, since the interaction between R^{3+} and Fe^{3+} moments is weak, the contribution of the R^{3+} moment decreases rapidly with increasing temperature, and is compensated by the Fe^{3+} moment at a particular temperature called the compensation point. In other words, the spontaneous magnetization of most garnet-type ferrites exhibits the N-type temperature dependence (see Fig. 9.9).

The magnetocrystalline anisotropy of the garnet-type ferrites is composed of two parts: a contribution from Fe^{3+} ions, and a contribution from R^{3+} ions. The former contribution can be found experimentally by measuring the anisotropy of yttrium–iron–garnet or YIG, because Y has no magnetic moment. Figure 12.45 shows

Table 12.5. Magnetocrystalline anisotropy constants of garnet-type ferrites, $R_3Fe_5O_{12}$, measured at various temperatures.[70]

R	T (K)	K_1 ($10^4\,J\,m^{-3}$), ($10^5\,erg\,cm^{-3}$)	K_2 ($10^4\,J\,m^{-3}$), ($10^5\,erg\,cm^{-3}$)
Y	4.2	−0.25	−0.023
	77	−0.22	−0.021
	295	−0.06	−0.0005
Sm	4.2	−400	
	77	−14.3	21
	295	−0.17	
Eu	4.2	−4	−1.56
	293	−0.38	
Gd	4.2	−2.41	−0.037
	80	−0.44	−0.035
	320	−0.041	−0.010
Tb	80	−7.6	−76
	300	−0.082	
Dy	80	−9.7	2.14
	300	< -0.05	
Ho	4.2	−120	
	78	−8	−2.7
	300	< -0.05	
Er	4.2	+90	+500
	77	+0.36	
	300	−0.06	
Tm	77	−0.30	
	293	−0.06	~0
Yb	4.2	−67	+80
	80	−0.39	+0.06
	300	−0.06	

the temperature dependence of K_1 and K_2 of YIG. The origin of this anisotropy is the same as that of Fe^{3+} ions in the spinel-type ferrites, so that they exhibit a monotonic temperature dependence similar to that in Fig. 12.44. In contrast to this, the contribution of R^{3+}, which has an orbital angular momentum L, is quite large and has a strong temperature dependence, as shown in Fig. 12.46. The reason for this large contribution of R^{3+} is that the orbital moment becomes aligned with decreasing temperature, thus giving rise to a large interaction with the crystalline field. Table 12.5 lists the values of K_1 and K_2 at a number of temperatures for various garnet-type ferrites. In some cases, the magnetostriction constants (see Chapter 14) are also very large,[71] and affect these anisotropy constants to a great extent (see Section 14.6).

PROBLEMS

12.1 When the magnetization is rotated in the (210) plane in a cubic crystal, express the anisotropy energy as a function of the angle θ between the magnetization and the [001] axis. Ignore terms higher than K_2.

12.2 Knowing the values for $I_s = 1.79\,\mathrm{T}$ and $K_{u1} = 4.1 \times 10^5\,\mathrm{J\,m^{-3}}$, calculate the anisotropy field in the c-axis for cobalt.

12.3 Assuming $q = 3 \times 10^{-25}\,\mathrm{J}$, calculate the magnetocrystalline anisotropy constant K_1 for a ferromagnetic metal with body-centered cubic structure ($a = 3\,\mathrm{Å}$), using the spin-pair model.

12.4 If the temperature dependence of an anisotropy constant were the same as that of spontaneous magnetization, what functional form should be expected for angular dependence of the anisotropy energy? Use the Zener–Carr formulation.

REFERENCES

1. R. Pauthenet, Y. Barnier, and G. Rimet, *J. Phys. Soc. Japan*, **17** Suppl. B-I (1962), 309.
2. H. Gengnagel and U. Hofmann, *Phys. Stat. Sol.*, **29** (1968), 91.
3. J. J. M. Franse and G. de Vries, *Physica*, **39** (1968), 477.
4. T. Yamada and S. Chikazumi, *Series in experimental physics* (in Japanese), **17** (Kyoritsu Publishing Co., 1968), No. 13, pp. 293–325.
5. T. Wakiyama and S. Chikazumi, *J. Phys. Soc. Japan*, **15** (1960), 1975; p. 306 in volume quoted in Ref. 4 above.
6. R. F. Pearson, *J. Phys. Radium*, **20** (1959), 409; p. 321 in volume quoted in Ref. 4 above.
7. K. Abe and S. Chikazumi, *Japan. J. Appl. Phys.*, **15** (1976), 623.
8. R. F. Penoyer, *Rev. Sci. Instr.*, **30** (1959), 711; in volume quoted in Ref. 4. above.
9. S. Chikazumi, *Physics of Magnetism* (1st edn) (Wiley, New York, 1964), pp. 132–3.
10. S. Chikazumi, Kotai Buturi (Solid State Phys.) (in Japanese), **9** (9), (1974), 495.
11. S. Chikazumi, K. Tajima, and K. Toyama, *Proc. Grenoble Conf.* (1970), **C-I**, 179.
12. J. S. Kouvel and C. D. Graham, Jr., *J. Appl. Phys.*, **28** (1957), 340.
13. S. Chikazumi, *J. Phys. Soc. Japan*, **11** (1956), 718.

14. E. A. Nesbitt, H. J. Williams, and R. M. Bozorth, *J. Appl. Phys.*, **25** (1954), 1014.
15. E. A. Nesbitt and H. J. Williams, *J. Appl. Phys.*, **26** (1955), 1217.
16. H. J. Williams, R. D. Heidenrich, and E. A. Nesbitt, *J. Appl. Phys.*, **27** (1956), 85.
17. L. R. Bickford, *Phys. Rev.*, **78** (1950), 449.
18. W. Heisenberg, *Z. Phys.*, **49**, (1928), 619.
19. L. W. McKeehan, *Phys. Rev.*, **43** (1933), 1025.
20. For instance, H. Kamimura, S. Sugano, and Y. Tanabe, *Ligand field theory and its applications* (in Japanese) (Syokabo Publishing Co., Tokyo, 1970).
21. J. C. Slonczewski, *J. Appl. Phys.*, **29** (1958), 448; *Phys. Rev.*, **110** (1958), 1341.
22. M. Tachiki, *Prog. Theor. Phys.*, **23** (1960), 1055.
23. K. Yosida and M. Tachiki, *Prog. Theor. Phys.*, **17** (1957), 331.
24. J. C. Slonczewski, *J. Appl. Phys.*, **32** (1961), 253S.
25. J. Kanamori, *Magnetism* (ed. by Rado and Suhl) (Academic Press, N.Y., 1963), Vol. 1, 127.
26. C. Zener, *Phys. Rev.*, **96** (1954), 1335.
27. W. J. Carr, Jr., *J. Appl. Phys.*, **29** (1958), 436.
28. H. E. Nigh, S. Legvold, and F. H. Spedding, *Phys. Rev.*, **132** (1963), 1092.
29. D. E. Hegland, S. Legvold, and F. H. Spedding, *Phys. Rev.*, **131** (1963), 158.
30. D. R. Behrendt, S. Legvold, and F. H. Spedding, *Phys. Rev.*, **109** (1958), 1544.
31. D. L. Strandburg, S. Legvold, and F. H. Spedding, *Phys. Rev.*, **127** (1962), 2046.
32. R. W. Green, S. Legvold, and F. H. Spedding, *Phys. Rev.*, **122** (1961), 827.
33. B. L. Rhodes, S. Legvold, and F. H. Spedding, *Phys. Rev.*, **109** (1958), 1547.
34. S. Chikazumi, S. Tanuma, I. Oguro, F. Ono, and K. Tajima, *IEEE Trans. Mag.*, **MAG-5** (1965), 265.
35. H. H. Liebermann and C. D. Graham, Jr., *AIP Conf. Proc.*, **No. 29** (1975), 598.
36. K. W. H. Stevens, *Proc. Phys. Soc. London*, **A65** (1962), 2058.
37. S. Chikazumi, T. Wakiyama, and K. Yosida, *Proc. Int. Conf. Mag.* (Nottingham, 1964), 756.
38. D. S. Rodbell, *J. Phys. Soc. Japan*, **17B-1** (1962), 313.
39. K. Yosida, A. Okiji, and S. Chikazumi, *Proc. Int. Conf. Mag.* (Nottingham, 1964), 22.
40. E. Tatsumoto, T. Okamoto, H. Fujii, and C. Inoue, *Proc. Int. Conf. Mag.* (Grenoble, 1970), CI-551.
41. R. S. Perkins and S. Strässler, *Phys. Rev.*, **B15** (1977), 477.
42. Y. Otani, H. Miyajima, and S. Chikazumi, *J. Appl. Phys.*, **61** (1987), 3436.
43. J. Smit and H. P. J. Wijn, *Ferrites* (Wiley, New York, 1959), p. 204.
44. S. Krupicka, *Physik der Ferrite* (Academic Press, 1973), 348.
45. J. Smit and H. P. J. Wijn, *Ferrites* (J. Wiley, New York, 1959), p. 118.
46. H. B. G. Casimir, J. Smit, U. Enz, J. F. Fast, H. P. J. Wijn, E. W. Gorter, A. J. W. Dryvesteyn, J. D. Fast, and J. J. de Jong, *J. de Phys. et Rad.*, **20** (1959), 360.
47. J. Kondo, *Prog. Theor. Phys.* (Kyoto), **23** (1960), 106.
48. L. R. Bickford, *J. Appl. Phys.*, **31** (1960), 259S; *Phys. Rev.*, **119** (1960), 1000.
49. K. Adachi, *J. de Phys.*, **24** (1963), 725.
50. H. J. Williams, R. C. Sherwood, and O. L. Boothby, *J. Appl. Phys.*, **28** (1957), 445.
51. H. J. Williams, R. C. Sherwood, F. G. Foster, and E. M. Kelley, *J. Appl. Phys.*, **28** (1957), 1181.
52. R. C. Hall, *J. Appl. Phys.*, **30** (1959), 816.
53. W. S. Chan, K. Mitsuoka, H. Miyajima, and S. Chikazumi, *J. Phys. Soc. Japan*, **48** (1980), 822.
54. E. Kneller, *Ferromagnetismus* (Springer-Verlag, Berlin, 1962), p. 197.
55. S. Kaya, *J. Fac. Sci. Hokkaido Imp. Univ.*, **2** (1938), 39.
56. R. M. Bozorth, *Rev. Mod. Phys.*, **25** (1953), 42.

57. H. Kagawa and S. Chikazumi, *J. Phys. Soc. Japan*, **48** (1980), 1476.
58. J. W. Shih, *Phys. Rev.*, **50** (1936), 376.
59. L. W. McKeehan, *Phys. Rev.*, **51** (1937), 136.
60. T. Wakiyama and S. Chikazumi, *J. Phys. Soc. Japan*, **15** (1960), 1975.
61. K. Abe, Y. Miyamoto, and S. Chikazumi, *J. Phys. Soc. Japan*, **41** (1976), 1894.
62. S. Chikazumi, *AIP Conf. Proc.*, **No. 29** (1975), 382.
63. E. J. W. Verwey and P. W. Haayman, *Physica*, **8** (1941), 979.
64. G. Shirane, S. Chikazumi, J. Akimitsu, K. Chiba, M. Matsui and Y. Fujii, *J. Phys. Soc. Japan*, **39** (1975), 947.
65. K. Shiratori and Y. Kino, *J. Mag. Mag. Mat.*, **20** (1980), 87.
66. N. Miyata, *J. Phys. Soc. Japan*, **16** (1961), 1291.
67. R. F. Penoyer and M. W. Shafer, *J. Appl. Phys.*, **30** (1959), 315S.
68. Y. Syono, *Jap. J. Geophys.*, **4** (1965), 71.
69. P. Hansen, *Proc. Int. School Phys. Enrico Fermi*, **LXX** (1978), 56.
70. R. F. Pearson, *J. App. Phys.*, **33** (1962), 1236S.
71. K. P. Belov, A. K. Gapeev, R. Z. Levitin, A. S. Markosyan, and Yu. F. Popov, *Sov. Phys. JETP*, **41** (1975), 117.

13

INDUCED MAGNETIC ANISOTROPY

The induced magnetic anisotropy differs from the ordinary magnetic anisotropy discussed in Chapter 12, in that the induced anisotropy is produced by a treatment (often an annealing treatment) that has directional characteristics. Not only the magnitude but also the easy axis of the anisotropy can be altered by appropriate treatments. We shall discuss various types of induced anisotropy in this section.

13.1 MAGNETIC ANNEALING EFFECT

The magnetic annealing effect is obtained by heating or annealing a magnetic material in an applied field. The effect was first observed by Kelsall[1] in Fe–Ni alloys, and was investigated by Dillinger and Bozorth[2] in detail. Figure 13.1 shows the effect of magnetic annealing on the shape of the magnetization curve of a 21.5% Fe–Ni alloy. Curves A and C are measured after annealing in a magnetic field applied parallel and perpendicular, respectively, to the direction along which the magnetization curve is measured. This behavior can be attributed to an induced anisotropy whose easy axis is parallel to the direction of the annealing field. In the case of curve B, no field was applied during annealing, but local easy axes were produced parallel to

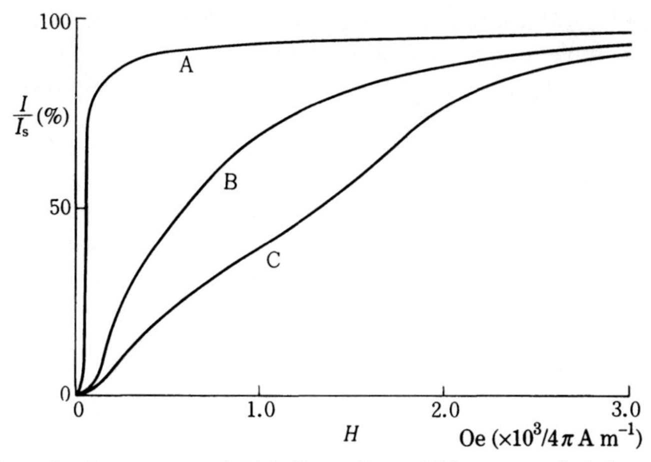

Fig. 13.1. Magnetization curves of 21.5 Permalloy which was cooled from 600°C (A) in a longitudinal magnetic field, (B) in the absence of magnetic field, and (C) with perpendicular (or circular) magnetic field.[7]

the direction of spontaneous magnetization in each magnetic domain during annealing. From the average susceptibility of curve C, the uniaxial anisotropy constant of the induced anisotropy is estimated to be $1 \times 10^2\,\mathrm{J\,m^{-3}}$.

Before going into details of this phenomenon, let us describe the history of this alloy. It was discovered by Arnold and Elmen[3] in 1923 that Ni-rich Ni–Fe alloys exhibit very high permeability, reaching a maximum at a composition of 21.5wt% Fe–Ni. This alloy, called Permalloy, has unusual annealing behavior: high permeability is attained only when the alloy is quenched from high temperatures. Slow cooling or annealing at intermediate temperatures greatly lowers the permeability. This behavior is different from other magnetic alloys, most of which show improved soft magnetic properties after slow cooling or annealing. This puzzle, known as the 'Permalloy problem', was investigated by many researchers with inconclusive results. Later Dahl[4] proved the presence of a superlattice in Ni–Fe alloys, and Kaya[5] investigated the process of formation of the superlattice in relation to changes in magnetic properties. The structure of this superlattice is shown in Fig. 12.33. It was found that the order–disorder transformation takes place at about 490°C, and a long annealing time (about one week) at this temperature is required to attain the equilibrium state. Tomono[6] discovered that magnetic annealing is effective at 500°C, where the ordering process starts. Chikazumi[7] found that the induced anisotropy tends to disappear as the alloy approaches the fully ordered state. He tried to explain this phenomenon in terms of 'directional order', or an anisotropic distribution of different atomic pairs such as Ni–Ni, Fe–Fe, or Ni–Fe. Based on the fact that complete ordering decreases the specific volume of the alloy by 5×10^{-4}, he assumed that the length of an Ni–Fe pair is smaller than the length of other possible atomic pairs. Then directional ordering causes lattice distortion, which can produce an induced anisotropy through the magnetoelastic energy. On the other hand, Kaya[8] assumed that ordering occurred by the growth of distinct volumes of the ordered phase, and explained the induced anisotropy as the result of shape anisotropy of the second phase (see (12.24)). Both models explained the fact that the induced anisotropy disappears when complete ordering is developed.

The open circles in Fig. 13.2 represent the maximum values of induced anisotropy constant observed after field cooling from 600°C at various cooling rates, as a function of Ni content in Fe.[9] The curve form is quite monotonic, and nearly quadratic with respect to the Fe composition. This result conflicts with both the models, because the strain model predicts a minimum at about 80% Ni where the magnetostriction constants go through zero (see Fig. 14.11), while the ordered phase model predicts a maximum at the Ni_3Fe composition, where the degree of order is maximum. The directional order model was then reinterpreted in terms of atomic pair interactions whose magnitudes are different for different pairs.[9–11] We discuss the problem using the formulation first given by Néel.

In a binary A–B alloy there are three kinds of atomic pairs, AA, BB, and AB. We assume that the coefficients of the dipole–dipole interactions (see (12.59)) of AA, BB, and AB pairs, given by l_{AA}, l_{BB}, and l_{AB}, are different from one another. We ignore pair interaction terms higher than the quadrupole interaction. Then the anisotropy

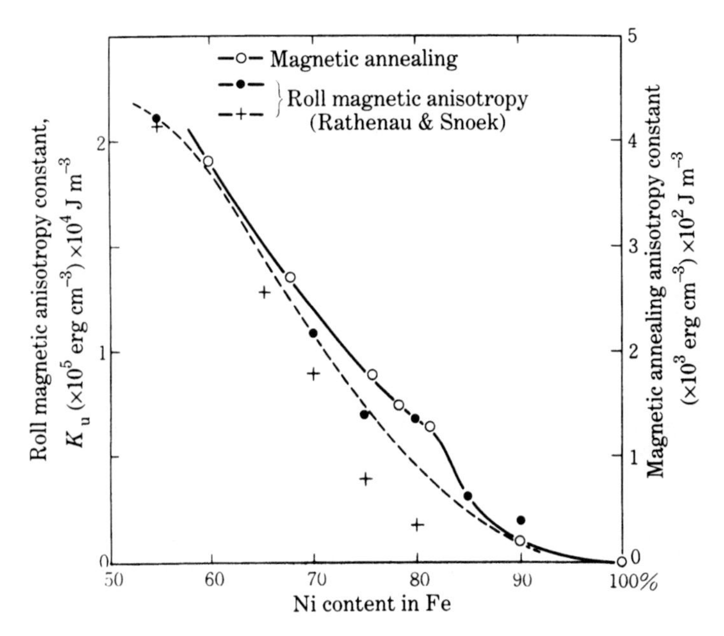

Fig. 13.2. Induced anisotropy of Fe–Ni alloys due to magnetic annealing and cold-rolling.[9]

energy due to an unbalanced distribution of three kinds of atomic pairs over differently oriented pair directions is

$$E_a = \sum_i (N_{AAi} l_{AA} + N_{BBi} l_{BB} + N_{ABi} l_{AB})\left(\cos^2 \varphi_i - \tfrac{1}{3}\right), \tag{13.1}$$

where i denotes the pair directions, N_{AAi}, N_{BBi}, and N_{ABi} are the number of AA, BB and AB pairs directed parallel to the ith direction, and φ_i is the angle between the magnetization and the ith bond direction. Note that N_{AAi}, N_{BBi}, and N_{ABi} are not independent variables, for the following reasons (see Fig. 13.3): if we divide each atom into z sectors, where z is the number of nearest neighbors, and attach each of these sectors to one nearest-neighbor bond, an AA pair has two A sectors, whereas an AB pair has one A sector. Thus the total number of A sectors is given by $2N_{AA} + N_{AB}$, where N_{AA} and N_{AB} are the total number of AA and BB pairs. The total number of A sectors is also given by zN_A, where N_A is the total number of A atoms. Thus the total number of A sectors attached to the atomic pairs with the ith bond direction is

$$2N_{AAi} + N_{ABi} = 2N_A, \tag{13.2}$$

where z is replaced by 2, because each A atom has just two sectors attached to bonds in a single direction. Similarly the number of B sectors attached to atomic pairs with the ith bond direction is

$$2N_{BBi} + N_{ABi} = 2N_B, \tag{13.3}$$

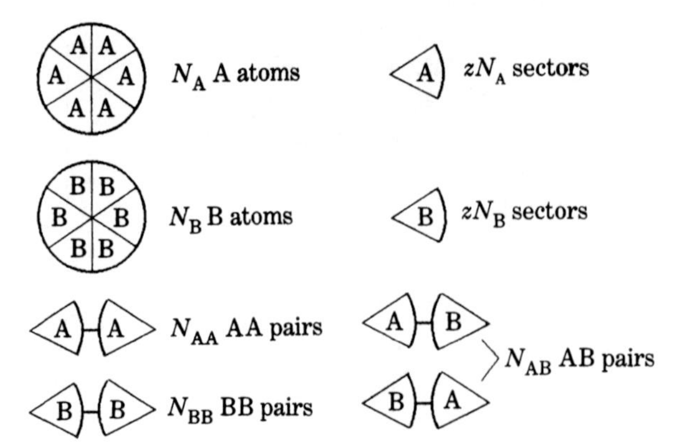

Fig. 13.3. Counting the number of A and B sectors in terms of the number of A and B atoms, and also in terms of the number of AA, BB and AB atom pairs.

where N_B is the total number of B atoms. Using (13.2) and (13.3), we can express the anisotropy energy (13.1) in terms of the number of BB pairs as

$$E_a = \sum_i N_{BBi} l_0 \left(\cos^2 \varphi_i - \tfrac{1}{3}\right) + \text{const.}, \tag{13.4}$$

where

$$l_0 = l_{AA} + l_{BB} - 2l_{AB}. \tag{13.5}$$

The anisotropy energy (13.3) can be expressed in terms of the direction cosines of the pair directions as

$$E_a = \sum_i N_{BBi} l_0 (\alpha_1 \gamma_{1i} + \alpha_2 \gamma_{2i} + \alpha_3 \gamma_{3i})^2, \tag{13.6}$$

where $(\alpha_1, \alpha_2, \alpha_3)$ and $(\gamma_{1i}, \gamma_{2i}, \gamma_{3i})$ are the direction cosines of the magnetization and of the ith bond directions respectively.

Now suppose that this alloy is annealed at T K, where the migration of atoms takes place easily, and at the same time a saturating field is applied parallel to the direction $(\beta_1, \beta_2, \beta_3)$. Then the BB pairs tend to align themselves parallel to the direction of magnetization, provided $l_0 < 0$. The energy change for one BB pair is given by $l_0' \cos^2 \varphi_i'$, where l_0' is the value of l_0 at T K. The number of BB pairs found in the ith bond direction at thermal equilibrium is proportional to the Boltzmann factor $\exp(-l_0' \cos^2 \varphi_i'/kT)$, and we have

$$N_{BBi} = N_{BB} \frac{\exp(-l_0' \cos^2 \varphi_i'/kT)}{\sum_i \exp(-l_0' \cos^2 \varphi_i'/kT)}, \tag{13.7}$$

where N_{BB} is the total number of BB pairs. If $l_0' \ll kT$, when we expand the exponential function, (13.7) becomes

$$N_{BBi} = \frac{2N_{BB}}{z}\left(1 - \frac{l_0' \cos^2 \varphi_i'}{kT}\right)$$
$$= \frac{2N_{BB}}{z}\left(1 - \frac{l_0'}{kT}(\beta_i\gamma_{1i} + \beta_2\gamma_{2i} + \beta_3\gamma_{3i})^2\right). \quad (13.8)$$

After the alloy has been quenched, these atom pairs are locked in place, and they give rise to magnetic anisotropy which is calculated by putting (13.8) into (13.6), as

$$E_a = \frac{2N_{BB}l_0}{z}\sum_i(\alpha_i\gamma_{1i} + \alpha_2\gamma_{2i} + \alpha_3\gamma_{3i})^2\left(1 - \frac{l_0'}{kT}(\beta_1\gamma_{1i} + \beta_2\gamma_{2i} + \beta_3\gamma_{3i})^2\right)$$
$$= -\frac{2N_{BB}l_0l_0'}{zkT}\sum_i(\alpha_1\gamma_{1i} + \alpha_2\gamma_{2i} + \alpha_3\gamma_{3i})^2(\beta_1\gamma_{1i} + \beta_2\gamma_{2i} + \beta_3\gamma_{3i})^2$$
$$+ \text{const.} \qquad \text{(for cubic crystal)}, \quad (13.9)$$

in which

$$\sum_i(\alpha_1\gamma_{1i} + \alpha_2\gamma_{2i} + \alpha_3\gamma_{3i})^2(\beta_1\gamma_{1i} + \beta_2\gamma_{2i} + \beta_3\gamma_{3i})^2$$
$$= (\alpha_1^2\beta_1^2 + \alpha_2^2\beta_2^2 + \alpha_3^2\beta_3^2)\sum_i \gamma_{1i}^4$$
$$+ (\alpha_1^2\beta_2^2 + \alpha_1^2\beta_3^2 + \alpha_2^2\beta_1^2 + \alpha_2^2\beta_3^2 + \alpha_3^2\beta_1^2 + \alpha_3^2\beta_2^2)\sum_i \gamma_{1i}^2\gamma_{2i}^2$$
$$+ 4(\alpha_1\alpha_2\beta_1\beta_2 + \alpha_2\alpha_3\beta_2\beta_3 + \alpha_3\alpha_1\beta_3\beta_1)\sum_i \gamma_{1i}^2\gamma_{2i}^2$$
$$= (\alpha_1^2\beta_1^2 + \alpha_2^2\beta_2^2 + \alpha_3^2\beta_3^2)\left(\sum_i \gamma_{1i}^4 - \sum_i \gamma_{1i}^2\gamma_{2i}^2\right)$$
$$+ 4(\alpha_1\alpha_2\beta_1\beta_2 + \alpha_2\alpha_3\beta_2\beta_3 + \alpha_3\alpha_1\beta_3\beta_1)\sum_i \gamma_{1i}^2\gamma_{2i}^2 + \sum_i \gamma_{1i}^2\gamma_{2i}^2.$$

Putting

$$\left.\begin{aligned} \frac{2}{z}\left(\sum_i \gamma_{1i}^4 - \sum_i \gamma_{1i}^2\gamma_{2i}^2\right) &= k_1 \\ \frac{8}{z}\sum_i \gamma_{1i}^2\gamma_{2i}^2 &= k_2 \end{aligned}\right\}, \quad (13.10)$$

we have the constants k_1 and k_2, which depend on the crystal structure; their numerical values are listed in Table 13.1. Using these constants, (13.9) is expressed as

$$E_a = -\frac{N_{BB}l_0l_0'}{kT}\left(k_1\sum_i \alpha_i^2\beta_i^2 + k_2\sum_{i>j}\alpha_i\alpha_j\beta_i\beta_j\right). \quad (13.11)$$

Table 13.1. Values of k_1 and k_2 for various crystal types (cf. (13.10)).

Crystal type	k_1	k_2
Isotropic	$\frac{2}{15}$	$\frac{4}{15}$
Simple cubic	$\frac{1}{3}$	0
Body-centered cubic	0	$\frac{4}{9}$
Face-centered cubic	$\frac{1}{12}$	$\frac{4}{12}$

For an isotropic material, (13.11) becomes

$$E_a = -\frac{2N_{BB}l_0 l_0'}{15kT}\left(\sum_i \alpha_i^2\beta_i^2 + 2\sum_{i>j}\alpha_i\alpha_j\beta_i\beta_j\right)$$
$$= -\frac{2N_{BB}l_0 l_0'}{15kT}\left(\sum_i \alpha_i\beta_i\right)^2$$
$$= -\frac{2N_{BB}l_0 l_0'}{15kT}\cos^2\varphi, \qquad (13.12)$$

where φ is the angle between the magnetization and the direction of the annealing field. If l_0 has the same sign as l_0', the anisotropy constant becomes negative, and the easy axis develops parallel to the annealing field. For a dilute and disordered solution of B atoms in a matrix of A, the probability of finding BB pairs is proportional to C_B^2, where C_B is the concentration of B atoms; hence

$$N_{BB} = \frac{zN}{2}C_B^2, \qquad (13.13)$$

where N is the total number of atoms per unit volume. On using (13.13), we can express the anisotropy constant as

$$K_u = \frac{zNl_0 l_0'}{15kT}C_B^2. \qquad (13.14)$$

The composition dependence of this expression explains the experimental result shown in Fig. 13.2.

Néel estimated the values of l_{AA}, l_{BB} and l_{AB} from the composition dependence of magnetostriction constants, which will be treated in Chapter 14. Since the numbers of AA, BB, and AB pairs in disordered alloys are proportional to C_A^2, C_B^2 and $2C_AC_B$, respectively, the average value of l is

$$l = C_A^2 l_{AA} + 2C_AC_B l_{AB} + C_B^2 l_{BB}. \qquad (13.15)$$

As we discuss in Chapter 14, the magnetostriction constants λ_{100} and λ_{111} are related to the dipole interaction, so that these magnetostriction constants of A–B alloys are also expected to change in a way similar to (13.15). As a matter of fact, the composition dependence of these constants, shown in Fig. 14.11, can be fitted with parabolic curves. Denoting Ni as A, and Fe as B, we get the expression

$$\left.\begin{aligned}\lambda_{100} \times 10^6 &= -55C_A^2 + 340C_AC_B - 245C_B^2 \\ \lambda_{111} \times 10^6 &= -27C_A^2 + 134C_AC_B + 13C_B^2\end{aligned}\right\}. \tag{13.16}$$

Using (14.50), we can estimate the values of l_{NiNi}, l_{FeFe} and l_{NiFe}, so that we have

$$\begin{aligned}Nl_0 &= N(l_{NiNi} + l_{FeFe} - 2l_{NiFe}) \\ &= 3.0 \times 10^7 \,\mathrm{J\,m^{-3}}.\end{aligned} \tag{13.17}$$

This formula will be described in detail in Section 14.2.

At the temperature at which the alloy is normally annealed, the value of l_0, or l_0', is much smaller than the value at room temperature. Using the I_s^3 law (see (12.73)), the value l_0' at 700 K, where I_s is 0.50 T, is estimated from the value at room temperature given by (13.17), where I_s is 1.13 T, as $Nl_0' = 3.0 \times 10^7 \times (0.50/1.13)^3 = 2.7 \times 10^6 \,\mathrm{J\,m^{-3}}$.

Using $z = 12$, $N = 9.17 \times 10^{28} \,\mathrm{m^{-3}}$, $k = 1.38 \times 10^{-23} \,\mathrm{J\,K^{-1}}$, K_u is calculated by (13.14) as

$$K_u = 7.5 \times 10^4 C_B^2 \,\mathrm{J\,m^{-3}}. \tag{13.18}$$

Since the experimental curve shown in Fig. 13.2 is expressed as

$$K_u = 2.9 \times 10^3 \, C_B^2 \,\mathrm{J\,m^{-3}}, \tag{13.19}$$

the coefficient given by (13.18) is large enough to explain the experimental result. The reasons why quantitative agreement is not obtained may lie in the approximate estimates of various quantities and in failure to attain complete thermal equilibrium.

The relationship (13.11) shows that the magnitude of the induced anisotropy depends on the crystallographic direction of the annealing field. Figure 13.4 shows the anisotropy energy curve observed for a (110) oblate single crystal of Ni_3Fe after cooling from 600°C at a rate of $14 \,\mathrm{K\,min^{-1}}$ in a magnetic field whose direction could be rotated in the (110) plane.[12] These energy curves are obtained by integrating torque curves with respect to the angle, and are then separated by Fourier analysis into the cubic crystal anisotropy and the induced uniaxial anisotropy energy curves. The arrows in the figure indicate the direction of the annealing field and the dots show for each curve the minimum value of the uniaxial anisotropy energy. It is seen from these curves that magnetic annealing is most effective for ⟨111⟩ annealing, less effective for ⟨110⟩, and least effective for ⟨100⟩. This figure also shows that the minimum point coincides with the direction of the annealing field for the three principal axes, whereas it tends to deviate toward ⟨111⟩ for intermediate directions of the annealing field.

To compare experiment with theory, we express the anisotropy energy as

$$E_a = -K_u \cos^2(\theta - \theta_0), \tag{13.20}$$

where θ is the angle of magnetization as measured in the $(1\bar{1}0)$ plane from the [001] direction (so that we have $\alpha_1 = \alpha_2 = (1/\sqrt{2})\sin\theta$ and $\alpha_3 = \cos\theta$) and θ_0 represents

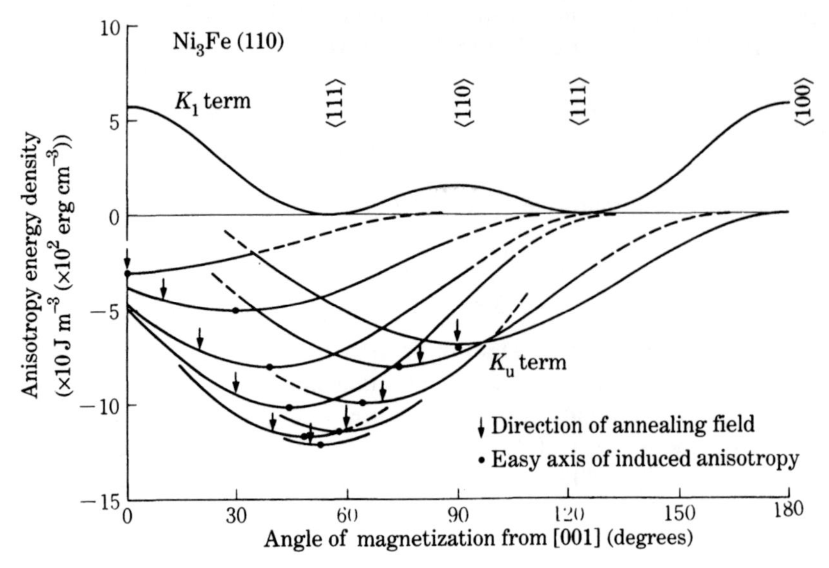

Fig. 13.4. Magnetic anisotropy energy induced by magnetic annealing of a (110) disk of Ni_3Fe.[12]

the minimum point. On comparing (13.20) and (13.11) we have, for three principal directions of the annealing field,

$$\langle 100\rangle \qquad K_u = \frac{N_{BB} l_0 l_0'}{kT} k_1, \tag{13.21}$$

$$\langle 110\rangle \qquad K_u = \frac{N_{BB} l_0 l_0'}{kT}\left(\frac{k_1}{2} + \frac{k_2}{4}\right), \tag{13.22}$$

$$\langle 111\rangle \qquad K_u = \frac{N_{BB} l_0 l_0'}{kT}\,\frac{k_2}{2}. \tag{13.23}$$

On putting into these formula the theoretical values of k_1 and k_2 listed in Table 13.1, we find that the ratio of the K_u should be 2:3:4 for the three principal axes. This agrees qualitatively with the experimental results. Quantitatively, however, the experiment shows a larger directional dependence, as shown in Fig. 13.5. Curve A is the theoretical curve deduced from (13.11) with the parameters deduced from the best fit to the experimental curve B. The deviation of the minimum point of the energy curve, $\Delta\theta$, from the direction of the annealing field, θ_t,

$$\Delta\theta = \theta_0 - \theta_t, \tag{13.24}$$

is plotted as a function of θ_t in Fig. 13.6. Curve A is the theoretical curve deduced from (13.11) and qualitatively explains the experimental curve B. The agreement is, however, not satisfactory. In order to fit the theoretical curve to the experiment quantitatively, we find that $k_1 : k_2 = 1:8.5$. The solid curves B in Figs. 13.4 and 13.5

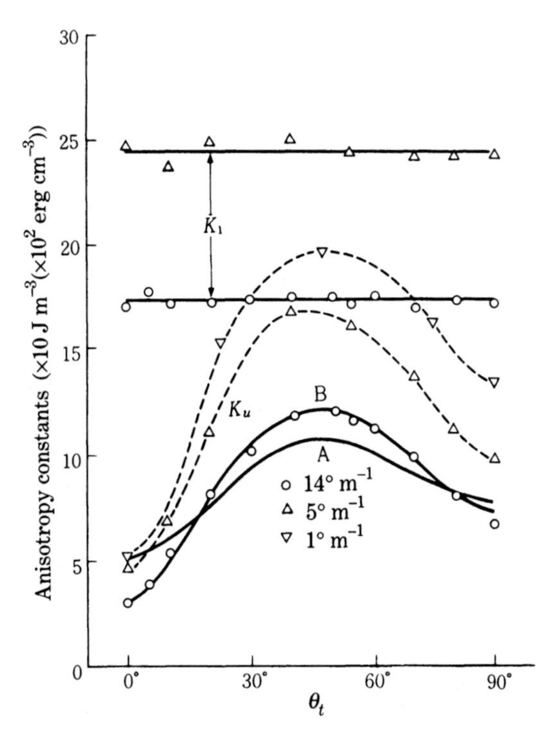

Fig. 13.5. Induced anisotropy constant as a function of crystallographic orientation of the annealing field.[12]

are drawn using this ratio. The values of the ratio $k_1 : k_2$ have been measured by several investigators[12-16] for a number of alloys; their results are listed in Table 13.2. The values range from 1.9 to 9.6, even for face-centered cubic alloys. The values for body-centered cubic Fe–Al alloys are also quite different from the theoretical ratio of 0:1. One of the reasons for this discrepancy may be a contribution from the interaction of second nearest neighbor atoms. For further details refer to Ferguson[17] and Iwata.[18]

Magnetic annealing has been found to be effective also for some ferrites, such as Co ferrites, which have been used as oxide permanent (O.P.) magnets.[19] This annealing has been used to improve the $(BH)_{\max}$ or energy product, which measures the quality of a permanent magnet material. The induced anisotropy was measured for Fe–Co ferrites by Iida *et al.*,[20] who discovered that the effect is sensitive to the partial pressure of oxygen during cooling (Fig. 13.7). They found that the material responds to magnetic annealing only when it is more or less oxidized. In other words, the magnetic annealing of ferrites is effective only in the presence of lattice vacancies. Single crystal experiments were made for the same system by Penoyer and Bickford,[21] who found that the coefficients of the first and second terms, F and G, or k_1 and k_2 times the proportionality factor outside the parentheses in (13.11), have different composition dependences, as shown in Fig. 13.8. That is, F increases as the square of

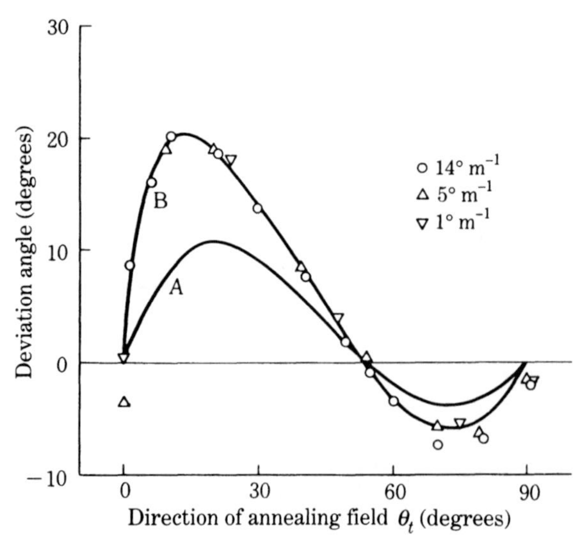

Fig. 13.6. Angle of difference, $\Delta\theta$, between the direction of annealing field and the easy axis of the induced anisotropy as a function of crystallographic orientation of the annealing field.[12]

the Co content, while G increases linearly with the Co content. Slonczewski[22] explained this behavior in terms of the one-ion anisotropy of the Co^{2+} ion on the octahedral sites, whose energy levels are split as shown in Fig. 12.20(c). The doublet thus produced gives rise to a uniaxial anisotropy with its axis parallel to $\langle 111 \rangle$ (see (12.69)). If all the Co^{2+} ions are distributed equally along the four equivalent $\langle 111 \rangle$s, the uniaxial anisotropies cancel out because of the cubic symmetry. When this ferrite is cooled in a magnetic field, the Co^{2+} ions tend to occupy the octahedral sites whose $\langle 111 \rangle$ axis is nearest to the magnetic field, so as to lower the anisotropy energy. After cooling, this unbalanced distribution of Co^{2+} ions results in an induced anisotropy. Since the axis of the uniaxial anisotropy of Co^{2+} ions is parallel to $\langle 111 \rangle$, the F or k_1

Table 13.2. Theoretical and experimental values of the ratio $k_1 : k_2$.

	$k_1 : k_2$		
Material	Theory	Experiment	Investigator
Polycrystal	1:2	1:2	
Ni_3Fe	1:4	1:8.5	Chikazumi[12]
20%Co–Fe	1:4	1:3.0	Aoyagi *et al.*[13]
12.5%Co–Fe	1:4	1:2.3–2.6	Aoyagi *et al.*[14]
54%Ni–Fe	1:4	1:1.9–2.4	Aoyagi *et al.*[14]
83%Ni–Fe	1:4	1:8.3–9.6	Aoyagi *et al.*[14]
$Fe_{3.35}Al$	0:1	1:3.4	Suzuki[15]
$Fe_{4.7}Al$	0:1	1:1	Chikazumi and Wakiyama[16]

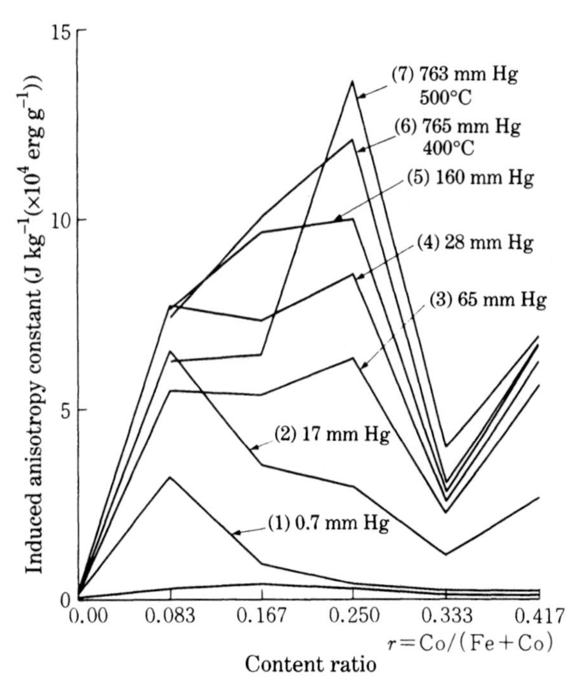

Fig. 13.7. Effect of magnetic annealing of Co–Fe ferrites under various partial pressures of oxygen.[20]

term should vanish just like the directional order in the body-centered cubic lattice. Actually the F term is very small compared with the G or k_2 term, as seen in Fig. 13.8. This one-ion induced anisotropy should be proportional to the available number of Co^{2+} ions. This explains the linear dependence of the G term on the composition. The quadratic composition dependence of the F term was ascribed to directional ordering of the Co^{2+}–Co^{2+} pairs. The presence of lattice vacancies is assumed to speed the diffusion of ions.[23–25] Oxidation was also found to be necessary for the magnetic cooling of Ni-ferrites.[26]

13.2 ROLL MAGNETIC ANISOTROPY

It was discovered by Six *et al.*[27] in 1934 that a large magnetic anisotropy is created during the process of cold-rolling iron–nickel alloys. Utilizing this effect, they produced a new magnetic material called Isoperm, which has constant permeability over a wide range of applied fields. This alloy, whose composition is 50% Fe–50% Ni, is first strongly cold-rolled, then recrystallized to give a (001)[100] crystallographic texture, and finally cold-rolled again to about 50% reduction in thickness. The sheet thus manufactured exhibits a large uniaxial magnetic anisotropy with its easy axis in the plane of the sheet but perpendicular to the rolling direction. Magnetization

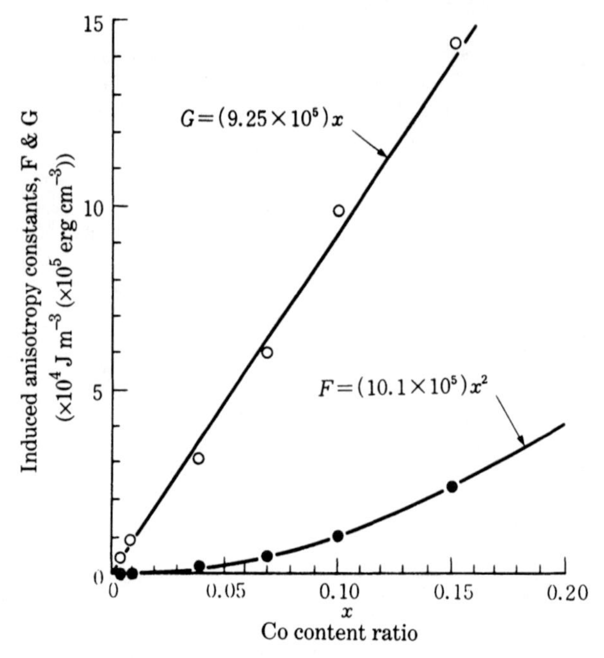

Fig. 13.8. Variation of coefficients F and G (which are proportional to k_1 and k_2, respectively) with change in Co content in Co–Fe ferrite.[21]

parallel to the rolling direction takes place exclusively through domain rotation, giving rise to a linear magnetization curve (Fig. 13.9).

Detailed investigations of this phenomenon were made by Conradt *et al.*[28] in 1940, and by Rathenau and Snoek[29] in 1941. They concluded that this anisotropy cannot be explained in terms of the coupling between magnetostriction and internal stresses.

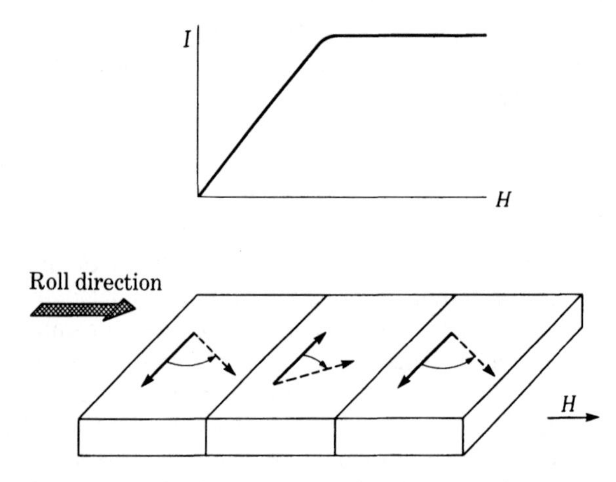

Fig. 13.9. Magnetization curve and domain structure of Isoperm.

Néel[30,10] and Taniguchi and Yamamoto[11] extended their interpretation of magnetic annealing to this phenomenon, and suggested that A and B atoms migrate to produce a stable directional order during the process of cold-rolling. They considered that the plastic deformation played a role in bringing atoms into stable positions. There is no doubt that directional order is the main origin of this anisotropy, because its magnitude is proportional to C_B^2, as is directional order anisotropy (shown by the dashed curve in Fig. 13.2). The magnitude of the anisotropy induced by rolling is, however, about 50 times larger than that induced by magnetic annealing.

However, roll magnetic anisotropy is not so simple as magnetic annealing, because single crystal measurements revealed that both the magnitude and the easy axis of the induced anisotropy are quite dependent on the crystallographic plane and direction of rolling. These phenomena were interpreted in terms of 'slip-induced anisotropy' by Chikazumi *et al.*[31] as will be explained below.

Generally speaking, when a crystal is deformed plastically, one part of the crystal slips relative to another part along a specific crystal plane, called the slip plane, and in a specific crystallographic direction, called the slip direction. Figure 13.10 shows a slip deformation in a face-centered cubic crystal. In this crystal structure, the slip plane is parallel to {111} and the slip direction is parallel to ⟨110⟩. Such a slip deformation usually takes place through the motion of a dislocation, a kind of line lattice defect, along the slip plane. The passage of one dislocation results in a displacement of one part of the crystal by one atomic distance (see Fig. 13.10). If such a slip deformation takes place in a perfectly ordered crystal, such as a A_3B-type superlattice, many BB

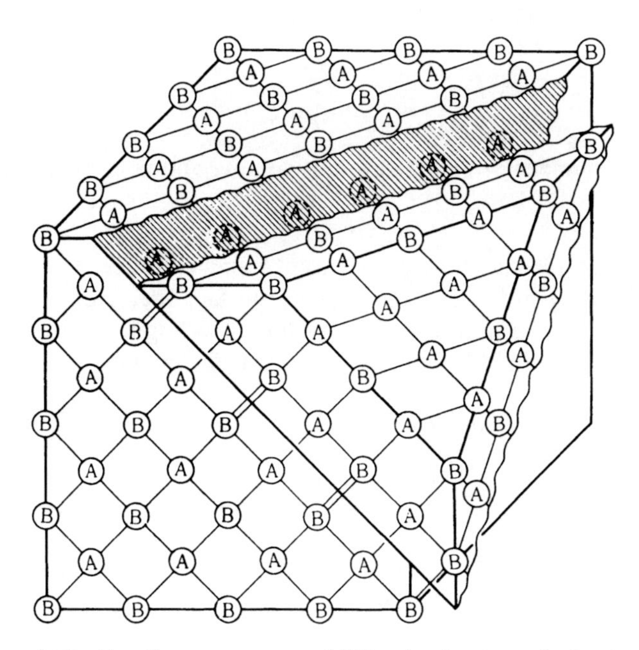

Fig. 13.10. Diagram indicating the appearance of BB pairs due to a single step slip along the (111) plane in the [0$\bar{1}$1] direction in an A_3B-type superlattice.[31]

pairs appear across the slipped plane, as indicated by double lines in Fig. 13.10. Since there are no BB pairs in the unslipped part of the crystal, the distribution of BB pairs becomes anisotropic, thus producing directional order. We call such directional order produced by slip deformation the *slip-induced directional order*.

The number of BB pairs thus produced depends on the degree of order S and on the probability p_0 of creating an isolated dislocation, because the motion of a pair of dislocations produces no BB pairs. As a matter of fact, there is a tendency for dislocations to exist and move in pairs in an ordered crystal, so as to avoid the creation of disordered atomic pairs. The number of BB pairs depends also on the number of slipped planes. We define the quantity p' as the probability with which dislocations are created on new atomic planes. Now we consider that ns dislocations run through the crystal per n atomic layers (s is called the slip density hereafter). Then the number of atomic planes upon which BB pairs are created in the ith slip system is given by

$$A_i = p_0 p' n s_i. \tag{13.25}$$

If we denote by α those lattice sites that are occupied by A atoms in the ordered state, and by β those occupied by B atoms, the number of $\beta\beta$ pairs which are created in a unit area of the slipped plane is given by $1/(\sqrt{3}\,a^2)$, where a is the lattice constant. Suppose that the alloy is partially ordered, with long-range order parameter S. The probability of finding B atoms in a β-site is S times larger than in other sites, and the number of BB pairs created in a unit area of the slipped plane is given by $S^2/(\sqrt{3}\,a^2)$. Since there are $n = \sqrt{3}/a$ atomic planes in a unit length perpendicular to the slip plane, the number of BB pairs created in a unit volume is given by

$$\begin{aligned} N_{\mathrm{BB}i} &= A_i S^2 n/(2\sqrt{3}\,a^2) \\ &= \tfrac{1}{8} N p S^2 \, |s_i|, \end{aligned} \tag{13.26}$$

where N is the number of atoms in a unit volume and is equal to $4/a^3$ and

$$p = p_0 p'. \tag{13.27}$$

Many dislocations tend to be created from a single source, such as a Frank–Read source. The factor $\frac{1}{2}$ in (13.26) results from the fact that even if all the dislocations are isolated from one another, the probability of having a disordered plane is still $\frac{1}{2}$.

Putting (13.26) into (13.6), we have the anisotropy energy

$$\begin{aligned} E_{\mathrm{a}} &= \tfrac{1}{8} N l_0 p S^2 \sum |s_i| (\alpha_i \gamma_{1i} + \alpha_2 \gamma_{2i} + \alpha_3 \gamma_{3i})^2 \\ &= \tfrac{1}{8} N l_0 p S^2 \sum |s_i| f_i(\alpha_1, \alpha_2, \alpha_3), \end{aligned} \tag{13.28}$$

where

$$\begin{aligned} f_i(\alpha_1, \alpha_2, \alpha_2) = \gamma_{1i}^2 \alpha_1^2 + \gamma_{2i}^2 \alpha_2^2 + \gamma_{3i}^2 \alpha_3^2 + 2\gamma_{2i}\gamma_{3i}\alpha_2\alpha_3 \\ + 2\gamma_{3i}\gamma_{1i}\alpha_3\alpha_1 + 2\gamma_{1i}\gamma_{2i}\alpha_1\alpha_2. \end{aligned} \tag{13.29}$$

The possible slip systems in a face-centered cubic lattice consist of four slip planes parallel to $\{111\}$ and three slip directions parallel to $\langle 110 \rangle$ in each slip plane.

Table 13.3. Direction cosines of BB pairs and the coefficients of each term in $f_i(\alpha_1, \alpha_2, \alpha_3)$ in the expression for long-range order-fine slip type (L.F. type) of anisotropy for each slip system (see (13.29)).

No. of slip system	Slip plane	Slip direction	γ_1	γ_2	γ_3	Coefficients in $f(\alpha_1, \alpha_2, \alpha_3)$					
						α_1^2	α_2^2	α_3^2	$\alpha_2\alpha_3$	$\alpha_3\alpha_1$	$\alpha_1\alpha_2$
1	(111)	$[01\bar{1}]$	0	$\frac{1}{\sqrt{2}}$	$\frac{1}{\sqrt{2}}$	0	$\frac{1}{2}$	$\frac{1}{2}$	1	0	0
2		$[10\bar{1}]$	$\frac{1}{\sqrt{2}}$	0	$\frac{1}{\sqrt{2}}$	$\frac{1}{2}$	0	$\frac{1}{2}$	0	1	0
3		$[1\bar{1}0]$	$\frac{1}{\sqrt{2}}$	$\frac{1}{\sqrt{2}}$	0	$\frac{1}{2}$	$\frac{1}{2}$	0	0	0	1
4	$(11\bar{1})$	[101]	$\frac{1}{\sqrt{2}}$	0	$-\frac{1}{\sqrt{2}}$	$\frac{1}{2}$	0	$\frac{1}{2}$	0	−1	0
5		[011]	0	$\frac{1}{\sqrt{2}}$	$-\frac{1}{\sqrt{2}}$	0	$\frac{1}{2}$	$\frac{1}{2}$	−1	0	0
6		$[1\bar{1}0]$	$\frac{1}{\sqrt{2}}$	$\frac{1}{\sqrt{2}}$	0	$\frac{1}{2}$	$\frac{1}{2}$	0	0	0	1
7	$(1\bar{1}1)$	[110]	$\frac{1}{\sqrt{2}}$	$-\frac{1}{\sqrt{2}}$	0	$\frac{1}{2}$	$\frac{1}{2}$	0	0	0	−1
8		$[10\bar{1}]$	$\frac{1}{\sqrt{2}}$	0	$\frac{1}{\sqrt{2}}$	$\frac{1}{2}$	0	$\frac{1}{2}$	0	1	0
9		[011]	0	$\frac{1}{\sqrt{2}}$	$-\frac{1}{\sqrt{2}}$	0	$\frac{1}{2}$	$\frac{1}{2}$	−1	0	0
10	$(\bar{1}11)$	$[01\bar{1}]$	0	$\frac{1}{\sqrt{2}}$	$\frac{1}{\sqrt{2}}$	0	$\frac{1}{2}$	$\frac{1}{2}$	1	0	0
11		[101]	$\frac{1}{\sqrt{2}}$	0	$-\frac{1}{\sqrt{2}}$	$\frac{1}{2}$	0	$\frac{1}{2}$	0	−1	0
12		[110]	$\frac{1}{\sqrt{2}}$	$-\frac{1}{\sqrt{2}}$	0	$\frac{1}{2}$	$\frac{1}{2}$	0	0	0	−1

Therefore the total number of slip systems is $4 \times 3 = 12$. The direction cosines of the induced BB pairs, $(\gamma_1, \gamma_2, \gamma_3)$, and the coefficients of the terms in $f_i(\alpha_1, \alpha_2, \alpha_3)$ are listed in Table 13.3. This type is called *long-range order (fine slip) type (L.F. type)*.

We can expect the appearance of another type of anisotropy. Suppose that the upper part of the crystal in Fig. 13.10 continues to travel over an out-of-step boundary to bring itself onto the neighboring order domain, in which one of the three kinds of α-sites is replaced by a β-site. Then the direction of BB pairs created on the slipped plane is changed from [011] to [110] or [101]. Thus when coarse slip takes place, BB pairs are expected to be distributed equally among the [011], [110], and [101] directions. The same thing is expected when the size of ordered domains is small, or in other words, when only short-range order prevails in the crystal. The average anisotropy of these three BB pairs is given by

$$\begin{aligned} w &= \frac{l_0}{3}\left(\cos^2 \varphi_{[011]} + \cos^2 \varphi_{[101]} + \cos^2 \varphi_{[110]}\right) \\ &= \frac{l_0}{3} \sum_{k=1,2,3} (\gamma_{1k}\alpha_1 + \gamma_{2k}\alpha_2 + \gamma_{3k}\alpha_3)^2 \\ &= l_0(n_{2i}n_{3i}\alpha_2\alpha_3 + n_{3i}n_{1i}\alpha_3\alpha_1 + n_{1i}n_{2i}\alpha_1\alpha_2) + \text{const.}, \end{aligned} \tag{13.30}$$

where (n_1, n_2, n_3) are the direction cosines of the normal to slip plane of the ith system. The anisotropy given by (13.30) is a uniaxial anisotropy with the axis normal to the slipped plane. The comprehensive explanation for this is that, if slip takes place in a crystal with short-range order, the atomic relation between the two sides of the slipped plane becomes disordered, signifying an abundance of BB pairs and thus rendering the hard axis normal to the slip plane, provided that $l_0 > 0$.

The number of BB pairs produced in a unit area of the slipped plane is

$$N_{BB} = \tfrac{1}{16}Np'\sigma\,|s_i|, \tag{13.31}$$

where σ is the short-range order parameter. The anisotropy energy is expressed as

$$E_a = \tfrac{1}{16}Nl_0 p'\sigma \sum |s_i| g_i(\alpha_1, \alpha_2, \alpha_3), \tag{13.32}$$

where

$$g_i(\alpha_1, \alpha_2, \alpha_3) = n_{2i}n_{3i}\alpha_2\alpha_3 + n_{3i}n_{1i}\alpha_3\alpha_1 + n_{1i}n_{2i}\alpha_1\alpha_2. \tag{13.33}$$

Table 13.4 lists the values of (n_{1i}, n_{2i}, n_{3i}) and the coefficients in each term of $g_i(\alpha_1, \alpha_2, \alpha_3)$ for all the slip systems. We call this type of anisotropy *short-range order (fine-slip) type (S.C. type).*

Assuming these two kinds of anisotropy, we can explain all the experimental facts for rolled single crystals. In the case of rolling on the (110) plane in the [001] direction, we observe slip bands as shown in Fig. 13.11(a), which proves that the slip systems 1, 2, 4, 5 contribute to the roll deformation (see Fig. 13.11(b)). By a simple calculation, we can deduce the relationship between rolling reduction r and the slip densities s_i, which is

$$\left.\begin{aligned} s_1 = s_4 &= -\frac{r}{2} \\ s_2 = s_5 &= \frac{r}{2} \end{aligned}\right\}. \tag{13.34}$$

Using (13.34), the roll magnetic anisotropy in this case is calculated as

$$\text{L.F. type} \qquad E_a = \tfrac{1}{16}Nl_0 pS^2 r\alpha_3^2, \tag{13.35}$$

and

$$\text{S.C. type} \qquad E_a = \tfrac{1}{24}Nl_0 p'\sigma r\alpha_1\alpha_2. \tag{13.36}$$

Table 13.4. Direction cosines of the normal to the slip plane and the coefficients of each term in $g_i(\alpha_1, \alpha_2, \alpha_3)$ in the expression for short-range order-coarse slip type (S.C. type) of anisotropy for each slip system (see (13.33)).

No. of slip system	n_1	n_2	n_3	Coefficients in $g(\alpha_1, \alpha_2, \alpha_3)$		
				$\alpha_2\alpha_3$	$\alpha_3\alpha_1$	$\alpha_1\alpha_2$
1	$\frac{1}{\sqrt{3}}$	$\frac{1}{\sqrt{3}}$	$\frac{1}{\sqrt{3}}$	$\frac{1}{3}$	$\frac{1}{3}$	$\frac{1}{3}$
2	$\frac{1}{\sqrt{3}}$	$\frac{1}{\sqrt{3}}$	$\frac{1}{\sqrt{3}}$	$\frac{1}{3}$	$\frac{1}{3}$	$\frac{1}{3}$
3	$\frac{1}{\sqrt{3}}$	$\frac{1}{\sqrt{3}}$	$\frac{1}{\sqrt{3}}$	$\frac{1}{3}$	$\frac{1}{3}$	$\frac{1}{3}$
4	$\frac{1}{\sqrt{3}}$	$\frac{1}{\sqrt{3}}$	$-\frac{1}{\sqrt{3}}$	$-\frac{1}{3}$	$-\frac{1}{3}$	$\frac{1}{3}$
5	$\frac{1}{\sqrt{3}}$	$\frac{1}{\sqrt{3}}$	$-\frac{1}{\sqrt{3}}$	$-\frac{1}{3}$	$-\frac{1}{3}$	$\frac{1}{3}$
6	$\frac{1}{\sqrt{3}}$	$\frac{1}{\sqrt{3}}$	$-\frac{1}{\sqrt{3}}$	$-\frac{1}{3}$	$-\frac{1}{3}$	$\frac{1}{3}$
7	$\frac{1}{\sqrt{3}}$	$-\frac{1}{\sqrt{3}}$	$\frac{1}{\sqrt{3}}$	$-\frac{1}{3}$	$\frac{1}{3}$	$-\frac{1}{3}$
8	$\frac{1}{\sqrt{3}}$	$-\frac{1}{\sqrt{3}}$	$\frac{1}{\sqrt{3}}$	$-\frac{1}{3}$	$\frac{1}{3}$	$-\frac{1}{3}$
9	$\frac{1}{\sqrt{3}}$	$-\frac{1}{\sqrt{3}}$	$\frac{1}{\sqrt{3}}$	$-\frac{1}{3}$	$\frac{1}{3}$	$-\frac{1}{3}$
10	$-\frac{1}{\sqrt{3}}$	$\frac{1}{\sqrt{3}}$	$\frac{1}{\sqrt{3}}$	$\frac{1}{3}$	$-\frac{1}{3}$	$-\frac{1}{3}$
11	$-\frac{1}{\sqrt{3}}$	$\frac{1}{\sqrt{3}}$	$\frac{1}{\sqrt{3}}$	$\frac{1}{3}$	$-\frac{1}{3}$	$-\frac{1}{3}$
12	$-\frac{1}{\sqrt{3}}$	$\frac{1}{\sqrt{2}}$	$\frac{1}{\sqrt{3}}$	$\frac{1}{3}$	$-\frac{1}{3}$	$-\frac{1}{3}$

In the case of Fe–Ni alloys, which have a positive l_0 as given by (13.17), (13.35) shows that the L.F. type has the hard axis parallel to the z-axis. In other words, the plane perpendicular to the rolling direction becomes an easy plane. Equation (13.36) shows that the S.C. type has the easy axis parallel to $[1\bar{1}0]$. If we measure the anisotropy in the rolling plane, or (110), both types result in an easy axis perpendicular to the rolling direction. Figure 13.12 shows the variation of K_u produced by (110)[001] rolling. As expected, the observed K_u is negative. With increasing reduction in thickness by rolling, the anisotropy grows monotonically, but at high reductions it tends to decrease again. This is probably due to the destruction of ordering by severe rolling. The effects of heat treatment will be explained later.

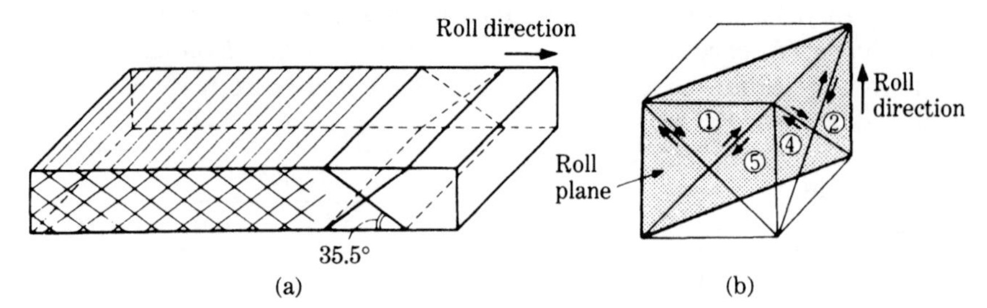

Fig. 13.11. (a) Slip bands and (b) slip systems for (110)[001] rolling.

In the case of (001)[110] rolling, it is possible that four slip systems contribute to the deformation, the same as for (110)[001] rolling. Actually, however, only two of the slip systems, 1 and 2, or 4 and 5 (see Fig. 13.13(b)), operate, as verified by the observation of slip bands on the side surface of the rolled slab (see Fig. 13.13(a)). The reason is as follows: The rolling process is equivalent to a tension in the rolling direction and a compression of the same strength perpendicular to the rolling plane. Then the maximum shear stress acts on a slip plane tilted 45° from the rolling plane. The slip plane in this case makes an angle greater than 45° from the rolling plane (see Fig. 13.13(a)). As a result of the rolling deformation, the slip plane rotates and its tilt angle approaches 45°, which encourages continued slip on the active slip system. This kind of slip is called *easy glide*. In contrast, in the case of (110)[001] rolling, the slip plane makes an angle of 35.5° with the rolling plane, so that the slip deformation along this slip plane causes a rotation of the crystal which encourages slip on the different slip plane (see Fig. 13.11(a)). This kind of slip is called *cross slip*.

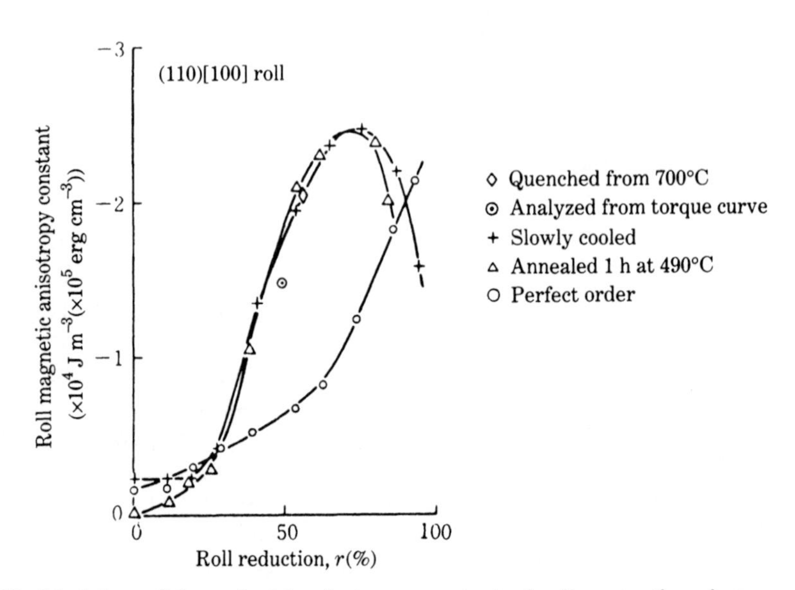

Fig. 13.12. Variation of the uniaxial anisotropy constant of roll magnetic anisotropy with the progress of (110)[001] rolling Ni_3Fe crystal.[31]

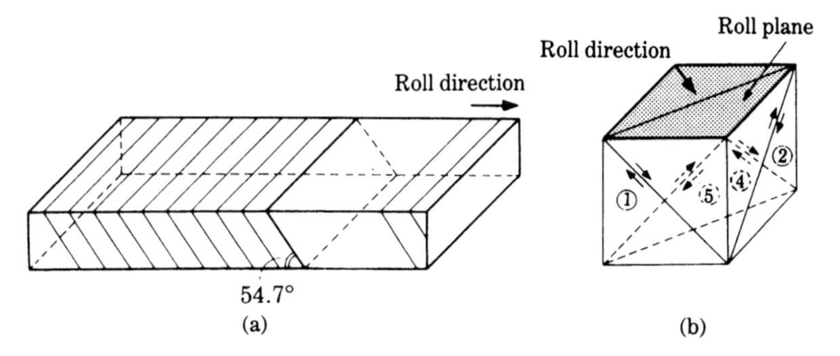

Fig. 13.13. (a) Slip bands and (b) slip systems for (001)[110] rolling.

In the case of (001)[110] rolling, therefore, the slip densities are given by

$$s_i = r, \qquad s_2 = -r, \tag{13.37}$$

from which the anisotropy energy leads to

$$\text{L.F. type} \qquad E_a = \tfrac{1}{16} N l_0 p S^2 r \alpha_3^2 (2\alpha_1 + 2\alpha_2 + \alpha_3), \tag{13.38}$$

and

$$\text{S.C. type} \qquad E_a = \tfrac{1}{24} N l_0 p' \sigma r (\alpha_1\alpha_2 + \alpha_2\alpha_3 + \alpha_3\alpha_1). \tag{13.39}$$

This L.F. type does not contribute to the anisotropy in the rolling plane (001), because $\alpha_3 = 0$ so that $E_a = 0$, whereas the S.C. type produces an easy axis perpendicular to the rolling direction in the rolling plane. Therefore, in the initial stage of rolling, the easy axis in the rolling plane is perpendicular to the rolling direction. As the rolling progresses, however, the crystal rotates by easy glide so as to tilt the slip plane, (111), closer to the rolling plane. Then the real easy axis of the L.F. type, [111], which is the intersection of the two easy planes (011) and (101), also tilts towards the rolling plane, thus producing an easy axis in the rolling plane parallel to the rolling direction.

Figure 13.14 shows the variation of the anisotropy constant, K_u, measured in the rolling plane for (001)[110] rolling, as a function of the roll reduction, r. As expected from the theory, K_u increases in a negative sense in the initial stage of rolling, but changes sign and increases in a positive sense in the later stage of rolling.

In the case of (001)[100] rolling, as in Isoperm, all four slip systems 1, 2, 4 and 5 contribute in a complex manner. The easy axis perpendicular to the rolling direction seems to be produced by crystal rotation.

Finally we shall discuss the influence of heat treatment of the alloy in advance of the rolling. We see in Fig. 13.12 that a quenched specimen develops almost the same roll magnetic anisotropy as a slowly cooled or annealed specimen. The reader may note that this behavior appears to be contradictory to the theory, which assumes the presence of some ordering in the alloy when rolling takes place. Moreover, a perfectly ordered specimen develops the anisotropy rather slowly as compared to a less-ordered specimen. These facts can be interpreted in terms of the pair-creation of dislocations in an ordered crystal. As explained before, the disordering created after one dislocation passes can be cancelled to recover the original state of order by the passage of a

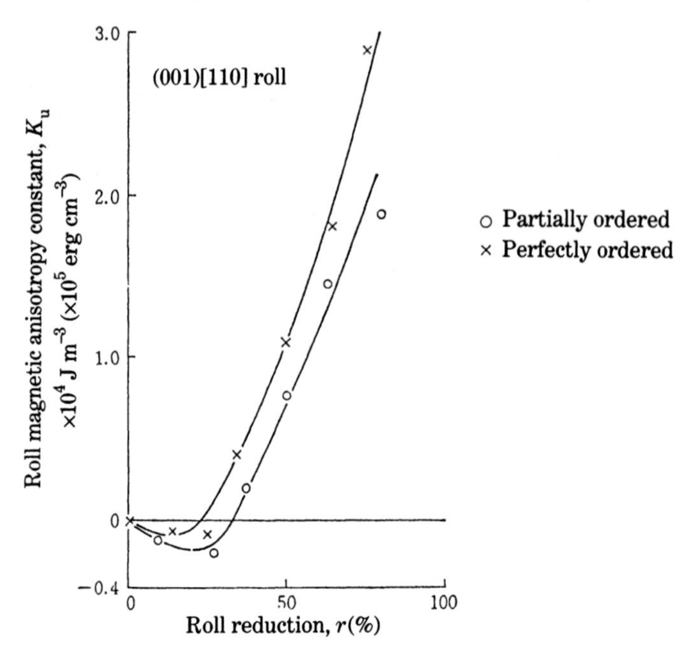

Fig. 13.14. Variation of the uniaxial anisotropy constant due to roll magnetic anisotropy with the progress of (001)[110] rolling Ni_3Fe crystal.[31]

second following dislocation. Therefore dislocations tend to occur in pairs in an ordered alloy, so as to make the stacking disorder between the pairs as short as possible. In other words, p_0 and therefore p in (13.27) decreases as S increases, keeping Sp_0 nearly constant.

The roll magnetic anisotropy was measured also for body-centered cubic alloys.[32] This kind of investigation has been extended to other kinds of cold-working, such as wire-drawing, by Chin.[33] Takahashi[34] used electron microscopy to observe dislocations produced by compression, and investigated their relationship to the created magnetic anisotropy.

13.3 INDUCED MAGNETIC ANISOTROPY ASSOCIATED WITH CRYSTALLOGRAPHIC TRANSFORMATIONS

Materials with uniaxial magnetocrystalline anisotropy will exhibit no net magnetic anisotropy if they are in the form of polycrystalline samples with the easy axes of the crystals (grains) distributed randomly. If, however, crystal growth or a phase transformation takes place at a temperature below the Curie temperature, the easy axes of the resulting grains may be aligned to some extent parallel to the magnetic field. This results in a net macroscopic induced anisotropy in the material.

First we treat the induced magnetic anisotropy of precipitation alloys, in which elongated precipitate particles are aligned by a magnetic field during the precipitation process. This treatment can be applied to Alnico to improve its permanent magnet characteristics. Another example is the neutron irradiation of a 50% Fe–50% Ni alloy in the presence of a field. This alloy transforms to a superlattice with tetragonal crystal structure during neutron irradiation, and a uniaxial anisotropy is induced if a magnetic field is present during irradiation. A third case is the phase transformation in cobalt metal from fcc to hcp (with uniaxial anisotropy) at about 400°C; an induced anisotropy is observed when polycrystalline cobalt is cooled through the transformation temperature in a magnetic field. A similar phenomenon is observed for magnetite, or Fe_3O_4, whose crystal structure changes from cubic to a lower symmetry at about 125 K. The difference between the last two cases is that the cobalt transformation requires the physical displacement of atoms, at least over small distances, while the magnetite transformation takes place purely by the transfer of electrons from one ion to another (electron hopping), changing the ionization state of both ions.

It is also observed that ferrimagnetic garnets exhibit growth-induced anisotropy with the easy axis normal to the surface, when the surface is a {111} plane.

These phenomena will be discussed below.

13.3.1 Induced magnetic anisotropy in precipitation alloys

In 1932 Mishima[35] invented a very strong magnet called MK steel, which is known as Alnico 5 in Europe and the United States. As suggested by its name, this alloy consists of Al, Ni, and Co, in addition to Fe. Above 1300°C it consists of a homogeneous solid solution with bcc crystal structure, but it separates into two phases by precipitation below 900°C. In 1938 Oliver and Shedden[36] discovered for a similar alloy that the magnetic properties could be greatly improved by cooling it in a magnetic field. Jones and Emden[37] applied this procedure to Alnico and obtained a rectangular hysteresis loop with a large value of $(BH)_{max}$. This is apparently due to the appearance of an induced magnetic anisotropy with its easy axis parallel to the direction of the magnetic field applied during cooling.

The electron micrographs of Fig. 13.15 show elongated precipitates in field-cooled Alnico. Photograph (a) shows a cross-sectional view and (b) a side view of the elongated precipitates. It is observed that the precipitates are always elongated parallel to the ⟨100⟩ direction which is nearest to the field applied during cooling. Both the precipitates and the matrix have bcc structures, but the precipitates contain more Fe and Co, and are ferromagnetic, while the matrix contains more Ni and Al, and is weakly magnetic or nonmagnetic. Therefore the spontaneous magnetization is different in the two phases, and a shape anisotropy as given by (12.24) is produced. When the precipitate particles are densely packed as in Fig. 13.15, we cannot ignore the magnetic interaction between particles. Let v be the volume fraction of precipitates. Then if v becomes 1, meaning that the entire sample consists of the precipitate phase, the shape anisotropy should disappear. At the opposite extreme, if v is very small, the anisotropy should be proportional to the precipitate volume. Therefore v in (12.24) should be replaced by a function which has the general behavior of $v(1-v)$. The anisotropy constant in this case is given by

$$K_u = \frac{1}{4\mu_0}(I_s - I_s')^2 v(1-v)(1-3N_z), \tag{13.40}$$

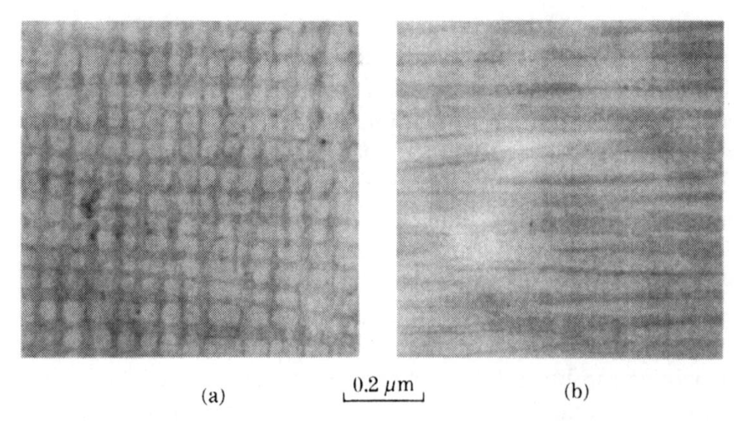

Fig. 13.15. Electron micrographs of Alnico 8 (36 wt% Co, 14 wt% Ni, 6.6 wt% Al, 5 wt% Ti, 3 wt% Cu, 0.1 wt% Si, balance Fe) annealed at 810°C for 10 min in a magnetic field of 0.13 MA m^{-1} (1600 Oe): (a) cross-sectional view; (b) side view of the elongated precipitate particles.[39]

where I_s and I'_s are the saturation magnetization of the matrix and the precipitate phases, respectively, and N_z is the demagnetization factor (along the long axis) of an isolated precipitate particle.

The mechanism of growth of the precipitate particles was explained by Cahn[38] as spinodal decomposition. In general, precipitation occurs if the free energy of the system decreases when an atom is transferred from the matrix phase to the precipitate phase. Therefore in order to initiate precipitation, the second derivative of free energy with respect to concentration must be negative. If, however, the precipitation results in a change in lattice constant, an increase in elastic energy will oppose the process of precipitation. This can result in a special kind of precipitation, in which the composition changes continuously from a homogeneous single phase to a two-phase structure by the gradual build-up of a sinusoidal composition variation. The wavelength of the fluctuation in concentration is determined by a balance between the chemical energy difference of the two phases and the elastic strain energy produced by the composition change. This precipitation process is called *spinodal decomposition*, and results in a very regular, small-scale precipitate structure.

If spinodal decomposition occurs near the Curie point of an alloy, and if the Curie point depends strongly on the composition, the local saturation magnetization will vary strongly with the local composition and therefore with position. In this case the magnetostatic energy is increased by the fluctuation in composition if the fluctuation occurs along the direction of magnetization. However, a fluctuation in composition perpendicular to the direction of magnetization will not affect the magnetostatic energy. For this reason the precipitate particles are elongated only along the direction of magnetization.

In the commercial heat treatment of Alnico 5, the alloy is first annealed for 30 min at 1250°C to produce a homogeneous single phase structure, quenched to 950°C to prevent the precipitation of the fcc gamma phase, and then cooled slowly from 900 to

700°C at a rate of 0.1–0.3 K s^{-1} in a magnetic field stronger than 0.12 MA m^{-1} (1500 Oe). For Alnico with higher Co content, the magnetic heat treatment is performed at 800–820°C. During these magnetic heat treatments many small precipitate nuclei are produced. Further heat treatment between 580 and 600°C for an appropriate time causes the elongated precipitates to grow to a size at which it is reasonable to regard them as single domain particles. This heat treatment is called aging. The final size of the particles is somewhat less than about 1000 Å long and 100 Å in width.

The long axis of these precipitates is parallel to the ⟨100⟩ direction which is nearest to the direction of the applied magnetic field. The reason for this is that the elastic modulus is smaller along ⟨100⟩ than along ⟨111⟩, so that the increase in elastic energy is smaller when the precipitates elongate parallel to ⟨100⟩. Once the precipitate nuclei have formed, the precipitates continue to grow in an anisotropic way even without the presence of a magnetic field. This formation of precipitate nuclei is most effective at a temperature just below the Curie point of the precipitate phase.[39]

The magnetic anisotropy thus produced is hardly changed by further aging with a magnetic field applied perpendicular to the easy axis, because any change in particle shape requires diffusion of atoms over long distances corresponding to the particle size, and also requires an increase in elastic energy.

Iwama *et al.*[40] observed that a slower cooling rate gave larger induced anisotropy and also large dimensional ratio (length l/diameter d) of the precipitate particles. In one case, $d = 200$–600 Å, $l/d = 3$–5, $K_u = (1.2 - 1.6) \times 10^5\,\mathrm{J\,m^{-3}}$ for v = volume fraction of precipitate = 0.7. In this work it was also observed that when the magnetic field is applied in a direction which deviates by the angle β ($\beta < 45°$) from [100] towards [010] in a (001) disk, the easy axis of the induced anisotropy deviates by an angle α which is always less than β as shown in Fig. 13.16. This results from the fact that the long axis of the precipitates approaches the nearest ⟨100⟩. The magnitude of K_u decreases as β increases from 0 to 45°. This is partly due to the development of an irregular geometry of the precipitate particles.

13.3.2 Induced anisotropy produced by neutron irradiation

It was discovered by Pauleve *et al.*[41] that neutron irradiation of 50 at% Fe–Ni alloy results in the formation of a tetragonal CuAu-type superlattice. A single crystal disk cut parallel to (011) was irradiated at 295°C by 1.5×10^{20} neutrons (14% of which had energy higher than 1 MeV) in a magnetic field of 200 kA m^{-1} (= 2500 Oe) applied parallel to [100]. It was discovered by X-ray diffraction that the crystal had divided into three regions, each with its tetragonal c-axis parallel to one of the original cubic [100], [010], or [001] axes, and each occupying a different volume fraction v_i, where i represents x, y or z. The degree of order, S_i, also may vary between the three regions. The values of $v_i S_i$ for $i = x$, y and z, determined by X-ray diffraction, were 0.32 ± 0.01, 0.21 ± 0.05 and 0.21 ± 0.05, respectively. This means that neutron irradiation in a magnetic field applied parallel to [100] resulted in a larger value of the product Sv for the region with its c-axis parallel to [100].

The magnetocrystalline anisotropy of this material determined by magnetization

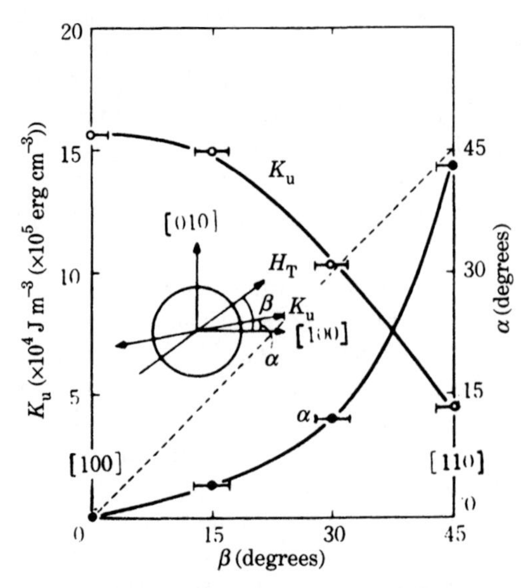

Fig. 13.16. The uniaxial anisotropy constant, K_u, and the angle, α, of the easy axis from [100] as functions of the angle β between the magnetic field applied during cooling and [100]. The anisotropy was determined from torque measurements on a (100) disk of Alnico 5 which was homogenized for 10 min at 900°C, cooled in a magnetic field of 0.64 MA m^{-1} (8 kOe) at a rate of 0.5 K s^{-1} in a direction β from [100], and finally annealed for 16 h at 600°C.[40]

curves was found to have the easy axis parallel to [100], and the hard axes parallel to [110]. The anisotropy constants were determined to be $K_{u1} = 3.2 \times 10^5$ J m^{-3} ($= 3.2 \times 10^6$ erg cm^{-3}) and $K_{u2} = 2.3 \times 10^5$ J m^{-3} ($= 2.3 \times 10^6$ erg cm^{-3}). The reason why the magnitude of K_{u2} is comparable with that of K_{u1} was explained as follows:[42,43] The size of the ordered regions is much smaller than that of the magnetic domains, so that the spin distribution is determined by a balance of the local magnetocrystalline anisotropy, the external field energy, the exchange energy, and the magnetostatic energy. When the external field is parallel to [100], all the spins are also aligned parallel to [100]; but when the magnetic field is in some other direction, the spins will take a transitional distribution, which makes [110] a hard axis. This reasoning is similar to that of the torque reversal discussed in Section 12.2.

As a result of this analysis, it was concluded that the size of the ordered regions is about fifty lattice constants, and that the volume fractions of the [100], [010], and [001] regions are 60%, 20%, and 20%, respectively. The anisotropy constant, $K_{u1[100]}$, of the ordered region with its c-axis parallel to [100] is larger than those of other regions, $K_{u1[010]}$ and $K_{u1[001]}$. These values, which are deduced from magnetization curves, are slightly different from those deduced from torque curves. The values from magnetization curves are

$$\left.\begin{aligned} K_{u1[100]} &= 1.37 \times 10^6 \text{ J m}^{-3} \quad (= 1.37 \times 10^7 \text{ erg cm}^{-3}) \\ K_{u1[010]} &= K_{u1[001]} = 1.03 \times 10^6 \text{ J m}^{-3} \quad (= 1.03 \times 10^7 \text{ erg cm}^{-3}) \end{aligned}\right\}, \tag{13.41}$$

while those from torque curves are

$$\left.\begin{array}{ll} K_{u1[100]} = 1.16\times10^{6}\,\mathrm{J\,m^{-3}} & (=1.16\times10^{7}\,\mathrm{erg\,cm^{-3}}) \\ K_{u1[010]} = K_{u1[001]} = 1.0\times10^{6}\,\mathrm{J\,m^{-3}} & (=1.0\times10^{7}\,\mathrm{erg\,cm^{-3}}) \end{array}\right\}. \quad (13.42)$$

In conclusion, the major effect of irradiation in a magnetic field is to alter the volume fractions of the three kinds of ordered regions, rather than to alter the anisotropy constants or the directional order in each region. Since, however, the induced anisotropy continues to increase with irradiation, the observed induced anisotropy may be far from the final value.[42]

Recently it was reported that the ordered phase of 50% Fe–Ni is found in iron meteorites.[44–47] The ordering may have been produced by annealing over astronomical time periods, or by a strong irradiation in space. In either case it is an interesting story.

13.3.3 Induced anisotropy associated with crystallographic transformation of Co and Co–Ni alloys

It was found by Takahashi and Kono[48] that polycrystalline Co and Co–Ni alloys exhibit induced magnetic anisotropy after they are cooled in a magnetic field through the phase transformation from fcc to hcp. The anisotropy constant was determined by the difference in magnetization curves between the two cooling processes: cooling with and without a magnetic field through 400°C, at which the phase transformation takes place. Figure 13.17 shows the dependence of K_u determined in this way, as a function of Co content in Ni. The sign of K_u is negative from 100 to 95% Co, because K_u of the hcp phase changes sign between the phase transformation temperature and room temperature. The sign of K_u is positive from 95 to 70% Co, because in this composition range K_u of the hcp phase keeps the same sign at all temperatures above room temperature. The maximum value of the induced anisotropy constant is $2.3\times10^{4}\,\mathrm{J\,m^{-3}}$ $(=2.3\times10^{5}\,\mathrm{erg\,cm^{-3}})$, which is only 7% of the estimated value, $3.3\times10^{5}\,\mathrm{J\,m^{-3}}$ $(=3.3\times10^{6}\,\mathrm{erg\,cm^{-3}})$, of the single crystal magnetocrystalline anisotropy of the alloy with the same composition. For compositions below 80% Co, the phase transformation occurs below room temperature, so that K_u is extremely small, and depends on composition as $C(\mathrm{Co})^2C(\mathrm{Ni})^2$, which is interpreted in terms of the induced anisotropy due to directional order in the binary alloy.

A large induced anisotropy constant, $-2\times10^{4}\,\mathrm{J\,m^{-3}}$, was observed by Graham[49] for polycrystalline pure Co after it was cooled in a magnetic field of $0.24\,\mathrm{MA\,m^{-1}}$ $(=3000\,\mathrm{Oe})$ through the phase transformation point. It was also observed that the temperature dependence of the induced anisotropy is proportional to that of the magnetocrystalline anisotropy of pure Co. Based on this experiment, he proposed the following mechanism: fcc Co transforms to hcp Co in such a way that the *c*-axis of an hcp grain is parallel to a $\langle 111\rangle$ axis of fcc Co. Since there are four $\langle 111\rangle$ axes in an fcc grain, there are four possible orientations for the *c*-axis of the transformed hcp Co. Thus the *c*-axis of hcp Co should be distributed equally among the four possible directions if the transformation takes place in the absence of a magnetic field.

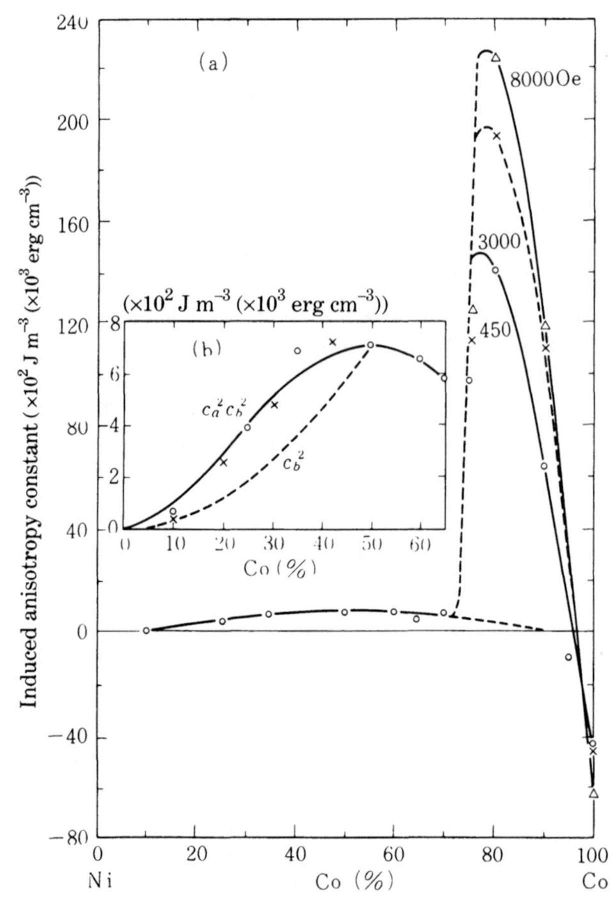

Fig. 13.17. The anisotropy constant, K_u, induced by cooling Ni–Co alloys in a magnetic field from 1000°C at a rate of $3.3\,\mathrm{K\,min^{-1}}$, as a function of Co content. The numerical values give the intensity of the magnetic field applied during cooling ($1\,\mathrm{Oe} = 80\,\mathrm{A\,m^{-1}}$).[48]

However, if a magnetic field strong enough to rotate the magnetization out of the easy axis of each fcc grain is applied during transformation, the anisotropy energy will cause an unbalanced volume distribution of these grains. Thus a preferred orientation of the *c*-axes results, producing an induced anisotropy.

Sambongi and Mitui[50] observed that the magnitude of K_u thus induced for Co is considerably influenced by cold working in advance of the heat treatment. No induced anisotropy due to magnetic cooling was observed for a specimen melted in an argon atmosphere, shaped, hot-worked, and annealed. It was found that the higher the annealing temperature, the more effective the magnetic cooling; and that the induced anisotropy cannot be removed by annealing below 700°C. These facts suggest that the presence of some lattice defects may play a role in the mechanism of this induced anisotropy.

Wakiyama *et al.*[51] observed the same effect for Fe–Co alloys with the double hcp structure (ABAC stacking sequence; see Fig. 12.28) and obtained the record value of $K_u = 1.3 \times 10^5\,\mathrm{J\,m^{-3}}$ ($= 1.3 \times 10^6\,\mathrm{erg\,cm^{-3}}$) for an alloy of Co-1.6 at% Fe.

13.3.4 Induced anisotropy associated with the low temperature transition of magnetite (Fe_3O_4)

The crystal structure of magnetite or Fe_3O_4 transforms from cubic to lower symmetry at about 125 K (see Section 12.4(b)). The magnetocrystalline anisotropy of this low temperature phase is given by (12.75). The anisotropy constants observed[52] at 120 K, which is just below the transition point, are given by

$$\left.\begin{array}{ll} K_0 = 6.7 \times 10^4\,\mathrm{J\,m^{-3}} & (= 6.7 \times 10^5\,\mathrm{erg\,cm^{-3}}) \\ K_a = 23 \times 10^4\,\mathrm{J\,m^{-3}} & (= 23 \times 10^5\,\mathrm{erg\,cm^{-3}}) \\ K_b = 7 \times 10^4\,\mathrm{J\,m^{-3}} & (= 7 \times 10^5\,\mathrm{erg\,cm^{-3}}) \\ K_u = -2.2 \times 10^4\,\mathrm{J\,m^{-3}} & (= -2.2 \times 10^5\,\mathrm{erg\,cm^{-3}}) \\ K_{aa} = 1.6 \times 10^4\,\mathrm{J\,m^{-3}} & (= 1.6 \times 10^5\,\mathrm{erg\,cm^{-3}}) \\ K_{bb}^* = 1.8 \times 10^4\,\mathrm{J\,m^{-3}} & (= 1.8 \times 10^5\,\mathrm{erg\,cm^{-3}}) \\ K_{ab}^* = 7.0 \times 10^4\,\mathrm{J\,m^{-3}} & (= 7.0 \times 10^5\,\mathrm{erg\,cm^{-3}}) \end{array}\right\}, \qquad (13.43)$$

where the symbol * indicates values at 4.2 K. Since $K_a > K_b > 0$, we find that the *a*-, *b*- and *c*-axes are the hard, intermediate, and easy axes, respectively. It was found that when a magnetic field is applied parallel to one of the $\langle 100 \rangle$ axes during cooling through the transition point, the *c*-axis of the low temperature phase is established parallel to this $\langle 100 \rangle$.[53] The crystal axis can be established at such a low temperature because the transition is caused by quenching of the electron hopping between Fe^{2+} and Fe^{3+} on the octahedral or B sites of the spinel lattice, so that no atomic migration is required. Verwey *et al.*[54] considered that as a result of this quenching of the electron hopping, Fe^{2+} and Fe^{3+} ions form an ordered arrangement on the B sites. However, this proposed ordering scheme has been disproved by later experiments.[55] No final conclusion has been reached on this point.

If the magnetic field applied during cooling is accurately parallel to one of the $\langle 100 \rangle$ directions, the *c*-axis is established parallel to this $\langle 100 \rangle$ as mentioned above, but the *b*-axis becomes parallel to one of the two possible $\langle 110 \rangle$ directions, both of which are perpendicular to the *c*-axis. The resulting mixture of two kinds of twinned regions will result in a uniaxial anisotropy with the easy axis parallel to the *c*-axis. In order to establish a unique *b*-axis, the magnetic field during cooling should be parallel to the direction in which the magnetocrystalline anisotropy of one twin is considerably lower than that of the others.

The magnetocrystalline anisotropy energies in the $(1\bar{1}0)$ plane are plotted in Fig. 13.18 for all the possible twins. The crystal axes or *a*-, *b*-, and *c*-axes of each twin with respect to the cubic axes are listed underneath the graph. In addition to the twins with different *a*-, *b*-, and *c*-axes, there is another class of twins, or sub-twins, in which

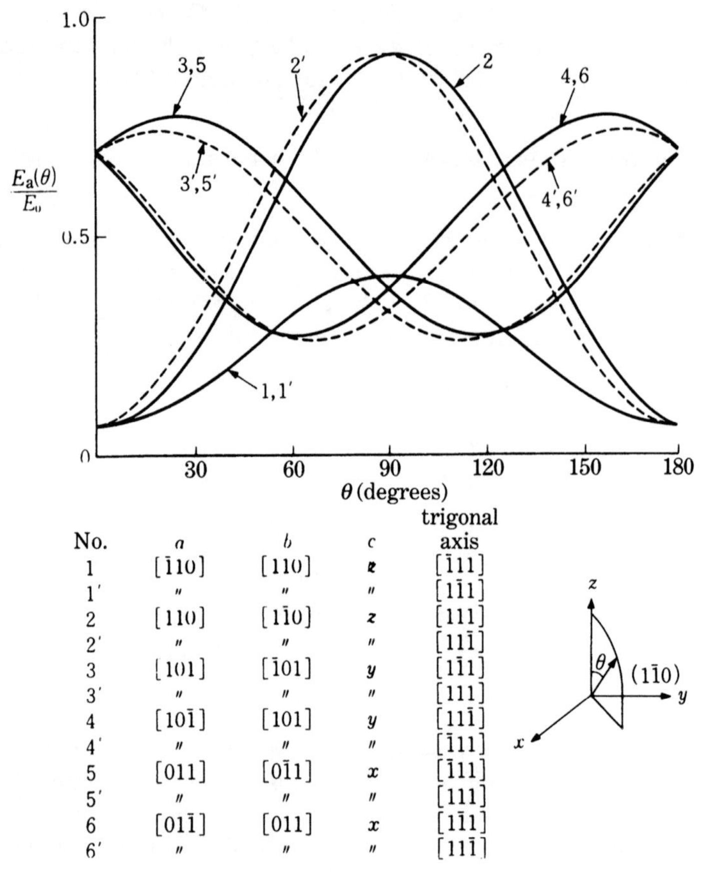

No.	*a*	*b*	*c*	trigonal axis
1	$[\bar{1}10]$	$[110]$	*z*	$[\bar{1}11]$
1′	″	″	″	$[1\bar{1}1]$
2	$[110]$	$[1\bar{1}0]$	*z*	$[111]$
2′	″	″	″	$[11\bar{1}]$
3	$[101]$	$[\bar{1}01]$	*y*	$[1\bar{1}1]$
3′	″	″	″	$[111]$
4	$[10\bar{1}]$	$[101]$	*y*	$[11\bar{1}]$
4′	″	″	″	$[\bar{1}11]$
5	$[011]$	$[0\bar{1}1]$	*x*	$[\bar{1}11]$
5′	″	″	″	$[111]$
6	$[01\bar{1}]$	$[011]$	*x*	$[1\bar{1}1]$
6′	″	″	″	$[11\bar{1}]$

Fig. 13.18. Variation of magnetocrystalline anisotropy in $(1\bar{1}0)$ of all possible twins of the low-temperature phase of magnetite as a function of the angle of magnetization measured from [001], where all the indices are referred to the cubic axes. The numbers on the curves refer to twins whose axes are shown in the table underneath the graph. The dashed curves with dashed numbers refer to sub-twins which have uniaxial anisotropy with different easy axis.

$$E_0 = 6.67 \times 10^4\,\mathrm{J\,m^{-3}}\ (= 6.67 \times 10^5\,\mathrm{erg\,cm^{-3}})$$

the c-axis tilts by 0.23° or −0.23° out of the z-axis in the z–a plane.[55] These twins have a K_u term (see (12.75)) with a different axis and are indicated by a dashed number under the graph. Because of a small difference due to the K_u term, the anisotropy energies of the twins with dashed numbers (shown by dashed curves) differ slightly from the solid curves for the non-dashed numbers. If a magnetic field is applied in the direction $\theta = 40°$ out of the ⟨100⟩ during cooling through the Verwey point, the anisotropy energy for twins 1 and 1′ is considerably lower than for the other twins, so that we expect the twins 1 and 1′ will develop uniquely. This is actually the case. The twins thus developed have the c-axis parallel to the ⟨100⟩ which is nearest to the magnetic field during cooling, and the b-axis at $\theta = 90°$ in the $\{1\bar{1}0\}$ plane

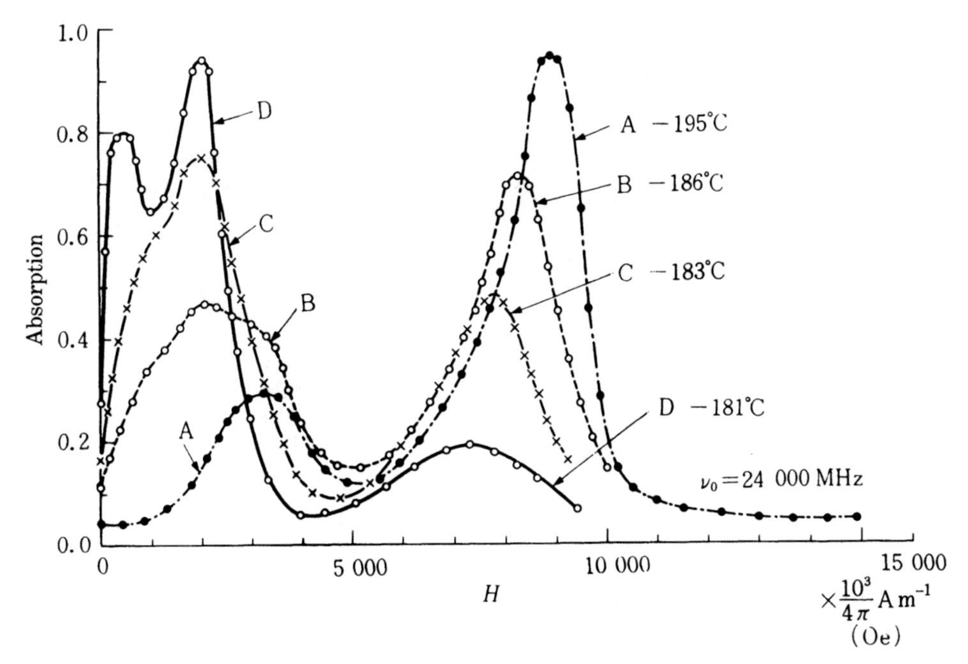

Fig. 13.19. Ferromagnetic resonance absorption curves observed for magnetite at temperatures below the Verwey transition point.[56]

containing the c-axis and the magnetic field during cooling.

In order to remove the twin 1′, we can utilize the difference in deformation of the two twins: twin 1 elongates by 0.4% parallel to $[\bar{1}11]$, while twin 1′ elongates parallel to $[1\bar{1}1]$. If a cylindrical specimen with its long axis parallel to $\langle 111\rangle$ is inserted (at room temperature) into tightly fitting aluminum rings and cooled through the Verwey point with a magnetic field applied as described above, the difference in thermal expansion coefficient between aluminum and the specimen exerts a uniform compressive stress in the plane perpendicular to the $[\bar{1}11]$ axis of the cylinder, and twin 1 is uniquely selected. The anisotropy constants given by (13.43) were determined for such a squeezed specimen. The effect of a compressive stress on K_u was estimated by changing the magnitude of the stress, and was found to be negligible.[52] Even if the stress is negligible, the space between the specimen and the aluminum rings is very small (less than 0.4%), so that the generation of more than one twin was impossible.

In a freely supported specimen, it is possible to switch a once-established c-axis to another $\langle 100\rangle$ by applying a magnetic field parallel to a new $\langle 100\rangle$ at a temperature below the Verwey point. Bickford[56] observed this switching phenomenon by means of ferromagnetic resonance. He observed a shift of the resonance line of lower fields by applying a magnetic field parallel to a new $\langle 100\rangle$, while heating the specimen from −195°C to −181°C (Fig. 13.19). The peaks at the higher field are due to the original twin, which has its easy axis perpendicular to the field and contributes a negative anisotropy field, so that the resonance occurs at a higher field, corresponding to the

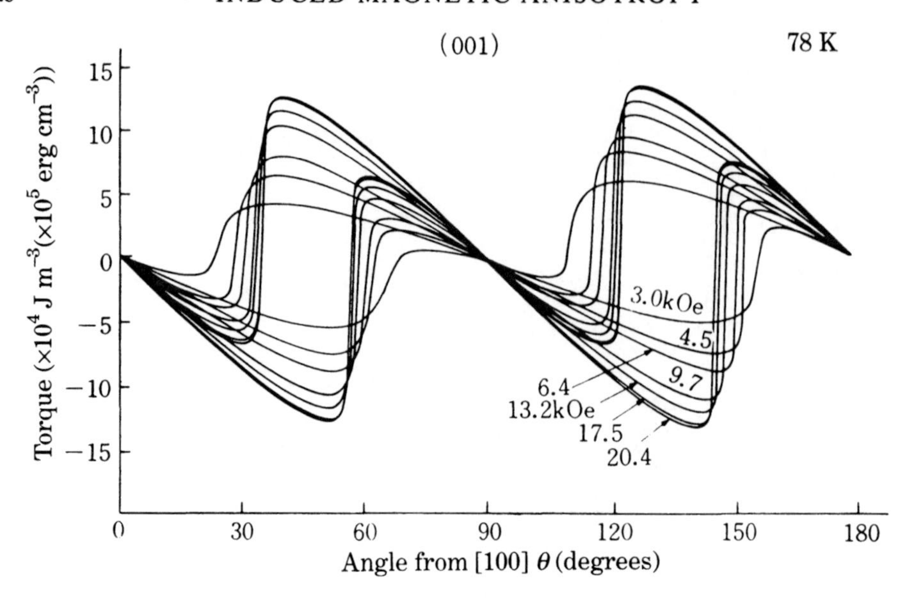

Fig. 13.20. Switching of the c-axis of the low-temperature phase caused by rotation of magnetization in the (001) plane of magnetite, as observed through a change in torque curves.[5]

resonance field plus the anisotropy field. The peaks at the lower field are due to a new twin, which has its easy axis parallel to the applied field and contributes a positive anisotropy field, so that the resonance occurs at a lower field, corresponding to the resonance field minus the anisotropy field. It should be noted that the resonance peaks change not by shifting position along the abscissa but by a decrease in the height of the higher field peak and an increase in the height of the lower field peak. This means that the change is not caused by changing the magnitude of the anisotropy in each twin, but by the growth of a new twin at the expense of the old twin.

Yamada[57] observed the switching phenomenon by means of a torque magnetometer. The torque changes abruptly when the switching occurs, as shown in Fig. 13.20. He also found that switching occurs between 77 and 120 K if the nucleus of a new twin is present. The numerical values on the torque curves give the magnetic field in which the torque was measured. Yamada showed that the switching occurs at the angle where the anisotropy energies of the two twins become equal. He explained this phenomenon in terms of the displacement of twin boundaries.

13.3.5 Growth-induced anisotropy

The rare earth iron garnets (see Section 9.3) containing more than two kinds of rare earth ions sometimes exhibit a uniaxial anisotropy in addition to cubic magnetocrystalline anisotropy. This uniaxial anisotropy has its axis parallel to the direction of the crystal growth,[58] and is therefore called the *growth-induced magnetic anisotropy*.

Two possible interpretations have been proposed: one considers that the anisotropy is caused by directional order of rare earth ions on dodecahedral sites of garnets,[59,60] while the other invokes a one-ion anisotropy caused by some anisotropic rare earth ions which select particular octahedral sites during crystal growth.[61,62] This anisotropy is rather complex: the direction of crystal growth becomes the easy or hard axis depending on the crystallographic plane of the crystal growth. For instance, it was observed by torque measurement that when a thin film of $Sm_{0.4}Y_{2.6}Ga_{1.2}Fe_{3.8}O_{12}$ (Sm-YIGG), 6–7 μm thick, is grown epitaxially onto gadolinium–gallium–garnet (GGG), the growth direction becomes the easy axis if the plane of the thin film is parallel to {111}, while the growth direction becomes the hard axis if the film plane is parallel to {100}.[63] If was also observed that in the case of growth plane parallel to {111}, the larger the difference in size between the two kinds of rare earth ions, the larger the growth-induced anisotropy.[64]

In general, when there is a large misfit of lattice parameters between the substrate and the garnet film, a fairly large uniaxial anisotropy is induced by stresses through magnetostriction.

13.4 OTHER INDUCED MAGNETIC ANISOTROPIES

Here we shall discuss induced magnetic anisotropies other than the categories already described in Sections 13.1–13.3. These are:

(1) exchange or unidirectional anisotropy;
(2) photo-induced anisotropy;
(3) rotatable anisotropy;
(4) mictomagnetic anisotropy;
(5) induced anisotropy in amorphous magnetic alloys.

13.4.1 Exchange anisotropy

Meiklejohn and Bean[65] discovered that when a slightly oxidized Co powder with particles of 100–1000 Å diameter is cooled from room temperature to 77 K in a magnetic field, it exhibits a *unidirectional* (as opposed to a uniaxial) *anisotropy*, which tends to hold the magnetization in the direction of the field applied during cooling.

Cobalt metal undergoes no crystal transformation in this temperature range, but the cobalt monoxide CoO, which covers the surface of Co particles, is antiferromagnetic with a Néel point at about 300 K. If a positive exchange interaction acts between a Co spin in the particle and a neighboring Co^{2+} spin in the CoO layer at the particle surface, the Co^{2+} spin at the surface must be aligned parallel to the Co spin in the particle when the antiferromagnetic spin structure is established in CoO at the Néel point during cooling (see Fig. 13.21(a)). After cooling in a magnetic field down to a temperature well below the Néel point, the spin structure of CoO is firmly fixed to the lattice through a strong anisotropy of the order of $5 \times 10^5\,\mathrm{J\,m^{-3}}$, so that the magnetization of the particle is forced to align parallel to the direction of the field applied during cooling. When an external magnetic field is applied out of the easy direction as

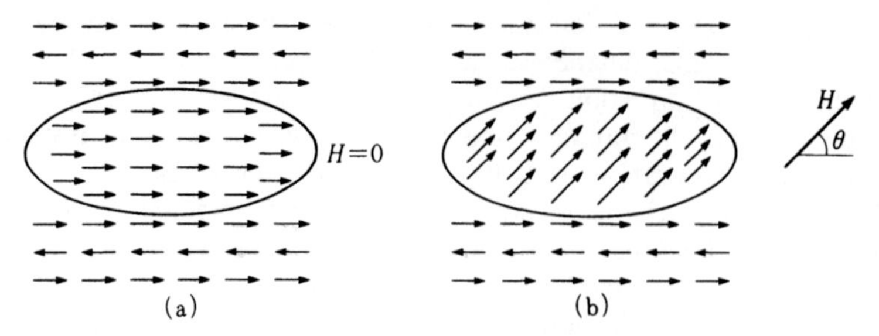

Fig. 13.21. Spin arrangement in a Co particle coated with CoO.

shown in Fig. 13.21(b), the magnetization of the particle rotates and makes an angle θ with the spin axis of CoO, thus increasing the exchange energy at the surface.

The anisotropy energy thus produced has the form

$$E_a = -K_d \cos\theta. \tag{13.44}$$

The value of K_d is of order $1 \times 10^5\,\mathrm{J\,m^{-3}}$, depending on the total surface area of the particles and therefore on the average particle size. The magnetic anisotropy thus produced is called the *exchange anisotropy*.

As a result of this anisotropy, the hysteresis curve shifts to the left along the abscissa as shown by the solid curve in Fig. 13.22. This curve was observed at 77 K after cooling through the Néel point with a magnetic field of $10^6\,\mathrm{A\,m^{-1}}$ (= 10^4 Oe) applied in the + direction. The dashed curve was observed for the same specimen cooled without magnetic field. The reason for this shift is that the magnetization of

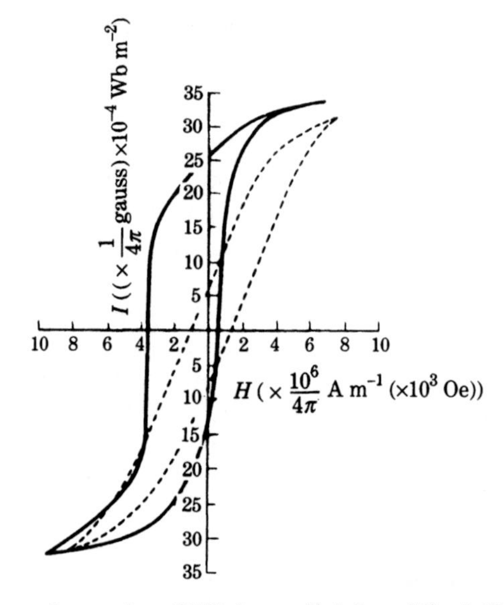

Fig. 13.22. Hysteresis loop observed at 77 K for a slightly oxidized Co powder. Solid curve: cooled in magnetic field; dashed curve: cooled in absence of magnetic field.[65]

the Co particle tends to point in the + direction due to the exchange anisotropy, so that an excess field must be applied in the − direction in order to reverse the magnetization into the − direction. If such a shifted hysteresis curve can be obtained at room temperature, it becomes possible to have a material in which the remanent magnetization always points in one specific direction, independent of the direction of the magnetizing field. This phenomenon may be useful for some magnetic logic system.

It was also observed that a rotational hysteresis (see Sections 12.2 and 18.6) is associated with the torque curve of this material, and that the hysteresis does not disappear even in a high magnetic field. The reason is thought to be that some of the antiferromagnetic spins in CoO are irreversibly flipped over when the magnetization in the Co particles is rotated.[65]

Exchange anisotropy was also observed in an Fe–FeO system[66] and for many ferromagnetic alloys such as Ni–Mn,[67] Fe–Al,[68] and $Fe_{65}(Ni_{1-x}Mn_x)$,[69] in all of which some antiferromagnetic interactions exist.

13.4.2 Photoinduced magnetic anisotropy

By illuminating some transparent ferromagnets in a magnetic field, a magnetic anisotropy can be induced. This induced anisotropy is called the *photoinduced magnetic anisotropy*. This phenomenon was first observed in yttrium–iron–garnet (YIG) by means of ferromagnetic resonance by Teale and Temple.[70]

The energy quantum of light of frequency ν is given by $h\nu$, so that $h\nu$ for visible light with a wavelength of 600 nm is calculated to be 3.3×10^{-19} J or 2.1 eV. Equating this energy with kT, we find the corresponding temperature to be 24 000 K. Therefore if an electron absorbs this light, it has more than enough energy to overcome the normal binding energy that holds the electron in the atom.

One of the photomagnetic effects is *photomagnetic annealing*, in which the spontaneous magnetization is stabilized by new ion distributions produced by illuminating an appropriate magnetic material with non-polarized light.[70-72] Another effect is the *polarization-dependent photoinduced effect*, in which polarized light can selectively excite electrons from the ions on some lattice sites.

As described in Section 9.3, the magnetic Fe^{3+} ions in YIG occupy $24d$ and $16a$ lattice sites in the ratio 3 : 2, and their spins are aligned antiparallel, thus exhibiting ferrimagnetism. Nonmagnetic Y^{3+} ions occupy $24c$ sites. This distribution can be described by the following formula:

$$\{Y_3\}[Fe_2^{3+}](Fe_3^{3+})O_{12}$$

where { }, [ ] and ( ) signify the $24c$, $16a$ and $24d$ sites, respectively. When Si ions are introduced into YIG, they occupy $24d$ sites selectively, so that the ion distribution is changed to

$$\{Y_3\}[Fe_{2-x}^{3+}Fe_x^{2+}](Fe_{3-x}^{3+}Si_x^{4+})O_{12},$$

where x is the number of Si atoms per formula unit.

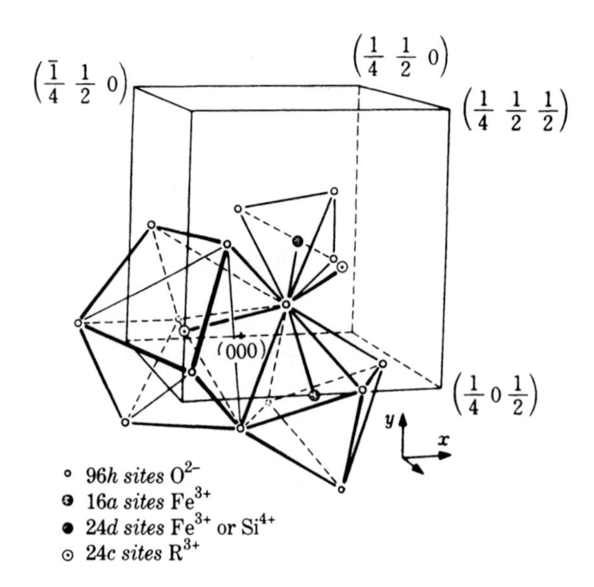

Fig. 13.23. Position of each sublattice site in the garnet structure.[73]

As explained in Chapter 12, the Fe^{2+} ion is anisotropic. At high temperatures, the distribution of Fe^{2+} and Fe^{3+} ions on 16*a* sites is random, but as the temperature is lowered, Fe^{2+} ions tend to be attracted by electrostatic forces to Si^{4+} ions on the 24*d* sites.

Figure 13.23 shows the positions of all the lattice sites in a garnet crystal.[73] Suppose that an Si^{4+} ion occupies the 24*d* site at the center of the cube. The Fe^{2+} ion tends to occupy the 16*a* site which is the nearest to Si^{4+} and located on the other side of the O^{2-} ion. There are four equivalent 16*a* sites. As seen in the figure, a 16*a* site is surrounded by six O^{2-} ions octahedrally. The shape of this octahedron is considerably distorted: the length of the edge of the face parallel to (111) is 2.68 Å, while the length of the other edges is 2.99 Å. Therefore the [111] axis is a special axis for this particular 16*a* site. It is clear from experiments that this special ⟨111⟩ is the easy axis of the Fe^{2+} ion on this site.[73] The four 16*a* sites which are the nearest to the Si^{4+} ion have their special axis parallel to different ⟨111⟩ directions, so that if the magnetization is parallel to, say, [111], the 16*a* site which has its special axis parallel to [111] is the most stable site for an Fe^{2+} ion.

In photomagnetic annealing, the electrons excited from Fe^{2+} ions surrounding Si^{4+} will collect in the most stable site, thus increasing the number of Fe^{2+} ions which have their easy axis parallel to the magnetization. In other words, anisotropy is induced in the direction of the magnetization during illumination. The lattice site from which the electrons are excited is called the *photomagnetic center*.

Figure 13.24 shows an experimental arrangement for observing the polarization-dependent photomagnetic effect. A spherical specimen of YIG with added Si is mounted on a torque magnetometer and cooled to 4.2 K in liquid helium. The specimen can be irradiated by polarized infra-red light in a magnetic field applied by

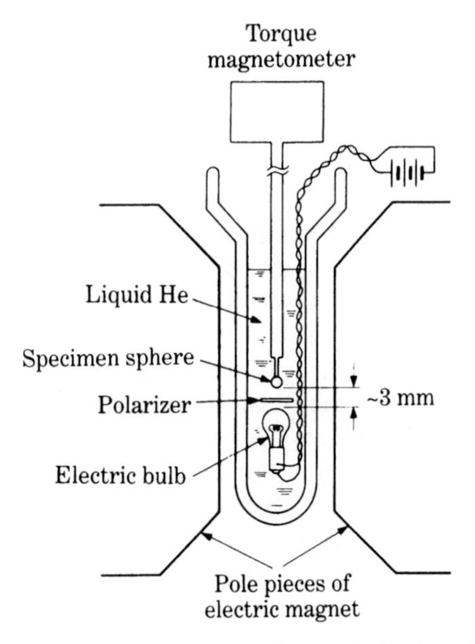

Fig. 13.24. Experimental arrangement for observing polarization-dependent photomagnetic effect.[72]

an electromagnet. The torque curve measured in the $(1\bar{1}0)$ plane is given by the general expression

$$L = L_s \sin 2\theta + L_c \cos 2\theta, \tag{13.45}$$

where θ is the angle between the magnetization and the [001] axis. When a specimen of YIG with Si content $x = 0.034$ was irradiated for four minutes with nonpolarized infra-red light in a magnetic field of $1.2\,\mathrm{MA\,m^{-1}}$ (= 15 kOe) parallel to [001] or [110], the torque curve was given by only the first term in (13.45). In other words, $L_c = 0$.

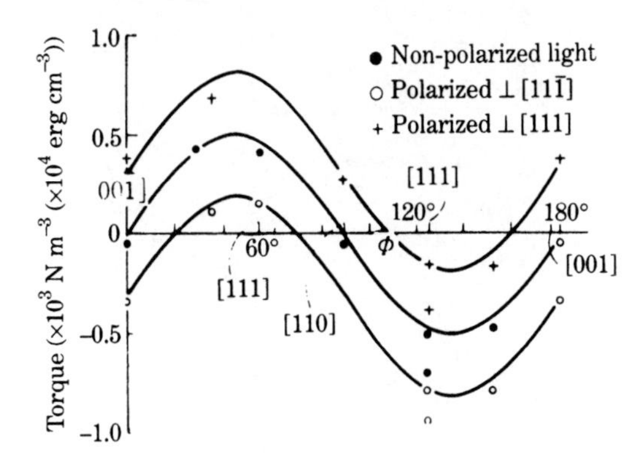

Fig. 13.25. Anisotropy constant, L_c in equation (13.45), of polarization-dependent photomagnetic induced anisotropy as a function of the direction of magnetization during illumination, Φ, measured from [001].[72]

This is reasonable, because there are two ⟨111⟩ axes in the (110) plane, which are symmetrical with respect to [001] or [110]. When the magnetic field was applied in a direction making the angle Φ with [001], L_c was nonzero, because the two ⟨111⟩ axes in the $(1\bar{1}0)$ plane are no longer symmetrical with respect to the applied field. The sinusoidal curve shown with solid circles in Fig. 13.25 is a plot of L_c as a function of Φ.

The same experiment was repeated with plane of polarization of the infra-red light perpendicular to $[11\bar{1}]$. The resulting L_c vs. Φ curve is shifted downwards, as shown by the open circles in the figure. The interpretation is that electrons are excited from the photomagnetic centers with the special axis parallel to [111] more effectively, resulting in a decrease in the number of Fe^{2+} ions on this site. On the other hand, when the light is polarized with the polarization perpendicular to [111], the curve is shifted upwards, as shown by the crosses in the figure. This is due to a decrease in Fe^{2+} on the site whose special axis is parallel to [111].

The mechanism of the excitation of electrons from the photomagnetic center is discussed from a more microscopic point of view by Alben *et al.*[74] Tucciarone[75] calculated the distribution of electrons on the basis of the symmetry of octahedral and tetrahedral sites and also discussed the associated dichroism together with the photomagnetic effect.

13.4.3 Rotatable magnetic anisotropy in anomalous magnetic thin films

In magnetic thin films, various magnetic anisotropies can be produced by varying the preparation procedures.

One of these is the anisotropy produced by control of the angle of incidence during evaporation;[76] that is, the metal vapor impinges onto the substrate in a direction tilted from the normal to the surface. The uniaxial anisotropy of a film produced in this way

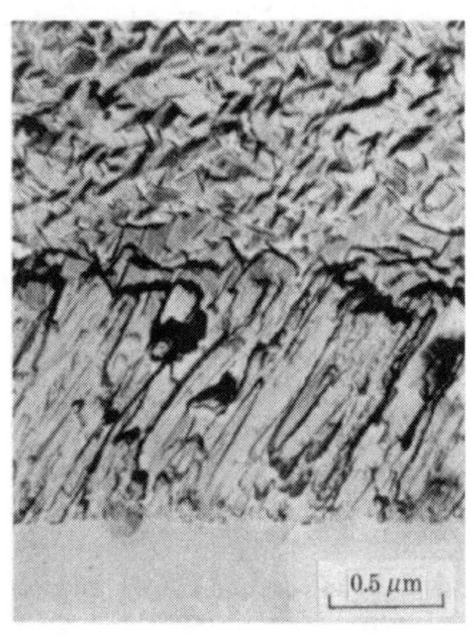

Fig. 13.26. Columnar structure produced by incident-angle evaporation.[78]

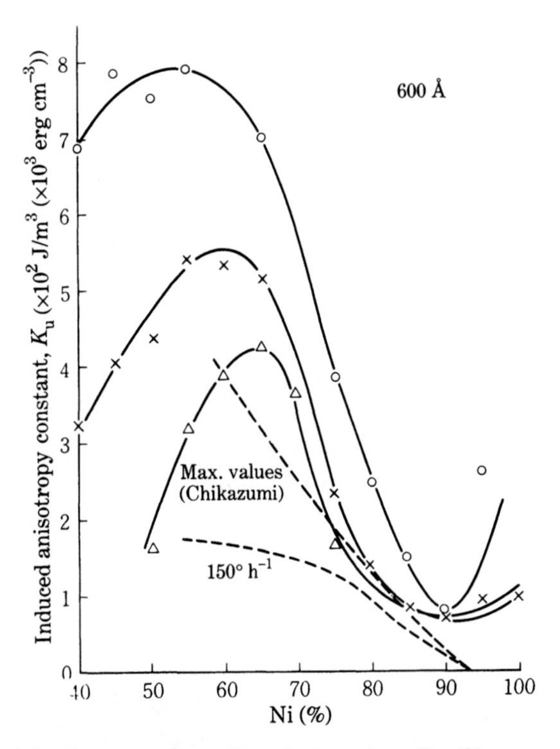

Fig. 13.27. Induced anisotropy constant, K_u, of anomalous Fe–Ni magnetic thin films evaporated under various conditions, as a function of Ni content (dashed curves show the anisotropy produced by directional order (see Fig. 13.2)). ○ Evaporated onto the substrate kept at 20°C in a magnetic field. × Evaporated onto the substrate kept at 300°C in a magnetic field. △ Evaporated onto the substrate kept at 300°C and annealed at 450°C in a magnetic field. In all cases the field strength is $20\,\mathrm{kA\,m^{-1}}$ (= 250 Oe).[79]

has its easy axis parallel to the direction of inclination of the incident vapor. The anisotropy is apparently caused by the shape anisotropy of a tilted columnar structure,[77] which is clearly observed by electron-microscopy as shown in Fig. 13.26.[78]

Even when the direction of evaporation is accurately normal to the substrate surface, if the film is magnetized parallel to its surface during the evaporation an in-plane anisotropy is produced. The anisotropy constant, K_u, of this anisotropy is plotted in Fig. 13.27[79] as a function of Ni content in Fe–Ni alloy films for various evaporation conditions. The dependence of this anisotropy on alloy composition is different from that of directional order, which is shown by dashed curves, in several ways: the magnitude is much greater and there is a non-zero value in pure Ni. Moreover, the easy axis can be rotated by applying a strong magnetic field. This anisotropy is called the *rotatable magnetic anisotropy*.[80]

It was found that the rotatable anisotropy is caused by *stripe domains*.[81,82] The structure of this domain will be described in detail in Section 17.4, so that now we simply glance at a conceptual view as shown in Fig. 13.28. The magnetization in each

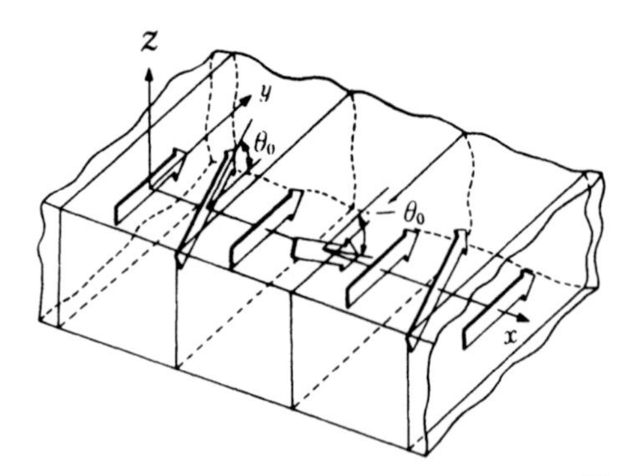

Fig. 13.28. Spin structure of stripe domains.[82]

domain deviates from the plane of the film, alternatively upwards and downwards. This deviation is caused by presence of an anisotropy with its easy axis perpendicular to the film surface, which is produced by the columnar structure. The spin structure of the stripe domain is determined by a balance of the perpendicular anisotropy energy, the magnetostatic energy due to surface free poles, and the exchange

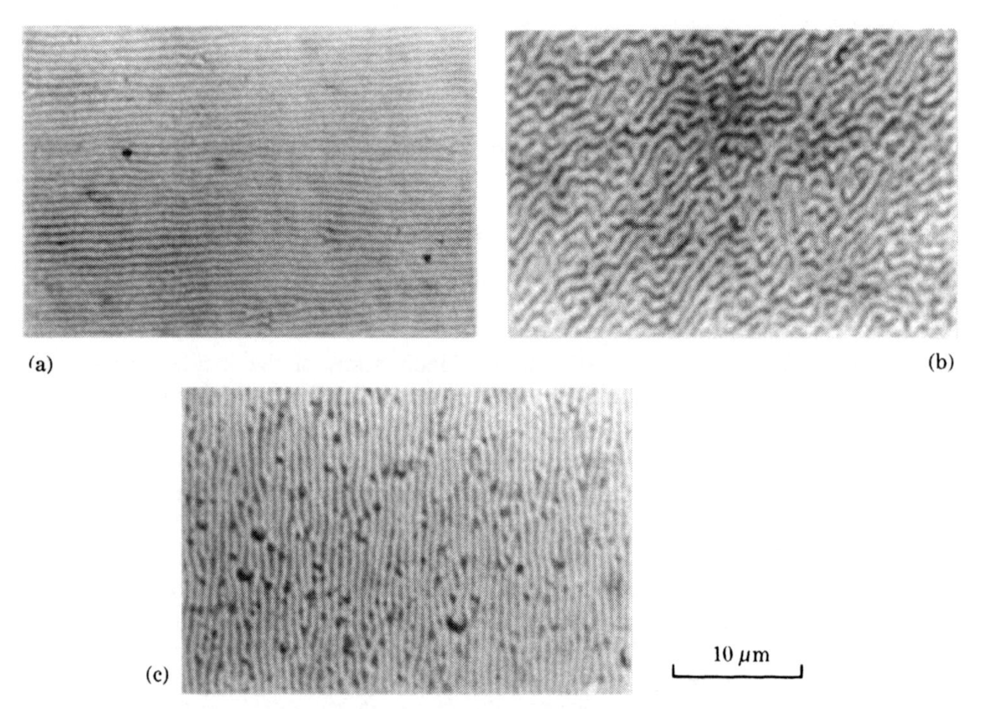

Fig. 13.29. Powder patterns of stripe domains in a 10Fe–90Ni alloy thin film: (a) observed after removal of a magnetic field applied in the horizontal direction; (b) observed after removal of a relatively strong magnetic field perpendicular to the original stripes; (c) observed after removal of a very strong magnetic field perpendicular to the original stripes.[82]

energy.[83] Figure 13.29(a) shows a photograph of stripe domains as observed by the powder-pattern technique.[82] When a weak magnetic field is rotated in the film plane, the magnetization rotates with the field without changing the direction of the stripes. Such a rotation of magnetization relative to the stripes changes the spin structure, thus changing the total energy. This is the origin of the anisotropy caused by stripe domains. When a magnetic field that is strong enough to decrease the deviation angle of the magnetization out of the film plane is applied perpendicular to the stripes, the stripes are disturbed (see the photo (b)), and finally rotate perpendicular to the original direction after removal of a very strong magnetic field (see (c)). This explains the mechanism of rotation of the easy axis.

The rotatable anisotropy is observed only for anomalous magnetic films, which are prepared by evaporation in relatively poor vacuum (less than 10^{-4} Torr) and onto a cold substrate (between -120°C and 100°C).[84] The appearance of this anisotropy has interfered considerably with the development of magnetic thin film computer memories.

13.4.4 Induced anisotropy associated with mictomagnetism

It was found by Satoh *et al.*[85,86] that when a partially ordered Ni_3Mn is cooled from room temperature to 77 K in a magnetic field, a weak uniaxial anisotropy of the order of $10^3\,J\,m^{-3}$ is induced.

In this alloy, the exchange coupling of the Mn–Ni and the Ni–Ni pairs is positive, so that their spins tend to align ferromagnetically, while the coupling of the Mn–Mn pairs is negative, so that their spins tend to align antiferromagnetically. When the alloy is perfectly ordered, Ni and Mn atoms form a Cu_3Au-type superlattice as shown in Fig. 12.33, in which no Mn–Mn nearest neighbor pairs exist. Therefore all the interactions between spins are positive, and the alloy is ferromagnetic. On the other hand, if the alloy is partially ordered, Mn–Mn pairs exist, and their negative interaction disturbs the ferromagnetic spin arrangement and produces mictomagnetism (see Section 7.5). A neutron scattering experiment showed directly that some of the Mn spins are antiparallel to the spontaneous magnetization.[87] Since no atomic migration can occur below room temperature, the origin of this induced anisotropy is to be attributed to a spin rearrangement rather than to an atomic rearrangement.

According to the experiment by Satoh *et al.*[88] this induced anisotropy contains unidirectional anisotropy, K_d, as well as uniaxial anisotropy, K_u, and magnetocrystalline anisotropy, K_1. The appearance of the unidirectional anisotropy suggests the presence of some antiferromagnetic interaction. The value of K_1 depends strongly on the intensity of the magnetic field used in the measurement of the anisotropy, and continues to increase up to a field of $1.6\,MA\,m^{-1}$ (20 kOe). Measurements on a single crystal show that this induced anisotropy is largest when the cooling field is applied in the $\langle 100 \rangle$ direction, intermediate for $\langle 110 \rangle$, and smallest for $\langle 111 \rangle$. This anisotropy also depends on the pre-annealing treatment that determines the degree of ordering, and tends to vanish as the degree of order approaches unity. This fact also suggests that the phenomenon depends strongly on the antiferromagnetic interaction of

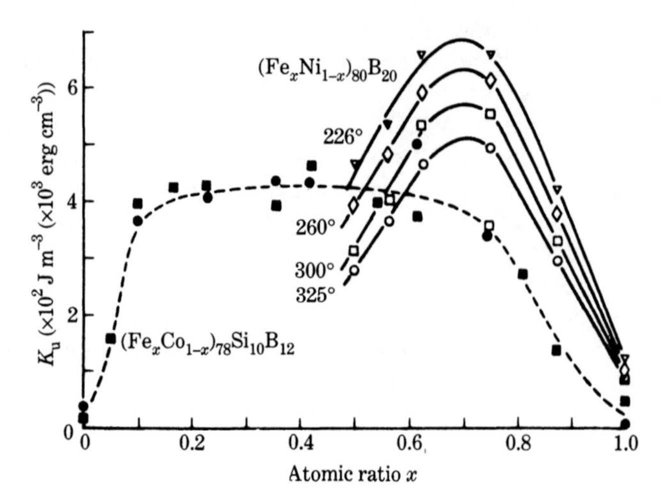

Fig. 13.30. Uniaxial anisotropy constant, K_u, induced in amorphous materials of composition $(Fe_xNi_{1-x})_{80}B_{20}$[91] and $(Fe_xCo_{1-x})_{78}Si_{10}B_{12}$[92] as a function of x. Numbers in the figure are the temperatures in °C at which the magnetic annealing was performed.

Mn–Mn pairs. Polycrystalline samples were also measured over a wider temperature range down to 4.2 K.[89]

This induced anisotropy is considered to be caused by rearrangement of the mictomagnetic spin system, in which ferromagnetic clusters are coupled antiferromagnetically with one another to minimize the pseudodipolar interaction between clusters during cooling in a magnetic field. However, no detailed theoretical treatment has been published.

13.4.5 Induced anisotropy of amorphous alloys

In amorphous alloys, which are treated in Chapter 11, a uniaxial anisotropy can be induced during preparation or by magnetic annealing.

First we discuss the anisotropy induced in amorphous alloys composed of 3*d* transition metals containing 10–20 at% of metalloid elements such as B, N, Si or P, prepared by quenching from the melt. Figure 13.30 shows the uniaxial anisotropy constant, K_u, induced by magnetic annealing of the alloys with compositions $(Fe_xNi_{1-x})_{80}B_{20}$[91] and $(Fe_xCo_{1-x})_{78}Si_{10}B_{12}$,[92] as a function of x. As in the case of magnetic annealing of crystalline alloys, the anisotropy constant, K_u, increases as x decreases from $x = 1$, suggesting that a major part of this induced anisotropy is caused by directional order. The only difference is that a nonzero anisotropy is found even at $x = 1$. This may be attributed to directional ordering of the nonmagnetic metalloid atoms. However, some mechanism specific to the amorphous structure can also be considered.

In the concept of directional order discussed in Section 13.1, we assumed that the pseudodipolar interactions of A–A, B–B, and A–B pairs are constants. This is the case for crystals, because every A–A pair has identical length and symmetry, as do all A–B and B–B pairs. However, in amorphous materials the distances between nearest neighbors are variable, so that the pseudodipolar interaction is not the same for every A–A pair (similarly for every A–B and B–B pair). For $x = 1$, only one kind of magnetic atom pair exists, but a slight change in the length of the pairs to decrease the total pseudodipolar interactions can cause the induced anisotropy.

Amorphous materials prepared by quenching will crystallize if heated above some critical temperature. However, it was found that even below this crystallization temperature a long annealing causes a change in the average atomic volume (structural relaxation).[93] It was pointed out that not only compositional short-range ordering but also topological short-range ordering may be responsible for the anisotropy induced in amorphous materials.[94] The relationship between this microscopic structure and the macroscopic magnetic properties has also been discussed.[95] Stress-induced anisotropy[96] and exchange anisotropy[97] have also been found in amorphous materials.

Evaporated or sputtered Gd–Co or Gd–Fe films exhibit a uniaxial anisotropy with the easy axis perpendicular to the film surface. However, not only the magnitude but also the sign of K_u is different for different preparation conditions. This phenomenon has been interpreted in terms of a columnar structure as well as by directional order.[98] The mechanism is, however, not so simple, because the sign of K_u depends on the presence of a DC bias voltage during sputtering. The reader may refer to the review by Eschenfelder[99] on these topics.

PROBLEMS

13.1 After a single crystal of a body-centered cubic alloy is cooled in a magnetic field applied parallel to [001], [110] or [111], the induced anisotropy is measured by rotating the magnetization in the $(1\bar{1}0)$ plane. Calculate the ratio of the induced anisotropy constants for the three cases, assuming only nearest neighbor interactions.

13.2 After a binary alloy with a cubic lattice was annealed in a magnetic field applied parallel to [110], the induced anisotropy was measured in the (001) and $(1\bar{1}0)$ planes. The anisotropy constants determined for two planes are generally not the same. Calculate the ratio between the two values for a simple cubic, a body-centered cubic and a face-centered cubic lattice.

13.3 A partially ordered single crystal of a binary alloy with the face-centered cubic lattice is deformed by slip deformation only on a single slip system, $(111)[1\bar{1}0]$. Calculate the induced anisotropy energy expressed in terms of the angle of rotation of magnetization in the (111) plane.

REFERENCES

1. G. A. Kelsall, *Physics*, **5** (1934), 169.
2. J. F. Dillinger and R. M. Bozorth, *Physics*, **6** (1935), 279, 285.
3. H. D. Arnold and G. W. Elmen, *J. Franklin Inst.*, **195** (1923), 621.

4. O. Dahl, *Z. Metallk.*, **28** (1936), 133.
5. S. Kaya, *J. Fac. Sci. Hokkaido Imp. Univ.*, **2** (1938), 39.
6. Y. Tomono, *J. Phys. Soc. Japan*, **4** (1948), 298.
7. S. Chikazumi, *J. Phys. Soc. Japan*, **5** (1950), 327, 333.
8. S. Kaya, *Rev. Mod. Phys.*, **25** (1953), 49.
9. S. Chikazumi and T. Oomura, *J. Phys. Soc. Japan*, **10** (1955), 842.
10. L. Néel, *Comp. Rend.*, **237** (1953), 1613; *J. Phys. Radium*, **15** (1954), 225.
11. S. Taniguchi and M. Yamamoto, *Sci. Rept. Res. Inst. Tohoku Univ.*, **A6** (1954), 330; S. Taniguchi, *Sci. Rept. Res. Inst. Tohoku Univ.*, **A7** (1955), 269.
12. S. Chikazumi, *J. Phys. Soc. Japan*, **11** (1956), 551.
13. K. Aoyagi, S. Taniguchi, and M. Yamamoto, *J. Phys. Soc. Japan*, **13** (1958), 532.
14. M. Yamamoto, S. Taniguchi, and K. Aoyagi, *Sci. Rept. Res. Inst. Tohoku Univ.*, **A13** (1961), 117; K. Aoyagi, *Sci. Rept. Res. Inst. Tohoku Univ.*, **A13** (1961), 137.
15. K. Suzuki, *J. Phys. Soc. Japan*, **13** (1958), 756.
16. S. Chikazumi and T. Wakiyama, *J. Phys. Soc. Japan*, **17**, Suppl. B-I (1962), 325; T. Wakiyama, Thesis (Tokyo, Univ. 1965), ISSP Rept. **A147** (1965), referenced by J. C. Slonczewski, *Magnetism I* (Academic Press, 1963), p. 234.
17. E. T. Ferguson, *J. Appl. Phys.*, **29** (1958), 252; *J. Phys. Radium*, **20** (1959), 251.
18. T. Iwata, *Sci. Rept. Res. Inst. Tohoku Univ.*, **A10** (1958), 34; **A13** (1961), 337, 356; *Trans. Japan Inst. Metals*, **2** (1961), 86, 92.
19. Y. Kato and T. Takei, *J. Inst. Elec. Eng. Japan*, **53** (1933), 408.
20. S. Iida, H. Sekizawa, and Y. Aiyama, *J. Phys. Soc. Japan*, **10** (1955), 907; **13** (1958), 58.
21. R. F. Penoyer and L. R. Bickford, Jr., *Phys. Rev.*, **108** (1957), 271.
22. J. C. Slonczewski, *Phys. Rev.*, **110** (1958), 1341.
23. L. R. Bickford, Jr., J. M. Brownlow, and R. F. Penoyer, *J. Appl. Phys.*, **29** (1958), 441.
24. T. Inoue, H. Mizuta, and S. Iida, *J. Phys. Soc. Japan*, **15** (1960), 1899.
25. S. Iida, *J. Appl. Phys.*, **31** (1961), 251S.
26. Y. Aiyama, H. Sekizawa, and S. Iida, *J. Phys. Soc. Japan*, **12** (1957), 742.
27. W. Six, J. L. Snoek, and W. G. Burgers, *Ingenieur*, **49** (1934), E195.
28. H. W. Conradt, O. Dahl, and K. J. Sixtus, *Z. Metallkd*, **32** (1940), 231.
29. G. W. Rathenau and J. L. Snoek, *Physica*, **8** (1941), 555.
30. L. Néel, *Compt. Rend.*, **238** (1954), 305.
31. S. Chikazumi and K. Suzuki, *Phys. Rev.*, **98** (1955), 1130; S. Chikazumi, K. Suzuki, and H. Iwata, *J. Phys. Soc. Japan*, **12** (1957), 1259; S. Chikazumi, *J. Appl. Phys.*, **29** (1958), 346.
32. S. Chikazumi, K. Suzuki, and H. Iwata, *J. Phys. Soc. Japan*, **15** (1960), 250; S. Chikazumi, *J. Appl. Phys.*, **31** (1960), 158S.
33. G. Y. Chin, *J. Appl. Phys.*, **36** (1965), 2915; *Adv. Mat. Res.*, **5** (Wiley, N.Y., 1971), 217.
34. S. Takahashi, *Phys. Stat. Solidi* (a) **42** (1977), 201, 529.
35. T. Mishima, *Ohm*, **19** (1932), 353.
36. D. A. Oliver and J. W. Shedden, *Nature*, **142** (1938), 209.
37. B. Jones, H. J. M. v. Emden, *Philips Tech. Rev.*, **6** (1941), 8.
38. J. W. Cahn, *J. App. Phys.*, **34** (1963), 3581.
39. Y. Iwama and M. Takeuchi, *Trans. Japan Inst. Metals*, **15** (1974), 371.
40. Y. Iwama, M. Takeuchi, and M. Iwata, *Trans. Japan Inst. Metals*, **17** (1976), 481.
41. J. Pauleve, D. Dautreppe, J. Laugier, and L. Néel, *Comptes Rendus*, **254** (1962), 965; **23** (1962), 841.
42. L. Néel, J. Pauleve, R. Pauthenet, J. Laugier, and D. Dautreppe, *J. Appl. Phys.*, **35** (1964), 873.
43. L. Néel, *J. Phys. Rad.*, **12** (1951), 431.

44. J. F. Petersen, M. Aydin, and J. M. Knudsen, *Phys. Lett.*, **62A** (1977), 192.
45. J. F. Albertsen, G. B. Jensen, and J. M. Knudsen, *Nature*, **273** (1978), 453.
46. E. R. D. Scott and R. S. Clarke, Jr., *Nature*, **281** (1979), 360.
47. J. Danon, R. B. Scorzelli, I. S. Azevedo, and M. Christophe-Michel-Levy, *Nature*, **281** (1979), 469.
48. M. Takahashi and T. Kono, *J. Phys. Soc. Japan*, **15** (1960), 936.
49. C. D. Graham, Jr., *J. Phys. Soc. Japan*, **16** (1961), 1481.
50. T. Sambongi and T. Mitui, *J. Phys. Soc. Japan*, **16** (1961), 1478; **18** (1963), 1253.
51. T. Wakiyama, G. Y. Chin, M. Robbins, R. C. Sherwood, and J. E. Bernard, *AIP Conf. Proc.*, **No. 29** (1975), 560.
52. K. Abe, Y. Miyamoto, and S. Chikazumi, *J. Phys. Soc. Japan*, **41** (1976), 1894.
53. C. H. Li, *Phys. Rev.*, **40** (1932), 1002.
54. E. J. W. Verwey and P. W. Haayman, *Physica*, **8** (1941), 979; E. J. W. Verwey, P. W. Haayman, and F. C. Romeijin, *J. Chem. Phys.*, **15** (1947), 181.
55. S. Chikazumi, *AIP Conf. Proc.*, **No. 29** (1975), 382.
56. L. R. Bickford, Jr., *Phys. Rev.*, **83** (1950), 449.
57. T. Yamada, Thesis (Gakushuin Univ. 1968); *Lectures on Experimental Physics* (Kyoritsu Publishing Co., Tokyo), **Vol. 17** (*Magnetic Measurement*, 1968), p. 315 (in Japanese).
58. A. H. Bobeck, E. G. Spencer, L. G. Van Uitert, S. C. Abraham, B. L. Barns, W. H. Grodiewicz, R. C. Sherwood, P. H. Smidt, D. H. Smith, and E. M. Walters, *Appl. Phys. Lett.*, **17** (1970), 131.
59. A. Rosencwaig and W. J. Tabor, *J. Appl. Phys.*, **42** (1971), 1643.
60. A. Rosencwaig, W. J. Tabor, F. B. Hagedorn, and L. G. Van Uitert, *Phys. Rev. Lett.*, **26** (1971), 775.
61. A. Rosencwaig, W. J. Tabor, and R. D. Pierce, *Phys. Rev. Lett.*, **26** (1971), 779.
62. H. Callen, *Appl. Phys. Lett.*, **18** (1971), 311.
63. F. B. Hagedorn and B. S. Hewitt, *J. Appl. Phys.*, **45** (1974), 925.
64. A. H. Eschenfelder, *J. Appl. Phys.*, **49** (1977), 1891.
65. W. H. Meiklejohn and C. P. Bean, *Phys. Rev.*, **102** (1956), 1413; **105** (1957), 904.
66. W. H. Meiklejohn, *J. Appl. Phys.*, **29** (1958), 454.
67. J. S. Kouvel, C. D. Graham, Jr., and I. S. Jacobs, *J. Phys. Radium*, **20** (1959), 198; J. S. Kouvel and C. D. Graham, Jr., *J. Appl. Phys.*, **30** (1959), 312S.
68. J. S. Kouvel, *J. Appl. Phys.*, **30** (1959), 313S.
69. Y. Nakamura and K. Miyata, *J. Phys. Soc. Japan*, **23** (1967), 223.
70. R. W. Teale and D. W. Temple, *Phys. Rev. Lett.*, **19** (1967), 223.
71. U. Enz, R. Metselaar, and P. J. Riinierse, *J. de Phys.*, **32C1** (1971), 703.
72. R. F. Pearson, A. D. Anin, and P. Kompfner, *Phys. Rev. Lett.*, **21** (1968), 1805.
73. S. C. Abrahams and S. Geller, *Acta Crystal.*, **11** (1958), 437.
74. R. Alben, E. M. Gyorgy, J. F. Dillon, Jr., and J. P. Remeika, *Phys. Rev.*, **5** (1972), 2560.
75. A. Tucciarone, Physics of magnetic garnets, *Proc. Int. School of Phys. E. Fermi LXX*, 1978, Italian Phys. Soc., p. 320.
76. D. O. Smith, *J. Appl. Phys.*, **30** (1959), 264S.
77. R. J. Prosen, J. O. Holmen, B. E. Gran, and T. J. Cebulla, *J. Phys. Soc. Japan*, **17**, Suppl. B-1 (1963), 580.
78. H. Fujiwara, K. Hara, K. Okamoto, and T. Hashimoto, *Trans. Japan Inst. Metals*, **20** (1979), 337.
79. M. Takahashi, D. Watanabe, T. Kono, and S. Ogawa, *J. Phys. Soc. Japan*, **15** (1961), 1351.
80. R. J. Prosen, J. O. Holmen, and B. E. Gran, *J. Appl. Phys.*, **32** (1961), 91S.
81. R. J. Spain, *Appl. Phys. Lett.*, **3** (1963), 208.

82. N. Saito, H. Fujiwara, and Y. Sugita, *J. Phys. Soc. Japan*, **19** (1964), 421, 1116.
83. Y. Murayama, *J. Phys. Soc. Japan*, **21** (1966), 2253; **23** (1967), 510.
84. T. Koikeda, S. Fujiwara, and S. Chikazumi, *J. Phys. Soc. Japan*, **21** (1966), 1914.
85. T. Satoh, Y. Yokoyama, and T. Nagashima, *J. Phys. Soc. Japan*, **22** (1967), 1296.
86. T. Satoh and T. Shimura, *J. Phys. Soc. Japan*, **29** (1970), 517.
87. J. W. Cable and H. R. Child, *J. Physique*, **32** (Colloq. No. 1-1971), C1-67.
88. T. Satoh, Y. Yokoyama, and I. Oguro, *J. Mag. Mag. Mat.*, **5** (1977), 18.
89. C. E. Patton and S. Chikazumi, *J. Phys. Soc. Japan*, **29** (1970), 1960; *J. Phys. Rad.*, **32** (1971), 99.
90. F. E. Luborsky, *Handbook of ferromagnetic materials*, Vol. 1 (ed. by E. P. Wohlfarth) (North-Holland, Amsterdam, 1980) Chap. 6, p. 512.
91. F. E. Luborsky and J. L. Walter, *IEEE Trans. Magnetics*, **MAG-13** (1977), 953.
92. H. Fujimori, H. Morita, Y. Obi, and S. Ohta, *Amorphous magnetism II* (eds R. A. Levy and R. Hasagawa) (Plenum Press, N.Y., 1977), p. 393.
93. T. Egami, *J. Mat. Sci. Eng.*, **13** (1978), 2587.
94. C. D. Graham, Jr., and T. Egami, *Ann. Rev. Mat. Sci.*, **8** (1978), 423.
95. K. Kronmüller, *J. Appl. Phys.*, **52** (1981), 1859.
96. B. S. Berry and W. C. Prichet, *Phys. Rev. Lett.*, **34** (1975), 1022; *AIP Conf. Proc.*, **34** (1976), 292.
97. R. C. Sherwood, E. M. Gyorgy, H. J. Leamy, and H. S. Chen, *AIP Conf. Proc.*, **34** (1976), 325.
98. G. S. Cargill and T. Mizoguchi, *J. Appl. Phys.*, **49** (1978), 1753.
99. A. H. Eschenfelder, *Handbook of ferromagnetic materials*, Vol. 2 (ed. by E. P. Wohlfarth) (North-Holland, Amsterdam, 1980), Chap. 6, p. 345.

14
MAGNETOSTRICTION

14.1 PHENOMENOLOGY OF MAGNETOSTRICTION

Magnetostriction is the phenomenon whereby the shape of a ferromagnetic specimen changes during the process of magnetization. The deformation $\delta l/l$ due to magnetostriction is usually very small, in the range 10^{-5} to 10^{-6}. A deformation of this magnitude can be conveniently measured by means of a strain-gauge technique, the details of which are described later. Although the deformation is small, magnetostriction is an important factor in controlling domain structure and the process of technical magnetization.

The magnetostrictive strain changes with increasing magnetic field as shown in Fig. 14.1, and finally reaches a saturation value λ. The reason for this behavior is that the crystal lattice inside each domain is spontaneously deformed in the direction of domain magnetization and its strain axis rotates with the rotation of the domain magnetization, thus resulting in a deformation of the specimen as a whole (Fig. 14.2). In order to calculate the dependence of the strain on the direction of magnetization, we consider a ferromagnetic sphere, which is an exact sphere with radius 1 when it is nonmagnetic but is elongated by $\delta l/l = e$ in the direction of magnetization, or along the x-axis, when it is magnetized to saturation (Fig. 14.3). Suppose that the elongation of the radius OP is measured along the direction AB that makes an angle φ with the direction of magnetization. The point P is displaced in the x-direction by PP$' = e\cos\varphi$, so that the elongation of the radius PP$''$ in the direction AB is given by

$$\frac{\delta l}{l} = e\cos^2\varphi. \tag{14.1}$$

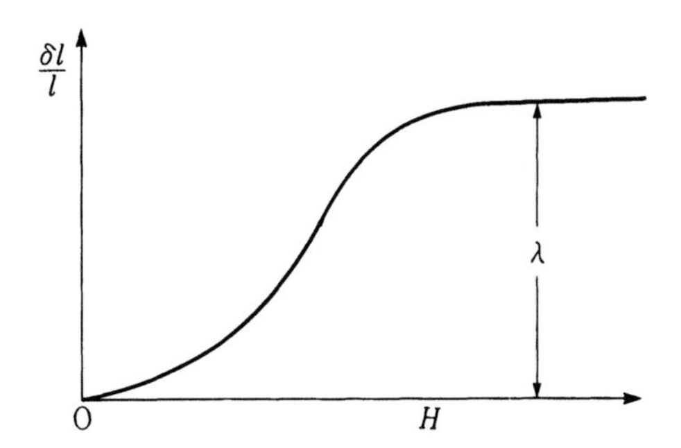

Fig. 14.1. Magnetostrictive elongation as a function of applied field.

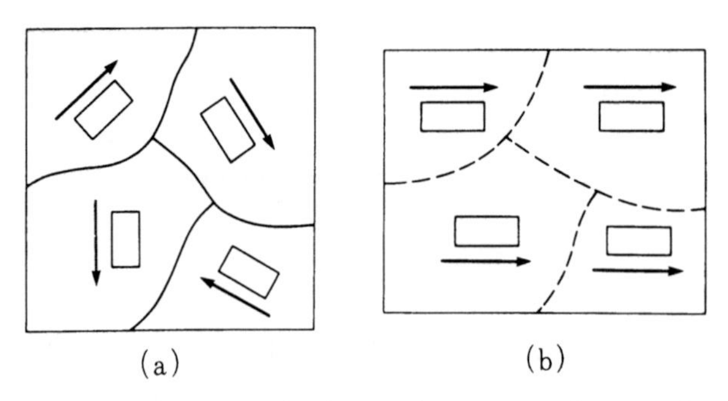

Fig. 14.2. Rotation of domain magnetization and accompanying rotation of the axis of spontaneous strain.

When the domain magnetization is distributed at random in a demagnetized state as shown in Fig. 14.2(a), the average deformation is given by the average of (14.1); thus

$$\begin{aligned}\left(\frac{\delta l}{l}\right)_{\text{dem}} &= \int_0^{\pi/2} e\cos^2\varphi\sin\varphi\,\mathrm{d}\varphi\\ &= e\int_0^1 t^2\,\mathrm{d}t \qquad (t=\cos\varphi)\\ &= e\left|\frac{t^3}{3}\right|_0^1\\ &= \frac{e}{3}.\end{aligned} \tag{14.2}$$

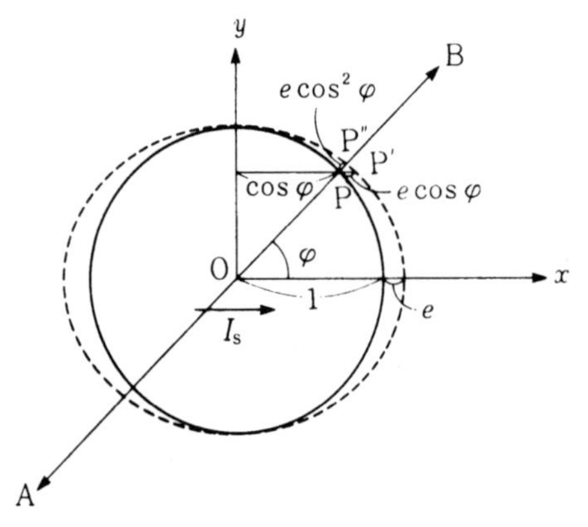

Fig. 14.3. Elongation of the radius of a sphere with unit radius in the direction making angle φ with the axis of spontaneous strain.

Since in the saturated state, as shown in Fig. 14.2(b),

$$\left(\frac{\delta l}{l}\right)_{\text{sat}} = e, \tag{14.3}$$

the saturation magnetostriction is given by

$$\lambda = \left(\frac{\delta l}{l}\right)_{\text{sat}} - \left(\frac{\delta l}{l}\right)_{\text{dem}} = \frac{2}{3}e. \tag{14.4}$$

Thus the spontaneous strain in the domain can be expressed in terms of λ as

$$e = \tfrac{3}{2}\lambda. \tag{14.5}$$

The factor $\frac{3}{2}$ which appears frequently in the following equations thus results from the definition of λ as a deformation from the demagnetized state. We have assumed here that the spontaneous strain $\frac{3}{2}\lambda$ is constant, irrespective of the crystallographic direction of the spontaneous magnetization. We call this quantity the *isotropic magnetostriction.*

Assuming isotropic magnetostriction, let us consider how the magnetostrictive elongation changes as a function of intensity of magnetization. First we consider a ferromagnetic material with uniaxial anisotropy, such as cobalt. If the magnetic field H makes an angle ψ with the easy axis (Fig. 14.4), the magnetization takes place by the displacement of 180° walls, that is, the domain walls separating anti-parallel domain magnetizations, until the magnetization reaches the value $I_s \cos\psi$. During this process no magnetostrictive elongation occurs, because the magnetization is everywhere (except within the domain walls) parallel to the easy axis. At higher fields, the domain magnetization rotates towards the direction of the applied field; during this process the elongation changes by

$$\Delta\left(\frac{\delta l}{l}\right) = \tfrac{3}{2}\lambda(1 - \cos^2\psi). \tag{14.6}$$

If H is parallel to the easy axis, that is, if $\psi = 0$, (14.6) gives $\Delta(\delta l/l) = 0$; in other

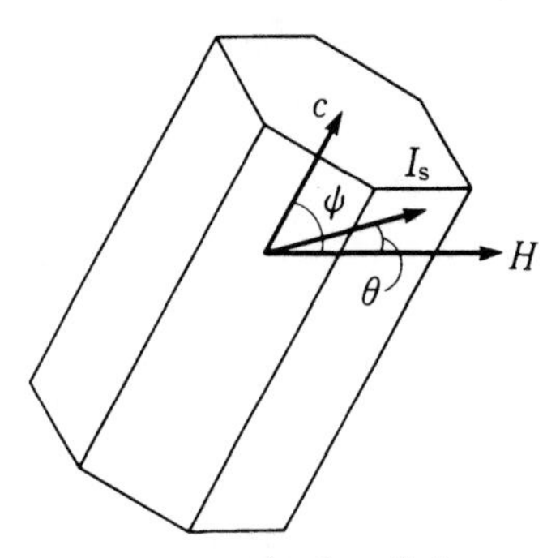

Fig. 14.4. Rotation of the spontaneous magnetization I_s by an applied field H in a uniaxial crystal.

words, there is no change in the length of the specimen from the demagnetized state to saturation. On the other hand, if H is perpendicular to the easy axis, that is, $\psi = \pi/2$, (14.6) gives $\Delta(\delta l/l) = \frac{3}{2}\lambda$. Since, in this case magnetization takes place entirely by rotation, we put $I = I_s \cos\varphi$ in (14.1); thus

$$\frac{\lambda l}{l} = \frac{3}{2}\lambda\left(\frac{I}{I_s}\right)^2. \tag{14.7}$$

If ψ has some value between 0 and $\pi/2$, the magnetizations at which the displacement of 180° walls is complete, and the corresponding elongation changes, are given by

$$\psi = 30° \qquad I = \frac{\sqrt{3}}{2} I_s, \qquad \Delta\left(\frac{\delta l}{l}\right) = \frac{1}{4}\left(\frac{3}{2}\lambda\right), \tag{14.8}$$

$$\psi = 45° \qquad I = \frac{1}{\sqrt{2}} I_s, \qquad \Delta\left(\frac{\delta l}{l}\right) = \frac{1}{2}\left(\frac{3}{2}\lambda\right), \tag{14.9}$$

$$\psi = 60° \qquad I = \frac{1}{2} I_s, \qquad \Delta\left(\frac{\delta l}{l}\right) = \frac{3}{4}\left(\frac{3}{2}\lambda\right). \tag{14.10}$$

The changes in $\delta l/l$ are shown graphically in Fig. 14.5 as a function of magnetization for various values of ψ. For a polycrystal, assuming that all wall displacements are finished before the onset of rotation magnetization, we have, simply by averaging the above values,

$$\psi = \overline{\psi}, \qquad I = \frac{I_s}{2}, \qquad \Delta\left(\frac{\delta l}{l}\right) = \lambda. \tag{14.11}$$

The $\delta l/l$ vs I relation for this crystal is shown by the dashed curve in Fig. 14.5.

For a cubic crystal with $K_1 > 0$, the magnetization is parallel to one of the ⟨100⟩ directions in each domain in the demagnetized state, so that the average elongation is

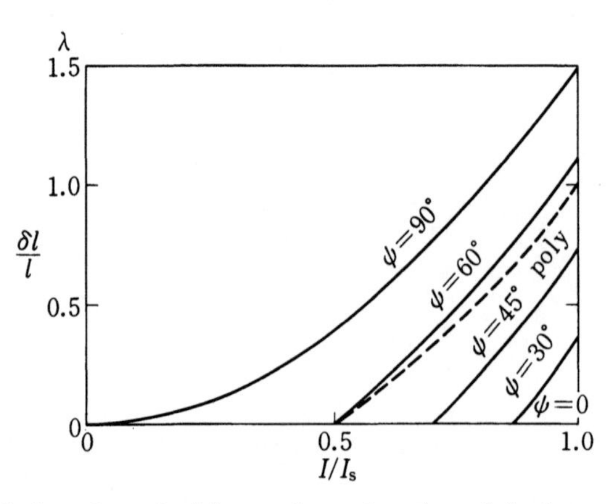

Fig. 14.5. Magnetostriction of a uniaxial crystal as a function of the intensity of magnetization.

given by $(\delta l/l)_{\text{demag}} = \lambda/2$, regardless of the direction of observation. If this crystal is magnetized to saturation parallel to [100] $(\delta l/l)_{\text{sat}} = \frac{3}{2}\lambda$, so that

$$\Delta\left(\frac{\delta l}{l}\right) = \left(\frac{3}{2}\lambda\right) - \frac{\lambda}{2} = \lambda. \tag{14.12}$$

In this example, the entire magnetization takes place by the displacement of domain walls. There are two kinds of walls: the 180° wall separating domains magnetized in opposite directions, and the 90° wall, separating domains magnetized in perpendicular directions. When magnetization changes by displacement of these domain walls, only 90° wall motion contributes to the elongation. Thus the $\delta l/l$ vs I/I_s curve depends on the ease of displacement of 90° walls relative to that of 180° walls. If 180° walls are very easily displaced, I should increase to $I_s/3$ without changing the length of the specimen, after which elongation due to 90° wall motion begins. Thus

$$\begin{aligned} \frac{\delta l}{l} &= 0 && \text{for} \quad \frac{I}{I_s} \le \tfrac{1}{3}, \\ \frac{\delta l}{l} &= \tfrac{3}{2}\lambda\left(\frac{I}{I_s} - \tfrac{1}{3}\right) && \text{for} \quad \frac{I}{I_s} \ge \tfrac{1}{3}. \end{aligned} \tag{14.13}$$

On the other hand, if the displacement of 90° and 180° walls takes place at the same time, the elongation should be given by

$$\frac{\delta l}{l} = \lambda \frac{I}{I_s}. \tag{14.14}$$

Both cases are shown in Fig. 14.6. The experiment points observed by Webster[1] for iron agree fairly well with line 1, which represents (14.13). A famous textbook on

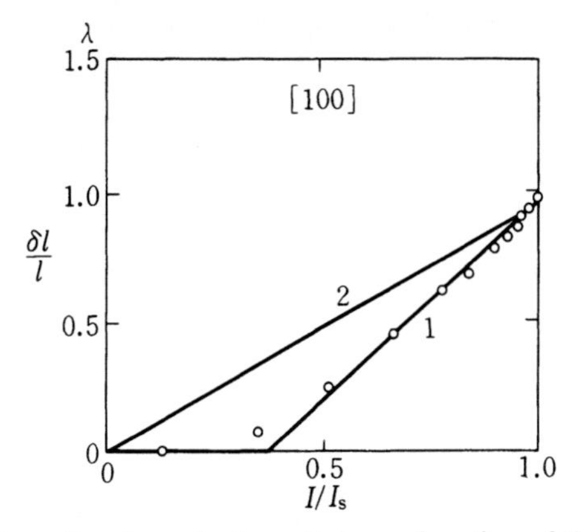

Fig. 14.6. Magnetostriction of an iron single crystal as a function of the intensity of magnetization in the [100] direction. The lines are drawn by assuming (1) preferential motion of 180° walls before 90° wall motion, (2) simultaneous displacement of 180° and 90° walls. Experimental points according to Webster.[1]

ferromagnetism written by Becker and Döring[2] describes the story of the interpretation of this phenomenon: Heisenberg[3] tried to explain this observation by statistical treatment of domain distribution using an old concept of magnetic domains, and fitted the experimental points to a theoretically deduced parabolic curve; while Akulov[4] fitted the points to a straight line, assuming the displacement of domain walls.

When the crystal is magnetized parallel to [111], the domains are first reduced to [100], [010] and [001] domains by the displacement of 180° walls at a sacrifice of $[\bar{1}00]$, $[0\bar{1}0]$ and $[00\bar{1}]$ domains. At the end of this stage $I = I_s/\sqrt{3} = 0.577I_s$. Then domain magnetization will rotate towards the direction of H; during this process $I = I_s \cos\theta$ and also $\delta l/l = \frac{3}{2}\lambda(\cos^2\theta - \frac{1}{3})$, where θ is the angle between I_s and H, so that

$$\begin{aligned} \frac{\delta l}{l} &= 0 & &\text{for} \quad \frac{I}{I_s} \le \frac{1}{\sqrt{3}}, \\ \frac{\delta l}{l} &= \frac{3}{2}\lambda\left[\left(\frac{I}{I_s}\right)^2 - \frac{1}{3}\right] & &\text{for} \quad \frac{I}{I_s} \ge \frac{1}{\sqrt{3}}. \end{aligned} \tag{14.15}$$

This relationship is shown graphically in Fig. 14.7. The experiment for iron reveals similar behavior, but the sign of the elongation is negative (see Fig. 14.8), just opposite to the case of [100] magnetization. Thus the sign of λ as well as its magnitude depends on the crystallographic direction of magnetization. We call this *anisotropic magnetostriction.* For [100] magnetization the effect of anisotropic magnetostriction is not observed, because I_s is always parallel to one of the $\langle 100 \rangle$ directions throughout the entire magnetization process.

Figure 14.9 shows the experimental result for [110] magnetization. A slight positive elongation due to 90° wall displacement occurs in the early stage of magnetization, while a fairly large contraction is observed later during the rotation process.

In order to discover the origin of the anisotropic magnetostriction, we must understand the physical origin of the magnetostriction.

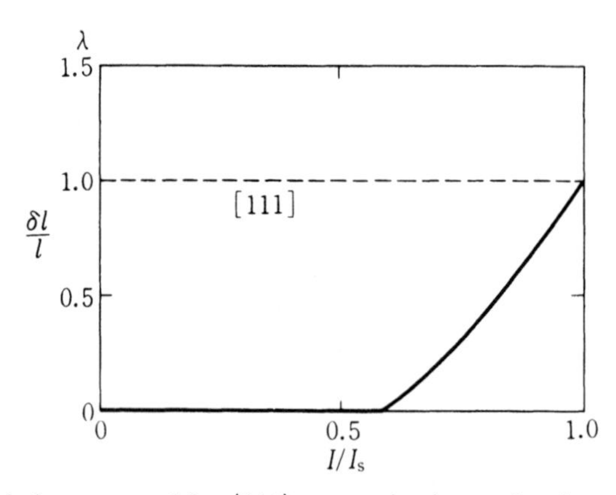

Fig. 14.7. Magnetostriction expected for $\langle 111 \rangle$ magnetization under the assumption of isotropic magnetostriction.

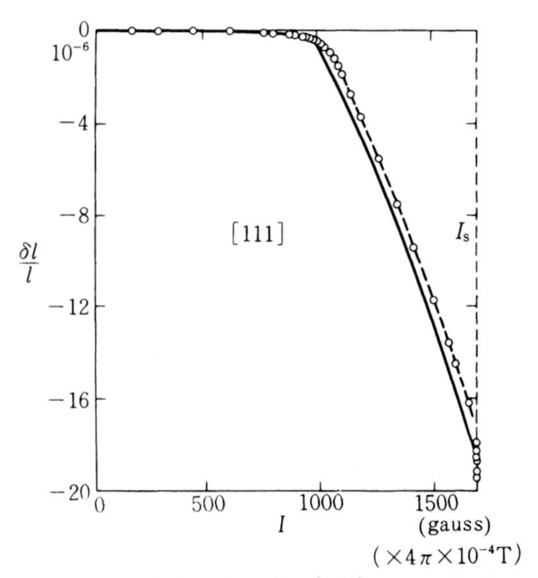

Fig. 14.8. Experimental magnetostriction data for ⟨111⟩ magnetization of iron. (Figure after Becker and Döring;[2] Experimental points according to Kaya and Takaki[5])

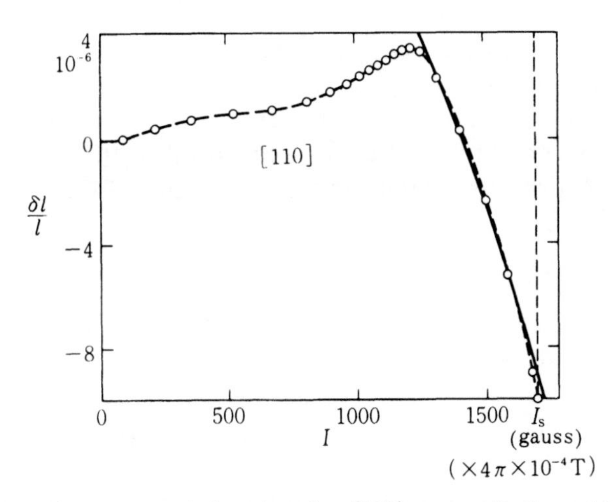

Fig. 14.9. Experimental magnetostriction data for ⟨110⟩ magnetization of iron. (Figure after Becker and Döring;[2] Experimental points according to Kaya and Takaki[5])

14.2 MECHANISM OF MAGNETOSTRICTION

Magnetostriction originates in the interaction between the atomic magnetic moments, as in magnetic anisotropy. We discuss the origin of magnetostriction following the treatment of Néel,[6] which was developed in his paper on magnetic annealing and

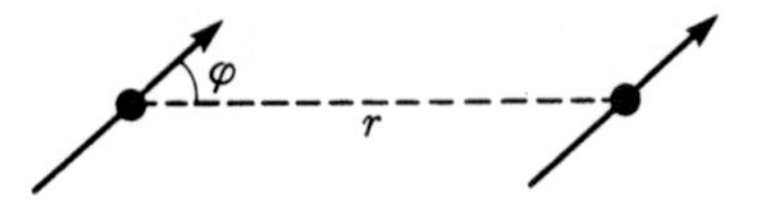

Fig. 14.10. A spin pair with a variable bond length, r, and angle, φ, between parallel spins and the bond.

surface anisotropy. When the distance between the atomic magnetic moments is variable, the interaction energy (12.59) may be expressed as

$$w(r, \cos\varphi) = g(r) + l(r)\left(\cos^2\varphi - \tfrac{1}{3}\right) + q(r)\left(\cos^4\varphi - \tfrac{6}{7}\cos^2\varphi + \tfrac{3}{35}\right) + \cdots, \tag{14.16}$$

where r is the interatomic distance (Fig. 14.10). If the interaction energy is a function of r, the crystal lattice will be deformed when a ferromagnetic moment arises, because such an interaction tends to change the bond length in a different way depending on the bond direction. The first term, $g(r)$, is the exchange interaction term; it is independent of the direction of magnetization. Thus the crystal deformation caused by the first term does not contribute to the usual magnetostriction, but it does play an important role in the volume magnetostriction, which is discussed in Section 14.5.

The second term represents the dipole–dipole interaction, which depends on the direction of magnetization, and can be regarded as the principal origin of the usual magnetostriction. The following terms also contribute to the usual magnetostriction, but normally their contributions are small compared to those of the second term. Neglecting these higher-order terms, we express the pair energy as

$$w(r, \varphi) = l(r)\left(\cos^2\varphi - \tfrac{1}{3}\right). \tag{14.17}$$

Let $(\alpha_1, \alpha_2, \alpha_3)$ denote the direction cosines of domain magnetization and $(\beta_1, \beta_2, \beta_3)$ those of the bond direction. Then (14.17) becomes

$$w(r, \varphi) = l(r)\left\{(\alpha_1\beta_1 + \alpha_2\beta_2 + \alpha_3\beta_3)^2 - \tfrac{1}{3}\right\}. \tag{14.18}$$

Now let us consider a deformed simple cubic lattice whose strain tensor components are given by e_{xx}, e_{yy}, e_{zz}, e_{xy}, e_{yz}, and e_{zx}. When the crystal is strained, each pair changes its bond direction as well as its bond length. For instance, a spin pair with its bond direction parallel to the x-axis has an energy in the unstrained state given by (14.18) with $\beta_1 = 1$, $\beta_2 = \beta_3 = 0$; that is

$$w_x(r, \varphi) = l(r_0)\left(\alpha_1^2 - \tfrac{1}{3}\right), \tag{14.19}$$

whereas, if the crystal is strained, its bond length r_0 will be changed to $r_0(1 + e_{xx})$ and the direction cosines of the bond direction will be changed to ($\beta_1 \simeq 1$, $\beta_2 = \frac{1}{2}e_{xy}$, $\beta_3 = \frac{1}{2}e_{zx}$). Then the pair energy (14.18) will be changed by an amount

$$\Delta w_x = \left(\frac{\partial l}{\partial r}\right) r_0 e_{xx}\left(\alpha_1^2 - \tfrac{1}{3}\right) + l\alpha_1\alpha_2 e_{xy} + l\alpha_3\alpha_1 e_{zx}. \tag{14.20}$$

Similarly, for the y and z pairs,

$$\Delta w_y = \left(\frac{\partial l}{\partial r}\right) r_0 e_{yy}\left(\alpha_2^2 - \tfrac{1}{3}\right) + l\alpha_2\alpha_3 e_{yz} + l\alpha_1\alpha_2 e_{xy}, \tag{14.21}$$

$$\Delta w_z = \left(\frac{\partial l}{\partial r}\right) r_0 e_{zz}\left(\alpha_3^2 - \tfrac{1}{3}\right) + l\alpha_3\alpha_1 e_{zx} + l\alpha_2\alpha_3 e_{yz}. \tag{14.22}$$

Adding these for all nearest neighbor pairs in a unit volume of a simple cubic lattice, we have

$$\begin{aligned} E_{\text{magel}} = B_1\left\{e_{xx}\left(\alpha_1^2 - \tfrac{1}{3}\right) + e_{yy}\left(\alpha_2^2 - \tfrac{1}{3}\right) + e_{zz}\left(\alpha_3^2 - \tfrac{1}{3}\right)\right\} \\ + B_2(e_{xy}\alpha_1\alpha_2 + e_{yz}\alpha_2\alpha_3 + e_{zx}\alpha_3\alpha_1), \end{aligned} \tag{14.23}$$

where

$$B_1 = N\left(\frac{\partial l}{\partial r}\right) r_0, \qquad B_2 = 2Nl. \tag{14.24}$$

The energy thus expressed in terms of lattice strain and the direction of domain magnetization is called the *magnetoelastic energy*. Similar calculations for the body-centered cubic and face-centered cubic lattice yield the same expression (14.23) with

$$B_1 = \tfrac{8}{3}Nl, \qquad B_2 = \tfrac{8}{9}N\left\{l + \left(\frac{\partial l}{\partial r}\right) r_0\right\}, \tag{14.25}$$

for the body-centered cubic lattice and

$$B_1 = \tfrac{1}{2}N\left\{6l + \left(\frac{\partial l}{\partial r}\right) r_0\right\}, \qquad B_2 = N\left\{2l + \left(\frac{\partial l}{\partial r}\right) r_0\right\}, \tag{14.26}$$

for the face-centered cubic lattice.

Since the magnetoelastic energy (14.23) is a linear function with respect to $e_{xx}, e_{yy}, \ldots, e_{zx}$, the crystal will be deformed without limit unless it is counterbalanced by the elastic energy which, for a cubic crystal, is given by

$$\begin{aligned} E_{\text{el}} = \tfrac{1}{2}c_{11}\left(e_{xx}^2 + e_{yy}^2 + e_{zz}^2\right) + \tfrac{1}{2}c_{44}\left(e_{xy}^2 + e_{yz}^2 + e_{zx}^2\right) \\ + c_{12}(e_{yy}e_{zz} + e_{zz}e_{xx} + e_{xx}e_{yy}), \end{aligned} \tag{14.27}$$

where c_{11}, c_{44}, and c_{12} are the elastic moduli. Since the elastic energy is a quadratic function of the strain, it increases rapidly with increasing strain and equilibrium is attained at some finite strain. The condition for equilibrium is obtained by minimizing the total energy,

$$E = E_{\text{magel}} + E_{\text{el}}. \tag{14.28}$$

That is,

$$\left.\begin{aligned}
\frac{\partial E}{\partial e_{xx}} &= B_1\left(\alpha_1^2 - \tfrac{1}{3}\right) + c_{11}e_{xx} + c_{12}(e_{yy} + e_{zz}) = 0 \\
\frac{\partial E}{\partial e_{yy}} &= B_1\left(\alpha_2^2 - \tfrac{1}{3}\right) + c_{11}e_{yy} + c_{12}(e_{zz} + e_{xx}) = 0 \\
\frac{\partial E}{\partial e_{zz}} &= B_1\left(\alpha_3^2 - \tfrac{1}{3}\right) + c_{11}e_{zz} + c_{12}(e_{xx} + e_{yy}) = 0 \\
\frac{\partial E}{\partial e_{xy}} &= B_2\alpha_1\alpha_2 + c_{44}e_{xy} = 0 \\
\frac{\partial E}{\partial e_{yz}} &= B_2\alpha_2\alpha_3 + c_{44}e_{yz} = 0 \\
\frac{\partial E}{\partial e_{zx}} &= B_2\alpha_3\alpha_1 + c_{44}e_{zx} = 0
\end{aligned}\right\}. \tag{14.29}$$

Solving these equations, we have the equilibrium strain

$$\left.\begin{aligned}
e_{xx} &= \frac{B_1}{c_{12} - c_{11}}\left(\alpha_1^2 - \tfrac{1}{3}\right) \\
e_{yy} &= \frac{B_1}{c_{12} - c_{11}}\left(\alpha_2^2 - \tfrac{1}{3}\right) \\
e_{zz} &= \frac{B_1}{c_{12} - c_{11}}\left(\alpha_3^2 - \tfrac{1}{3}\right) \\
e_{xy} &= -\frac{B_2}{c_{44}}\alpha_1\alpha_2 \\
e_{yx} &= -\frac{B_2}{c_{44}}\alpha_2\alpha_3 \\
e_{zx} &= -\frac{B_2}{c_{44}}\alpha_3\alpha_1
\end{aligned}\right\}. \tag{14.30}$$

The elongation observed in the direction $(\beta_1, \beta_2, \beta_3)$, which is given by

$$\frac{\delta l}{l} = e_{xx}\beta_1^2 + e_{yy}\beta_2^2 + e_{zz}\beta_3^2 + e_{xy}\beta_1\beta_2 + e_{yz}\beta_2\beta_3 + e_{zx}\beta_3\beta_1, \tag{14.31}$$

becomes

$$\frac{\delta l}{l} = \frac{B_1}{c_{12} - c_{11}}\left(\alpha_1^2\beta_1^2 + \alpha_2^2\beta_2^2 + \alpha_3^2\beta_3^2 - \tfrac{1}{3}\right)$$
$$- \frac{B_2}{c_{44}}(\alpha_1\alpha_2\beta_1\beta_2 + \alpha_2\alpha_3\beta_2\beta_3 + \alpha_3\alpha_1\beta_3\beta_1). \tag{14.32}$$

If the domain magnetization is along [100], the elongation in the same direction is obtained by putting $\alpha_1 = \beta_1 = 1$, $\alpha_2 = \alpha_3 = \beta_2 = \beta_3 = 0$ in (14.32); thus

$$\lambda_{100} = \frac{2}{3}\frac{B_1}{c_{12} - c_{11}}. \tag{14.33}$$

Similarly, when the magnetization is along [111], the elongation is calculated to be

$$\lambda_{111} = -\frac{1}{3}\frac{B_2}{c_{44}}, \tag{14.34}$$

by putting $\alpha_i = \beta_i = 1/\sqrt{3}$ ($i = 1, 2$, and 3) in (14.32). By using λ_{100} and λ_{111}, (14.32) can be expressed as

$$\frac{\delta l}{l} = \tfrac{3}{2}\lambda_{100}\left(\alpha_1^2\beta_1^2 + \alpha_2^2\beta_2^2 + \alpha_3^2\beta_3^2 - \tfrac{1}{3}\right)$$
$$+ 3\lambda_{111}(\alpha_1\alpha_2\beta_1\beta_2 + \alpha_2\alpha_3\beta_2\beta_3 + \alpha_3\alpha_1\beta_3\beta_1). \tag{14.35}$$

Thus the magnetostriction of a cubic crystal may be expressed in terms of λ_{100} and λ_{111}. The elongation in [110] is not independent of λ_{100} and λ_{111}, but is related to them by

$$\lambda_{110} = \tfrac{1}{4}\lambda_{100} + \tfrac{3}{4}\lambda_{111} \tag{14.36}$$

as will be found by putting $\alpha_1 = \beta_1 = \alpha_2 = \beta_2 = 1/\sqrt{2}$, $\alpha_3 = \beta_3 = 0$ in (14.35). Putting (14.24), (14.25) and (14.26) into (14.32) into (14.33) and (14.34), we can express λ_{100} and λ_{111} in term of the coefficients of the pair energy:

$$\lambda_{100} = \frac{2}{3}\frac{N}{c_{12} - c_{11}}\left(\frac{\partial l}{\partial r}\right)r_0, \quad \lambda_{111} = -\frac{2}{3}\frac{N}{c_{44}}l \quad \text{for sc,} \tag{14.37}$$

$$\lambda_{100} = \frac{16}{9}\frac{N}{c_{12} - c_{11}}l, \quad \lambda_{111} = -\frac{8}{27}\frac{N}{c_{44}}\left\{1 + \left(\frac{\partial l}{\partial r}\right)r_0\right\} \quad \text{for bcc,} \tag{14.38}$$

$$\left.\begin{aligned} \lambda_{100} &= \frac{1}{3}\frac{N}{c_{12} - c_{11}}\left\{6l + \left(\frac{\partial l}{\partial r}\right)r_0\right\}, \\ \lambda_{111} &= -\frac{1}{3}\frac{N}{c_{44}}\left\{2l + \left(\frac{\partial l}{\partial r}\right)r_0\right\} \end{aligned}\right\} \quad \text{for fcc} \tag{14.39}$$

If

$$\lambda_{100} = \lambda_{111} = \lambda, \tag{14.40}$$

(14.35) becomes

$$\frac{\delta l}{l} = \frac{3}{2}\lambda\left\{(\alpha_1\beta_1 + \alpha_2\beta_2 + \alpha_3\beta_3)^2 - \frac{1}{3}\right\}$$
$$= \frac{3}{2}\lambda\left(\cos^2\theta - \frac{1}{3}\right), \tag{14.41}$$

where θ is the angle between the direction of domain magnetization and that of the measured strain. The final form is the same as that of the isotropic magnetostriction given by (14.1). The condition for isotropic magnetostriction is thus expressed by equating the expressions for λ_{100} given by (14.37), (14.38), and (14.39) to those for λ_{111}. Since the coefficients of the pair energy and the elastic moduli included in these expressions are entirely independent of each other, it is hard to attribute any significant physical meaning to the isotropic magnetostriction.

For isotropic magnetostriction, it can be seen by putting (14.40) into (14.36) that $\lambda = \lambda_{100} = \lambda_{111} = \lambda_{110}$. Figure 14.11 shows the dependence of magnetostriction constants on the alloy composition for iron–nickel alloys, where the magnetostriction is isotropic at 15% and at 40% Fe–Ni.

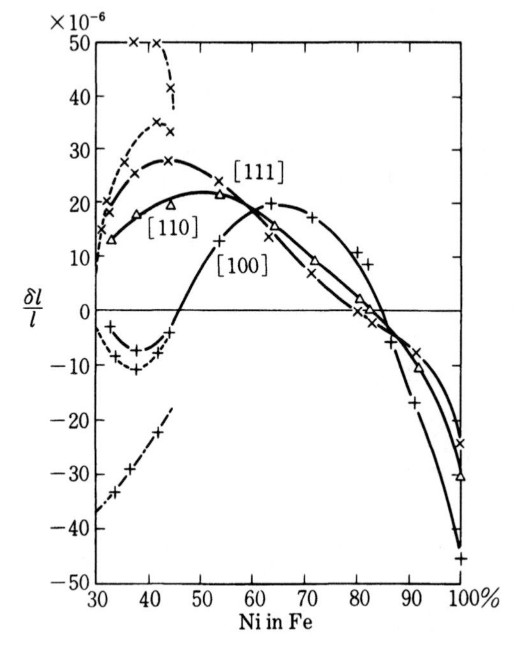

Fig. 14.11. Composition dependence of the magnetostriction constants λ_{100}, λ_{110}, and λ_{111} for Ni–Fe alloys. (After Lichtenberger[7] for solid curves and Kim *et al.*[8] for dashed curves measured at room temperature and dot and dashed curves measured at 4.2 K for less than 44% Ni.)

For polycrystalline materials, the magnetostriction is isotropic, because the deformation in each grain is averaged to produce the overall magnetostriction, even if $\lambda_{100} \neq \lambda_{111}$. The average longitudinal magnetostriction is calculated by averaging (14.35) for different crystal orientations, assuming $\alpha_i = \beta_i$ ($i = 1$, 2, and 3); thus

$$\bar{\lambda} = \tfrac{2}{5}\lambda_{100} + \tfrac{3}{5}\lambda_{111}. \tag{14.42}$$

In the preceding discussion we considered only the dipole–dipole interaction term in (14.16). If we take into account the third term of (14.16), the expression for the magnetostriction of a cubic crystal becomes more complicated than (14.35):

$$\begin{aligned}\frac{\delta l}{l} &= h_1\left(\alpha_1^2\beta_1^2 + \alpha_2^2\beta_2^2 + \alpha_3^2\beta_3^2 - \tfrac{1}{3}\right)\\ &+ h_2(2\alpha_1\alpha_2\beta_1\beta_2 + 2\alpha_2\alpha_3\beta_2\beta_3 + 2\alpha_3\alpha_1\beta_3\beta_1)\\ &+ h_4\left(\alpha_1^4\beta_1^2 + \alpha_2^4\beta_2^2 + \alpha_3^4\beta_3^2 + \tfrac{2}{3}s - \tfrac{1}{3}\right)\\ &+ h_5(2\alpha_1\alpha_2\alpha_3^2\beta_1\beta_2 + 2\alpha_2\alpha_3\alpha_1^2\beta_2\beta_3 + 2\alpha_3\alpha_1\alpha_2^2\beta_3\beta_1)\\ &+ h_3\left(s - \tfrac{1}{3}\right) \qquad \text{for } K_1 < 0\\ &+ h_3 s \qquad \text{for } K_1 > 0,\end{aligned} \tag{14.43}$$

where

$$s = \alpha_1^2\alpha_2^2 + \alpha_2^2\alpha_3^2 + \alpha_3^2\alpha_1^2, \tag{14.44}$$

and h_1 and h_2 are related to λ_{100} and λ_{111} by $h_1 = (3/2)\lambda_{100}$ and $h_2 = (3/2)\lambda_{111}$, respectively.

In a hexagonal crystal, by setting the z-axis parallel to the c-axis, the deformation from the state with the magnetization parallel to the c-axis is given by[9]

$$\begin{aligned}\frac{\delta l}{l} &= \lambda_A\left[(\alpha_1\beta_1 + \alpha_2\beta_2)^2 - (\alpha_1\beta_1 + \alpha_2\beta_2)\alpha_3\beta_3\right]\\ &+ \lambda_B\left[(1-\alpha_3^2)(1-\beta_3^2) - (\alpha_1\beta_1 + \alpha_2\beta_2)^2\right]\\ &+ \lambda_C\left[(1-\alpha_3^2)\beta_3^2 - (\alpha_1\beta_1 + \alpha_2\beta_2)\alpha_3\beta_3\right]\\ &+ 4\lambda_D(\alpha_1\beta_1 + \alpha_2\beta_2)\alpha_3\beta_3.\end{aligned} \tag{14.45}$$

A more precise expression[10] is given by

$$\begin{aligned}\frac{\delta l}{l} &= \lambda_0 + \lambda_0\beta_3^2 + (\lambda_1 + \lambda_2\beta_3^2)\alpha_3^2 + \lambda_3(\beta_1\alpha_1 + \beta_2\alpha_2)^2\\ &+ \lambda_4(\beta_1\alpha_1 + \beta_2\alpha_2)\beta_3\alpha_3 + (\lambda_5 + \lambda_6\beta_3^2)\alpha_3^4 + \lambda_7(\beta_1\alpha_1 + \beta_2\alpha_2)^2\alpha_3^2\\ &+ \lambda_8(\beta_1\alpha_1 + \beta_2\alpha_2)\beta_3\alpha_3^3 + \lambda_9(2\alpha_1\alpha_2\beta_1 + \alpha_1^2\beta_2 - \alpha_2^2\beta_2)^2.\end{aligned} \tag{14.46}$$

Since the magnetostriction constants are related to the dipole–dipole interaction, it is possible to deduce its coefficient, l, from the composition dependence of λ_{100} and λ_{111} observed for binary alloys.

As seen in Fig. 14.11, the composition dependences of λ_{100} and λ_{111} can be approximated by quadratic functions. In Ni–Fe alloys, the composition dependences in the fcc region are expressed by

$$\lambda_{100} \times 10^6 = -55C_{Ni}^2 + 340C_{Ni}C_{Fe} - 245C_{Ni}^2$$
$$\lambda_{111} \times 10^6 = -27C_{Ni}^2 + 134C_{Ni}C_{Fe} + 13C_{Ni}^2, \tag{14.47}$$

where C_{Ni} and C_{Fe} are the fractions of Ni and Fe, respectively. Using these relationships, Néel[6] deduced the coefficients, the ls, for all three possible kinds of atomic pairs. If the atomic distribution of A and B atoms is completely random, the probabilities of finding AA, AB and BB pairs are proportional to C_A^2, $2C_AC_B$, and C_B^2, respectively, where C_A and C_B are the fractions of A and B atoms. Therefore the average l is given by

$$l = C_A^2 l_{AA} + 2C_AC_B l_{AB} + C_B^2 l_{BB}. \tag{14.48}$$

Similarly the average $\partial l/\partial r$ is given by

$$\frac{\partial l}{\partial r} = C_A^2 \frac{\partial l_{AA}}{\partial r} + 2C_AC_B \frac{\partial l_{AB}}{\partial r} + C_B^2 \frac{\partial l_{BB}}{\partial r}. \tag{14.49}$$

Solving (14.39) for an fcc lattice, we have

$$l = \frac{3}{4N}\{(c_{12} - c_{11})\lambda_{100} + c_{44}\lambda_{111}\}, \tag{14.50}$$

$$\left(\frac{\partial l}{\partial r}\right)_{r_0} = -\frac{3}{2N}\{(c_{12} - c_{11})\lambda_{100} + 3c_{44}\lambda_{111}\}. \tag{14.51}$$

If we assume that the elastic moduli of Ni–Fe alloys are the same as those of Ni, as given in Table 14.3 in Section 14.4, we have

$$\begin{aligned}
Nl &= \tfrac{3}{4}(c_{12} - c_{11})\lambda_{100} + \tfrac{3}{4}c_{44}\lambda_{111} \\
&= -6.75 \times 10^{10}\lambda_{100} + 8.85 \times 10^{10}\lambda_{111} \\
&= (6.75 \times 55 - 8.85 \times 27) \times 10^4 C_{Ni}^2 \\
&\quad + (-6.75 \times 340 + 8.85 \times 134) \times 10^4 C_{Ni}C_{Fe} \\
&\quad + (6.75 \times 245 + 8.85 \times 13) \times 10^4 C_{Fe}^2 \\
&= 0.132 \times 10^7 C_{Ni}^2 - 1.11 \times 10^7 C_{Ni}C_{Fe} + 1.77 \times 10^7 C_{Fe}^2.
\end{aligned} \tag{14.52}$$

Comparing (14.52) with (14.48), we can determine

$$Nl_{NiNi} = 0.13 \times 10^7, \quad 2Nl_{NiFe} = -1.11 \times 10^7, \quad Nl_{FeFe} = 1.77 \times 10^7, \tag{14.53}$$

from which we have

$$Nl_0 = Nl_{NiNi} + Nl_{FeFe} - 2Nl_{NiFe} = 3.0 \times 10^7\ \mathrm{J\,m^{-3}} \tag{14.54}$$

for Ni–Fe alloys. This value was used for estimating the induced anisotropy due to directional order in Ni–Fe alloys in Chapter 13.

As discussed above, the origin of magnetostriction is the magnetoelastic energy, which is the magnetocrystalline anisotropy of the deformed crystal. Therefore the microscopic origin of magnetostriction is the same as that of magnetocrystalline anisotropy.

A calculation of magnetostriction was made by Tsuya[11] for ferrites using a one-ion model. However, the agreement with experiment was not satisfactory. Slonczewski[12] calculated magnetostriction in Co ferrites using the one-ion model discussed in Section 12.3. The calculated λ_{100} agreed with experiment, but the value for λ_{111} and its temperature dependence were not satisfactory. For more details the reader may refer to the review by Kanamori.[13]

14.3 MEASURING TECHNIQUE

A convenient means for measuring magnetostriction is the *strain gauge technique*.[14,15] A strain gauge, sometimes called a resistance strain gauge, is made of a small piece of paper or polymer sheet, on which a thin serpentine resistance wire or foil is cemented as shown in Fig. 14.12. When a specimen to which this strain gauge is cemented is elongated, the resistance wire in the strain gauge is also elongated, thus changing its resistance. The proportionality factor between $\Delta l/l$ and $\Delta R/R$, where R is the resistance of the strain gauge, is called the *gauge factor*. If we assume that Poisson's ratio is 0.5, which means that the volume of the gauge material does not change during elongation, and also that the effect of strain is only geometrical, an elongation of $\Delta l/l$, and a decrease in cross-sectional area, S, by $\Delta S/S = \Delta l/l$, should result in a gauge factor of 2. Usually the value of the gauge factor ranges from 1.8 to 2.2.

In order to determine λ_{100} and λ_{111} in a cubic material, we cut a disk parallel to $(1\bar{1}0)$ from a single crystal and cement a strain gauge parallel to [100] or [111]; then rotate a magnetic field in the plane of the disk, and measure the strain parallel to [100] or [111]. If we denote the angle between [001] and the magnetization by θ, as shown in Fig. 12.7, the direction cosines of the magnetization are given by (12.12), so that the elongation given by (14.35) is reduced to

$$\frac{\delta l}{l} = \tfrac{3}{2}\lambda_{100}\left(\tfrac{1}{2}\sin^2\theta\,(\beta_1^2+\beta_2^2)+\cos^2\theta\,\beta_3^2-\tfrac{1}{3}\right) + 3\lambda_{111}\left(\tfrac{1}{2}\sin^2\theta\,\beta_1\beta_2+\frac{1}{\sqrt{2}}\sin\theta\cos\theta\,(\beta_2\beta_3+\beta_3\beta_1)\right). \tag{14.55}$$

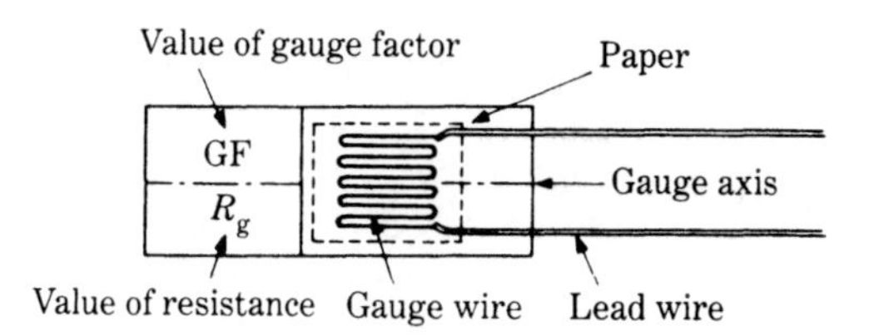

Fig. 14.12. Paper strain gauge.[14]

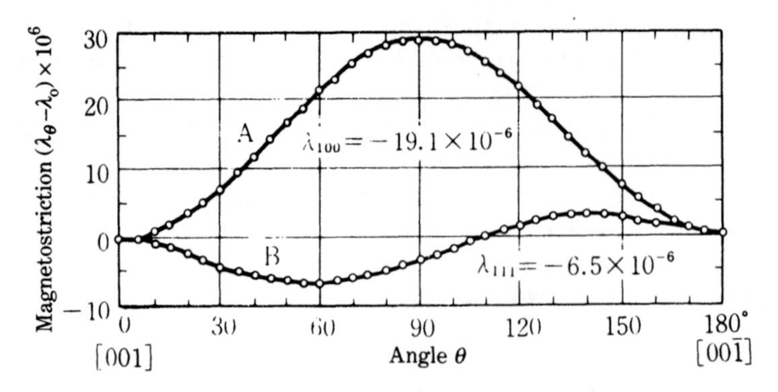

Fig. 14.13. Magnetostriction measured as a function of the angle of magnetization from [001] for 3.93% V–Ni (1$\bar{1}$0) disk: (A) elongation parallel to [001]; (B) parallel to [111].[14]

When the strain gauge is parallel to [001], we put $\beta_1 = \beta_2 = 0$, and $\beta_3 = 1$, so that (14.55) gives

$$\frac{\delta l}{l} = \tfrac{3}{2}\lambda_{100}\left(\cos^2\theta - \tfrac{1}{3}\right) = \tfrac{3}{4}\lambda_{100}(\cos 2\theta + \tfrac{1}{3}). \tag{14.56}$$

Figure 14.13 shows the experimental result of magnetostriction measured for 3.93 at% V–Ni alloy. Curve A shows the elongation parallel to [001] as a function of the angle θ. The functional form of curve A is well fitted by $\cos 2\theta$, in agreement with (14.56). Equating the amplitude of this term to $\frac{3}{4}\lambda_{100}$, we can determine λ_{100}. For the strain gauge parallel to [111], we put $\beta_1 = \beta_2 = \beta_3 = 1/\sqrt{3}$ in (14.55), so that we have

$$\begin{aligned}\frac{\delta l}{l} &= \lambda_{111}\left(\tfrac{1}{2}\sin^2\theta + \sqrt{2}\sin\theta\cos\theta\right)\\ &= \tfrac{1}{4}\lambda_{111}(1 - \cos 2\theta + 2\sqrt{2}\sin 2\theta)\\ &= \tfrac{3}{4}\lambda_{111}(\cos 2(\theta - \theta_0) + \tfrac{1}{3}),\end{aligned} \tag{14.57}$$

where $\theta_0 = 54.7°$. The data in this case is shown as curve B in Fig. 14.13. This curve has the functional form $\cos 2\theta$, which has a minimum at 54.7° in agreement with (14.57). Equating the amplitude of this curve to $\frac{3}{4}\lambda_{111}$, we can determine λ_{111}.

It should be noted that if the cement applied to attach the strain gauge to the specimen is too thick or the specimen is too thin, the rigidity of the specimen is altered by the presence of the strain gauge, so that the sensitivity of the strain gauge is reduced. On the other hand, if the specimen is too thick, the demagnetizing factor becomes so large that it may be difficult to reach magnetic saturation. Moreover, the volume magnetostriction is increased, which makes the determination of saturation magnetostriction difficult (see Section 14.5).

Changes in gauge resistance are measured in a bridge circuit. To compensate for resistance changes due to temperature and to magnetic field, three dummy gauges completing the bridge may be cemented onto a nonmagnetic disk placed near and parallel to the specimen, so as to experience the same magnetic field and the same temperature as the specimen. Even in this case, the gauge factor is influenced by the magnetic field and by temperature, so that a calibration is needed. However, some strain gauges are made with the gauge factor quite insensitive to both magnetic field and temperature. Details are given by Wakiyama and Chikazumi.[14]

Strain gauges made of semiconductor materials are also available; they have very high sensitivity (gauge factor over 100), but correspondingly high values of magnetoresistance and temperature coefficient of resistivity.

Magnetostriction measurements can also be made by a capacitance technique, in which the magnetostrictive strain changes the gap between two conducting plates and hence changes the capacitance. For wire and ribbon samples, which may be too small to mount a strain gauge, a sensitive length-measuring device called a linear variable differential transformer (LVDT) may be useful.

14.4 EXPERIMENTAL DATA

14.4.1 Magnetostriction of uniaxial crystals

(a) Magnetostriction of rare earth metals

Heavy rare-earth metals with more than seven $4f$ electrons such as Gd, Tb, Cy, Ho, Er, and Tm exhibit ferromagnetism at low temperatures (Section 8.3). Except for Gd, these metals have unquenched orbital magnetic moments which give rise to large magnetocrystalline anisotropies, so that they can be magnetized to saturation only in very high fields (see Section 12.4.1.(a)). Gadolinium is an exceptional case, because it

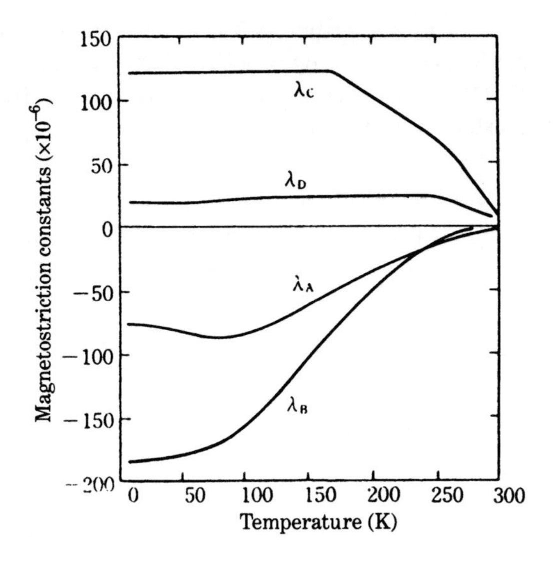

Fig. 14.14. Temperature dependences of magnetostriction constants λ_A, λ_B, λ_C, and λ_D observed for gadolinium single crystal.[16]

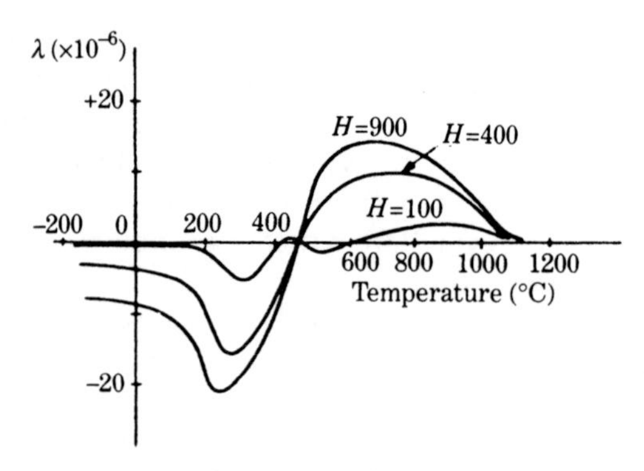

Fig. 14.15. Temperature dependence of magnetostriction constant observed for polycrystalline cobalt in various magnetic fields[20] (numerical values of H are in Oe ($= 10^3/4\pi\,\mathrm{A\,m^{-1}}$)).

has no orbital moment ($L = 0$), and is easily magnetized to saturation in any crystallographic direction. Therefore magnetostriction can be measured without any difficulty. Figure 14.14 shows the temperature dependences of magnetostriction constants (see (14.45)) of Gd.[16,17] Note that the magnitudes of some of these constants are as large as 100 to 200×10^{-6}, even though there is no orbital magnetic moment. For other rare earth metals (with nonzero orbital magnetic moments) it is hard to determine accurate values of magnetostriction constants. However, extrapolations to 0 K from the small values near the Néel points give very large values, some thousands times 10^{-6} for Dy,[17] Ho[18], Er.[17,19]

(b) Magnetostriction of cobalt

Cobalt transforms from fcc to hcp at about 420°C on cooling. The magnetostriction constant of polycrystalline cobalt changes its sign at this transformation point as shown in Fig. 14.15.[20] The magnetostriction constants and elastic moduli measured for single crystals are listed in Table 14.1.

(c) Magnetostriction of hexagonal oxides and compounds

Magnetostriction constants of a hexagonal ferrite $BaFe_{18}O_{27}$ (see Section 9.4) and MnBi, a NiAs-type compound, (see Section 10.3) are listed in Table 14.2. The

Table 14.1. Magnetostriction constants and elastic moduli of hexagonal cobalt.[21]

t(°C)	λ_A	λ_B	λ_C	λ_D
400	-16.5×10^{-6}	-70.5×10^{-6}	105×10^{-6}	-52×10^{-6}
200	-32.5	-88.5	120	-82
0	-52	-109	126	-108
-200	-66	-123	126	-128
-200	$c_{11} = 3.07$, $c_{12} = 1.65$,	$c_{13} = 1.03$,	$c_{33} = 3.58$,	$c_{44} = 0.76 \times 10^{11}\,\mathrm{N\,m^{-2}}$ ($\times 10^{12}\,\mathrm{dyn\,cm^{-2}}$)

Table 14.2. Magnetostriction constants of hexagonal compounds.

Compounds	λ_A	λ_B	λ_C	λ_D	Ref.
$BaFe_{18}O_{27}$	13×10^{-6}	3×10^{-6}	-23×10^{-6}	3×10^{-6}	22
MnBi	-800	-210	640		23

constants of MnBi are two orders of magnitude larger than those of $BaFe_{18}O_{27}$, although the magnetocrystalline anisotropy constants of these materials are almost the same (see Sections 12.4.1(d) and 12.4.1(e)).

14.4.2 Magnetostriction of cubic crystals

(a) Magnetostriction of 3d transition metals and alloys

The magnetostriction constants and elastic moduli observed for iron at room temperature are listed in Table 14.3, together with those of nickel. Temperature dependences of five constants (see (14.43)) are shown in Fig. 14.16 for Fe. The effect of dilution with nonmagnetic elements such as Si, Al, and Ti on λ_{100} and λ_{111} of Fe are shown in Figs 14.17, 14.18, and 14.19, respectively. It is interesting that the values of λ_{100} for Fe–Si and Fe–Al increase with a dilution with these nonmagnetic elements, in spite of the fact that the saturation magnetization of both the alloys decreases as if the magnetic moments of Fe atoms are diluted by nonmagnetic atoms (see Section 8.2). Since the Fe–Al alloy develops the superlattice Fe_3Al, the magnetostriction constants are also influenced by heat treatments. In the case of Fe–Ti alloys, the saturation magnetization stays almost constant, while the Curie point increases, with addition of Ti.[29] The magnetostriction constants of Fe–Ti alloys exhibit fairly complicated composition dependences (Fig. 14.19).

The magnetostriction constants and elastic moduli observed for nickel at room temperature are listed in Table 14.3. Contrary to iron, both the constants λ_{100} and λ_{111} are negative. The temperature dependences of five constants are shown in Fig. 14.20. Effects of dilution with nonmagnetic elements such as Cu, Cr, V, and Ti are shown in Figs 14.21–14.24, respectively. In all these cases, the magnitude of the magnetostriction decreases monotonically with dilution, reflecting a monotonic decrease of the saturation magnetization.

Magnetostriction constants for Fe–Ni alloy are already shown as a function of

Table 14.3. Magnetostriction constants and elastic moduli of iron and nickel (room temperature).

			c_{11}	c_{12}	c_{44}	
Metal	λ_{100}	λ_{111}	$\times10^{11}\,N\,m^{-2}$ ($\times10^{12}$ dyn cm^{-2})			Ref.
Fe	20.7×10^{-6}	-21.2×10^{-6}	2.41	1.46	1.12	24
Ni	-45.9	-24.3	2.50	1.60	1.18	24

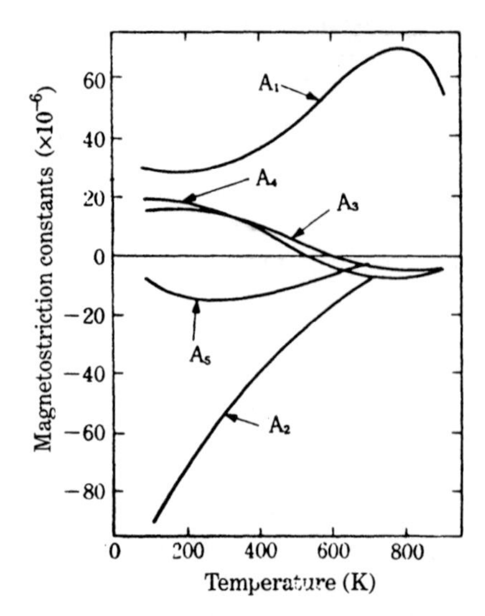

Fig. 14.16. Temperature dependences of five magnetostriction constants observed for iron in a magnetic field of 0.64 MA m^{-1} (= 8 kOe).[25] (The constants in the figure are related to those in (14.43) by $A_1 = h_1$, $A_2 = 2h_2$, $A_3 = h_3$, $A_4 = h_4$, and $A_5 = 2h_5$.)

composition in Fig. 14.11. The isotropic magnetostrictions are realized at 60 and 85% Ni. In the vicinity of 75% Ni, λ_{100} decreases and λ_{111} increases with the formation of the superlattice Ni_3Fe, and isotropic magnetostriction occurs.[33,34]

Magnetostriction constants of Ni–Co and Fe–Co alloys are shown as a function of composition in Figs 14.25 and 14.26, respectively. Despite the formation of the superlattice FeCo[37] (see Section 8.4), the magnetostriction constants are insensitive to heat treatments.

(b) Magnetostriction of spinel-type ferrites

The magnetostriction constants λ_{100} and λ_{111} are listed for various spinel-type ferrites in Table 14.4. Most ferrites have negative λ_{100}, except for some ferrites containing Fe^{2+} ions. Most of the constants are of the order of magnitude of 10^{-5}, except for very large values for ferrites containing Co or Ti. In the case of Co, the one-ion anisotropy of the Co^{2+} ion discussed in Section 12.3 may contribute to these large values. In the case of Ti, the introduction of Ti^{4+} may give rise to a change in valence from Fe^{3+} to Fe^{2+}, which is an anisotropic ion and can be responsible for the large magnetostriction. Particularly in those ferrites containing more than 50% Ti, a fraction of Fe^{3+} ions on the A sites change to Fe^{2+} ions, which contribute to large values of magnetocrystalline anisotropy and magnetostriction.[48] Also the Fe^{2+} ions on the A sites tend to produce the Jahn–Teller effect (see Section 14.7) and the accompanying *lattice softening* may also contribute to the large magnetostriction. Figure 14.27 shows the composition dependence for $Mn_xFe_{3-x}O_4$. Similar to the composition dependence of K_1 of the same system, both constants change abruptly as x becomes smaller than 1.0. This is also due to the appearance of Fe^{2+} ions.

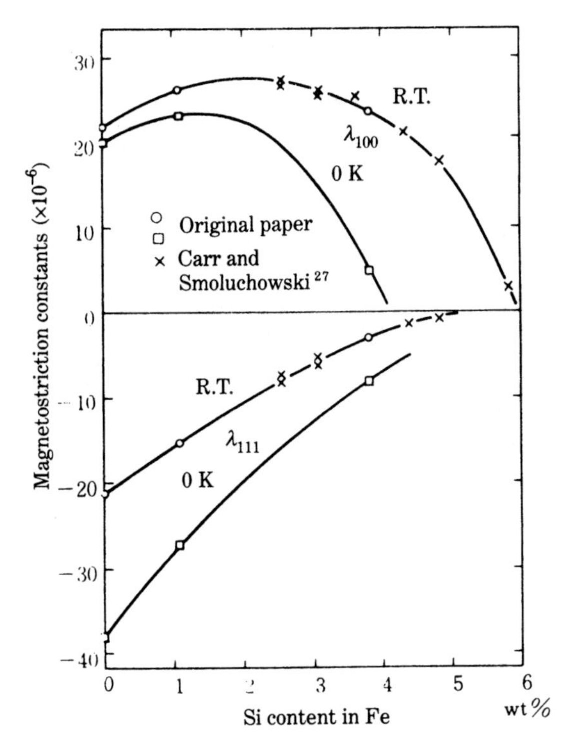

Fig. 14.17. Composition dependence of magnetostriction constants of Fe–Si alloys.[26] (Data after Tatsumoto and Kamoto[26] and Carr and Smoluchowski[27])

(c) Magnetostriction of garnet-type ferrites

The magnetostriction constants measured at various temperatures for pure iron garnets are listed in Table 14.5. The values at room temperature are of the order of magnitude of 10^{-6}–10^{-5}, but for those garnets containing rare earth ions with nonzero L the values are very large at low temperatures. This is due to the fact that the paramagnetic rare earth ions are strongly magnetized by the exchange field at low temperatures. Figure 14.28 shows the temperature dependence of magnetostriction constants of YIG containing only nonmagnetic yttrium. With the addition of anisotropic Tb, the values at low temperatures become very large as seen in Fig. 14.29 which shows the temperature dependences of magnetostriction constants. A change in magnetostriction by replacing Fe^{3+} with Ga^{3+} in YIG or by an increase in x for $Y_3Fe_{5-x}Ga_xO_{12}$ was also measured.[59]

14.5 VOLUME MAGNETOSTRICTION AND ANOMALOUS THERMAL EXPANSION

In the preceding sections we assumed that the volume of a ferromagnet remains

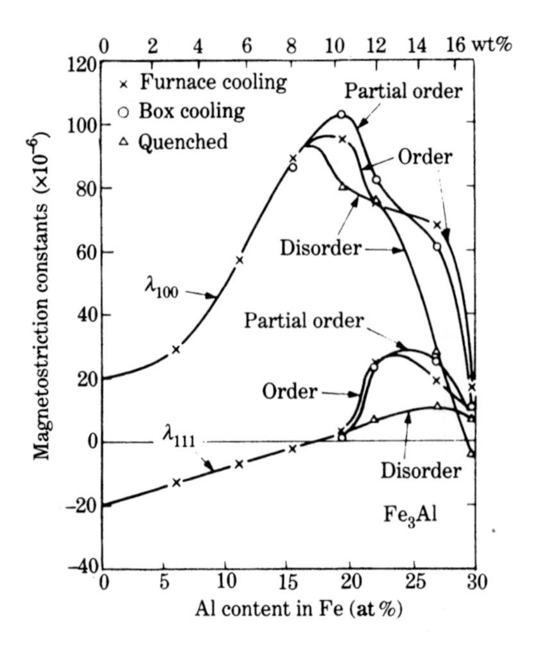

Fig. 14.18. Composition dependence of magnetostriction constants of Fe–Al alloys (room temperature).[28]

constant during the process of technical magnetization. In terms of strain tensors, the fractional volume change is given by

$$\frac{\delta v}{v} = e_{xx} + e_{yy} + e_{zz}. \tag{14.58}$$

Using the spontaneous strain tensors given by (14.30), we find from (14.58) that

$$\frac{\delta v}{v} = 0. \tag{14.59}$$

However, we have ignored the first term and all terms higher than the second term in (14.16). These terms can produce a small but nonzero change in volume with magnetization. We call this phenomenon the *volume magnetostriction.*

First we discuss the effect of the first term, $g(r)$, in (14.16). This term describes the exchange interaction between spins, and is independent of the direction of the spins relative to the bond direction. However, this term depends on the length of the bond, or the distance between the paired spins, r. For the spin pair whose bond direction is given by the direction cosines $(\beta_1, \beta_2, \beta_3)$, a change in the pair energy is given in terms of the lattice strain by

$$\Delta w = \left(\frac{\partial g}{\partial r}\right) r_0 \left(e_{xx}\beta_1^2 + e_{yy}\beta_2^2 + e_{zz}\beta_3^2 + e_{xy}\beta_1\beta_2 + e_{yz}\beta_2\beta_3 + e_{zx}\beta_3\beta_1\right). \tag{14.60}$$

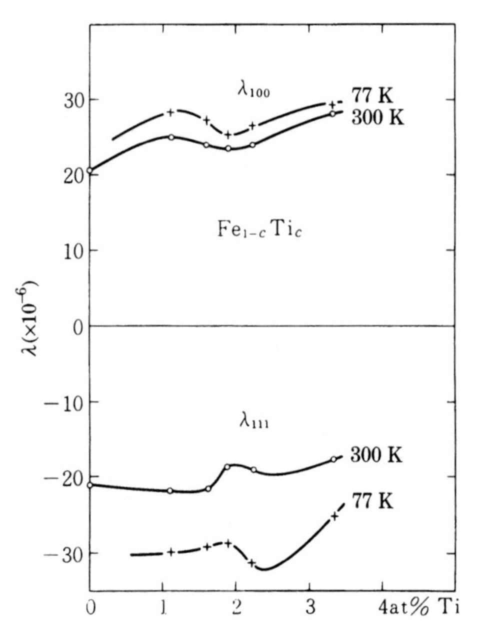

Fig. 14.19. Composition dependence of magnetostriction constants of Fe–Ti alloys at 77 K and 300 K.[29]

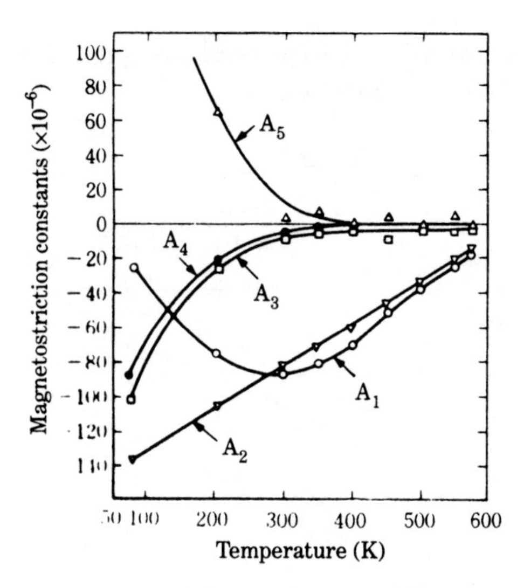

Fig. 14.20. Temperature dependence of five magnetostriction constants observed for nickel.[30] (The constants in the figure are related to those in (14.43) by $A_1 = h_1$, $A_2 = 2h_2$, $A_3 = h_3$, $A_4 = h_4$, and $A_5 = 2h_5$.)

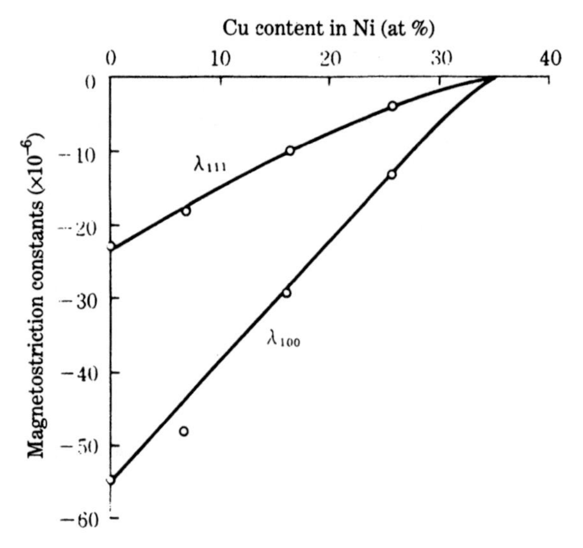

Fig. 14.21. Composition dependence of magnetostriction constants of Ni–Cu alloys (room temperature).[31]

Summing up the pair energies for all the nearest neighbor pairs in a simple cubic lattice, we have

$$E_{\text{volmag}} = N\left(\frac{\partial g}{\partial r}\right) r_0 (e_{xx} + e_{yy} + e_{zz}), \tag{14.61}$$

which is called the *magneto-volume energy*. Minimizing the total energy, or the sum of the magneto-volume and elastic energies,

$$E = E_{\text{volmag}} + E_{\text{el}}, \tag{14.62}$$

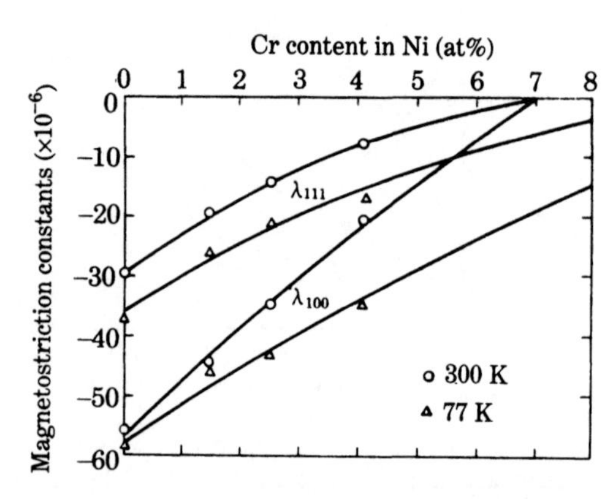

Fig. 14.22. Composition dependence of magnetostriction constants of Ni–Cr alloys.[32]

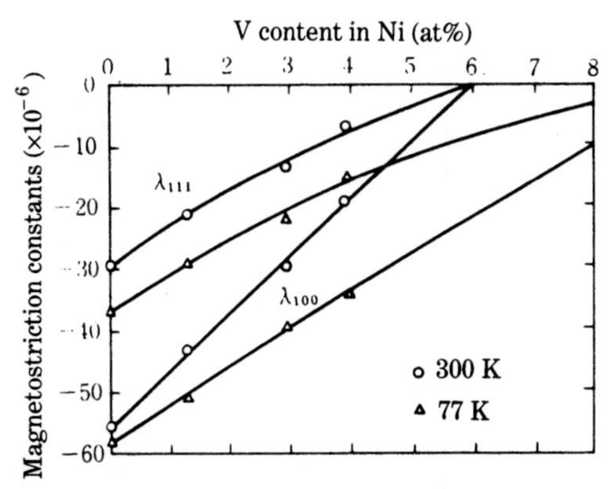

Fig. 14.23. Composition dependence of magnetostriction constants of Ni–V alloys.[32]

we have the relationships

$$\left.\begin{aligned}\frac{\partial E}{\partial e_{xx}} &= N\left(\frac{\partial g}{\partial r}\right)r_0 + c_{11}e_{xx} + c_{12}(e_{yy} + e_{zz}) = 0\\ \frac{\partial E}{\partial e_{yy}} &= N\left(\frac{\partial g}{\partial r}\right)r_0 + c_{11}e_{yy} + c_{12}(e_{zz} + e_{xx}) = 0\\ \frac{\partial E}{\partial e_{zz}} &= N\left(\frac{\partial g}{\partial r}\right)r_0 + c_{11}e_{zz} + c_{12}(e_{xx} + e_{yy}) = 0\end{aligned}\right\}. \quad (14.63)$$

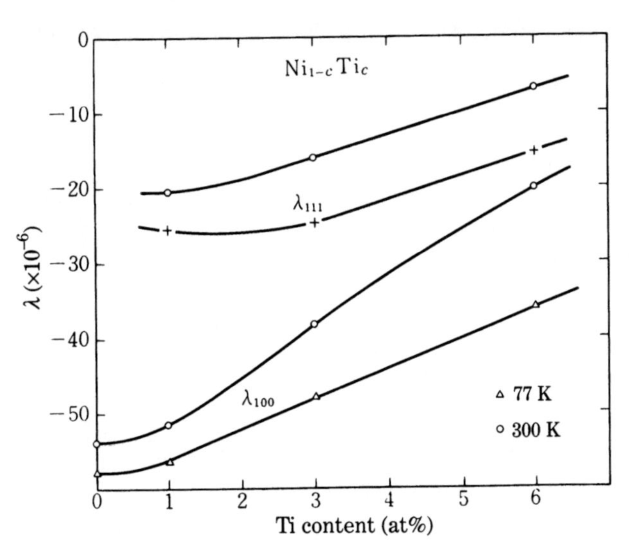

Fig. 14.24. Composition dependence of magnetostriction constants of Ni–Ti alloys.[29]

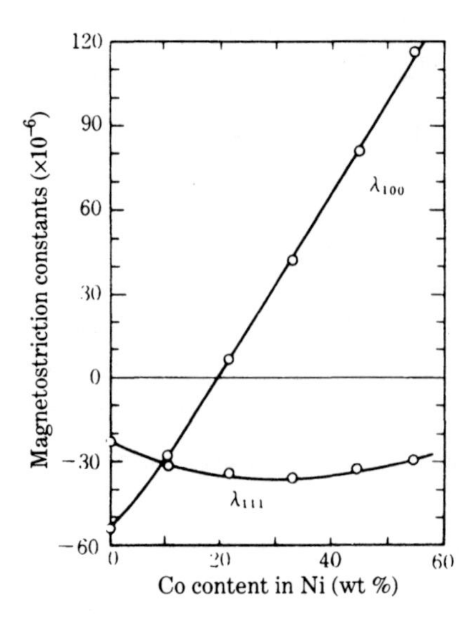

Fig. 14.25. Composition dependence of magnetostriction constants of Ni–Co alloys (room temperature).[35]

Adding up these terms, we have the volume change

$$\frac{\delta v}{v} = e_{xx} + e_{yy} + e_{zz} = -\frac{3Nr_0}{c_{11} + 2c_{12}}\left(\frac{\partial g}{\partial r}\right) \quad \text{for sc} \tag{14.62}$$

This is the volume magnetostriction produced by the appearance of spontaneous

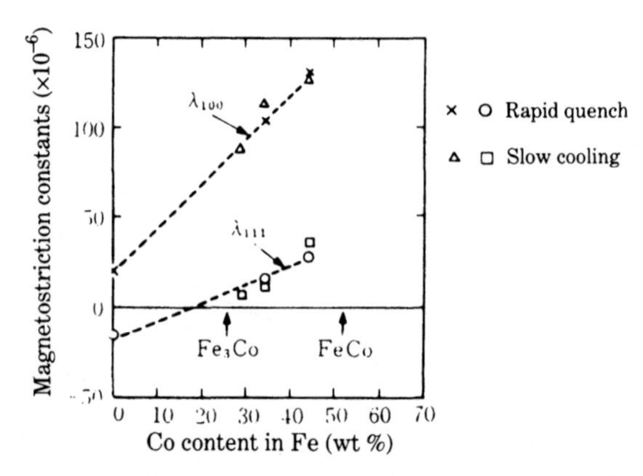

Fig. 14.26. Composition dependence of magnetostriction constants of Fe–Co alloys (room temperature).[36]

Table 14.4. Magnetostriction constants of spinel-type ferrites (mainly after Tsuya[49]).

Composition	λ_{100}	λ_{111}	Temp.	Ref.
$MnFe_2O_4$	-31×10^{-6}	6.5×10^{-6}		38
Fe_3O_4	-20	78	20°C	39
$Co_{0.8}Fe_{2.2}O_4$	-590	120	20°C	40
$NiFe_2O_4$	-42	-14		41
$CuFe_2O_4$	-57.5	4.7		42
$MgFe_2O_4$	-10.5	1.7		41
$Li_{0.5}Fe_{2.5}O_4$	-26	-3.8		38
$Mn_{0.6}Fe_{2.4}O_4$	-5	45	20°C	43
$Mn_{0.28}Zn_{0.16}Fe_{2.37}O_4$	-0.5	36		44
$Mn_{1.04}Zn_{0.22}Fe_{1.82}O_4$	-22	3		44
$Mg_{0.63}Fe_{1.26}Mn_{1.11}O_4$	49.5	-2.6		45
$Co_{0.32}Zn_{0.22}Fe_{2.2}O_4$	-210	110		40
$Co_{0.1}Ni_{0.9}Fe_2O_4$	-109	-38.6		46
$Li_{0.43}Zn_{0.14}Fe_{2.07}O_4$	-27.1	3.2		41
$Li_{0.5}Al_{0.35}Fe_{2.15}O_4$	-19.1	0.2		47
$Li_{0.56}Ti_{0.10}Fe_{2.35}O_4$	-16.0	4.3		47
$Li_{0.5}Ga_{1.4}Fe_{1.1}O_4$	-12.3	2.9		47
$Ti_{0.18}Fe_{2.82}O_4$	47	109	290 K	48
	142	86	80 K	48
$Ti_{0.56}Fe_{2.44}O_4$	170	92	290 K	48
	990	(330)	80 K	48

magnetization in a simple cubic lattice. Similarly we have

$$\frac{\partial v}{v} = -\frac{4Nr_0}{c_{11}+2c_{12}}\left(\frac{\partial g}{\partial r}\right) \quad \text{for bcc,} \tag{14.65}$$

and

$$\frac{\delta v}{v} = -\frac{6Nr_0}{c_{11}+2c_{12}}\left(\frac{\partial g}{\partial r}\right) \quad \text{for fcc.} \tag{14.66}$$

These are the volume changes produced by the appearance of ferromagnetism. The coefficient, g, is independent of α_i, but is nonzero when the material is ferromagnetic. In other words, the coefficient, g, is regarded as the average exchange interaction energy. Then the internal energy associated with the spontaneous magnetization is given by

$$E = \tfrac{1}{2}Nzg, \tag{14.67}$$

where z is the number of nearest neighbors.

Using this relationship, the coefficient, g, in (14.51), (14.52), and (14.53) is replaced

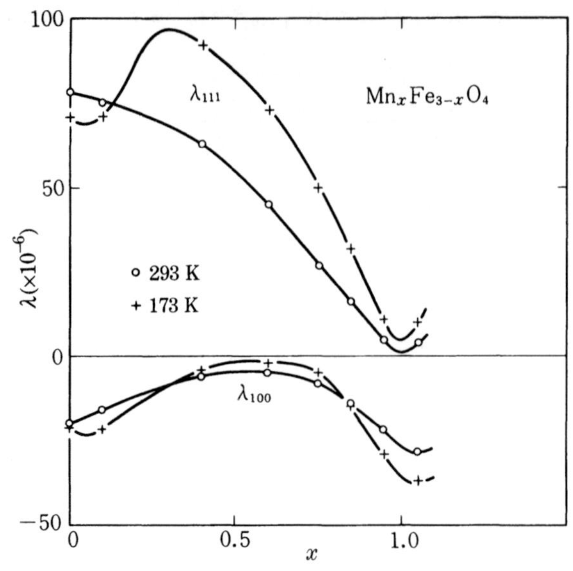

Fig. 14.27. Dependence of magnetostriction constants on the composition x in $Mn_xFe_{3-x}O_4$. (Data after Miyata and Funatogawa[43])

by E. Also, using the relationship between the bulk modulus, c, and the elastic moduli, c_{11} and c_{12},

$$c = \frac{1}{3}(c_{11} + 2c_{12}), \tag{14.68}$$

we have for all crystal types

$$\frac{\delta v}{v} = -\frac{1}{c}\frac{\partial E}{\partial \omega} = \frac{\partial E}{\partial p}, \tag{14.69}$$

where $\omega = \delta v/v$. This relationship can also be deduced from thermodynamic considerations.

Since the magnetic internal energy depends on temperature through a change in spontaneous magnetization, the volume change given by (14.69) results in an anomalous thermal expansion coefficient, or

$$\alpha = \frac{1}{3}\frac{\partial}{\partial T}\left(\frac{\delta v}{v}\right) = \frac{1}{3}\frac{\partial^2 E}{\partial T \partial p} = \frac{1}{3}\frac{\partial}{\partial p}\left(\frac{\partial E}{\partial T}\right) = \frac{1}{3}\frac{\partial C_p}{\partial p}, \tag{14.70}$$

where C_p is the anomalous heat capacity per unit volume due to the appearance of ferromagnetism. In fact, the anomalous thermal expansion coefficient of Ni shown in the inset of Fig. 14.30 is quite similar to its anomalous heat capacity.

The main graph in Fig. 14.30 shows the temperature dependence of thermal expansion coefficient for Invar, which is a 34 at% Ni–Fe alloy used as a low thermal expansion material (see Section 8.2). The anomalous region shown by the hatched area in Fig. 14.30 is much larger than that of Ni, covers a wider temperature range, and has no sharp peak at the Curie point. This behavior can be interpreted in terms

Table 14.5. Magnetostriction constants of garnet-type ferrites (after Tsuya[57]; Hansen[58]).

RIG	λ_{100}	λ_{111}	Temp. (K)	Ref.
YIG	-1.0×10^{-6}	-3.6×10^{-6}	78	50
	-1.1	-3.9	196	50
	-1.4	-2.4	300	50
SmIG	159	-183	78	50
	49	-28.1	196	50
	21	-8.5	300	50
EuIG	110	20	4.2	51
	86	9.7	78	50
	51	5.3	196	50
	21	1.8	300	50
GdIG	7.5	-4.1	4.2	52
	4.0	-5.1	78	50
	1.7	-4.5	196	50
	0	-3.1	300	53
TbIG	1200	2420	4.2	54
	67	560	78	50
	-10.3	65	196	50
	-3.3	12	300	50
DyIG	-1400	-550	4.2	55
	-169	-145	78	50
	-46.6	-21.6	196	50
	-12.5	-5.9	300	50
HoIG	-1400	-330	0	55
	-82.2	-56.3	78	50
	-10.6	-7.4	196	50
	-3.4	-4.0	300	50
ErIG	630	-450	0	55
	10.7	-19.4	78	50
	4.1	-8.8	196	50
	2.0	-4.9	300	50
TmIG	25	-31.2	78	50
	4.9	-11.3	196	50
	1.4	-5.2	300	50
YbIG	97	-31	4.2	56
	18.3	-14.4	78	50
	5.0	-7.1	196	50
	1.4	-4.5	300	50

of a model which assumes two spin states for Fe atoms: a low-spin state with a small magnetic moment and small atomic volume, and a high-spin state with a large magnetic moment and large atomic volume.[61–3] A theory based on the spin-fluctuation model can also explain this phenomenon.[64]

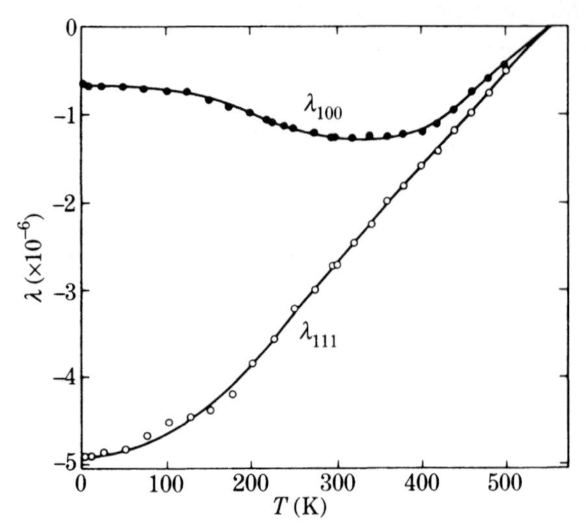

Fig. 14.28. Temperature dependence of magnetostriction constants of YIG.[58] (Data after Hansen[59])

The exchange energy, g, can also be changed by the application of an external field, which causes an increase in spontaneous magnetization. In the presence of a high magnetic field, H, the energy of spontaneous magnetization is given by

$$E = -\tfrac{1}{2}wI_s^2 - I_sH \tag{14.71}$$

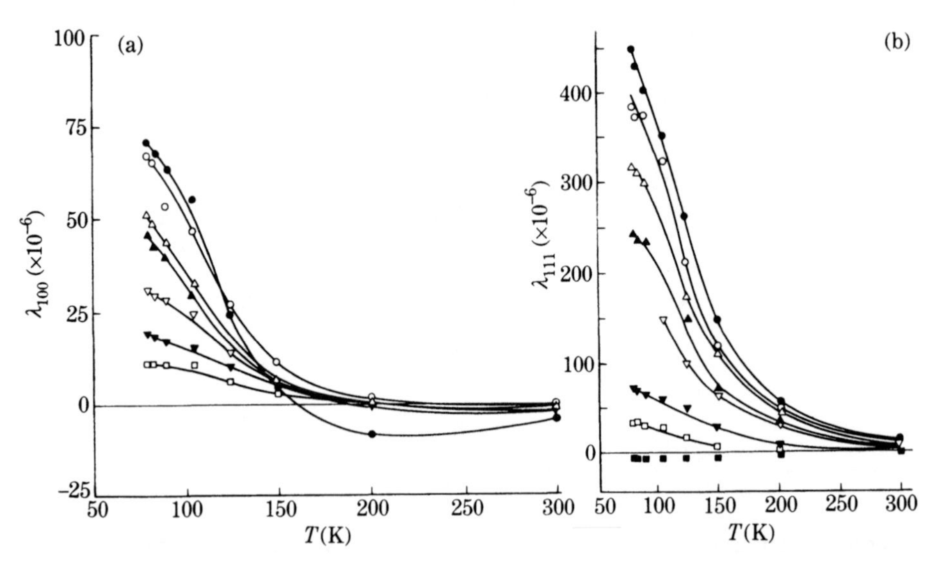

Fig. 14.29. Temperature dependence of magnetostriction constants of $Tb_xY_{3-x}Fe_5O_{12}$ (x-values are from up to down: 3.00, 2.54, 2.12, 1.65, 1.17, 0.50, 0.26 and 0.00)[58] (Data after Belov *et al.*[60])

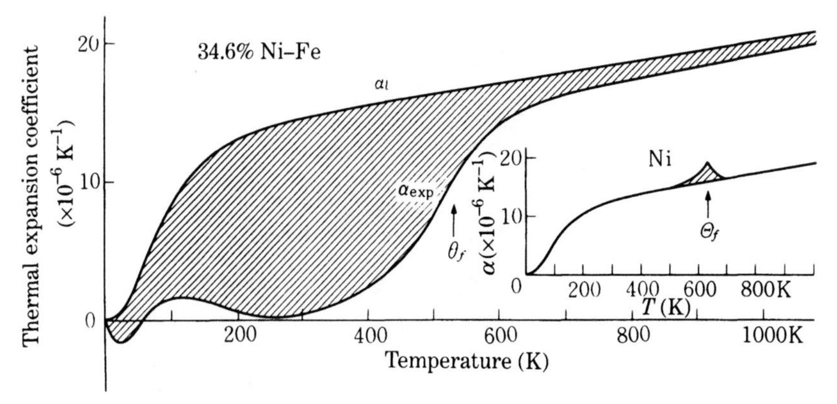

Fig. 14.30. Temperature dependence of the anomalous thermal expansion coefficient for Ni and Invar.[61]

(see (6.23)), so that, referring to (14.69), we have a volume change produced by the application of a field

$$\frac{\delta v}{v} = \frac{-1}{c}\frac{\partial}{\partial H}\left(\frac{\partial E}{\partial \omega}\right)H = \frac{-1}{c}\frac{\partial}{\partial \omega}\left(\frac{\partial E}{\partial H}\right)H = \frac{1}{c}\frac{\partial I_s}{\partial \omega}H. \tag{14.72}$$

We call this volume magnetostriction caused by the forced alignment of spins by high magnetic field the *forced magnetostriction*. The anomalous thermal expansion and the forced magnetostriction are essentially isotropic, because they are caused by the isotropic exchange interaction.

Let us proceed to the volume magnetostriction that arises from the quadrupole interaction $q(r)$ in (14.16). For the simple cubic lattice, this term produces a volume magnetostriction given by

$$\frac{\delta v}{v} = e_{xx} + e_{yy} + e_{zz} = \frac{2N(\partial q/\partial r)r_0}{c_{11}+2c_{12}}\left(\alpha_1^2\alpha_2^2 + \alpha_2^2\alpha_3^2 + \alpha_3^2\alpha_1^2 - \tfrac{1}{5}\right). \tag{14.73}$$

Using the relationships (12.62) and (14.68), the main term in (14.73) (other than the constant term) is

$$\frac{\delta v}{v} = -\frac{1}{c}\frac{\partial K_1}{\partial \omega}(\alpha_1^2\alpha_2^2 + \alpha_2^2\alpha_3^2 + \alpha_3^2\alpha_1^2). \tag{14.74}$$

Comparing with (14.69), we see that this effect is caused by a rotation of the spontaneous magnetization against the magnetocrystalline anisotropy; it is called the *crystal effect*. This effect is isotropic: in other words, the volume change is spherical, because (14.74) does not contain any β_i, which are the direction cosines of the observation direction. The h_3 term in (14.43), which does not contain any β_i, corresponds to this effect.

The volume magnetostriction depends also on the shape of the specimen. Consider a ferromagnetic specimen of nearly unit volume, $(1+\omega)$, where ω is much smaller than 1, and with demagnetizing factor, N. When this sample is magnetized to an intensity of magnetization, I, the magnetostatic energy is given by

$$U=\frac{1}{2}\frac{N}{\mu_0}\frac{M^2}{1+\omega}, \tag{14.75}$$

where M is the magnetic moment of the specimen as a whole. Substituting U in (14.75) for E in (14.69), and considering $M \approx I$, we have

$$\frac{\delta v}{v}=\frac{NI^2}{2c\mu_0}. \tag{14.76}$$

We call this effect the *form effect*. When the specimen is magnetized to saturation, this volume change also reaches a saturation value

$$\left(\frac{\delta v}{v}\right)_s=\frac{NI_s^2}{2c\mu_0}. \tag{14.77}$$

The feature of this effect is that the volume change is proportional to I^2, or to H^2 if the magnetization curve is linear.

In summary, the volume magnetostriction of a specimen with a finite size is expected to change as a function of external field as shown in Fig. 14.31. First the volume increases proportionally to H^2 as a result of the form effect, and then the crystal effect appears in the region of rotation magnetization. The sign of the crystal effect can be either positive or negative, depending on the sign of q (or of K_1) and that of $\partial q/\partial r$. The sign is negative for iron and nickel. In this case, the volume change decreases and then increases linearly due to the forced magnetostriction. Generally speaking, the volume magnetostriction is small when the magnetic field is reasonably weak, but it becomes comparable to the linear magnetostriction when the magnetic field is stronger than $1\,\mathrm{MA\,m^{-1}}$ (about 10^4 Oe), because the forced magnetostriction is linear with H (see (14.72)). Fig. 14.32 shows the volume magnetostriction observed for two iron samples with different demagnetizing factors.

It should be noted that the form effect produces not only volume magnetostriction but also normal linear magnetostriction. For a specimen in the form of a rotational

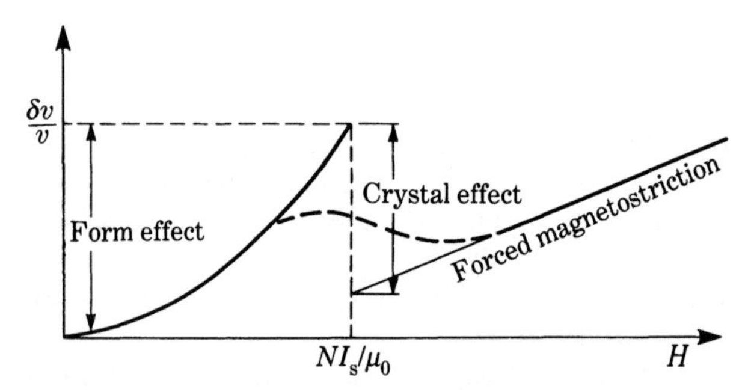

Fig. 14.31. Schematic variation of the volume magnetostriction with increasing magnetic field.[65]

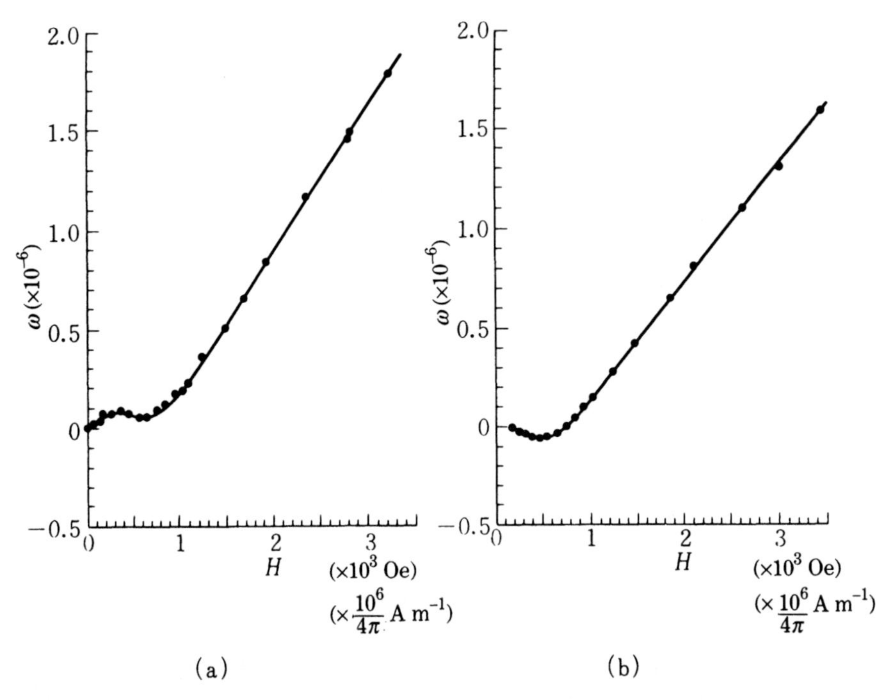

Fig. 14.32. Experimental data on the field dependence of the volume magnetostriction for two iron specimens with different dimensional ratios; (a) $k = 16.9$; (b) $k = 41.6$. (Data after Kornetzki[66])

ellipsoid with the long axis parallel to the x-axis, the demagnetizing factor depends on the strain as

$$N = N_0 + \frac{\mathrm{d}N}{\mathrm{d}k}\left(\frac{\partial k}{\partial e_{xx}} e_{xx} + \frac{\partial k}{\partial e_{yy}} e_{yy} + \frac{\partial k}{\partial e_{zz}} e_{zz}\right), \tag{14.78}$$

where k is the dimensional ratio or aspect ratio (length/diameter). Let the length of the specimen be l and the diameter d. Then we have

$$\frac{\partial k}{\partial e_{xx}} = \frac{l}{d} = k, \qquad \frac{\partial k}{\partial e_{yy}} = \frac{\partial k}{\partial e_{zz}} = -\frac{l}{2d} = -\frac{k}{2}. \tag{14.79}$$

Since the demagnetizing factor of the elongated ellipsoid is given by (1.43), we have

$$\frac{\mathrm{d}N}{\mathrm{d}k} = -\frac{1}{k^3}(2\ln 2k - 3). \tag{14.80}$$

Using (14.78) and (14.79), (14.78) becomes

$$N = N_0\left\{1 - a\left(e_{xx} - \tfrac{1}{2}(e_{yy} + e_{zz})\right)\right\}, \tag{14.81}$$

where a is given by

$$a = \frac{N'k}{N_0} = \frac{2\ln 2k - 3}{\ln 2k - 1} \tag{14.82}$$

($N' = \mathrm{d}N/\mathrm{d}k$). Ignoring the volume magnetostriction, the magnetostatic energy is expressed in terms of strain tensors as

$$E_N = \frac{1}{2\mu_0} NI^2 = \frac{1}{2\mu_0} N_0 I^2 \left\{1 - a\left(e_{xx} - \tfrac{1}{2}(e_{yy} + e_{zz})\right)\right\}. \tag{14.83}$$

Therefore in order to decrease the magnetostatic energy, e_{xx} should become more positive, and e_{yy} and e_{zz} should become more negative. The resultant strain can be found by minimizing the total energy

$$E = E_N + E_{\mathrm{el}}, \tag{14.84}$$

or

$$\left.\begin{aligned} \frac{\partial E}{\partial e_{xx}} &= -\frac{a}{2\mu_0} N_0 I^2 + c_{11}e_{xx} + c_{12}(e_{yy} + e_{zz}) = 0 \\ \frac{\partial E}{\partial e_{yy}} &= \frac{a}{4\mu_0} N_0 I^2 + c_{11}e_{yy} + c_{12}(e_{zz} + e_{xx}) = 0 \\ \frac{\partial E}{\partial e_{zz}} &= \frac{a}{4\mu_0} N_0 I^2 + c_{11}e_{xx} + c_{12}(e_{xx} + e_{yy}) = 0 \end{aligned}\right\}. \tag{14.85}$$

Solving (14.85), we have

$$e_{xx} = \frac{1}{2\mu_0} \frac{aN_0 I^2}{c_{11} - c_{12}}, \qquad e_{yy} = e_{zz} = -\frac{1}{4\mu_0} \frac{aN_0 I^2}{c_{11} - c_{12}}, \tag{14.86}$$

where e_{xx} is the longitudinal magnetostriction, and e_{yy} or e_{zz} is the transverse magnetostriction. In order to avoid this effect in measuring the ordinary linear magnetostriction, we must use a specimen with a small demagnetizing factor. The measurement of the anisotropic form effect is discussed by Gersdorf.[67]

14.6 MAGNETIC ANISOTROPY CAUSED BY MAGNETOSTRICTION

Since a magnetostrictive elongation is caused by magnetization, a mechanical stress is expected to have some effect on magnetization. Such an effect is called the *inverse effect of magnetostriction.* Magnetostriction plays an important role in determining domain structures through this effect.

Suppose that a stress, σ ($\mathrm{N\,m^{-2}}$), is acting on a ferromagnetic body. Let the direction cosines of this tension be $(\gamma_1, \gamma_2, \gamma_3)$. Then the tensor components are given by $\sigma_{ij} = \sigma\gamma_i\gamma_j$, which results in a strain with components

$$e_{xx} = \sigma\left\{s_{11}\gamma_1^2 + s_{12}(\gamma_2^2 + \gamma_3^2)\right\}\dots, \qquad e_{xy} = \sigma s_{44}\gamma_1\gamma_2 \dots, \tag{14.87}$$

where s_{11}, s_{12}, s_{44} are the elastic constants. Using these relationships in (14.23) we have the magnetoelastic energy

$$\begin{aligned} E_\sigma = {} & B_1\sigma(s_{11} - s_{12})\left(\alpha_1^2\gamma_1^2 + \alpha_2^2\gamma_2^2 + \alpha_3^2\gamma_3^2 - \tfrac{1}{3}\right) \\ & + B_2\sigma s_{44}(\gamma_1\gamma_2\alpha_1\alpha_2 + \gamma_2\gamma_3\alpha_2\alpha_3 + \gamma_3\gamma_1\alpha_3\alpha_1). \end{aligned} \tag{14.88}$$

The coefficients B_1 and B_2 are related to λ_{100} and λ_{111} by (14.33) and (14.34). Using this relationship between c_{11}, c_{12}, and c_{44} and s_{11}, s_{12}, and s_{44},* (14.88) can be rewritten as

$$E_\sigma = -\tfrac{3}{2}\lambda_{100}\sigma\left(\alpha_1^2\gamma_1^2 + \alpha_2^2\gamma_2^2 + \alpha_3^2\gamma_3^2 - \tfrac{1}{3}\right) - 3\lambda_{111}\sigma(\alpha_1\alpha_2\gamma_1\gamma_2 + \alpha_2\alpha_3\gamma_2\gamma_3 + \alpha_3\alpha_1\gamma_3\gamma_1). \tag{14.89}$$

For a ferromagnet with $K_1 > 0$ such as Fe, a domain with magnetization parallel to [100] has energy

$$E_\sigma = -\tfrac{3}{2}\lambda_{100}\sigma\gamma_1^2. \tag{14.90}$$

Similarly, for the [010] and [001] domains we have

$$E_\sigma = -\tfrac{3}{2}\lambda_{100}\sigma\gamma_2^2 \tag{14.91}$$

and

$$E_\sigma = -\tfrac{3}{2}\lambda_{100}\sigma\gamma_3^2, \tag{14.92}$$

respectively. This difference in energy between domains results in a force acting on the 90° domain walls between them.

When the easy axis is parallel to [111], (14.89) leads to

$$E_\sigma = -\lambda_{111}\sigma(\gamma_1\gamma_2 + \gamma_2\gamma_3 + \gamma_3\gamma_1). \tag{14.93}$$

Let φ be the angle between [111] and σ. Then we have the relationship

$$\cos\varphi = \frac{1}{\sqrt{3}}(\gamma_1 + \gamma_2 + \gamma_3), \tag{14.94}$$

so that (14.93) becomes

$$E_\sigma = -\tfrac{3}{2}\lambda_{111}\cos^2\varphi. \tag{14.95}$$

This relationship holds always if we define φ as the angle between the magnetization and σ.

If instead of fixing the direction of magnetization, we fix the axis of stress, σ, the energy changes as the magnetization rotates, according to (14.89). In other words, a magnetic anisotropy is produced. For simplicity, we assume isotropic magnetostriction, $\lambda_{100} = \lambda_{111} = \lambda$. Then, (14.89) becomes

$$E_\sigma = -\tfrac{3}{2}\lambda\sigma(\alpha_1\gamma_1 + \alpha_2\gamma_2 + \alpha_3\gamma_3)^2 = -\tfrac{3}{2}\lambda\sigma\cos^2\varphi, \tag{14.96}$$

where φ is the angle of the magnetization measured from the axis of stress. This is a kind of uniaxial magnetic anisotropy. Hereafter we use (14.96) when we discuss the effect of stress on magnetization.

Magnetostriction can also influence the cubic anisotropy, because as the magnetization rotates, the lattice distorts, thus producing changes in magnetoelastic and elastic

* $c_{11} = \dfrac{s_{11}+s_{12}}{(s_{11}-s_{12})(s_{11}+2s_{12})}$, $c_{12} = \dfrac{-s_{12}}{(s_{11}-s_{12})(s_{11}+2s_{12})}$, $c_{44} = \dfrac{1}{s_{44}}$.

energies. Substituting the strain tensor in the magnetoelastic energy (14.23) using the strain tensor components in (14.30), we have

$$\Delta E_{\text{magel}} = \frac{B_1^2}{c_{12}-c_{11}}\left\{\left(\alpha_1^2-\tfrac{1}{3}\right)^2+\left(\alpha_2^2-\tfrac{1}{3}\right)^2+\left(\alpha_3^2-\tfrac{1}{3}\right)^2\right\}$$
$$-\frac{B_2^2}{c_{44}}(\alpha_1^2\alpha_2^2+\alpha_2^2\alpha_3^2+\alpha_3^2\alpha_1^2)$$
$$= -\left(\frac{2B_1^2}{c_{12}-c_{11}}+\frac{B_2^2}{c_{44}}\right)(\alpha_1^2\alpha_2^2+\alpha_2^2\alpha_3^2+\alpha_3^2\alpha_1^2). \tag{14.97}$$

We have also a change in elastic energy from (14.27):

$$\Delta E_{\text{el}} = \tfrac{1}{2}c_{11}\frac{B_1^2}{(c_{12}-c_{11})^2}\left\{\left(\alpha_1^2-\tfrac{1}{3}\right)^2+\left(\alpha_2^2-\tfrac{1}{3}\right)^2+\left(\alpha_3^2-\tfrac{1}{3}\right)^2\right\}$$
$$+c_{12}\frac{B_1^2}{(c_{12}-c_{11})^2}\left\{\left(\alpha_1^2-\tfrac{1}{3}\right)\left(\alpha_2^2-\tfrac{1}{3}\right)\right.$$
$$\left.+\left(\alpha_2^2-\tfrac{1}{3}\right)\left(\alpha_3^2-\tfrac{1}{3}\right)+\left(\alpha_3^2-\tfrac{1}{3}\right)\left(\alpha_1^2-\tfrac{1}{3}\right)\right\}$$
$$+\tfrac{1}{2}c_{44}\frac{B_2^2}{c_{44}^2}(\alpha_1^2\alpha_2^2+\alpha_2^2\alpha_3^2+\alpha_3^2\alpha_1^2)$$
$$= \left(\frac{B_1^2}{c_{12}-c_{11}}+\frac{1}{2}\frac{B_2^2}{c_{44}}\right)(\alpha_1^2\alpha_2^2+\alpha_2^2\alpha_3^2+\alpha_3^2\alpha_1^2). \tag{14.98}$$

Thus a change in the total energy (14.28) becomes

$$\Delta E = \Delta K_1(\alpha_1^2\alpha_2^2+\alpha_2^2\alpha_3^2+\alpha_3^2\alpha_1^2), \tag{14.99}$$

where ΔK_1 is given by

$$\Delta K_1 = -\left(\frac{B_1^2}{c_{12}-c_{11}}+\frac{1}{2}\frac{B_2^2}{c_{44}}\right)$$
$$= -\tfrac{9}{4}\{(c_{12}-c_{11})\lambda_{100}^2+2c_{44}\lambda_{111}^2\} \tag{14.100}$$

from the relationships (14.33) and (14.34). Therefore the observed magnetocrystalline anisotropy constant must be compared with the theoretical value using the correction given by (14.100). Using the values of Table 14.3 for Ni, (14.100) gives

$$\Delta K_1 = -\tfrac{9}{4}\{-(2.50-1.60)\times 45.9^2+2\times 1.18\times 24.3^2\}\times 10^{-1}$$
$$= 1.13\times 10^2\ (\mathrm{J\,m^{-3}}). \tag{14.101}$$

This value is small compared with the observed value of $-5.7\times 10^3\ \mathrm{J\,m^{-3}}$ (see (12.7)), but is not negligible.

14.7 ELASTIC ANOMALY AND MAGNETOSTRICTION

In this section, we discuss several topics on the relationship between elasticity and magnetostriction of magnetic materials.

As noted in Section 14.2, the magnetostriction is produced by a balance between a decrease in the magnetoelastic energy and an increase in the elastic energy. Therefore if the elastic moduli are decreased for some reason, the magnetostriction must increase.

One of the mechanisms that can cause a decrease in the elastic moduli is a structural phase transition. When a cubic phase at high temperatures transforms to phase with lower symmetry at some transition temperature, the elastic modulus corresponding to this deformation decreases with a decrease of temperature towards this transition point. A typical example is the *Jahn–Teller distortion*.[68] Intuitively this phenomenon is interpreted as a deformation of the lattice by anisotropic electron clouds of $3d$ electrons, as shown in Section 3.4. For example, suppose that the Cu^{2+} ion, which has nine $3d$ electrons, occupies an octahedral site in a cubic crystal. Then nine electrons distribute on five energy levels according to Hund's rule as shown in Fig. 14.33. As discussed in Chapter 12, three d levels are lowered and two d levels are raised by cubic crystalline fields. One of the $d\gamma$ functions has the form x^2-y^2 and stretches along the x- and y-axes, while the other $d\gamma$ function has the form $2z^2-x^2-y^2$ and stretches along the z-axis (see Fig. 3.20). If the cubic crystal is elongated along the z-axis ($c>a$), the $3d$ ions under consideration are more separated from the O^{2-} ions on the same z-axis than from those on the same x- or y-axis. Then the $d\gamma$ function with $2z^2-x^2-y^2$ has a lower electrostatic energy than the other. Therefore the ninth electron occupies this level, so that the total energy of the occupied levels is lowered as the lattice is elongated along the z-axis. Since such a deformation of the crystal increases the elastic energy, the balance of the two energies results in a nonzero distortion of the crystal. This is the Jahn–Teller distortion.

The tetragonal distortion of copper ferrite (see Section 9.2) observed at room temperature is due to this effect. At high temperatures this material is cubic, because

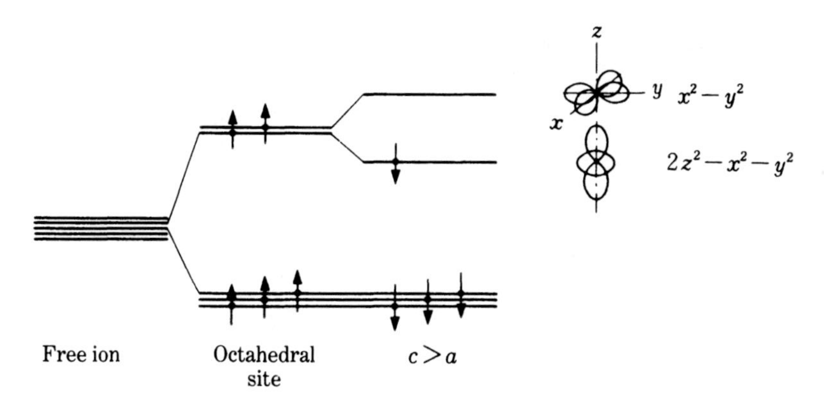

Fig. 14.33. Splitting of the $d\gamma$ levels of the Cu^{2+} ion on the octahedral site.

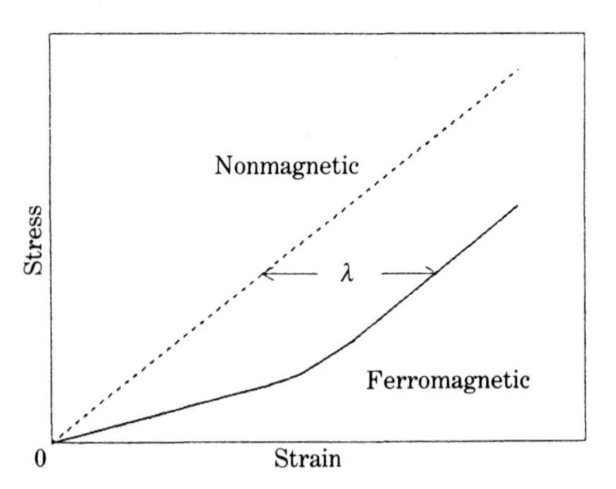

Fig. 14.34. Schematic diagram of stress–strain relationships for ferro and nonferromagnetic materials.

the thermal excitation to the excited $d\gamma$ level lowers the total energy levels by reducing the separation of the two levels. However, there is still a tendency for the lattice to deform to a tetragonal structure, so that the elastic modulus c_{11} is reduced. This phenomenon is called the *lattice softening*. A value of magnetostriction observed for Ti-ferrite (see Section 14.4.2(b)) is considered to be due to lattice softening caused by Fe^{2+} ions on the A sites.

The elastic moduli of ferromagnetic materials are reduced to some extent due to magnetostriction. Figure 14.34 shows the stress–strain curves for ferromagnetic and

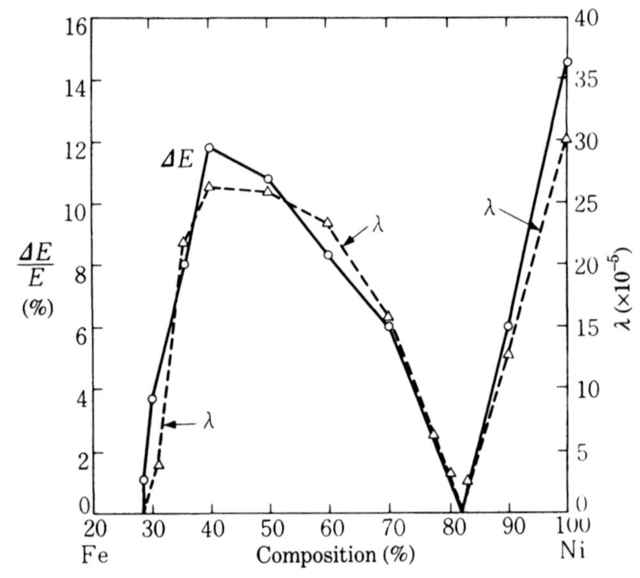

Fig. 14.35. Comparison between the ΔE effect and λ for Fe–Ni alloys.[69] (Data of ΔE effect from Köster[70])

nonferromagnetic materials. At moderate stresses, where magnetization can rotate freely, the magnetostriction contributes an additional strain irrespective to the sign of λ, thus resulting in a decrease of the Young's modulus, E. This phenomenon is called the ΔE *effect*. It is naturally proportional to the magnitude of λ as shown in Fig. 14.35, which demonstrates the proportionality between $\Delta E/E$ and λ in Fe–Ni alloys.

In ferromagnetic metals, which exhibit the ΔE effect, a mechanical vibration produces eddy currents through the rotational vibration of magnetization, which results in an additional internal friction. Turbine blades are usually made from magnetic 13% Cr stainless steel, instead of nonmagnetic 18–8 stainless steel, because vibrational energy is more strongly absorbed in a magnetic material because of this effect.[71,72]

PROBLEMS

14.1 Consider a cubic ferromagnetic crystal with a large positive magnetocrystalline anisotropy constant ($K_1 > 0$) which contains only [100] and [$\bar{1}$00] domains in the demagnetized state and is magnetized parallel to [010]. Calculate the elongation measured in the [010] direction as a function of the intensity of magnetization in that direction.

14.2 Knowing that the saturation value of magnetostriction is in the order of 10^{-5} and also that the elastic modulus is in the order of $10^{11}\,\mathrm{N\,m^{-2}}$, estimate the order of magnitude of the dipole–dipole interaction, Nl.

14.3 When a single crystal sphere with magnetostriction constants h_1, h_2, and h_3 is magnetized to its saturation by a constant magnetic field which makes an angle θ with [100] in the (001) plane, how do the elongations parallel to [100] and [010] change as a function of θ?

14.4 When a tensile stress σ is applied parallel to [123] in a ferromagnetic crystal with positive K_1, how much energy can be stored in the x, y, and z domains? Assume $\lambda_{100}\sigma \ll K_1$.

14.5 Calculate the value of ΔK_1 for iron at room temperature.

REFERENCES

1. W. L. Webster, *Proc. Roy. Soc.* (London), **A109** (1925), 570.
2. R. Becker and W. Döring, *Ferromagnetismus* (Springer, Berlin, 1939), S284.
3. W. Heisenberg, *Z. Phys.*, **69** (1931), 287.
4. N. Akulov, *Z. Phys.*, **69** (1931), 78.
5. S. Kaya and H. Takaki, *J. Fac. Sci. Hokkaido Univ.*, **2** (1935), 227.
6. L. Néel, *J. Phys. Radium*, **15** (1954), 225.
7. F. Lichtenberger, *Ann. Physik.*, **10** (1932), 45.
8. C. D. Kim, M. Matsui, and S. Chikazumi, *J. Phys. Soc. Japan*, **44** (1978), 1152.
9. W. P. Mason, *Phys. Rev.*, **96** (1954), 302.
10. W. J. Carr, *Magnetic properties of metals and alloys* (American Society of Metals, Cleveland, 1959) Chap. 10; *Handbuch der Physik* (Springer-Verlag, Berlin, 1966), **XIII/2**, 274.

11. N. Tsuya, *J. Appl. Phys.*, **29** (1958), 449: *Sci. Repts. Res. Inst. Tohoku Univ.*, **B8** (1957), 161.
12. J. C. Slonczewski, *J. Appl. Phys.*, **30** (1959), 310S; *Phys. Chem. Solids*, **15** (1960), 335.
13. J. Kanamori, *Magnetism* (Academic Press, Inc. N.Y., 1963), Vol. 1, p. 127.
14. T. Wakiyama and S. Chikazumi, *Experimental Physics Series* (Kyoritsu Publishing Co., Tokyo, 1968), Vol. 17, *Magnetism*, Chap. 14, p. 328 (in Japanese).
15. H. Zijlstra, *Experimental methods in magnetism*, Vol. 2 (Wiley-Interscience, New York, 1967), Chap. 4.
16. R. M. Bozorth and T. Wakiyama, *J. Phys. Soc. Japan*, **18** (1963), 97.
17. A. E. Clark, B. F. DeSavage, and R. M. Borzorth, *Phys. Rev.*, **138** (1965), A216.
18. S. Legvold, J. Alstad, and J. Rhyne, *Phys. Rev. Lett.*, **10** (1963), 509.
19. J. Rhyne and S. Legvold, *Phys. Rev.*, **140** (1965), A1243.
20. K. Honda and S. Shimizu, *Phil. Mag.*, **6** (1903), 392.
21. H. J. McSkimin, *J. Appl. Phys.*, **26** (1955), 406.
22. S. S. Fonton and A. V. Zalesskii, *Soviet Phys. JETP*, **20** (1965), 1138.
23. H. J. Williams, R. C. Sherwood, and O. L. Boothby, *J. Appl. Phys.*, **28** (1957), 445.
24. E. W. Lee, *Rept. Prog. Phys.*, **18** (1955), 184.
25. G. M. Williams and A. S. Pavlovic, *J. Appl. Phys.*, **39** (1968), 571.
26. E. Tatsumoto and T. Okamoto, *J. Phys. Soc. Japan*, **14** (1959), 1588.
27. W. J. Carr and R. Smoluchowski, *Phys. Rev.*, **83** (1951), 1236.
28. R. C. Hall, *J. Appl. Phys.*, **28** (1957), 707.
29. W. C. Chan, K. Mitsuoka, H. Miyajima, and S. Chikazumi, *J. Phys. Soc. Japan*, **48** (1980), 822.
30. G. N. Benninger and A. S. Pavlovic, *J. Appl. Phys.*, **38** (1957), 1325.
31. M. Yamamoto and T. Nakamichi, *Sci. Rept. Res. Inst. Tohoku Univ.*, **11** (1959), 168.
32. T. Wakiyama and S. Chikazumi, *J. Phys. Soc. Japan*, **15** (1960), 1975.
33. R. M. Bozorth and J. G. Walker, *Phys. Rev.*, **89** (1953), 624.
34. R. M. Bozorth, *Rev. Mod. Phys.*, **25** (1953), 42.
35. M. Yamamoto and T. Nakamichi, *J. Phys. Soc. Japan*, **13** (1958), 228.
36. R. C. Hall, *J. Appl. Phys.*, **30** (1959), 816.
37. J. E. Goldman, *Phys. Rev.*, **80** (1950), 301.
38. Y. N. Kotyukov, *Soviet Phys. Solid State*, **9** (1967), 899.
39. L. R. Bickford, Jr., J. Pappis, and J. L. Stull, *Phys. Rev.*, **99** (1955), 1210.
40. R. M. Bozorth, E. F. Tilden, and A. J. Williams, *Phys. Rev.*, **99** (1955), 1788.
41. K. I. Arai and N. Tsuya, *Proc. Int. Conf. Ferrites* (1970), 51; *J. Appl. Phys.*, **42** (1971), 1673; *Japanese J. Appl. Phys.*, **11** (1972), 1303; *J. Phys. Chem. Solids*, **34** (1973), 431.
42. G. A. Petrakovski, *Soviet Phys. Solid State*, **10** (1968), 2544.
43. N. Miyata and Z. Funatogawa, *J. Phys. Soc. Japan*, **17** (1962), 279.
44. K. Ohta, *J. Appl. Phys.*, **3** (1964), 576.
45. S. Kainuma, K. I. Arai, K. Ishizumi, and N. Tsuya, *Spring Meeting of Phys. Soc. Japan* (1973), 8.
46. A. B. Smith and R. V. Jones, *J. Appl. Phys.*, **34** (1963), 1283; **37** (1966), 1001.
47. G. F. Donne, *J. Appl. Phys.*, **40** (1969), 4486.
48. Y. Syono, *Japanese J. Geophys.*, **4** (1965), **71**.
49. N. Tsuya, *Handbook on magnetic materials*, ed. by S. Chikazumi *et al.* (Asakura Publishing Co., 1975), p. 852, Table 13.9 (in Japanese).
50. S. Iida, *Phys. Lett.*, **6** (1963), 165.
51. W. G. Nilsen, R. L. Comstock, and R. L. Walker, *Phys. Rev.*, **139** (1965), A472.
52. A. E. Clark, J. J. Rhyne, and E. R. Callen, *J. Appl. Phys.*, **39** (1968), 573.
53. T. G. Phillips and R. L. White, *Phys. Rev.*, **153** (1967), 616.

54. V. I. Sokolov and T. D. Hien, *Sov. Phys. JETP*, **25** (1967), 986.
55. A. E. Clark, B. F. DeSavage, N. Tsuya, and S. Kawakami, *J. Appl. Phys.*, **37** (1966), 1324.
56. R. L. Comstock and J. J. Raymond, *J. Appl. Phys.*, **38** (1967), 3737.
57. N. Tsuya, *Handbook on magnetic materials*, ed. by S. Chikazumi *et al.* (Asakura Publishing Co., 1975), p. 855, Table 13.12 (in Japanese).
58. P. Hansen, *Proc. Int. School Phys., Enrico Fermi*, **LXX** (1978), 56.
59. P. Hansen, *J. Appl. Phys.*, **45** (1974), 3638.
60. K. P. Belov, A. K. Gapeev, R. Z. Levitin, A. S. Markosyan, and Yu. F. Popov, *Sov. Phys. JETP*, **41** (1975), 117; Landolt-Börnstein (Springer-Verlag, Berlin, 1978), **III/12a**, Mag. Oxides & Related Comp. Part a, p. 157.
61. S. Chikazumi, *J. Mag. Mag. Mat.*, **15–18** (1980), 1130.
62. R. J. Weiss, *Proc. Phys. Soc.* (London), **82** (1963), 281.
63. M. Matsui and S. Chikazumi, *J. Phys. Soc. Japan*, **45** (1978), 458.
64. H. Hasegawa, *J. Phys.*, **C14** (1981), 2793.
65. R. Becker and W. Döring, *Ferromagnetismus* (Springer-Verlag, Berlin, 1939).
66. M. Kornetzki, *Z. Physik*, **87** (1933), 560.
67. R. Gersdorf, J. H. M. Stoelinga, and G. W. Rathenau, *Physica*, **27** (1961), 381.
68. H. A. Jahn and E. Teller, *Proc. Roy. Soc.* (London), **A161** (1937), 220.
69. R. M. Bozorth, *Ferromagnetism* (Van Nostrand, Princeton, N.J., 1951), p. 131.
70. W. Köster, *Z. Metalk*, **35** (1943), 194.
71. L. Lazan and L. J. Demer, *Proc. ASTM*, **51** (1951), 611.
72. A. W. Cochardt, *J. Appl. Mech.*, **20** (1953), 196.

Part VI
DOMAIN STRUCTURES

Ferro- or ferrimagnetic materials have *domain structures* in which the spontaneous magnetization takes different directions in different domains. In this Part we shall discuss various aspects of domain structure.

15

OBSERVATION OF DOMAIN STRUCTURES

15.1 HISTORY OF DOMAIN OBSERVATIONS AND POWDER-PATTERN METHOD

Weiss pointed out in his famous paper[1] on spontaneous magnetization in 1907 that ferromagnetic materials are not necessarily magnetized to saturation, because the spontaneous magnetization takes different directions in different domains.

In 1919 Barkhausen[2] discovered that the magnetization process in ferromagnetic materials takes place in many small discontinuous steps. This effect is called the *Barkhausen effect*. Barkhausen was a famous vacuum tube engineer, who used his skill to amplify the noise signals which are produced in a search coil wound on a ferromagnetic specimen when the magnetization of the specimen is changed (see Fig. 15.1). He connected a loud speaker to the circuit and heard a roaring noise like the sound of waves on the sea shore. This is called the *Barkhausen noise*. It means that the magnetization of a ferromagnetization specimen takes place in many small discontinuous steps. At the time, it was thought that each step in magnetization corresponds to the flip of a complete domain. From the magnitude of a single step, the volume of a single domain was estimated to be about 10^{-8} cm^3. Later this volume was found to be simply the volume traversed by a domain wall released from some constraint, but people at that time tended to believe that ferromagnetic domains were small enough to be regarded as a mesoscopic feature.

In 1931 Sixtus and Tonks[3] succeeded in producing a large domain in an elastically stretched Permalloy wire. Permalloy is a 21.5% Fe–Ni alloy with a very small magnetocrystalline anisotropy and a positive magnetostriction, so that the axis of tension becomes an easy axis. After removing the domain structure by magnetizing the whole wire to saturation in a long solenoid, C_1 (see Fig. 15.2), they reduced the field to zero and applied a small reverse field. Then they applied a strong local field

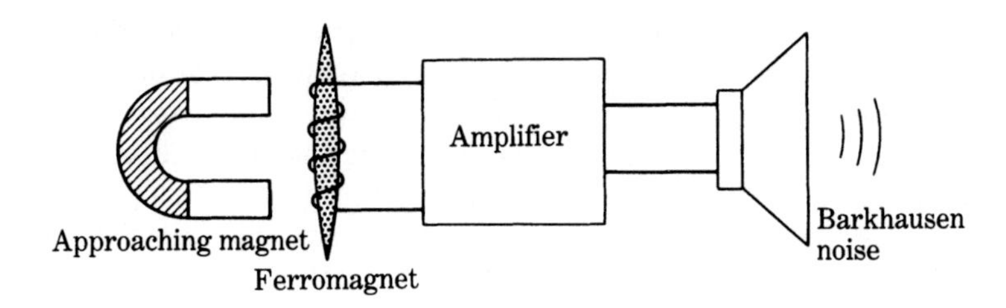

Fig. 15.1. Barkhausen effect.

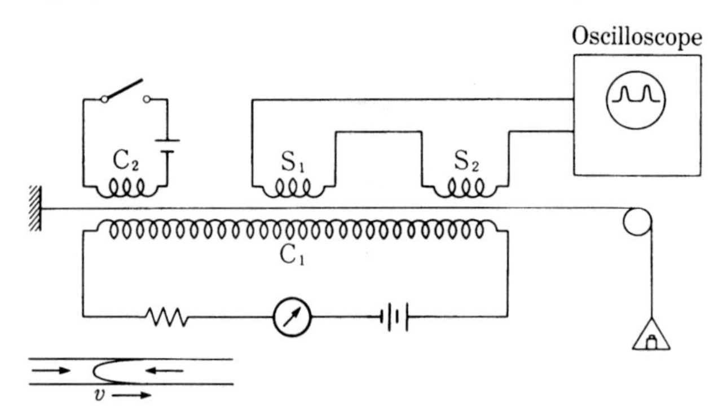

Fig. 15.2. Large Barkhausen effect.

with a small coil, C_2, to reverse the magnetization. They observed that a reverse domain nucleated at C_2, and a single domain wall travelled along the wire. The speed of the wall displacement was determined by measuring the time interval between the signals arising in two separate search coils, S_1 and S_2. This phenomenon can be considered as a *large Barkhausen effect*. Sixtus and Tonks also observed that the shape of the domain wall is concave, as shown in the inset of Fig. 15.2. This is because the magnetization reversal toward the center of the wire is damped or retarded by local eddy currents. This experiment was the first measurement of the speed of domain wall motion.

The first attempt to observe ferromagnetic domains directly under a microscope was made by Bitter[4] in 1931 and independently by Hamos and Thiessen[5] in 1932. The experiment consisted of placing colloidal ferromagnetic particles onto the polished surface of a ferromagnetic crystal and observing the image of domains outlined by the magnetic particles using a reflecting (metallurgical) optical microscope. The idea is similar to the observation of the magnetic lines of force around a permanent magnet by scattering iron filings on a sheet of paper above the magnet. This technique of observing domain structure is called the *powder-pattern method*. Although some micrographs in Bitter's paper revealed domain walls, he hesitated to conclude that these were real images of domain walls. The reason may be that the sizes of the observed domains were too large according to the then-current concept of domain structures. Many investigations were made thereafter using the same technique, but no definite conclusion was drawn for about seventeen years. The main reason may be that the surfaces of the single crystals used were not oriented so that they contained a magnetic easy axis; in this case a complicated layer of surface domains prevents the observation of the underlying simple domain structure. Some improvements were made by Elmore,[6] both in the techniques for preparing the collidal ferromagnetic suspension and for electropolishing the crystal surface. During this period, one thing that was confusing the investigators was the appearance of the *maze domain pattern* as shown in Fig. 15.3(a). One block of the maze pattern is about 10^{-2} to 10^{-3} mm in width, so that the volume of a single block was estimated to be about 10^{-8} cm^3; this is in good agreement with the domain volume estimated from the Barkhausen

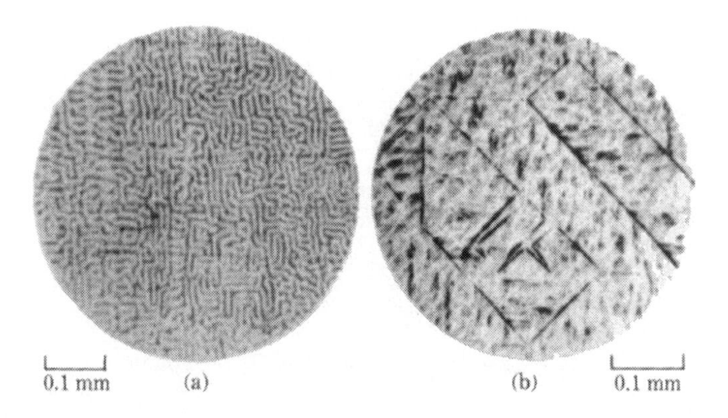

Fig. 15.3. Ferromagnetic domain pattern observed on (001) surface of 4% Si–Fe single crystal: (a) maze surface domain; (b) true domain observed after removal of 28 μm of material from the surface. (After Chikazumi and Suzuki[7])

effect. This agreement seemed to support the 'small block concept' of ferromagnetic domains. However Kaya[8] showed in 1934 that the maze pattern is caused by stresses introduced during polishing the surface. Nevertheless, the true domain size remained unclarified.

A theoretical prediction of domain structures was made by Landau and Lifshitz[9] in 1935; they took into consideration the effect of magnetostatic energy. In 1944 Néel[10] made a detailed calculation of a particular type of domain structure. In 1949 Williams *et al.*[11] succeeded in observing well-defined domain structures by an improved powder-pattern method on a precisely cut, stress-free surface of an Si–Fe single crystal, as shown in Fig. 15.3(b). Quite unlike the maze domain shown in Fig. 15.3(a), the true domains are much larger in size, and more geometrical, bounded by planar domain walls, as predicted by Néel and others. The structure of the maze domain will be described in Section 16.4.

In the following we describe some details of the powder-pattern method. Colloidal particles of Fe_3O_4 are prepared by mixing a solution containing ferrous chloride, $FeCl_2$, and hydrated ferric chloride, $FeCl_3$, with a solution of sodium hydroxide, NaOH (Fig. 15.4). Black magnetite precipitate is deposited as a result of the reactions

$$FeCl_2 + 2FeCl_3 + 8NaOH = Fe(OH)_2 + 2Fe(OH)_3 + 8NaCl$$
$$Fe(OH)_2 + 2Fe(OH)_3 = Fe_3O_4 + 4H_2O.$$

The precipitate particles are removed by filtration, and rinsed repeatedly with distilled water until the wash water shows no trace of Cl^{1-} ions. The particles are then added to a 0.3% solution of a high-quality soap which is free of Cl^{1-} ions, and mixed using a mechanical stirrer or an ultrasonic blender. A high-quality colloidal suspension is reddish-brown and transparent. Similar suspensions, called ferro-fluids, are commercially available.

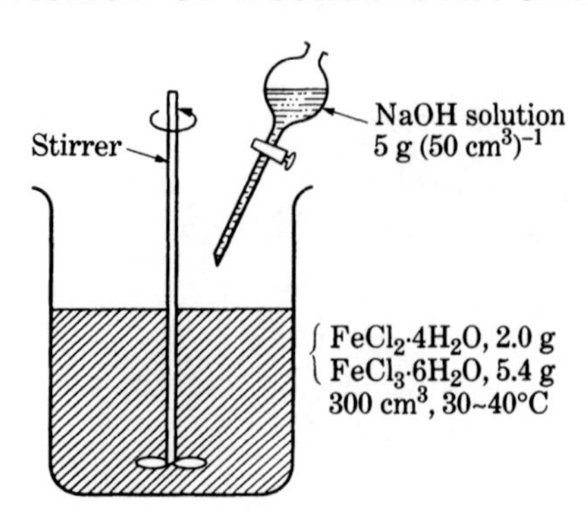

Fig. 15.4. Preparation of Fe_3O_4 precipitate.

In order to observe a well-defined domain pattern, we must cut a crystal along a crystallographic plane that contains at least one of the easy axes. Figure 15.5 shows an example of a complicated domain pattern observed on a tilted crystal plane; many small surface domains appear.

It is also necessary to remove residual stresses introduced during the process of cutting. For this purpose, the best means is electrolytic polishing using high current density. An electrolytic solution is made by mixing phosphoric acid (85%) and solid chromic acid in the ratio 9 to 1 (by weight). The solution is contained in a beaker of 100–600 ml capacity, with a large copper plate serving as the cathode of the electrolytic cell (Fig. 15.6). Fairly heavy currents in the range 10–20 $A\,cm^{-2}$ are used to polish the specimen, which is held in the electrolyte by tweezers or a clamp connected to the anode of the power supply. The area of the cathode should be fairly large compared to that of the specimen, so as to produce a large potential drop near the specimen. Before observation of domain patterns, the electrolyte should be completely removed from the specimen by washing with distilled water.

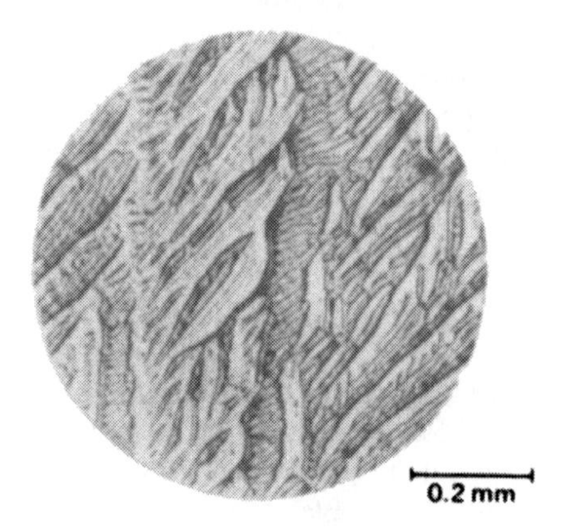

Fig. 15.5. Complicated domain pattern observed on a crystal surface which makes a fairly large angle with an easy axis (4% Si–Fe).

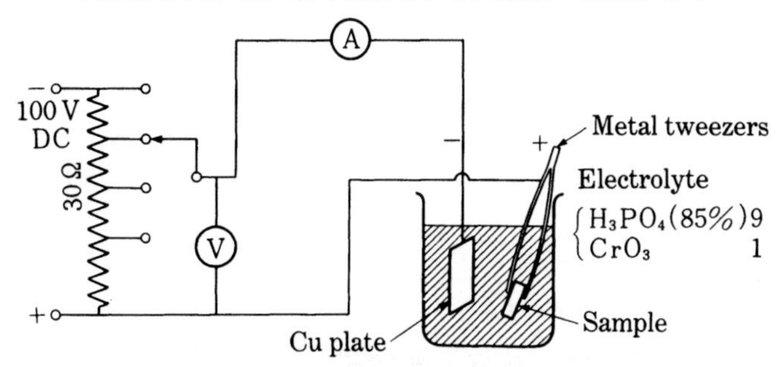

Fig. 15.6. Method of electrolytic polishing.

To observe the domain structure, we use a reflecting or metallurgical microscope, at a magnification of 70–150 times. Sometimes a dark-field illumination system, in which the light strikes the sample surface at a small angle, increases the optical contrast and makes the domain walls more easily visible. A small electromagnet as shown in Fig. 15.7 is convenient for applying a horizontal or a vertical magnetic field to the specimen. First the specimen is placed on the magnet, which is set under the microscope; then a drop of colloidal suspension is applied to the electropolished surface and a thin microscope cover glass is placed on the liquid drop to spread the colloidal suspension uniformly on the crystal surface (Fig. 15.8). The ferromagnetic colloidal particles are attracted to the domain walls where the gradient of the stray field is maximum; thus the domain walls are visible as black lines under normal illumination, or light lines under dark-field illumination. Figure 15.9 shows an example of the domain pattern observed on a (001) surface of a 4% Si–Fe crystal. The black straight lines are the domain walls and the arrows show the direction of spontaneous magnetization in each domain. The most convenient means to determine the axis of magnetization is to observe the striations which appear perpendicular to

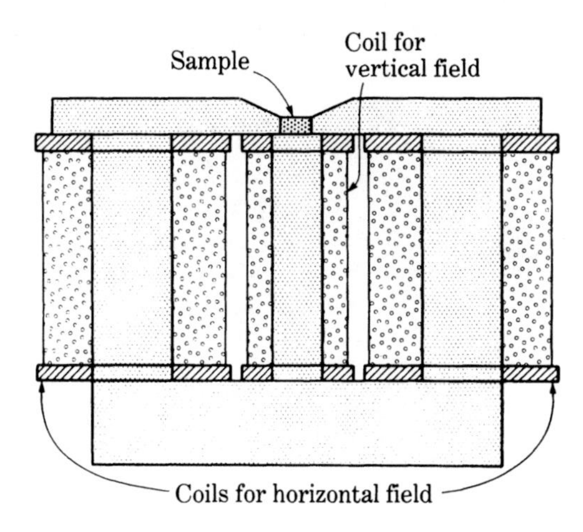

Fig. 15.7. Small electromagnet for the observation of domain patterns.

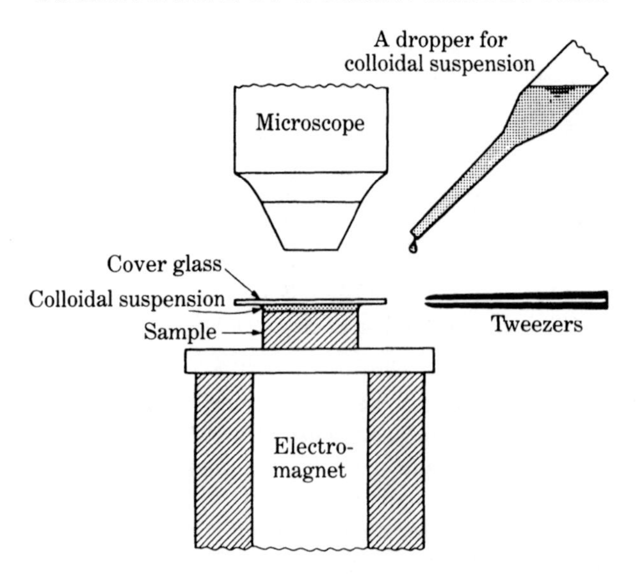

Fig. 15.8. Powder-pattern method.

the domain magnetization. The reason is that magnetic free poles appear at irregularities which are perpendicular to the local magnetization. The irregularities which cause the striations are thought to be non-flatness of the electropolished surface and inhomogeneity in alloy composition. The same thing happens at a groove made deliberately by scratching the electropolished surface with a fine glass fiber. If the groove is perpendicular to the domain magnetization, the magnetic flux emerges from the groove as shown in Fig. 15.10(a), thus collecting the colloidal particles; if the scratch is parallel to the domain magnetization, it induces no free poles, does not attract colloid, and is not visible (Fig. 15.10(b)). This behavior of a scratch is shown schematically in Fig. 15.9.

In order to find the sense of the local magnetization, Williams *et al.*[11] devised a

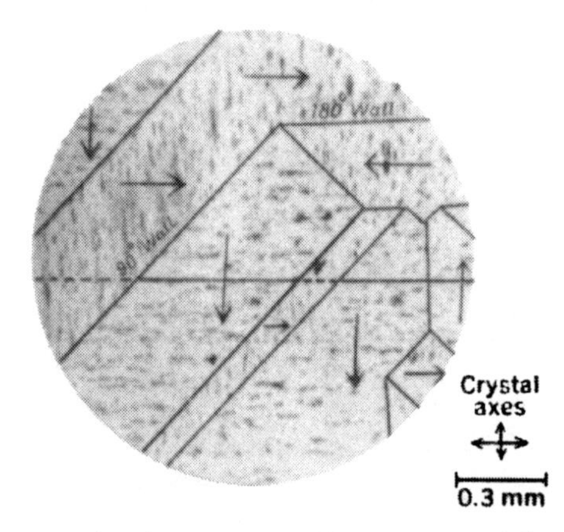

Fig. 15.9. Domain pattern on (001) surface of 4% Si–Fe crystal (retouched). The black line at the center is drawn as a schematic illustration of the appearance of a mechanical scratch.

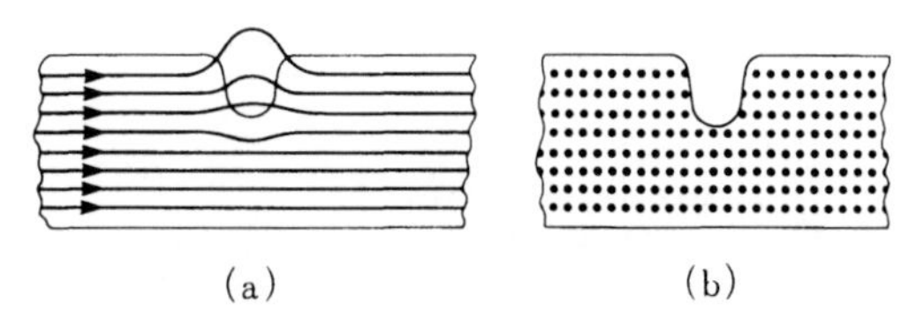

Fig. 15.10. Magnetic flux line around a mechanical scratch: (a) scratch perpendicular to the domain magnetization; (b) scratch parallel to the domain magnetization.

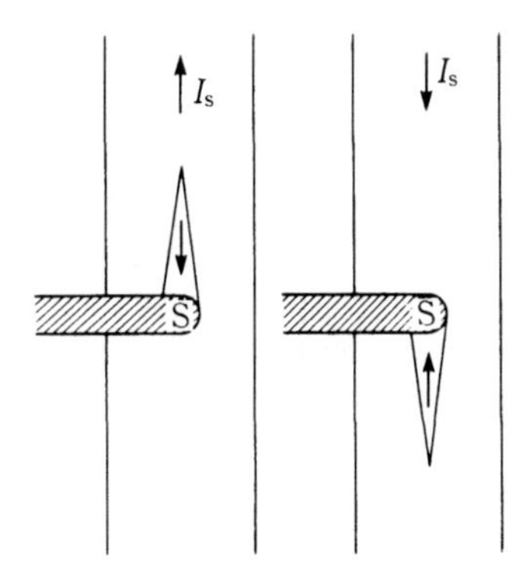

Fig. 15.11. Small spike domain induced by a thin magnet wire, indicating the sense of domain magnetization.

clever technique using a thin permanent magnet needle: if the S pole at the end of the needle approaches a domain surface, the field produced by this S pole induces a spike-like reverse domain which points in the direction of domain magnetization (Fig. 15.11).

The powder pattern technique was adopted for electron microscopy by making a replica of the surface after the powder dried.[12] High-temperature observation using a dry ferromagnetic powder[13,14] and low-temperature observation using solid oxygen powder[15] (oxygen is strongly paramagnetic) have also been reported.

15.2 MAGNETO-OPTICAL METHOD

Magneto-optical effects such as the magnetic Kerr effect and the Faraday effect can be used for observing magnetic domain structures. These effects do not make use of a colloidal magnetic suspension, so that they can be used at any temperature.

The magnetic Kerr effect is the rotation of the plane of polarization of light on reflection from the surface of a magnetized material. Figure 15.12 illustrates a domain observation system using the Kerr effect. Figure 15.12(a) shows a polarizing microscope as used for the observation of magnetic domain structures. The light from the lamp is polarized by a polarizer, reflected by a half-silvered mirror, and projected onto the surface of the specimen. The light reflected by the specimen reaches the observer after going through the half-silvered mirror, analyzer, and eyepiece. Figure 15.12(b)

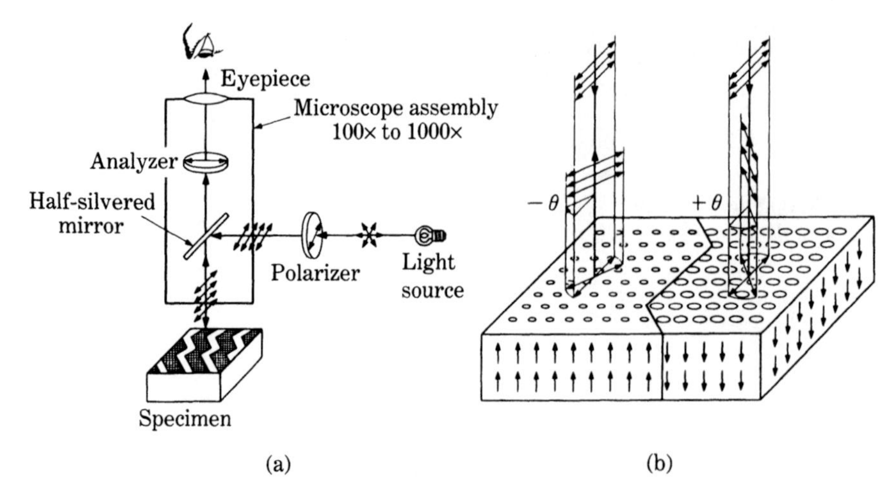

Fig. 15.12. (a) Optical microscope for the observation of domain structures by means of the magnetic Kerr effect and (b) rotation of polarization axis upon reflection of polarized light from the surface of domains with magnetizations perpendicular to the surface.

shows rotation of the polarization axis in opposite senses on reflection from domains magnetized in opposite directions, for the case of magnetization perpendicular to the surface. If the analyzer is adjusted for maximum transmission of light reflected from domains of one sign, domains of the other sign will appear darker. Alternatively, and for stronger contrast, the analyzer can be adjusted for minimum transmission from one set of domains, so that the other set appears relatively light. Figure 15.13 shows the domain structures observed by this device on the *c*-plane of MnBi. As explained in Section 12.4.1(e) MnBi has its easy axis parallel to the *c*-axis, and its magnetocrystalline anisotropy is enormously large, so that the magnetization can lie perpendicular to the *c*-plane, in spite of a large demagnetizing field. Figures 15.13(a)–(c) show the domains observed for specimens of different thickness; the size and shape of domains vary with sample thickness. We shall see the reason in Chapter 17.

If the magnetization lies parallel to the surface, the arrangement shown in Fig. 15.12(a) will not reveal the domain structure. In this case we must use a polarized incident beam making a small incident angle with the surface, so that the magnetization has a nonzero component parallel to the beam. In this apparatus the polarizer and the analyzer must be arranged so as to satisfy the law of reflection.

The Faraday effect is the phenomenon by which the polarization axis rotates during the propagation of light through the material. Domain structures in transparent ferromagnetic materials such as iron garnets can be observed by this method. The construction of the apparatus is the same as that for the Kerr effect, except that the incident polarized light passes through the specimen as in a biological microscope.

15.3 LORENTZ ELECTRON MICROSCOPY

In magnetic thin films which are thin enough to allow the transmission of an electron

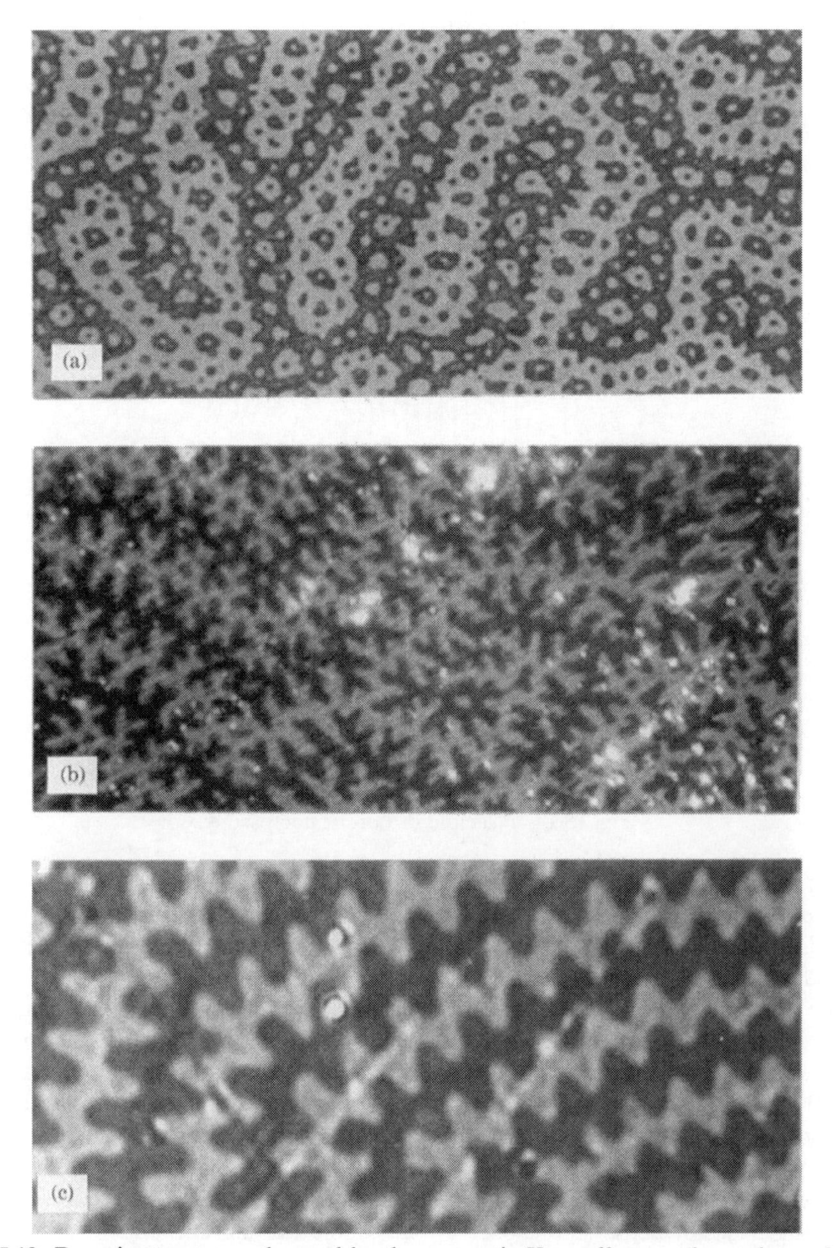

Fig. 15.13. Domain structures observed by the magnetic Kerr effect on the c-plane of MnBi plates: (a) the thickest plate; (b) intermediate; (c) the thinnest. (Due to Roberts and Bean[16])

beam, the domain structure can be observed using the electron microscope. The principle is to use the deflection of the electron beam by the Lorentz force which acts on the moving electrons because of the spontaneous magnetization.[17] If the objective electron lens is defocused slightly from the specimen film, the domain walls appear as

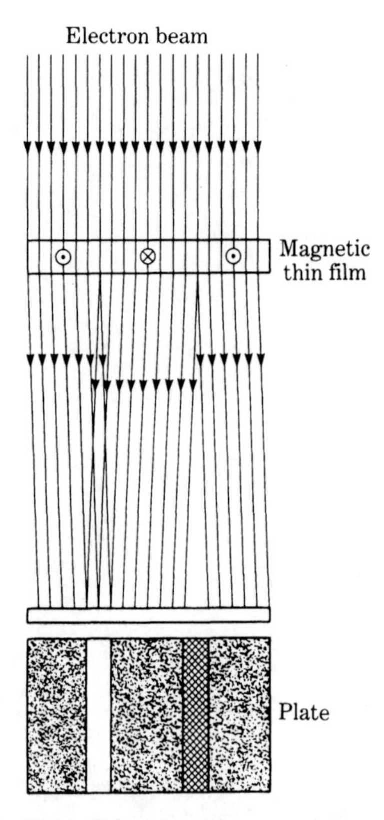

Fig. 15.14. Principle of Lorenz microscopy.

back or white lines, as illustrated in Fig. 15.14. This method is called *Lorentz microscopy*. Figure 15.15 shows an example of domain structures observed in a Permalloy thin film of thickness 600 Å.[18] The spin structures of the cross-ties perpendicular to the main domain walls will be explained in Section 16.5.

If the objective lens is focused at infinity, a parallel electron beam focuses to a spot (a diffraction spot), from which we can deduce the direction of magnetization. Figure 15.16 shows a Lorentz micrograph observed for a single crystal film of iron parallel to the (001) plane.[19] We see many domain walls appearing as black and white lines. The inset shows the diffraction pattern, from which we know that the magnetization lies in four equivalent crystallographic directions.

The strong point of this method is its high magnification, but because of the defocusing it is not possible to observe well-defined crystalline and magnetic structures at the same time. However, the method known as *transmission Lorentz* scanning electron microscopy (SEM) overcomes this difficulty, as will be described in Section 15.4.

15.4 SCANNING ELECTRON MICROSCOPY

Scanning electron microscopy (SEM) enables the observation of domain structures in

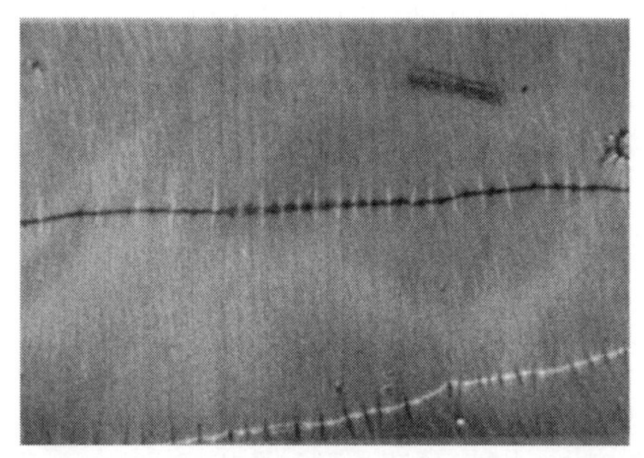

Fig. 15.15. Domain pattern observed in a Permalloy film of thickness 600 Å.[18]

thick specimens by using the deflection of electron beams. As shown in Fig. 15.17, when electrons penetrate into a tilted ferromagnetic specimen, they are deflected by the Lorentz force in directions which are different in different domains. Some of the domains deflect the electrons towards the surface of the specimen, while other domains deflect the electrons more deeply into the specimen. The electron beam is scanned over the specimen surface and the deflected electrons are collected by the electron detector. The electron beam in the display tube is synchronized with the scanning electron beam in the electron microscope, and its intensity is modulated by the number of collected electrons. Then the domain structure is reproduced as a black and white pattern as shown in Fig. 15.18. This method is called *reflection Lorentz SEM*.

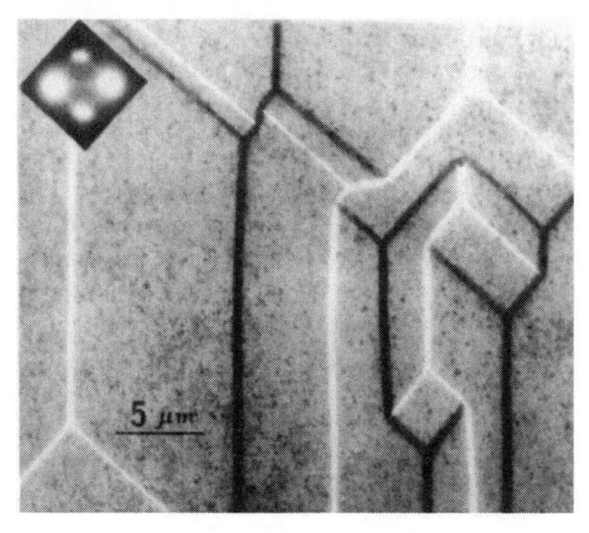

Fig. 15.16. Domain pattern observed in a (001) iron single crystal film of thickness 2500 Å, and its diffraction pattern (inset).[19]

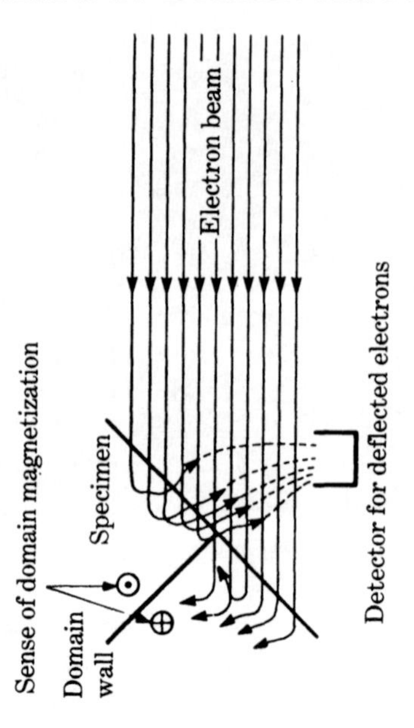

Fig. 15.17. The principle of the observation of domain walls in a thick specimen using scanning electron microscopy.

By a refinement of this method it is possible to observe the dynamic behavior of domain walls. An AC magnetic field is applied to the sample at a frequency which is a small multiple of the scanning electron beam sweep frequency, and the domain walls appear as sinusoidal waves as shown in Fig. 15.19. From such a wave form we can, if the wall motion is sufficiently repeatable, see the nucleation and motion of the domain walls.

Instead of reflecting electrons from a bulk specimen, we can utilize the Lorentz deflection[22] from a magnetic thin film (see Section 15.3). In this method, an electron

Fig. 15.18. Domain pattern observed on a Si–Fe plate.[20]

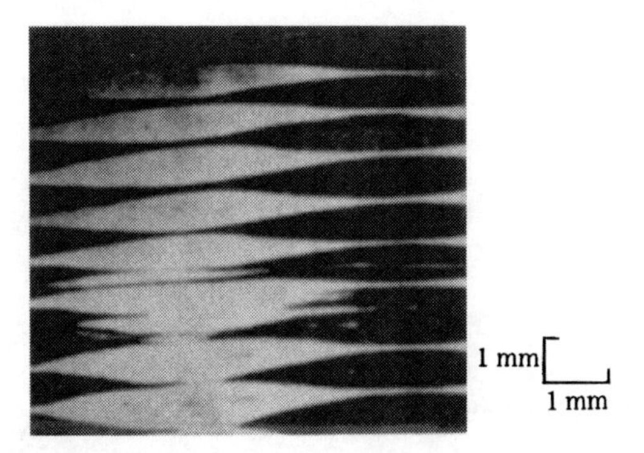

Fig. 15.19. Dynamical behavior of the domain wall observed by the application of an AC magnetic field to a Si–Fe plate.[21]

beam is scanned over the specimen film; the brightness of the electron beam in the display tube is modulated according to the diffraction angle to make the domain pattern visible. As compared with the usual Lorentz microscopy, in which the electron beam is slightly defocused from the film, the strong point of this method is that the electron beam can be sharply focused on the film. Thus well-defined structural and micro-magnetic structures can be observed on the same pattern.[23] We call this method *transmission Lorentz SEM.*

Another method for observing domain patterns on bulk specimens is *spin-polarized SEM*. In this method the electron beam is scanned over the surface of a ferromagnetic specimen and the polarization of the scattered secondary electrons from domains, the sense of which is different for different domains, is detected by a Mott detector.[24] Figure 15.20 illustrates the principle of the Mott detector. The secondary electrons are collected and accelerated towards a thin-film target made of gold, under a potential gradient that produces an electron wave with a wavelength of the order of the atomic radius of gold. As a result of interference scattering, the cross-section of the scattering beam depends strongly on the scattering angle at some specific

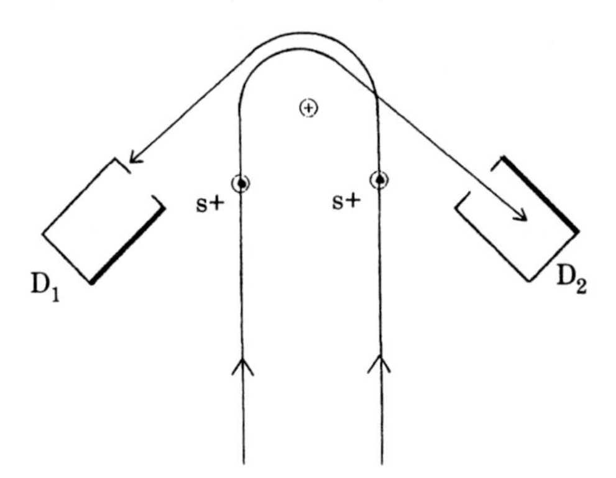

Fig. 15.20. The principle of the Mott detector.

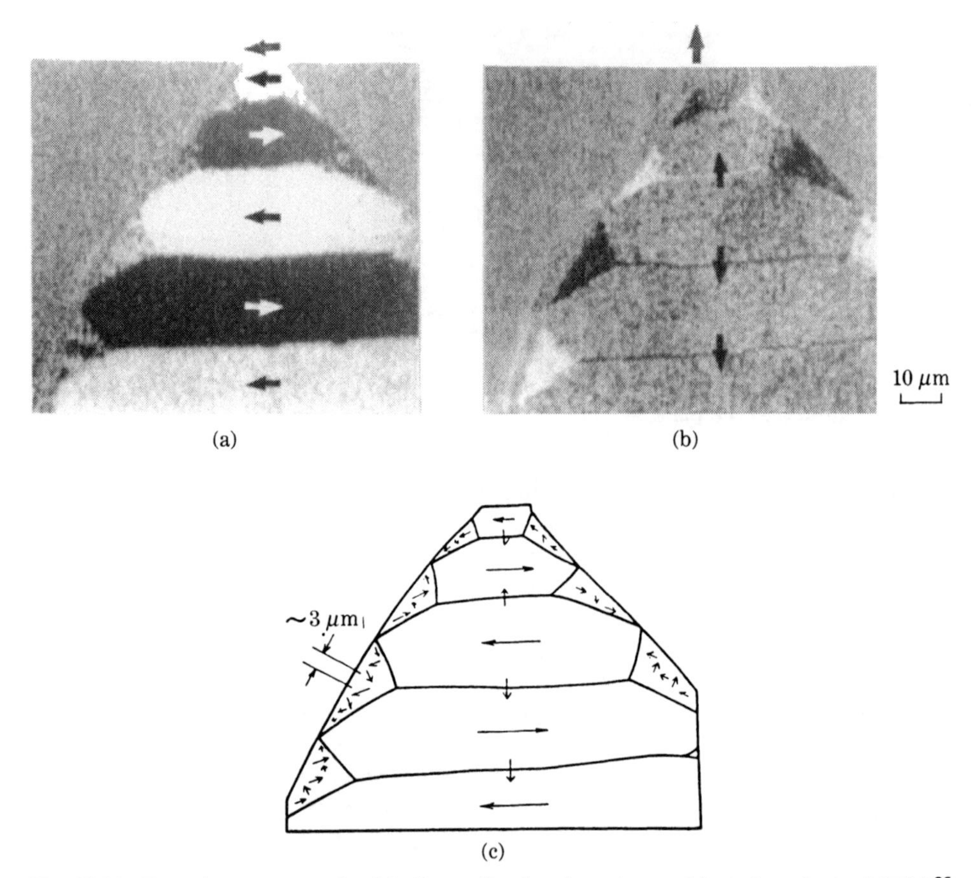

Fig. 15.21. Domain structure of a thin Permalloy head as observed by spin-polarized SEM.[25]

scattering angle. Two electron detectors, D_1 and D_2, are set at this angle symmetrically on both sides of the direction of acceleration. Suppose that the spin angular moment, s, of the electron points upwards perpendicular to the paper (denoted as s_+ in the figure). Its magnetic moment tends to depress the clockwise orbital motion about the nucleus because of the spin–orbit interaction (see Section 3.2), so that the orbital radius becomes smaller than that of the anticlockwise orbital motion of the electron with the same spin. Therefore the unbalanced signal between D_1 and D_2 becomes proportional to the spin polarization. For electrons with opposite spin, the sign of the unbalanced signal is opposite.

Figure 15.21 shows the domain structure of a Permalloy thin-film head used for high-density magnetic recording, as observed by *spin-polarized SEM*. Picture (a) was taken with the Mott detector adjusted to detect the horizontal spin directions: the black and white zones indicate that the spin directions point right and left, respectively. Picture (b) was taken with the detector adjusted to detect the vertical spin directions: not only are spin directions in the closure domains distinguishable, but

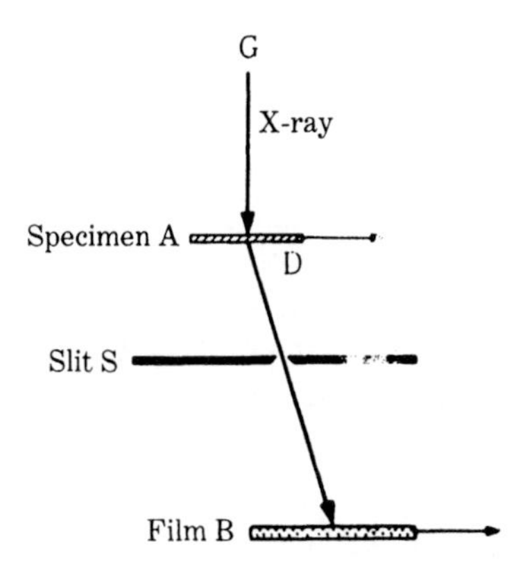

Fig. 15.22. Principle of X-ray topography.

also we can see the spins inside the 180 degree walls. Figure (c) illustrates the spin map as deduced from these experiments.

15.5 X-RAY TOPOGRAPHY

X-ray topography was invented by Lang[26] to investigate the distribution of internal strains in a crystal. The principle is to utilize the change in the diffraction angle of the diffracted X-rays due to a strain in the crystal. As shown in Fig. 15.22, the X-rays

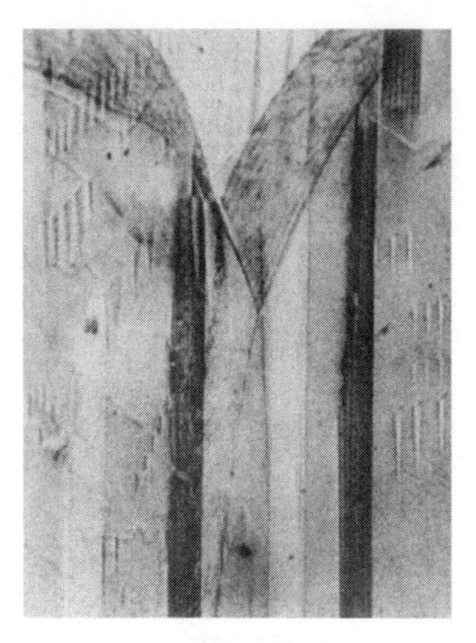

Fig. 15.23. Domain pattern observed by means of X-ray topography on a Si–Fe (001) plate 120 μm thick.[27,28]

diffracted from the specimen A go through the slit S and strike the film B unless the crystal lattice is distorted. If the crystal is distorted, the diffraction angle is changed, so that the diffracted X-rays are blocked by the slit S and cannot reach the film B. If the specimen A and the film B are displaced at the same velocity, the distribution of the internal strain in the specimen appears as a black and white pattern in the film B. Figure 15.23 shows the domain pattern observed by this method for a (001) single crystal plate of Si–Fe, 120 μm thick. This is not a magnetic image, but a strain image caused by magnetostriction. The strong point of this method is that it allows the observation of the interior domain structure.

15.6 ELECTRON HOLOGRAPHY

Electron holography enables the direct observation of lines of magnetic flux not only in a magnetic thin film but also outside the film.[29] The electron microscope used for this purpose is provided with an electron biprism which overlaps a reference beam with the electron beam going through the specimen to produce an electron hologram, as shown in Fig. 15.24. The actual construction of the electron biprism consists of two earthed parallel electrodes with a positively charged thin gold wire inserted between them.[30] The electrons tend to be attracted towards the gold wire from both the electrodes, thus being deflected in the same way as a beam of light passing through an optical biprism. In order to produce a good coherent electron beam, this electron

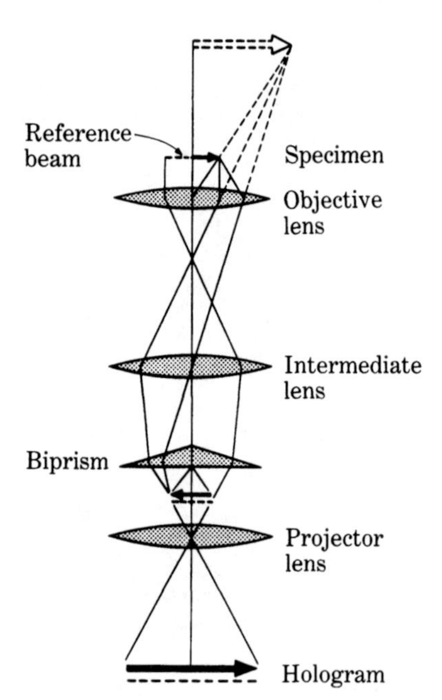

Fig. 15.24. Schematic diagram of electron-hologram formation.[29]

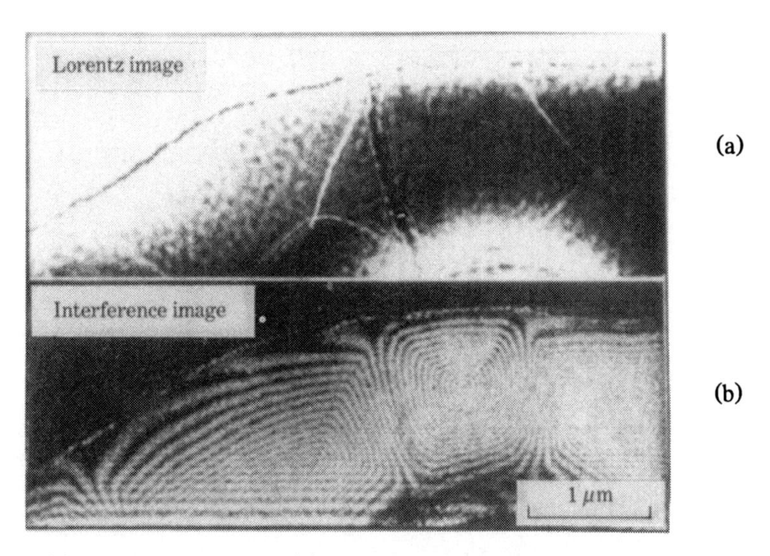

Fig. 15.25. (a) Lorentz image and (b) electron interference fringes as observed by electron holography for an amorphous film of $(Co_{0.94}Fe_{0.06})_{79}Si_{10}B_{11}$.[31]

microscope uses field emission from a sharp cold tungsten cathode. The hologram exposed on the photographic plate is developed and transferred to an optical system to reproduce a real image of the specimen. A laser beam with the same wavelength as that of the electron beam used to produce the hologram is passed through the hologram plate, and after overlapping with the reference optical beam reproduces a real image of the specimen.

Figure 15.25(a) shows a Lorentz micrograph produced in this way from an amorphous film of composition $(Co_{0.94}Fe_{0.06})_{79}Si_{10}B_{11}$. Several domain walls can be observed.

The holograph contains information not only of the real image but also of the phase difference of the electron waves. Therefore when the reference optical beam is directed almost parallel to the beam passed through the hologram, interference fringes as seen in (b) are formed. These fringes run parallel to the magnetization in the film. The reason for this will be explained below. The magnetic structure in these photographs will be discussed in Chapter 16.

According to quantum mechanics, when an electron passes through a space where the vector potential $\boldsymbol{A}$ exists, the electron wave undergoes a phase shift given by

$$\frac{S}{\hbar} = -\frac{e}{\hbar}\int \boldsymbol{A}\cdot \mathrm{d}\boldsymbol{s}, \tag{15.1}$$

where $\hbar$ is Planck's constant divided by 2π, and the integration is along the electron path.[32]

Suppose that the electron waves P and Q in Fig. 15.26, which have the same phase

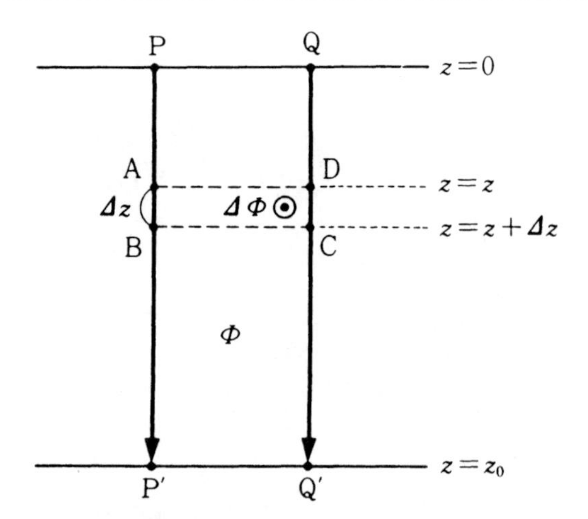

Fig. 15.26. Illustration of phase difference between two electron waves.

at $z=0$, travel through the magnetic field and reach P′ and Q′ at $z=z_0$. From (15.1), the phase difference between the two electron beams at $z=z_0$ is given by

$$\frac{S}{\hbar}=-\frac{e}{\hbar}\int_{\mathrm{P}}^{\mathrm{P'}}\boldsymbol{A}\cdot\mathrm{d}\boldsymbol{s}+\frac{e}{\hbar}\int_{\mathrm{Q}}^{\mathrm{Q'}}\boldsymbol{A}\cdot\mathrm{d}\boldsymbol{s}=\frac{e}{\hbar}\oint_{\mathrm{QQ'P'P}}\boldsymbol{A}\cdot\mathrm{d}\boldsymbol{s}=\frac{e}{\hbar}\Phi,^{*} \tag{15.2}$$

where Φ is the magnetic flux penetrating the area PP′Q′Q.

If this phase difference were 2π, we see by putting the final term in (15.2) equal to 2π,

$$\Phi=\frac{h}{e}=4.1357\times10^{-15}\ \mathrm{Wb}, \tag{15.3}$$

which is just twice the flux quantum $h/2e$. In other words, if neighboring interference fringes correspond to P′ and Q′, Φ in (15.3) must be twice the flux quantum.

Suppose that in Fig. 15.27 two electron waves PP′ and QQ′ passing through a magnetic thin film of thickness d form two neighboring interference fringes. If the saturation flux density B_s in the film is the only magnetic flux penetrating the area PP′Q′Q, we have

$$\Phi=\frac{h}{e}=ldB_s, \tag{15.4}$$

where l is the separation between the two fringes. From (15.4), we have

$$l=\frac{h}{edB_s}. \tag{15.5}$$

* Since $\boldsymbol{B}=\operatorname{curl}\boldsymbol{A}$, the integration in (15.1) along ABCD is given by

$$\oint\boldsymbol{A}\cdot\mathrm{d}\boldsymbol{s}=\iint\operatorname{curl}_n\boldsymbol{A}\,\mathrm{d}S=\iint B_n\,\mathrm{d}S=\Delta\Phi,$$

where $\Delta\Phi$ is the magnetic flux penetrating the area ABCD.

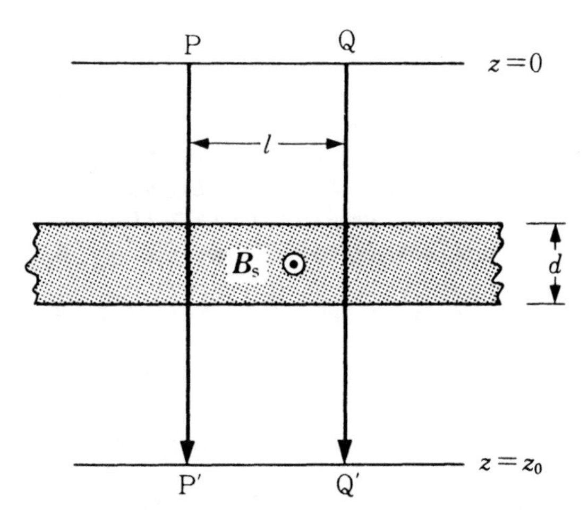

Fig. 15.27. Magnetic flux in a magnetic thin film.

Therefore it the thickness of the film, d, is uniform, the fringes run parallel to the magnetization, and the separation of the fringes is inversely proportional to B_s. However, if there is leakage flux in the area PP′Q′Q, (15.5) does not hold. For instance, if the flux caused by the magnetization in the film leaks into space and comes back through the area PP′Q′Q, Φ in (15.4) decreases, so that the separation l becomes larger than (15.5).

On the other hand, when we observe the flux in space, the interference fringes run along the lines of flux density. This method is useful for finding the location of free magnetic poles, which exist where the lines of flux emerge.

REFERENCES

1. P. Weiss, *J. Phys.*, **6** (1907), 661.
2. H. Barkhausen, *Phys. Z.*, **20** (1919), 401.
3. K. J. Sixtus and L. Tonks, *Phys. Rev.*, **37** (1931), 930; **39** (1932), 357; **42** (1932), 419; **43** (1933), 70, 931.
4. F. Bitter, *Phys. Rev.*, **38** (1931), 1903; **410** (1932), 507.
5. L. V. Hamos and P. A. Thiessen, *Z. Phys.*, **71** (1932), 442.
6. W. C. Elmore, *Phys. Rev.*, **51** (1937), 982; **53** (1938), 757; **54** (1938), 309; **62** (1942), 486.
7. S. Chikazumi and K. Suzuki, *J. Phys. Soc. Japan*, **10** (1955), 523; *IEEE Trans. Mag.*, **MAG-15** (1979), 1291.
8. S. Kaya, *Z. f. Phys.*, **89** (1934), 796; **90** (1934), 551; S. Kaya and J. Sekiya, *Z. f. Phys.*, **96** (1935), 53.
9. L. Landau and E. Lifshitz, *Phys. Z. Sowjet U.*, **8** (1935), 153; E. Lifshitz, *J. Phys. USSR*, **8** (1944), 337.
10. L. Néel, *J. Phys. Rad.*, **5** (1944), 241. 265.
11. H. J. Williams, R. M. Bozorth, and W. Shockley, *Phys. Rev.*, **75** (1949), 155.
12. D. J. Craik and P. M. Griffiths, *British J. Appl. Phys.*, **9** (1958), 279.

13. W. Andrä, *Ann. Phys.* [7] **3** (1959), 334.
14. W. Andrä and E. Schwabe, *Ann. Physik*, **17** (1955), 55.
15. K. Piotrowski, A. Szewczyk, and Szymczak, *J. Mag. Mag. Mat.*, **31-34**, Part II (1983), 979.
16. B. W. Roberts and C. P. Bean, *Phys. Rev.*, **96** (1954), 1494.
17. M. E. Hale, H. W. Fuller, and H. Rubinstein, *J. Appl. Phys.*, **30** (1959), 789.
18. T. Ichinokawa, *Mem. Sci. Eng. Waseda Univ.*, **25** (1961), 80.
19. S. Tsukahara and H. Kawakatsu, *J. Phys. Soc. Japan*, **32** (1972), 72.
20. T. Nozawa, T. Yamamoto, Y. Matsuo, and Y. Ohya, *IEEE Trans. Mag.*, **MAG-15** (1979), 972.
21. T. Nozawa, T. Yamamoto, Y. Matsuo, and Y. Ohya, *IEEE Trans. Mag.*, **MAG-14** (1978), 252.
22. J. N. Chapman, P. E. Batson, E. M. Waddel, and R. P. Ferrier, *Ultramicroscopy*, **3** (1978), 203.
23. Y. Yajima, Y. Takahashi, M. Takeshita, T. Kobayashi, M. Ichikawa, Y. Hosoe, Y. Shiroishi, and Y. Sugita, *J. Appl. Phys.*, **73** (1993), 5811.
24. For instance, J. Kessler, *Polarized electrons* (2nd edn.), (Springer-Verlag, 1985).
25. K. Mitsuoka, S. Sudo, N. Narishige, M. Hanazono, Y. Sugita, K. Koike, H. Matsuyama, and K. Hayakawa, *IEEE Trans. Mag.*, **MAG-23** (1987), 2155.
26. A. R. Lang, *Acta Cryst.*, **12** (1959), 249.
27. M. Polcarova and A. R. Lang, *Phys. Lett.*, **1** (1962), 13.
28. M. Polcarova, *IEEE Trans. Mag.*, **MAG-5** (1969), 536.
29. A. Tonomura, T. Matsuda, H. Tanabe, N. Osakabe, I. Endo, A. Fukahara, K. Shinagawa, and H. Fujiwara, *Phys. Rev.*, **B25** (1982), 6799.
30. G. Mollenstedt and H. Ducker, *Z. Phys.*, **145** (1956), 377.
31. S. Takayama, T. Matsuda, A. Tonomura, N. Osakabe, and H. Fujiwara, *1982 TMS-AIME Fall Meeting Abs.* (St. Louis) (1982), 66.
32. A. Aharonov and D. Bohm, *Phys. Rev.*, **115** (1959), 485.

16

SPIN DISTRIBUTION AND DOMAIN WALLS

16.1 MICROMAGNETICS

Micromagnetics is the name coined by Brown[1,2] for the procedure by which the distribution of spins in a ferromagnet of finite size is solved from first principles. In an infinitely long, thin ferromagnet, the spontaneous magnetization is aligned parallel to the long axis, thus forming a single domain. In a ferromagnet of finite size, such a single domain produces surface free poles which give rise to *magnetostatic energy*, U_{mag}. In order to reduce U_{mag}, the spin distribution must be altered, which modifies the complete parallel spin alignment. As a result, *exchange energy*, U_{ex}, *magnetocrystalline energy*, U_a, or *magnetoelastic energy*, U_λ, are increased. The stable spin distribution is determined by minimizing the total energy

$$U = U_{mag} + U_{ex} + U_a + U_\lambda. \tag{16.1}$$

Consider a ferromagnetic disk of radius r and thickness d. If this disk is uniformly magnetized to saturation along a diameter, as shown in Fig. 16.1, free magnetic poles N and S appear at the edges, and the demagnetizing field NI_s/μ_0 appears opposite to the spontaneous magnetization I_s, so that there is a magnetostatic energy (1.99) given by

$$U_{mag} = \frac{1}{2\mu_0} NI_s^2 v, \tag{16.2}$$

where v is the volume of the disk given by

$$v = \pi r^2 d. \tag{16.3}$$

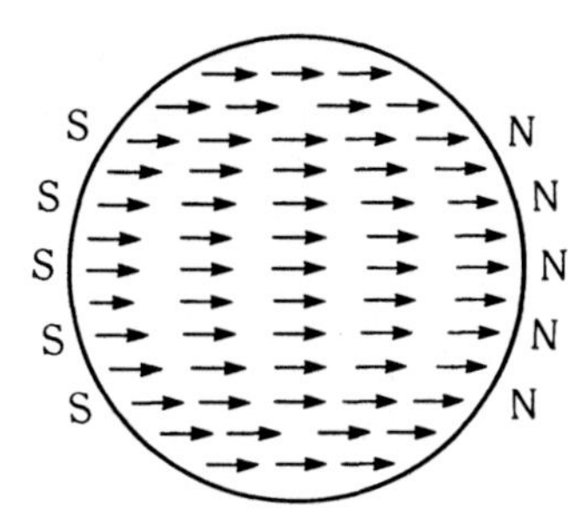

Fig. 16.1. Uniformly magnetized disk (single domain structure).

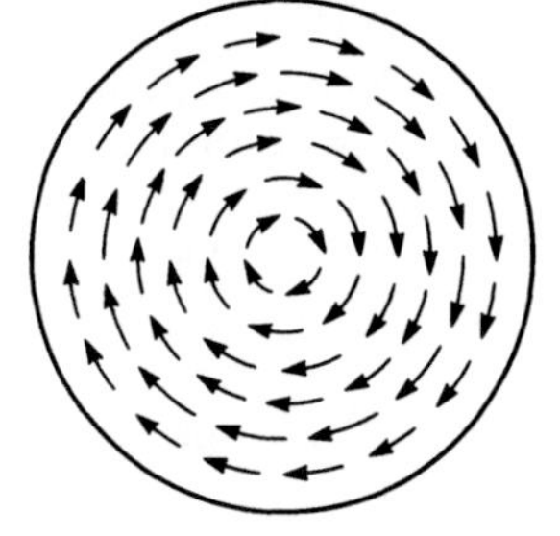

Fig. 16.2. Circularly magnetized disk (no free poles).

The demagnetizing factor N can be approximated by that of a thin oblate spheroid, and can be calculated using equation (1.44) or found from Table 1.1. A general discussion of magnetostatic energy is given in Section 1.5.

One possible spin configuration that eliminates magnetostatic energy is the circular configuration shown in Fig. 16.2. There is no divergence of magnetization and therefore no free poles. In other words, $U_{\text{mag}} = 0$. Instead, neighboring spins make some non-zero angle, so that some exchange energy is stored. As discussed in Section 6.2, the exchange energy stored in a pair of spins $\boldsymbol{S}_i$ and $\boldsymbol{S}_j$ is given by

$$w_{ij} = -2J\boldsymbol{S}_i \cdot \boldsymbol{S}_j, \tag{16.4}$$

where J is the exchange integral. In a ferromagnet, $J > 0$. The value of J is related to the Curie point as listed in Table 6.1 for various statistical approximations. According to a detailed calculation by Weiss,[3] the relationship between J and the Curie point is given by

$$\begin{aligned} J &= 0.54k\Theta_f \quad \text{for a simple cubic lattice } (S = \tfrac{1}{2}), \\ J &= 0.34k\Theta_f \quad \text{for a body-centered cubic lattice } (S = \tfrac{1}{2}), \\ J &= 0.15k\Theta_f \quad \text{for a body-centered cubic lattice } (S = 1). \end{aligned} \tag{16.5}$$

For instance, if we assume $S = 1$ for iron, it follows that

$$\begin{aligned} J &= 0.15 \times 1043 \times 1.38 \times 10^{-23} \\ &= 2.16 \times 10^{-21}\ \text{J}. \end{aligned} \tag{16.6}$$

When the angle φ between spins $\boldsymbol{S}_i$ and $\boldsymbol{S}_j$ is small, the exchange pair energy given by (16.4) reduces to

$$w = -2JS^2 \cos\varphi \simeq JS^2\varphi^2 + \text{const.}, \tag{16.7}$$

where S is the magnitude of the spin. This formula shows that the exchange increases as the square of the angle φ. This is analogous to the elastic energy, which increases as the square of the strain in the lattice.

The elastic energy density can be expressed in terms of the strain tensor as shown in (14.27). Similarly the exchange energy density can be expressed in terms of the spin distribution in the lattice.[4] Let a unit vector parallel to the spin $\boldsymbol{S}_i$ at P be $\boldsymbol{\alpha}$, and that of $\boldsymbol{S}_j$ at Q, which is separated from $\boldsymbol{S}_i$ by the distance r_j, be $\boldsymbol{\alpha}'$. If we assume the variation of $\boldsymbol{\alpha}$ is smooth and continuous, $\boldsymbol{\alpha}'$ can be expanded as

$$\boldsymbol{\alpha}' = \boldsymbol{\alpha} + \frac{\partial \boldsymbol{\alpha}}{\partial x} x_j + \frac{\partial \boldsymbol{\alpha}}{\partial y} y_j + \frac{\partial \boldsymbol{\alpha}}{\partial z} z_j + \frac{1}{2}\left(\frac{\partial^2 \boldsymbol{\alpha}}{\partial x^2} x_j^2 + \frac{\partial^2 \boldsymbol{\alpha}}{\partial y^2} y_j^2 + \frac{\partial^2 \boldsymbol{\alpha}}{\partial z^2} z_j^2 \right) + \cdots, \tag{16.8}$$

where (x_j, y_j, z_j) are the components of the distance r_j. Since $\cos\varphi$ in (16.7) can be expressed in terms of $\boldsymbol{\alpha}$ and $\boldsymbol{\alpha}'$ as $\cos\varphi = \boldsymbol{\alpha} \cdot \boldsymbol{\alpha}'$, the exchange pair energy (16.7) can be expressed in terms of $\boldsymbol{\alpha}$, using the relationship (16.8). Let us calculate the sum of the exchange pair energies between $\boldsymbol{S}_i$ and its z nearest neighbor $\boldsymbol{S}_j$s. In a cubic lattice, in addition to the spin $\boldsymbol{S}_j$ at Q, we find another spin at Q′ which is located $-r$

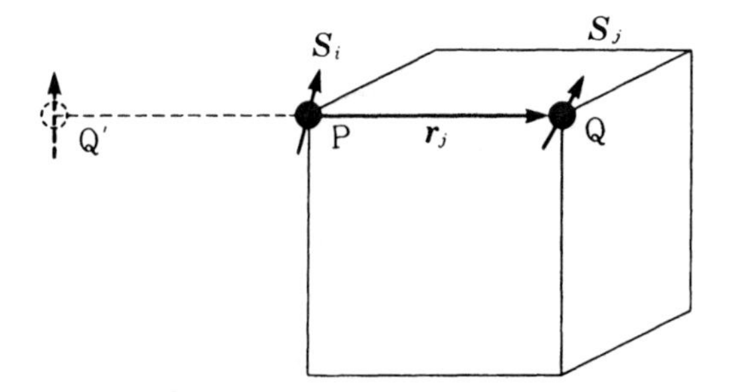

Fig. 16.3. Calculation of exchange energy in a given spin distribution.

from P (see Fig. 16.3), so that the summations of the second terms in (16.8) over Q and Q′ vanish. Therefore the summation of the exchange pair energy over the nearest z neighbors is given by

$$\sum_{j=1}^{z} w_{ij} = -JS^2 \sum_{j=1}^{z} \left(\boldsymbol{\alpha} \cdot \frac{\partial^2 \boldsymbol{\alpha}}{\partial x^2} x_j^2 + \boldsymbol{\alpha} \cdot \frac{\partial^2 \boldsymbol{\alpha}}{\partial y^2} y_j^2 + \boldsymbol{\alpha} \cdot \frac{\partial^2 \boldsymbol{\alpha}}{\partial z^2} z_j^2 \right). \tag{16.9}$$

In a cubic lattice

$$\sum_{j=1}^{z} x_j^2 = \sum_{j=1}^{z} y_j^2 = \sum_{j=1}^{z} z_j^2, \tag{16.10}$$

which gives the value $2a^2$ for the simple cubic, body-centered cubic, and the face-centered cubic lattices, where a is the lattice constant or the length of the edge of a unit cell. The number of atoms in a unit cell, n, is given by

$$\left. \begin{aligned} n &= 1 \quad \text{for a simple cubic lattice} \\ &= 2 \quad \text{for a body-centered cubic lattice} \\ &= 4 \quad \text{for a face-centered cubic lattice} \end{aligned} \right\}. \tag{16.11}$$

Summing (16.9) over all the atomic pairs in a unit volume, the exchange energy density becomes

$$E_{\text{ex}} = -\frac{nJS^2}{a} \left(\boldsymbol{\alpha} \cdot \frac{\partial^2 \boldsymbol{\alpha}}{\partial x^2} + \boldsymbol{\alpha} \cdot \frac{\partial^2 \boldsymbol{\alpha}}{\partial y^2} + \boldsymbol{\alpha} \cdot \frac{\partial^2 \boldsymbol{\alpha}}{\partial z^2} \right). \tag{16.12}$$

On the other hand, differentiating $(\boldsymbol{\alpha} \cdot \boldsymbol{\alpha}) = 1$ twice with respect to x, we have

$$\boldsymbol{\alpha} \cdot \frac{\partial^2 \boldsymbol{\alpha}}{\partial x^2} = -\left(\frac{\partial \boldsymbol{\alpha}}{\partial x} \right)^2. \tag{16.13}$$

Using (16.13), (16.12) becomes

$$E_{\text{ex}} = A \left\{ \left(\frac{\partial \boldsymbol{\alpha}}{\partial x} \right)^2 + \left(\frac{\partial \boldsymbol{\alpha}}{\partial y} \right)^2 + \left(\frac{\partial \boldsymbol{\alpha}}{\partial z} \right)^2 \right\}, \tag{16.14}$$

where A is given by

$$A = \frac{nJS^2}{a}. \tag{16.15}$$

The terms such as $\partial\boldsymbol{\alpha}/\partial x$ correspond to the strain in an elastic body, and the expression (16.14) is similar to that for the elastic energy density which is expressed as a quadratic function of strain tensors. In this sense the coefficient A in (16.15) is called the *exchange stiffness constant*.

Let us calculate the exchange energy of the spin configuration shown in Fig. 16.2. If we rewrite (16.14) using cylindrical coordinates (r, θ, z) with the origin at the center of the disk and the z-axis perpendicular to the plane of the disk, we have

$$E_{ex} = A\left\{\left(\frac{\partial\boldsymbol{\alpha}}{\partial r}\right)^2 + \frac{1}{r^2}\left(\frac{\partial\boldsymbol{\alpha}}{\partial\theta}\right)^2 + \left(\frac{\partial\boldsymbol{\alpha}}{\partial z}\right)^2\right\}. \tag{16.16}$$

Since

$$\boldsymbol{\alpha} = \sin\theta\boldsymbol{i} - \cos\theta\boldsymbol{j}, \tag{16.17}$$

where $\boldsymbol{i}$ and $\boldsymbol{j}$ are unit vectors parallel to the x- and y-axes, respectively, we have

$$\frac{\partial\boldsymbol{\alpha}}{\partial\theta} = \cos\theta\boldsymbol{i} + \sin\theta\boldsymbol{j}, \tag{16.18}$$

so that

$$\left(\frac{\partial\boldsymbol{\alpha}}{\partial\theta}\right)^2 = \cos^2\theta + \sin^2\theta = 1. \tag{16.19}$$

In addition to this, $\partial\boldsymbol{\alpha}/\partial r = 0$, $\partial\boldsymbol{\alpha}/\partial z = 0$, so that from (16.16) we have

$$E_{ex} = \frac{A}{r^2}. \tag{16.20}$$

Therefore the exchange energy of a disk with radius r and thickness d is given by

$$\begin{aligned} U_{ex} &= 2\pi d \int_0^r rE_{ex}\,dr \\ &= 2\pi Ad \ln r = \left(\frac{2Av}{r^2}\right)\ln r. \end{aligned} \tag{16.21}$$

We see from (16.21) that the average exchange energy density U_{ex}/v increases as the radius r decreases. This is because the exchange energy density E_{ex} given by (16.20) is large where r is small.

The actual domain structure in an amorphous film as seen in Fig. 15.25(b) shows such a circular spin configuration at the center. At the edge of the film, however, the spin direction is parallel to the edge, so as to reduce the magnetostatic energy. The spin configuration in the intermediate region is not circular but rather rectangular, with areas where the spin direction changes sharply. In such regions there appear black or white lines in a Lorentz image, as shown in Fig. 15.25(a), depending on whether the sense of spin rotation is clockwise or anticlockwise (see Section 15.3).

When the magnetocrystalline anisotropy is large, the spins are forced to align parallel to one of the easy axes. As a result, the domain structure in a disk with cubic

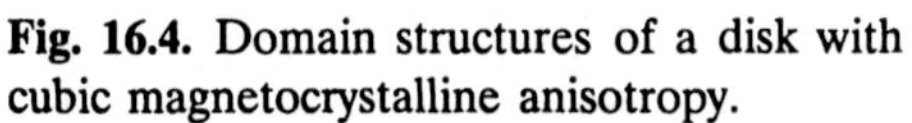
Fig. 16.4. Domain structures of a disk with cubic magnetocrystalline anisotropy.

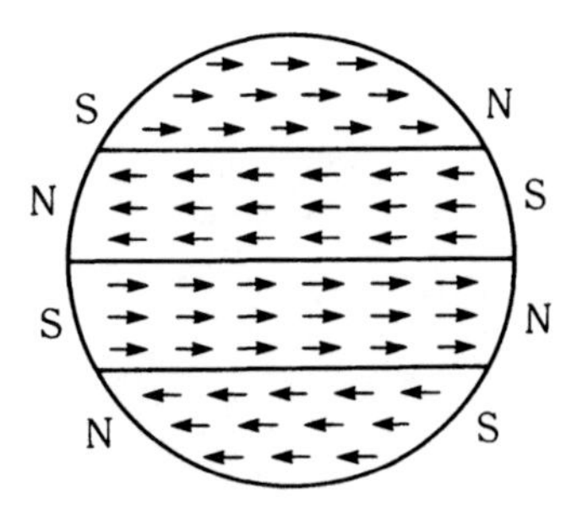

Fig. 16.5. Domain structures of a disk with uniaxial magnetocrystalline anisotropy.

crystal structure and large crystal anisotropy becomes as shown in Fig. 16.4, while that of a uniaxial crystal becomes as shown in Fig. 16.5. Because of the free magnetic poles appearing at the edges of the disk, some magnetostatic energy is stored. In addition to this, the domain walls which separate the neighboring domains store energy which is called *domain wall energy*. We shall discuss this problem in Section 16.2.

Consider next the case of a ferromagnet with a large magnetostriction ($\lambda > 0$). If each domain elongates as required by the magnetostrictive strain, the crystal will separate at the domain boundaries as shown in Fig. 16.6. Since the magnitude of λ is generally small, the magnetostrictive strain does not actually cause gaps to appear. In order to keep the domains in contact with each other, however, a considerable amount of elastic energy must be stored in the crystal. One possible way to avoid or reduce this energy is to increase the volume of the main domains with magnetization parallel to one of the easy axes and reduce the volume of domains magnetized along the other axes. (Domains with antiparallel magnetization have the same strain.) In this case the deformation of the bulk sample will be determined by the magnetostriction of the main domains, and the elastic energy will be concentrated into the small flux-closure domains, which are forced to strain so as to fit the deformation of the main domains (Fig. 16.7).

We have discussed here various domain structures and the various kinds of associated energies. Real domain structures will be determined by minimizing the total energies given by (16.1). In the equilibrium state, the total energy is distributed among these associated energies in appropriate ratios. We shall treat this problem in Chapter 17.

16.2 DOMAIN WALLS

At domain boundaries there are transition layers where spins gradually change their direction from one domain to the other. These transition layers are called *domain walls*. Sometimes they are referred to as *Bloch walls*, after Bloch[5] who first investigated the spin structure of these transition layers in detail. However, this term is also

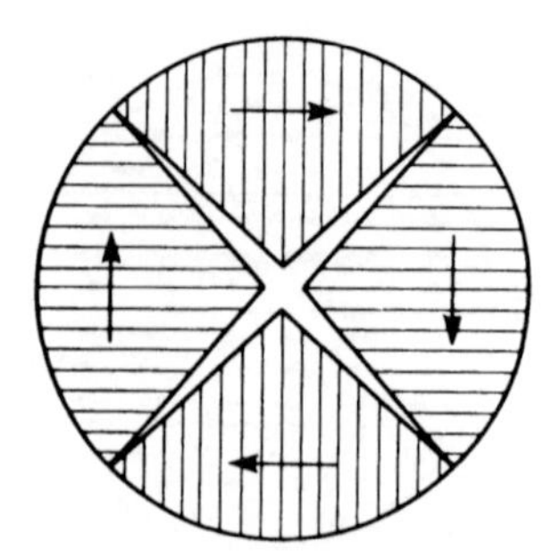

Fig. 16.6. Postulated domain structures of a disk with a large positive magnetostriction (unrestrained magnetostrictive deformation leads to the formation of gaps at the domain boundaries).

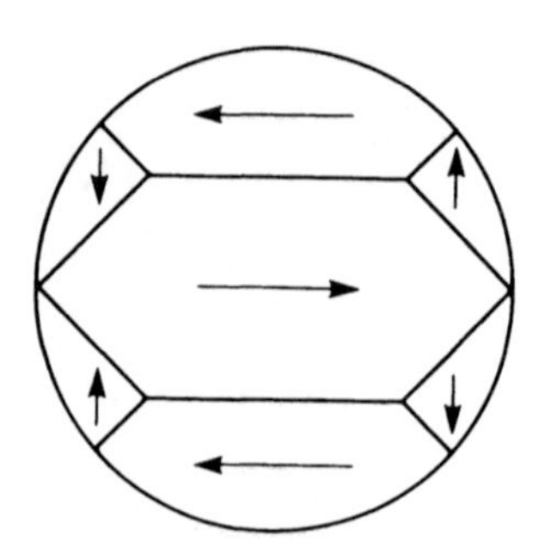

Fig. 16.7. Expected domain structure of a disk with normal magnetostriction (elastic energy is concentrated in the small flux-closure domains).

sometimes used to describe one particular type of domain wall, to distinguish it from another type called a Néel wall (see Section 16.5).

The reason why spins change their direction gradually is that the exchange energy of spin pairs increases as the square of the angle φ between neighboring spins (see (16.7)), so that an abrupt change in the angle φ increases exchange energy to a great extent.

Let us calculate the domain wall energy using a simplified model: as shown in Fig. 16.8, the spins change their direction in steps of equal angle from $\varphi = 0$ to 180° over N atomic layers. Therefore the angle φ_{ij} between spins on the neighboring two layers is given by π/N. Let us consider a simple cubic lattice with lattice constant a. Since the number of atoms in a unit area of one layer is given by $1/a^2$, the number of nearest neighbor spin pairs in a unit area of the wall is given by N/a^2, so that the exchange energy stored in a unit area of the domain wall is

$$\gamma_{ex} = \frac{N}{a^2} w_{ij} = \frac{JS^2\pi^2}{a^2 N}. \tag{16.22}$$

This formula shows that the exchange energy is inversely proportional to the thickness of the wall. In other words, the thicker the wall, the smaller the exchange energy.

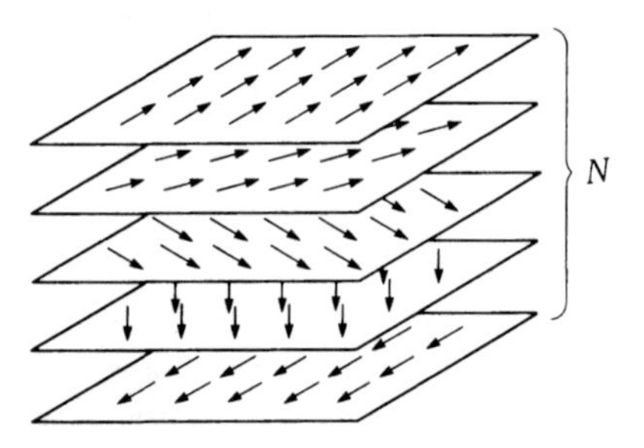

Fig. 16.8. Rotation of spins in the domain wall.

On the other hand, the direction of each spin in the wall deviates from the easy axis, so that anisotropy energy is stored in the wall. Roughly speaking, the anisotropy energy density is increased by the anisotropy constant K when a spin rotates from the easy axis to the hard axis (see Chapter 12). In this simple model, the volume of the wall per unit area is given by $(N/a^2)\times a^3 = Na$, so that the anisotropy energy stored in a unit area of the domain wall is

$$\gamma_a = KNa. \tag{16.23}$$

This formula shows that the anisotropy energy is proportional to the thickness of the wall. In other words, the thicker the wall, the larger the anisotropy energy.

The actual thickness of the wall is determined by a balance of these competing energies. The equilibrium thickness is obtained by minimizing the total energy

$$\begin{aligned}\gamma &= \gamma_{ex} + \gamma_a \\ &= \frac{JS^2\pi^2}{a^2N} + KNa,\end{aligned} \tag{16.24}$$

or

$$\frac{\partial\gamma}{\partial N} = -\frac{JS^2\pi^2}{a^2N^2} + Ka = 0. \tag{16.25}$$

Solving (16.25), we have

$$N = \pi\sqrt{\frac{JS^2}{Ka^3}}. \tag{16.26}$$

Therefore the thickness δ of the wall is given by

$$\delta = Na = \pi\sqrt{\frac{JS^2}{Ka}}. \tag{16.27}$$

For iron, $J = 2.16\times10^{-21}$ (see (16.6)), $S = 1$, $K = 4.2\times10^4$ and $a = 2.86\times10^{-10}$, so that

$$\begin{aligned}\delta &= 3.14\times\sqrt{\frac{2.16\times10^{-21}\times1^2}{4.2\times10^4\times2.86\times10^{-10}}} = 4.2\times10^{-8}\ (\text{m}) \\ &= 150 \text{ lattice constants.}\end{aligned} \tag{16.28}$$

The total energy of the wall per unit area in this equilibrium state is obtained by substituting the values of N given by (16.26) into (16.22) and (16.23), giving

$$\gamma = \pi\sqrt{\frac{JS^2K}{a}} + \pi\sqrt{\frac{JS^2K}{a}} = 2\pi\sqrt{\frac{JS^2K}{a}}. \tag{16.29}$$

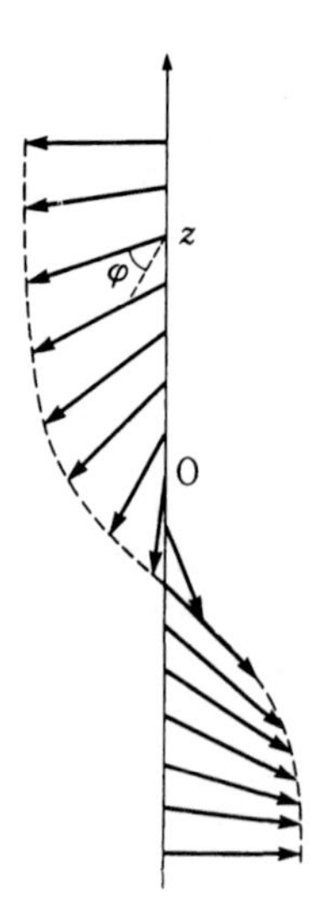

Fig. 16.9. Azimuthal angle of spin rotation in the wall.

It is interesting to note that γ_{ex} and γ_a are equal in the equilibrium state.* For iron, the energy of the wall per unit area is calculated as

$$\gamma = 2 \times 3.14 \times \sqrt{\frac{2.16 \times 10^{-21} \times 1^2 \times 4.2 \times 10^4}{2.86 \times 10^{-10}}}$$
$$= 1.1_2 \times 10^{-3}\ \mathrm{J\,m^{-2}}\ (= 1.1_2\ \mathrm{erg\,cm^{-2}}). \tag{16.30}$$

In the above model, we assumed that the rotation of spin occurs at a constant rate. Actually, however, the orientation of each spin in a wall is determined so as to minimize the total energy. Let us solve this problem using the variational method.

Set the z-axis perpendicular to the wall surface. The azimuthal angle φ of the spin about the z-axis is measured from the direction of the spin at $z = 0$ (the center of the wall) (Fig. 16.9). The angle of spins between two neighboring layers is given by $(\partial\varphi/\partial z)a$, so that the exchange energy is expressed as $JS^2a^2(\partial\varphi/\partial z)^2$. Accordingly the total exchange energy stored in a unit area of the wall is given by

$$\gamma_{ex} = \frac{JS^2}{a} \int_{-\infty}^{\infty} \left(\frac{\partial\varphi}{\partial z}\right)^2 \mathrm{d}z. \tag{16.31}$$

On the other hand, if the anisotropy energy measured as a function of the angle of spin, $\varphi(z)$, is denoted by $g(\varphi)$, the total anisotropy energy stored in a unit area of the wall is given by

$$\gamma_a = \int_{-\infty}^{\infty} g(\varphi)\,\mathrm{d}z. \tag{16.32}$$

* When the energy is given by a term proportional to x plus a term inversely proportional to x, the two terms are numerically equal in the equilibrium state with respect to changes in x.

Therefore the total energy becomes

$$\gamma = \gamma_{ex} + \gamma_a = \int_{-\infty}^{\infty} \left\{ g(\varphi) + A\left(\frac{\partial \varphi}{\partial z}\right)^2 \right\} dz, \tag{16.33}$$

where A is the exchange stiffness constant given by (16.15). Since the exchange interaction is isotropic, the value of A is the same for any wall, regardless of its orientation in the lattice. For iron, $J = 2.16 \times 10^{-21}$ (see (16.6)), $S = 1$, $a = 2.9 \times 10^{-10}$ and $n = 2$, so that

$$A = 1.49 \times 10^{-11}\,(\mathrm{J\,m^{-1}}). \tag{16.34}$$

Next the stable spin configuration can be obtained by minimizing the total energy (16.33) for a small variation of the spin arrangement inside the wall. When the angle φ of the spin at z is varied by $\delta\varphi$, the total energy of the wall is changed by

$$\delta\gamma = \int_{-\infty}^{\infty} \left[\frac{\partial g(\varphi)}{\partial \varphi}\,\delta\varphi + 2A\left(\frac{\partial \varphi}{\partial z}\right)\left(\frac{\partial \delta\varphi}{\partial z}\right) \right] dz, \tag{16.35}$$

which should vanish for the stable spin arrangement. The second term of the integrand is treated by integrating by parts:

$$\begin{aligned}\int_{-\infty}^{\infty} 2A\left(\frac{\partial \varphi}{\partial z}\right)\left(\frac{\partial \delta\varphi}{\partial z}\right) dz &= \left| 2A\left(\frac{\partial \varphi}{\partial z}\right) \delta\varphi \right|_{-\infty}^{\infty} - \int_{-\infty}^{\infty} 2A\left(\frac{\partial^2 \varphi}{\partial z^2}\right) \delta\varphi\, dz \\ &= -\int_{-\infty}^{\infty} 2A\left(\frac{\partial^2 \varphi}{\partial z^2}\right) \delta\varphi\, dz,\end{aligned} \tag{16.36}$$

where the first term vanishes because $\delta\varphi = 0$ at $z = \infty$ or $-\infty$. Then the vanishing of (16.35) requires that

$$\int_{-\infty}^{\infty} \left[\frac{\partial g(\varphi)}{\partial \varphi} - 2A\left(\frac{\partial^2 \varphi}{\partial z^2}\right) \right] \delta\varphi\, dz = 0. \tag{16.37}$$

In order that this condition be satisfied for any selection of $\delta\varphi(z)$, the integrand must always be zero; that is

$$\frac{\partial g(\varphi)}{\partial \varphi} - 2A\left(\frac{\partial^2 \varphi}{\partial z^2}\right) = 0. \tag{16.38}$$

This equation is usually called the Euler equation of the variational problem. Multiplying by $(\partial\varphi/\partial z)$ and integrating from $z = -\infty$ to $z = z$, we have

$$g(\varphi) = A\left(\frac{\partial \varphi}{\partial z}\right)^2. \tag{16.39}$$

Here the origin of $g(\varphi)$ is defined to be the value at $\varphi = \pm\pi/2$, so that $g = 0$ inside the domains at $z = \pm\infty$. Modifying (16.39), we have

$$\mathrm{d}z = \sqrt{A}\,\frac{\mathrm{d}\varphi}{\sqrt{g(\varphi)}}. \tag{16.40}$$

On integrating, this becomes

$$z = \sqrt{A}\int_0^{\varphi}\frac{\mathrm{d}\varphi}{\sqrt{g(\varphi)}}. \tag{16.41}$$

We shall calculate this function for several examples later.

The relationship (16.39) tells us that the anisotropy energy density is equal to the exchange energy density at any part of the wall. From this fact we can see that the spin rotation is more rapid at any position where the spin has a higher anisotropy energy, and vice versa. Referring to (16.40), we obtain, for the total surface energy of the wall given by (16.33),

$$\gamma = 2\sqrt{A}\int_{-\pi/2}^{\pi/2}\sqrt{g(\varphi)}\,\mathrm{d}\varphi. \tag{16.42}$$

Let us treat the domain wall of a ferromagnet which has uniaxial anisotropy. The anisotropy energy in this case is expressed as

$$g(\varphi) = K_{\mathrm{u}}\cos^2\varphi. \tag{16.43}$$

This expression differs from the usual one such as the first term in (12.1), because the origin of φ is defined to be the hard axis, so that g must be zero for $\varphi = 90°$. Then (16.41) becomes

$$\begin{aligned} z &= \sqrt{\frac{A}{K_{\mathrm{u}}}}\int_0^{\varphi}\frac{\mathrm{d}\varphi}{\cos\varphi} \\ &= \sqrt{\frac{A}{K_{\mathrm{u}}}}\ln\tan\left(\frac{\varphi}{2}+\frac{\pi}{4}\right). \end{aligned} \tag{16.44}$$

The relationship between φ and z given by (16.44) is graphically shown in Fig. 16.10. As discussed above, the spin rotation is most rapid at the center of the wall, where the anisotropy energy is the highest, and tends to vanish far from the center of the wall, where the anisotropy energy approaches zero. The thickness of the wall is strictly speaking infinite. An effective wall thickness is conventionally defined as that of a wall in which the spin rotation at the center remains constant throughout the wall, as indicated by the dashed line in Fig. 16.10. The rate of spin rotation at $z = 0$ is given by

$$\left(\frac{\partial z}{\partial\varphi}\right)_{z=0} = \frac{1}{2}\sqrt{\frac{A}{K_{\mathrm{u}}}}\left(\sec^2\left(\frac{\varphi}{2}+\frac{\pi}{4}\right)\Big/\tan\left(\frac{\varphi}{2}+\frac{\pi}{4}\right)\right)_{\varphi=0} = \sqrt{\frac{A}{K_{\mathrm{u}}}}, \tag{16.45}$$

so that the thickness of the wall is given by

$$\delta = \pi\left(\frac{\partial z}{\partial\varphi}\right)_{z=0} = \pi\sqrt{\frac{A}{K_{\mathrm{u}}}}. \tag{16.46}$$

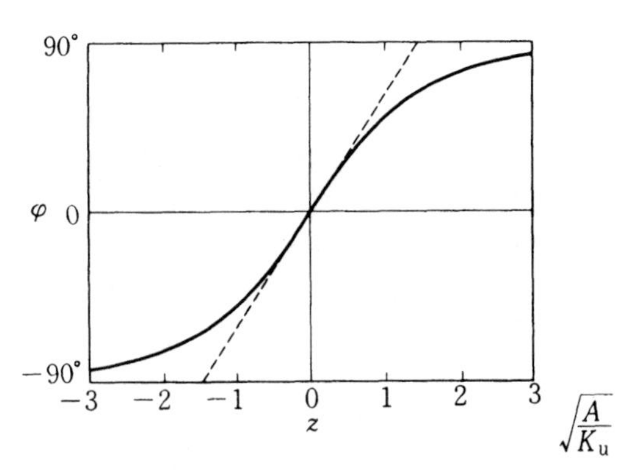

Fig. 16.10. Variation of spin direction within a 180° wall in a ferromagnetic crystal with uniaxial anisotropy.

This expression is the same as (16.27), which was deduced from the simplified model shown in Fig. 16.8. The energy of the wall can be calculated by using (16.43) in (16.42):

$$\gamma = 2\sqrt{AK_u}\int_{-\pi/2}^{\pi/2}\cos\varphi\, d\varphi = 4\sqrt{AK_u}, \tag{16.47}$$

which is also nearly equal to (16.29).

Similar results are obtained for domain walls in materials with cubic anisotropy (see Problems 16.2 and 16.3).

16.3 180° WALLS

Magnetic domain walls are classified into two categories: 180° walls separating two oppositely magnetized domains, and 90° walls separating two domains whose magnetizations make a 90° angle. Consider a cubic ferromagnet having its easy axes parallel to ⟨100⟩, as in iron. There are six possible directions of domain magnetization: [100], [$\bar{1}$00], [010], [0$\bar{1}$0], [001], and [00$\bar{1}$]. A domain wall between [100] and [$\bar{1}$00] domains is a 180° wall, while a wall between [100] and [010] is a 90° wall. In the case of a ferromagnet having its easy axes parallel to ⟨111⟩, such as nickel, there are three kinds of domain walls: 180°, 109°, and 71°, as seen in the domain pattern observed on the (110) surface of Permalloy (Fig. 16.11). Commonly, all domain walls other than 180° walls are classified as 90° walls; they share the property of being sensitive to mechanical stresses. In this section we treat 180° walls.

Let us consider the geometry of a 180° wall separating [100] and [$\bar{1}$00] domains in a cubic ferromagnet with $K_1 > 0$. If we observe such a wall on the (001) surface, it appears as a straight line, as seen in Fig. 15.3(b). The reason is that if the wall were curved as shown in Fig. 16.12(b), magnetic free poles would appear along the curved portion, giving rise to a demagnetizing field opposite to the magnetization

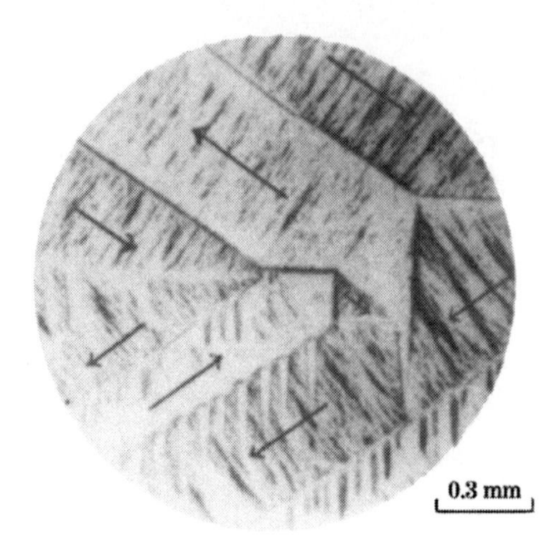

Fig. 16.11. Domain pattern observed on the (110) surface of ordered Permalloy.[6]

in the swollen part of the domain. Therefore the wall is straightened as shown in Fig. 16.12(a).

However, when the wall is observed in a direction parallel to the domain magnetization, the cross-sectional view of the wall can be curved as shown in Fig. 16.13, because this curvature does not result in the appearance of any magnetic free poles

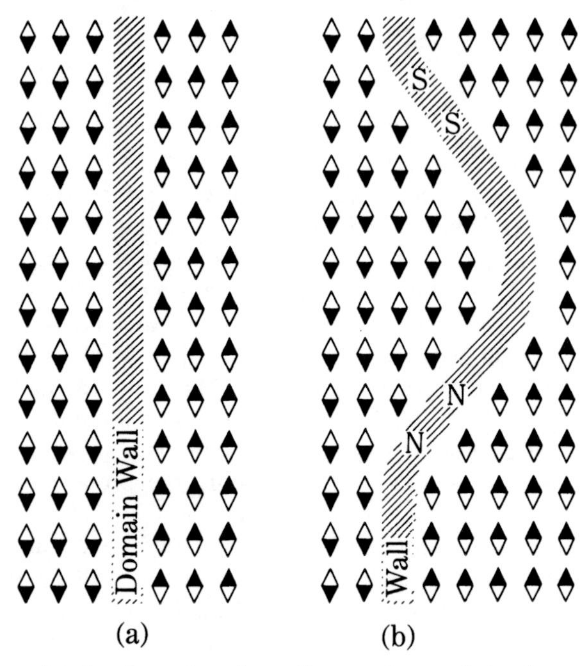

Fig. 16.12. Diagrams showing how domain walls are flattened to decrease magnetostatic energy.

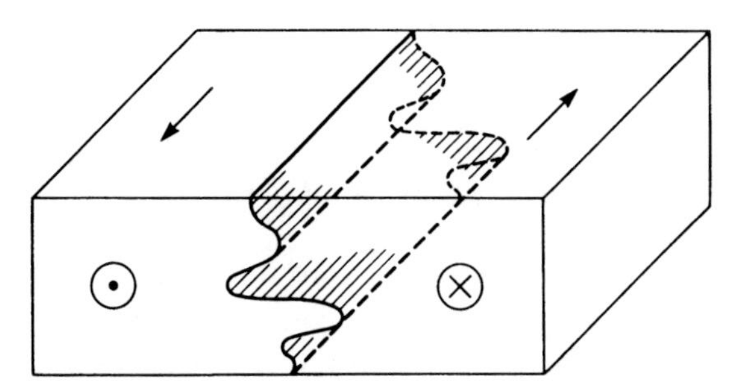

Fig. 16.13. Possible curvature of a domain wall.

on the wall. This curvature does increase the total area of the domain wall, and accordingly the total wall energy, so the wall tends to decrease its surface area unless there is some reason to sustain the curvature. Possible mechanisms for sustaining curvature in the wall result from the presence of inclusions or voids, irregular distribution of internal stresses or alloy composition, and the dependence of the wall energy on its crystallographic orientation.

Imagine a cylindrical 180° wall between [100] and [$\bar{1}$00] domains as shown in Fig. 16.14, with the x-axis parallel to the magnetization, and consider how the wall energy depends on the crystallographic direction of the wall normal, $\boldsymbol{n}$. As explained above, the normal $\boldsymbol{n}$ must lie in the y–z plane. Let the angle between $\boldsymbol{n}$ and y-axis be ψ. The spins in the wall rotate about $\boldsymbol{n}$ as shown in Fig. 16.9 to avoid the appearance of magnetic free poles in the wall.

The magnetocrystalline anisotropy energy is given by

$$E_a = K_1(\alpha_1^2\alpha_2^2 + \alpha_2^2\alpha_3^2 + \alpha_3^2\alpha_1^2), \tag{16.48}$$

as shown in (12.5). We set new coordinates (x', y', z'), in which the x'-axis is parallel

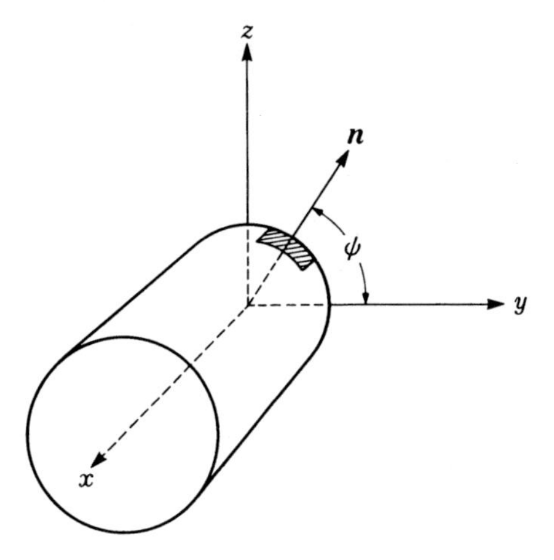

Fig. 16.14. A cylindrical domain wall. The angle ψ indicates the direction of the normal to the wall surface.

to the x-axis. Using direction cosines $(\alpha_1', \alpha_2', \alpha_3')$ with respect to the new coordinates, the direction cosines with respect to (x, y, z) are expressed as

$$\left.\begin{aligned} \alpha_1 &= \alpha_1' \\ \alpha_2 &= \alpha_2' \sin\psi + \alpha_3' \cos\psi \\ \alpha_3 &= -\alpha_2' \cos\psi + \alpha_3' \sin\psi \end{aligned}\right\}. \tag{16.49}$$

Considering that $\alpha_3' = 0$ in the wall, the anisotropy energy (16.48) is expressed, through the use of (16.49), in terms of new direction cosines as

$$E_a = K_1\left(\alpha_1'^2 + \alpha_2'^2 \sin^2\psi \cos^2\psi\right)\alpha_2'^2. \tag{16.50}$$

Using the angle, φ, as indicated in Fig. 16.9, we have

$$\left.\begin{aligned} \alpha_1' &= \sin\varphi \\ \alpha_2' &= \cos\varphi \end{aligned}\right\}, \tag{16.51}$$

so that the anisotropy energy in the wall is expressed as

$$g(\varphi) = K_1(\sin^2\varphi + \cos^2\varphi \sin^2\psi \cos^2\psi)\cos^2\varphi. \tag{16.52}$$

Using this expression, the wall energy (16.42) becomes

$$\gamma = 2\sqrt{AK_1} \int_{\pi/2}^{\pi/2} \sqrt{\sin^2\varphi + \cos^2\varphi \sin^2\psi \cos^2\psi}\, \cos\varphi\, \mathrm{d}\varphi. \tag{16.53}$$

Putting $\sin\psi\cos\psi = s$, $\sin\varphi = t$, or $\cos\varphi\,\mathrm{d}\varphi = \mathrm{d}t$, we have from (16.53)

$$\begin{aligned} \gamma &= 2\sqrt{AK_1} \int_{-1}^{+1} \sqrt{(1-s^2)t^2 + s^2}\, \mathrm{d}t \\ &= \sqrt{AK_1} \left| t\sqrt{(1-s^2)t^2+s^2} + \frac{s^2}{\sqrt{1-s^2}} \ln\left(t + \sqrt{t^2 + \frac{s^2}{1-s^2}}\right) \right|_{-1}^{1} \\ &= 2\sqrt{AK_1}\left(1 + \frac{s^2}{\sqrt{1-s^2}} \ln\frac{1+\sqrt{1-s^2}}{s}\right). \end{aligned} \tag{16.54}$$

If the plane of the wall is parallel to the x–y plane ($\psi = \pi/2$), or parallel to the x–z plane ($\psi = 0$), it follows that $s = 0$, so that

$$\gamma_{100} = 2\sqrt{AK_1}. \tag{16.55}$$

The change in γ during the rotation of the wall from $\psi = 0$ to 90° can be calculated using (16.54) and is shown by the solid line in Fig. 16.15. As seen in this graph, the wall energy is maximum at $\psi = 45°$, where the wall is parallel to (011). The maximum value is

$$\gamma_{011} = 2.76\sqrt{AK_1}. \tag{16.56}$$

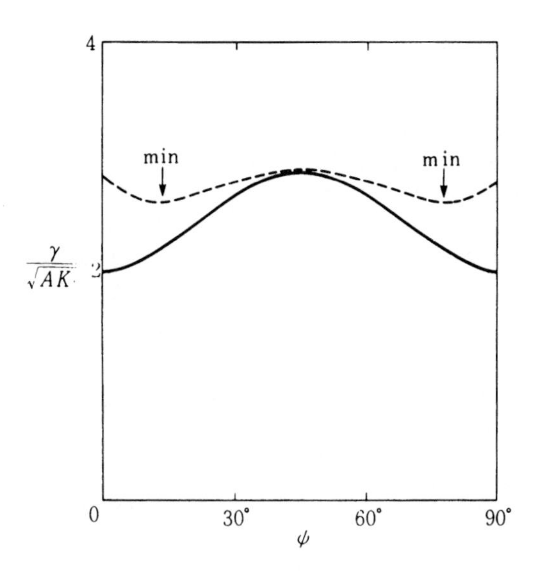

Fig. 16.15. Dependence of 180° wall energy on the orientation of the normal to the wall surface. Solid line, wall energy per unit area; dashed line, total wall energy (energy per unit area × total area) for a wall in a $(01\bar{1})$ single crystal plate.

For iron, using values of A given by (16.34) and K_1 given by (12.6), we have

$$\gamma_{100} = 1.7 \times 10^{-3}\,(\mathrm{J\,m^{-2}}), \tag{16.57}$$

and

$$\gamma_{110} = 2.3 \times 10^{-3}\,(\mathrm{J\,m^{-2}}). \tag{16.58}$$

Therefore the 180° wall tends to become parallel to {100}. However, in the case of a single crystal plate cut parallel to $(01\bar{1})$, the 180° wall tends to tilt so as to minimize the total energy (= wall energy × total area) as shown in Fig. 16.16. The total energy is given by

$$U = ld\,\frac{\gamma(\psi)}{\cos(\psi - 45°)}, \tag{16.59}$$

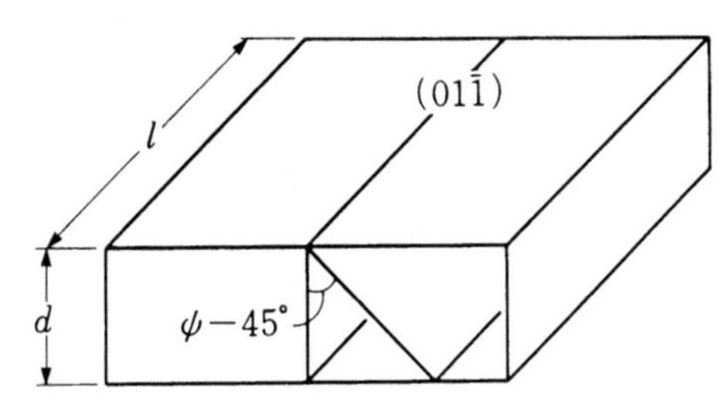

Fig. 16.16. Orientation of a 180° wall in a $(01\bar{1})$ single crystal plate.

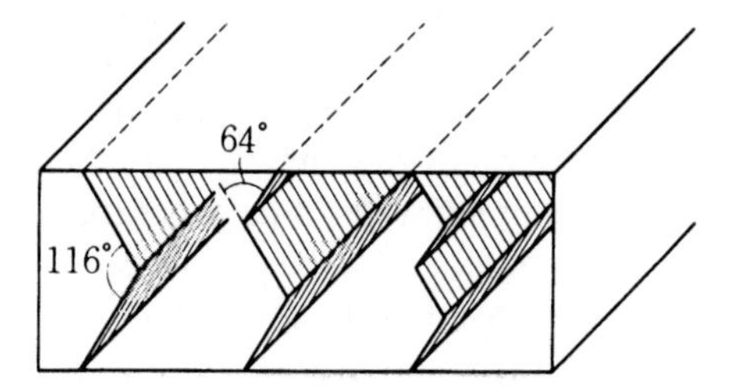

Fig. 16.17. Configuration of 180° walls in a (011) single crystal plate.

where d is the thickness of the crystal and l is the length measured along the direction of domain magnetization. In Fig. 16.15 the function $U/(ld)$ is shown as a dashed curve, which has two minima at $\psi \simeq 13°$ and $\psi \simeq 77°$. Accordingly the wall has two stable orientations, thus making possible the zig-zag shapes shown in Fig. 16.17. It is almost impossible to observe such an end-on or cross-sectional view of the zig-zag wall, because of the appearance of strong magnetic free poles on a plane perpendicular to the direction of magnetization. However, the presence of a tilt was confirmed by observing the position of 180° walls on the top and bottom surfaces of a thin crystal.[7]

The geometry of spin rotation in a 180° wall is solved by using (16.52) in (16.41) to give

$$\begin{aligned} z &= \sqrt{\frac{A}{K_1}} \int_0^{\varphi} \frac{\mathrm{d}\varphi}{\sqrt{\sin^2\varphi + \cos^2\varphi \sin^2\psi \cos^2\psi \cos\varphi}} \\ &= \sqrt{\frac{A}{K_1}} \int_0^{t} \frac{\mathrm{d}t}{\sqrt{(1-s^2)t^2 + s^2(1-t^2)}} \qquad \begin{cases} s = \sin\psi\cos\psi \\ t = \sin\varphi \end{cases} \\ &= \frac{1}{2}\sqrt{\frac{A}{K_1}} \left| \ln \frac{\sqrt{(1-s^2)t^2+s^2} + t}{\sqrt{(1-s^2)t^2+s^2} - t} \right|_0^{\sin\varphi} \\ &= \frac{1}{2}\sqrt{\frac{A}{K_1}} \ln\left(\frac{\sqrt{(1-s^2)\sin^2\varphi + s^2} + \sin\varphi}{\sqrt{(1-s^2)\sin^2\varphi + s^2} - \sin\varphi} \right). \end{aligned} \tag{16.60}$$

Figure 16.18 shows the angle of spin rotation φ as a function of z (normalized) for $\psi = 45°$ and $\psi = 2.87°$. In the case of $\psi = 45°$, the spin rotates smoothly, whereas in the case of $\psi = 2.87°$, the spin rotation is separated into two stages. In this case the wall is nearly parallel to (010), so that the spin direction approaches another easy axis, [001], where the rotation rate should be very small. In the case of $\psi = 0$, the wall tends to separate into two 90° walls. In an actual ferromagnet with nonzero magnetostriction, the domain between the two 90° walls is highly strained, so that it must store a magnetoelastic energy density $\frac{1}{2}\{(\frac{3}{2})\lambda_{100}\}^2 \times (c_{11} - c_{12})$. In order to reduce this energy, the two 90° walls cannot move very far apart.[4]

The magnitude of this effect depends, of course, on the magnitude of the magnetostriction constants of the material. For a material with small magnetostriction, a 180° wall is expected to split into two 90° walls. Such a double wall behaves in the same way as a single 180° wall in fields parallel to the direction of magnetization on either side of the double wall. When, however, the field is applied parallel to the magnetization of the intermediate domain, the double wall may be separated into two independent 90° walls, thus generating a new domain which is magnetized perpendicular to the original domains. This is one possible mechanism for nucleation of 90° walls in the process of demagnetization.

16.4 90° WALLS

A 90° domain wall separates domains whose magnetizations make an angle of 90°

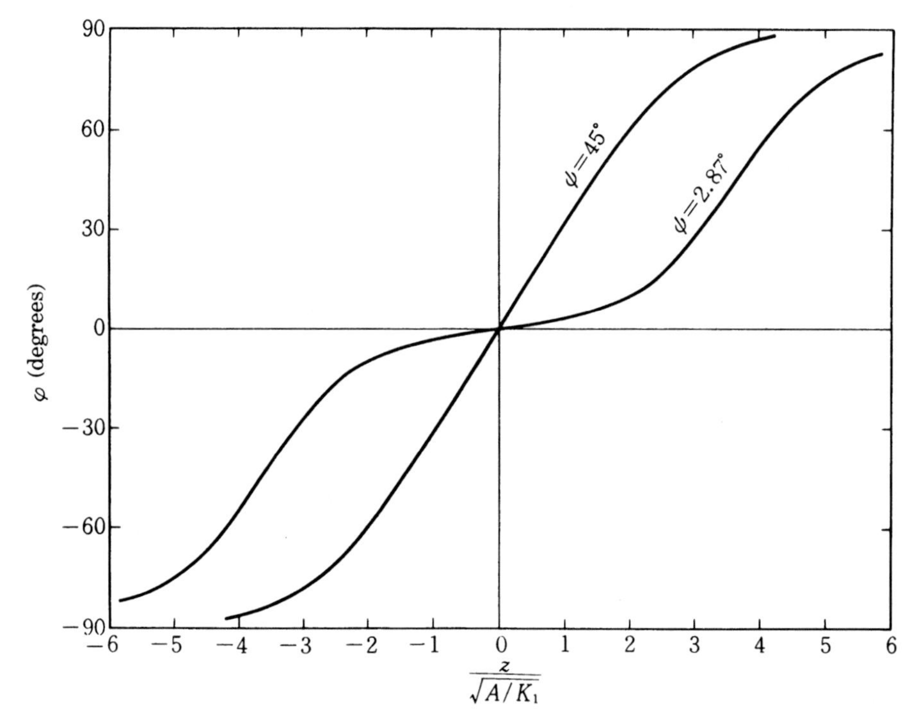

Fig. 16.18. Geometry of spin rotation in a 180° wall making angles $\psi = 45°$ and $\psi = 2.87°$ with the (001) plane ($K_1 > 0$).

(or near 90°). The fundamental properties of 90° walls are similar to those of 180° walls. However, the details are somewhat different. We shall discuss these matters in this section.

The 90° wall is oriented so as to make the normal component of magnetization the same on both sides of the wall, to avoid the appearance of magnetic free poles on the wall surface. This condition does not uniquely specify the orientation of the wall. Let the two magnetizations on both sides of the 90° wall by I_1 and I_2. It is easily verified that the plane of the wall which satisfies this condition must contain the bisector of the angle between I_1 and $-I_2$. But the 90° wall has the freedom to change its orientation about this bisector (Fig. 16.19). Let the angle between the normal to the wall, $\boldsymbol{n}$, and the normal to the plane which contains the two magnetizations be ψ, and the angle between the normal, $\boldsymbol{n}$, and the domain magnetizations be θ. The two angles are related by*

$$\sin \psi = \sqrt{2} \cos \theta. \tag{16.61}$$

Now let us investigate how the wall energy changes as the orientation of the wall varies by changing the angle ψ. In order to see the rotation of spins in the wall more

* Let the direction cosines of the normal to the wall, $\boldsymbol{n}$, with respect to (x', y', z') (see Fig. 16.19) be $(\alpha_1, \alpha_2, \alpha_3)$, and those of the magnetization of one domain be $(\beta_1, \beta_2, \beta_3)$. Then we have $\cos \theta = \alpha_1 \beta_1 + \alpha_2 \beta_2 + \alpha_3 \beta_3$, where $\alpha_1 = 0$, $\alpha_2 = \sin \psi$, $\alpha_3 = \cos \psi$, $\beta_1 = \beta_2 = 1/\sqrt{2}$, $\beta_3 = 0$, so that $\cos \theta = (1/\sqrt{2}) \sin \psi$, from which we have (16.61).

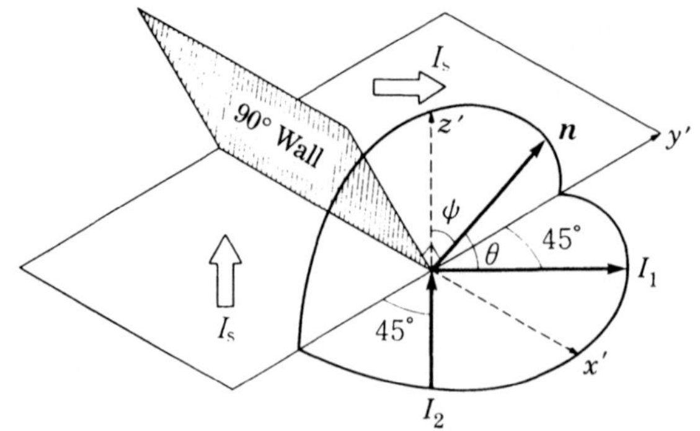

Fig. 16.19. Diagram showing possible orientation of a 90° wall which has no magnetic free poles.

easily, let us choose new coordinate axes so that the x–y plane is parallel to the plane of the wall as shown in Fig. 16.20. As defined above, the magnetization makes the angle θ with the normal to the plane wall, z. Let us assume that the azimuthal angle of the magnetization about the z-axis changes from ϕ_1 to ϕ_2 ($\phi_1 = -\phi_2$). From a consideration of the geometry,* we have

$$\sin\theta = \frac{1}{\sqrt{2}\sin\phi_1} = \frac{1}{\sqrt{2}\sin\phi_2}. \tag{16.62}$$

The angle between the spins in neighboring atomic layers in the 90° wall is given by

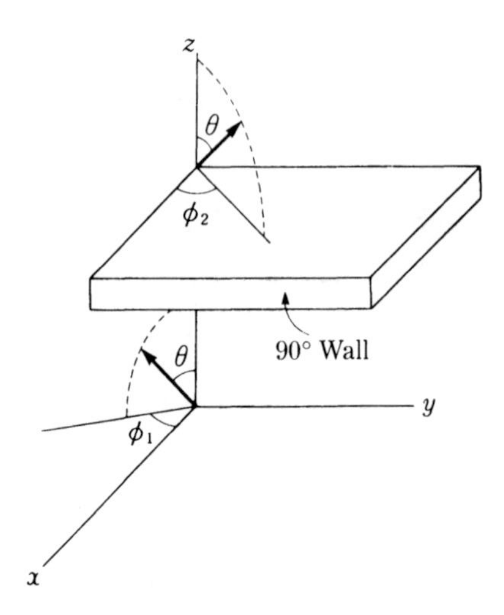

Fig. 16.20. Directions of magnetization on both sides of a 90° domain wall.

* Let the direction cosines of the two magnetizations with respect to (x, y, z) (see Fig. 16.20) be $(\alpha_1, \beta_1, \gamma_1)$ and $(\alpha_2, \beta_2, \gamma_2)$. Since the two magnetizations are perpendicular, $\alpha_1\alpha_2 + \beta_1\beta_2 + \gamma_1\gamma_2 = 1$, where $\alpha_1 = \alpha_2 = \sin\theta\cos\phi_1$, $\beta_1 = -\beta_2 = \sin\theta\sin\phi_1$, and $\gamma_1 = \gamma_2 = \cos\theta$, so that $\sin^2\theta(\cos^2\phi_1 - \sin^2\phi_1) + \cos^2\theta = 0$, from which we have (16.62).

$\sin\theta(\partial\phi/\partial z)a$, where a is the spacing of the atomic layers. Therefore the exchange energy stored in a unit area of the wall is given by

$$\gamma_{ex} = A\sin^2\theta\int_{-\infty}^{\infty}\left(\frac{\partial\phi}{\partial z}\right)^2 dz, \tag{16.63}$$

where $\sin^2\theta$ can be taken outside the integral because the angle must be the same for all the spins to avoid the appearance of magnetic free poles in the wall. On the other hand, if the anisotropy energy density is expressed as a function of the two angles θ and ϕ, $g(\theta,\phi)$, then the anisotropy energy per unit area is given by

$$\gamma_{ani} = \int_{-\infty}^{\infty} g(\theta,\phi)\,dz. \tag{16.64}$$

The spin rotation in the wall can be found by the variational method, minimizing the sum of the energies (16.63) and (16.64). By using this result, the total energy of the wall is given [similar to (16.42) for the 180° wall], by

$$\gamma = 2\sqrt{A}\sin\theta\int_{\phi_1}^{\phi_2}\sqrt{g(\theta,\phi)}\,d\phi. \tag{16.65}$$

By the numerical calculation of (16.65), using the first term of the magnetocrystalline anisotropy given by (12.5), we get the wall energy as a function of the angle ψ as shown in Fig. 16.21. As seen in this graph, the wall energy per unit area, γ, is minimum for $\psi=0$, where the plane of the wall is parallel to the magnetization of the domains on either side. The reason for this is that the rotation of the spin in the wall is only 90°. On the contrary, for $\psi=90°$, the total change in the azimuthal angle is 180°, so that the larger range of integration makes the wall energy larger in spite of the smaller value of $\sin^2\theta$.

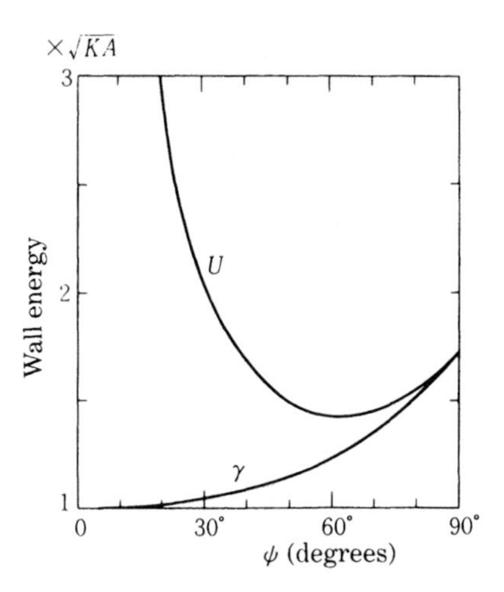

Fig. 16.21. The 90° wall energy, γ, as a function of orientation of the wall. The energy per unit area parallel to the average zigzag wall, U, is minimum at $\psi = 62°$.

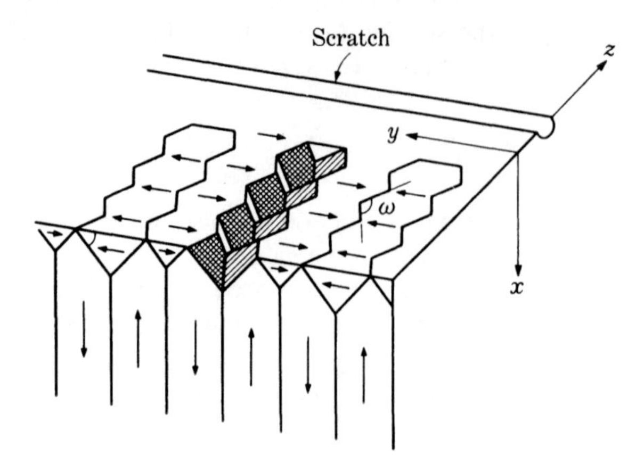

Fig. 16.22. Domain structure induced by the application of a mechanical scratch on a (100) surface of 4% Si–Fe crystal.

Let us consider the actual shape of 90° walls which are produced by a mechanical scratch on the (100) surface of a silicon iron single crystal. It was discovered that the mechanical stresses induced by the scratch result in the appearance of many fine domains which have their magnetizations perpendicular to the crystal surface (Fig. 16.22). These domains are connected by closure domains at the crystal surface to avoid the appearance of magnetic free poles. The 90° walls between the fundamental domains and the closure domains tend to rotate their segments toward an orientation parallel to (001) (the x–y plane in Fig. 16.22) in order to attain the minimum total wall energy. Since in this case 90° walls must stretch along the z-axis, each segment must be out of the (001) plane, thus resulting in zigzag-shaped walls. The energy of the zigzag wall per unit area of the average wall plane containing the z-axis is given by

$$U = \frac{\gamma}{\sin \psi}, \tag{16.66}$$

which is graphically shown in Fig. 16.21 as a function of ψ. As seen in this graph the energy, U, is minimum at $\psi = 62°$. It is concluded, therefore, that the stable configuration of the 90° wall is a zigzag shape, composed of alternate segments each making an angle $\psi = \pm 62°$ with the x–y plane. When we observe these zigzag walls on the (100) crystal surface (the y–z plane in Fig. 16.22), the angle ω between the zigzag line segments, which we call the zigzag angle, is 106°.

Figure 16.23 shows zigzag walls observed on the scratched (100) surface of a 4% Si–Fe single crystal. The thick groove seen on the right-hand edge of the photograph was made by a weighted ball point pen drawn parallel to [010] (the y-axis). Many zigzag lines run perpendicular to the scratch. It is observed that the zigzag angle is nearly 106° at places far from the scratch, but becomes smaller at places nearer to the scratch. The observed zigzag angle is plotted as a function of the distance from the scratch in Fig. 16.24. The reason for the variation of the zigzag angle, ω, may be understood as follows: a tension stress is produced in the region around the scratch,

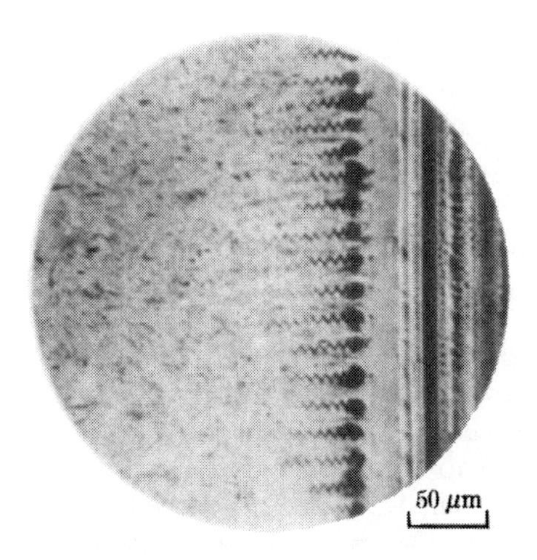

Fig. 16.23. Zigzag domain walls produced by a mechanical scratch on a (100) surface of 4% Si–Fe crystal (Chikazumi and Suzuki[8]).

which has only the component T_{xx} underneath the maze domain. This tension is the main origin of the underlying domains parallel to the x-axis, because the magnetostriction constant parallel to ⟨100⟩ is positive. This tension, however, discourages the closure domains, because their magnetizations are perpendicular to the tension, T_{xx}. Therefore the closure domains tend to become shallower, so that the normal components of magnetization across the zigzag walls are no longer continuous, thus

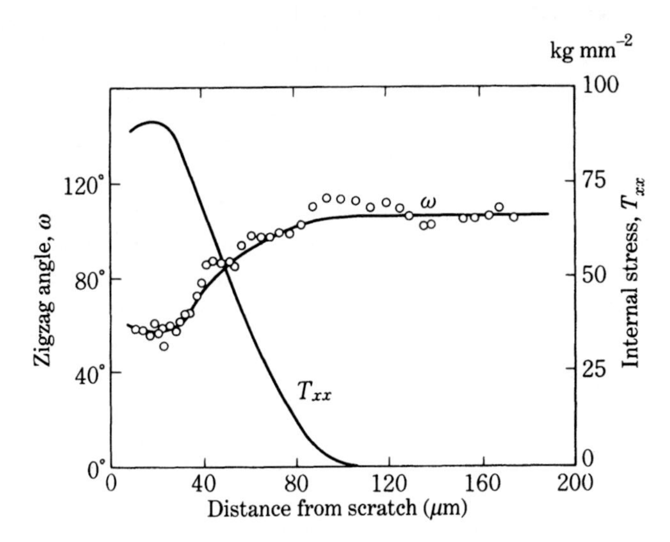

Fig. 16.24. Variation of zigzag angle and calculated internal stress as a function of a distance from the scratch.[8]

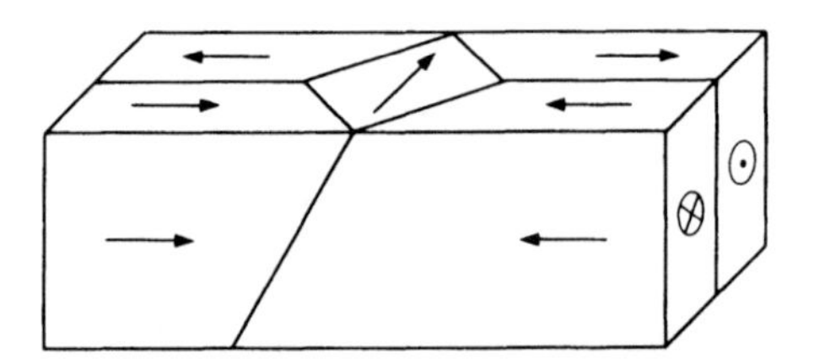

Fig. 16.25. A tilted 90° wall as observed for an iron whisker.[9]

producing magnetic free poles on the zigzag walls. The total magnetic free pole on a zigzag segment is independent of the angle ψ, so that a decrease in the angle results in the same number of free poles spread over a larger wall area, thus reducing the magnetostatic energy. Since T_{xx} increases nearer to the scratch, the zigzag angle is expected to decrease close to the scratch. The zigzag angle was theoretically calculated as a function of T_{xx}.[8] The value of T_{xx} deduced from the observed value of the zigzag angle is also plotted in Fig. 16.24. The result shows that the maximum value of the internal tension is $90\,\mathrm{kg\,mm^{-2}}$, which is close to the yield strength of the material.

This experiment was performed for the purpose of clarifying the nature of maze domains (see Fig. 15.3(a)) observed on mechanically polished crystal surfaces. It was found by a careful observation of the powder pattern using a dilute colloidal suspension that the walls of maze domains, which normally appear to be poorly defined and thick, are in fact composed of fine zigzag lines.

The tilt of 90° walls was also observed on iron whisker crystals, grown by the reduction of ferrous bromide by hydrogen gas at about 700°C.[9] The crystal is a few millimeters in length and has a rectangular cross-section. The domain patterns observed on each surface of the crystal revealed that the 90° wall tilts as shown in Fig. 16.25, in good agreement with the calculation.

16.5 SPECIAL-TYPE DOMAIN WALLS

We have assumed in the preceding discussion that the normal component of magnetization is continuous across a domain wall so that no magnetic free poles exist in the wall. However, such a spin rotation must produce free poles on the sample surface where the domain wall terminates. In thick samples, these surface free poles have little effect, but in thin samples their contribution to the total energy of the system may not be negligible. Néel pointed out that in the case of very thin films, a rotation of the spins in a plane parallel to the thin film surface has less magnetostatic energy than rotation in a plane parallel to the wall.[10] We call the former type of wall a *Néel wall*, and the latter a *Bloch wall* (see Fig. 16.26).

Let us calculate the energy of the Néel wall for a uniaxial ferromagnet. In contrast to the spin rotation in the Bloch wall (Fig. 16.9), the spin rotation in the Néel wall occurs in a plane containing the normal to the wall (Fig. 16.27). The exchange energy stored in a unit area of this wall is given, similarly to (16.31), by

$$\gamma_{\mathrm{ex}} = \frac{JS^2}{a}\int_{-\infty}^{\infty}\left(\frac{\partial\theta}{\partial z}\right)^2 \mathrm{d}z. \qquad (16.67)$$

(a)

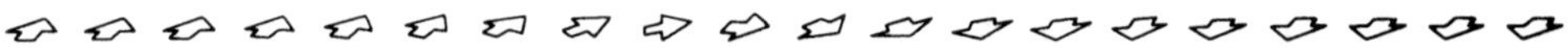

(b)

Fig. 16.26. Two types of spin rotation across a magnetic domain wall, for uniaxial anisotropy: (a) Bloch wall; (b) Néel wall (Chikazumi).

The anisotropy energy per unit area of the wall is given, similarly to (16.32), by

$$\gamma_a = \int_{-\infty}^{\infty} g(\theta)\,\mathrm{d}z, \tag{16.68}$$

where $g(\theta)$ is given, similarly to (16.43), by

$$g(\theta) = K_\mathrm{u} \cos^2 \theta. \tag{16.69}$$

In addition, the Néel wall stores magnetostatic energy due to the demagnetizing field

$$\gamma_m = \frac{I_\mathrm{s}^2}{2\mu_0} \int_{-\infty}^{\infty} \cos^2 \theta\,\mathrm{d}z. \tag{16.70}$$

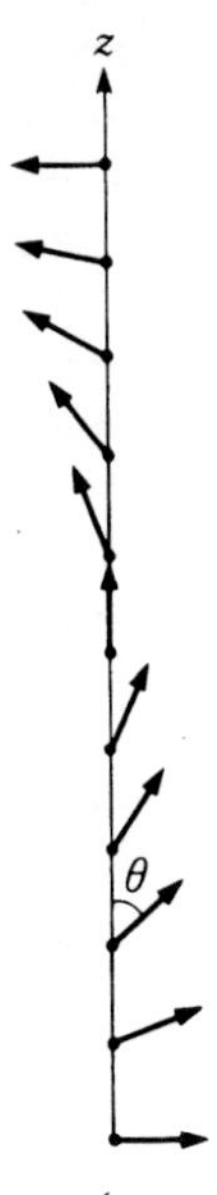

Fig. 16.27. Spin rotation across a Néel wall (the rotation occurs in the plane of the page).

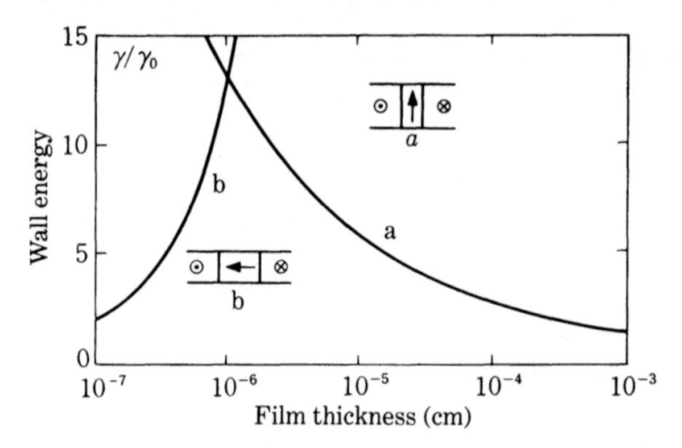

Fig. 16.28. Dependence of the wall energy on film thickness (a) Bloch wall; (b) Néel wall.[10]

This term has the same functional form as (16.68) and (16.69), so that all the equations from (16.33) to (16.47) hold in this case by using the substitution

$$K_u \to K_u + \frac{I_s^2}{2\mu_0}. \tag{16.71}$$

Therefore, referring to (16.47), the wall energy of the Néel wall is given by

$$\gamma = 4\sqrt{A\left(K_u + \frac{I_s^2}{2\mu_0}\right)}. \tag{16.72}$$

Comparing (16.72) with (16.47), we find that the Néel wall always has higher energy than the Bloch wall.

In the case of magnetic thin films, the Bloch wall has an additional magnetostatic energy due to free poles at the film surfaces, which increases with a decrease of film thickness as shown by the curve a in Fig. 16.28. On the other hand, the magnetostatic energy of the Néel wall decreases with a decrease of film thickness, because the demagnetizing factor, N, which was assumed to be 1 in (16.70), decreases with a

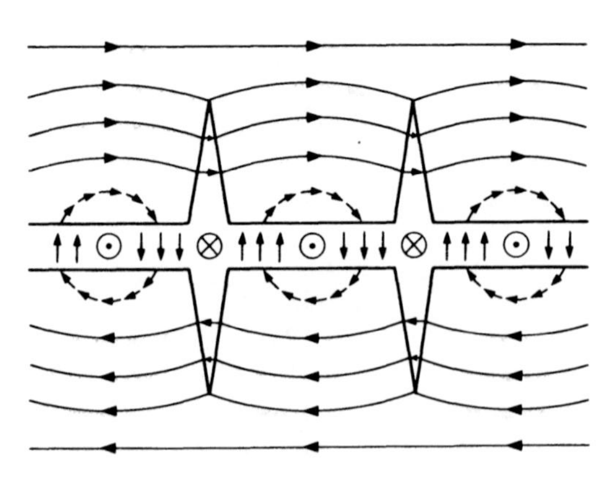

Fig. 16.29. Spin configurations inside and outside the cross-tie wall.[11]

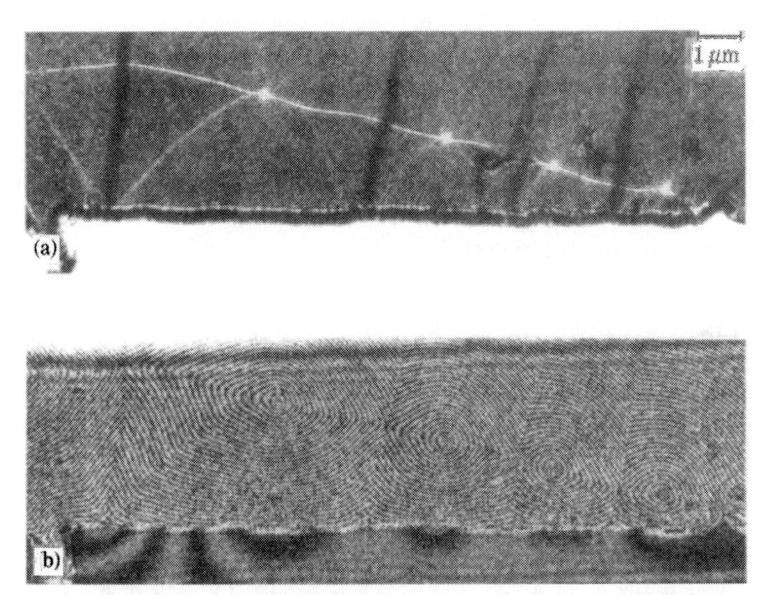

Fig. 16.30. Electron interference micrographs of a cross-tie wall as observed for a Permalloy film: (a) Lorentz image; (b) interference fringes.[12]

decrease of film thickness as shown by the curve b in Fig. 16.28. Therefore the Bloch wall is stable for relatively thick films, while the Néel wall is stable for relatively thin films.

For films of intermediate thickness, where the energies of both types of domain wall are comparable, a cross-tie wall as shown in Fig. 16.29 appears.[11] In this wall, Bloch-type and Néel-type spin configurations appear alternatively along the wall. The spin arrangement on both sides of the wall is modified and results in cross-ties, where the direction of spin changes discontinuously. Figure 16.30(a) shows a Lorentz image of a cross-tie wall observed for a Permalloy film, while (b) shows interference fringes

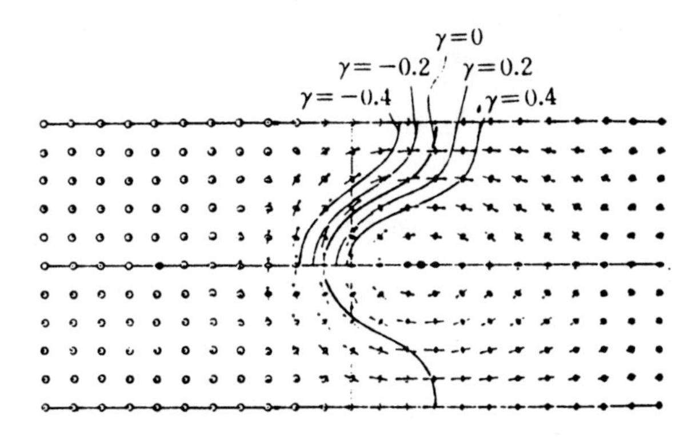

Fig. 16.31. Leakage field-free spin configuration of a Bloch wall. The value of γ is the direction cosine of spins with respect to the normal to the page.[13]

as observed by electron interference microscopy which show the same spin configuration as the postulated structure in Fig. 16.29. The line in either a Bloch or a Néel wall where the spin rotation changes from right-handed to left-handed, or vice versa, is called a *Bloch line*. It is a line boundary in a domain wall. In Fig. 16.30(a), Bloch lines appear as bright spots or dark spots from which cross-ties emerge.

Hubert[13] theoretically predicted a leakage field-free spin configuration of a Bloch wall, as shown in Fig. 16.31. Such an asymmetric spin configuration was experimentally verified by electron microscopy.[14]

PROBLEMS

16.1 Assuming that a ferromagnetic crystal with isotropic magnetostriction constant and with negligibly small magnetocrystalline anisotropy energy is subjected to a uniform tension σ, formulate the energy of a 180° wall which is present in this crystal.

16.2 Calculate the energy of 180° walls which are parallel to the (100) and (110) planes in a cubic crystal with $K_1 = 0$ and $K_2 > 0$.

16.3 Calculate the energy of a 180° wall which is parallel to (100) of a bcc ferromagnetic crystal with S (spin) $= \frac{1}{2}$, a (lattice constant) $= 3\,\text{Å}$, Θ_f (Curie point) $= 527°\text{C}$, and $K_1 = 6 \times 10^4\,\text{J m}^{-3}$.

16.4 In a single crystal plate parallel to the (001) plane, there is an infinitely long x-domain touching both crystal surfaces with a quadrilateral cross-section surrounded by y-domains. Calculate the inner angles of the cross-section of the x-domain perpendicular to its length. Assume that the effect of magnetostriction can be ignored, and the magnetocrystalline anisotropy is expressed only by the first term.

REFERENCES

1. W. F. Brown, Jr, *J. Appl. Phys.*, **11** (1940), 160; *Phys. Rev.*, **58** (1940), 736.
2. W. F. Brown, Jr., *Micromagnetics* (Wiley-Interscience, N.Y., 1963).
3. P. R. Weiss, *Phys. Rev.*, **74** (1948), 1493.
4. C. Kittel, *Rev. Mod. Phys.*, **21** (1949), 541.
5. F. Bloch, *Z. f. Phys.*, **74** (1932), 295.
6. S. Chikazumi, *Phys. Rev.*, **85** (1952), 918.
7. C. D. Graham, Jr. and P. W. Neurath, *J. Appl. Phys.*, **28** (1957), 888.
8. S. Chikazumi and K. Suzuki, *J. Phys. Soc. Japan*, **10** (1955), 523. This paper ignored the μ^*-correction, which is described in a later paper: S. Chikazumi and K. Suzuki, *IEEE Trans. Mag.*, **MAG-15** (1979), 1291.
9. R. W. DeBlois and C. D. Graham, Jr., *J. Appl. Phys.*, **29** (1958), 525.
10. L. Néel, *Compt. Rend.*, **241** (1955), 533.
11. E. E. Huber, D. O. Smith, and J. B. Goodenough, *J. Appl. Phys.*, **29** (1958), 294.
12. A. Tonumura, T. Matsuda, H. Tanabe, N. Osakabe, I. Endo, A. Fukuhara, K. Shinagawa, and H. Fujiwara, *Phys. Rev.*, **B25** (1982), 6799.
13. A. Hubert, *Phys. Stat. Solidi*, **32** (1962); **38** (1970), 699.
14. T. Suzuki, *Jap. J. Appl. Phys.*, **17** (1978), 141.

17

MAGNETIC DOMAIN STRUCTURES

17.1 MAGNETOSTATIC ENERGY OF DOMAIN STRUCTURES

Ferro- and ferrimagnetic materials are divided into multiple domains, in which the spontaneous magnetization takes different orientations. The reason for the existence of domains is that the magnetostatic energy is greatly reduced. In this section we discuss how the magnetostatic energy is reduced by the appearance of domains.

First we calculate the magnetostatic energy of the domain structure as shown in Fig. 17.1. We assume a crystal with a surface parallel to the x–y plane at $z = 0$, in which there are many plate-like domains with magnetization I_s along the $+z$ or $-z$ direction. The domains are separated by 180° walls lying parallel to the y–z plane. For simplicity we assume that the crystal extends indefinitely in the $-z$ and also in the $\pm x$- and $\pm y$-directions. Let the thickness of each domain be d. Then the surface density of magnetic free poles, ω, is given by

$$\omega = \begin{cases} I_s & 2md < x < (2m+1)d \\ -I_s & (2m+1)d < x < 2(m+1)d, \end{cases} \tag{17.1}$$

where m is an index that numbers the domains. As shown by (1.77), the magnetostatic energy is given by

$$U = \tfrac{1}{2} \sum_i m_i \varphi_i, \tag{17.2}$$

where φ is the magnetic potential. The magnetic potential must satisfy the Laplace equation

$$\frac{\partial^2 \varphi}{\partial x^2} + \frac{\partial^2 \varphi}{\partial z^2} = 0, \tag{17.3}$$

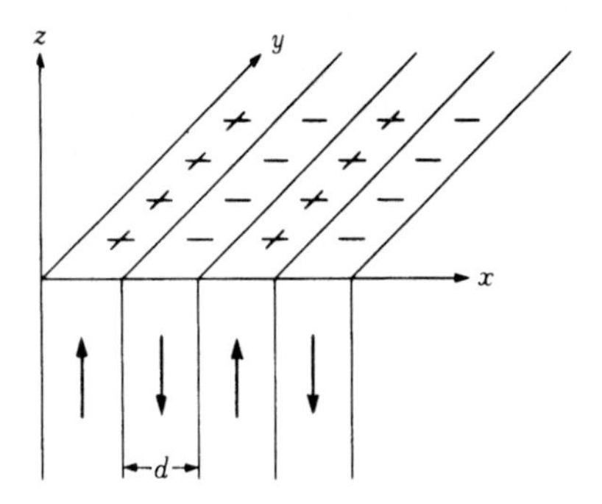

Fig. 17.1. Free poles on the crystal surface produced by the plate-like or laminated domains.

in all space other than at the crystal surface, where magnetic free poles exist. It must satisfy the boundary condition

$$\left(\frac{\partial\varphi}{\partial z}\right)_{z=+0}-\left(\frac{\partial\varphi}{\partial z}\right)_{z=-0}=-\frac{\omega}{\mu_0}, \tag{17.4}$$

at $z=0$. In the present case, the potential must be symmetrical with respect to the plane $z=0$, because the free poles are confined to this plane, so that

$$\left(\frac{\partial\varphi}{\partial z}\right)_{z=+0}=-\left(\frac{\partial\varphi}{\partial z}\right)_{z=-0}. \tag{17.5}$$

Using this relationship in (17.4), we have

$$\left(\frac{\partial\varphi}{\partial z}\right)_{z=-0}=\frac{\omega}{2\mu_0}. \tag{17.6}$$

The solution of the Laplace equation (17.3) must have the form

$$\varphi=\sum_{n=1}^{\infty}A_n\sin n\left(\frac{\pi}{d}\right)x\cdot e^{n(\pi/d)z}, \tag{17.7}$$

where n is an odd number. In order to prove that (17.7) is a solution of (17.3), we may use (17.7) for φ in (17.3) and find that (17.3) holds. The coefficients A_n are determined to satisfy the boundary condition (17.6). Using (17.7) and (17.1) in (17.6), we have

$$\left(\frac{\pi}{d}\right)\sum_{n=1}^{\infty}nA_n\sin n\left(\frac{\pi}{d}\right)x=\begin{cases}\dfrac{I_s}{2\mu_0} & \text{for } 2md<x<(2m+1)d\\[2ex] -\dfrac{I_s}{2\mu_0} & \text{for } (2m+1)d<x<2(m+1)d.\end{cases}$$

The coefficients A_n are obtained by integrating the above formula from $2md$ to $2(m+1)d$ after multiplying by $\sin n(\pi/d)x$ to give

$$\begin{aligned}A_n&=\frac{I_s}{2\pi\mu_0 n}\left\{\int_{2md}^{(2m+1)d}\sin n\left(\frac{\pi}{d}\right)x\,dx-\int_{(2m+1)d}^{2(m+1)d}\sin n\left(\frac{\pi}{d}\right)x\,dx\right\}\\&=\frac{2I_s d}{\pi^2\mu_0 n^2},\end{aligned} \tag{17.8}$$

where n is an odd integer. The value of φ at $z=0$ is obtained from (17.8) as

$$\varphi_{(z=0)}=\frac{2I_s d}{\pi^2\mu_0}\sum_{\substack{n=1\\ \text{odd}}}^{\infty}\frac{1}{n^2}\sin n\left(\frac{\pi}{d}\right)x. \tag{17.9}$$

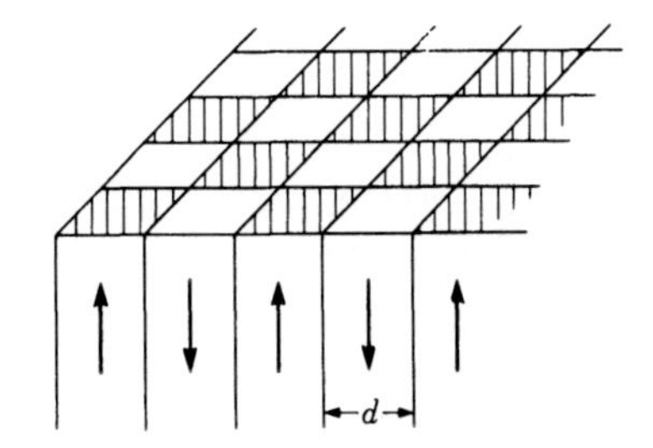

Fig. 17.2. Checkerboard domain pattern.

Using (17.2), we can calculate the magnetostatic energy per unit area of the crystal surface as

$$\begin{aligned}
\varepsilon_m &= \frac{I_s^2 d}{\pi^2 \mu_0} \sum_{n=1}^{\infty} \frac{1}{n^2 d} \int_0^d \sin n\left(\frac{\pi}{d}\right) x \, \mathrm{d}x \\
&= \frac{2 I_s^2 d}{\pi^3 \mu_0} \sum_{n \,\text{odd}}^{\infty} \frac{1}{n^3} \\
&= \frac{2 I_s^2 d}{\pi^3 \mu_0} \times (1.0517) \\
&= 5.40 \times 10^4 I_s^2 d \qquad (= 0.85_2 I_s^2 d \text{ in CGS}).
\end{aligned} \tag{17.10}$$

Thus we find that the magnetostatic energy is proportional to the width of the domains, d.

The above calculation was carried out by Kittel,[1] who also extended the treatment to the checkerboard pattern shown in Fig. 17.2 and the circular pattern as shown in Fig. 17.3, obtaining

$$\varepsilon_m = 3.36 \times 10^4 I_s^2 d \; (= 0.53 I_s^2 d \text{ in CGS}) \tag{17.11}$$

for the checkerboard pattern and

$$\varepsilon_m = 2.37 \times 10^4 I_s^2 d \; (= 0.37_4 I_s^2 d \text{ in CGS}) \tag{17.12}$$

for the circular pattern. Comparing (17.11) and (17.12) with (17.10), we see that the magnetostatic energy is greatly reduced by mixing N and S poles more closely. All three formulas (17.10), (17.11) and (17.12) show that the magnetostatic energy is proportional to the size of the domains, d. A decrease in d, therefore, results in a decrease in the magnetostatic energy, because the N and S poles are mixed more densely. Goodenough[2] showed that a wavy pattern as shown in Fig. 17.4 has a much lower magnetostatic energy than the plate-like or laminated domain structure of Fig. 17.1. The energy is rather closer to that of the checkerboard pattern (17.11). Such wavy domains were observed in MnBi, as shown in Fig. 15.13(c).

In all the cases considered above, the spontaneous magnetization lies perpendicular to the crystal surface. In these cases, magnetic free poles are confined in the crystal

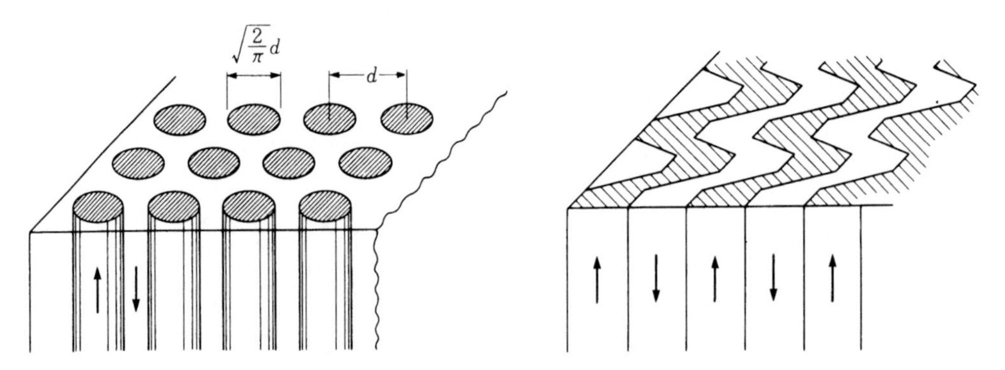

Fig. 17.3. Circular domain pattern.

Fig. 17.4. Wavy domain pattern.

surface, where there is a discontinuous change in magnetization. However, if the easy axis makes an angle other than a right angle with the crystal surface, as shown in Fig. 17.5, the demagnetizing field in each domain will rotate the magnetization out of the easy axis, thus producing a volume distribution of free poles. This results in a reduction of magnetostatic energy.

Let the angle of inclination of the easy axis from the crystal surface be θ. Then the surface density of free poles is given by

$$\omega = \pm I_s \sin\theta. \tag{17.13}$$

If the magnetization is fixed in the easy axis (Fig. 17.6(a)), the surface density of magnetic free poles is obtained by the substitution $I_s \to I_s \sin\theta$ in (17.10), giving the magnetostatic energy per unit area

$$\varepsilon_m = 5.40 \times 10^4 I_s^2 d \sin^2\theta \qquad (= 0.85_2 I_s^2 d \sin^2\theta \text{ in CGS}). \tag{17.14}$$

However, in the crystal each domain magnetization is rotated non-uniformly by the demagnetizing field resulting from the free poles at the crystal surface, as shown in

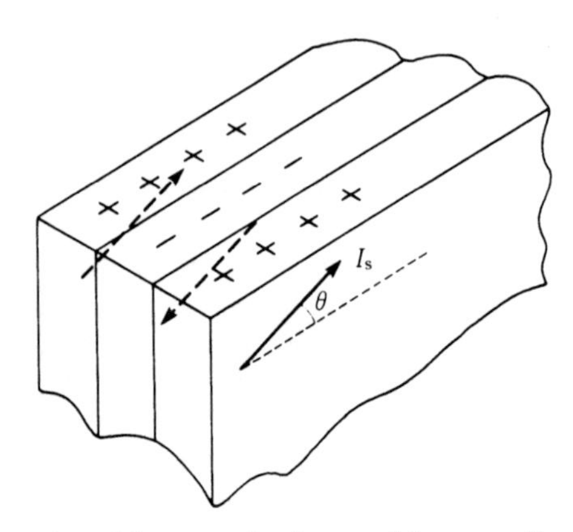

Fig. 17.5. Laminated domains with magnetizations making a small angle θ with the surface.

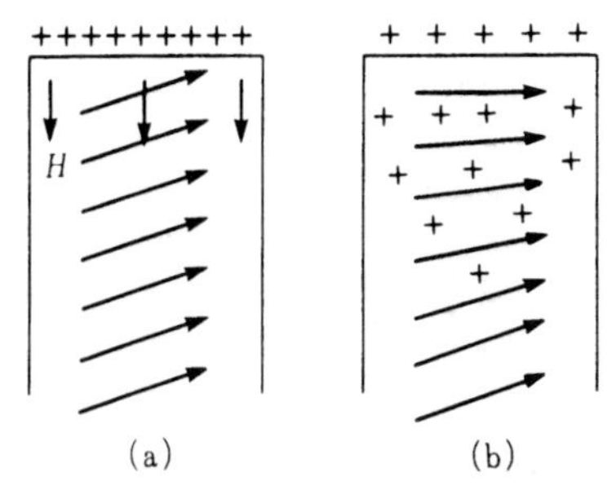

Fig. 17.6. Distribution of magnetization vectors in a domain in Fig. 17.5: (a) The direction of magnetization is fixed; (b) the local magnetization is rotated out of the easy axis by the demagnetizing field.

Fig. 17.6(b). This produces a volume distribution of magnetic free poles instead of a planar distribution. It is easily shown that the total number of free poles is the same in Figs. 17.6(a) and (b). It is easy to see that the magnetostatic energy is much lower for (b) than for (a), because free poles with the same sign tend to repel one another. It must be noted that there is some increase in anisotropy energy in case (b), because the magnetization is rotated away from the easy axis. One way to solve this problem is to replace the ferromagnetic crystal by a homogeneous magnetic material with a modified permeability μ^* and to assume the same free pole distribution as in (a) (Fig. 17.7). As mentioned in Section 1.5, the magnetostatic energy in such a system is calculated in terms of magnetic potential at the permanent free poles: the polarization free charge in the soft magnetic material is automatically taken into consideration.

The permeability μ^* is not the real permeability, which results from domain wall motion, but a special permeability resulting from rotation of the magnetization. Then the boundary condition (17.4) must be replaced by

$$\mu_0\left(\frac{\partial\varphi}{\partial z}\right)_{z=+0} - \mu^*\left(\frac{\partial\varphi}{\partial z}\right)_{z=-0} = -\omega. \tag{17.15}$$

This means that the permanent free poles give a divergence of $\boldsymbol{B}$ which does not include the permanent magnetization.

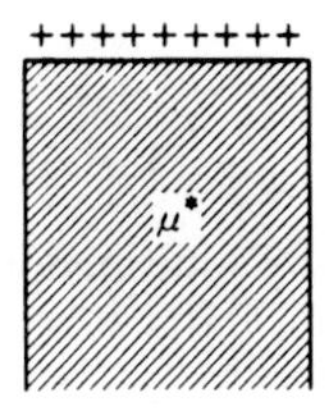

Fig. 17.7. Model for μ^* correction, equivalent to Fig. 17.6(b).

In this case the magnetic potential, φ, is no longer symmetrical with respect to the plane $z=0$, so that the relationship (17.5) does not hold. Let the solution for $\mu^* = \mu_0$ be

$$\varphi = f(x, z). \tag{17.16}$$

The solution for $\mu^* = \mu^*$ is no longer symmetrical for $z > 0$ and $z < 0$, so that we can assume the form

$$\left.\begin{aligned} \varphi &= Af(x, \alpha z), \qquad z > 0 \\ \varphi &= Af(x, \beta z), \qquad z < 0 \end{aligned}\right\}. \tag{17.17}$$

Using this formula in (17.15), we have

$$\mu_0 A \alpha f_z(x, +0) - \mu^* A \beta f_z(x, -0) = -\omega. \tag{17.18}$$

Since (17.16) must satisfy the conditions (17.4) and (17.5), we have

$$f_z(x, +0) - f_z(x, -0) = -\frac{\omega}{\mu_0} \tag{17.19}$$

and

$$f_z(x, +0) = -f_z(x, -0), \tag{17.20}$$

where f_z signifies $\partial f(x, z)/\partial z$. Using these relationships in (17.18), we have

$$A(\mu_0 \alpha + \mu^* \beta) = 2\mu_0. \tag{17.21}$$

On the other hand, the potential must satisfy the Laplace equation for $z > 0$, so that

$$\Delta \varphi = \frac{\partial^2 \varphi}{\partial x^2} + \frac{\partial^2 \varphi}{\partial z^2} = Af_{xx} + A\alpha^2 f_{zz} = 0. \tag{17.22}$$

Similarly in the case $\mu^* = \mu_0$, we have

$$\Delta \varphi = f_{xx} + f_{zz} = 0. \tag{17.23}$$

Comparing (17.22) and (17.23), we find

$$\alpha = 1. \tag{17.24}$$

Similarly from the Laplace equation for $z < 0$, we have

$$\beta = 1. \tag{17.25}$$

Therefore (17.21) becomes

$$A = \frac{2}{1 + \bar{\mu}^*}. \tag{17.26}$$

The potential at $z = 0$ is obtained by multiplying A by (17.9), thus

$$\varphi_{(z=0)} = \frac{4 I_s d \sin\theta}{(1 + \bar{\mu}^*)\pi^2 \mu_0} \sum_{n=1}^{\infty} \frac{1}{n^2} \sin n\left(\frac{\pi}{d}\right) x. \tag{17.27}$$

Therefore the magnetostatic energy (17.14) must also be multiplied by A, so that we have

$$\varepsilon_m = \frac{2}{1+\bar{\mu}^*}(5.40 \times 10^4) I_s^2 d \sin^2 \theta$$

$$\left(= \frac{2}{1+\bar{\mu}^*} 0.852 I_s^2 d \sin^2 \theta \text{ in CGS} \right). \tag{17.28}$$

The value of $\bar{\mu}^*$ is determined by the rotation of the magnetization out of the easy axis (see (18.82) in which $\theta_0 \approx \pi/2$), so that

$$\bar{\mu}^* = 1 + \frac{I_s^2}{2\mu_0 K_1}. \tag{17.29}$$

When the magnetization makes 45° with the crystal surface, by putting $\theta_0 \approx \pi/4$ in (18.82) we have

$$\bar{\mu}^* = 1 + \frac{I_s^2}{4\mu_0 K_1}. \tag{17.30}$$

In the case of the 90° wall on which magnetic free poles appear, such as the zigzag 90° domain walls in Fig. 16.22, the crystal on both sides of the wall has μ^*, and (17.26) becomes

$$A = \frac{1}{\bar{\mu}^*}. \tag{17.31}$$

This situation can be understood by considering that the magnetic field is reduced by a factor $1/\bar{\mu}$ due to the presence of the soft magnetic material (see (1.103)).

Let us calculate the value of $\bar{\mu}^*$ for iron, which has magnetocrystalline anisotropy $K_1 = 4.2 \times 10^4\,\mathrm{J\,m^{-3}}$ and spontaneous magnetization $I_s = 2.15$ (T). Using these values, (17.29) gives $\bar{\mu}^* = 45$, so that the factor A given by (17.26) is 0.0435, a very small value. Therefore such a correction can never be ignored when the fields produced by free poles cause a rotation of the magnetization. However, when the magnetization is perpendicular to the crystal surface as in Figs. 17.1–17.4, $\bar{\mu}^* = 1$, and (17.26) gives $A = 1$ and there is no such correction. This correction is called the *mu-star correction*; its importance was first pointed out by Kittel.[1]

17.2 SIZE OF MAGNETIC DOMAINS

As mentioned previously, a ferromagnetic body is divided into many domains in order to reduce the magnetostatic energy. As seen from the preceding calculations, the magnetostatic energy is always proportional to the width, d, of the domains, irrespective of the geometry of the domain structure. Therefore the magnetostatic energy is reduced by reducing the width of the domains. On the other hand, the total domain wall area per unit volume is increased by reducing the width of the domains. The equilibrium size of the domains is determined by minimizing the sum of the magnetostatic energy and the total wall energy.

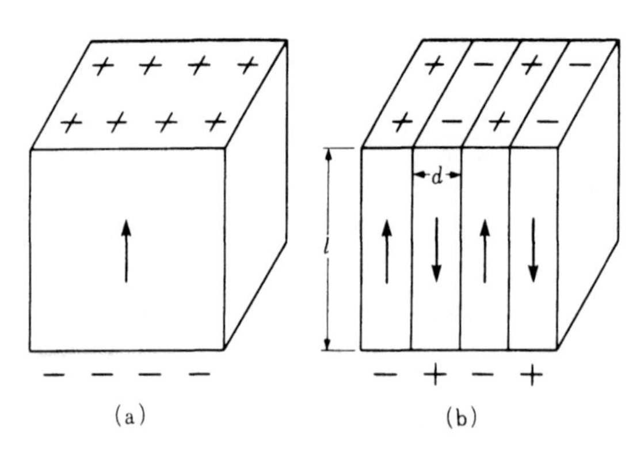

Fig. 17.8. (a) Single domain structure; (b) multi-domain structure.

Consider a ferromagnetic plate with thickness l having an easy axis perpendicular to the plate. If the entire plate consists of a single domain as shown in Fig. 17.8(a), N poles appear on the top and S poles appear on the bottom surfaces, both of which produce demagnetizing fields $-I_s/\mu_0$ in the plate. The magnetostatic energy stored per unit area is given by

$$\varepsilon_m = \frac{I_s^2}{2\mu_0} l. \tag{17.32}$$

The magnetostatic energy is lowered if the plate is divided into many laminated domains of thickness d, as shown in Fig. 17.8(b). The magnetostatic energy of such a free pole distribution is given by (17.10). In the present case, free poles exist on the top as well as on the bottom surfaces. If $l \gg d$, we can ignore the interaction between free poles on the two surfaces, so that the magnetostatic energy per unit area of the plate is given by two times the results of (17.10), or

$$\varepsilon_m = 1.08 \times 10^5 I_s^2 d. \tag{17.33}$$

On the other hand, the total domain wall area is l/d, so that the total wall energy per unit area of the plate is given by

$$\varepsilon_w = \frac{\gamma l}{d}, \tag{17.34}$$

where γ is the wall energy. The equilibrium domain width, d, is given by minimizing the total energy

$$\varepsilon = \varepsilon_m + \varepsilon_w = 1.08 \times 10^5 I_s^2 d + \frac{\gamma l}{d}, \tag{17.35}$$

or

$$\frac{\partial \varepsilon}{\partial d} = 1.08 \times 10^5 I_s^2 - \frac{rl}{d^2} = 0. \tag{17.36}$$

Solving (17.36), we have

$$d = 3.04 \times 10^{-3} \frac{\sqrt{\gamma l}}{I_s}. \tag{17.37}$$

In the case of iron, $\gamma_{100} = 1.6 \times 10^{-3}$, $I_s = 2.15$, so that the domain width in a plate with $l = 1\,\mathrm{cm} = 0.01$ (m) is calculated to be

$$d = 3.04 \times 10^{-3} \frac{\sqrt{1.6 \times 10^{-5}}}{2.15} = 5.6 \times 10^{-6}\ (\mathrm{m}). \tag{17.38}$$

Thus the magnetic domain width is small, of the order of 0.01 mm. The total energy of this domain structure is evaluated by using (17.37) in (17.35), to give

$$\varepsilon = 6.56 \times 10^2 I_s \sqrt{\gamma l} \approx 5.63\ (\mathrm{J\,m^{-2}}). \tag{17.39}$$

The magnetostatic energy of the single domain given by (17.32) is evaluated as

$$\varepsilon = \frac{I_s^2}{2\mu_0} l = \frac{2.15^2 \times 10^{-2}}{2 \times 4\pi \times 10^{-7}} \approx 1.8 \times 10^4\ (\mathrm{J\,m^{-2}}). \tag{17.40}$$

Therefore the energy of the equilibrium domain structure is about $1/10\,000$ of that of a single domain.

In the case of the checkerboard pattern shown in Fig. 17.2, the magnetostatic energy of the free poles on the top and bottom surfaces is given by

$$\varepsilon_m = 6.72 \times 10^4 I_s^2 d, \tag{17.41}$$

which is smaller that that of the laminated pattern, given by (17.33). On the other hand, the total domain wall area is twice as large as that of the laminated domains, or

$$\varepsilon_w = \frac{2\gamma l}{d}. \tag{17.42}$$

Minimizing the total energy

$$\frac{\partial \varepsilon}{\partial d} = 6.72 \times 10^4 I_s^2 - \frac{2\gamma l}{d^2} = 0 \tag{17.43}$$

leads to an equilibrium domain width

$$d = 5.47 \times 10^{-3} \frac{\sqrt{\gamma l}}{I_s}, \tag{17.44}$$

so that the total energy becomes

$$\varepsilon = 7.37 \times 10^2 I_s \sqrt{\gamma l}. \tag{17.45}$$

Comparing (17.45) with (17.39), we find that (17.45) is larger than (17.39) for any value of I_s, γ, or l. We therefore conclude that the checkerboard domain structure will never exist. Actually, however, we observe a checkerboard-like domain pattern on the c-plane of a cobalt single crystal, as shown in Fig. 17.9(a). This contradiction between theory and experiment can be understood by observing the domain pattern on the surface parallel to the c-axis of the same crystal (Fig. 17.9(b)). We see that the width of the domains is small at the surface perpendicular to the c-axis, but increases

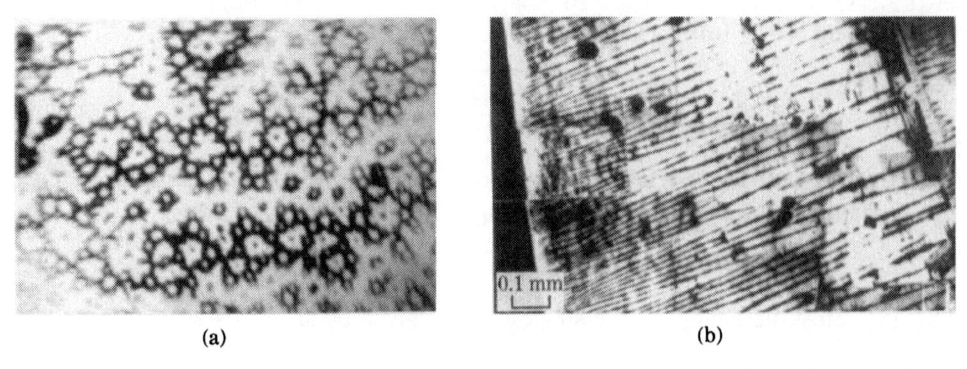

Fig. 17.9. Domain pattern observed on cobalt crystals with the crystal surface (a) perpendicular to the c-plane and (b) parallel to the c-axis. (After Y. Takata)

gradually with increasing the distance from this surface. Figure 17.10 is a sketch of the domain structure at the crystal surface. The surface domains exist only to a limited depth, so that the total domain wall area is not as large as that assumed in the above calculation.

In a cubic crystal such an iron or nickel where the number of easy axes is more than one, the formation of magnetic free poles can be avoided by generating *closure domains* with magnetizations parallel to the crystal surface, as shown in Fig. 17.11. In this case no free poles appear, and no magnetostatic energy is stored.

In general, however, ferromagnetic materials exhibit magnetostriction, so that if the magnetostriction constant is positive, the closure domains tend to elongate parallel to the crystal surface, as shown in Fig. 17.12. Since the closure domain must be compressed into a triangular form which matches the major domains, some magnetoelastic energy is stored. In this case, the width of the domains will be determined by a balance between the magnetoelastic energy and the wall energy. The strain in the closure domain is given by $e_{xx} = 3\lambda_{100}/2$, which means that work amounting to $c_{11}e_{xx}^2/2$ per unit volume must be done in order to squeeze the closure domain to fit

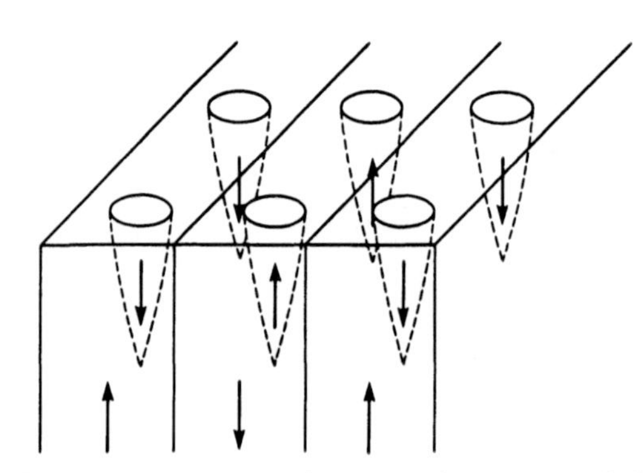

Fig. 17.10. Schematic illustration of the surface domain structure below the crystal surface perpendicular to the uniaxial easy axis.

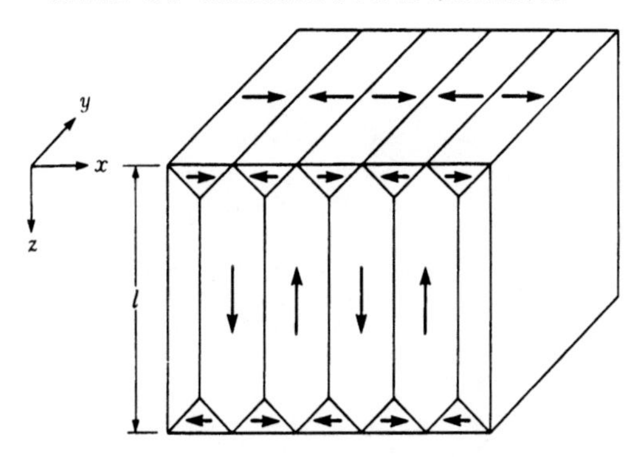

Fig. 17.11. Closure domain structure of a cubic single crystal.

with the neighboring major domains. Here c_{11} is the elastic modulus of the crystal. The magnetoelastic energy stored per unit area of crystal surface is given by

$$\epsilon_{el} = \frac{c_{11}}{2}\left(\tfrac{9}{4}\lambda_{100}^2\right)\frac{d}{2} = \tfrac{9}{16}\lambda_{100}^2 c_{11} d. \tag{17.46}$$

The total energy is given by

$$\varepsilon = \varepsilon_{el} + \varepsilon_w = \tfrac{9}{16}\lambda_{100}^2 c_{11} d + \frac{\gamma l}{d}. \tag{17.47}$$

The condition for minimizing the total energy is

$$\frac{\partial \varepsilon}{\partial d} = \tfrac{9}{16}\lambda_{100}^2 c_{11} - \frac{\gamma l}{d^2} = 0, \tag{17.48}$$

from which we have

$$d = \frac{4}{3}\sqrt{\frac{\gamma l}{\lambda_{100}^2 c_{11}}}. \tag{17.49}$$

For an iron single crystal with thickness $l = 1$ cm, $\gamma = 1.6 \times 10^{-3}$, $\lambda_{100} = 2.07 \times 10^{-5}$, $c_{11} = 2.41 \times 10^{11}$, so that

$$d = \frac{4}{3}\sqrt{\frac{1.6 \times 1 \times 10^{-5}}{2.07^2 \times 2.41 \times 10}} = 5.3 \times 10^{-4}\ (\mathrm{m}). \tag{17.50}$$

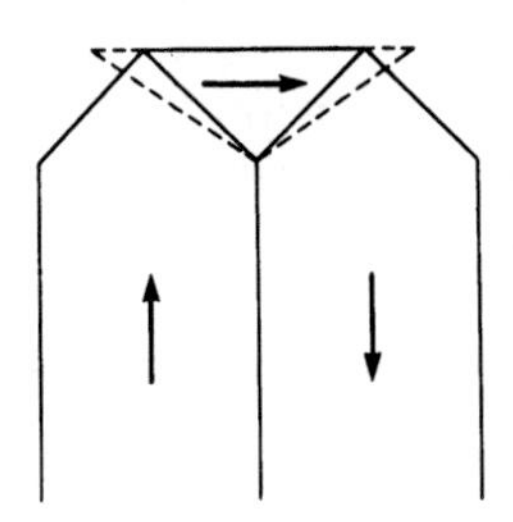

Fig. 17.12. Hypothetical magnetostrictive elongation of a closure domain.

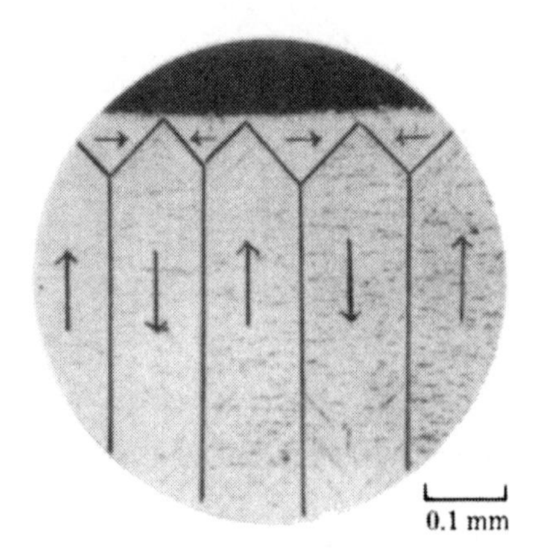

Fig. 17.13. Powder pattern of closure domains observed on the (100) surface of an Fe + 4% Si crystal (Chikazumi and Suzuki).

This value is of the order of 0.5 mm, and is much larger than the case of (17.38). The total energy is given by

$$\varepsilon = \tfrac{3}{2}\sqrt{\lambda_{100}^2 c_{11}\gamma l} \approx 6.1 \times 10^{-2}\ (\mathrm{J\,m^{-2}}), \tag{17.51}$$

which is much lower than (17.39). Such closure domains are actually observed, as shown in Fig. 17.13, for the (001) surface of an iron single crystal containing a small amount of silicon.

In a uniaxial crystal such as cobalt, it is also possible to produce closure domains which are magnetized in hard directions. The closure domain in this case stores magnetic anisotropy energy, thus

$$K_{\mathrm{u}} = K_{\mathrm{u1}} + K_{\mathrm{u2}}, \tag{17.52}$$

where K_{u1} and K_{u2} are the anisotropy constants defined by (12.1). The anisotropy energy stored per unit area of the crystal surface is given by

$$\varepsilon_a = \tfrac{1}{2}K_{\mathrm{u}}d. \tag{17.53}$$

For cobalt $K_{\mathrm{u}} = 6.0 \times 10^5$ (see (12.3)), so that

$$\varepsilon_a = 3.0 \times 10^5 d. \tag{17.54}$$

On the other hand, the magnetostatic energy of the domain structure without closure domains is calculated by putting $I_{\mathrm{s}} = 1.79$ in (17.33), giving

$$\varepsilon_m = (1.08 \times 10^5) \times 1.79^2 d \approx 3.5 \times 10^5 d. \tag{17.55}$$

Comparing (17.55) with (17.54), we find that the structure with closure domains has lower energy.

When the easy axis tilts from the normal to the crystal surface, the shape of the closure domains is greatly modified. For instance, Fig. 17.14 shows a fir-tree pattern,

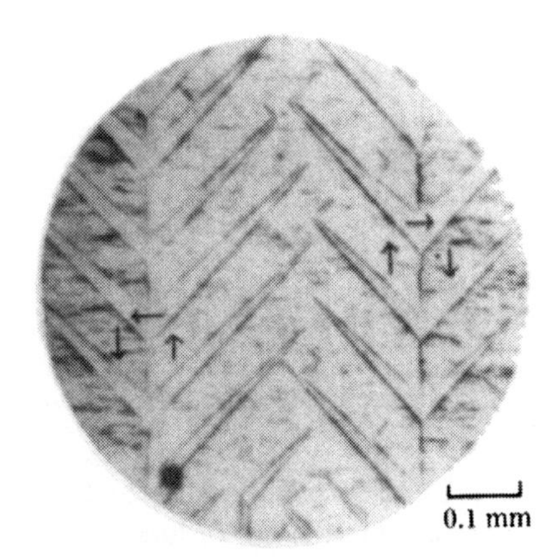

Fig. 17.14. Powder pattern of fir-tree domains (a kind of closure domain structure) observed on a tilted (001) surface of an Fe + 4% Si crystal (Chikazumi and Suzuki).

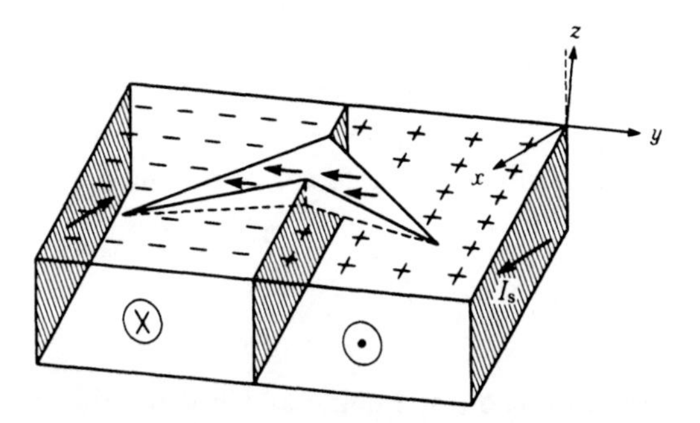

Fig. 17.15. Schematic illustration of fir-tree domains.

which is the closure domain structure observed on a tilted (001) surface of an Fe or Fe + Si crystal. The structure is illustrated in Fig. 7.15. The tree-like domain serves to transport a part of the free poles appearing on the crystal surface from one domain to the neighboring domain. The domain boundaries of the fir-tree pattern tilt from the main domain wall by some angle, which is determined by the condition that the normal component of magnetization is continuous across the 90° boundary underneath the crystal surface (see Fig. 16.19).

The observation of well-defined domain patterns and the interpretation mentioned in this section were developed by Williams *et al.*[3]

17.3 BUBBLE DOMAINS

A *bubble domain* is a cylindrical domain created in a ferromagnetic plate with an easy axis perpendicular to the plate in the presence of a magnetic field normal to the plate. Bobeck,[4] who discovered the bubble domain, suggested the possibility of utilizing it as

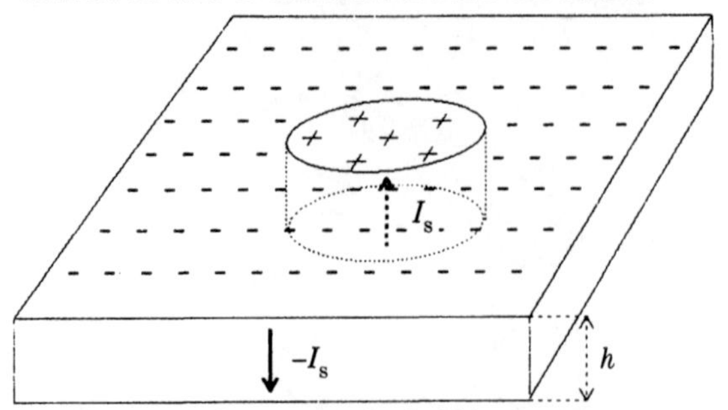

Fig. 17.16. Structure of a bubble domain.

a digital memory device. Figure 17.16 shows the structure of the bubble domain. The surface tension of the domain wall tends to reduce the bubble radius, while the magnetostatic energy of the magnetic free poles tends to increase the radius. The latter effect is due to the fact that the internal magnetic field in the bubble domain is parallel to its magnetization. The reason is as follows. As shown in Fig. 17.17, the magnetization and free pole distribution of the bubble domain (a) can be regarded as a superposition of those of the magnetic plate magnetized downwards without any bubble domain (b) and those of a single bubble domain with a double magnetization (c). The demagnetizing field in (b) is given by I_s/μ_0, whereas that in (c) is given by $-2NI_s/\mu_0$, where N is the demagnetizing factor of the bubble domain. The resultant demagnetizing field is given by

$$H_d = (1 - 2N)\frac{I_s}{\mu_0}. \tag{17.56}$$

If $r = h/2$, where r is the radius of the bubble and h is the thickness of the plate, the shape of the bubble may be regarded as a sphere, so that N is approximately $\frac{1}{3}$.

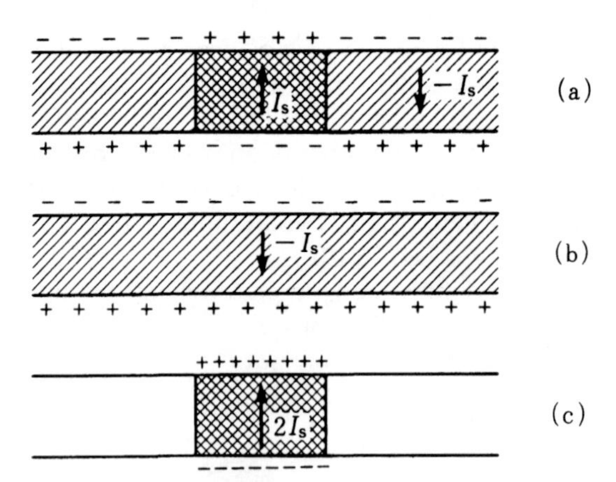

Fig. 17.17. Surface free poles and the demagnetizing field of a bubble domain: (a) a bubble domain, (b) a magnetic field without bubbles; (c) a hypothetical bubble domain with double magnetization. A superposition of (b) and (c) is equivalent to (a).

Therefore we find that the demagnetizing field given by (17.56) is positive, which favors the magnetization in the bubble. Therefore the demagnetizing field in the bubble tends to increase its radius. On the other hand, the domain wall energy γ ($\mathrm{J\,m^{-2}}$) gives rise to a surface tension, which tends to decrease the radius of the bubble. As in the case of a soap bubble, the inner pressure of the magnetic bubble is given by γ/r ($\mathrm{N\,m^{-2}}$). Comparing this with the pressure on the wall produced by a hypothetical field H_γ, which is $2I_sH_\gamma$ (see (18.4)), we have

$$H_\gamma = -\frac{\gamma}{2I_s r}, \tag{17.57}$$

where the negative sign means that the effect of the field is to reduce the radius of the bubble. In order to stabilize the bubble, we must apply a bias field $-H_b$. These three fields must be balanced, giving

$$H_d + H_\gamma - H_b = 0. \tag{17.58}$$

Using (17.56) and (17.57), (17.58) becomes

$$(1-2N)\frac{I_s}{\mu_0} = \frac{\gamma}{2I_s r} + H_b. \tag{17.59}$$

The demagnetizing factor, N, is a function of the radius of the bubble, being 0 for $r=0$, so that the left-hand side of (17.59) is I_s/μ_0. For $r = h/2$, $N = \frac{1}{3}$, so that the left-hand side of (17.59) is $I_s/3\mu_0$. For $r \to \infty$, $N \to 1$, so that this value approaches $-I_s/\mu_0$. Fig. 17.18 shows the left-hand and right-hand side values of (17.59) as functions of the reduced radius, r/h. The two values are equal at two cross-over points, P and Q. These points represent equilibrium states, where (17.59) is satisfied.

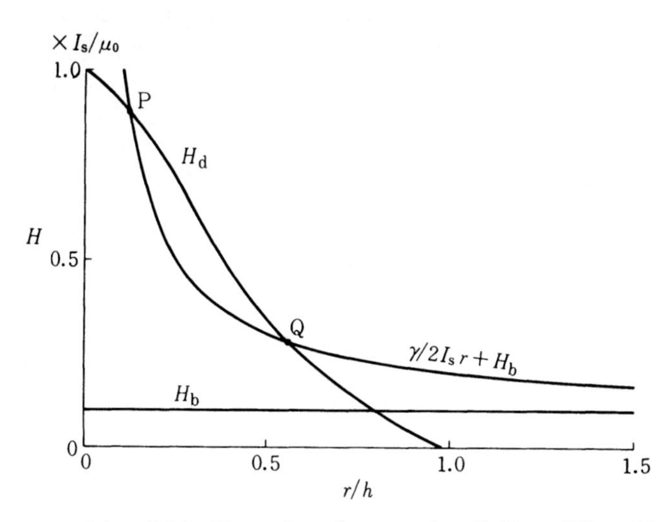

Fig. 17.18. Demagnetizing field, H_d, and surface tension field, $\gamma/2I_s r$ + bias field, H_b, as functions of the reduced radius, r/h.

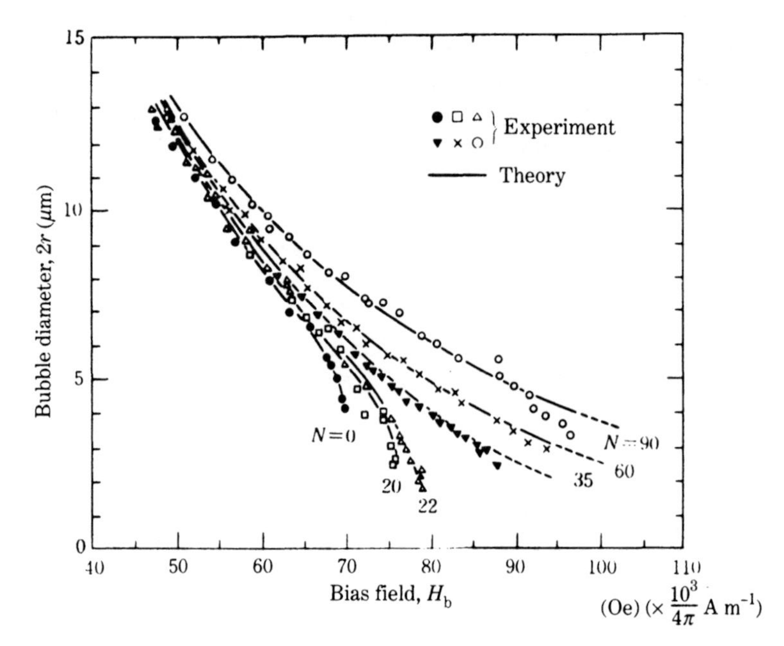

Fig. 17.19. Variation of the bubble diameter, $2r$, with increasing bias field, H_b, for a number of bubbles with different values of N.[5]

However, the point P represents an unstable equilibrium, because if r increases, H_d, which acts to increase r, becomes even larger; if r decreases, the term that decreases r becomes larger. On the other hand, the point Q represents a stable equilibrium, because any deviations of r tend to return r to its original value. When the bias field, H_b, is increased, the point Q shifts to the left, so that the bubble shrinks. This behavior is illustrated by the experimental curve denoted by $N = 0$ in Fig. 17.19, which corresponds to the physical situation described in this paragraph.

There is a special class of bubbles whose radius is rather insensitive to H_b. These bubbles contain many Bloch lines, as shown in Fig. 17.20(b), and are called *hard bubbles*.[5-7] Let the number of revolutions of the spin in the midplane of the bubble wall along the full circumference of the wall be N.[8] Therefore a normal bubble as shown in Fig. 17.20(a) has $N = 0$. For a hard bubble with large N, a decrease in the bubble radius, r, results in an increase in the density of Bloch lines, thus increasing the exchange energy and accordingly the wall energy, γ. Therefore the curve in Fig. 17.18 representing the right-hand side of (17.59) becomes steeper, so that the decrease in r caused by increasing H_b should become more gentle. Figure 17.19 shows the dependence of the bubble radius on the bias field, H_b, for a number of hard bubbles with different values of N. We see a beautiful agreement between theory and experiment.

Figure 17.21 shows a Lorentz electron micrograph of the bubble domains observed

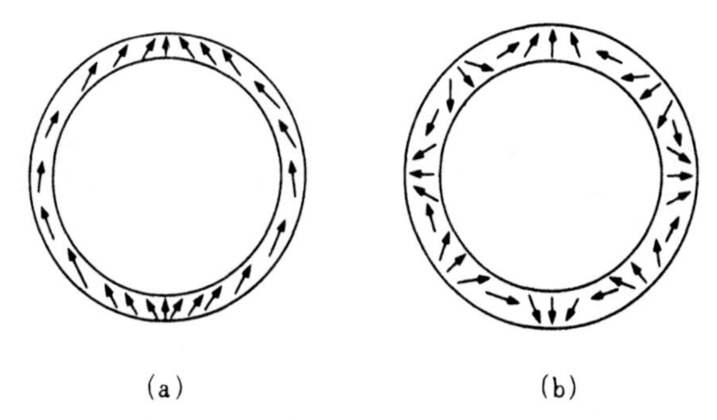

Fig. 17.20. Variation of orientation of spins in the midlayer of the bubble domain; (a) a normal bubble ($N = 0$), (b) a hard bubble ($N = 3$).

in a Co thin film with the c-axis perpendicular to the film surface, under the application of a bias field of $0.74\,\mathrm{MA\,m^{-1}}$ ($=9.2\,\mathrm{kOe}$). Three kinds of bubbles are seen in the photograph. The contrasts of these bubbles illustrated in (a), (b), and (c) can be interpreted assuming the spin structure as shown in (a)′, (b)′, and (c)′, corresponding to $N = 1$, -1, and 0, respectively.

Bubble domains can be utilized as memory or logic devices in which the bubbles are transferred along a circuit patterned on a single-crystal plate (see Section 22.3). However, hard bubbles are undesirable for such devices, because their movement is abnormal. In order to suppress the appearance of hard bubbles, the perpendicular anisotropy of a surface layer of the crystal is erased by ion bombardment,[9] or a thin Permalloy film is evaporated on top of the crystal.[10] The soft magnetic layer is effective in removing the complicated spin structure of the hard bubbles, because some normal wall nuclei are created in the soft layer and propagate downwards through the wall to destroy the unstable spin structure.

The necessary condition to create a bubble domain is that a large perpendicular anisotropy exists in the crystal. Specifically, the anisotropy field $2K_u/I_s$ must be larger than the demagnetizing field I_s/μ_0, or

$$\frac{2K_u}{I_s} \geqq \frac{I_s}{\mu_0}$$

$$\therefore\ K_u \geqq \frac{1}{2}\frac{I_s^2}{\mu_0}. \tag{17.60}$$

One crystal that meets this condition is orthoferrite (see Section 9.5), which exhibits a feeble spontaneous magnetization by spin-canted magnetism. Bubble domains can also be produced in rare-earth iron garnets with a small spontaneous magnetization made by replacing some of Fe ions on the $24d$ lattice sites by nonmagnetic Ga ions. In this case a growth-induced anisotropy (see Section 13.3.5) is used to create the bubble domains.

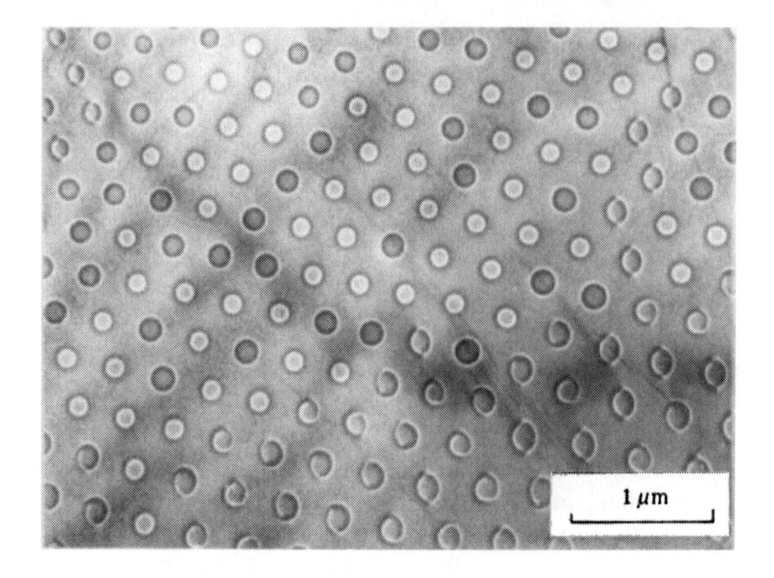

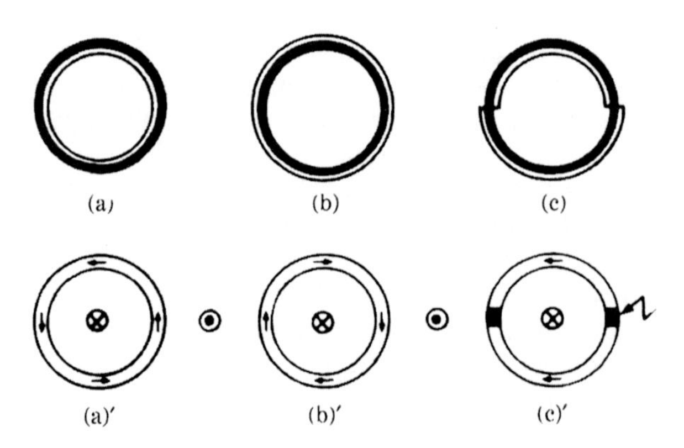

Fig. 17.21. Lorentz electron micrograph of bubble domains observed in a Co film with the c-axis perpendicular to the film surface in the presence of the bias field of $0.74\,\mathrm{MA\,m^{-1}}$ ($=9.2$ kOe). Three kinds of bubbles (a), (b) and (c) are interpreted by assuming the spin structure (a)′, (b)′ and (c)′, respectively. The broken arrow indicates a Bloch line. (Courtesy of D. Watanabe).

17.4 STRIPE DOMAINS

Stripe domains are associated with a rotatable magnetic anisotropy discovered in anomalous magnetic thin films (see Section 13.4.3).[11] Figure 17.22 shows a schematic illustration of the spin distribution in stripe domains. Stripe domains appear when the perpendicular anisotropy is not greater than the value that satisfies (17.60). This spin

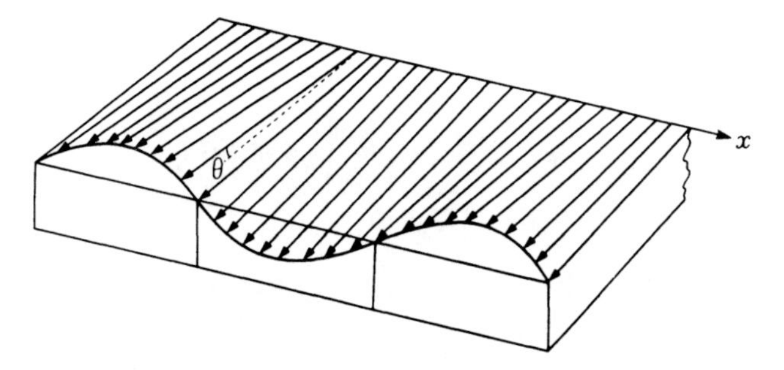

Fig. 17.22. Schematic illustration of the spin structure of stripe domains.

structure is nature's clever solution to the problem of simultaneously minimizing the perpendicular anisotropy and the magnetostatic energy.

We assume that the angle of deviation of the spins from the surface of the thin film can be approximated by a sinusoidal function,

$$\theta = \theta_0 \sin\left(2\pi \frac{x}{\lambda}\right), \tag{17.61}$$

where x is the coordinate axis perpendicular to the average spin axis and λ is the wavelength of the spin variation or the width of a stripe domain. When $\theta \ll \pi/2$, the surface density of magnetic free poles is given by

$$\begin{aligned} \omega &= I_s \sin\theta \simeq I_s \theta \\ &= I_s \theta_0 \sin\left(2\pi \frac{x}{\lambda}\right). \end{aligned} \tag{17.62}$$

The magnetostatic energy of such a free pole distribution is easily calculated by adopting the first term ($n = 1$) in (17.7) and averaging over one wavelength, as in (17.10):

$$\begin{aligned} \varepsilon_m &= \frac{I_s^2 \lambda \theta_0^2}{4\pi\mu_0} \left\langle \sin^2\left(2\pi \frac{x}{\lambda}\right) \right\rangle \\ &= \frac{I_s^2 \lambda \theta_0^2}{8\pi\mu_0}, \end{aligned} \tag{17.63}$$

where the value is doubled by considering the effect of free poles on both sides of the thin film. The magnetic anisotropy energy is expressed in terms of the angle, θ, as

$$\begin{aligned} E_a &= -K_u \cos^2\left(\frac{\pi}{2} - \theta\right) \\ &= -K_u \sin^2\theta \\ &\simeq -K_u \theta^2. \end{aligned} \tag{17.64}$$

Using (17.61) and averaging over one wavelength of the spin oscillation, we have the anisotropy energy per unit area as

$$\varepsilon_a = -K_u\theta_0^2\left\langle \sin^2\left(2\pi\frac{\pi}{\lambda}\right)\right\rangle h$$
$$= -\tfrac{1}{2}K_u\theta_0^2 h, \tag{17.65}$$

where h is the thickness of the film. The exchange energy per unit area stored in this spin system is calculated, using (17.61) in (16.14), to be

$$\varepsilon_{ex} = A\left\langle\left(\frac{\partial\theta}{\partial x}\right)^2\right\rangle h$$
$$= A\left(\frac{2\pi}{\lambda}\right)^2\theta_0^2\left\langle\cos^2\left(2\pi\frac{x}{\lambda}\right)\right\rangle h$$
$$= \frac{2\pi^2}{\lambda^2}\theta_0^2 Ah. \tag{17.66}$$

The total energy is given by

$$\varepsilon = \varepsilon_m + \varepsilon_a + \varepsilon_{ex}$$
$$= \left(\frac{I_s^2\lambda}{8\pi\mu_0} - \frac{K_u h}{2} + \frac{2\pi^2 Ah}{\lambda^2}\right)\theta_0^2$$
$$= -w\theta_0^2, \tag{17.67}$$

where $-w$ represents the function in parentheses. If $w > 0$, the stripe domain appears. The wavelength, λ, is obtained by minimizing w with respect to λ, giving

$$\frac{\partial w}{\partial\lambda} = 0$$
$$\therefore \frac{I_s^2}{8\pi\mu_0} - \frac{4\pi^2 Ah}{\lambda^3} = 0 \tag{17.68}$$
$$\therefore \lambda = 4\pi\sqrt[3]{\frac{\pi_0 Ah}{2I_s^2}}.$$

Using this value, the condition $w > 0$ reduces to

$$K_u > \frac{3I_s}{2}\sqrt[3]{\frac{AI_s}{2\pi_0^2 h^2}}. \tag{17.69}$$

When a rotating field of moderate intensity is applied to the stripe domains, the configuration of stripe domains remains unchanged, whereas the spontaneous magnetization rotates with the external field. In this process some magnetostatic energy is

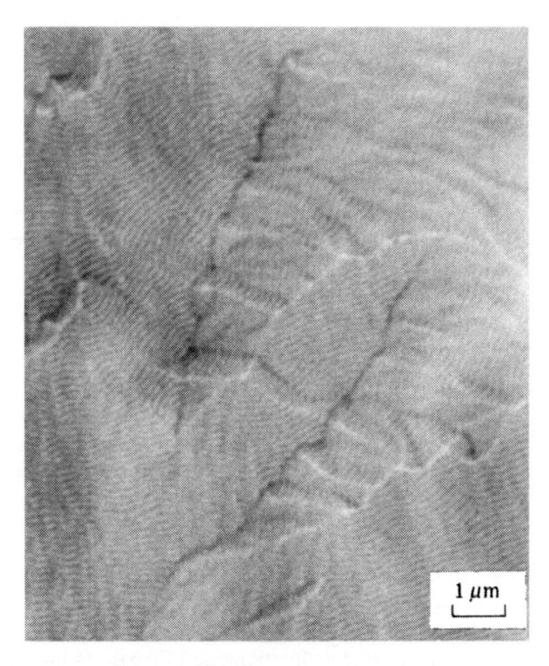

Fig. 17.23. Lorentz electron micrograph of stripe domains as observed on a 95Fe–5Ni alloy thin film, 1200 Å thick.[13]

added to the energy given by (17.67), because some magnetic free poles appear at the domain boundaries. In other words, an additional magnetic anisotropy energy is created. If the field exceeds some critical value, the stripes rotate, which means that the easy axis of this anisotropy is changed. This is the rotatable anisotropy discussed in Section 13.4.3.

We have assumed that the variation of the spin-canting angle is sinusoidal. Murayama[12] calculated this spin configuration more precisely by using a variational method.

Figure 17.23 shows a Lorentz electron micrograph of stripe domains as observed on an Fe–Ni thin film. Stripe domains observed by the powder pattern method were shown in Fig. 13.29.

17.5 DOMAIN STRUCTURE OF FINE PARTICLES

As discussed in Section 17.2, the domain width is related to the thickness of the crystal, l. For instance, as seen in (17.37), (17.44), and (17.49), the domain width is always proportional to $\sqrt{l}$. Therefore when the size of the particle is decreased, keeping its shape unchanged, the domain width decreases more gradually than the size of the particle, thus finally resulting in a *single domain structure*.

Consider a spherical ferromagnetic particle with radius r. If this sphere is divided

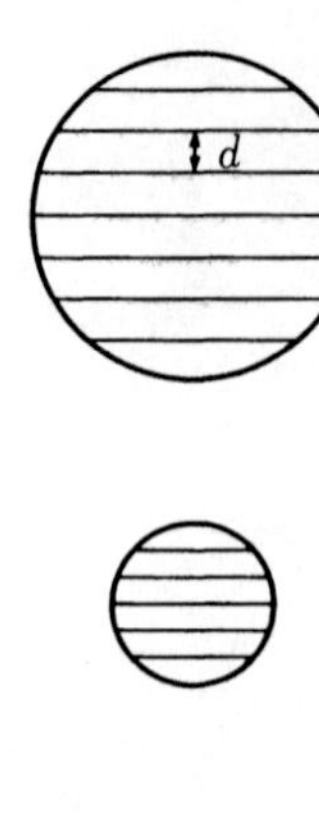

Fig. 17.24. Relationship between the size of ferromagnetic particles and the width of magnetic domains.

into multiple domains, each of width d (see Fig. 17.24), the total wall energy is roughly estimated as

$$U_w = \gamma \times (\pi r^2) \frac{2r}{d} = \frac{2\pi\gamma r^3}{d}. \tag{17.70}$$

On the other hand, the magnetostatic energy is roughly estimated to be $d/2r$ times the value for a single domain particle, giving

$$U_m = \frac{I_s^2}{6\mu_0} \frac{4}{3} \pi r^3 \frac{d}{2r} = \frac{\pi I_s^2 r^2}{9\mu_0} d. \tag{17.71}$$

The total energy of the particle

$$U = U_w + U_m \tag{17.72}$$

is minimized by the condition

$$\frac{\partial U}{\partial d} = -\frac{2\pi\gamma r^3}{d^2} + \frac{\pi I_s^2 r^2}{9\mu_0} = 0, \tag{17.73}$$

giving

$$d = \sqrt{\frac{18\gamma\mu_0 r}{I_s^2}}. \tag{17.74}$$

As the radius r of the particle is decreased, the width of the domains is decreased proportional to $\sqrt{r}$, so the number of domains in the sphere decreases as shown in Fig. 17.24, attaining finally the single domain structure below a critical radius, r_c. At this critical radius, $d = 2r_c$, so that from (17.74) we have

$$2r_c = \sqrt{\frac{18\gamma\mu_0 r_c}{I_s^2}}$$

$$\therefore r_c = \frac{9\gamma\mu_0}{2I_s^2}. \tag{17.75}$$

For iron, $I_s = 2.15$, $\gamma = 1.6 \times 10^{-3}$, so that

$$r_c = 2 \times 10^{-9}\ \mathrm{m} = 20\,\text{Å}. \tag{17.76}$$

The presence of a single domain structure was first predicted by Frenkel and Dorfman[14] in 1930, and more than ten years later, further detailed calculations were made by Kittel,[15] Néel,[16] and Stoner and Wohlfarth.[17] It was pointed out by Kittel[15] that single domain structures are also expected for ferromagnetic fine wires and thin films.

Figure 17.25 shows an experimental verification of single domain structure as observed for Ba-ferrite fine particles by means of a powder pattern method using a

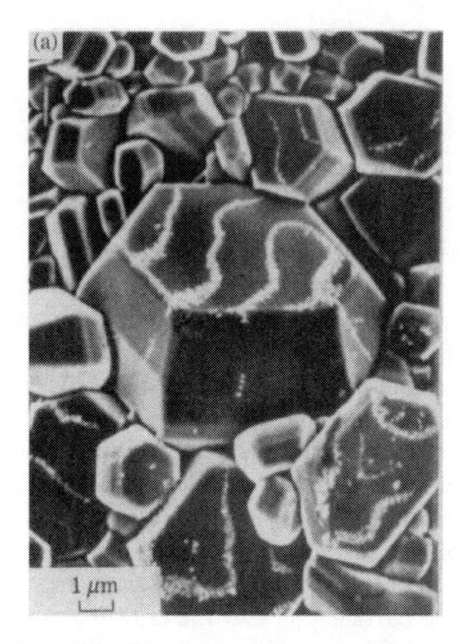

Fig. 17.25. Electron micrographs of powder patterns (Colloid SEM) of Ba-ferrite: (a) for relatively large particles showing a multi-domain structure; (b) for relatively small particles showing a single domain structure.[19] (Courtesy of K. Goto)

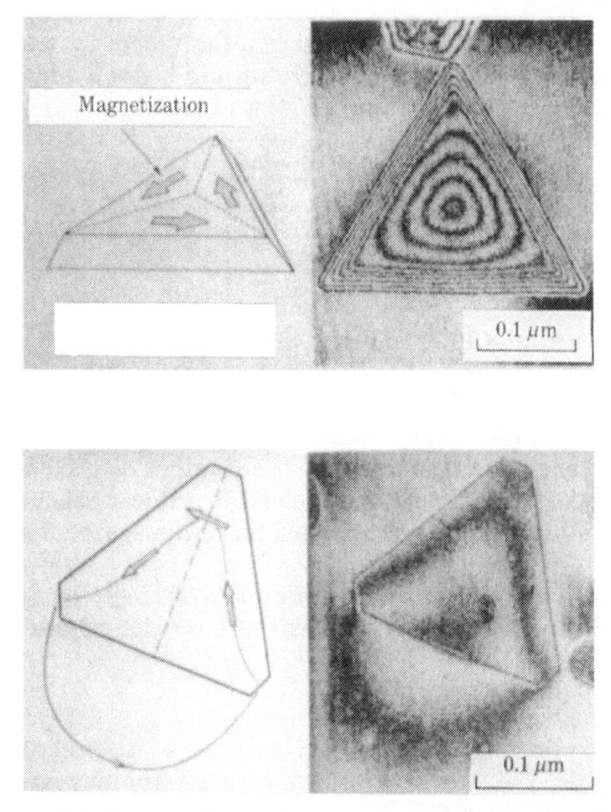

Fig. 17.26. Electron interference fringes observed for Co platelets showing lines of magnetic flux inside and outside the platelets. The thickness of the platelet in (b) is about one quarter of that of (a).[20,21] (Courtesy of A. Tonomura)

scanning electron microscope (Colloid SEM).[18] We see many domain walls on a relatively large particle, showing a multi-domain structure (see photo (a)), but only a cluster of magnetic powder on a small particle, showing a single domain structure (see photo (b)).[19]

Figure 17.26 shows electron interference micrographs of Co platelets with different thicknesses, showing the magnetic flux lines inside and outside of the platelet. In photo (a) we see that the lines of magnetic flux are confined inside a relatively thick platelet, while in photo (b) a flux line emerges outside a thin platelet, showing that this platelet tends to have a single domain structure.[20,21]

In the preceding discussion we treated an isolated particle. In an assembly of single domain particles, separated by a nonmagnetic matrix, the magnetostatic energy is reduced by the magnetostatic interaction between particles, so that the critical size for a single domain structure is more or less increased. The situation is also the same for a ferromagnet that contains many nonmagnetic inclusions or voids. Even in a purely magnetic material, if the direction of the easy axis changes from grain to grain, the individual grains will have a single domain structure if the grains are sufficiently small.

The domain structure of such a non-uniform magnetic material will be discussed in Section 17.6.

17.6 DOMAIN STRUCTURES IN NON-IDEAL FERROMAGNETS

In the preceding section, we learned that ideal ferromagnets have quite regular domain structures. Such a regular domain structure can be actually attained for carefully prepared single crystals. In contrast, the domain structures of ordinary ferromagnetic materials are influenced by irregularities such as voids, nonmagnetic inclusions, internal stresses, and grain boundaries. In magnetically hard materials such as permanent magnets, such irregularities are the main factors which govern the size and distribution of the ferromagnetic domains.

First we discuss the domain structure of polycrystals. The domain magnetization is more or less continuous across grain boundaries, to decrease the magnetostatic energy. This continuity is, however, not perfect,[22] except in the special case where the grain boundary has a particular orientation with respect to the directions of the axes of easy magnetization on both sides of the boundary. Let us calculate the magnetostatic energy of the free poles which appear at an average grain boundary. First we treat a uniaxial crystal with positive K_u, such as cobalt, that has only one easy axis. Suppose that the domain magnetizations on the two sides of the grain boundary make angles θ_1 and θ_2 with the normal to the boundary, as shown in Fig. 17.27. The surface density of magnetic free poles is given by

$$\omega = I_s(\cos\theta_1 - \cos\theta_2). \tag{17.77}$$

As seen in previous calculations, the magnetostatic energy is always proportional to ω^2; hence we only need to calculate the average value of ω^2 over all possible combinations of θ_1 and θ_2. If θ_1 is in the range $-\pi/2 < \theta_1 < \pi/2$, θ_2 must be also in the range $-\pi/2 < \theta_2 < \pi/2$, because otherwise the two magnetizations have components in opposite directions. This would lead to a large magnetic free pole density that could be lowered if one of the magnetizations reversed its direction. Thus

$$\begin{aligned}
\overline{\omega^2} &= I_s^2 \int_0^{\pi/2}\int_0^{\pi/2} (\cos\theta_1 - \cos\theta_2)^2 \sin\theta_1 \sin\theta_2 \, d\theta_1 \, d\theta_2 \\
&= I_s^2 \left\{ \int_0^{\pi/2} \cos^2\theta_1 \sin\theta_1 \, d\theta_1 + \int_0^{\pi/2} \cos^2\theta_2 \sin\theta_2 \, d\theta_2 \right. \\
&\qquad \left. -2\int_0^{\pi/2}\int_0^{\pi/2} \cos\theta_1 \cos\theta_2 \sin\theta_1 \sin\theta_2 \, d\theta_1 \, d\theta_2 \right\} \\
&= I_s^2 \left\{ \frac{1}{3} + \frac{1}{3} - \frac{1}{2} \right\} = \frac{I_s^2}{6}.
\end{aligned} \tag{17.78}$$

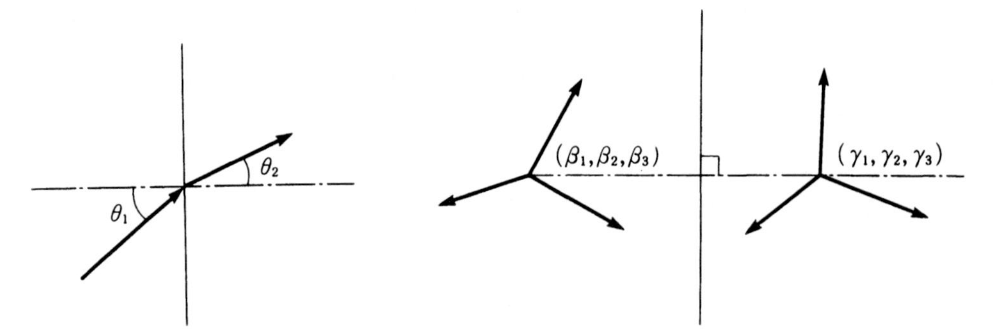

Fig. 17.27. Angles of magnetizations of the neighboring grains as measured from the normal to the grain boundary.

Fig. 17.28. Crystal orientations on both sides of a grain boundary.

If the individual grains are separated from one another, it follows that

$$\overline{\omega^2} = I_s^2 \int_0^{\pi/2} \cos^2\theta_1 \sin\theta_1 \, d\theta_1 = \frac{I_s^2}{3}. \quad (17.79)$$

Thus we find that the magnetostatic energy of a uniaxial polycrystal is about one-half that of the isolated grains. Therefore from the calculation in (b) above, we find that the domain width in a polycrystal is about $\sqrt{2}$ times larger than the domain width in isolated grains, and that the total domain energy is lower by a factor $1/\sqrt{2}$.

For a cubic crystal, which has three or four axes of easy magnetization (depending on the sign of the anisotropy), the probability of having a good continuity of magnetization across the grain boundary is much greater. Let the direction cosines of the normal to the grain boundary be $(\beta_1, \beta_2, \beta_3)$ with respect to the cubic axes of the left-hand grain, and $(\gamma_1, \gamma_2, \gamma_3)$ with respect to the right-hand grain (Fig. 17.28). For crystals with positive K_1 such as iron, the axes of easy magnetization are $\langle 100 \rangle$; hence the magnetization in either grain can lie along the x-, y-, or z-direction (i or $j = 1, 2,$ or 3). When the magnetization is in the ith easy axis in the left-hand grain and the jth easy axis in the right-hand grain, the surface density of magnetic free pole is given by

$$\omega = I_s(\beta_i - \gamma_j). \quad (17.80)$$

Now let us find the magnetization directions in the two grains that minimize the surface density of magnetic free poles given by (17.80). Figure 17.29 shows the three crystal axes (x, y, z) of the right-hand grain, the normal $\boldsymbol{n}$ to the grain boundary, and the direction of magnetization I_1 in the left-hand grain. If $\gamma_3 > \gamma_2 > \gamma_1$ and $\beta_i > \gamma_3$, the direction of magnetization, I_2, in the right-hand grain should be parallel to the z-axis, because this is the easy axis closest to I_1. The free pole density on the grain boundary is then given by

$$\omega = I_s(\beta_i - \gamma_3). \quad (17.81)$$

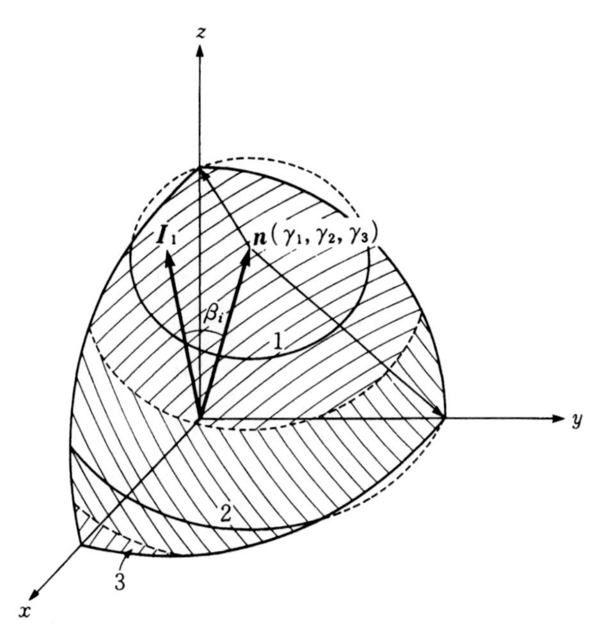

Fig. 17.29. If the domain magnetization I_1 in the first grain is in the region indicated by 1, 2, and 3, the domain magnetization I_2 in the second grain must be parallel to the z-, y-, and x-direction, respectively, to minimize the surface density of free poles appearing on the grain boundary.

If β_i is less than γ_3, then ω changes its sign. If $\beta_i < (\gamma_3 + \gamma_2)/2$, ω becomes smaller when the magnetization in the right-hand grain, I_2, is parallel to the y-axis. Then

$$\omega = I_s(\beta_i - \gamma_2). \tag{17.82}$$

If β_i is further decreased and it happens that $\beta_i < (\gamma_2 + \gamma_1)/2$, the x-direction becomes the most favorable for I_2; thus

$$\omega = I_s(\beta_i - \gamma_1). \tag{17.83}$$

On averaging ω^2 with respect to β_i, we have

$$\begin{aligned}\int_0^1 \omega^2 \, d\beta_i = I_s^2 \bigg\{ & \int_{(\gamma_3-\gamma_2)/2}^1 (\beta_i - \gamma_3)^2 \, d\beta_i + \int_{(\gamma_2+\gamma_1)/2}^{(\gamma_3+\gamma_2)/2} (\beta_i - \gamma_2)^2 \, d\beta_i \\ & + \int_0^{(\gamma_2+\gamma_1)/2} (\beta_i - \gamma_1)^2 \, d\beta_i \bigg\} \\ = I_s^2 \big(& \tfrac{1}{3} - \gamma_3 + \gamma_3^2 - \tfrac{1}{4}\gamma_3^3 - \tfrac{1}{4}\gamma_3^2\gamma_2 + \tfrac{1}{4}\gamma_3\gamma_2^2 \\ & - \tfrac{1}{4}\gamma_2^2\gamma_1 + \tfrac{1}{4}\gamma_2\gamma_1^2 + \tfrac{1}{4}\gamma_1^3 \big).\end{aligned} \tag{17.84}$$

On averaging each term in (17.84) over all possible orientations of the right-hand

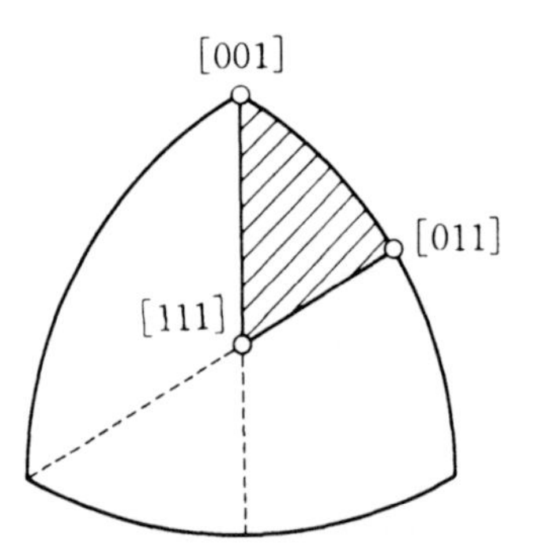

Fig. 17.30. Illustration indicating the range of integration to calculate the average values in (17.85).

grain in the range $\gamma_3 > \gamma_2 > \gamma_1$, which is shown as a shaded area in Fig. 17.30, we have

$$\left.\begin{aligned} &\bar{\gamma}_3 = 0.832,\ \overline{\gamma_3^2} = 0.764,\ \overline{\gamma_3^3} = 0.596 \\ &\overline{\gamma_3^2\gamma_2} = 0.405,\ \overline{\gamma_3\gamma_2^2} = 0.202,\ \overline{\gamma_2^2\gamma_1} = 0.054 \\ &\overline{\gamma_2\gamma_1^2} = 0.041,\ \overline{\gamma_1^3} = 0.020 \end{aligned}\right\}. \tag{17.85}$$

Then (17.84) becomes

$$\begin{aligned} \overline{\omega^2} &= I_s^2\left(\tfrac{1}{3} - 0.832 + 0.764 - \tfrac{0.596}{4} - \tfrac{0.405}{4} + \tfrac{0.202}{4} - \tfrac{0.054}{4} + \tfrac{0.041}{4} + \tfrac{0.020}{4}\right) \\ &= 0.067 I_s^2. \end{aligned} \tag{17.86}$$

On comparing (17.86) with the average value (17.79) for an isolated grain, we see that the magnetostatic energy of a cubic polycrystal is about 0.20 that of the separated grains, so that its domain width will be $1/\sqrt{0.20} = 2.2$ times larger, and the total energy will be $\sqrt{0.20} = 0.45$ that of the separated grains. Thus the magnetostatic energy of the magnetic free poles appearing on the grain boundaries is small but still nonzero, so that the domain structures of polycrystalline materials are essentially determined by the size of the grains.

Next let us discuss the domain structure of a strongly stressed crystal. When a tension stress σ exists in crystal, there will be an induced magnetic anisotropy, given by (14.96), which rotates the local magnetization towards the axis of tension if $\lambda > 0$. Figure 17.31 shows the domain pattern observed on the (001) surface of an Fe_3Al crystal containing a mechanical indentation. Since this alloy has a large magnetostriction ($\lambda = 3.7 \times 10^{-5}$) and a small magnetocrystalline anisotropy, the direction of domain magnetization is mainly determined by the distribution of residual stresses in the crystal. If we assume that $\sigma = 100\,\mathrm{kg\,mm^{-2}} = 10^9\,\mathrm{N\,m^{-2}}$, the anisotropy constant becomes

$$\tfrac{3}{2}\lambda\sigma = (1.5) \times (3.7 \times 10^{-5}) \times 10^9 \approx 5.6 \times 10^4\ (\mathrm{J\,m^{-3}}),$$

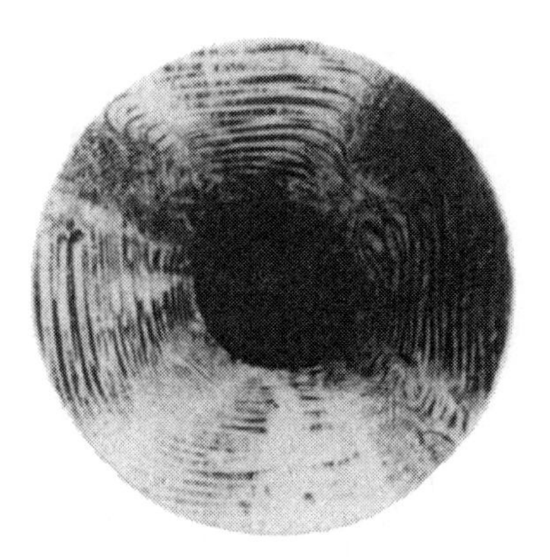

Fig. 17.31. Magnetic domains induced by strong stresses produced by an indentation on (001) surface of Fe_3Al (Chikazumi, Suzuki, and Shimizu).

which is almost of the same order of magnitude as the magnetocrystalline anisotropy constant of iron. The domain structure of such a strongly stressed crystal is similar to the uniaxial polycrystal, except that the easy axis changes its orientation from place to place. Domain structures of extremely soft materials such as Permalloy and Supermalloy, which have a small magnetocrystalline anisotropy, are considered to belong to this category.[20]

Next we discuss the domain structure of a material which includes voids, inclusions, and precipitates.

Figure 17.32(a) shows a spherical void or nonmagnetic inclusion existing in a magnetic domain. Since the surface free poles on a sphere are equivalent to a dipole, the magnetostatic energy associated with the void is given by

$$U = \frac{1}{2}\frac{I_s^2}{3\mu_0}\frac{4\pi r^3}{3} = 5.6 \times 10^5 I_s^2 r^3, \tag{17.87}$$

where r is the radius of the sphere. If a domain wall bisects this void as shown in (b), the free pole areas are divided into regions of opposite polarity, thus reducing the magnetostatic energy to about half the value given by (17.87). Therefore the magnetostatic energy is reduced by

$$\Delta U \approx 2.8 \times 10^5 I_s^2 r^3. \tag{17.88}$$

Suppose that N spherical voids are distributed in a unit volume of the crystal. We assume that the crystal is composed of $\pm y$ domains separated by 180° walls which are almost parallel to the y–z plane (Fig. 17.33). The total wall energy per unit volume of the crystal is given by γ/d, where d is the average separation of the neighboring walls. On the other hand, an individual wall tends to pass through a void to decrease the

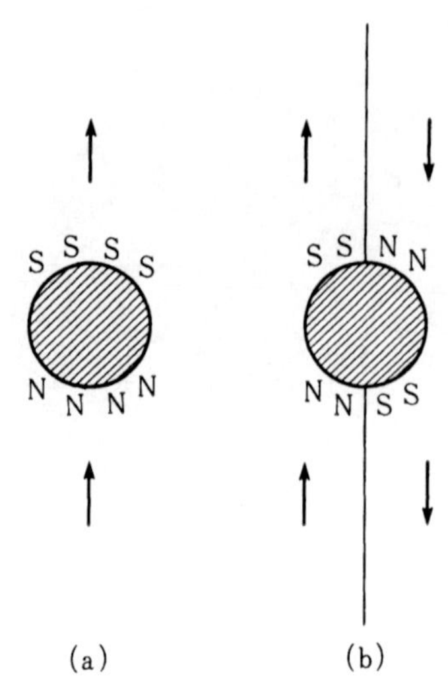

Fig. 17.32. Free magnetic poles appearing on a surface of a spherical void or a nonmagnetic inclusion: (a) a void isolated in a magnetic domain; (b) a void bisected by a domain wall.

magnetostatic energy by ΔU, given by (17.88). The number of voids which are intersected by a single wall is roughly estimated to be $N^{2/3}$, so that there are $N^{2/3}/d$ voids in a unit volume, leading to a decrease in the magnetostatic energy. Therefore the decrease in total magnetostatic energy is given by $N^{2/3}\Delta U/d$. If

$$N^{2/3}\Delta U \geqq \gamma, \tag{17.89}$$

a wall is generated, irrespective of the outside shape of the crystal. In the case of iron,

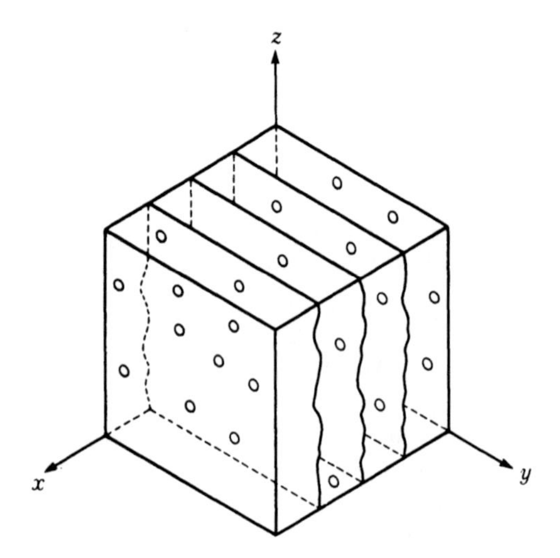

Fig. 17.33. Domain walls stabilized by a high density of distributed voids in a crystal.

which contains voids of radius $r = 0.01$ mm, we can estimate ΔU taking $I_s = 2.15$, obtaining

$$\Delta U = (3 \times 10^5) \times (2.15)^2 \times (1 \times 10^{-5})^3 = 1.4 \times 10^{-9}\ (\mathrm{J}). \tag{17.90}$$

From condition (17.89) and using $\gamma = 1.6 \times 10^{-3}$, we find that

$$N \geqq \left(\frac{\gamma}{\Delta U}\right)^{3/2} = \left(\frac{1.6 \times 10^{-3}}{1.4 \times 10^{-9}}\right)^{3/2} = 1.2 \times 10^9\ (\mathrm{m}^{-3}). \tag{17.91}$$

In other words, if the separation of the voids is less than $N^{-1/3} \cong 10^{-3}$ (m) or 1 mm, these voids can create a domain wall. Since even carefully prepared single crystals can contain such a number of voids or nonmagnetic particles, the domain structures in these crystals are determined by the distribution of voids, rather than by the external shape of the crystal.

PROBLEMS

17.1 Assume an array of ferromagnetic domains with their magnetizations parallel to the $\pm x$-axis and also parallel to the $\pm y$-axis separated by a 90° wall and many 180° walls parallel to the x–y plane at an interval d (see the figure). When the angle between the surface of the 90° wall and the x–z plane, ϕ, is varied, how does the magnetostatic energy per unit area of the 90° wall change? Let the intensity of domain magnetization be I_s, the magnetocrystalline anisotropy constant be K_1, and assume that the crystal is infinitely large.

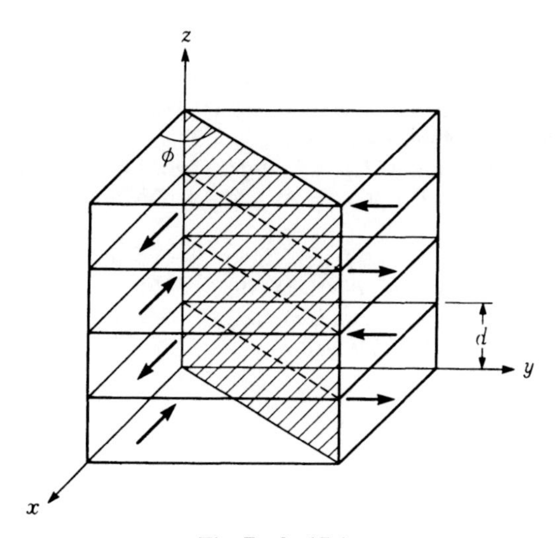

Fig. Prob. 17.1

17.2 Consider a single crystal plate of a ferromagnet with uniaxial anisotropy described by an anisotropy constant K_u. How does the width of parallelepiped domains change with a change in K_u in the following cases: (i) the easy axis is perpendicular to the crystal surface, and (ii) the easy axis makes a small angle θ with the crystal surface. Assume that $I_s^2/\mu_0 \gg K_u$.

17.3 Calculate the width of parallelepiped domains for a 1 cm thick cobalt crystal with the c-axis (easy axis) making the angle $\theta = 0.1$ radian with the crystal surface. Assume that $K_u = 4.1 \times 10^5 \, \mathrm{J\,m^{-3}}$, $I_s = 1.8\,\mathrm{T}$, and $\gamma = 1.5 \times 10^{-2} \, \mathrm{J\,m^{-2}}$.

17.4 Estimate the critical size of a grain of polycrystalline Fe in which the individual grain is composed of a single domain. Assume that the shape of the grain is a sphere.

REFERENCES

1. C. Kittel, *Rev. Mod. Phys.*, **21** (1949), 541.
2. J. B. Goodenough, *Phys. Rev.*, **102** (1956), 356.
3. H. J. Williams, R. M. Bozorth, and W. Shockley, *Phys. Rev.*, **75** (1949), 155.
4. A. H. Bobeck, *IEEE Trans. Mag.*, **MAG-5** (1969), 554.
5. T. Kobayashi, H. Nishida, and Y. Sugita, *J. Phys. Soc. Japan*, **34** (1973), 555.
6. W. J. Tabor, A. H. Bobeck, G. P. Vella-Coleiro, and A. Rosencweig, *Bell. Sys. Tech. J.*, **51** (1972), 1427.
7. A. P. Malozemoff, *Appl. Phys. Lett.*, **21** (1972), 142.
8. J. C. Slonczewski, A. P. Malozemoff, and O. Voegili, *AIP Conf. Proc.*, **10** (1973), 458.
9. R. Wolfe and J. C. North, *Bell Sys. Tech. J.*, **51** (1972), 1436.
10. M. Takahashi, N. Nishida, T. Kobayashi, and Y. Sugita, *J. Phys. Soc. Japan*, **34** (1973), 1416.
11. N. Saito, H. Fujiwara, and Y. Sugita, *J. Phys. Soc. Japan*, **19** (1964), 421, 1116.
12. Y. Murayama, *J. Phys. Soc. Japan*, **21** (1966), 2253; **23** (1967), 510.
13. T. Koikeda, K. Suzuki, and S. Chikazumi, *Appl. Phys. Lett.*, **4** (1964), 160.
14. J. Frenkel and J. Dorfman, *Nature*, **126** (1930), 274.
15. C. Kittel, *Phys. Rev.*, **70** (1946), 965.
16. L. Néel, *Comp. Rend.*, **224** (1947), 1488.
17. E. C. Stoner and E. P. Wohlfarth, *Nature*, **160** (1947), 650; *Phil. Trans.*, **A240** (1948), 599.
18. K. Goto and T. Sakurai, *Appl. Phys. Letter*, **30** (1977), 355.
19. K. Goto, M. Ito, and T. Sakurai, *Japan J. Appl. Phys.*, **19** (1980), 1339.
20. A. Tonomura, T. Matsuda, J. Endo, T. Arii, and K. Mihama, *Phys. Rev. Lett.*, **44** (1980), 1430.
21. T. Matsuda, A. Tonomura, K. Suzuki, J. Endo, N. Osakabe, H. Umezaki, H. Tanabe, Y. Sugita, and H. Fujiwara, *J. Appl. Phys.*, **53** (1982), 5444.
22. W. S. Paxton and T. G. Nilan, *J. Appl. Phys.*, **26** (1955), 65.
23. S. Chikazumi, *Phys. Rev.*, **85** (1952), 918.

Part VII
MAGNETIZATION PROCESS

In this Part, we discuss how a ferromagnet increases its magnetization upon the application of an external magnetic field. In Chapter 18 we consider technical magnetization, in which the net magnetization is changed by domain wall displacement and by rotation of domain magnetization. In Chapter 19, we see how the intrinsic magnetization responds to very high magnetic fields, where the phenomena are described in terms of a spin phase diagram. Finally, dynamical magnetization processes or time-dependent magnetization will be discussed in Chapter 20.

18
TECHNICAL MAGNETIZATION

18.1 MAGNETIZATION CURVE AND DOMAIN DISTRIBUTION

When a magnetic material is subjected to an increasing magnetic field, its magnetization is increased and finally reaches a limiting value called the *saturation magnetization*. This process is called *technical magnetization*, because it is essentially achieved by a change in the direction of domain magnetization and can be distinguished from a change in the intensity of spontaneous magnetization.

The technical magnetization process is composed of domain wall displacements and rotation of the domain magnetization. Suppose that a magnetic field, H, is applied parallel to the magnetization of one of two domains separated by a domain wall (Fig. 18.1). When the domain wall is displaced as shown in (b), there is an increase in the volume of the domain magnetized parallel to H, and an equal decrease in the volume of the domain magnetized opposite to H. The net or resultant magnetization is therefore increased. This process is called *domain wall displacement*.

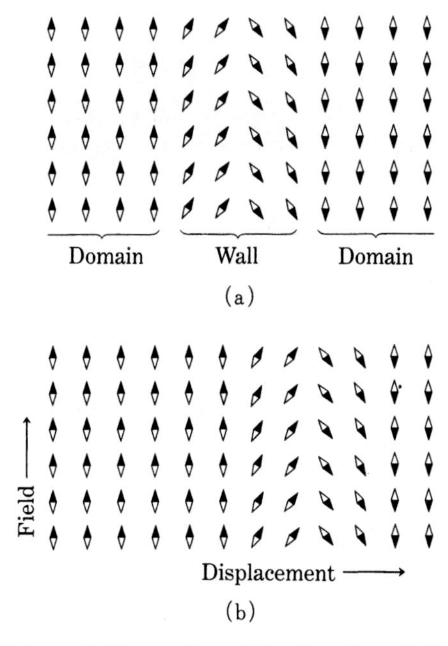

Fig. 18.1. Schematic illustration of the domain wall displacement.

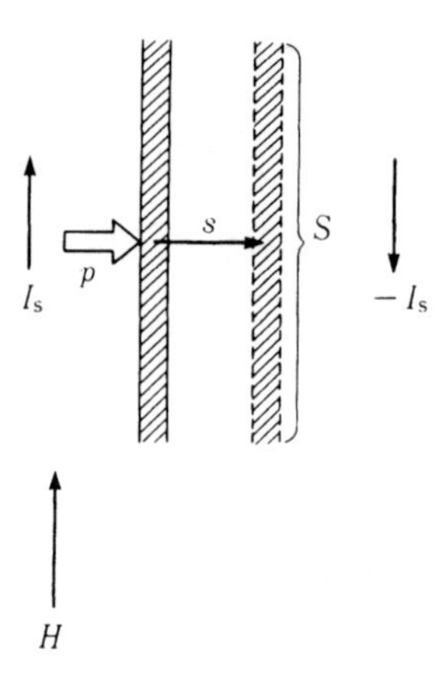

Fig. 18.2. Hypothetical pressure acting on a domain wall.

The effect of the magnetic field H on the domain wall can be replaced by a hypothetical pressure, p. Suppose that a 180° domain wall of area S is displaced by a distance s by the application of a magnetic field H parallel to the magnetization of one of the domains (see Fig. 18.2). Since the magnetization in a volume of magnitude Ss is reversed, the magnetic moment M is increased by

$$M = 2I_s Ss, \tag{18.1}$$

where I_s is the spontaneous magnetization. Therefore the work W done by the field is given by

$$W = MH = 2I_s SsH. \tag{18.2}$$

The hypothetical pressure p acting perpendicular to the domain wall gives a force pS on the wall, and the work done by this force in moving the domain wall a distance s is given by

$$W = pSs. \tag{18.3}$$

Comparing (18.3) with (18.2), we have

$$p = 2I_s H. \tag{18.4}$$

Thus the effect of the magnetic field acting on a domain wall is equivalent to the hypothetical pressure p given by (18.4). When the field makes an angle θ with I_s, only the component of the field $H \cos \theta$ is effective in moving the wall, so that

$$p = 2I_s H \cos \theta. \tag{18.5}$$

In the case of a 90° wall, in which the domain magnetizations on either side of the wall make the angles θ_1 and θ_2 with the magnetic field, the hypothetical pressure is given by

$$p = I_s H(\cos \theta_1 - \cos \theta_2). \tag{18.6}$$

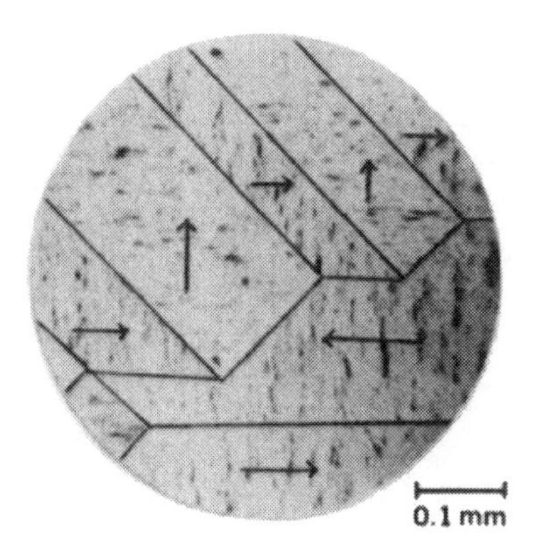

Fig. 18.3. Domain pattern observed on the (001) surface of Si–Fe crystal.

Figure 18.3 shows a typical domain structure that includes several 90° walls in addition to 180° walls, as observed on the (001) surface of a Si–Fe crystal with positive K_1. Suppose that an external magnetic field is applied parallel to one of the easy axes as shown in Fig. 18.4(a). Then a pressure is exerted on every domain wall as the domains magnetized parallel to the field (shown as shaded areas) try to expand into the neighboring domains. This results in an increase in the volume of the shaded domains, until finally the shaded area expands to cover the entire sample. This state corresponds to *saturation magnetization.*

If there is no hindrance to domain wall displacement, this process is achieved in a weak magnetic field, so that the magnetization curve rises parallel to the ordinate until it reaches the saturation magnetization (see the curve labelled [100] in Fig. 18.5).

When a magnetic field is applied parallel to the [110] axis as shown in Fig. 18.4(b), the domain walls enclosing the domains which have their direction of magnetization nearest to the field (shown by the shaded areas) expand so as to increase the shaded area. Finally there remain two kinds of domains, with magnetizations parallel to [100] or [010]. Further increase in the external field causes rotation of the domain magnetizations away from the easy axes. This process is called *magnetization rotation.* The intensity of magnetization, I_r, at which the magnetization rotation starts is given in this case by $I_s \cos 45°$, so that

$$I_r = \frac{I_s}{\sqrt{2}} = 0.71 I_s. \tag{18.7}$$

The value of I_r is called the *residual magnetization.*

When the field is applied parallel to [111], all the domain magnetizations lie along [100], [010], or [001] after the wall displacement is completed, so the residual magnetization becomes

$$I_r = \frac{I_s}{\sqrt{3}} = 0.58 I_s. \tag{18.8}$$

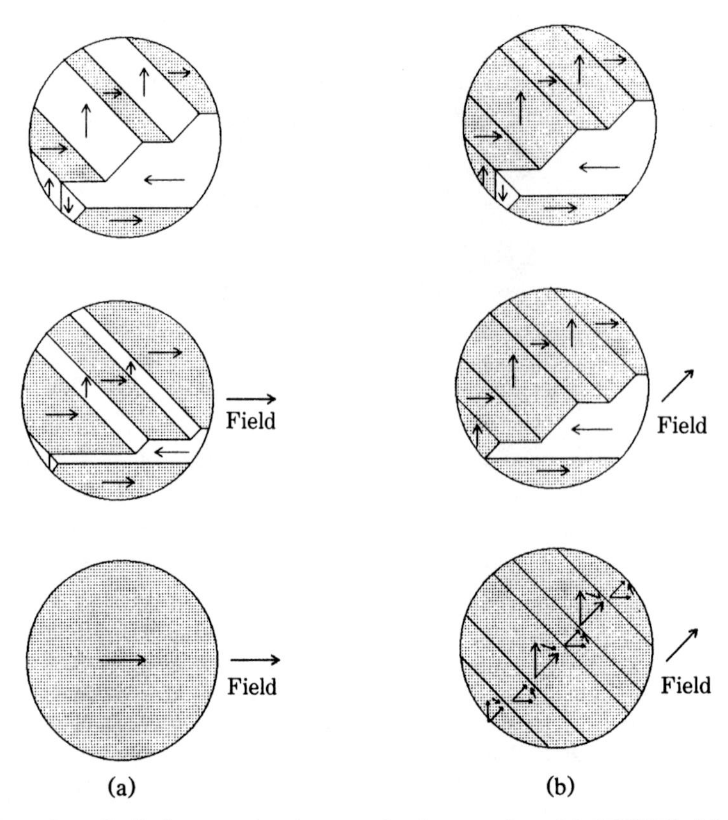

Fig. 18.4. Domain wall displacement and magnetization rotation: (a) $H \parallel [100]$; (b) $H \parallel [110]$.

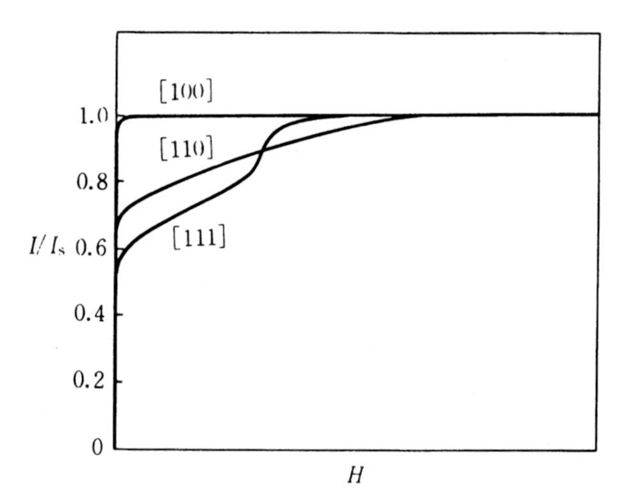

Fig. 18.5. Magnetization curves measured for a cubic ferromagnet with positive K_1 along three principal crystal axes.

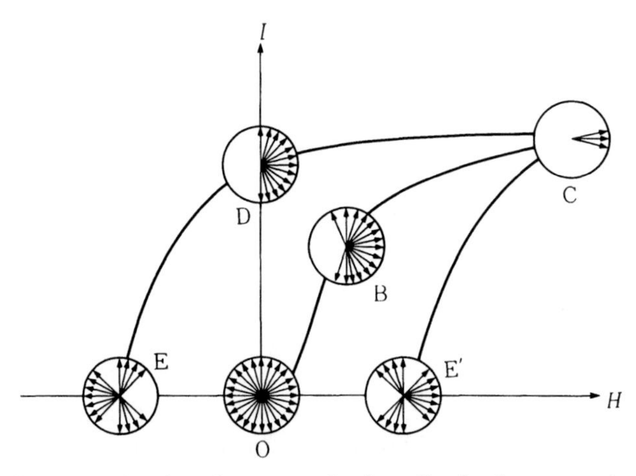

Fig. 18.6. Polar diagrams showing the magnetization distribution at various points on the magnetization curve.

In fact, the magnetization curves measured for a single crystal of iron (Fig. 18.5) break away from the y-axis at the magnetization levels given by (18.7) and (18.8) when the field is applied parallel to [110] or [111].

In polycrystalline uniaxial ferromagnets, the situation is not so simple. Figure 18.6 shows the changes in the polar distribution of domain magnetizations on the magnetization curve. Starting from the demagnetized state at O where the distribution is isotropic, the magnetizations pointing in the negative direction are reversed and increase the population in the positive hemisphere as the field increases in the positive direction, as shown by the distribution at B. In this case the magnetization in the negative direction is reversed first, because the strongest pressure acts on the walls which contribute to this domain reversal. At a sufficiently strong field, all magnetizations are lined up nearly parallel to the field direction, as shown at C. This is the *saturated state*. As the field is reduced from point C, each domain magnetization rotates back to the nearest positive easy direction and thus covers a hemisphere at the residual magnetization point D. When the field is then increased in the negative direction, the most unstable magnetization, which is the magnetization in the positive direction, is reversed first. This results in the distribution shown at point E, where the net magnetization is zero. The intensity of the field at point E is called the *coercive force* or *coercive field*, and is a measure of the stability of the residual magnetization. Further increase in the field in the negative direction results in negative saturation. Figure 18.6 shows the equivalent magnetization process towards positive saturation from E′ to C.

Such a change in the distribution of domain magnetizations is different in different materials. Figure 18.6 shows the case of a uniaxial material such as polycrystalline cobalt or randomly stressed nickel. In this case, the local magnetization in zero or small fields lies parallel to the easy axis, in either the positive or negative direction. At the point of residual magnetization, the domain magnetization lies everywhere in the

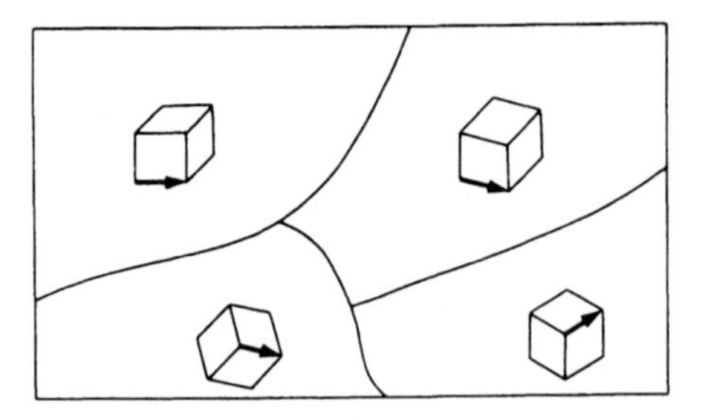

Fig. 18.7. Domain magnetizations at the residual magnetization of polycrystalline ferromagnet with positive K_1.

positive direction of the local easy axis and the distribution covers the positive hemisphere. The value of the residual magnetization is given by

$$I_r = \int_0^{\pi/2} I_s \cos\theta \sin\theta \, d\theta = \frac{I_s}{2}, \tag{18.9}$$

where θ is the angle between the magnetization and the previously applied field. In fact, severely stressed ferromagnets exhibit an I_r/I_s value of about 50%.

For cubic ferromagnets, the situation is different. If $K_1 > 0$, the easy axes are $\langle 100 \rangle$. Therefore at residual magnetization each domain magnetization lies parallel to the $\langle 100 \rangle$ which is the nearest to the previously applied field direction, as shown in Fig. 18.7. The largest deviation of the magnetization from the direction of the previously applied field occurs when the field was applied parallel to $\langle 111 \rangle$. The deviation angle in this case is $\cos^{-1}(1/\sqrt{3}) \simeq 55°$. The polar distribution of domain magnetization in this case is shown in Fig. 18.8.

In order to calculate the value of the residual magnetization for any direction of the applied field, we fix the coordinate axes and vary the direction of the applied magnetic field, as shown in Fig. 18.9. When the previously applied field direction is in the area marked with concentric circles, the magnetization should remain parallel to

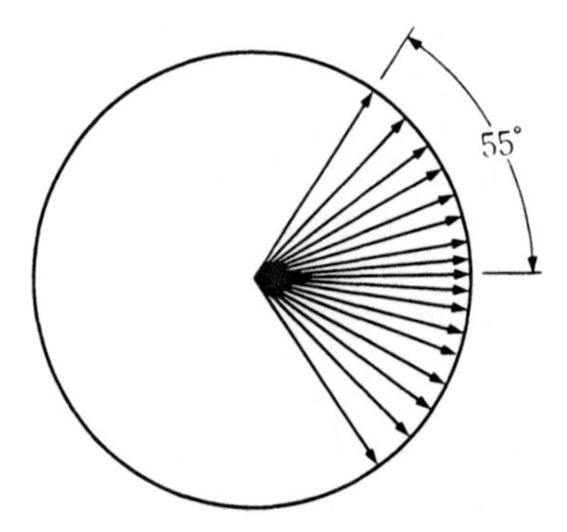

Fig. 18.8. Polar diagram of magnetization distribution at the residual magnetization for a cubic ferromagnet with positive K_1.

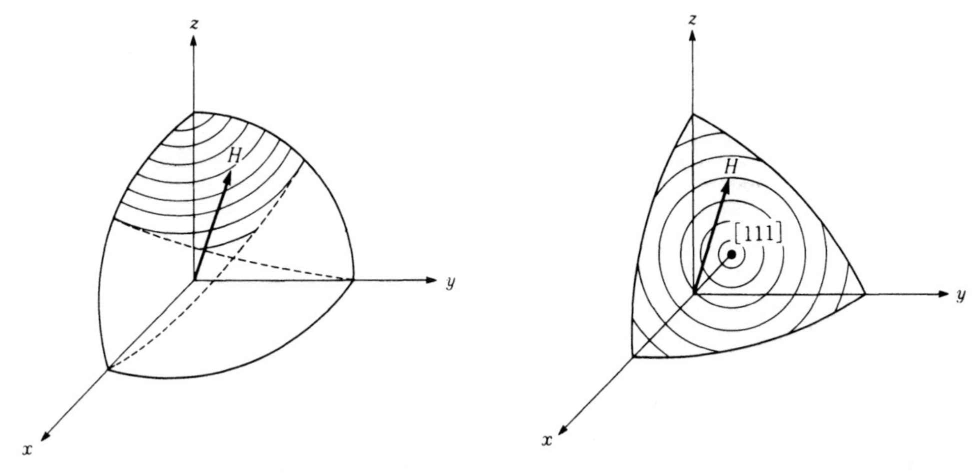

Fig. 18.9. The range of integration in (18.10).

Fig. 18.10. The range of integration in (18.12).

the z-axis. Let the angle between the z-axis and the direction of the applied field be θ. The average z-component of magnetization, which is the residual magnetization, is given by

$$I_r = I_s \overline{\cos\theta}$$
$$= \frac{6I_s}{\pi}\left(\int_0^{\pi/4}\int_{-\pi/4}^{\pi/4}\cos\theta\sin\theta\,\mathrm{d}\varphi\,\mathrm{d}\theta + \int_{\pi/4}^{\cos^{-1}(1/\sqrt{3})}\int_{-\varphi_0}^{\varphi_0}\cos\theta\sin\theta\,\mathrm{d}\varphi\,\mathrm{d}\theta\right)$$
$$= 0.832 I_s, \tag{18.10}$$

where φ is the azimuthal angle of the field measured from the $(1\bar{1}0)$ plane, and φ_0 is given by

$$\varphi_0 = \cos^{-1}\cot\theta. \tag{18.11}$$

Similarly, in the case of $K_1 < 0$ (easy axes parallel to $\langle 111\rangle$), the magnetization remains parallel to [111] when the applied field was in the first quadrant (the area marked with circles) in Fig. 18.10. If we let the angle between [111] and the applied field be θ, and average over the circled area, we find the residual magnetization to be

$$I_r = I_s \overline{\cos\theta}$$
$$= \frac{6I_s}{\pi}\left(\int_0^{\cos^{-1}(\sqrt{2/3})}\int_{-\pi/3}^{\pi/3}\cos\theta\sin\theta\,\mathrm{d}\varphi\,\mathrm{d}\theta + \int_{\cos^{-1}(\sqrt{2}/\sqrt{3})}^{\cos^{-1}(1/\sqrt{3})}\int_{-\varphi}^{\varphi_0}\cos\theta\sin\theta\,\mathrm{d}\varphi\,\mathrm{d}\theta\right)$$
$$= 0.886 I_s, \tag{18.12}$$

where φ is the azimuthal angle about [111] measured from the $(0\bar{1}1)$ plane, and φ_0 is given by

$$\varphi_0 = \frac{\pi}{3} - \cos^{-1}\left(\frac{1}{\sqrt{2}}\cot\theta\right). \tag{18.13}$$

This value is larger than that given by (18.10), in spite of the fact that the largest deviation angle is 55° in both cases. The reason is that in the latter case the number of easy axes is four, while in the former case it is three. The probability of selecting the nearest easy axis is larger when there are more easy axes. Thus the value of I_r/I_s is in the range 70–90% for cubic ferromagnets, if the cubic anisotropy is predominant. Accordingly the shape of the hysteresis loop is more rectangular than in materials with uniaxial anisotropy.

In the above treatment, we ignored the influence of free poles which may appear on the grain boundaries. In real polycrystals, sometimes the free poles which appear at residual magnetization reverse a part of the magnetization to reduce the magnetostatic energy. As we discussed in Section 17.6, the magnetostatic energy of the free poles which appear on grain boundaries is 20–50% of that of an isolated particle. Therefore, unless the grain size is smaller than the critical size for a single domain, domain walls will be generated to create domains of reverse magnetization, thus reducing the magnetostatic energy. Actually, however, because of the difficulty of nucleating reverse domains, or the lack of mobility of the domain walls, the reduction of the residual magnetization is incomplete. In any case, the residual magnetization is reduced by this effect. It is reported[1] that Permalloy has residual magnetization as small as 7% of the saturation magnetization. The reason is that this alloy is magnetically very soft, so that the magnetization is easily reversed.

The value of the residual magnetization is influenced by the shape of the specimen. Kaya[2] measured the residual magnetizations of variously oriented cylindrical single crystals of iron, and discovered a rule called the *lmn rule* or *Kaya's rule*: The residual magnetization, I_r, is given by the formula

$$I_r = \frac{I_s}{l+m+n}, \tag{18.14}$$

where I_s is the saturation magnetization, and l, m, and n are the direction cosines of the long axis of the cylinder with respect to the cubic axes. Figure 18.11 shows the experimental plot I_r as a function of $1/(l+m+n)$, where we see that the experiment is well expressed by (18.14). Kaya explained this rule by assuming that the vector sum of the x-, y-, and z-magnetizations at the residual point is parallel to the long axis of the cylinder (Fig. 18.12). Otherwise nonzero free poles appear on the side surface of the cylinder, thus increasing the magnetostatic energy. Let the magnetization components parallel to the x-, y-, and z-axis be I_x, I_y, and I_z, respectively. Then we have the relationships

$$\left.\begin{aligned} I_x + I_y + I_z &= I_s \\ I_x &= lI_r \\ I_y &= mI_r \\ I_z &= nI_r \end{aligned}\right\}, \tag{18.15}$$

from which we can deduce (18.14).

In ferromagnets with a special anisotropy, unusual values of I_r/I_s may be obtained. Isoperm, which is made by rolling a grain-oriented Ni–Fe alloy (see Section 13.2), has

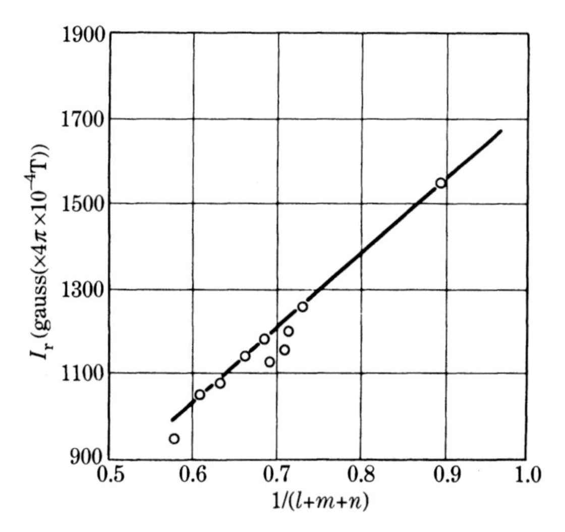

Fig. 18.11. The relationship between residual magnetization and crystal orientation of cylindrical single crystals of iron. The orientation is expressed by $1/(l+m+n)$, where l, m, and n are the direction cosines of the cylinder axis with respect to the cubic crystal axes. (After Kaya[2])

its easy axis perpendicular to the rolling direction, so that the residual magnetization (measured after magnetizing parallel to the rolling direction) is almost zero. On the other hand, grain-oriented Si-steel (see Section 22.1.1) has its easy axis parallel to the rolling direction, so that the value of I_r/I_s is almost 100%.

Permalloy annealed at 490°C exhibits an I_r/I_s value of about 30%. Bozorth[3] explained this phenomenon as follows: This alloy exhibits an induced uniaxial anisotropy as a result of the annealing (see Section 13.1), in addition to some cubic anisotropy with positive K_1. As the field is reduced from a high value, the domain magnetization rotates back to the nearest cubic easy axis and stays in this direction, if the easy axis of the induced anisotropy coincides with this cubic easy axis. However, if the easy axis of the induced anisotropy is parallel to one of the other cubic easy axes,

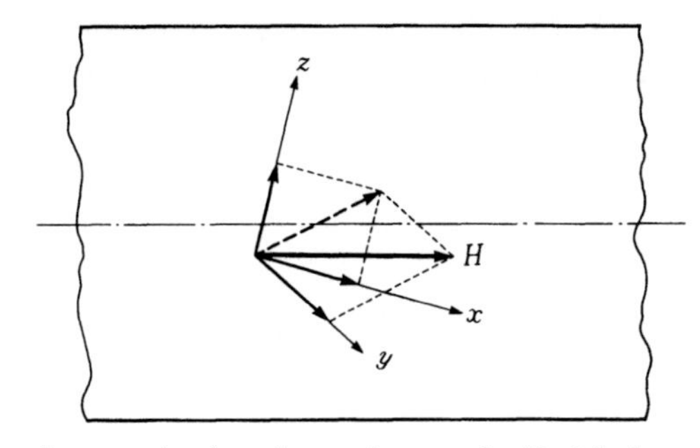

Fig. 18.12. Distribution of magnetizations in an elongated cylindrical specimen with positive K_1 (Kaya's rule).

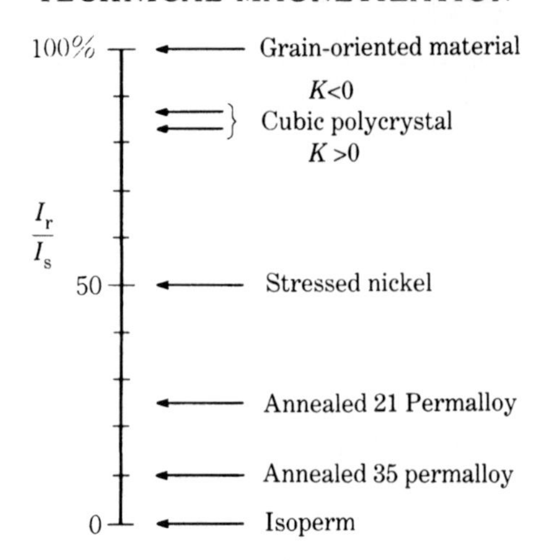

Fig. 18.13. Values of I_r/I_s for various categories of ferromagnetic materials.

the magnetization will split into plus or minus directions along this preferred easy axis, thus giving no contribution to the residual magnetization. Since the probability that the easy axis of the induced uniaxial anisotropy coincides with one of the cubic axes is $\frac{1}{3}$, the I_r/I_s value becomes

$$\frac{I_r}{I_s} = \frac{83.2}{3}\% = 27.7\%. \tag{18.16}$$

We expect the same situation for a sample of iron under elastic stress. Figure 18.13 summarizes the predicted values of I_r/I_s for various magnetic materials.

Next we shall discuss the polar distribution of domain magnetizations at the coercive field point. As seen in Fig. 18.6, when the field is increased in the negative direction from the residual magnetization, the most unstable magnetization is reversed first. If the easy axis of the uniaxial anisotropy is distributed at random and the magnetization reversal occurs by wall displacement, the domain magnetization which points in the positive direction is most unstable. Then the distribution of magnetization is shown at point E in Fig. 18.6. Let the half-polar angle of the cone of the reversed magnetization be θ_0. Since the reversed magnetization is equal to the unreversed magnetization at the coercive field, we have the relationship

$$I_s \int_0^{\theta_0} \cos\theta \sin\theta \, d\theta = \frac{I_s}{4}, \tag{18.17}$$

so that

$$\cos\theta_0 = \frac{1}{\sqrt{2}};$$

therefore

$$\theta_0 = 45°. \tag{18.18}$$

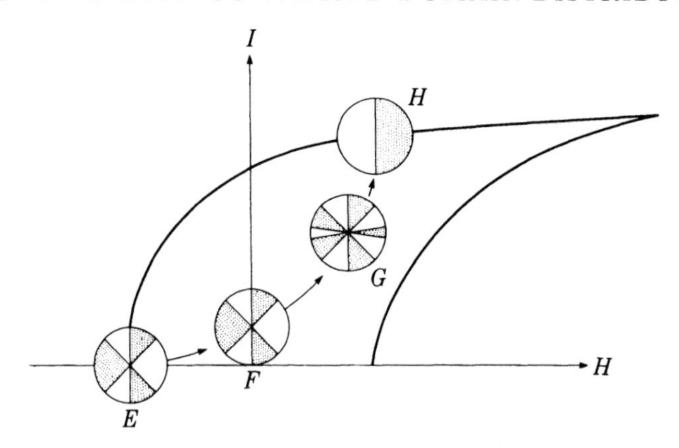

Fig. 18.14. Change in domain distribution during the magnetization process, starting from the coercive field, assuming uniaxial anisotropy and magnetization occurring by domain wall displacement.

Thus the half-polar angle of the cone is 45°, and the coercive field is the field needed to move a 180° wall making 45° with the positive direction.

Let us consider the magnetization process which may occur when the field is increased in the positive direction from the coercive field point E. As shown in Fig. 18.14, only reversible magnetization rotation occurs from the point E to F (where $H = 0$). Further increase in H in the positive direction causes a reversal of the most unstable magnetization, which is the magnetization pointing in the negative direction (point G in the diagram). Finally the sample reaches point H, whose distribution is almost the same as that of the residual magnetization. If we try to attain the same domain distribution from the demagnetized state, we need a fairly strong field to reverse the directions of domain magnetizations which are almost perpendicular to the field direction.

A permanent magnet is often magnetized after it is assembled into a device, which requires the application of a large field. But if the magnet is first magnetized and then brought to the coercive field point by applying an appropriate reverse field, it can be easily remagnetized by a relatively weak field. For example, a Ba-ferrite magnet with coercive field $H_c = 0.12\,\mathrm{MA\,m^{-1}}$ $(= 1500\,\mathrm{Oe})$ requires about $0.80\,\mathrm{MA\,m^{-1}}$ $(= 10\,000\,\mathrm{Oe})$ to magnetize to saturation from the demagnetized state. However, once it has been brought to the coercive field point, only $0.24\,\mathrm{MA\,m^{-1}}(= 3000\,\mathrm{Oe})$ is required to magnetize to residual magnetization. This field is about $0.12\,\mathrm{MA\,m^{-1}}(= 1500\,\mathrm{Oe})$ larger than H_c, because the field must overcome the demagnetizating field.[4] It should be noted that this process of remagnetizing permanent magnets is effective, irrespective of magnetization mechanism, except for a grain-oriented uniaxial magnet.

Next we discuss the case of a ferromagnet in which a cubic magnetocrystalline anisotropy is predominant. When the field is increased in the negative direction from the residual magnetization to the coercive field, the most unstable domain magnetizations, which make small angles with the positive field direction, are reversed first by

180° wall displacement. If $K_1 > 0$, the coercive field state is attained when half of the residual magnetization $0.832I_s$, or $0.416I_s$, is reversed. Thus we have the relationship

$$3I_s \int_{\theta}^{\theta_0} \cos\theta \sin\theta \, d\theta = 0.416I_s, \tag{18.19}$$

from which we have

$$\cos\theta_0 = 0.850$$
$$\therefore \quad \theta_0 = 31.8°. \tag{18.20}$$

Equation (18.19) does not include a complicated integration with respect to the azimuthal angle as in (18.10), because θ_0 is less than 45° (see Fig. 18.9).

Similarly, if $K_1 < 0$, the residual magnetization is $0.866I_s$, so that half of this value, or $0.433I_s$, must be reversed to get to the coercive field point. Therefore we have the relationship

$$4I_s \int_{\theta}^{\theta_0} \cos\theta \sin\theta \, d\theta = 0.433I_s, \tag{18.21}$$

from which we have

$$\cos\theta_0 = 0.885$$
$$\therefore \quad \theta_0 = 27.7°. \tag{18.22}$$

Again the equation (18.21) includes no complicated integration with respect to the azimuthal angle, because $\theta_0 < 35°$ (see Fig. 18.10).

In the case of an assembly of single domain particles, the magnetization reversal is exclusively performed by magnetization rotation. As will be discussed in Section 18.3, in the case of uniaxial crystal anisotropy, or an elongated particle with uniaxial shape anisotropy, magnetization at an angle of 45° to H is most unstable, and is reversed first. If we assume that domain magnetizations in the range $\theta_1 = \pi/4 - \varepsilon$ to $\theta_2 = \pi/4 + \varepsilon$ are reversed until the net magnetization goes to zero, we have the relationship

$$I_s \int_{\theta_1}^{\theta_2} \cos\theta \sin\theta \, d\theta = \frac{I_s}{4}, \tag{18.23}$$

from which we have

$$\cos^2\theta_1 - \cos^2\theta_2 = \tfrac{1}{2}$$
$$\therefore \quad \sin 2\varepsilon = \tfrac{1}{2}$$
$$\therefore \quad \varepsilon = 15°. \tag{18.24}$$

Therefore the domain magnetizations in the angular range $\theta_1 = 30°$ to $\theta_2 = 60°$ are reversed to reach the coercive field point.

Figure 18.15 summarizes polar diagrams of magnetization distribution at residual magnetization and at the coercive field for various categories of magnetic materials.

Finally we discuss the process of *demagnetization* or of obtaining zero net magnetization, and its physical meaning. There are two methods of demagnetization: *thermal demagnetization* and *AC demagnetization*. The former method consists in heating the

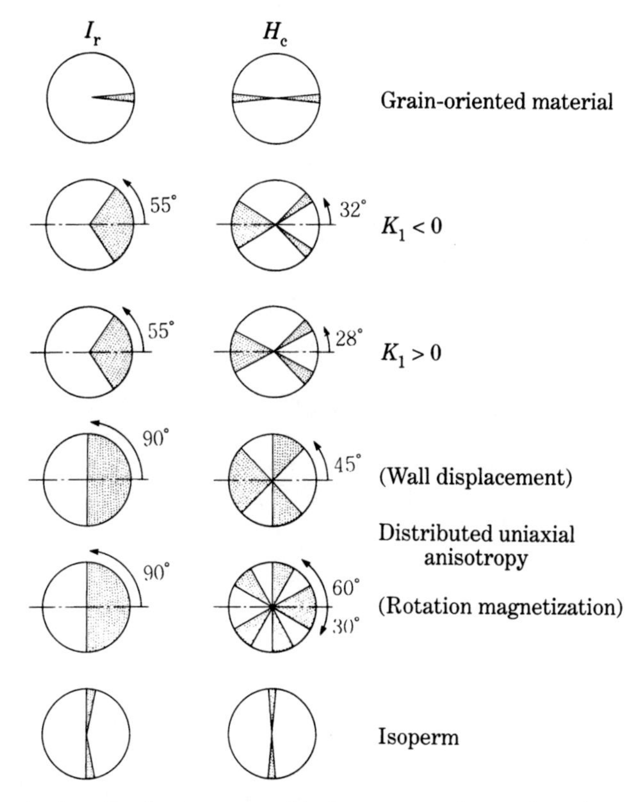

Fig. 18.15. Magnetization distributions at residual magnetization and at the coercive field for various categories of ferromagnetic materials.

specimen above its Curie point and cooling it to room temperature in the absence of a magnetic field. The latter method consists in magnetizing the specimen with an AC (alternating current) field with a sufficiently large amplitude and then decreasing the amplitude of the AC field to zero in the absence of a DC field. Since the thermal demagnetization requires a heat treatment and accordingly is time-consuming, usually AC demagnetization is used.

Let us consider what happens during AC demagnetization. Figure 18.16 shows the changes in domain distribution during AC demagnetization for a material with uniaxial anisotropy, in which the magnetization reversal takes place by wall displacement. As the amplitude of the AC field is decreased, the domain magnetizations that make large angles with the field direction are settled first. The final state is an isotropic distribution as shown by the diagram at the origin. In order to realize a completely isotropic angular distribution, the rate of decrease in the amplitude of the AC field should be small.

It should be noted that in the case of a cubic ferromagnet, AC demagnetization results in a domain distribution in which the local magnetizations are confined to a cone with a half angle of 55°. In order to realize an isotropic domain distribution in

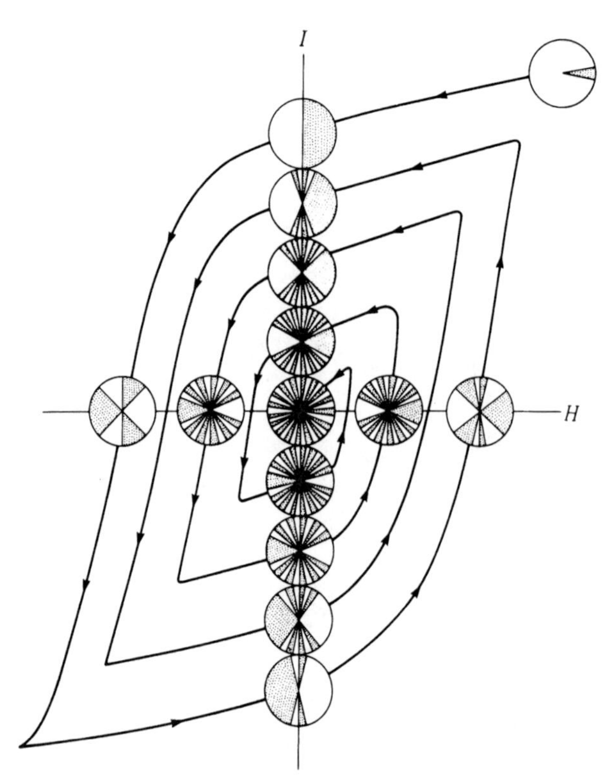

Fig. 18.16. Variation of angular distribution of magnetization during AC demagnetization, assuming random uniaxial anisotropy and magnetization occurring by domain wall displacement.

this case, AC demagnetization must be performed by using a rotating field of decreasing magnitude instead of an AC field along a fixed axis.

18.2 DOMAIN WALL DISPLACEMENT

When a domain wall moves in a large plate of a homogeneous magnetic material, the wall energy is independent of its position. The wall therefore does not return to its original position after the field is removed. In other words, the wall is in neutral equilibrium. This situation corresponds to a sphere placed on a level surface, which has no stable position. In order that the sphere have a position of stable equilibrium, the surface of the plate must be uneven; then the sphere will settle in a concave depression, about which position its displacement becomes reversible for small displacements. Similarly the displacement of the wall becomes reversible only when the wall energy varies from place to place.

Suppose that a plane wall moves, keeping its shape unchanged. Let us assume that the wall energy per unit area, ε_w, changes as a function of position, s, as shown in Fig. 18.17. When there is no external magnetic field, the wall must settle at the position

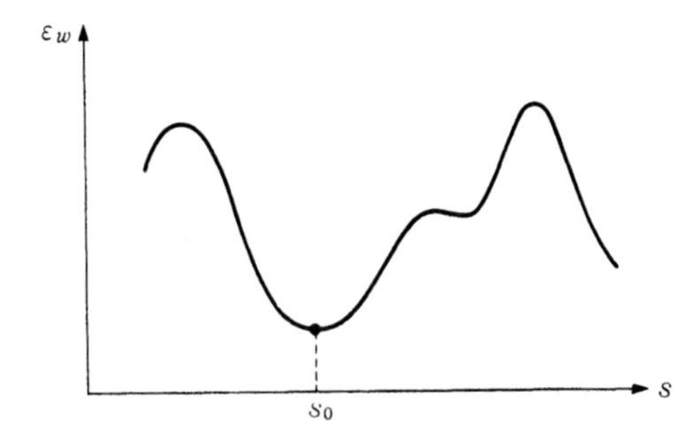

Fig. 18.17. Variation of domain wall energy with position (schematic).

where $\partial \varepsilon_w / \partial s = 0$, or the wall energy is minimum. The wall energy can be expressed approximately by

$$\varepsilon_w = \tfrac{1}{2} \alpha s^2. \tag{18.25}$$

If a magnetic field H is applied in the direction which makes an angle θ with the domain magnetization, I_s, this field gives a pressure p given by (18.5) on the wall, so that the energy supplied by the field is given by

$$\varepsilon_H = -2I_s H \cos\theta\, s. \tag{18.26}$$

The total energy is given by

$$\varepsilon = \varepsilon_w + \varepsilon_H = \tfrac{1}{2} \alpha s^2 - 2I_s H \cos\theta\, s, \tag{18.27}$$

and we find its minimum from

$$\frac{\partial \varepsilon}{\partial s} = \alpha s - 2I_s H \cos\theta = 0, \tag{18.28}$$

giving

$$s = \frac{2I_s \cos\theta}{\alpha} H. \tag{18.29}$$

As a result of displacement, the magnetization component parallel to H is increased by $2I_s \cos\theta\, s$, so that the magnetization of the specimen is increased by

$$I = \frac{4I_s^2 \cos^2\theta}{\alpha} SH, \tag{18.30}$$

where S is the total area of the domain wall in a unit volume. Therefore the initial susceptibility is given by

$$\chi_a = \frac{4I_s^2 \cos^2\theta}{\alpha} S. \tag{18.31}$$

If the easy axes are $\langle 100 \rangle$ as in the case of iron, and the field is applied parallel to the direction $[\alpha_1, \alpha_2, \alpha_3]$, the value of $\cos^2\theta$ averaged over all the 180° walls is given by

$$\overline{\cos^2\theta} = \tfrac{1}{3}(\alpha_1^2 + \alpha_2^2 + \alpha_3^2) = \tfrac{1}{3}. \tag{18.32}$$

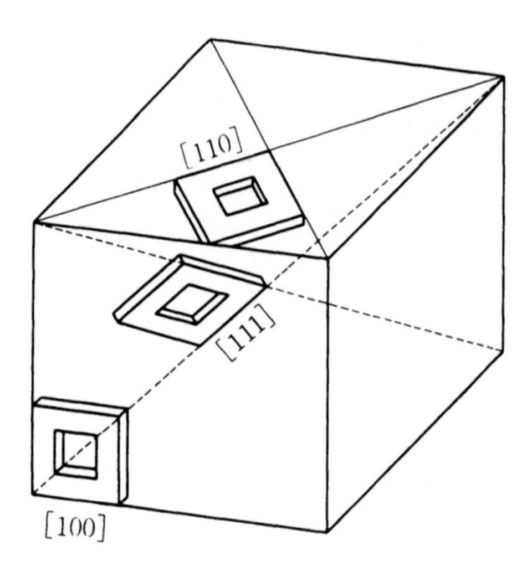

Fig. 18.18. Method of cutting picture-frame single crystals with sides parallel to ⟨100⟩, ⟨110⟩ or ⟨111⟩ from a bulk single crystal.

Therefore we expect that the initial susceptibility is constant, irrespective of direction of the applied field. In order to check this, Williams[5] measured initial permeability parallel to ⟨100⟩, ⟨110⟩ and ⟨111⟩, using picture-frame single crystal specimens cut out of a bulk single crystal of Si–Fe as shown in Fig. 18.18. He discovered that the initial susceptibility was not constant, but varied as

$$\bar{\mu}_{100} : \bar{\mu}_{110} : \bar{\mu}_{111} = 1 : \tfrac{1}{2} : \tfrac{1}{3}. \tag{18.33}$$

This result, however, is not necessarily in disagreement with the theoretical treatment given above, because the domain structure is not the same for the three picture-frame specimens. Becker and Döring[6] explained the ratio given by (18.33) as follows: in the case of the ⟨100⟩ crystal, a 180° wall runs parallel to the edge, so that $\overline{\cos^2\theta} = 1$. In the ⟨110⟩ crystal, domain walls are parallel to either one of the two easy axes which are closest to ⟨110⟩ edges, so that

$$\overline{\cos^2\theta} = \tfrac{1}{2}\left((1/\sqrt{2})^2 + (1/\sqrt{2})^2\right) = \tfrac{1}{2}.$$

In the ⟨111⟩ crystal, domain walls are parallel to either one of the three easy axes, so that

$$\overline{\cos^2\theta} = \tfrac{1}{3}\left((1/\sqrt{3})^2 + (1/\sqrt{3})^2 + (1/\sqrt{3})^2\right) = \tfrac{1}{3}.$$

Thus we have the relationship (18.33) from (18.31), provided the total domain wall area S is the same. However, the last assumption is quite doubtful.

In the case that $K_1 < 0$, easy axes are parallel to $\langle 111 \rangle$, so that

$$\overline{\cos^2\theta} = \frac{1}{4}\left\{\left(\frac{1}{\sqrt{3}}\alpha_1 + \frac{1}{\sqrt{3}}\alpha_2 + \frac{1}{\sqrt{3}}\alpha_3\right)^2 + \left(\frac{1}{\sqrt{3}}\alpha_1 + \frac{1}{\sqrt{3}}\alpha_2 - \frac{1}{\sqrt{3}}\alpha_3\right)^2\right.$$
$$\left. + \left(\frac{1}{\sqrt{3}}\alpha_1 - \frac{1}{\sqrt{3}}\alpha_2 + \frac{1}{\sqrt{3}}\alpha_3\right)^2 + \left(-\frac{1}{\sqrt{3}}\alpha_1 + \frac{1}{\sqrt{3}}\alpha_2 + \frac{1}{\sqrt{3}}\alpha_3\right)^2\right\}$$
$$= \frac{1}{3}. \tag{18.34}$$

Thus the initial susceptibility is again constant, irrespective of the field direction.

When the easy axes are randomly distributed, we have

$$\overline{\cos^2\theta} = \int_0^{\frac{\pi}{2}} \cos^2\theta \sin\theta \, d\theta = \tfrac{1}{3}. \tag{18.35}$$

Thus in all the cases, (18.31) becomes

$$\chi_{a,180°} = \frac{4I_s^2}{3\alpha} S. \tag{18.36}$$

One mechanism for generating variations in wall energy as shown in Fig. 18.17 is the presence of randomly distributed internal stresses, as first pointed out by Kondorsky.[7] This idea was developed by Kersten,[8] whose treatment may be described as follows: he assumed that the internal stress varies as a function of displacement, s, as

$$\sigma = \sigma_0 \cos 2\pi \frac{s}{l}. \tag{18.37}$$

The anisotropy constant therefore varies as

$$K = K_1 - \tfrac{3}{2}\lambda\sigma_0 \cos 2\pi \frac{s}{l}. \tag{18.38}$$

This form is assumed for simplicity, although the functional form is generally different between anisotropy energy and the magnetoelastic energy. If the wavelength of the stress variation is sufficiently larger than the wall thickness, we can regard the anisotropy given by (18.38) as remaining constant throughout the wall. Then the wall energy is deduced from (16.55) as

$$\gamma = 2\sqrt{A\left(K_1 - \tfrac{3}{2}\lambda\sigma_0 \cos 2\pi \frac{s}{l}\right)}. \tag{18.39}$$

Assuming that $K_1 \gg \lambda\sigma$, (18.39) can be expanded near $s = 0$ as

$$\gamma = 2\sqrt{AK_1}\left(1 - \frac{3\lambda\sigma_0}{4K_1}\cos 2\pi \frac{s}{l} + \cdots\right)$$
$$= 2\sqrt{AK_1}\left(1 - \frac{3\lambda\sigma_0}{4K_1}\left(1 - \frac{2\pi^2 s^2}{l^2}\right) + \cdots\right). \tag{18.40}$$

Comparing (18.40) with (18.25), we have

$$\alpha = \frac{\partial^2 \gamma}{\partial s^2} = 6\pi^2 \sqrt{\frac{A}{K_1}} \frac{\lambda \sigma_0}{l^2}. \tag{18.41}$$

Based on (16.46) or Fig. 16.18, we can express the thickness of the domain wall approximately as

$$\delta = 3\sqrt{\frac{A}{K_1}}, \tag{18.42}$$

so that (18.41) is expressed in terms of δ as

$$\alpha = 2\pi^2 \frac{\lambda \sigma_0}{l^2} \delta. \tag{18.43}$$

Moreover, if we assume that a domain wall exists at every energy minimum, and also that the wall energy varies similarly in the x-, y-, and z-directions, it follows that

$$S = \frac{3}{l}. \tag{18.44}$$

Using (18.43) and (18.44), we have from (18.36)

$$\chi_{a,180°} = \frac{2I_s^2}{\pi^2 \lambda \sigma_0} \frac{l}{\delta}. \tag{18.45}$$

According to this expression, the susceptibility is large when the wavelength of stress variation, l, is large. This is because the wall is easily displaced when the wall energy changes slowly with position. When the wavelength, l, is smaller than the thickness of the wall, δ, the above calculation is not applicable. If $l < \delta$, the effect of the stress will be smoothed over by the effect of the exchange interaction, so that the wall becomes easily movable again. Therefore the minimum susceptibility is realized when $l \approx \delta$, so that

$$\chi_{a_{\min}} = \frac{2I_s^2}{\pi^2 \lambda \sigma_o}. \tag{18.46}$$

We can estimate the value of the initial susceptibility for iron from (18.46) by assuming that $\lambda_{100} = 2.07 \times 10^{-5}$, $I_s = 2.16$, $\sigma \approx 50\,\mathrm{kg\,mm^{-2}} = 5 \times 10^8\,\mathrm{N\,m^{-2}}$, giving

$$\bar{\mu}_a \approx \bar{\chi}_a = \frac{2 \times 2.16^2}{(3.14)^2 \times 2.07 \times 10^{-5} \times 5 \times 10^8 \times 4 \times 3.14 \times 10^{-7}} \approx 73. \tag{18.47}$$

Since there are contributions from other mechanisms such as 90° domain wall displacement and magnetization rotation, we can estimate that $\bar{\mu}_a \approx 200$. Ordinary soft steel or iron exhibits $\bar{\mu}_a \approx 100 \sim 200$, so that this estimate seems reasonable. However, our assumptions of internal stress and total domain wall area are fairly arbitrary, so we cannot have great faith in the calculation.

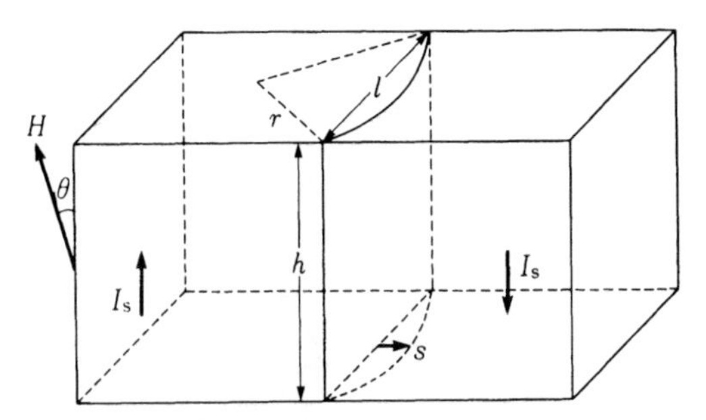

Fig. 18.19. A 180° wall bulging under the action of a magnetic field.[10]

We assumed above that the wall is rigid and undeformable. Actually, however, the wall is deformable, as shown in Fig. 16.13. This was first pointed out by Néel,[9] and Kersten[10] proposed a new mechanism of reversible wall displacement as shown in Fig. 18.19. Let us assume a domain wall pinned from place to place as shown in Fig. 17.33. The wall shown in Fig. 18.19 is pinned at the side edges but is free to bulge under the action of the field pressure. When a field H is applied in a direction which makes the angle θ with the magnetization I_s, a pressure given by (18.5) acts on the wall, so that the wall bulges in a cylindrical form as shown in Fig. 18.19. If the radius of curvature is r, the relationship between the pressure and r leads to the formula

$$\frac{\gamma}{r} = 2I_s H \cos\theta. \tag{18.48}$$

The volume increase of the upward-magnetized domain is given by

$$\delta V = \tfrac{2}{3} lhs, \tag{18.49}$$

where s is the displacement of the wall at the center. The increase in magnetization produced by this wall bulging is

$$I = \tfrac{4}{3} I_s \cos\theta\, Ss, \tag{18.50}$$

where S is the total domain wall area per unit volume. Since s is approximately related to r and l by

$$s = \frac{l^2}{8r}, \tag{18.51}$$

(18.50) is modified using (18.48) and (18.51) to give

$$I = \frac{Sl^2 I_s^2}{3\gamma} \cos^2\theta\, H. \tag{18.52}$$

In many cases we have $\overline{\cos^2\theta} = \frac{1}{3}$ (see (18.32), (18.34) and (18.35)), so that (18.52) leads to the susceptibility

$$\chi_{a,180^\circ} = \frac{Sl^2 I_s^2}{9\gamma}. \tag{18.53}$$

If we assume that voids or nonmagnetic inclusions (pinning sites) are distributed in a simple cubic array with spacing l, we have

$$S = \frac{2}{l}. \tag{18.54}$$

Then (18.53) leads to

$$\chi_a = \frac{2I_s^2 l}{9\gamma}. \tag{18.55}$$

In the case of iron, $I_s = 2.15$ (Wb m^{-2}), $\gamma = 1.6 \times 10^{-3}$, and we can assume from the observation of domain patterns that $l = 10^{-4}$ (m), so that we have

$$\bar{\mu}_a \approx \bar{\chi}_a = \frac{2 \times (2.15)^2 \times 10^{-4}}{9 \times 1.6 \times 10^{-3} \times 4\pi \times 10^{-7}} = 51\,000. \tag{18.56}$$

Well-annealed pure iron exhibits $\bar{\chi}_a \approx 10\,000$. The mechanism described here is considered to contribute to this high initial susceptibility.

Generally speaking, the initial susceptibility increases with increasing temperature and exhibits a sharp maximum just below the Curie temperature. This phenomenon is called the *Hopkinson effect*. Kersten[10] treated this phenomenon. In order to describe the temperature dependence of the wall energy, which is given by (16.47) for a 180° wall, he assumed that the exchange stiffness constant, A, is proportional to I_s^2, or

$$A \propto I_s^2, \tag{18.57}$$

because A is proportional to S^2 (see (16.15)). Therefore the wall energy changes with temperature as

$$\gamma \propto I_s \sqrt{K}. \tag{18.58}$$

Using this expression in (18.53), we have the relationship

$$\chi_a \propto \frac{I_s}{\sqrt{K}}. \tag{18.59}$$

Figure 18.20 shows the temperature dependence of χ_a for Fe, Ni, and Co fitted to formula (18.59). Excellent agreement is seen.

When the field becomes strong enough to drive the domain wall out of the stable pinned region, the wall displaces irreversibly and does not return to its original position even after the field is removed. For simplicity, let us treat a rigid plane wall, which displaces without any deformation. Suppose that the wall energy, ε_w, varies with the displacement s as shown in Fig. 18.21. When a field is applied, a pressure acts on the wall (see (18.5)), which displaces until the restoring force becomes equal to the pressure, or

$$\frac{\partial \varepsilon_w}{\partial s} = p = 2I_s H \cos\theta. \tag{18.60}$$

If the wall reaches the point s_1 where the gradient is maximum, further increase in

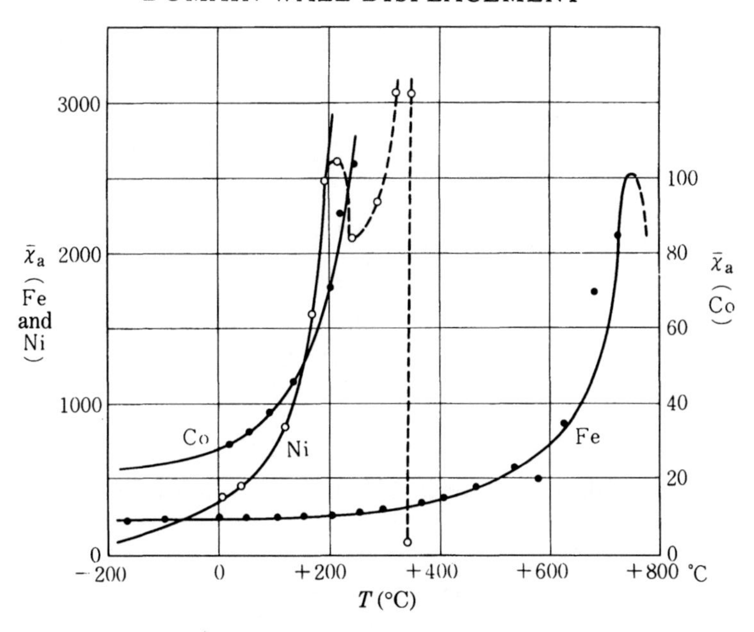

Fig. 18.20. Temperature dependence of the relative initial susceptibility for iron, cobalt and nickel. Solid curves represent the theory (see (18.59)), where $K = K_1 + K_2$ for cobalt.[10]

field intensity causes an irreversible displacement of the wall to the next equilibrium position, s_2, where the condition (18.60) is satisfied for a new field. If the field is reduced to zero, the wall comes back to the nearest stable point, but not to the original position, s_0. Further increase of the field will bring the wall to the next equilibrium position, s_3. Thus the critical field, H_0, at which the wall is swept out of the crystal is determined by the maximum gradient of the curve in Fig. 18.21, so that we have

$$H_0 = \frac{1}{2I_s \cos\theta}\left(\frac{\partial \varepsilon_w}{\partial s}\right)_{\max}. \tag{18.61}$$

If we assume that the variation of the wall energy is caused by variations in the

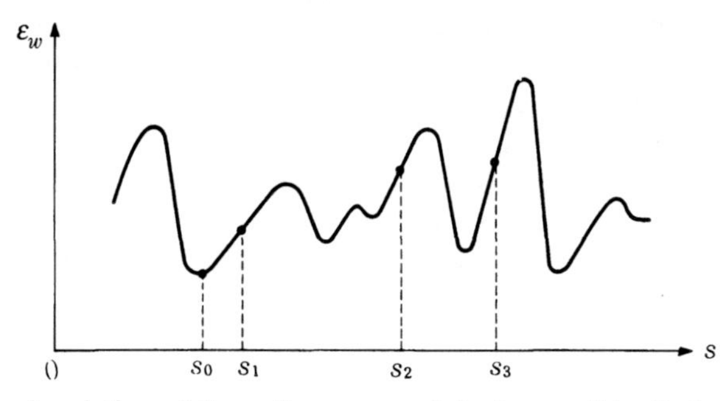

Fig. 18.21. Positional variation of the wall energy and the irreversible displacement of the wall.

internal stress as shown by (18.39), the maximum gradient of the wall energy is given by

$$\left(\frac{\partial \gamma}{\partial s}\right)_{\max} = \frac{3\pi\lambda\sigma_0}{l}\left(\sqrt{\frac{A}{K_1 - \frac{3}{2}\lambda\sigma_0\cos(2\pi s/l)}} \times \sin(2\pi s/l)\right)_{\max}$$

$$= \frac{3\pi\lambda\sigma_0}{l}\sqrt{\frac{A}{K_1}} = \frac{\pi\lambda\sigma_0}{l}\,\delta, \tag{18.62}$$

where δ is the wall thickness (see (18.42)). Using this expression in (18.61), where $\varepsilon_w = \gamma$, we have

$$H_0 = \frac{\pi\lambda\sigma_0}{2I_s\cos\theta}\,\frac{\delta}{l}. \tag{18.63}$$

As already mentioned, the wall energy is maximum when $l \approx \delta$, so that H_0 (18.63) also becomes maximum. Therefore we have

$$(H_0)_{\max} = \frac{\pi\lambda\sigma_0}{2I_s\cos\theta}. \tag{18.64}$$

For ordinary ferromagnets, we assume that $\lambda \simeq 10^{-5}$, $I_s = 1\,\mathrm{T}$, $\cos\theta \sim 1$, and also we adopt a maximum internal stress $\sigma_0 = 100\,\mathrm{kg\,mm^{-2}} \simeq 10^9\,\mathrm{N\,m^{-2}}$, so that the numerical value of (18.64) becomes

$$(H_0)_{\max} = \frac{\pi \times 10^{-5} \times 10^9}{2 \times 1 \times 1} \approx 1.5 \times 10^4(\mathrm{A\,m^{-1}}) \quad (\approx 200\,\mathrm{Oe}), \tag{18.65}$$

which is about the average coercive field of alloy permanent magnets.

Kersten[11] measured the coercive field of Ni as a function of the internal stress (see Fig. 18.22). Curve a shows the result for a Ni wire which was stretched plastically by various amounts, while curve b is for a cold-rolled Ni plate that was annealed for various lengths of time to release the internal stress. Both curves agree well with (18.64).

Next we shall discuss the critical field for a pinned domain wall. Figure 18.23 shows the cross-sectional view of a domain wall with both ends pinned. For $H = 0$, the wall is plane (curve a). As the field, H, is increased, the wall bulges under the action of the pressure (18.5) as shown by curves b and c. The radius of curvature of the cylindrical wall, r, is inversely proportional to H as shown in (18.48), so that the radius of the wall decreases as H is increased, up to curve c where the radius r is equal to half the distance between the two pinned points. Further increase in H causes an increase in r, which, however, does not satisfy the equilibrium condition (18.48), so that the wall must expand irreversibly to the next pinned points. Therefore the critical field is obtained by putting $r = l/2$ in (18.48) to give

$$H_{0,180°} = \frac{\gamma}{I_s l \cos\theta}. \tag{18.66}$$

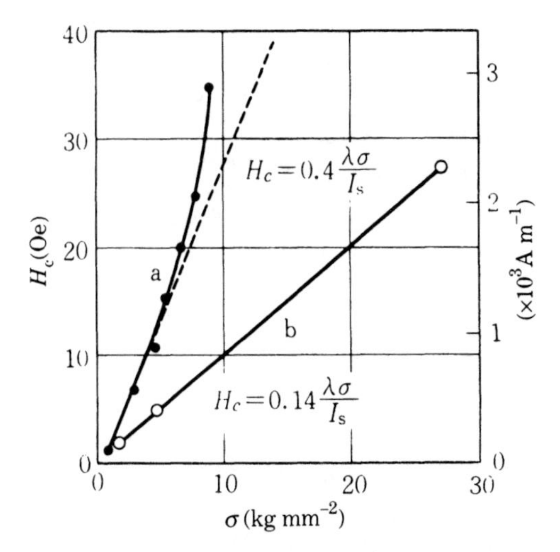

Fig. 18.22. Relationship between coercive field and internal stress for nickel:[11] (a) Ni wire plastically elongated by various amounts; (b) cold-rolled Ni plate annealed for various lengths of time.

Assuming that $l = 10^{-5} (= 10^{-3}$ cm), Kersten[10] explained the composition dependence of the coercive field in Fe–Ni alloys and also the temperature dependence of H_c of Fe.

In this model, the pinning mechanism is considered to be voids or nonmagnetic inclusions, as shown in Fig. 17.32. Kersten assumed in this paper that a dislocation may be a possible pinning site. In any case, if the pinning is strong enough, the critical field is exclusively determined by the distance between neighboring pinning points, and is independent on the pinning strength. However, if the pinning is not so strong, the wall will pull free of the pinning site before reaching position c in Fig. 18.23 and expanding irreversibly. Suppose that the pinning points are arranged in a simple cubic array with spacing l, and a wall is trapped in a (100) plane. Under the action of the

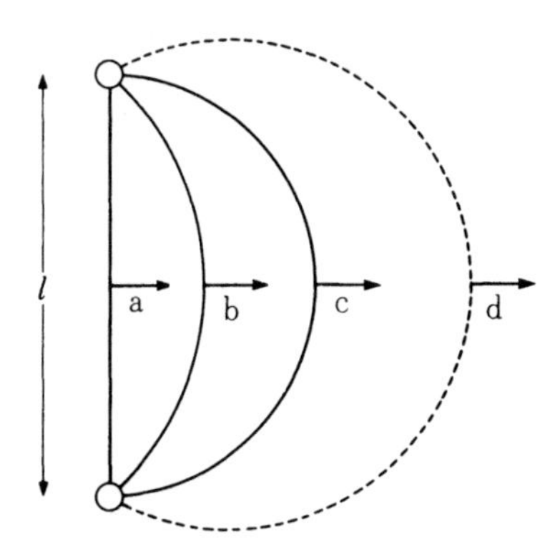

Fig. 18.23. Reversible and irreversible expansion of a wall with both ends pinned by voids.

field, H, the wall bulges in a cylindrical form. Let the energy of pinning be ΔU at each pinning site. Suppose that the wall is displaced by the radius r of the pinning point under the action of the field H. Then the energy of the field is reduced by $2I_s H \cos\theta\, l^2 r$, and if this value exceeds ΔU, the wall will leave the pinning point. The critical field in this case is given by the condition

$$2I_s H_0 \cos\theta\, l^2 r = \Delta U,$$

or

$$H_0 = \frac{\Delta U}{2I_s \cos\theta\, l^2 r}. \tag{18.67}$$

If we use (17.88) for ΔU, (18.67) becomes

$$H_0 = 1.4 \times 10^5 \frac{I_s r^2}{\cos\theta\, l^2}. \tag{18.68}$$

For iron $I_s = 2.15\,\mathrm{T}$, $\cos\theta \sim 1$, $r = 10^{-5}$ ($= 0.01\,\mathrm{mm}$), and $l = 10^{-4}$ ($= 0.1\,\mathrm{mm}$), so that we have

$$H_0 = (1.4 \times 10^5) \times (2.15) \times 10^{-2} \approx 3 \times 10^3\,\mathrm{A\,m^{-1}} \quad (\simeq 38\,\mathrm{Oe}). \tag{18.69}$$

On the other hand, since $\gamma = 1.6 \times 10^{-3}$, (18.66) gives the value

$$H_0 = \frac{1.6 \times 10^{-3}}{(2.15) \times 10^{-4}} = 7.4\,\mathrm{A\,m^{-1}} \quad (\simeq 0.1\,\mathrm{Oe}), \tag{18.70}$$

which is much smaller than (18.69). Therefore the irreversible expansion of the wall will occur before the wall leaves the pinning points. However, if the interval between

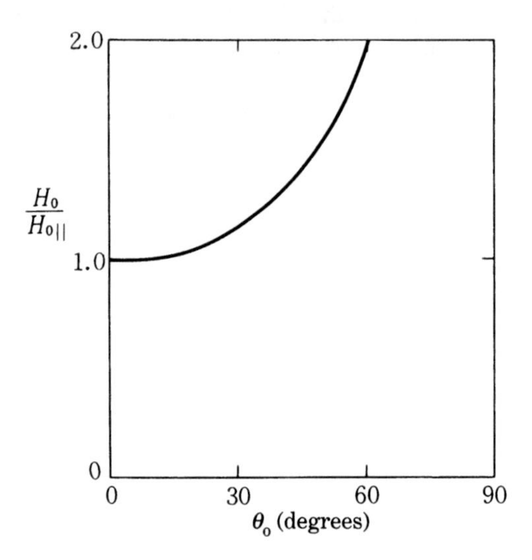

Fig. 18.24. Dependence of the critical field, H_0 on the angle, θ_0, between the field and easy axis of the uniaxial anisotropy (see (18.71)).

the pinning points is smaller by two orders of magnitude, the wall leaves the pinning points before the irreversible expansion point is reached.

Generally speaking, for domain wall displacement, the critical field is proportional to $1/\cos\theta$. For instance, for uniaxial anisotropy, H_0 is smallest when the field is applied parallel to the easy axis. Let this value be $H_{0\parallel}$, then when the field makes an angle θ_0 with the easy axis, the critical field is given by

$$H_0 = \frac{H_{0\parallel}}{\cos\theta_0}, \tag{18.71}$$

which is shown graphically in Fig. 18.24. Thus the critical field is larger for larger θ_0.

18.3 MAGNETIZATION ROTATION

If a ferromagnetic medium contains no domain walls or only immobile domain walls, it can be magnetized only by rotation of the domain magnetization. This mechanism is called *magnetization rotation.*

First we discuss reversible magnetization rotation. As shown in Fig. 18.25, the domain magnetization I_s rotates from the easy axis by the application of a magnetic field H, and rotates back to the easy axis reversibly upon removal of the magnetic field.

If the magnetic anisotropy is uniaxial, and the applied magnetic field H makes an angle θ_0 with the easy axis, the domain magnetization I_s rotates from the easy axis to a direction which makes the angle θ with the field (Fig. 18.25). The energy density in this case is given by

$$E = -K_u \cos^2(\theta - \theta_0) - I_s H \cos\theta, \tag{18.72}$$

where K_u is the uniaxial anisotropy constant. This is equal to K_{u1} for uniaxial magnetocrystalline anisotropy, and equal to $\frac{3}{2}\lambda\sigma$ for magnetoelastic energy due to stress σ. Moreover, a uniaxial anisotropy can arise from various induced anisotropies such as magnetic annealing, roll magnetic anisotropy, and shape magnetic anisotropy of elongated precipitates.

The equilibrium direction of magnetization can be found by minimizing the energy (18.72) with respect to θ, giving

$$\frac{\partial E}{\partial\theta} = K_u \sin 2(\theta - \theta_0) + I_s H \sin\theta = 0. \tag{18.73}$$

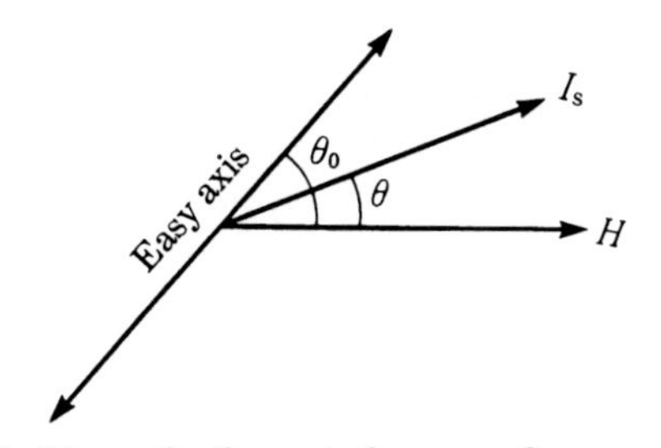

Fig. 18.25. Magnetization rotation away from an easy axis.

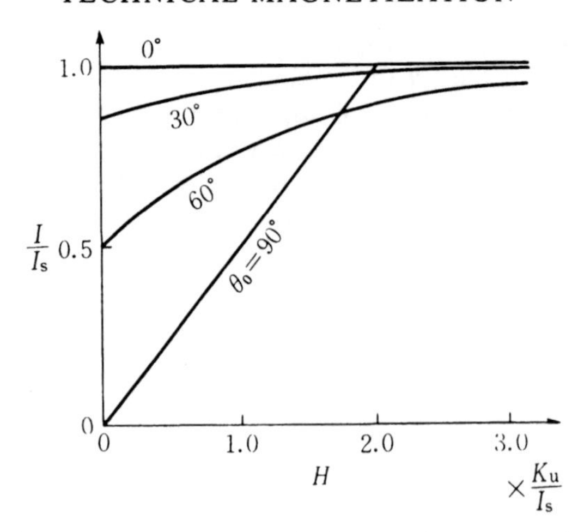

Fig. 18.26. Magnetization curves due to magnetization rotation against the uniaxial anisotropy. Numerical values are the angles between the field and the easy axis.

Putting $\cos\theta = x$, we have from (18.73)

$$4x^4 + 4p\cos 2\theta_0\, x^3 - (4 - p^2)x^2 - 4p\cos 2\theta_0\, x + \sin^2 2\theta_0 - p^2 = 0, \quad (18.74)$$

where $p = I_s H/K_u$. Solving for x in (18.74), we have

$$I = I_s \cos\theta = I_s x. \quad (18.75)$$

The magnetization curve can be obtained by plotting x as a function of p as shown in Fig. 18.26. For the case in which $\theta_0 = 90°$, we have from (18.73)

$$\cos\theta = \frac{I_s H}{2K_u}. \quad (18.76)$$

Putting this in (18.75), we have

$$I = \frac{I_s^2}{2K_u} H, \quad (18.77)$$

which is a linear magnetization curve (see Fig. 18.26). Such linear magnetization curves are observed for a stretched nickel wire ($\lambda < 0$); for a Permalloy wire annealed with an electric current flowing through itself (see curve c in Fig. 13.1); for Isoperm, which is a cold-rolled grain-orientated Fe–Ni alloy (see Fig. 13.9); and for a Co single crystal with the c-axis perpendicular to the magnetic field. When the magnetic field is removed, the magnetization decreases along the same linear curve, with almost no residual magnetization. As θ_0 is reduced, the residual magnetization increases, while the slope of the magnetization curve, or the susceptibility, decreases as seen in the magnetization curves calculated from (18.74) in Fig. 18.26. It is interesting that for $\theta_0 = 30°$ and $60°$, the magnetization does not saturate, in contrast to the case of $\theta_0 = 90°$, for which the magnetization saturates completely at $H = 2K_u/I_s$.

In very weak magnetic fields ($H \ll K_u/I_s$), $\theta \approx \theta_0$. If we put

$$\Delta\theta = \theta_0 - \theta, \tag{18.78}$$

the anisotropy energy in (18.72) is expressed as $K_u \Delta\theta^2$, so that condition (18.73) becomes

$$2K_u \Delta\theta = I_s H \sin\theta_0,$$

or

$$\Delta\theta = \frac{I_s H}{2K_u} \sin\theta_0. \tag{18.79}$$

The initial susceptibility is obtained from (18.75) as

$$\chi_a = \left(\frac{\partial I}{\partial H}\right)_{H=0} = -I_s \sin\theta_0 \frac{\partial\theta}{\partial H}, \tag{18.80}$$

so that from (18.79) we have

$$\chi_a = \frac{I_s^2 \sin^2\theta_0}{2K_u}. \tag{18.81}$$

In the case of cubic anisotropy with positive K_1, the anisotropy energy is expressed as $K_1\Delta\theta^2$ as shown in (12.52), so that the initial susceptibility is given by

$$\chi_a = \frac{I_s^2 \sin^2\theta_0}{2K_1}. \tag{18.82}$$

In the case of negative K_1 the anisotropy energy is expressed as $-(2K_1/3)\Delta\theta^2$ as shown in (12.55), so that the initial susceptibility is given by

$$\chi_a = -\frac{3I_s^2 \sin^2\theta_0}{4K_1}. \tag{18.83}$$

In the case of a random distribution of the easy axes, $\overline{\sin^2\theta_0} = \frac{2}{3}$, so that (18.82) becomes

$$\bar{\mu}_a \approx \bar{\chi}_a = \frac{I_s^2}{3K_1 \mu_0}. \tag{18.84}$$

For iron in which $I_s = 2.15\,T$ and $K_1 = 4.2 \times 10^4\,J\,m^{-3}$, (18.84) gives

$$\bar{\mu}_a = \frac{2.15^2}{3 \times 4.2 \times 4\pi \times 10^{-3}} \simeq 29. \tag{18.85}$$

Since iron usually exhibits relative initial permeability of 100–200, much larger than the value given by (18.85), there is no doubt that domain wall displacement contributes to the mechanism of magnetization at low fields. As mentioned in the previous section, the susceptibility due to wall displacement is quite structure-

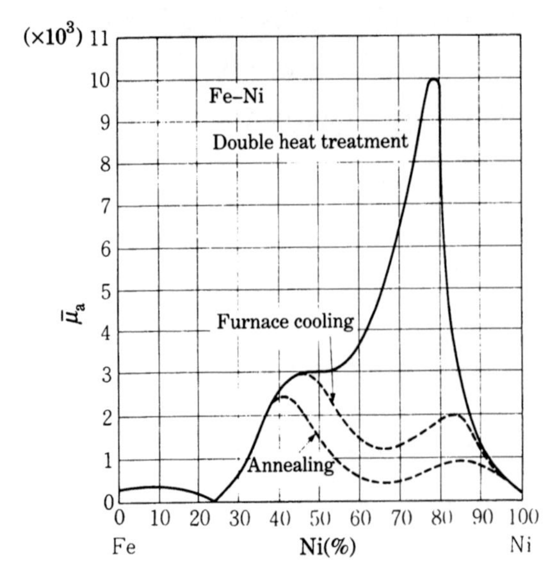

Fig. 18.27. Initial permeability of Fe–Ni alloys (After Bozorth[12]). Double heat treatment means that the specimen was annealed for 30 min at 1000°C, cooled slowly to 600°C, and then quenched on a copper plate kept at room temperature. Furnace cooling means cooling from 1000°C in the furnace. Annealing means prolonged annealing below 500°C after annealing at 1000°C.

sensitive, so that it can increase indefinitely by improving the homogeneity of the material. On the other hand, the susceptibility due to magnetization rotation depends only on the magnitude of magnetic anisotropy and is insensitive to the homogeneity of the material. Thus we can say that the susceptibility due to magnetization rotation is the lower limit of the initial susceptibility.

As shown in Fig. 18.27, the relative permeability of Fe–Ni alloys depends upon the heat treatment: The maximum value of about 10 000 is observed at 22% Fe, where K_1 is almost zero. Assuming that this value is due to magnetization rotation, we can calculate the value of K using (18.84) with $\bar{\mu}_a = 10\,000$, and $I_s = 1.1$, as follows:

$$K = \frac{I_s^2}{3\mu_0 \bar{\mu}_a} = \frac{1.1^2}{3 \times 4\pi \times 10^{-7} \times 10\,000} \approx 32. \tag{18.86}$$

In Fe–Ni alloys K_1 is zero at 24% Fe, and λ is zero at 19% Fe. At 19% Fe the magnetoelastic anisotropy is zero irrespective of stress, while $K_1 = 420\,\mathrm{J\,m^{-3}}$ (see the curve for the disordered state in Fig. 12.35). This value is too large to permit such a high permeability. At 24% Fe, $K_1 = 0$, while $\lambda = 7 \times 10^{-6}$ (see Fig. 14.11) which means a stress σ less than $3 \times 10^6\,\mathrm{N\,m^{-2}} = 310\,\mathrm{g\,mm^{-2}}$ is needed in order to produce a magnetoelastic anisotropy less than $32\,\mathrm{J\,m^{-3}}$. This value of stress is about 1% of ordinary working stresses. Thus the high permeability is not due to magnetization rotation. Therefore we must consider wall displacement, as discussed previously.

However, there is another possibility involving magnetization rotation: As explained in Chapter 13, directional order develops in this alloy system, creating an induced magnetic anisotropy K_{uf} which stabilizes the magnetization parallel to its direction during annealing. In order to obtain high permeability, the alloy must be annealed at 1000°C for 30 min, cooled slowly down to 600°C, and then quenched at a moderate rate to room temperature (double heat treatment). By this treatment, an induced anisotropy K_{uf} with the easy axis randomly oriented is produced. This local anisotropy cancels the cubic anisotropy from place to place in a composition range where $K_1 \approx K_{uf}$, thus resulting in easy rotation of magnetization locally. Further annealing or slow cooling results in the development of directional order, thus creating a uniaxial anisotropy much larger than the cubic anisotropy. Therefore the high permeability is reduced, as seen in the dashed curves in Fig. 18.27. This picture can explain why such a high permeability is obtained reproducibly in a definite composition range.

Next we discuss irreversible magnetization rotation. In ordinary ferromagnetic materials, irreversible magnetization takes place by domain wall displacement, while in single domain particles, irreversible magnetization occurs by magnetization rotation.

First we treat the case of uniaxial anisotropy. Figure 18.28 shows an elongated single domain particle whose long axis makes an angle θ_0 with the x-axis. For uniaxial anisotropies other than shape anisotropy, the treatment is the same, with the easy axis in the direction of the long axis of Fig. 18.28. When a magnetic field, H, is applied in the $-x$-direction, the spontaneous magnetization rotates to the direction which makes the angle, θ, with the magnetic field. The energy density of this system is given by

$$E = -K_u \cos^2(\theta - \theta_0) + I_s H \cos\theta. \tag{18.87}$$

The equilibrium direction of the magnetization is obtained by minimizing this energy, giving

$$\frac{\partial E}{\partial \theta} = K_u \sin 2(\theta - \theta_0) - I_s H \sin\theta = 0. \tag{18.88}$$

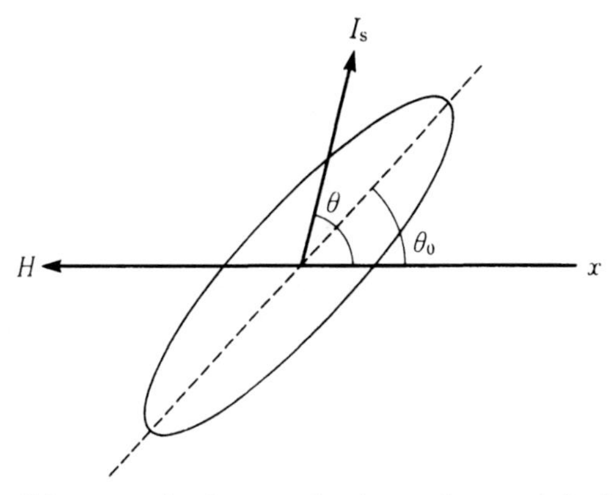

Fig. 18.28. Irreversible magnetization rotation in an elongated single domain particle.

When this is a stable equilibrium,

$$\frac{\partial^2 E}{\partial \theta^2} > 0,$$

and it is unstable when

$$\frac{\partial^2 E}{\partial \theta^2} < 0.$$

With increasing field H, the magnetization rotates, increasing the angle θ, and then suddenly rotates towards the $-x$-direction. At this instant, the magnetization changes from a stable to unstable equilibrium, so it must be that

$$\frac{\partial^2 E}{\partial \theta^2} = 0.$$

Differentiating (18.88), this condition gives

$$\frac{\partial^2 E}{\partial \theta^2} = 2K_u \cos 2(\theta - \theta_0) - I_s H_0 \cos \theta = 0. \tag{18.89}$$

Equations (18.88) and (18.89) can be expressed as

$$\left.\begin{aligned} \sin 2(\theta - \theta_0) &= p \sin \theta \\ \cos 2(\theta - \theta_0) &= \frac{p}{2} \cos \theta \end{aligned}\right\}, \tag{18.90}$$

where

$$p = \frac{I_s H_0}{K_u}. \tag{18.91}$$

Squaring both sides of the two equation in (18.90), and adding them together, we have an equation for $\sin^2 \theta$ from which we have

$$\sin \theta = \sqrt{\frac{4 - p^2}{3p^2}}, \qquad \cos \theta = \pm 2\sqrt{\frac{p^2 - 1}{3p^2}}. \tag{19.92}$$

Using (19.92) we can solve for $\sin 2\theta_0$ from (18.90), to get

$$\sin 2\theta_0 = \frac{1}{p^2}\left(\frac{4 - p^2}{3}\right)^{3/2}. \tag{18.93}$$

This relationship between p and θ_0 is shown graphically in Fig. 18.29. We see that p, or the critical field, H_0, is minimum when $\theta_0 = 45°$. In other words, when the field makes the angle 45° the magnetization reversal occurs at the smallest critical field, which is given by

$$H_0 = \frac{K_u}{I_s}. \tag{18.94}$$

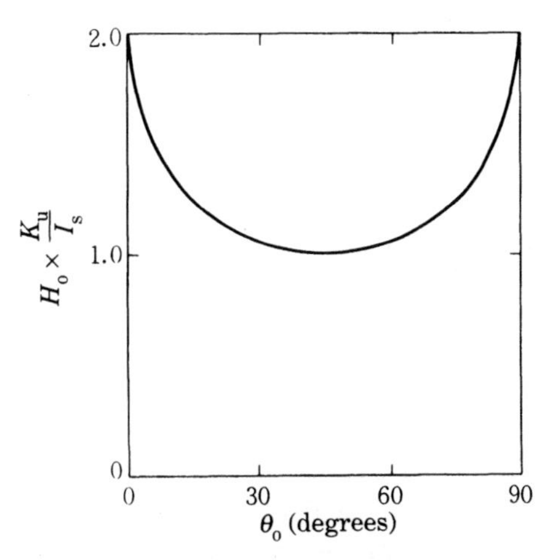

Fig. 18.29. Dependence of the critical field, H_0, on the orientation of the applied field (refer to (18.93)).

The critical field becomes larger as θ_0 deviates from 45°. At $\theta_0 = 0°$ or 90°, we have

$$H_0 = \frac{2K_u}{I_s}. \tag{18.95}$$

Next we shall consider the case of cubic anisotropy. Since the functional form of the anisotropy is more complex, the situation is not so simple. We discuss only the case in which the field is applied opposite to the magnetization pointing parallel to one of the easy axes. Since the anisotropy energy with positive K_1 is approximately expressed by (12.52), the total energy of the system is given by

$$E = K_1\theta^2 + I_s H \cos\theta. \tag{18.96}$$

Therefore the critical field can be solved by the conditions

$$\frac{\partial E}{\partial \theta} = 2K_1\theta - I_s H \sin\theta = 0, \tag{18.97}$$

and

$$\frac{\partial^2 E}{\partial \theta^2} = 2K_1 - I_s H \cos\theta = 0. \tag{18.98}$$

Solving these two equations, we have the critical field

$$H_0 = \frac{2K_1}{I_s}. \tag{18.99}$$

In the case of negative K_1, the anisotropy energy is given by (12.55) when I_s is close to an easy axis, so that the total energy is given by

$$E = -\tfrac{2}{3}K_1\theta^2 + I_s H \cos\theta. \tag{18.100}$$

From the conditions for neutral equilibrium

$$\frac{\partial E}{\partial \theta} = -\frac{4}{3} K_1 \theta - I_s H \sin\theta = 0, \tag{18.101}$$

and

$$\frac{\partial^2 E}{\partial \theta^2} = -\frac{4}{3} K_1 - I_s H \cos\theta = 0, \tag{18.102}$$

we have the critical field

$$H_0 = -\frac{4}{3}\frac{K_1}{I_s} \qquad (K_1 < 0). \tag{18.103}$$

Thus the magnitude of the critical field for irreversible magnetization rotation is the order of K/I_s. In the common ferromagnets $I_s = 1 \sim 2\,\mathrm{T}$, and $K_1 = 10^4 \sim 10^5\,\mathrm{J\,m^{-3}}$, so that $H_0 = 10^4 \sim 10^5\,\mathrm{A\,m^{-1}}$ ($= 10^2 \sim 10^3$ Oe). These values are in the range for permanent magnet materials. Thus if such an irreversible rotation magnetization is realized, we expect a reasonably good permanent magnet material.

18.4 RAYLEIGH LOOP

In or near the demagnetized state the magnetization caused by the application of a weak magnetic field can be described by

$$I = \chi_a H + \tfrac{1}{2}\eta H^2, \tag{18.104}$$

where χ_a is the initial susceptibility and η is the Rayleigh constant. This phenomenon was investigated by Lord Rayleigh[13] a long time ago. In the reversible magnetization region where H is small, the magnetization is well approximated by the first term. With increasing amplitude of H, the second term becomes significant because of the appearance of irreversible magnetization. When H is decreased, the second term changes its sign. Therefore if H is oscillated about the origin, the magnetization changes along a loop as shown in Fig. 18.30. In the lower half of the loop, which starts at point B, the magnetization changes approximately as given by (18.104), so that the magnetization is described by

$$I + I_1 = \chi_a (H + H_1) + \tfrac{1}{2}\eta (H + H_1)^2, \tag{18.105}$$

where I_1 and H_1 are the amplitudes of I and H, respectively. The upper half of the loop can be described by

$$I - I_1 = \chi_a (H - H_1) - \tfrac{1}{2}\eta (H - H_1)^2. \tag{18.106}$$

The value of I_1 can be obtained by putting $I = I_1$ at $H = H_1$, to give

$$I_1 = \chi_a H_1 + \eta H_1^2. \tag{18.107}$$

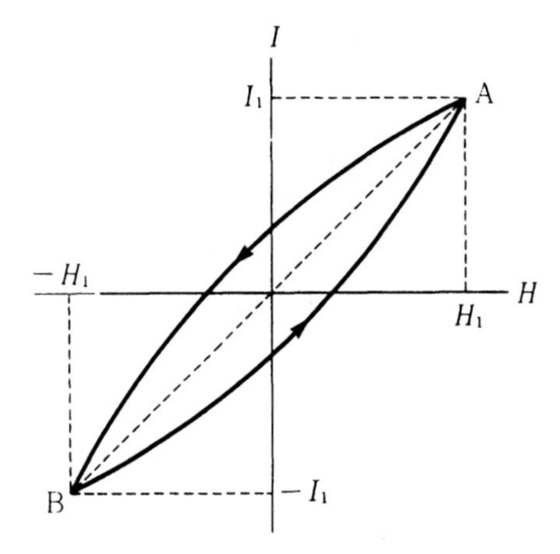

Fig. 18.30. Rayleigh loop.

Using (18.107), we can rewrite (18.105) and (18.106) as

$$\left.\begin{aligned} &\text{ascending part:} \quad I = (\chi_a + \eta H_1)H + \tfrac{1}{2}\eta(H^2 - H_1^2) \\ &\text{descending part:} \quad I = (\chi_a + \eta H_1)H - \tfrac{1}{2}\eta(H^2 - H_1^2) \end{aligned}\right\}. \tag{18.108}$$

These curves are called the *Rayleigh loop*. The hysteresis loss is calculated to be

$$W_\eta = \oint I\,\mathrm{d}H = \tfrac{4}{3}\eta H_1^3, \tag{18.109}$$

which shows that the loss increases rapidly with an increase of the amplitude H_1.

The appearance of this hysteresis loop is undesirable for magnetic cores, because it causes not only power losses but also a distortion of the waveform. Figure 18.31 shows how the waveform of I is distorted by the Rayleigh loop when a pure sine wave of H is applied. By analyzing this waveform, we find that the phase of the fundamental waveform of I is shifted by the angle δ, and also there appears a third harmonic wave $\sin 3\omega t$. Thus the flux density B is expressed by

$$B = B_\omega \sin(\omega t - \delta) + B_{3\omega} \sin 3\omega t + \cdots. \tag{18.110}$$

Calculation shows that

$$\tan\delta = \frac{4}{3\pi}\,\frac{\eta H_1}{\mu_a + \eta H_1} \tag{18.111}$$

and

$$k = \frac{3B_{3\omega}}{B_\omega} = \frac{4}{5\pi}\,\frac{\eta H_1}{\mu_a}. \tag{18.112}$$

The quantity $\tan\delta$ is called the *loss factor*, and has a nonzero value when there is hysteresis loss. The quantity k given by (18.112) is called the *Klirr factor*; it is a

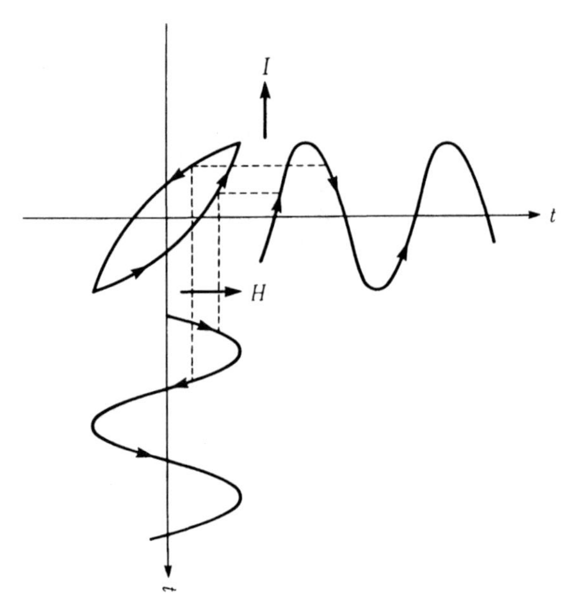

Fig. 18.31. Distortion of the wave form of magnetization by the Rayleigh loop when a pure sinusoidal field is applied.

measure of the third harmonic relative to the fundamental frequency. The presence of the third harmonic is harmful because it not only distorts the waveform but also gives rise to a loss if three-phase AC lines are delta-connected, for then the voltage induced by the third harmonics is shorted in the delta circuit.

The origin of the Rayleigh loop is the same as that of the ordinary hysteresis. The only difference is that the amplitude of the field is very small in the case of the Rayleigh loop. Let us consider the mechanism in terms of wall displacement. As discussed in Section 18.2, the wall is held at the place where condition (18.60) is satisfied. In Fig. 18.32, $\partial \varepsilon_\omega / \partial s$ is plotted as a function of the displacement s. At $H = 0$ the wall stays at a point, say A, where $\partial \varepsilon_\omega / \partial s = 0$. With an increase of the field, the point rises along the slope, and displaces discontinuously from the maximum to the next positive slope, say to B. With further increase of H, the point displaces from B to C → D → E. It should be noted that the stronger the field, the longer the jump in each discontinuous displacement. The reason is that the maxima which are lower than those that have already been surmounted by the wall are no longer effective in stopping the wall.

Now we can derive the second term in Rayleigh's equation from this model. The critical field to get over a maximum, H_0, is given by (18.61). Let the number of maxima that have critical fields between H_0 and $H_0 + \mathrm{d}H_0$ be $f(H_0)\,\mathrm{d}H_0$, where $f(H_0)$ is a function of H_0, similar to the Gaussian distribution function centered at $H_0 = 0$. In the present case, H_0 is limited to a small range, so that we can postulate that $f(H_0)$ is a constant, say f_0. If the field is increased from H to $H + \mathrm{d}H$, the number of walls which are released should be proportional to $f_0\,\mathrm{d}H$. On the other

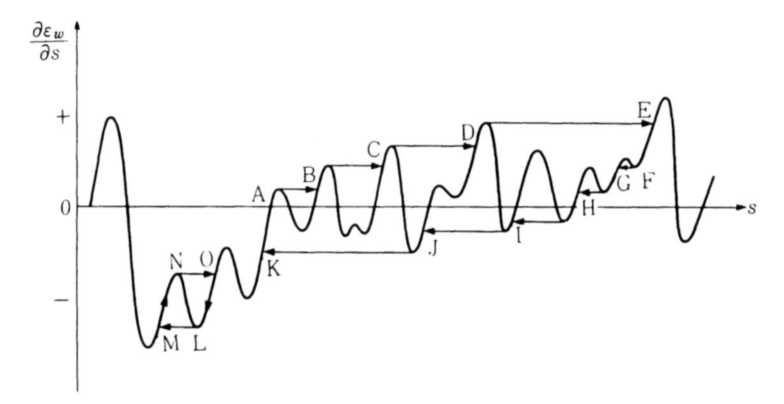

Fig. 18.32. Irreversible displacement of domain wall along the Rayleigh loop.

hand, each displacement jump is proportional to the number of maxima with H_0 less than the applied field H, or

$$\int_0^H f_0 \, \mathrm{d}H_0 = f_0 H. \tag{18.113}$$

The irreversible magnetization should then be proportional to (number of displaced walls) × (one jump of irreversible displacement) or

$$\mathrm{d}I_{\mathrm{irr}} = cf_0^2 H \, \mathrm{d}H, \tag{18.114}$$

where c is a proportionality factor. The irreversible magnetization caused by the change from 0 to H is thus given by

$$I_{\mathrm{irr}} = \int_0^{I_{\mathrm{irr}}} \mathrm{d}I_{\mathrm{irr}} = cf_0^2 \int_0^H H \, \mathrm{d}H = \tfrac{1}{2} cf_0^2 H^2, \tag{18.115}$$

which is Rayleigh's second term.

When the field is decreased from the point E in Fig. 18.32, the point displaces reversibly along the curve to F and then jumps to G, H, I, J, and K. The shape of the loop AEFK closely resembles the shape of a Rayleigh loop.

The abscissa of the graph in Fig. 18.32 is the displacement of the wall, so that it corresponds to the magnetization, while the ordinate corresponds to the magnetic field. Therefore the minor loop LMNO in Fig. 18.32 corresponds to the shifted rectangular hysteresis loop as shown in Fig. 18.33. Preisach[14] explained the Rayleigh loop by regarding the magnetic material as an assembly of units, each of which exhibits a shifted hysteresis loop with different values of half-width a and shift b.

Figure 18.34 shows the states of magnetization of these different units. Here + means that the magnetization of the unit points the + direction, while − means the magnetization points in the − direction. Figure (a) shows the demagnetized state, where in the region $b > a$ above the line OA, all the magnetizations are − (see Fig. 18.33), while in the region $b < a$ below the line OB, all the magnetizations are +.

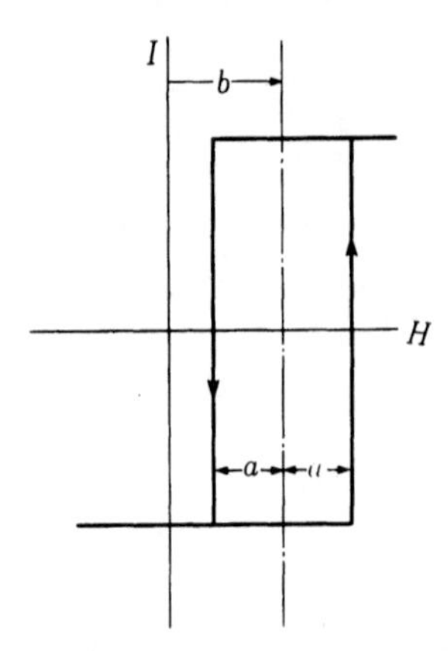

Fig. 18.33. A shifted rectangular hysteresis loop.

Now we apply the field H_1 in the + direction, so that the boundary lines AOB are shifted by H_1 to A′O′B′ (see Fig. 18.34 (b)), reversing the − magnetization below the

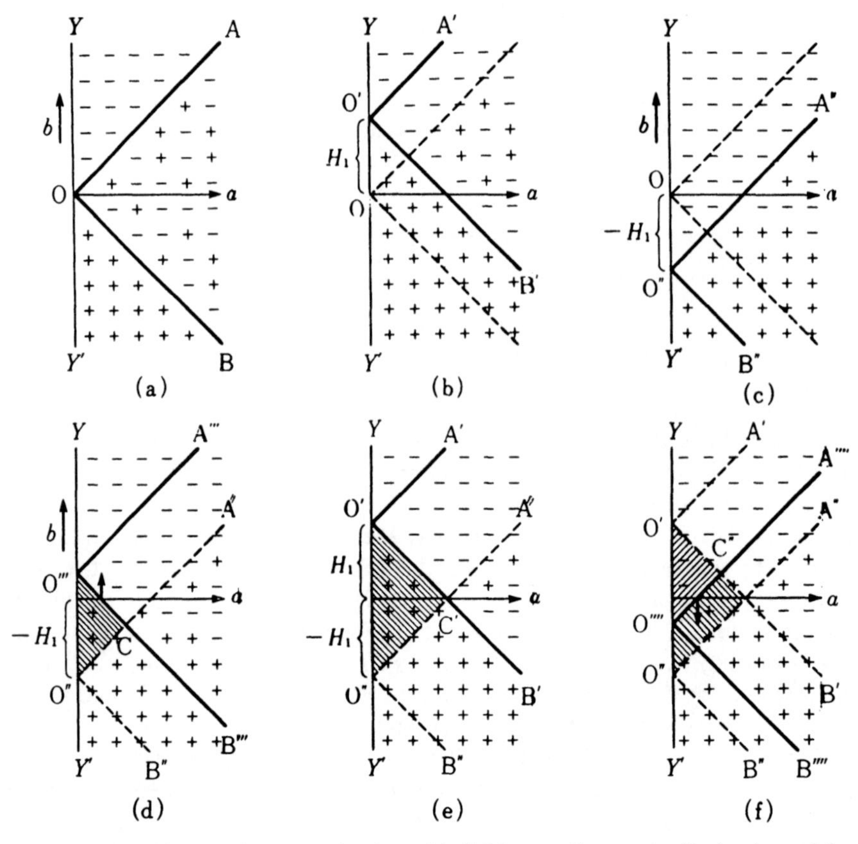

Fig. 18.34. Changes in magnetization with field according to the Preisach model.

Table 18.1. Parameters of Rayleigh loops[15]

Material	$\bar{\mu}_a$	$\bar{\eta}$*	W_h for $\Delta I = 10^{-4}$ T $= 1/4\pi$ gauss
		per A m^{-1}(per $\frac{10^3}{4\pi}$ Oe)	$\mu\text{J m}^{-3}(\times 10\ \mu\text{erg cm}^{-3})$
Iron	200	25	2.6
Fe powder	30	0.013	0.41
Cobalt	70	0.13	0.32
Nickel	220	3.1	0.25
45 Permalloy	2 300	200	0.014
4–79 Mo Permalloy	20 000	4 300	0.0005
Supermalloy	100 000	150 000	0.0001
45-25 Perminvar	400	0.0013	0.00002

* The value of $\bar{\eta}$ measured in units of μ_0. It is the rate of increase of μ_a by an increase in the amplitude of H.

line O′B′ to +. If the field is decreased to $-H_1$, the boundary lines are shifted to A″O″B″ (see Fig. 18.34 (c)), thus resulting in an increase in the number of − magnetized units. If the field is changed back and forth between H_1 and $-H_1$, magnetization changes only in the shaded triangular area O′C′O″ shown in Fig. 18.34 (c). The magnetization reversal occurs from one end of the triangular area, and the magnetization change is proportional to the area of ΔO′C″O‴, or to H^2 (see Fig. 18.34 (f)). This corresponds to the Rayleigh relationship given by the second term in (18.104).

The Preisach model is essentially the same as the explanation given above with respect to Fig. 18.32. It is also possible to regard the Preisach model as an assembly of small magnetic particles with various rectangular hysteresis loops. In this case, the shift b is caused by the magnetic interaction from the neighboring particles.

The parameters of Rayleigh loops measured in various magnetic materials are listed in Table 18.1.

18.5 LAW OF APPROACH TO SATURATION

Under a moderately strong field, ferromagnetic materials are generally magnetized to their saturated state, where all possible wall displacements have taken place and the magnetization is pointed almost parallel to the applied field.

In this section we examine the rotation of the magnetization under such circumstances. Let the angle between magnetization and the magnetic field be θ. Then the component of magnetization in the direction of the field is given by

$$I = I_s \cos\theta = I_s\left(1 - \frac{\theta^2}{2} + \cdots\right). \tag{18.116}$$

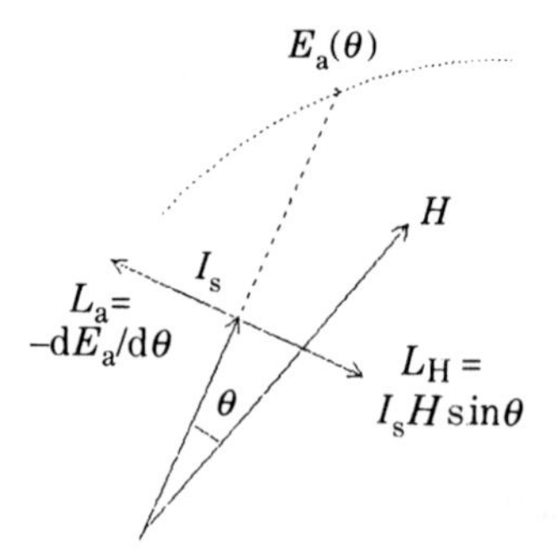

Fig. 18.35. Rotation of magnetization against the magnetic anisotropy.

The torque exerted by the magnetic field is counterbalanced by the torque resulting from the magnetic anisotropy (see Fig. 18.35):

$$I_s H \sin \theta = -\frac{\partial E_a}{\partial \theta}, \tag{18.117}$$

where E_a is the anisotropy energy. Since θ is very small, it can be found from (18.117) to be

$$\theta = -\frac{(\partial E_a/\partial\theta)_{\theta=0}}{I_s H} = \frac{C}{I_s}\frac{1}{H}, \tag{18.118}$$

where

$$C = -\left(\frac{\partial E_a}{\partial \theta}\right)_{\theta=0}. \tag{18.119}$$

On substituting (18.118) in (18.116), we have

$$I = I_s\left(1 - \frac{b}{H^2} - \cdots\right), \tag{18.120}$$

where

$$b = \frac{1}{2}\frac{C^2}{I_s^2}. \tag{18.121}$$

Now we solve for the actual form of C for cubic anisotropy. Since the magnetization rotates along the maximum gradient of the anisotropy energy in the vicinity of H,

$$C^2 = |\mathrm{grad}\, E_a|^2 = \left(\frac{\partial E_a}{\partial \theta}\right)^2 + \frac{1}{\sin^2\theta}\left(\frac{\partial E_a}{\partial \varphi}\right)^2, \tag{18.122}$$

where (θ, φ) are the polar coordinates of the magnetization. Since E_a is normally expressed as a function of the direction cosines $(\alpha_1, \alpha_2, \alpha_3)$ of magnetization, which are related to (θ, φ) by $\alpha_1 = \sin\theta\cos\varphi$, $\alpha_2 = \sin\theta\sin\varphi$, $\alpha_3 = \cos\theta$,

$$\left.\begin{aligned}\frac{\partial E_a}{\partial\theta} &= \left(\frac{\partial E_a}{\partial\alpha_1}\right)\frac{\partial\alpha_1}{\partial\theta} + \left(\frac{\partial E_a}{\partial\alpha_2}\right)\frac{\partial\alpha_2}{\partial\theta} + \left(\frac{\partial E_a}{\partial\alpha_3}\right)\frac{\partial\alpha_3}{\partial\theta}\\ &= \left(\frac{\partial E_a}{\partial\alpha_1}\right)\cos\theta\cos\varphi + \left(\frac{\partial E_a}{\partial\alpha_2}\right)\cos\theta\sin\varphi - \left(\frac{\partial E_a}{\partial\alpha_3}\right)\sin\theta\\ \text{and}\qquad \frac{1}{\sin\theta}\frac{\partial E_a}{\partial\varphi} &= \left(\frac{\partial E_a}{\partial\alpha_1}\right)\frac{\partial\alpha_1}{\sin\theta\,\partial\varphi} + \left(\frac{\partial E_a}{\partial\alpha_2}\right)\frac{\partial\alpha_2}{\sin\theta\,\partial\varphi} + \left(\frac{\partial E_a}{\partial\alpha_3}\right)\frac{\partial\alpha_3}{\sin\theta\,\partial\varphi}\\ &= -\left(\frac{\partial E_a}{\partial\alpha_1}\right)\sin\varphi + \left(\frac{\partial E_a}{\partial\alpha_2}\right)\cos\varphi.\end{aligned}\right\}\quad(18.123)$$

Then (18.122) becomes

$$\begin{aligned}C^2 &= \left(\frac{\partial E_a}{\partial\alpha_1}\right)^2(\cos^2\theta\cos^2\varphi + \sin^2\varphi) + \left(\frac{\partial E_a}{\partial\alpha_2}\right)^2(\cos^2\theta\sin^2\varphi + \cos^2\varphi)\\ &\quad + \left(\frac{\partial E_a}{\partial\alpha_3}\right)^2\sin^2\theta + 2\left(\frac{\partial E_a}{\partial\alpha_1}\right)\left(\frac{\partial E_a}{\partial\alpha_2}\right)(\cos^2\theta\sin\varphi\cos\varphi - \sin\varphi\cos\varphi)\\ &\quad - 2\left(\frac{\partial E_a}{\partial\alpha_1}\right)\left(\frac{\partial E_a}{\partial\alpha_3}\right)\sin\theta\cos\theta\cos\varphi - 2\left(\frac{\partial E_a}{\partial\alpha_2}\right)\left(\frac{\partial E_a}{\partial\alpha_3}\right)\sin\theta\cos\theta\sin\varphi\\ &= \left(\frac{\partial E_a}{\partial\alpha_1}\right)^2(\alpha_3^2 + \alpha_2^2) + \left(\frac{\partial E_a}{\partial\alpha_2}\right)^2(\alpha_3^2 + \alpha_1^2) + \left(\frac{\partial E_a}{\partial\alpha_3}\right)^2(\alpha_1^2 + \alpha_2^2)\\ &\quad - 2\left(\frac{\partial E_a}{\partial\alpha_1}\right)\left(\frac{\partial E_a}{\partial\alpha_2}\right)\alpha_1\alpha_2 - 2\left(\frac{\partial E_a}{\partial\alpha_1}\right)\left(\frac{\partial E_a}{\partial\alpha_3}\right)\alpha_1\alpha_3 - 2\left(\frac{\partial E_a}{\partial\alpha_2}\right)\left(\frac{\partial E_a}{\partial\alpha_3}\right)\alpha_2\alpha_3\\ &= \left(\frac{\partial E_a}{\partial\alpha_1}\right)^2 + \left(\frac{\partial E_a}{\partial\alpha_2}\right)^2 + \left(\frac{\partial E_\alpha}{\partial\alpha_3}\right)^2 - \left\{\left(\frac{\partial E_a}{\partial\alpha_1}\right)\alpha_1 + \left(\frac{\partial E_a}{\partial\alpha_2}\right)\alpha_2 + \left(\frac{\partial E_a}{\partial\alpha_3}\right)\alpha_3\right\}^2.\end{aligned}\quad(18.124)$$

If we adopt the first term of (12.5),

$$\frac{\partial E_a}{\partial\alpha_1} = 2K_1\alpha_1(1-\alpha_1^2),\ \frac{\partial E_a}{\partial\alpha_2} = 2K_1\alpha_2(1-\alpha_2^2),\ \frac{\partial E_a}{\partial\alpha_3} = 2K_1\alpha_3(1-\alpha_3^2).\quad(18.125)$$

Then (18.124) becomes

$$\begin{aligned} C^2 &= 4K_1^2\{1 - 2(\alpha_1^4 + \alpha_2^4 + \alpha_3^4) + (\alpha_1^6 + \alpha_2^6 + \alpha_3^6)\} \\ &\quad - 4K_1^2\{1 - (\alpha_1^4 + \alpha_2^4 + \alpha_3^4)\}^2 \\ &= 4K_1^2\{(\alpha_1^6 + \alpha_2^6 + \alpha_3^6) - (\alpha_1^8 + \alpha_2^8 + \alpha_3^8) \\ &\quad - 2(\alpha_1^4\alpha_2^4 + \alpha_2^4\alpha_3^4 + \alpha_3^4\alpha_1^4)\}. \end{aligned} \tag{18.126}$$

For a polycrystal, averaging over the all possible orientations of the individual crystallites, we have $\overline{\alpha_i^6} = \frac{1}{7}$, $\overline{\alpha_i^8} = \frac{1}{9}$, $\overline{\alpha_i^4\alpha_j^4} = \frac{1}{105}$; hence

$$\overline{C}^2 = 4K_1^2\{\tfrac{3}{7} - \tfrac{3}{9} - \tfrac{6}{105}\} = \tfrac{16}{105}K_1^2. \tag{18.127}$$

On putting this in (18.121), we obtain

$$b = \frac{8}{105}\frac{K_1^2}{I_s^2} = 0.0762\frac{K_1^2}{I_s^2}. \tag{18.128}$$

This result coincides with that obtained by Becker and Döring[16] by an entirely different method.

The law of approach to saturation, obtained by experiment, is

$$I = I_s\left(1 - \frac{a}{H} - \frac{b}{H^2} - \cdots\right) + \chi_0 H. \tag{18.129}$$

The last term, $\chi_0 H$, is caused by an increase of the spontaneous magnetization itself. The first term, a/H, has been discussed by many investigators, as we shall see later. Czerlinsky[17] measured $\mathrm{d}I/\mathrm{d}H$ in the high-field region, instead of measuring I itself, because in the latter method a small error in the measurement of I may have a fatal influence on the determination of a and b in (18.129). From (18.129), we have

$$\frac{\mathrm{d}I}{\mathrm{d}H} = I_s\left(\frac{a}{H^2} + \frac{2b}{H^3} + \cdots\right) + \chi_0. \tag{18.130}$$

Figures 18.36 and 18.37 show $\mathrm{d}I/\mathrm{d}H$ as a function of $1/H^3$ for iron and nickel. The experimental points are well fitted by straight lines. This means that the experiments are well described by the b/H^2 term in (18.129). Using the values of b thus determined we can calculate from (18.128) the values of K_1. We obtain (at room temperature)

$$\begin{aligned} &\text{Fe} \quad |K_1| = 4.14, \quad 3.98 \times 10^4\ (\mathrm{J\,m^{-3}}) \\ &\text{Ni} \quad |K_1| = 5.0, \quad 4.66 \times 10^3\ (\mathrm{J\,m^{-3}}). \end{aligned} \tag{18.131}$$

These values are in fair agreement with the values given in (12.6) and (12.7).

A close inspection of the graphs in Figs. 18.36 and 18.37 reveals that the experimental points deviate from the linear relationship near the origin. This means the presence of the a/H term. This term cannot be explained in terms of a uniform

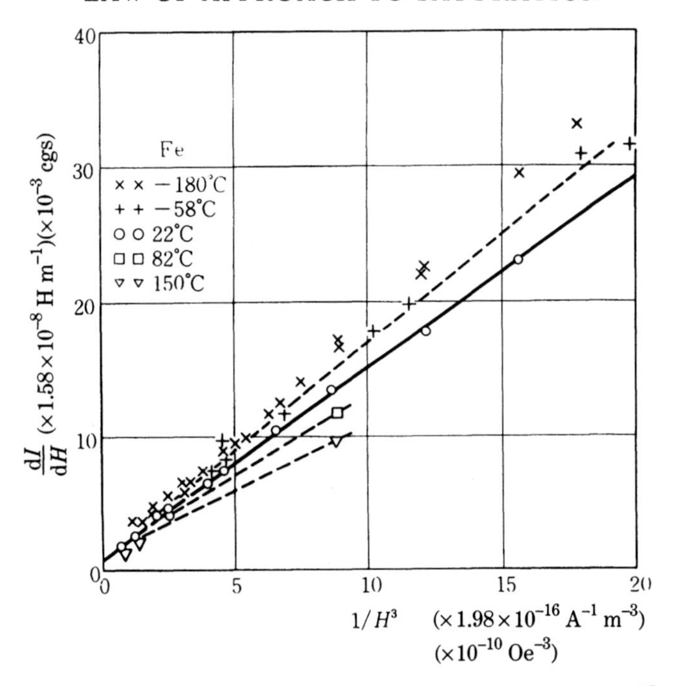

Fig. 18.36. Law of approach to saturation for iron (Czerlinsky[17]).

magnetization rotation, because this leads to the b/H^2 term, as already discussed in (18.120). Brown[18] explained the a/H term as the result of the stress field about dislocations, while Néel[19] interpreted it in terms of nonmagnetic inclusions and voids. Both calculations are very complicated, so that here we show only the basic ideas of the two interpretations. As pointed out by Néel, if the expression including the a/H term is valid to infinitely strong magnetic fields, the work necessary to magnetize the specimen to complete saturation diverges, as shown by

$$W = \int_I^{I_s} H \, dI = \int_H^{\infty} H\left(\frac{dI}{dH}\right) dH$$
$$= \int_H^{\infty} I_s\left(\frac{a}{H}\right) dH = I_s \, |a \ln H|_H^{\infty} = \infty. \tag{18.132}$$

Thus we must conclude that the term a/H is valid only within some finite range of field strength. Since a constant restoring force or torque leads to the b/H^2 term, we must assume the presence of a restoring force which increases as the magnetization approaches saturation. For instance, small spike domains, which remain around voids at high fields, will diminish in volume with increasing field strength up to a demagnetizing field of the order of I_s/μ_0. If local internal stresses caused by dislocations or lattice vacancies fix the direction of the local magnetization firmly, the magnetization surrounding these points will form transition layers which are similar to an ordinary domain wall. The same thing is also expected for the magnetization surrounding a special precipitated particle producing an exchange anisotropy (see Section 13.4.1).

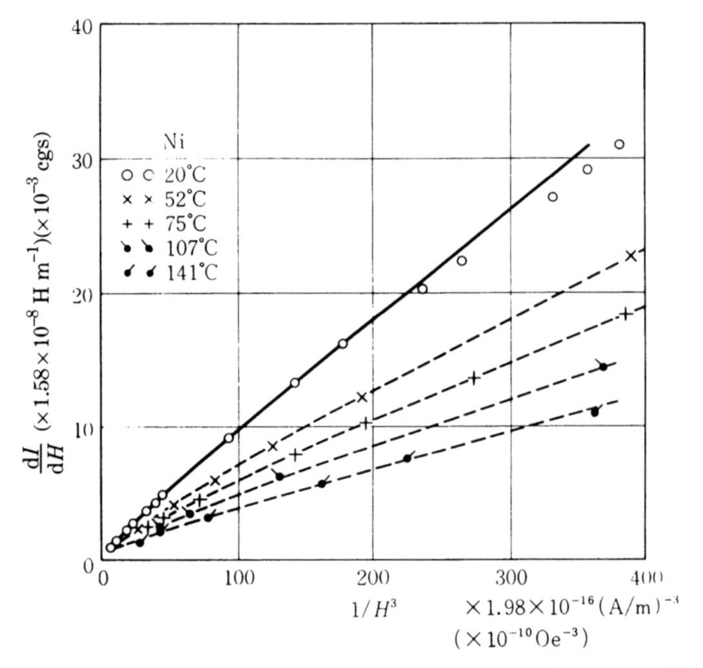

Fig. 18.37. Law of approach to saturation for nickel (Czerlinsky[17]).

The thickness of such a transition layer decreases with increasing field strength. In both cases, the change in magnetization is first proportional to $1/H^{1/2}$ and finally to $1/H^2$. It is naturally expected, therefore, that the change becomes proportional to $1/H$ in an intermediate field range. In any event, the magnitude of the term a/H is expected to be a good measure of the inhomogeneity of the magnetic material. Fähnle and Kronmüller[20] treated such a problem for amorphous ferromagnets where spatially random magnetostatic, magnetocrystalline, magnetostrictive and exchange fluctuations all are present.

Finally we discuss the last term, $\chi_0 H$, in (18.129). This term arises from an increase in spontaneous magnetization produced by the external magnetic field. According to the Weiss theory, the change in spontaneous magnetization by the external field is given by (6.14), so that the high-field susceptibility χ_0 is given by

$$\chi_0 = \frac{dI}{dH} = \frac{NM^2L'(\alpha)}{k(T-3\Theta L'(\alpha))}, \tag{18.133}$$

where $L(\alpha)$ is the Langevin function (see (5.20)). Experimentally, the value of χ_0 can be obtained from the extrapolation of the curve to the ordinate in Figs 18.36 and 18.37. The experimental values thus obtained are normally about ten times larger than the value calculated by (18.133). These values were found to be dependent on the impurity level of the materials. When the specimen is purified, the value is considerably reduced. Becker and Döring[21] considered the reason to be that the spins in the impurities or the spins of the matrix separated by impurities will be thermally agitated more than the spins in a pure material, giving rise to the large value of χ_0.

18.6 SHAPE OF HYSTERESIS LOOP

As discussed above, there are various mechanisms by which materials may be technically magnetized. The shape of the hysteresis loop depends on the predominant mechanism.

The simplest case may be an assembly of randomly oriented single-domain uniaxial ferromagnets, such as a random aggregate of elongated single-domain particles. Since there are no domain walls in the single-domain particles, all the technical magnetization must take place by magnetization rotation. Reversible magnetization rotation was discussed in Section 18.3, and the resulting magnetization curve is given in Fig. 18.26. When the magnetic field is reversed, an irreversible rotation of the magnetization occurs at a critical field given by Fig. 18.29. By combining the two magnetization processes, the hysteresis loop for a single-domain particle whose easy axis makes an angle θ_0 with the external field can be constructed, as shown by the curves in Fig. 18.38 for different values of θ_0. In these curves, the curved portions correspond to reversible magnetization and the vertical jumps correspond to irreversible magnetization rotation. It is interesting that for $\theta_0 > 45°$ the reversible magnetization is predominant, so that even after a fairly strong negative field brings the magnetization to a negative value, the magnetization returns to a positive value when the field is removed (see the curve denoted by 70° in Fig. 18.38).

If an array of such single-domain particles are oriented randomly, the hysteresis loop of the assembly is obtained by averaging the individual hysteresis loops in Fig. 18.38 to give Fig. 18.39. In this graph the solid curves represent reversible magnetization rotation, while the dashed curves represent irreversible magnetization rotation.

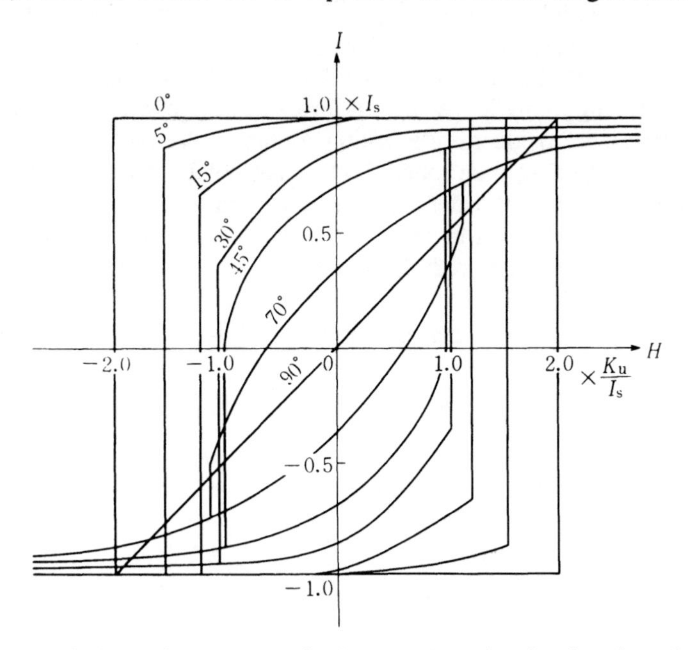

Fig. 18.38. Hysteresis loops due to magnetization rotation of variously oriented single domain particles with uniaxial anisotropy.

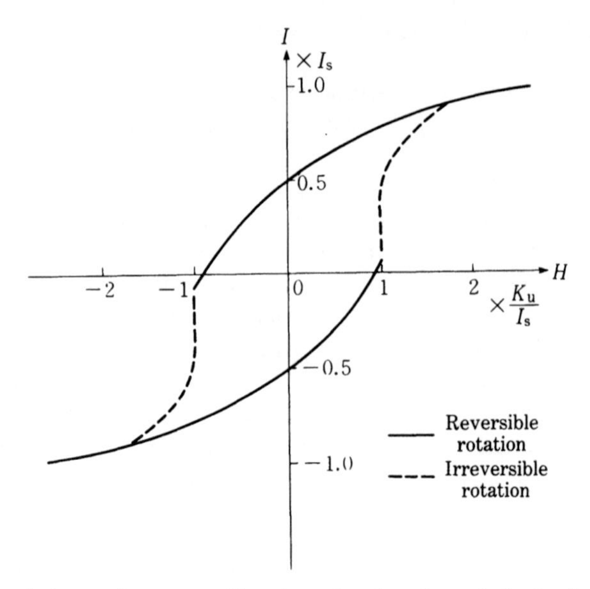

Fig. 18.39. Hysteresis loop of an assembly of randomly oriented single domain particles with uniaxial anisotropy, calculated by averaging the loops in Fig. 18.38.

In the field range $H > -K_u/I_s$, magnetization takes place exclusively by reversible magnetization rotation. Therefore, if the magnetic field is reduced from the coercive field to zero, the magnetization returns reversibly to the residual magnetization ($I_r = 0.5I_s$). In this model, the critical field H_0 is a unique function of the direction θ_0 of the elongated particles. In real materials, the critical field depends more or less on the nature of the particles; the hysteresis loop for real magnetic materials can be obtained by averaging many loops with different scale factors of the abscissa of Fig. 18.39. Thus some irreversible magnetization occurs in the second quadrant, and the ideal reversible rotation as mentioned above may not be observed. A difference in behavior between particles of different sizes may be the main reason why H_0 depends on the nature of the particles. Moreover if the size is too small, H_0 can be reduced by thermal fluctuations (see Section 20.1). Various parameters for the hysteresis loop in Fig. 18.39 are listed below:

$$\begin{aligned}
&\text{Residual magnetization} && I_r = 0.5I_s \\
&\text{Coercive field} && H_c \simeq 1.0\,\frac{K_u}{I_s} \\
&\text{Maximum susceptibility} && \chi_{\max} = 0.56\,\frac{I_s^2}{K_u} \\
&\text{Hysteresis loss} && W_h = 1.98K_u.
\end{aligned} \tag{18.134}$$

Next we discuss the hysteresis loops of uniaxial materials in which the magnetization reversal takes place exclusively by the displacement of domain walls. This magnetization mechanism was discussed in Section 18.2. When the magnetic field is applied exactly antiparallel to the direction of magnetization, we assume that the

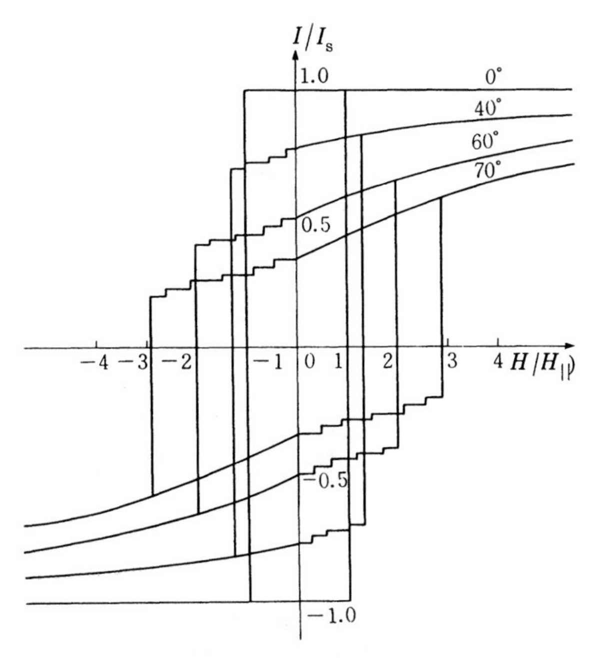

Fig. 18.40. Hysteresis loops due to reversible magnetization rotation and irreversible displacement of domain walls for variously oriented multi-domain particles with uniaxial anisotropy.

magnetization reversal takes place at the critical field $H_{0\parallel}$. The critical field when the easy axis makes an angle θ_0 with the magnetic field is given by (18.71), and the hysteresis loops for various θ_0 are shown in Fig. 18.40. The small irreversible displacements of domain walls prior to the main magnetization reversal appear as small vertical steps in the curves.

Assuming that the directions of the easy axes are randomly oriented, the resultant hysteresis loop is calculated to be as shown in Fig. 18.41. In this figure, the solid curves correspond to reversible rotation of the magnetization; the dashed curves correspond to irreversible magnetization; and the dotted curves correspond to small irreversible displacements of the domain walls. The unit of the scale of the abscissa is $H_{0\parallel}$. The coercive field of this average hysteresis loop is $1.3H_{0\parallel}$. The reversible magnetization rotation is determined by the value of K_u/I_s, irrespective of the value of $H_{0\parallel}$. The hysteresis loop in Fig. 18.41 is drawn by assuming $H_{0\parallel} = 0.2K_u/I_s$. In this graph, the curve in the second quadrant is due to irreversible displacement of the wall. Various parameters of this hysteresis loop are as follows:

$$\begin{aligned}
&\text{Residual magnetization} && I_r = 0.5I_s \\
&\text{Coercive field} && H_c = 1.3H_{0\parallel} \\
&\text{Maximum susceptibility} && \chi_{max} = 0.2\,\frac{I_s}{H_{0\parallel}} \\
&\text{Hysteresis loss} && W_h = 3.21 I_s H_{0\parallel}\,.
\end{aligned} \tag{18.135}$$

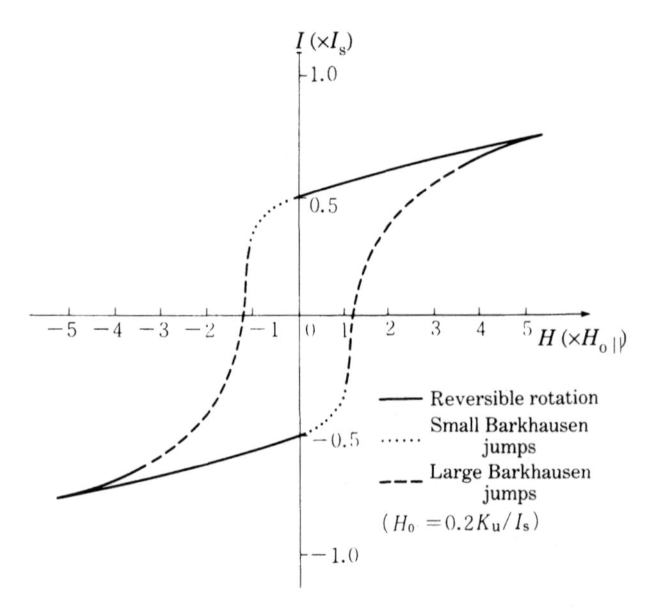

Fig. 18.41. Hysteresis loop of an assembly of randomly oriented multi-domain particles with uniaxial anisotropy, calculated by averaging the loops in Fig. 18.40.

Let us consider the magnitude of the residual magnetization after applying a magnetic field, H, to an assembly of elongated particles and reducing the field to zero. If the magnetization reversal occurs by domain wall displacement, the magnetization reversal occurs first in particles with $\theta_0 = 0$. But if the magnetization reversal occurs by rotation of the magnetization, the magnetization reversal occurs first in particles with $\theta_0 = 45°$. Figure 18.42 shows the dependence of residual magnetization (normalized to the saturation magnetization) on the previously applied field (normalized to the minimum critical field H_{min}) for two mechanisms of magnetization reversal. It is seen that the initial rise is steeper for magnetization rotation than for domain wall displacement. The reason is that the population of particles with $\theta_0 = 45°$

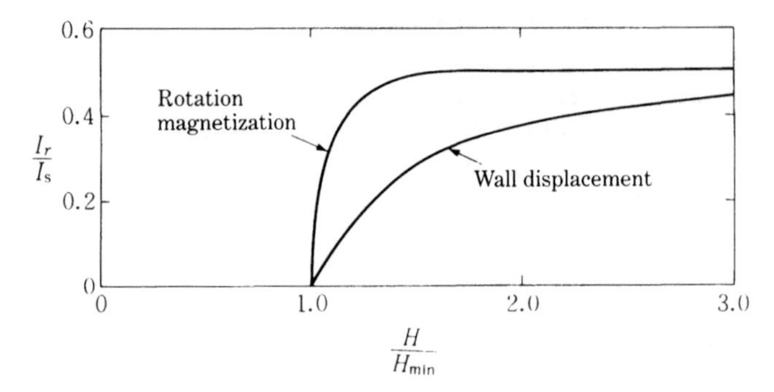

Fig. 18.42. Residual magnetization (normalized to saturation magnetization) as a function of the previously applied magnetic field (normalized to the minimum critical field), calculated for two magnetization mechanisms.

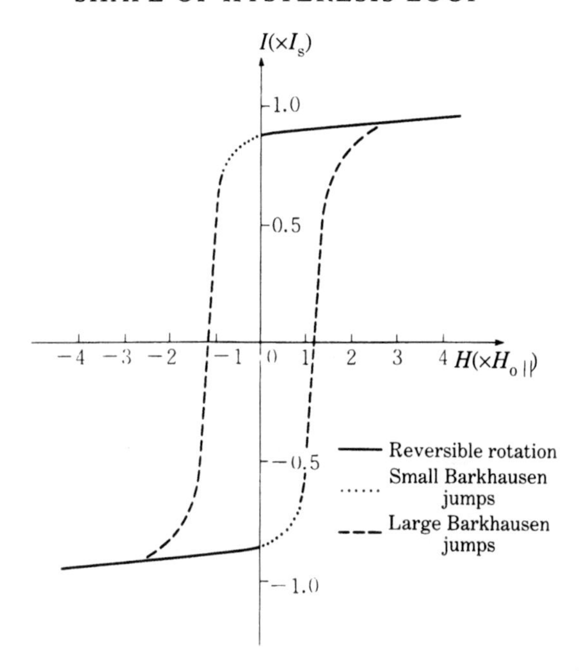

Fig. 18.43. Hysteresis loop calculated for cubic magnetic materials ($K_1 > 0$).

is much larger than that with $\theta_0 = 0$. Such a sharp rise in residual magnetization is desirable to increase the sensitivity in magnetic recording.

Next we consider the case in which the cubic anisotropy is predominant. As shown in Fig. 18.15, in the condition of residual magnetization the spins lie within an angular range $\theta \leqq 55°$. The critical field for domain wall displacement must be

$$H_0 \leqq \sqrt{3}\, H_{0\parallel}\,. \tag{18.136}$$

Therefore the magnetization reversal will take place more easily than in the case of uniaxial anisotropy. Accordingly the shape of the hysteresis loop is expected to be more rectangular. Assuming that the cubic crystallites are randomly oriented, and that all the magnetization reversals are performed by 180° wall displacement, we have the average hysteresis loop as shown in Fig. 18.43. This is the case where $K_1 > 0$, but the shape of the loop is not very different for $K_1 < 0$. We assume that H_0 is the same for all the crystallites, and also that none of the 90° walls contribute to the magnetization reversal. Even so, Fig. 18.43 well reproduces the common hysteresis loop of cubic metallic magnetic materials. Various parameters of this hysteresis loop are listed below:

$$\begin{aligned}
&\text{Residual magnetization} && I_r = 0.832 I_s \\
&\text{Coercive field} && H_c = 1.2 H_{0\parallel} \\
&\text{Maximum susceptibility} && \chi_{max} = 0.5\,\frac{I_s}{H_{0\parallel}} \\
&\text{Hysteresis loss} && W_h = 4.30 I_s H_{0\parallel}\,.
\end{aligned} \tag{18.137}$$

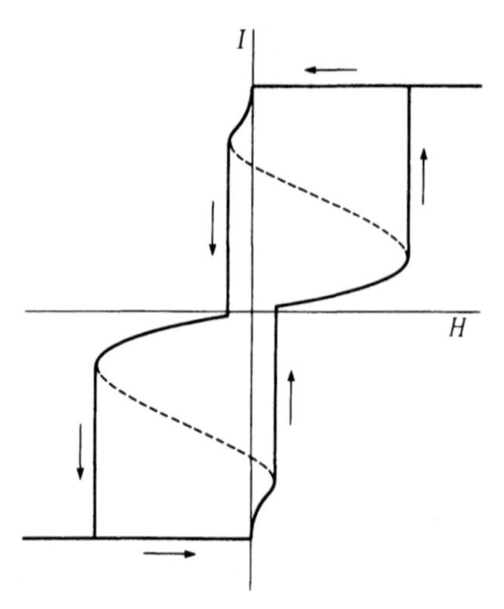

Fig. 18.44. A snake-shaped or constricted hysteresis loop. (After Taniguchi[22])

If the coercive field is determined by the internal stress, we have from (18.64)

$$H_{0\parallel} = \frac{\pi\lambda\sigma_0}{I_s}. \tag{18.138}$$

Using this relationship, we find the hysteresis loss in (18.137) is expressed as

$$W_h = 4.30\pi\lambda\sigma_0. \tag{18.139}$$

In spite of the assumption that the cubic anisotropy is predominant, the expression in (18.139) does not include K_1. This is because the coercive field in this case is determined by internal stresses and not by cubic anisotropy. On the other hand, the expression for W_h in (18.134) does include K_u, because the magnetization reversal in this case occurs by irreversible rotation magnetization against the uniaxial anisotropy.

Comparing Fig. 18.43 with Fig. 18.41, we see that the hysteresis loop for cubic materials is more rectangular than that for uniaxial materials. This results from the fact that the residual magnetization is larger for cubic materials than for uniaxial materials (see Fig. 18.13). From the I_r/I_s values, we can deduce the shape of the hysteresis loop.

In some special materials, the snake-shaped or constricted hysteresis loop shown in Fig. 18.44 appears. This phenomenon is often observed for alloys or mixed ferrites which are strongly affected by magnetic annealing (see Section 13.1). The real mechanism is as follows: when these materials are cooled in the absence of a magnetic field, a magnetic anisotropy with its easy axis parallel to the local magnetization is induced. This is also the case even inside the domain walls, where the direction of magnetization changes gradually. As a result of stabilizing the direction of the spins by the local induced anisotropy, the wall is stabilized in the position it occupied during the anneal. When the magnetic field is strong enough to drive the wall from its pinned position, the magnetization is driven irreversibly to saturation[22] (see Fig. 18.44). In order to avoid this effect, the material must be cooled in a strong magnetic

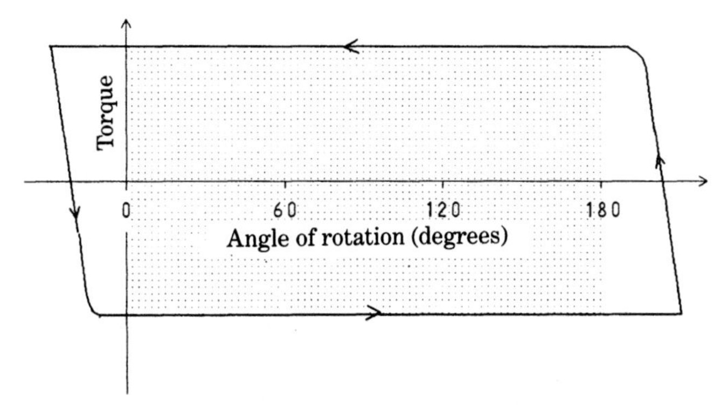

Fig. 18.45. Schematic illustration of the experimental determination of rotational hysteresis loss. The torque must be measured over an angular range of more than 180°, and the loss is determined from the area between two torque curves in the interval between 0 and 180° (see the shaded area).

field that removes all the domain walls, or in a rotating magnetic field that does not allow atoms or vacancies to occupy permanent preferential positions.

Hysteresis loss is caused by irreversible displacement of domain walls and/or by irreversible magnetization rotation. In order to separate these two effects, the measurement of rotational hysteresis loss is useful. When torque is measured as a function of angle of rotation by means of a torque magnetometer (see Fig. 12.9) for polycrystalline or powder specimens, the energy loss resulting from irreversible magnetization rotation appears as a torque opposing the rotation of the sample. That is, energy loss produces a negative torque for positive direction of rotation, and vice versa. Therefore, the integral of torque with respect to angle of rotation over one revolution,

$$W_r = -\int_0^{2\pi} L \, d\theta \tag{18.140}$$

becomes nonzero. This value is equal to the energy loss per unit volume of the specimen during one rotation and is called the *rotational hysteresis loss*. In practice, the value of W_r can be measured by rotating the field back and forth between the angles of $-40°$ or less and 220° or more, and measuring the area between the two torque curves between 0 and 180° (see the shaded area in Fig. 18.45). There is no contribution to this hysteresis loss from the irreversible displacement of domain walls, because no domain walls exit at the relatively high fields used in this experiment.

Rotational hysteresis appears only when the field intensity is moderately strong: in a weak field magnetization cannot follow the rotation of field, while in a field stronger than the anisotropy field (see Section 12.2) magnetization rotation occurs reversibly. The form of the W_r vs. H curve depends on the model of magnetization reversal, which is characterized by a quantity called the rotational hysteresis integral

$$I_{rh} = \int_0^{\infty} (W_r/I_s) \, d(1/H). \tag{18.141}$$

The experimental value of this integral indicates that in the case of electrodeposited elongated single-domain particles the magnetization occurs by a non-uniform rotation rather than a uniform rotation of magnetization.[23]

PROBLEMS

18.1 Calculate the value of I_r/I_s for an assembly of elongated single domain particles randomly oriented in a plane.

18.2 In a ferromagnet with cubic crystal structure and $K_1 > 0$, many voids are distributed in a face-centered cubic array with the cube edge length a. Find the critical field in [110] for a 180° wall separating x and $\bar{x}$ domains. Assume that the pinning by the voids is sufficiently large that the wall will expand irreversibly before it leaves the pinning points. Denote the wall energy by $\gamma_{180°}$, and the spontaneous magnetization by I_s.

18.3 The initial susceptibility measured parallel to the c-axis of a cobalt single crystal is $\bar{\chi}_a = 20$, and that measured perpendicular to the c-axis is $\bar{\chi}_a = 5$. What values can we expect for the directions 30°, 45°, and 60° away from the c-axis? What value can we expect for a polycrystalline sample made from the same material?

18.4 Suppose that elongated single domain particles are randomly distributed in the x–y plane (assume that all the particles are held parallel to this plane). A magnetic field slightly stronger than $H = I_s/4\mu_0 (= \pi I_s$ in CGS) is applied parallel to the $+x$-direction, then rotated towards y-, $-x$-, $-y$-, and finally back to the $+x$-direction, where its intensity is decreased to zero. How much residual magnetization is left and in which direction?

18.5 When a magnetic material is magnetized by a small AC magnetic field, how does the average susceptibility change with an increase in the amplitude of the AC field?

REFERENCES

1. R. M. Bozorth, *Ferromagnetism* (Van Nostrand, 1951), p. 502.
2. S. Kaya, *Z. Phys.*, **84** (1933), 705.
3. R. M. Bozorth, *Z. Phys.*, **84** (1933), 519.
4. Gerald, Hugs and Weber (N. B. Philips Fab.), Japan Patent S32-2125 (1957).
5. H. J. Williams, Phys. Rev., **52** (1937), 747, 1004.
6. R. Becker and W. Döring, *Ferromagnetismus* (Springer, Berlin, 1939), p. 153.
7. E. Kondorsky, *Phys. Z. Sowjet.*, **11** (1937), 597.
8. M. Kersten, *Phys. Z.*, **39** (1938), 860.
9. L. Néel, *Annal. Univ. Grenoble*, **22** (1946), 299.
10. M. Kersten, *Z. f. angew. Phys.*, **7** (1956), 313; **8** (1956), 382, 496.
11. M. Kersten, *Probleme der tech. Magn. Kurve* (Berlin, 1938); R. Becker and W. Döring, *Ferromagnetismus* (Springer, Berlin, 1939), p. 215.
12. R. M. Bozorth, *Ferromagnetism* (Van Nostrand, Princeton, N.J., 1951), p. 114.

13. Lord Rayleigh, *Phil. Mag.*, **23** (1887), 225.
14. F. Preisach, *Z. Physik*, **94** (1935), 277.
15. R. M. Bozorth, *Ferromagnetism* (Van Nostrand, Princeton, N.J., 1951), p. 494.
16. R. Becker and W. Döring, *Ferromagnetismus* (Springer, Berlin, 1939), p. 168.
17. E. Czerlinsky, *Ann. Physik*, **V13** (1932), 80.
18. W. F. Brown, *Phys. Rev.*, **60** (1941), 139.
19. L. Néel. *J. Phys. Rad.*, **9** (1948), 184.
20. M. Fähnle and M. Krönmüller, *J. Mag. Mag. Mat.*, **8** (1978), 149.
21. R. Becker and W. Döring, *Ferromagnetismus* (Springer, Berlin, 1939), p. 176.
22. S. Taniguchi, *Sci. Rept. Res. Inst. Tohoku Univ.*, **A8** (1956), 173.
23. I. S. Jacobs and F. E. Luborsky, *J. Appl. Phys.*, **28** (1957), 467.

19
SPIN PHASE TRANSITION

19.1 METAMAGNETIC MAGNETIZATION PROCESSES

Metamagnetism is defined as a transition from an antiferromagnetic to a ferromagnetic spin arrangement by applying a magnetic field or by changing temperature.[1] When a magnetic field is applied to an antiferromagnetic material with a small anisotropy, the spin axis flops to the direction perpendicular to the magnetic field, because susceptibility in this case, $\chi_\perp$, is larger than that, $\chi_\parallel$, in the case with the spin axis parallel to the magnetic field (see Section 7.1). This phenomenon is called *spin-axis flopping*. Further increase in the magnetic field results in the magnetization increasing with a constant susceptibility given by

$$\chi = -\frac{1}{w_2}, \tag{19.1}$$

where w_2 is the molecular field coefficient acting between the A and B sublattices (see (7.27)). The magnetization curve in this case is shown in Fig. 19.1. Along the linear magnetization curve, the two sublattice magnetizations rotate towards the direction of the field and reach a saturation magnetization, I_s, at a field given by

$$H_s = -w_2 I_s. \tag{19.2}$$

When the spin axis is held in the easy axis (z-axis) by a strong anisotropy, an increasing magnetic field applied parallel to the easy axis (z-axis) increases the sublattice moment parallel to the field gradually, with the susceptibility, $\chi_\parallel$, until the spin axis flops at a critical field H_c. The critical field H_c can be found as follows: The anisotropy energy density is assumed to be uniaxial, thus given by

$$E_a = -K_u \cos^2 \theta, \tag{19.3}$$

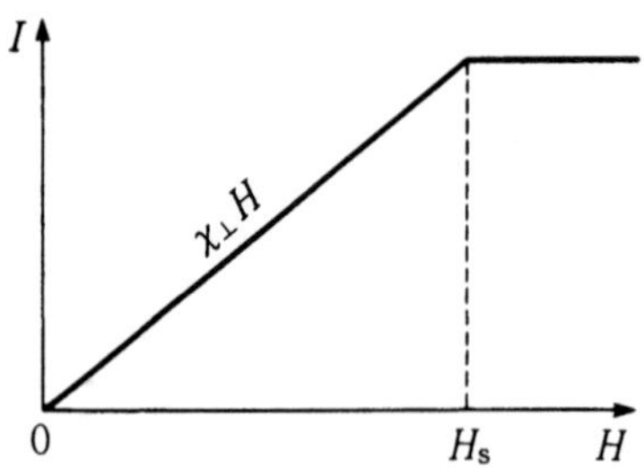

Fig. 19.1. Magnetization curve of antiferromagnetic material with magnetic field applied perpendicular to the spin-axis.

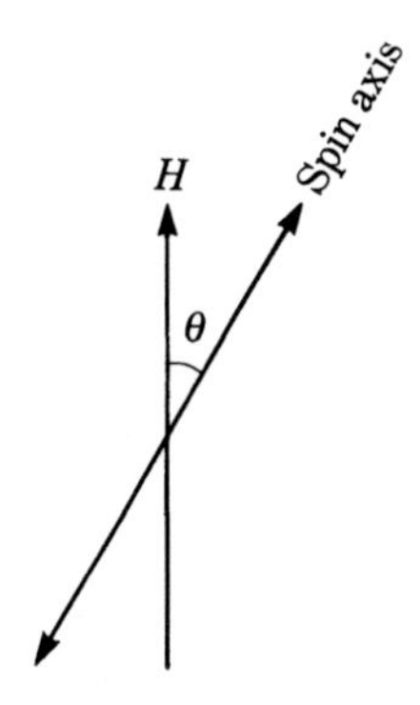

Fig. 19.2. Magnetic field applied at angle θ to the spin axis.

where K_u is the anisotropy constant, and θ is the angle between the spin axis and the z-axis. When the spin axis makes an angle, θ, with the z-axis (Fig. 19.2), the susceptibility is given by

$$\chi = \chi_\parallel \cos^2\theta + \chi_\perp \sin^2\theta. \tag{19.4}$$

Therefore the field energy density is given by

$$E_H = -\tfrac{1}{2}\chi H^2 = -\tfrac{1}{2}(\chi_\parallel \cos^2\theta + \chi_\perp \sin^2\theta)H^2. \tag{19.5}$$

Spin-axis flopping occurs when the total energy

$$E = E_a + E_H \tag{19.6}$$

reaches an unstable equilibrium, that is

$$\frac{\partial E}{\partial \theta} = \{2K_u + (\chi_\parallel - \chi_\perp)H^2\}\sin\theta\cos\theta = 0, \tag{19.7}$$

and

$$\frac{\partial^2 E}{\partial \theta^2} = \{2K_u + (\chi_\parallel - \chi_\perp)H^2\}\cos 2\theta \gtrless 0. \tag{19.8}$$

From (19.7) we have $\theta = 0°$ or $90°$. The condition (19.8) gives

$$H \lessgtr \sqrt{\frac{2K_u}{\chi_\perp - \chi_\parallel}} = H_c, \tag{19.9}$$

depending on the coefficient within braces in (19.8) $\gtrless 0$, where H_c is the critical field for the spin-axis flopping. When the field is increased beyond H_c, θ is changed from $0°$ to $90°$, and the susceptibility is changed discontinuously from $\chi_\parallel$ to $\chi_\perp$. If $H_c < H_s$ (see (19.2)), the magnetization curve should be as shown in Fig. 19.3. If $H_c > H_s$, the magnetization curve is as shown in Fig. 19.4. Such a discontinuous change in spin orientation is called *spin flopping*.

Figure 19.5 shows metamagnetic magnetization curves observed for MnF_2 at 4.2 K. This antiferromagnetic crystal has a body-centered tetragonal structure with the spin

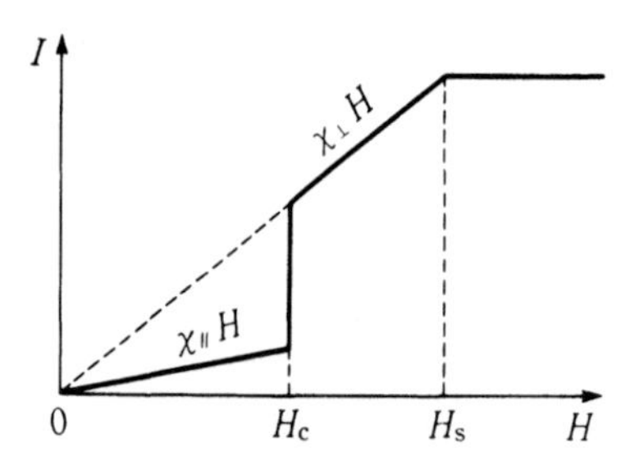

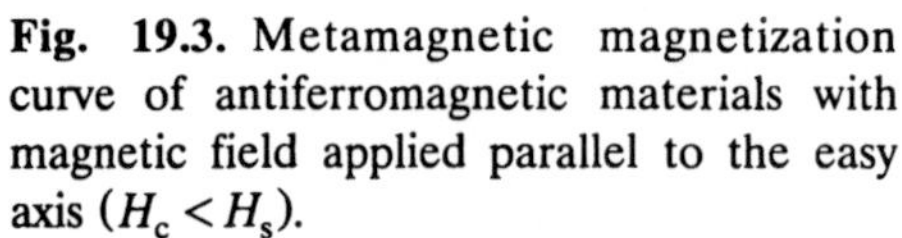

Fig. 19.3. Metamagnetic magnetization curve of antiferromagnetic materials with magnetic field applied parallel to the easy axis ($H_c < H_s$).

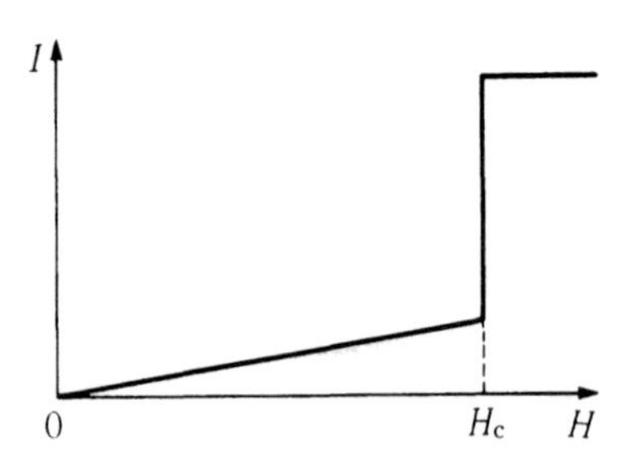

Fig. 19.4. Metamagnetic magnetization curve of antiferromagnetic materials with the magnetic field applied parallel to the easy axis ($H_c > H_s$).

axis parallel to the c-axis. The number for each curve denotes the angle between the direction of applied magnetic field and the c-axis. For 0°, we expect a sharp transition as indicated by a solid line. In the actual experiment, θ is nonzero, so that a spin-axis rotation takes place, before the spin flopping occurs. For the powder specimen, in which the spin axes are distributed at random, the average magnetization curve is expected to be as shown by the dashed curve.[2] A two-stage spin flopping was observed for $CoCl_2 \cdot 2H_2O$.[3] This phenomenon was interpreted by assuming the presence of an

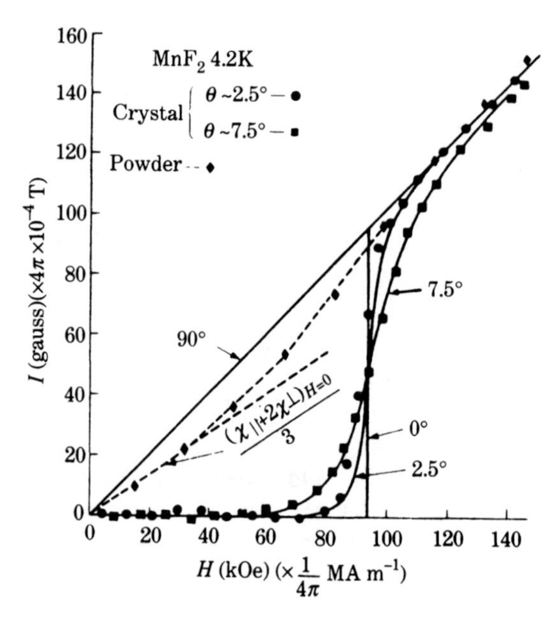

Fig. 19.5. Magnetization curves for MnF_2 measured at 4.2 K (θ is the angle between the c-axis and the applied magnetic field).[2]

exchange energy, J_2, acting between second nearest neighbor spins, in addition to the first neighbor exchange, J_1.[4] A theoretical treatment of metamagnetic magnetization processes in antiferromagnetic materials is given in a detailed review.[5] Experimental data on metamagnetic magnetization processes for various antiferromagnetic materials are given by Jacobs.[6]

19.2 SPIN FLOP IN FERRIMAGNETISM

Ferrimagnetism is the result of two uncompensated sublattice magnetizations, I_A and I_B, aligned antiparallel and having different magnitudes ($I_A > |I_B|$). The spontaneous magnetization is the difference between the two sublattice magnetizations ($I_s = I_A - |I_B|$) (see Section 7.2). The two sublattice magnetizations are aligned antiparallel by the superexchange interaction (see Section 7.1). This interaction is equivalent to a magnetic field of several hundred $\mathrm{MA\,m^{-1}}$ (several MOe), so that a ferrimagnet behaves like a normal ferromagnet in magnetic fields of moderate strength. Therefore the technical magnetization for ferrimagnetic materials behaves in the same way as that for ferromagnetic materials.

However, a ferrimagnet responds to a high magnetic field quite differently from a normal ferromagnet. A theoretical investigation of this behavior was made in 1968,[7] but experimental studies have been started only recently when ultra-high magnetic fields become available.

Let us consider a ferrimagnet consisting of two sublattice magnetizations, I_A and I_B. The molecular fields acting on these sublattices are expressed as

$$\left.\begin{aligned} \boldsymbol{H}_{mA} &= w_A \boldsymbol{I}_A - w \boldsymbol{I}_B \\ \boldsymbol{H}_{mB} &= -w \boldsymbol{I}_A + w_B \boldsymbol{I}_B \end{aligned}\right\}, \tag{19.10}$$

where w_A and w_B are the intra-sublattice molecular field coefficients, while $-w$ is the inter-sublattice molecular field coefficient ($w > 0$). If $|I_A| > |I_B|$, I_A is aligned parallel to the magnetic field, while I_B is aligned antiparallel, as long as the field is moderately weak. As the field is increased, I_B becomes unstable, and tilts from the $-H$-direction. At the same time I_A tilts from the $+H$-direction under the action of the molecular field from the B sublattice. Thus I_A and I_B form a *canted spin arrangement*. Further increase in H causes a rotation of I_A and I_B towards the field, resulting in a ferromagnetic arrangement (see Fig. 19.6).

In an external field H, the effective fields which I_A and I_B feel are given by

$$\left.\begin{aligned} \boldsymbol{H}_{A\,\mathrm{eff}} &= \boldsymbol{H} + \boldsymbol{H}_{mA} = \boldsymbol{H} + w_A \boldsymbol{I}_A - w \boldsymbol{I}_B \\ \boldsymbol{H}_{B\,\mathrm{eff}} &= \boldsymbol{H} + \boldsymbol{H}_{mB} = \boldsymbol{H} - w \boldsymbol{I}_A + w_B \boldsymbol{I}_B \end{aligned}\right\}. \tag{19.11}$$

Taking the difference on each side in (19.11), we have

$$\boldsymbol{H}_{A\,\mathrm{eff}} - \boldsymbol{H}_{B\,\mathrm{eff}} = (w_A + w)\boldsymbol{I}_A - (w_B + w)\boldsymbol{I}_B$$

therefore

$$\boldsymbol{H}_{A\,\mathrm{eff}} - (w_A + w)\boldsymbol{I}_A = \boldsymbol{H}_{B\,\mathrm{eff}} - (w_B + w)\boldsymbol{I}_B. \tag{19.12}$$

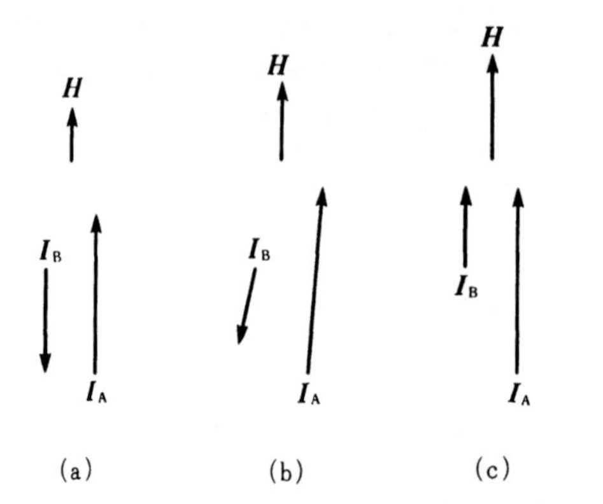

Fig. 19.6. Ferrimagnetic spin-flopping: (a) $H < H_{c1}$, (b) $H_{c1} < H < H_{c2}$; (c) $H_{c2} < H$.

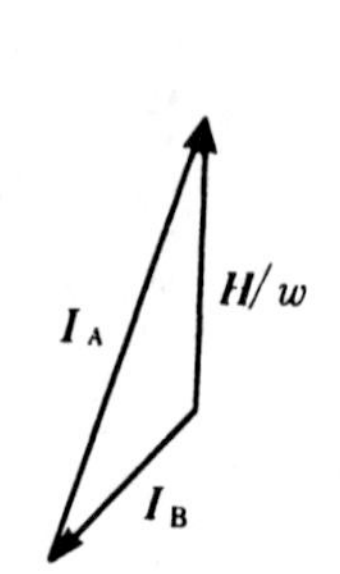

Fig. 19.7. Canted spin arrangement.

Since I_A and I_B must be parallel to $H_{A\,\mathrm{eff}}$ and $H_{B\,\mathrm{eff}}$ respectively, the left-hand side of (19.12) must be parallel to I_A, whereas the right-hand side must be parallel to I_B. In the canted-spin arrangement, these sublattice magnetizations have different directions, so (19.12) is satisfied only when both sides of the equation are zero. In other words,

$$\left.\begin{aligned} \boldsymbol{H}_{A\,\mathrm{eff}} &= (w_A + w)\boldsymbol{I}_A \\ \boldsymbol{H}_{B\,\mathrm{eff}} &= (w_B + w)\boldsymbol{I}_B \end{aligned}\right\}. \tag{19.13}$$

This means that $|I_A|$ and $|I_B|$ are determined by their own magnitudes, independent of the intensity of the external field. Thus in the canted-spin arrangement, $|I_A|$ and $|I_B|$ have constant magnitudes independent of the angles between I_A, I_B and H.

Using the relationship (19.13) in (19.11), we have

$$\boldsymbol{I}_A + \boldsymbol{I}_B = \frac{\boldsymbol{H}}{w}. \tag{19.14}$$

This relationship is shown in a vector diagram in Fig. 19.7. The magnetization composed of I_A and I_B is given by

$$\boldsymbol{I} = \boldsymbol{I}_A + \boldsymbol{I}_B = \frac{\boldsymbol{H}}{w}, \tag{19.15}$$

which is proportional to H, independent of temperature. Thus the magnetization curve of a ferrimagnet must be the same straight line with susceptibility, $1/w$, at any temperature, as long as the canted-spin arrangement exists. However, the field, H_{c1}, at which the linear magnetization curve begins and the field, H_{c2}, at which it ends, do depend on temperature. We can find H_{c1}, by putting $I = |I_A| - |I_B|$ to give

$$H_{c1} = w(|I_A| - |I_B|). \tag{19.16}$$

In the same way, we can find H_{c2}, by putting $I = |I_A| + |I_B|$, to give

$$H_{c2} = w(|I_A| + |I_B|). \tag{19.17}$$

Since $|I_A|$ and $|I_B|$ depend on temperature, these critical fields are also dependent of temperature. A computer simulation of such spin-flop behavior is shown below.

Yttrium iron garnet (YIG) is a ferrimagnet with the chemical formula $Y_3Fe_5O_{12}$, where Y is nonmagnetic while Fe is magnetic. Three Fe ions on the $24d$ sites and two Fe ions on the $16a$ sites are coupled antiferromagnetically, resulting in a spontaneous magnetization of 5 Bohr magnetons per formula unit associated with one Fe ion (see Section 9.3). The number of nearest neighbor j-sites of an i-site, z_{ij}, for the $16a$ and $24d$ sites are given by

$$z_{aa} = 8, \quad z_{ad} = 6, \quad z_{da} = 4, \quad z_{dd} = 4. \tag{19.18}$$

The exchange energy between the i-site spin S_i and the j site spin S_j is given by $-2J_{ij}S_iS_j$, where J_{ij} is the exchange integral. Using this formula, we can express the exchange energy density for a–a interaction as

$$E_{aa} = -\left(\tfrac{2}{5}N\frac{z_{aa}}{2}\right) \times (2J_{aa}S^2), \tag{19.19}$$

where N is the number of Fe ions per unit volume, S is the spin of an Fe ion ($S = \frac{5}{2}$). Using (19.18) and this S-value, we have

$$E_{aa} = -20NJ_{aa} \tag{19.20}$$

Similarly we have for a–d interaction

$$\begin{aligned} E_{ad} &= -\tfrac{1}{2}\left(\tfrac{2}{5}Nz_{ad} + \tfrac{3}{5}Nz_{da}\right)(2J_{ad}S^2) \\ &= -30NJ_{ad}, \end{aligned} \tag{19.21}$$

and for d–d interaction

$$\begin{aligned} E_{dd} &= -\left(\tfrac{3}{5}N\frac{z_{dd}}{2}\right) \times (2J_{dd}S^2) \\ &= -15NJ_{dd}. \end{aligned} \tag{19.22}$$

We can express these energy densities in terms of the molecular field in (19.10) by converting the sublattice names from A and B to d and a, respectively. Since

$$I_a = -\tfrac{2}{5}N \times 5M_B = -2NM_B, \qquad I_d = \tfrac{3}{5}N \times 5M_B = 3NM_B,$$

we have

$$\left.\begin{aligned} E_{aa} &= -\tfrac{1}{2}w_a I_a^2 = -2w_a N^2 M_B^2 \\ E_{ad} &= -wI_aI_d = 6wN^2M_B^2 \\ E_{dd} &= -\tfrac{1}{2}w_d I_d^2 = -\tfrac{9}{2}w_d N^2 M_B^2 \end{aligned}\right\} \tag{19.23}$$

where M_B is the Bohr magneton. Comparing this expression with (19.20), (19.21) and

(19.22), we have the molecular field coefficients in terms of the exchange integrals as

$$\left.\begin{aligned} w_a &= \frac{10J_{aa}}{NM_B^2} \\ w &= -\frac{5J_{ad}}{NM_B^2} \\ w_d &= \frac{10J_{dd}}{3NM_B^2} \end{aligned}\right\}. \tag{19.24}$$

These exchange integrals can be determined from the temperature dependence of spontaneous magnetization. The spontaneous magnetization of a ferrimagnetic garnet is given by

$$I_s = I_d + I_a, \tag{19.25}$$

where

$$\left.\begin{aligned} I_d &= (\tfrac{3}{5}N)(5M_B)B_{5/2}\left(\frac{5M_B H_{md}}{kT}\right) \\ I_a &= -(\tfrac{2}{5}N)(5M_B)B_{5/2}\left(\frac{5M_B H_{ma}}{kT}\right) \end{aligned}\right\}. \tag{19.26}$$

According to (19.10), H_{md} and H_{ma} are given by

$$\left.\begin{aligned} H_{md} &= w_d I_d - w I_a \\ H_{ma} &= -w I_d + w_a I_a \end{aligned}\right\}. \tag{19.27}$$

Starting from the initial values at 0 K, I_{d0} and I_{a0} for YIG, or

$$I_{d0} = 3NM_B, \qquad I_{a0} = -2NM_B, \tag{19.28}$$

the temperature dependence of I_d and I_a can be found by successive integrations of (19.26) and (19.27). Anderson[8] assumed the following two sets of values for the exchange integrals for YIG:

$$\left.\begin{aligned} J_{aa} &= -8.45\,\mathrm{cm}^{-1} \\ J_{dd} &= -11.86\,\mathrm{cm}^{-1} \\ J_{ad} &= -25.36\,\mathrm{cm}^{-1} \end{aligned}\right\} \quad (\text{set A}), \tag{19.29}$$

and

$$\left.\begin{aligned} J_{aa} &= 0 \\ J_{dd} &= 0 \\ J_{ad} &= -13.61\,\mathrm{cm}^{-1} \end{aligned}\right\} \quad (\text{set B}). \tag{19.30}$$

The results are plotted in Fig. 19.8, which shows the temperature dependence of I_d

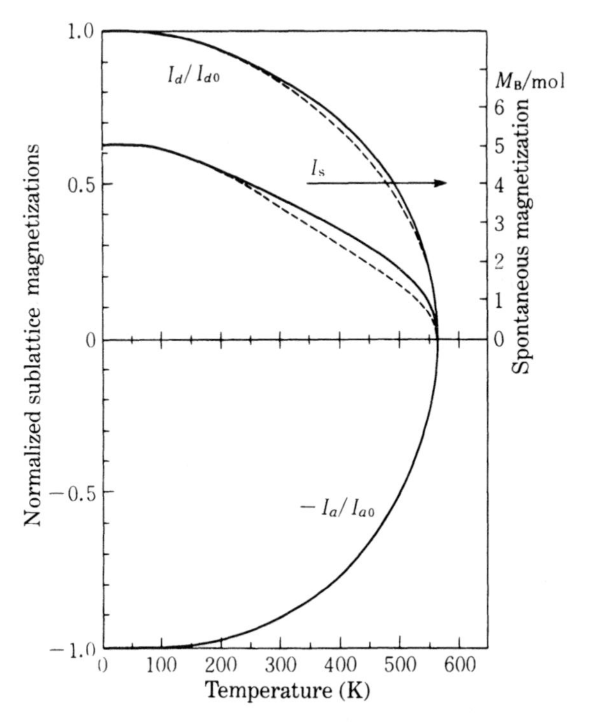

Fig. 19.8. Temperature dependence of normalized sublattice magnetization calculated by using two sets of exchange integrals, and temperature dependence of the spontaneous magnetization; the solid curves are calculated by using set A (see (19.29)) and the dashed curves are drawn using set B (see (19.30)).[9]

and I_a as well as the spontaneous magnetization I_s.[9] The solid curves assume the values of J_{ij} for set A. The temperature dependence of I_s thus obtained is in good agreement with the experimental values of Anderson.[8] The dashed curves assume the values of set B, which ignores J_{aa} and J_{dd}, and a different value for J_{ad}. Nevertheless the temperature dependence of I_s thus obtained is not so different from the solid curve. This means that values of the three J_{ij} cannot be accurately determined by such a procedure.

Another method for determining the value for w or J_{ad} is to observe spin flopping in a high magnetic field (see (19.15) or (19.16)). The magnetization curve in high magnetic fields can be simulated by using (19.26), in which the molecular fields H_m are replaced by the effective fields H_{eff} in (19.11), together with (19.27). Figures 19.9(a) and (b) show the results for YIG at 100 K and 200 K, respectively, calculated by using the values J_{ij} for set A (see (19.29)). These graphs show field dependences of normalized values of sublattice magnetizations, their z-components and the resultant magnetization. In the field range $H_{c1} < H < H_{c2}$, where the spin canting takes place, the magnitude of the sublattice magnetizations, and the inclination of the magnetization curve, stay constant as mentioned before. However, the values of H_{c1} and H_{c2}

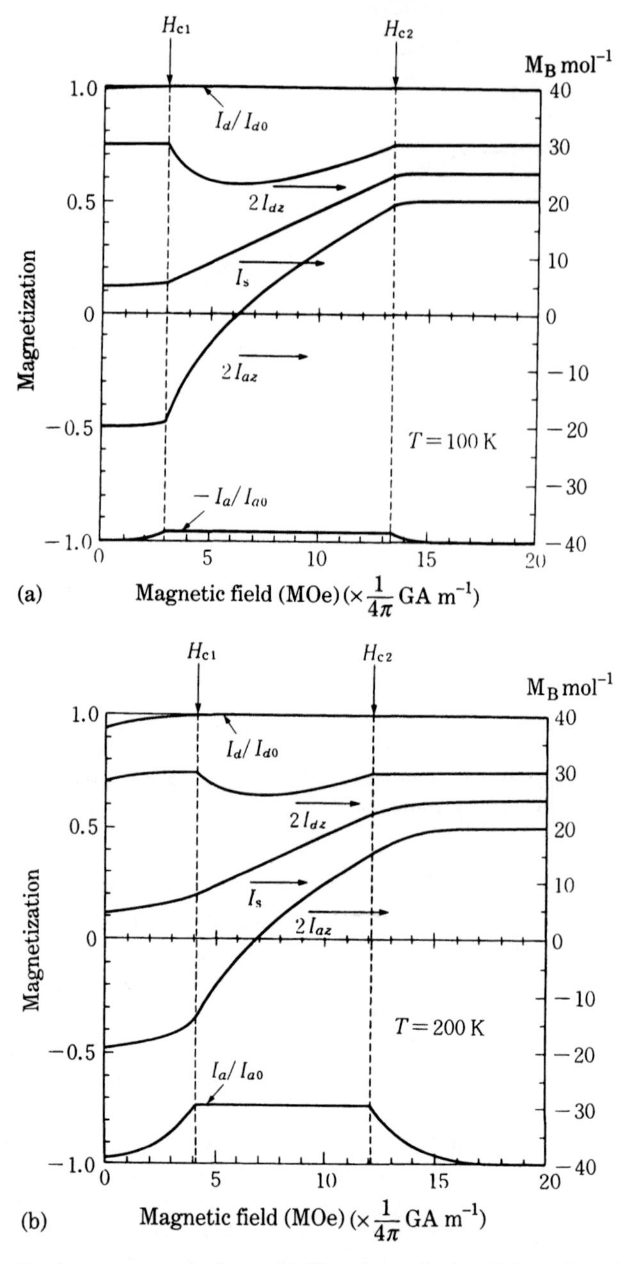

Fig. 19.9. Magnetization process during spin flopping calculated by using the exchange integrals (set A) in (19.29) for YIG: (a) at $T = 100$ K; (b) $T = 200$ K.[9]

are different at different temperatures. Figure 19.10 shows a *spin-phase diagram* in which the traces of H_{c1} and H_{c2} are drawn in the H–T plane.

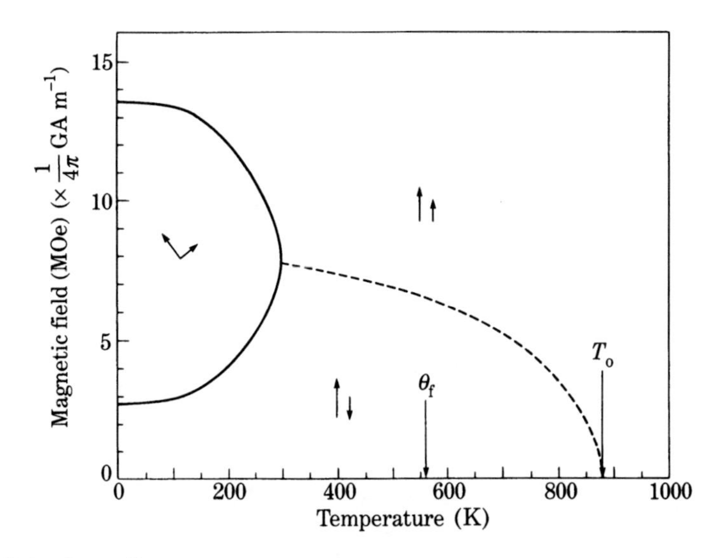

Fig. 19.10. Spin phase diagram constructed for YIG by using the exchange integrals (set A) in (19.29).[9]

This spin-phase diagram is difficult to verify experimentally, because it requires very high magnetic fields. The dashed curve in Fig. 19.10 is the phase boundary at which the ferrimagnetic spin arrangement changes to ferromagnetic. It is interesting to note that this boundary continues above the Curie point, Θ_f. This means that above the Curie point the material is paramagnetic in zero field, but becomes ferrimagnetic in an applied field up to the phase boundary (dashed curve), and

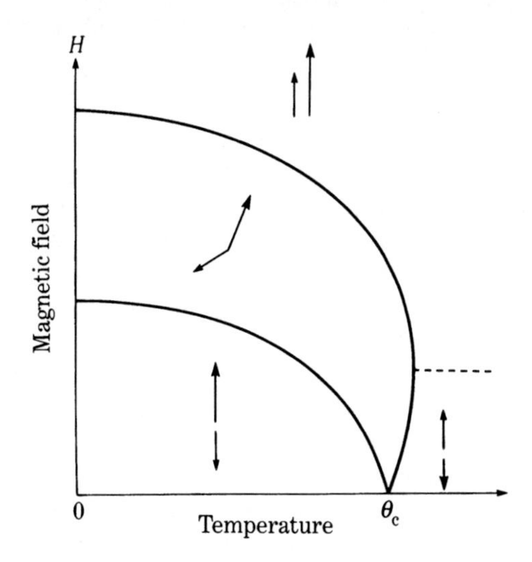

Fig. 19.11. Conceptual spin phase diagram for an N-type ferrimagnet.

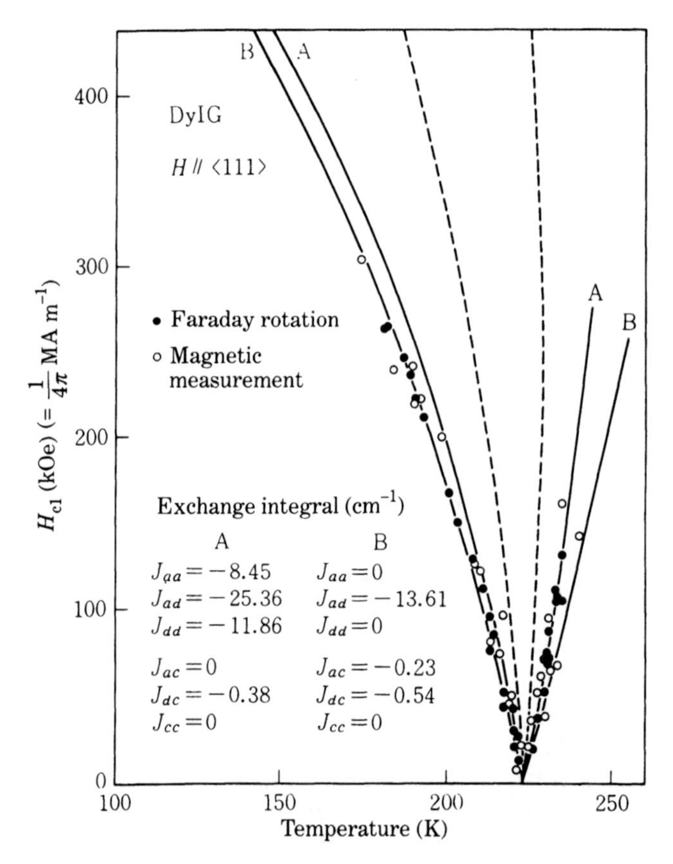

Fig. 19.12. A part of the spin phase diagram for DyIG constructed by using two sets of exchange integrals, and a comparison with experiments.[11] The unit of the exchange integrals is cm^{-1}.

ferromagnetic in higher fields. Above T_0, it is always ferromagnetic in a nonzero field.

When nonmagnetic Ga ions are added to YIG, they dilute the Fe ions on *d*- and *a*-sites, so that the exchange interaction between *a*- and *d*-sites is weakened. In this case spin-flopping can be observed in a relatively weak magnetic field such as $80\,\mathrm{MA\,m^{-1}}(=1$ MOe).[9,10] In an N-type ferrimagnet (see Section 7.2), $|I_A|-|I_B|=0$ at the compensation point, Θ_c, and $H_{c1}=0$ as found by (19.16). Therefore spin-flopping can be observed in a relatively weak field near the compensation point.

Figure 19.11 is a conceptual spin phase diagram of an N-type ferrimagnet. The critical fields, H_{c1} and H_{c2}, go to zero at the compensation point. Figure 19.12 shows a part of the spin phase diagram observed for DyIG, which is an N-type ferrimagnet. In this ferrimagnet a Dy^{3+} spin on the 24*c* site and an Fe^{3+} spin on the strongly coupled 16*a* and 24*d* sites are loosely coupled antiferromagnetically. The dashed critical curves are calculated on the basis of this two sublattice model. They are in poor agreement with experiments, shown by solid and open circles. The solid circles were measured by means of Faraday rotation[11] (see Section 21.3). If we assume the three sublattice model in which not only the Dy spin on 24c site but also Fe spins on 16a and 24d sites form a canted-spin arrangement, we obtain the solid curves in the Figure, which fit the experimental points reasonably well. Unlike sets A and B in (19.29) and (19.30), the number of parameters is increased, because interactions between Dy and other spins are introduced. Therefore by selecting 6 parameters (sets A and B) in the figure we can easily fit the experiment with the calculation. This means that such an experiment is not very satisfactory for measuring exchange integrals.

In order to determine exchange parameters, it is necessary to observe spin-flopping in pure YIG. This has been done by applying ultra-high magnetic fields[12] up to 350 T (see Section 2.1). By fitting a calculated H_{c1} versus temperature curve to the experiment, the exchange integrals $J_{ad} = -25.6\,\mathrm{cm}^{-1}$ and $J_{aa} = -6.2\,\mathrm{cm}^{-1}$ were determined. The value of J_{dd} cannot be determined, because the sublattice magnetization on the d-site, M_d, is saturated during the spin flopping at low temperatures.

19.3 HIGH-FIELD MAGNETIZATION PROCESS

As discussed in Section 18.5, the high-field susceptibility is given by (18.133), which shows $\chi_0 \to 0$ as $T \to 0$, because $L'(a)$ decreases as T^2. The reason for this is that we assumed the atomic magnetic moment to be independent of the field. In some real magnetic materials, however, this is not necessarily true.

For example, the atomic magnetic moment of rare earth atoms is generally well described by Hund's rule (see Section 3.2). In Sm^{3+} and Eu^{3+} ions, the energy difference is rather small between the ground state with $J = |L| - |S|$ and the first excited state with $J+1$, in which L and S make some angle. For instance, the ground state for Eu^{3+} has $|L| = |S| = 3$, so that $J = 0$, while the first excited state with $J = 1$ is higher than the ground state by 600 K or about $700\,\mathrm{MA\,m^{-1}}$ (= 9 MOe).[13] Therefore at temperatures above 600 K or in an ultra-high magnetic field above $700\,\mathrm{MA\,m^{-1}}$ we expect that the excited state with $J = 1$ will be mixed with the ground state.

In 3d transition metals, L is generally quenched by the crystalline field. Moreover, the atomic d-wave functions are split into $d\varepsilon$ and $d\gamma$ states under the influence of the cubic crystalline field (see Section 3.4). When such an ion is located on an octahedral site in an oxide, it may find a nearest neighbor O^{2-} ion on the x-, y- or z-axis. Therefore the $d\varepsilon$ orbits whose orbital wave functions are stretched between these cubic axes are more stable than the $d\gamma$ orbits whose orbital wave functions are stretched along the cubic axes (see Fig. 12.20). Suppose we place an Fe^{2+} ion that has six 3d electrons on the octahedral site. According to Hund's rule, all the + spin levels are filled first and then the lowest − spin level is occupied, as shown in Fig. 19.13(a). This results in a magnetic moment of $4M_B$. If, however, the level splitting due to the crystalline field is too large, the two + spins which occupied $d\gamma$ levels fall into the lowest $d\varepsilon$ levels (see Fig. 19.13(b)), thus resulting in a non-magnetic ion. The former state is called the *high-spin state*, while the latter is called the *low-spin state*. The situation is the same for the Co^{3+} ion.

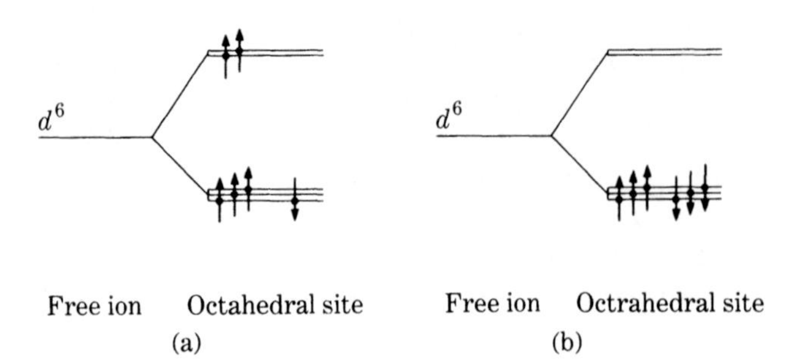

Fig. 19.13. Level splitting and electronic states of Fe^{2+} or $Co^{3+}(d^6)$ ions on octahedral site: (a) high-spin state; (b) low-spin state.

Saturation magnetization of these two-spin-state compounds changes considerably with temperature and applied magnetic field. For simplicity, let us assume that the magnetic moment of the low-spin state is zero, while that of the high-spin state is gJM_B. The energy separation between the two states is denoted by ΔU. Generally speaking, the high-spin states are split into $2J+1$ levels in a magnetic field H with $J_z = J, J-1, \ldots, -J$. Figure 19.14 shows the variation of the energy levels of the two-spin states for $J=1$; the high-spin states are split into three levels with $J_z = 1, 0, -1$, whose separations increase with increasing field H. Assuming the energy of the low-spin state to be zero, the energy levels of these states are given by

$$\left.\begin{array}{lll} \text{low-spin state} & U_0 & = 0 \\ \text{high spin state with } J_z = -1 & U_1 & = \Delta U - gJM_B H \\ \text{high spin state with } J_z = 0 & U_2 & = \Delta U \\ \text{high spin state with } J_z = 1 & U_3 & = \Delta U + gJM_B H \end{array}\right\}, \tag{19.31}$$

where the subscripts on the Us number the states in order of increasing energy levels. According to statistical mechanics, the probability of obtaining the state with energy U is proportional to the Boltzmann factor $\exp(-U/kT)$, so that the probability of

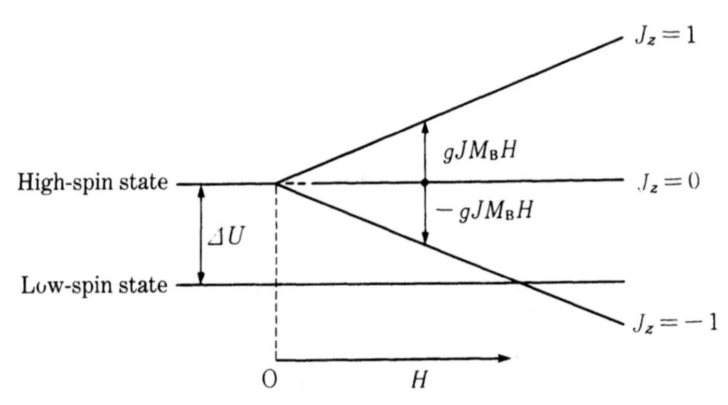

Fig. 19.14. Variation of level splitting between low-spin and high-spin ($J=1$) states with increasing magnetic field.

finding each of these states is given by

$$\left.\begin{aligned} p_0 &= \frac{1}{\sum_{i=0}^{3} \exp(-U_i/kT)} \\ p_1 &= \frac{\exp(-U_1/kT)}{\sum_{i=0}^{3} \exp(-U_i/kT)} \\ p_2 &= \frac{\exp(-U_2/kT)}{\sum_{i=0}^{3} \exp(-U_i/kT)} \\ p_3 &= \frac{\exp(-U_3/kT)}{\sum_{i=0}^{3} \exp(-U_i/kT)} \end{aligned}\right\} \tag{19.32}$$

Therefore the average atomic magnetic moment M_s is given by

$$M_s = gJM_B(p_1 - p_3), \tag{19.33}$$

since the levels for states 0 and 2 are nonmagnetic.

Assuming that $\Delta U = 0.0086\,\text{eV}$ (corresponding to 100 K), the field dependences of the saturation magnetic moment M_s calculated from (19.33) for $T = 4$, 40, 100 and 200 K are plotted in Fig. 19.15 up to 300 MA m^{-1} (= 3.75 MOe). At 4 K, the transition from low-spin state to high-spin state due to a crossover of the high-spin level with $J_z = -1$ and the low-spin level is clearly seen. However, at high temperatures such as

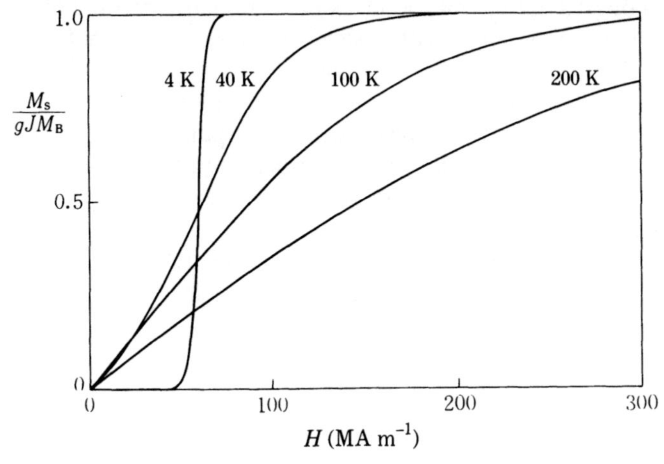

Fig. 19.15. Temperature dependence of saturation magnetic moment with $J = 1$ calculated by assuming $\Delta U = 0.0086\,\text{eV}$ (100 K).

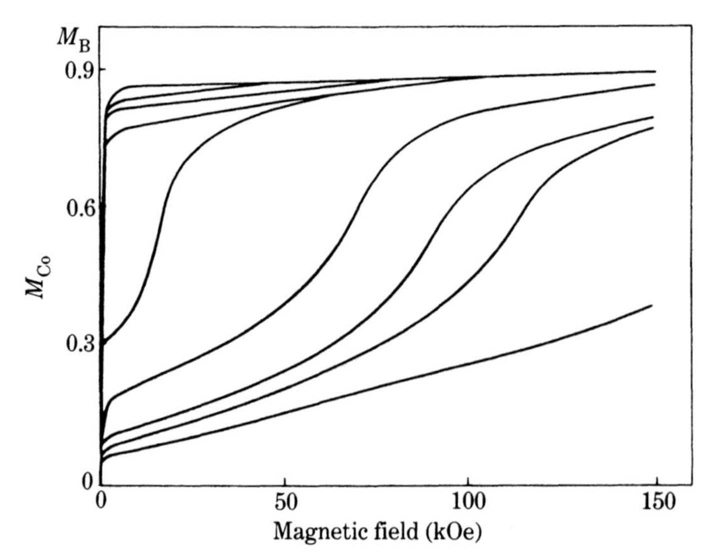

Fig. 19.16. Magnetization curves measured for $CoS_{2-x}Se_x$ at 4.2 K. (from top to bottom $x = 0.1$, 0.2, 0.25, 0.28, 0.30, 0.35, 0.375, 0.40 and 0.45).[14]

100 and 200 K the transitions are spread out and the curves resemble the Brillouin function.

If the magnetic material containing such two-spin-state ions is ferromagnetic, we must consider the effect of the molecular field and use $H + wI$ for H. Figure 19.16 shows the magnetization curves for $CoS_{2-x}Se_x$, which is a pyrite-type compound, measured at 4 K. The compound CoS_2 is ferromagnetic, with the Co ion having an atomic magnetic moment of $0.9M_B$. When Se is added, Co ions with many Se neighbors tend to take the low-spin state[14] with zero magnetic moment. Therefore a transition from the low-spin state to high-spin state occurs, as seen for the curves for $x > 0.28$. The nonzero atomic magnetic moment remains in low fields, because there are still some high-spin Co ions with a small number of Se neighbors.

Invar is an alloy of 35 at% Ni–Fe whose thermal expansion coefficient at room temperature is almost zero; it has a number of practical uses. The small thermal expansion has been interpreted in terms of a high-spin to low-spin-state transition by assuming that the high-spin state has a larger atomic volume than the low-spin state.[15,16] Recently this characteristic was alternately interpreted in terms of the temperature dependence of local magnetic moment in spin fluctuation theory.[17]

The high field susceptibility, χ_0, of 3d transition metals exhibits non-zero values even at 0 K. The values measured for Fe, Co, and Ni at 4.2 K are[18]

$$\left.\begin{array}{ll} \mathrm{Fe}\langle 100\rangle & \bar{\chi}_0 = 3.34\times 10^{-3} \\ \mathrm{Co} & \bar{\chi}_0 = 3.33\times 10^{-3} \\ \mathrm{Ni}\langle 111\rangle & \bar{\chi}_0 = 1.42\times 10^{-3} \end{array}\right\}. \tag{19.34}$$

This phenomenon is interpreted as an enhancement of the band-polarization by the external field. The composition dependence of the high field susceptibility could be

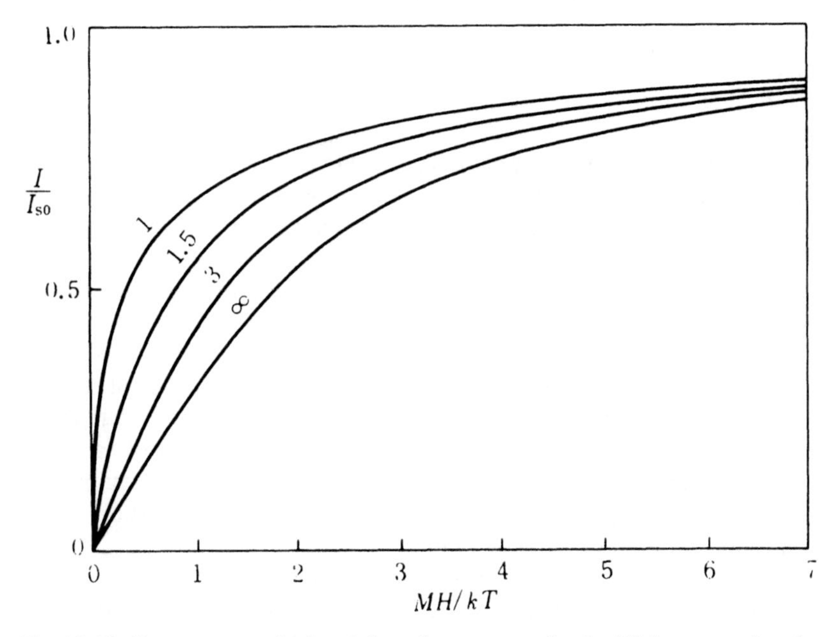

Fig. 19.17. Paraprocess calculated for a ferromagnet by the Weiss approximation.

well accounted for in this way.[19]

As discussed above, a high magnetic field influences the magnetic structure as the temperature does. In the statistical mechanics of spin systems, the populations of the spin states in a magnetic field H are described in terms of the parameter $\alpha = MH/kT$. Therefore one is tempted to consider that the field H and temperature T are not independent parameters to describe the magnetic state. However, this is not the case. For example, the magnetization curve of a ferromagnet above its Curie point, which is called the *paraprocess*, depends independently on H and T. Figure 19.17 shows the paraprocess of a ferromagnet calculated by the Weiss approximation. The parameter on each curve is the value of T/Θ_f. The abscissa is MH/kT. If the paraprocess were described by a unique function of H/T, all the curves would be identical. As seen in the figure, however, the shape of the curve changes with temperature.

In the usual statistical theory of ferromagnetism, temperature and magnetic field are two major parameters: Temperature ranges from 0 to 1000 K, while magnetic field ranges from 0 to $1\,\mathrm{MA\,m^{-1}}$ (= 13 kOe). Since, however, the energy of a magnetic moment of $1 M_B$ in a magnetic field of $80\,\mathrm{MA\,m^{-1}}$ (= 1 MOe) corresponds to 68 K, an ultra-high magnetic field of $800\,\mathrm{MA\,m^{-1}}$ (= 10 MOe) corresponds to 680 K. Therefore all magnetic phase transitions can be caused by applying an ultra-high magnetic field as well as by temperature. This additional experimental parameter may be useful for checking various statistical theories (see Section 6.2).

19.4 SPIN REORIENTATION

Spin reorientation is defined as a transition of the state of magnetic ordering between two of the possible magnetic structures (antiferromagnetic, canted, ferrimagnetic or ferromagnetic), or a change in the orientation of the spin axis.[20] In these phenomena, not only the exchange interaction but also the magnetic anisotropy plays an important role.

Rare earth orthoferrites are magnetic oxides with composition given by $RFeO_3$ (R: rare earth); they have orthorhombic crystal structure (see Section 9.3). Because this crystal structure is a distortion of the cubic structure, an antisymmetric exchange interaction (see Section 7.4) acts on the antiferromagnetically arranged Fe^{3+} spins, thus resulting in a canted-spin arrangement with a small spontaneous magnetization. In the case R = Sm, the spontaneous magnetization points parallel to the a-axis at 450 K and rotates gradually with increasing temperature until it reaches the c-axis at 500 K.[21] This is a spin reorientation corresponding to a second-order phase transition. Similar phenomena are observed for other magnetic orthoferrites, but their transitions occur at low temperatures.[20] This spin reorientation is interpreted as occurring because the sublattice magnetization comprising the Re^{3+} moment changes dramatically with temperature and this affects the spontaneous magnetization of the Fe^{3+} moment through the R–Fe exchange interaction.[22]

Many RCo_5-type intermetallic compounds (see Section 12.4) undergo second-order type spin reorientations. In the case R = Dy, Co- and Dy-spins are coupled ferrimagnetically on the a-axis below 325 K (T_{SR1}), and then the spontaneous magnetization rotates gradually until it reaches the c-axis at 367 K (T_{SR2}). Figure 19.18 shows the temperature dependences of I_s, the spontaneous magnetization; I_a, the a-component of I_s; and I_c, the c-component ($I_s = \sqrt{I_a^2 + I_c^2}$). In order to explain all the phenomena reasonably, we must assume that the Dy spins have taken a conical spin arrangement above T_{SR1}.[23]

Figure 19.19 shows the temperature dependence of spontaneous magnetization measured for a spherical specimen of $DyCo_5$ hung by a thin thread perpendicular to

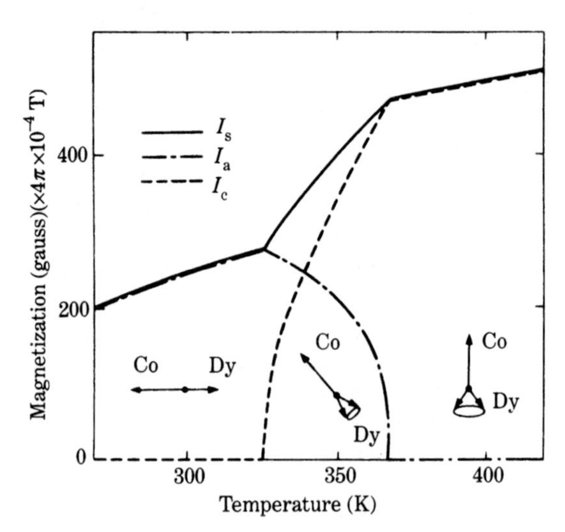

Fig. 19.18. Spin reorientation of $DyCo_5$ caused by temperature change.[23]

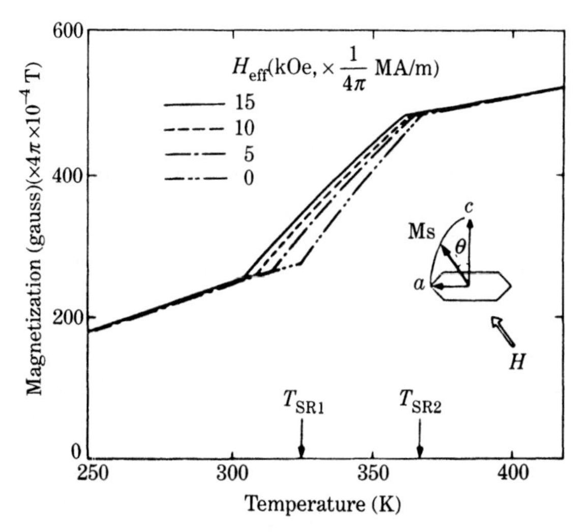

Fig. 19.19. Temperature dependence of I_s measured for a spherical specimen of $DyCo_5$ freely rotatable about an axis perpendicular to the a–c plane. The numbers identifying the various curves are the angle θ measured from the c-axis.[23]

the a–c plane. It was observed that the specimen rotates from the a- to the c-axis with increasing temperature from T_{SR1} to T_{SR2}. It is also interesting that $|I_s|$ depends on the intensity of the field in this temperature range. In the case R = Nd, the spontaneous magnetization is parallel to the b-axis below $T_{SR1} = 245$ K, and rotates with increasing temperature until it reaches the c-axis at $T_{SR2} = 285$ K.[24]

Some magnetic materials such as FeRh (see Section 8.4), $CrS_{1.17}$ (see Section 10.4) and MnAs (see Section 10.3) undergo a first-order type spin reorientation. There are a variety of spin reorientations, including reorientation of the easy axis, change in sign of the anisotropy constant, etc. However, we shall not go further into the details of such behavior.

PROBLEMS

19.1 Draw the spin-phase diagram for an antiferromagnet containing N magnetic ions each with magnetic moment M. Assume that the magnetic anisotropy is negligible, and the temperature dependence of $I_A = -I_B$ is the Q-type (see Fig. 7.12).

19.2 Suppose a ferrimagnet contains N magnetic ions with a magnetic moment M in a unit volume. The sublattice magnetizations of the A- and B-sublattices are given by $I_A = \frac{2}{3}NM$ and $I_B = -\frac{1}{3}NM$ at $T = 0$ and $H = 0$. Assuming the molecular field constant to be $-w$, draw a high field magnetization curve at 0 K. When I_B is perpendicular to H, how many degrees does I_A tilt from H? Find the magnitude of H in this case.

19.3 Consider a possible engineering application of spin reorientation.

REFERENCES

1. J. Becquerel and J. van den Handel, *J. Phys. Radium 10* (1939) **10**; L. Néel, *Nuovo Cimento*, **6**S (1957), 942.
2. I. S. Jacobs, *J. Appl. Phys.*, **32** (1961), 61S.
3. H. Kobayashi and T. Haseda, *J. Phys. Soc. Japan*, **19** (1964), 765.
4. J. Kanamori, *Prog. Theor. Phys.*, **35** (1966), 16.
5. T. Nagamiya, K. Yosida, and R. Kubo, *Adv. Phys.*, **4** (1955), 1.
6. I. S. Jacobs, GE Report No. 69-C-112 (1969); *American Institute of Physics Handbook*, 3rd edn (McGraw-Hill, New York, 1972), No. 5f-16, p. 5-242.
7. A. E. Clark and E. Callen, *J. Appl. Phys.*, **39** (1968), 5972.
8. E. E. Anderson, *Phys. Rev.*, **134** (1964), A1581.
9. N. Miura, I. Oguro, and S. Chikazumi, *J. Phys. Soc. Japan*, **45** (1978), 1534.
10. N. Miura, G. Kido, I. Oguro, K. Kawauchi, S. Chikazumi, J. F. Dillon, Jr., and L. G. Uitert, *Physica*, **86-88B** (1977), 1219.
11. T. Tanaka, K. Nakao, G. Kido, N. Miura, and S. Chikazumi, *Proc. ICM*, **82** (JMMM 31-34) (1983), 773.
12. T. Goto, K. Nakao, and N. Miura, *Physica* B, **155** (1989), 285.
13. J. H. Van Vleck, *Theory of electric and magnetic susceptibilities* (Oxford University Press, 1932), p. 246.
14. G. Krill, P. Panissod, M. Lahrichi, and M. F. Lapierre-Revet, *J. Phys. C: Solid State Phys.*, **12** (1979), 4269.
15. R. J. Weiss, *Proc. Phys. Soc.*, **82** (1963), 281.
16. M. Matsui and S. Chikazumi, *J. Phys. Soc. Japan*, **45** (1978), 458.
17. T. Moriya, *J. Mag. Mag. Mat.*, **14** (1979), 1.
18. J. P. Rebouillat, *IEEE Trans. Magnetics*, **MAG-8** (1972), 630.
19. F. Ono and S. Chikazumi, *J. Phys. Soc. Japan*, **37** (1974), 631.
20. R. L. White, *J. Appl. Phys.* **40** (1969), 1061.
21. E. M. Gyorgy, J. P. Remeika, and F. B. Hagedorn, *J. Appl. Phys.*, **39** (1968), 1369.
22. S. Washimiya and C. Satoko, *J. Phys. Soc. Japan*, **45** (1978), 1204.
23. M. Ohkoshi, H. Kobayashi, T. Katayama, M. Hirano, and T. Tsushima, *Physica*, **86–8** (1977), 195.
24. M. Ohkoshi, H. Kobayashi, T. Katayama, M. Hirano, and T. Tsushima, *IEEE Trans. Mag.*, **MAG13** (1977), 1158.

20

DYNAMIC MAGNETIZATION PROCESSES

In this section we treat time-dependent magnetization processes. These processes become particularly important when the magnetization must change rapidly, as in high-frequency or pulsed fields.

20.1 MAGNETIC AFTER-EFFECT

Magnetic after-effect is defined as the phenomenon in which a change in magnetization is partly delayed after the application of a magnetic field. This phenomenon is sometimes referred to as *magnetic viscosity*. Eddy currents (see Section 20.2) may cause a delay in magnetization, but this kind of purely electrical phenomenon is regarded here as distinct from magnetic after-effect. Magnetization may also be affected by purely metallurgical phenomena such as precipitation, diffusion, or crystal phase transition; this kind of magnetization change is also excluded from the definition of magnetic after-effect. Structural changes like those in the previous sentence, if they occur slowly at or near room temperature, are usually known as *aging*. The difference between magnetic after-effect and magnetization changes due to aging is that changes resulting from magnetic after-effect can be erased by applying an appropriate magnetic field, while changes due to aging are not recoverable by purely magnetic means.

Suppose that a magnetic field $H = H_1$ is applied to a magnetic material, and is suddenly changed to $H = H_2$ at $t = 0$. There is an immediate change in magnetization, I_i, but there is an additional change, I_n, occurring over time as shown in Fig. 20.1. The time-varying part I_n can be written generally as

$$I_n = I_n(t). \tag{20.1}$$

The magnitude of I_n is a function of I_i and also depends on the final magnetic state at H_2. For example, if the final state is in the field range for magnetization rotation, the value of I_n is very small, while if the final state is in a field range for irreversible magnetization, such as near the residual magnetization or the coercive field, I_n may be fairly large.

In the simplest case, $I_n(t)$ can be expressed by a single relaxation time τ, to give

$$I_n(t) = I_{n0}(1 - e^{-t/\tau}), \tag{20.2}$$

where I_{n0} is the total change in magnetization from $t = 0$ to ∞, not including I_i.

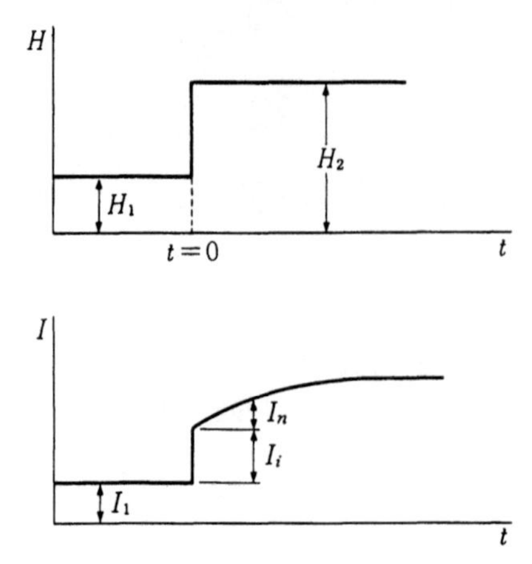

Fig. 20.1. Time change in a magnetic field and the associated change in magnetization.

Figure 20.2 shows a semi-log plot of $I_n(t)$ vs time measured for pure iron containing a small amount of carbon.[1] The curves measured at various temperatures can all be fitted with straight lines corresponding to a ($\log I_n$)–t relationship. This means that (20.2) describes the experiment well. The applied field in this experiment is in the initial permeability range, where the Rayleigh term (the second term in (18.104)) is

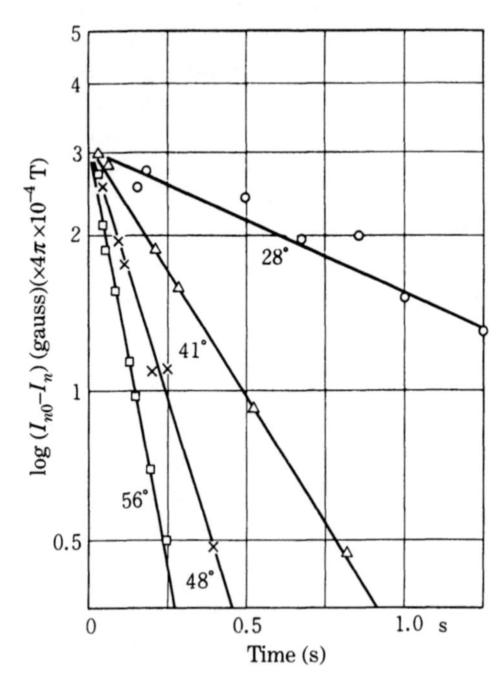

Fig. 20.2. Magnetic after-effect observed for a low-carbon iron.[1]

negligibly small. The value I_{n0}/I_i was 30%. If we denote this ratio as ζ, the magnetization can be expressed as

$$I = \chi_a H\{1 + \zeta(1 - e^{-t/\tau})\}. \tag{20.3}$$

The magnetic after-effect also causes a delay in magnetization in a material subjected to an AC magnetic field. In order to express this phenomenon mathematically, we consider a differential equation which leads to (20.3) for DC magnetization, to give

$$\frac{d}{dt}(I - \chi_a H) = -\frac{1}{\tau}\{I - \chi_a H(1 + \zeta)\}. \tag{20.4}$$

Suppose that an AC magnetic field expressed as

$$H = H_0 e^{i\omega t} \tag{20.5}$$

is applied to a magnetic material. Then changes in magnetization are delayed so that

$$I = I_0 e^{i(\omega t - \delta)}, \tag{20.6}$$

where δ is the delay expressed as a phase angle. In order to find the angle δ and the amplitude I_0, we use (20.4) and (20.5) in (20.6). Then we find

$$\tan \delta = \frac{\zeta \omega \tau}{(1 + \zeta) + \omega^2 \tau^2}, \tag{20.7}$$

and

$$I_0 = \frac{\omega \tau}{\omega \tau \cos \delta - \sin \delta} \chi_a H. \tag{20.8}$$

Since the appearance of a nonzero angle δ results in a power loss, we call this angle δ the *loss angle* and $\tan \delta$ the *loss factor*.

Figure 20.3 shows the temperature dependence of the loss factor measured at

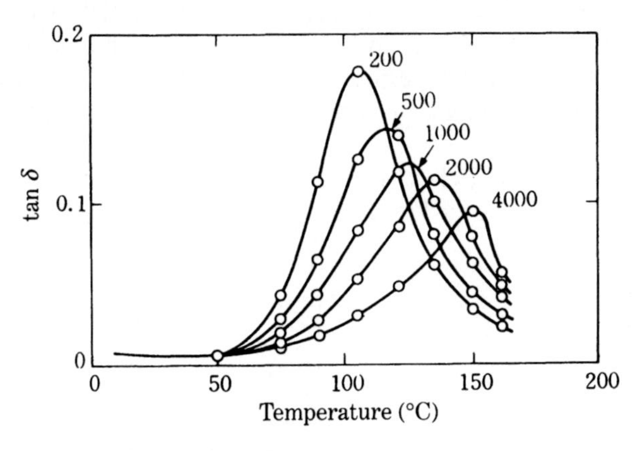

Fig. 20.3. Temperature dependence of the loss factor, $\tan \delta$, observed for a low carbon iron. The numerical values are the frequencies of the AC field in hertz.[1]

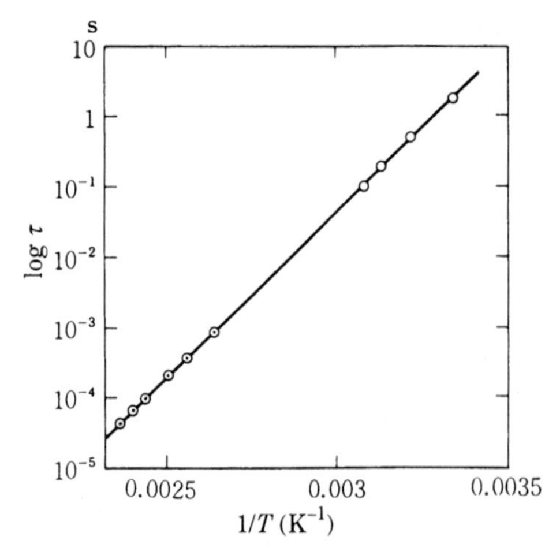

Fig. 20.4. Log τ vs $1/T$ curve obtained from semistatic (open circles) and high-frequency measurements (center-dot circles).[1]

various frequencies for low-carbon iron,[1] the same material used in the semistatic measurement shown in Fig. 20.2. Each curve shows a maximum at some temperature, because the relaxation time τ varies with temperature. If we consider (20.7) to be a function of τ, $\tan\delta$ becomes very small for large τ, because the denominator increases as τ^2, while the numerator increases only as τ. For small τ, $\tan\delta$ becomes also small, because the numerator becomes small, while the denominator stays almost constant. The maximum occurs for

$$\tau = \frac{\sqrt{1+\zeta}}{\omega}. \tag{20.9}$$

Therefore we can determine the relaxation time from the angular frequency of the maximum at a particular temperature. Figure 20.4 shows the logarithm of relaxation time, as determined from AC measurements as well as from semistatic measurements, as a function of reciprocal temperature. We see that both groups of experimental points fall on the same straight line, which means that both phenomena have the same origin.

In order to understand the nature of this phenomenon, a model proposed by Snoek[2] is helpful. As shown in Fig. 20.5, suppose a heavy ball is placed on a concave

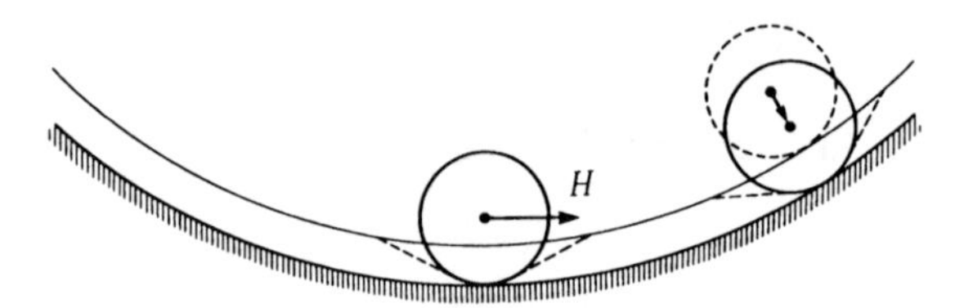

Fig. 20.5. Snoek's model of magnetic after-effect.

surface covered with a layer of thick mud. If the ball is displaced by a lateral force H, it will sink gradually into the layer of mud, changing its equilibrium position. This corresponds to the semistatic magnetic after-effect. AC magnetization corresponds to the case when the ball is oscillating back and forth around the minimum position of the concave surface under the action of an alternating force. As the viscosity of mud increases at low temperatures, the ball will move on the hard mud surface with very low loss. On the other hand, when the viscosity decreases at high temperatures, the ball will move through a low-viscosity mud layer, resting on the concave surface, again resulting in very low loss. At an intermediate temperature, the motion of the ball is most severely damped, and a very large loss results. This is the reason why the loss factor has a maximum at an intermediate temperature.

The magnetic after-effect is not necessarily described by a single relaxation time. In general the relaxation times are distributed over some finite range. If the relaxation times are distributed over a wide range, we can conveniently use $\ln \tau$ as a parameter instead of τ. Let the volume in which the logarithm of the relaxation time is in the range $\ln \tau$ to $\ln \tau + \mathrm{d}(\ln \tau)$ be $g(\tau)\mathrm{d}(\ln \tau)$. Since $g(\tau)\mathrm{d}(\ln \tau) = (g(\tau)/\tau)\mathrm{d}\tau$, the distribution function $g(\tau)$ can be normalized by

$$\int_0^\infty \frac{g(\tau)}{\tau}\,\mathrm{d}\tau = 1. \tag{20.10}$$

Then the time change of magnetization can be described by

$$I_n(t) = I_{n0}\left(1 - \int_0^\infty \frac{g(\tau)}{\tau}\,\mathrm{e}^{-t/\tau}\,\mathrm{d}\tau\right). \tag{20.11}$$

If we assume for simplicity that the distribution function is a constant g from τ_1 to τ_2 and zero outside this range, we have, from (20.10),

$$g = \frac{1}{\ln(\tau_2/\tau_1)}, \tag{20.12}$$

for $\tau_1 < \tau < \tau_2$. If we put $t/\tau = y$, the second term of (20.11) becomes

$$\Delta I_n = I_{n0} - I_n = \frac{I_{n0}}{\ln(\tau_2/\tau_1)} \int_{\tau_1}^{\tau_2} \frac{\mathrm{e}^{-t/\tau}}{\tau}\,\mathrm{d}\tau = \frac{I_{n0}}{\ln(\tau_2/\tau_1)} \int_{t/\tau_2}^{t/\tau_1} \frac{\mathrm{e}^{-y}}{y}\,\mathrm{d}y. \tag{20.13}$$

If we set

$$N(\alpha) = \int_\alpha^\infty \frac{\mathrm{e}^{-y}}{y}\,\mathrm{d}y, \tag{20.14}$$

(20.13) can be expressed as

$$\Delta I_n = \frac{I_{n0}}{\ln(\tau_2/\tau_1)}\left(N\left(\frac{t}{\tau_2}\right) - N\left(\frac{t}{\tau_1}\right)\right). \tag{20.15}$$

The function $N(\alpha)$ is expressed approximately by

$$\left.\begin{aligned} &\text{for } \alpha \ll 1 \qquad N(\alpha) = -0.577 - \ln\alpha + \alpha - \frac{1}{2}\frac{\alpha^2}{2!} + \frac{1}{3}\frac{\alpha^3}{3!} + \cdots \\ &\text{for } \alpha \gg 1 \qquad N(\alpha) = \frac{e^{-\alpha}}{\alpha}\left(1 - \frac{1!}{\alpha} + \frac{2!}{\alpha^2} - \cdots\right), \\ &\text{for } \alpha = 1 \qquad N(1) = 0.219 \end{aligned}\right\}. \tag{20.16}$$

Using these formulas, we see that ΔI_n decreases linearly with time t, or

$$\Delta I_n \cong I_{n0}\left(1 - \frac{(1/\tau_1) - (1/\tau_2)}{\ln(\tau_2/\tau_1)}t\right) \tag{20.17}$$

for $t \ll \tau_1 < \tau_2$, then ΔI_n varies proportionally to $\ln t$, or

$$\Delta I_n \cong \frac{I_{n0}}{\ln(\tau_2/\tau_1)}(\ln\tau_2 - 0.577 - \ln t) \tag{20.18}$$

for $\tau_1 \ll t \ll \tau_2$, and finally tends towards to zero according to

$$\Delta I_n \cong I_{n0}\frac{\tau_2}{\ln(\tau_2/\tau_1)}\frac{1}{t}e^{-t/\tau_2} \tag{20.19}$$

for $\tau_1 < \tau_2 \ll t$.

Figure 20.6 shows the time decrease of magnetization vs. $\log t$ measured at 55°C for well-annealed carbonyl iron. The solid curve represents the function (20.15) drawn by assuming $\tau_1 = 0.0048$ s, $\tau_2 = 0.14$ s, and it reproduces the experiment very well. If this material is magnetized by an AC magnetic field, the loss factor will be maximum at a frequency where $1/\omega$ lies between τ_1 and τ_2. This type of magnetic after-effect was investigated by Richter,[4] so that it is referred to as the *Richter-type magnetic after-effect*.

According to (20.7), we expect that the loss factor $\tan\delta$ will be zero when τ becomes very large at low temperatures. We see in Fig. 20.3, however, that some non-zero value remains even at 0 K. This magnetic loss is also independent of the angular frequency ω. This loss is referred to as the *Jordan-type magnetic after-effect*.

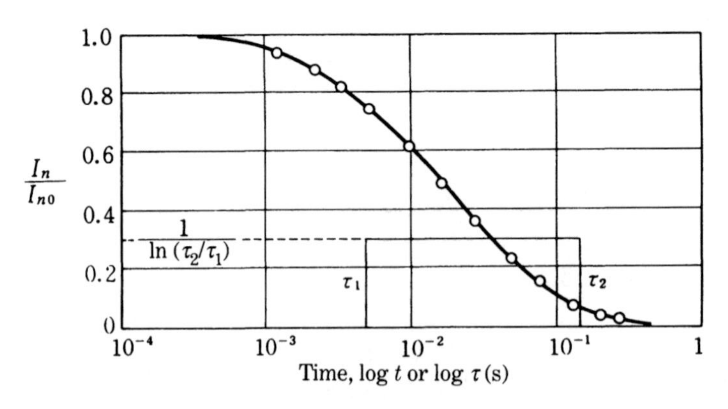

Fig. 20.6. Richter-type magnetic after-effect.

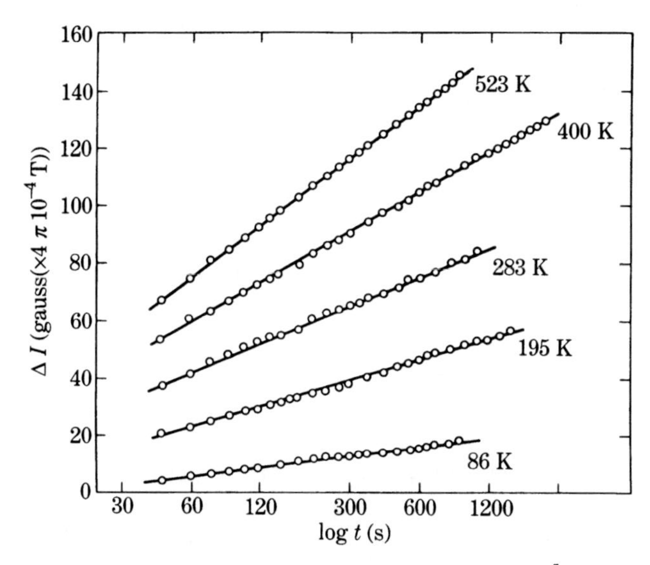

Fig. 20.7. Magnetic after-effect for Alnico V.[5]

The origin of this magnetic after-effect is that the distribution of relaxation times is very large: τ_1, is very small, while τ_2 is sufficiently large. In such a case, the volume in which τ is equal to $1/\omega$ is constant independent of ω, so that $\tan\delta$ becomes independent of ω. In this case (20.18) holds, and for $\tau_1 \ll t \ll \tau_2$, we have

$$\frac{\Delta I_n}{I_n} = \text{const.} - S \ln t, \tag{20.20}$$

where S is called the *magnetic viscosity parameter.*

This type of magnetic after-effect is often observed for permanent magnet materials. Figure 20.7 shows the time variation of magnetization as a function of $\log t$ observed for Alnico V. The experimental points lie very well on straight lines, verifying (20.20).

The physical mechanism of the magnetic after-effect was proposed by Snoek[2] and later corrected by Néel.[6] The carbon or nitrogen atoms, which are very small compared with an iron atom, occupy interstitial sites in the body-centered lattice of iron. There are three kinds of interstitial sites, identified as x-, y-, and z-sites in Fig. 20.8. If many carbon atoms occupy x-sites preferentially, the pseudodipolar interaction of the Fe–Fe pairs in the x-direction is changed, thus inducing a uniaxial magnetic anisotropy (see Section 13.1). Let the number of carbon atoms on the x-, y-, and z-sites be N_x, N_y, and N_z, respectively, in a unit volume of the iron lattice. Then the anisotropy energy is given by

$$E_a = \sum_{i,j} \omega_{ij} = N_x l_C\left(\alpha_1^2 - \tfrac{1}{3}\right) + N_y l_C\left(\alpha_2^2 - \tfrac{1}{3}\right) + N_z l_C\left(\alpha_3^2 - \tfrac{1}{3}\right), \tag{20.21}$$

where $(\alpha_1, \alpha_2, \alpha_3)$ are the direction cosines of the spontaneous magnetization and l_C is the change in dipolar interaction coefficient due to the insertion of the carbon

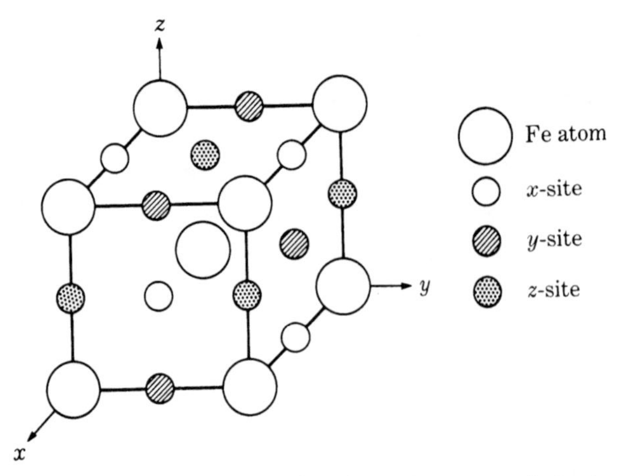

Fig. 20.8. Three kinds of interstitial lattice sites for carbon atoms in body-centered cubic iron.

atom. At thermal equilibrium at a temperature satisfying the condition $l_C \ll kT$, the distribution of carbon atoms is given by

$$\left.\begin{aligned} N_{x\infty} &= \frac{N_C}{3}\exp\left\{\left(\tfrac{1}{3}-\alpha_1^2\right)\frac{l_C}{kT}\right\} \\ N_{y\infty} &= \frac{N_C}{3}\exp\left\{\left(\tfrac{1}{3}-\alpha_2^2\right)\frac{l_C}{kT}\right\} \\ N_{z\infty} &= \frac{N_C}{3}\exp\left\{\left(\tfrac{1}{3}-\alpha_3^2\right)\frac{l_C}{kT}\right\} \end{aligned}\right\}, \tag{20.22}$$

where the subscript ∞ indicates an equilibrium value after a sufficiently long time.

Now we consider the process by which carbon atoms change their position from the given distribution N_x, N_y, and N_z at $t = 0$ to thermal equilibrium at $t = \infty$ given by (20.22). We postulate that a carbon atom must get over a potential barrier of height Q when transferring from one site to a neighboring site (Fig. 20.9). This height is also influenced by the orientation of the domain magnetization. It is given by $Q + (\frac{1}{3} - \alpha_1^2)l_C$, $Q + (\frac{1}{3} - \alpha_2^2)l_C$ and $Q + (\frac{1}{3} - \alpha_3^2)l_C$ for the three kinds of lattice sites.

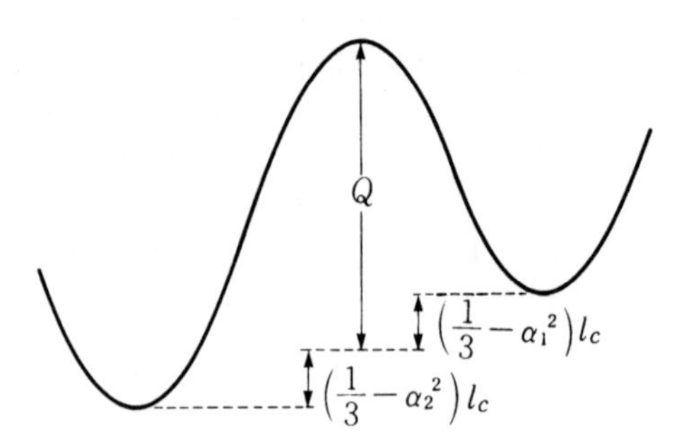

Fig. 20.9. The potential barrier between x- and y-sites.

The numbers of carbon atoms which transfer from one kind of site to another kind should be proportional to (the numbers of carbon atoms present in the same kind of sites) $\times$ exp($-$ height of barrier). Since the rate of increase in N_x is given by one-half of the carbon atoms which escape from y-sites and from z-sites less the number of atoms which escape from the x-sites, we have

$$\frac{\mathrm{d}N_x}{\mathrm{d}t} = \frac{1}{2} cN_y \exp\left(-\frac{Q + (\frac{1}{3} - \alpha_2^2)l_\mathrm{C}}{kT}\right) + \frac{1}{2} cN_z \exp\left(-\frac{Q + (\frac{1}{3} - \alpha_3^2)l_\mathrm{C}}{kT}\right) - cN_x \exp\left(-\frac{Q + (\frac{1}{3} - \alpha_1^2)l_\mathrm{C}}{kT}\right)$$

$$= \frac{c}{6} N_\mathrm{C} \mathrm{e}^{-Q/kT}\left(\frac{N_y}{N_{y\infty}} + \frac{N_z}{N_{z\infty}} - 2\frac{N_x}{N_{x\infty}}\right). \tag{20.23}$$

Assuming that $l_\mathrm{C} \ll kT$, we expand the exponential functions in (20.22) in power series of l_C/kT, and, neglecting the higher order terms and using the relation $N_x + N_y + N_z = N_\mathrm{C}$, we have

$$\frac{\mathrm{d}N_x}{\mathrm{d}t} = -\tfrac{3}{2}c\mathrm{e}^{-Q/kT}(N_x - N_{x\infty}). \tag{20.24}$$

Since N_y and N_z change with time in a similar way, the anisotropy energy given by (20.21) changes with time as

$$\frac{\mathrm{d}E_\mathrm{a}}{\mathrm{d}t} = -\tfrac{3}{2}c\mathrm{e}^{-Q/kT}(E_\mathrm{a} - E_{\mathrm{a}\infty}), \tag{20.25}$$

where c is a coefficient that includes the relaxation time.

First we discuss the effect of such diffusion of carbon atoms on magnetization rotation. If the magnetocrystalline anisotropy is much greater than anisotropy resulting from the carbon atoms, or $K \gg Nl_\mathrm{C}$, the domain magnetization rotates instantaneously to the equilibrium direction upon the application of a magnetic field, and gradually approaches the final direction. Since the angle of the final rotation is small, $E_{\mathrm{a}\infty}$ in (20.25) can be regarded as constant. Then (20.25) is readily solved as

$$E_\mathrm{a} = E_{\mathrm{a}\infty} + (E_{\mathrm{a}0} - E_{\mathrm{a}\infty})\mathrm{e}^{-t/\tau}, \tag{20.26}$$

where $E_{\mathrm{a}0}$ is the anisotropy energy at $t = 0$ and

$$\tau = \frac{2}{3c}\mathrm{e}^{Q/kT}. \tag{20.27}$$

Thus the time variation in the anisotropy is, in this case, described by a single relaxation time. This equation explains the relationship between $\log \tau$ and $1/T$ shown in Fig. 20.4. From the slope of the $\log \tau$ vs $1/T$ curve, we obtain the activation energy Q, which is equal to 0.99 eV in this experiment. This value agrees well with the activation energy for diffusion of carbon atoms in body-centered cubic iron.

The effect of diffusion of carbon atoms on the displacement of domain walls is more complicated. When a field drives a domain wall to a new place, the carbon atoms in the wall rearrange their lattice sites so as to stabilize the local orientation of spins inside the wall. Such a rearrangement of carbon atoms changes the wall energy through a change in the anisotropy energy, and the wall starts to move gradually. If the length of such a gradual displacement is larger than the wall thickness, the directions of spins in the wall are changed by an angle comparable to π, so that $E_{a\infty}$ in (20.25) must be regarded as a function of time. Néel[6] made a detailed calculation for such a gradual displacement of the wall. We refer to the magnetic after-effect caused by diffusion of carbon or nitrogen atoms as the *diffusion after-effect.*

Néel[7] proposed another type of magnetic after-effect called the *thermal fluctuation after-effect*. This phenomenon is caused by thermal fluctuation of the magnetization of an isolated single domain. Let us consider an elongated single domain particle which is magnetized first in the positive direction and then is subjected to a field applied in the negative direction. If the intensity of the field is less than the critical field H_0, the magnetization will stay in the positive direction, in which state the energy of the system is given by

$$U_{+} = v(-K_u + I_s H), \tag{20.28}$$

where v is the volume of the particle. If the magnetization is reversed to the negative direction, the energy becomes

$$U_{-} = v(-K_u - I_s H). \tag{20.29}$$

The potential barrier between the two states can be easily calculated from (18.87) and (18.88), by letting $\theta_0 = 0$, to give $U_{max} = vI_s^2H^2/4K_u$ at $\cos\theta = I_sH/2K_u$.

At temperature T each spin is subject to thermal agitation whose energy is $kT/2$ per degree of freedom of motion. The coherent rotation of all spins included in this particle is also thermally activated and has energy $kT/2$. Since usually the height of the potential barrier is much larger than $kT/2$, such coherent rotation is not able to overcome the potential barrier. If, however, the volume of the particle is so small that the height of the potential barrier vK_u at $H = 0$ is the same order of magnitude as $kT/2$, thermal activation will allow the domain magnetization to rotate over the potential barrier. At room temperature $T = 273\,\text{K}$, $kT = 3.77 \times 10^{-21}\,\text{J}$, so that for $K_u = 10^5\,\text{J/m}^3$, the critical volume becomes

$$v_0 \simeq \frac{kT}{2K_u} = \frac{3.77 \times 10^{-21}}{2 \times 10^5} = 1.9 \times 10^{-26}\,\text{m}^3. \tag{20.30}$$

If we assume that the particle is spherical in shape, its radius is

$$r_0 = \left(\frac{3}{4\pi} v_0\right)^{1/3} = 1.7 \times 10^{-9}\,\text{m} = 17\,\text{Å}. \tag{20.31}$$

In such particles the domain magnetization is always thermally activated, and oscillating. This phenomenon is similar to Langevin paramagnetism (see Section 5.2), so that it is called *superparamagnetism*.[8]

When a magnetic field of intensity H is applied in the negative direction, the height of the potential barrier as measured from U_+ and U_- is changed to $v(2K_u - I_sH)^2/4K_u$ and $v(2K_u + I_sH)^2/4K_u$, respectively. Since the former is less than the latter, the number of particles whose magnetization is activated from the plus direction towards the minus direction is greater than the number which activated in the opposite direction. Thus the rate of increase in the number of particles magnetized in the plus direction is given by

$$\frac{\mathrm{d}N_+}{\mathrm{d}t} = -c'\left\{N_+ \exp\left(-\frac{v(2K_\mathrm{u}-I_\mathrm{s}H)^2}{4K_\mathrm{u}kT}\right) - N_- \exp\left(-\frac{v(2K_\mathrm{u}+I_\mathrm{s}H)^2}{4K_\mathrm{u}kT}\right)\right\}, \tag{20.32}$$

where N_- is the number of particles magnetized in the minus direction. Néel considered c' to be determined by the precessional speed of coherent rotation of the spin system caused by a thermal distortion of the crystal lattice through the change in magnetostrictive anisotropy or in demagnetizing field.

If the field is increased to just below the critical field, $H_0 = 2K_\mathrm{u}/I_\mathrm{s}$, the first term of (20.32) becomes sufficiently large compared to the second term to permit us to neglect the second term, and we have

$$\frac{\mathrm{d}N_+}{\mathrm{d}t} = -\frac{1}{\tau}N_+, \tag{20.33}$$

where

$$\tau = \frac{1}{c'}\exp\left(\frac{v(2K_\mathrm{u}-I_\mathrm{s}H)^2}{4K_\mathrm{u}kT}\right). \tag{20.34}$$

Thus thermal activation of the flux reversal can occur even for particles having a volume larger than the critical volume v_0, if H is close enough to the critical field. Although the form of time change of N_+ given by (20.33) is quite similar to (20.24), the important difference between the two is that the activation energy in (20.34) includes H. If the particles volumes v, or the values of K_u, are scattered around average values, the value of τ given by (20.34) is expected to cover a very wide range. Néel[7] estimated that a particle of volume $1\times10^{-24}\,\mathrm{m}^3$ exhibits a relaxation time $\tau \approx 10^{-1}$ s at room temperature, whereas a particle of volume $2\times10^{-24}\,\mathrm{m}^3$ exhibits $\tau \approx 10^9$ s (several tens of years) under the same conditions. Thus the condition $\tau_1 \ll t \ll \tau_2$ is always valid for a practical duration of measurement, and the time change in magnetization in this case is expected to be proportional to $\log t$ as shown by (20.18). If we let $K_{\mathrm{u\,max}}$ and $v_{\max}$ denote the maximum values of K_u and v, it follows from (20.34)

$$\left.\begin{aligned} \tau_1 &= \frac{1}{c'} \\ \tau_2 &= \frac{1}{c'}\exp\left(\frac{v_{\max}(2K_{\mathrm{u\,max}}-I_\mathrm{s}H)^2}{4K_{\mathrm{u\,max}}kT}\right) \end{aligned}\right\}. \tag{20.35}$$

On putting these two expressions into (20.18), we have

$$\Delta I_n = \text{const} - \frac{4K_{\mathrm{u\,max}} I_{n0} kT}{v_{\max}(2K_{\mathrm{u\,max}} - I_{\mathrm{s}} H)^2} \ln t. \tag{20.36}$$

According to this conclusion, the slope of the $\Delta I_n - \log t$ curve, or the magnetic viscosity parameter, S in (20.20), is expected to be proportional to the absolute temperature T. This is actually the case for permanent magnets such as Alnico V (see Fig. 20.7). However, some materials exhibit a quite different temperature dependence of the parameter S. For instance, in a Pt-Co alloy the parameter S decreases rapidly with increasing temperature.[9] This phenomenon was interpreted in terms of Néel's disperse theory.[10,12]

A thermal fluctuation after-effect can also be considered for domain wall displacement. When a domain wall is pinned by a small obstacle, the wall can be released by a thermal disturbance. Néel[10] considered that the thermal agitation of the local spin system gives rise to a fluctuation in local magnetic fields which may cause irreversible displacement of weakly pinned domain walls. Such a slow displacement of domain walls was actually observed by means of the polar Kerr magneto-optical effect for sputter-deposited Tb-Fe thin films.[11] Under a fixed magnetic field of 0.4 $\mathrm{MA\,m^{-1}}$ (5 kOe) a domain of 10 μm in diameter was observed to increase to about 100 μm in a time of about 3 min. In this case the viscosity parameter S decreases with increasing temperature. The result was interpreted in terms of temperature dependence of the "activation volume".

The important difference between the diffusion after-effect and the thermal fluctuation after-effect is as follows: In the former type, the history of the previous magnetization distribution is retained by a non-magnetic mechanism, such as the distribution of carbon atoms; while in the latter type the history is retained only by magnetic means. Therefore in the former type, if the magnetization is changed from state A to state B, and then later from B to C, the magnetic after-effect corresponding to the change from A to C continues to occur in addition to that of the change from B to C. This phenomenon is called the *superposition principle*. In contrast, in the latter type any memory of the previous state would be destroyed by changing the magnetic state; hence the superposition principle is not valid for the latter type. Néel[12] called the former type the *reversible after-effect*, and the latter the *irreversible after-effect*.

It is commonly observed that the permeability of a magnetic material changes with time after application of a magnetic field or a mechanical stress. Figure 20.10 shows the time decrease of permeability measured for Mn–Zn ferrite. We see that the permeability changes significantly for a long period of time. This phenomenon was discovered from the fact that the resonance frequency of an L–C circuit used in a commercial radio receiver shifts with time from the designed frequency. In this sense Snoek[13] named the phenomenon *disaccommodation*.

Snoek used the model shown in Fig. 20.5 to explain this phenomenon. When the ball is displaced to a new place, it can be moved fairly easily on the surface of the mud layer by an external force. After some period of time the ball will sink into the mud layer and will lose its mobility. We can regard the ball in this model as representing a domain magnetization driven by an external field against the

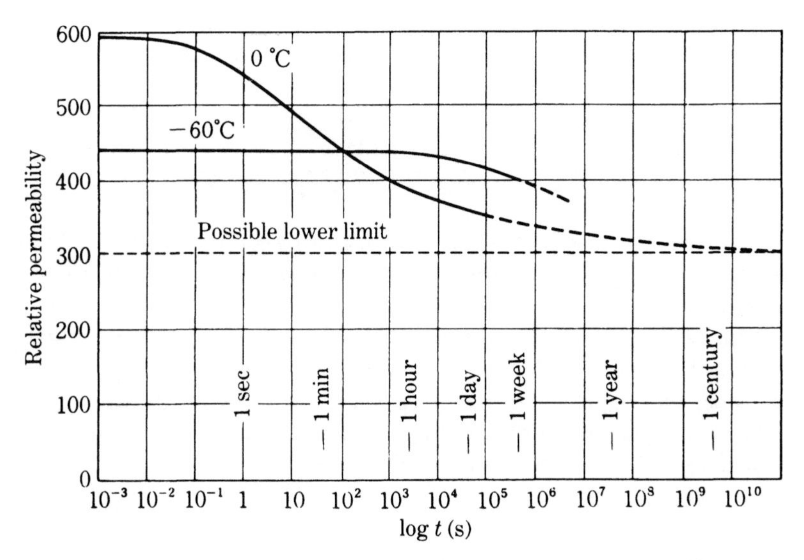

Fig. 20.10. Disaccommodation observed for Mn–Zn ferrite.[13]

magnetic anisotropy, a part of which can be changed by diffusion of C or some other structural change.

Let us consider this phenomenon in terms of diffusion of carbon atoms. Suppose that the magnetization lies parallel to [100] in a material with $K_1 > 0$. After some period of time, carbon atoms will diffuse into energetically favorable sites, stabilizing the domain magnetization as it exists. The final distribution of the carbon atoms is calculated by putting $\alpha_1 = 1$, $\alpha_2 = \alpha_3 = 0$ in (20.22) to give

$$\left.\begin{aligned} N_{x\infty} &= \frac{N_C}{3}\exp\left(-\frac{2l_C}{3kT}\right) \\ N_{y\infty} &= \frac{N_C}{3}\exp\left(\frac{l_C}{3kT}\right) \\ N_{z\infty} &= \frac{N_C}{3}\exp\left(\frac{l_C}{3kT}\right) \end{aligned}\right\} \tag{20.37}$$

Expanding these exponential functions and putting the results into (20.21), we have the anisotropy energy

$$E_a = \frac{N_C l_C^2}{9kT}(1 - 3\alpha_1^2) \simeq \text{const} + \frac{N_C l_C^2}{3kT}\Delta\theta^2, \tag{20.38}$$

where $\Delta\theta$ is the angle of deviation of magnetization from [100]. Therefore, similarly to (18.82), the susceptibility due to magnetization rotation is given by

$$\chi_a = \frac{I_s^2 \sin^2\theta_0}{2\left(K_1 + \dfrac{N_C l_C^2}{3kT}\right)}. \tag{20.39}$$

If we assume that the concentration of carbon atoms is 0.01 at%, then $N_C/N = 1 \times 10^{-4}$, where N is the total number of iron atoms in a unit volume $= 8.5 \times 10^{28}\,\mathrm{m}^{-3}$, and also $Nl_C \approx 10^7$ (see (14.53)), we can estimate the induced anisotropy at 300 K as

$$\frac{N_C l_C^2}{3kT} \approx \frac{10^{-4} \times (10^7)^2}{3 \times 8.5 \times 10^{28} \times 1.38 \times 10^{-23} \times 300} = 9.5, \tag{20.40}$$

which is lower than the magnetocrystalline anisotropy K_1 by a factor 10^{-3}. Therefore we cannot explain a large disaccommodation as shown in Fig. 20.10 by this magnetization rotation model.

Let us consider next the displacement of domain walls. This time we regard the ball as representing a domain wall. After the wall has remained for a prolonged time in the same place, each spin in the wall is stabilized by a local anisotropy given by (20.38). When the wall is displaced by a distance Δs, which is smaller than the thickness of the wall δ, the spin in the wall rotates by the angle

$$\Delta\theta = \frac{\pi \Delta s}{\delta}, \tag{20.41}$$

if we postulate uniform rotation of the spins across the wall. The local anisotropy energy is changed by

$$\Delta E_a = \frac{\pi^2 N_C l_C^2}{3\delta^2 kT} \Delta s^2, \tag{20.42}$$

so that the wall energy is changed by

$$\Delta\gamma = \Delta E_a \times \delta = \frac{\pi^2 N_C l_C^2}{3\delta kT} \Delta s^2. \tag{20.43}$$

Comparing (20.43) with (18.25), we find the second derivative of the wall energy with respect to the displacement s to be

$$\alpha = \alpha_0 + \frac{2\pi^2 N_C l_C^2}{3\delta kT}, \tag{20.44}$$

where α_0 is due to potential fluctuation of γ. If α_0 is caused by internal stress, its value is given by (18.43). Since the susceptibility is inversely proportional to α, as shown by (18.31), the susceptibility change is

$$\frac{\Delta\chi_a}{\chi_a} = 1 - \frac{3\lambda\sigma_0 kT}{N_C l_C^2}\left(\frac{\delta}{l}\right)^2, \tag{20.45}$$

provided

$$\alpha_0 \ll \frac{2\pi^2 N_C l_C^2}{3\delta kT}. \tag{20.46}$$

If we tentatively assume that $\lambda\sigma_0 \sim 10^4 J\,\mathrm{m}^{-3}$ and $\delta/l = 0.01$ for annealed iron, we have

$$\frac{\Delta\chi_a}{\chi_a} = 1 - \frac{3\times 10^4 \times 8.5\times 10^{28} \times 1.38\times 10^{-23} \times 300\times 10^{-4}}{10^{-4}\times (10^7)^2} = 89\%. \quad (20.47)$$

Thus we can explain the experimental values by this model.

Disaccommodation in ferrites was first observed by Snoek[13] for Mn–Zn ferrite. He suggested that electron hopping between the octahedral sites of the spinel lattice could be the cause of this phenomenon. However, considering that the activation energy for the electron hopping is 0.1 eV, while that for disaccommodation is 0.5–0.8 eV, and also that disaccommodation is larger for a specimen containing more lattice vacancies, Ohta[14] concluded that the selective distribution of lattice vacancies on B sites could be the real origin of this phenomenon. In fact, it was confirmed that the cooling of ferrites from high temperatures in a nitrogen atmosphere, which avoids oxidation and the resulting generation of lattice vacancies, is effective in suppressing disaccommodation. In order to prove this consideration, Ohta and Yamadaya[15] observed an induced anisotropy of $10^2\,\mathrm{J\,m}^3$ after cooling Mn–Zn ferrite in a magnetic field. They interpreted this result as due to the selective distribution of vacancies among four kinds of B sites which have different trigonal axes. In contrast, Yanáse[16] explained this phenomenon in terms of magnetic dipole interaction affected by lattice vacancies.

20.2 EDDY CURRENT LOSS

Eddy current loss is defined as the power loss resulting from the eddy currents induced by changing magnetization in magnetic metals and alloys. Let us consider first a long cylinder of radius r_0 made from ferromagnetic or ferrimagnetic material, magnetized parallel to its long axis (see Fig. 20.11(a)). Applying the integral form of the law of electromagnetic induction

$$\int E_s\,\mathrm{d}s = -\iint \frac{\mathrm{d}B_n}{\mathrm{d}t}\,\mathrm{d}S \quad (20.48)$$

to a circuit with radius r drawn about the center axis of the cylinder, and assuming uniform magnetization I, we have, for $r < r_0$,

$$2\pi r E(r) = -\pi r^2 \frac{\mathrm{d}I}{\mathrm{d}t},$$

or

$$E(r) = -\frac{r}{2}\frac{\mathrm{d}I}{\mathrm{d}t}. \quad (20.49)$$

Then the current density is given by

$$i(r) = -\frac{r}{2\rho}\frac{\mathrm{d}I}{\mathrm{d}t}, \quad (20.50)$$

where ρ is the resistivity. The power loss per unit volume is, therefore, given by

$$P = \frac{1}{\pi r_0^2} \int_0^{r_0} 2\pi E(r) i(r) r \, dr$$

$$= \frac{1}{2\rho r_0^2} \left(\frac{dI}{dt} \right)^2 \int_0^{r_0} r^3 \, dr = \frac{r_0^2}{8\rho} \left(\frac{dI}{dt} \right)^2. \tag{20.51}$$

Thus the eddy current loss is proportional to the square of the rate of the magnetization change. This is also true for an alternating magnetization. That is, the power loss increases proportional to the square of the frequency as long as the flux penetration is complete. It is also seen in (20.51) that the power loss is proportional to r_0^2; this means that the loss can be decreased by subdivision of the material into electrically isolated regions. It is natural that the loss is inversely proportional to the resistivity. This is the reason that ferrites are more useful at high frequencies than ferromagnetic metals and alloys.

If dI/dt is sufficiently large, the eddy currents become strong enough to give rise to a magnetic field that is comparable to the applied field. Since the eddy current that flows around a cylinder of radius r produces a magnetic field only inside this cylinder, the integrated magnetic field produced by the total eddy current is strongest at the center and decreases to zero at the surface of the cylindrical specimen. The magnetic field produced by the eddy current always opposes the change in magnetization, so that the magnetization is damped away inside the cylinder. The amplitude of magnetization change is decreased to $1/e$ of that at the specimen surface at a depth

$$s = \sqrt{\frac{2\rho}{\omega\mu}}, \tag{20.52}$$

where ω is the angular frequency of the alternating magnetic field and μ is the permeability, treated as a constant.[17] The depth s is called the *skin depth*. Its value is independent of the size or shape of the specimen, as long as the skin depth is sufficiently small compared to the diameter or thickness of the specimen. When a magnetic metal with $\rho = 1 \times 10^{-7}\,\Omega\,\text{m}$ and $\bar{\mu} = 500$ is magnetized by a 50 Hz AC magnetic field, the skin depth is calculated to be

$$s = \sqrt{\frac{2 \times 1 \times 10^{-7}}{2\pi \times 50 \times 500 \times 4\pi \times 10^{-7}}} \approx 1.0 \times 10^{-3}\,(\text{m}) = 1\,\text{mm}. \tag{20.53}$$

The magnetic cores of AC machines are usually made of laminated thin sheets of magnetic metal, each sheet thinner than the skin depth, and electrically isolated from its neighbors, so that the magnetic flux penetrates completely through each lamination.

In Fig. 20.11(a), we considered a homogeneous change in magnetization. In real ferromagnets, magnetization mostly occurs by domain wall displacement. In Fig. 20.11(b) a ferromagnetic cylinder is separated into two domains by a cylindrical 180° domain wall. Inside the wall, for $r < R$, where R is the radius of the wall, there is no flux change; hence

$$E(r) = 0 \qquad (r \leqq R). \tag{20.54}$$

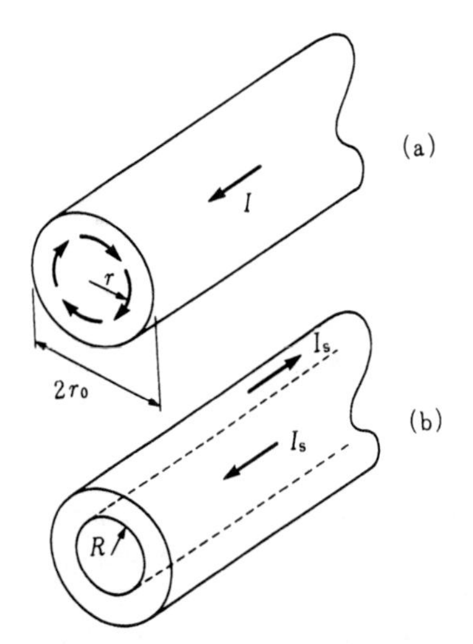

Fig. 20.11. Eddy current in a cylindrical specimen with magnetization changing (a) homogeneously, and (b) by wall displacement.

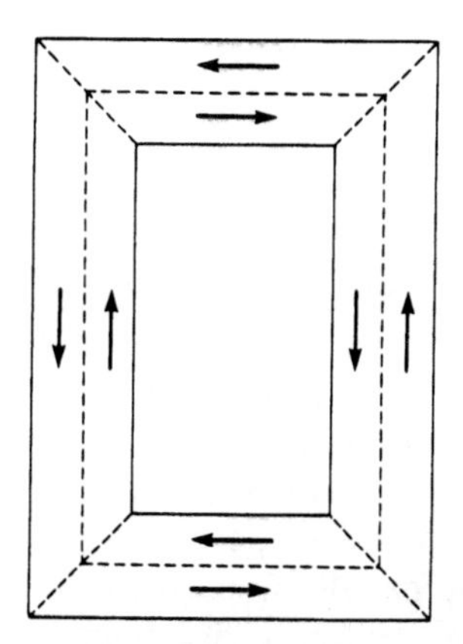

Fig. 20.12. A single crystal of Si–Fe cut into a picture-frame shape with ⟨100⟩ legs, and its domain structure.[18]

Outside the wall, for $r > R$, it follows from (20.48) that

$$2\pi r E(r) = -4\pi I_s R \frac{dR}{dt},$$

or

$$E(r) = -2I_s R \frac{dR}{dt} \frac{1}{r} \qquad (r > R). \tag{20.55}$$

Thus the average power loss per unit volume is given by

$$P = \frac{1}{\pi r_0^2} \int_R^{r_0} \frac{E^2(r)}{\rho} 2\pi r \, dr = \frac{8 I_s^2 R^2}{\rho r_0^2} \left(\frac{dR}{dt} \right)^2 \ln\left(\frac{r_0}{R} \right). \tag{20.56}$$

If we use the rate of magnetization change,

$$\frac{dI}{dt} = \frac{4I_s}{r_0^2} R \frac{dR}{dt} \tag{20.57}$$

for (20.56), we have

$$P = \frac{r_0^2}{2\rho} \ln\left(\frac{r_0}{R} \right) \left(\frac{dI}{dt} \right)^2. \tag{20.58}$$

Thus we find that the power loss depends on R, the position of the wall. On average,

$$\overline{\ln\left(\frac{r_0}{R}\right)} = \frac{1}{r_0}\int_0^{r_0} \ln\left(\frac{r_0}{R}\right) \mathrm{d}R = 1, \tag{20.59}$$

so that

$$\bar{P} = \frac{r_0^2}{2\rho}\left(\frac{\mathrm{d}I}{\mathrm{d}t}\right)^2. \tag{20.60}$$

Comparing this with (20.51), we find that the power loss is four times larger than that for a homogeneous magnetization change. The reason is that the eddy currents are localized at the wall and such localization gives rise to a larger power loss, because the power loss is proportional to i^2.

This fact was first experimentally verified by Williams *et al.*[18] They used a picture-frame single crystal specimen of 3% silicon iron which contains a 180° wall running parallel to each leg of the specimen (Fig. 20.12). The velocity of the domain wall was measured as a function of the applied field and found to be expressed by

$$v = c(H - H_0), \tag{20.61}$$

where c is a proportionality factor and H_0 is the critical field for displacement of the wall. The proportionality between v and H can easily be inferred from the model shown in Fig. 20.11(b). When a cylindrical wall with radius R is expanding with the velocity $\mathrm{d}R/\mathrm{d}t$, the power supplied by the magnetic field per unit volume is

$$P = H\frac{\mathrm{d}I}{\mathrm{d}t} = \frac{4I_s HR}{r_0^2}\left(\frac{\mathrm{d}R}{\mathrm{d}t}\right). \tag{20.62}$$

Since there is no change in the potential energy, this energy must appear as heat produced by the eddy current. Thus on equating (20.62) with (20.56), we have

$$v = \frac{\mathrm{d}R}{\mathrm{d}t} = \frac{\rho}{2I_s R \ln\left(\frac{r_0}{R}\right)} H, \tag{20.63}$$

which shows the proportionality between v and H. Williams *et al.*[18] calculated the distribution of eddy currents for a rectangular cross section in which a plane domain wall is moving as shown in Fig. 20.13(a), and they showed that the calculated power loss is in good agreement with the observed one. They also found that the wall shrinks into a cylinder as shown in Fig. 20.13(b) when a strong magnetic field is applied, because the displacement of the wall is strongly damped in the interior of the specimen. The cylindrical wall vanishes very rapidly, since the surface tension of the wall helps to diminish its area. There was very good agreement between the calculated and observed behavior of the wall in this cylindrical case also.

Before the success of this experiment, the calculation of eddy current losses made under the assumption of a homogeneous magnetization gave only one-half to one-third of the observed value, which was known as the *eddy current anomaly.* It is now known from this experiment that the anomaly results from ignoring the localization of eddy currents at the domain walls.

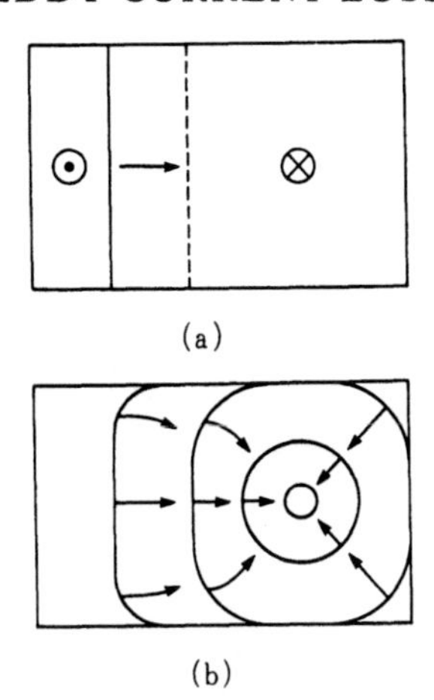

Fig. 20.13. Cross-sectional view of a domain wall moving (a) at low speed and (b) at high speed.

The calculation of the distribution of eddy currents is very difficult for an actual material which contains many walls, since it depends not only on the shape and the distribution of domain walls, but also on the dimensions and external shape of the specimen. Figure 20.14 shows a comparison of the eddy current distribution in the two cases shown in Fig. 20.11. It is seen that the eddy current changes discontinuously at the location of the domain wall. This discontinuity, Δi, is easily found by letting $r = R$ in (20.55):

$$\Delta i = \frac{2I_s}{\rho} v. \tag{20.64}$$

This relation is generally valid for a 180° wall moving at velocity v. Let us now assume

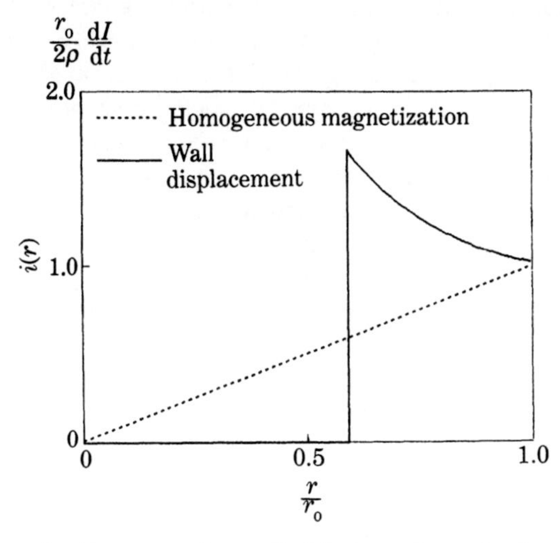

Fig. 20.14. Distribution of eddy current in a cylindrical specimen for homogeneous magnetization and for wall displacement.

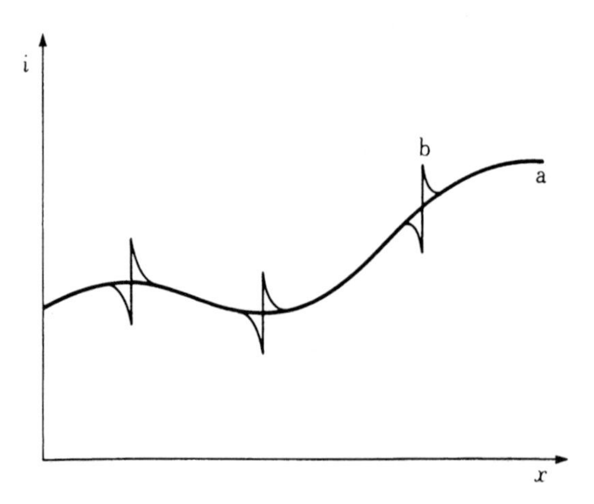

Fig. 20.15. Spatial distribution of macroscopic and microscopic eddy currents.

that the distribution of eddy currents is shown by curve a in Fig. 20.15 for homogeneous magnetization. Apparently the actual eddy current is expected to exhibit sharp changes at the locations of domain walls as shown by curve b in the figure. We call the eddy current due to homogeneous magnetization the *macroscopic eddy current*, i_{ma}, and the deviation from macroscopic eddy current due to wall displacement the *microscopic eddy current*, i_{mi}. The power loss is then given by

$$P = \rho \int (i_{ma} + i_{mi})^2 \, dv$$

$$= \rho \int i_{ma}^2 \, dv + \rho \int i_{mi}^2 \, dv + 2\rho \int i_{ma} I_{mi} \, dv, \tag{20.65}$$

where the integration should be made over a unit volume of the specimen. If the spatial variation of the macroscopic eddy current is gentle compared to that of the microscopic eddy current, as it is in Fig. 20.15, the third term should vanish, since plus and minus values of i_{mi} cancel each other. For such a case we can calculate the power loss as the sum of those of the macroscopic and microscopic eddy currents. If, however, the paths of eddy currents are complicated because of the presence of many non-conducting inclusions, or if the separation of domain walls is comparable to the size of the specimen, we cannot ignore the third term in (20.65), and also cannot distinguish the two categories of macroscopic and microscopic eddy currents. If there are a large number of domain walls, and accordingly the velocity of an individual wall is small, the individual microscopic eddy currents become small, as seen in (20.64). For this case we can approximate the power loss by a macroscopic eddy current. This is a natural conclusion because this presence of a large number of domain walls means that the magnetization is quite homogeneous.

20.3 HIGH-FREQUENCY CHARACTERISTICS OF MAGNETIZATION

In this section we summarize various losses and resonances that appear in high-frequency magnetization. Desirable properties for soft magnetic materials are high permeability and low loss. If a magnetic material is magnetized by an AC magnetic field $\boldsymbol{H} = H_0 e^{i\omega t}$, the magnetic flux density $\boldsymbol{B}$ is generally delayed by the phase angle δ because of the presence of loss, and is thus expressed as $\boldsymbol{B} = B_0 e^{i(\omega t - \delta)}$. The permeability is then

$$\mu = \frac{B}{H} = \frac{B_0 e^{i(\omega t - \delta)}}{H_0 e^{i\omega t}} = \frac{B_0}{H_0} e^{-i\delta}$$
$$= \frac{B_0}{H_0} \cos\delta - i \frac{B_0}{H_0} \sin\delta. \tag{20.66}$$

If we put

$$\left.\begin{aligned} \mu' &= \frac{B_0}{H_0} \cos\delta \\ \mu'' &= \frac{B_0}{H_0} \sin\delta \end{aligned}\right\}, \tag{20.67}$$

(20.66) becomes

$$\mu = \mu' - i\mu''. \tag{20.68}$$

In these expressions μ' expresses the component of $\boldsymbol{B}$ which is in phase with $\boldsymbol{H}$, so it corresponds to the normal permeability: if there are no losses, we should have $\mu = \mu'$. The permeability μ'' expresses the component of $\boldsymbol{B}$ which is delayed by the phase angle 90° from $\boldsymbol{H}$. The presence of such a component requires a supply of energy to maintain the alternating magnetization, regardless of the origin of the delay. The ratio μ'' to μ' is, from (20.67),

$$\frac{\mu''}{\mu'} = \frac{(B_0/H_0)\sin\delta}{(B_0/H_0)\cos\delta} = \tan\delta \tag{20.69}$$

and so $\tan\delta$ is also called the *loss factor*. The quality of soft magnetic materials is often measured by the factor $\mu/\tan\delta$.

Let us consider what kinds of losses appear as the angular frequency ω is increased.

In the low-frequency region, the most important loss is the hysteresis loss. If the amplitude of magnetization is very small, and accordingly in the Rayleigh region, the loss factor due to the hysteresis loss depends on the amplitude of magnetic field, as distinguished from the other types by reducing the amplitude of $\boldsymbol{H}$ towards zero. The hysteresis loss becomes less important in the high-frequency range, because the wall displacement, which is the main origin of the hysteresis, is mostly damped in this range and is replaced by magnetization rotation, as will be discussed later.

The next important loss for ferromagnetic metals and alloys is the eddy current loss. Since a power loss of this type increases in proportion to the square of the frequency, as discussed in the preceding section, it plays an important role in the high-frequency range. One means to reduce eddy current loss is to reduce the dimension of the material in one or both directions perpendicular to the axis of magnetization. For instance, very thin Permalloy sheets (approaching 10 μm) are produced by rolling to reduce eddy currents. Vacuum evaporation, electroplating and sputtering are also effective in preparing very thin metal films which are used in high-speed and high-density memory devices. Magnetic cores composed of fine metallic particles ('dust cores') are also made for the purpose of reducing eddy currents. The most effective means of avoiding eddy currents, however, is to use electrically insulating ferromagnetic materials such as ferrites and garnets. The resistivity of a typical ferrite is about $10^4\,\Omega\,\text{m}$ ($=10^6\,\Omega\,\text{cm}$), so that even bulk material can be used as a magnetic core. Since, however, the resistivity of magnetite (Fe_3O_4) is fairly low ($10^{-4}\,\Omega\,\text{m}=10^{-2}\,\Omega\,\text{cm}$), the presence of excess Fe^{2+} ions in various mixed ferrites results in a decrease in resistivity and accordingly in an increase in eddy current loss at very high frequencies. For instance, Mn-ferrites with excess Fe content are used to attain increased permeability, but the resistivity is of the order of $1\,\Omega\,\text{m}$ ($=10^2\,\Omega\,\text{cm}$). In this case the eddy current loss is not negligible at frequencies over 100 kHz. The Ni-ferrites have resistivities as high as $10^7\,\Omega\,\text{m}$ ($=10^9\,\Omega\,\text{cm}$), so that they can be and are used extensively at high frequencies (see Fig. 20.17). Even in this case, if the material contains excess Fe^{2+} ions, eddy current losses are observed. In contrast, the rare earth iron garnets (see Section 9.3) contain no divalent metal ions, and therefore no electron hopping occurs, so that they exhibit extraordinary low magnetic losses. Some of the garnet crystals are transparent to visible light.

Magnetic after-effect also gives rise to a magnetic loss, as discussed in Section 20.1. The relation between $\tan\delta$ and the relaxation time τ is given by (20.7), which shows a maximum at a certain value of ω. Since the relaxation time decreases with increasing temperature, $\tan\delta$ exhibits a temperature maximum when measured at a fixed frequency, as shown in Fig. 20.3 for a low-carbon iron. A similar phenomenon is also observed for Ni–Zn and Mn–Zn ferrites at $-150°C$ at a frequency of several kHz.[19] The origin of this phenomenon in ferrites is considered to be the diffusion or 'hopping' of electrons between Fe^{2+} and Fe^{3+}. It is estimated that this temperature maximum would be shifted to room temperature at a frequency of several hundred megaherz. However, as will be discussed later, this phenomenon cannot be distinguished from the natural resonance which occurs in this frequency range. The loss factor due to ionic diffusion exhibits a maximum at room temperature only for very low frequencies (as low as a few herz), so that it does not contribute to the high-frequency loss at room temperature.

Thus metallic cores tend to be replaced by ferrite cores for high-frequency applications. However, if ferrite cores are used for a large-scale machine, sometimes their permeability drops at a certain frequency. Figure 20.16 shows a drop of permeability between 1 and 2 MHz for an Mn–Zn ferrite core with cross-section dimensions of $1.25\times2.5\ \text{cm}^2$.[20] This drop in permeability is shifted to a higher frequency when the size of the cross-section is reduced. The origin of this phenomenon is considered to be the building up of an electromagnetic standing wave inside the core. The velocity of an electromagnetic wave is reduced by a factor $(\bar{\varepsilon}\bar{\mu})^{-1}$ as compared with that in vacuum, where $\bar{\varepsilon}$ and $\bar{\mu}$ are the relative permittivity and relative permeability, respectively. Hence the wavelength in the material is given by

$$\lambda=\frac{c}{f\sqrt{\bar{\varepsilon}\bar{\mu}}},\tag{20.70}$$

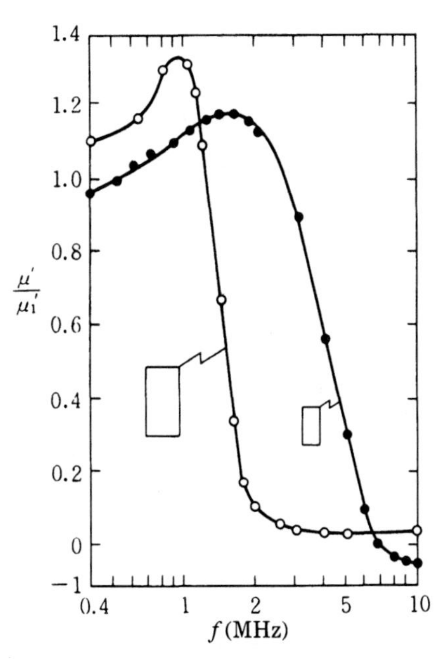

Fig. 20.16. Dimensional resonance observed for Mn–Zn ferrite cores with different cross-sections. The ordinate is normalized to the value at 1 kHz, μ_1'.[20]

where c is the velocity of light in vacuum and f is the frequency. For Mn–Zn ferrite, $\bar{\mu} \sim 10^3, \bar{\varepsilon} \sim 5 \times 10^4$; if we assume $f = 1.5$ MHz, the wavelength λ is found to be 2.6 cm. If, therefore, the dimension of the core is an integer multiple of the wavelength λ, the electromagnetic wave will resonate within the core, giving rise to a standing wave. It is commonly known that μ' and μ'' (or χ' and χ'') vary with frequency as shown later in Fig. 20.23, if some kind of resonance is induced at frequency f_r (H_z is replaced by f, and H_r by f_r). The form of the experimental curves in Fig. 20.16 is recognized as the resonance type. This phenomenon is called *dimensional resonance*.

Generally, permeability drops off and magnetic loss increases at very high frequencies because of the occurrence of a magnetic resonance. Figure 20.17 shows the frequency dependence of μ' and μ'' observed for Ni–Zn ferrites with various compositions.[21] The curve forms are also of the resonance type. It is seen in this graph that a ferrite with high permeability tends to have its permeability decrease at a relatively low frequency. Snoek[22] explained this fact in terms of the resonance of magnetization rotation under the action of the anisotropy field. The resonance frequency can be obtained by setting $H = 0$ in (12.49), to give

$$\omega = \nu H_a, \tag{20.71}$$

where ν is the gyromagnetic constant given by

$$\nu = g\frac{e\mu_0}{2m}$$
$$= 1.105 \times 10^5 g \ (\mathrm{m\,A^{-1}\,s^{-1}}). \tag{20.72}$$

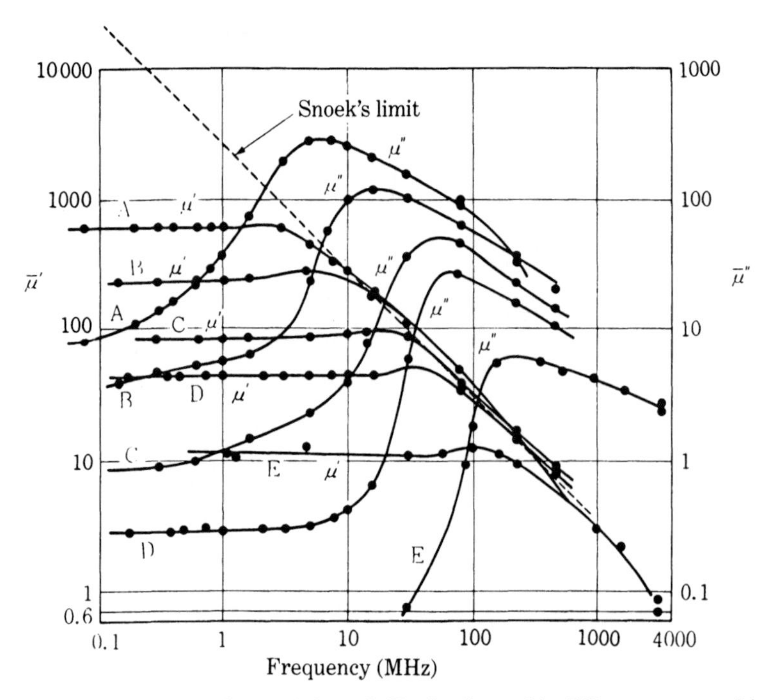

Fig. 20.17. Natural resonance observed for Ni–Zn ferrites with different compositions: Mole ratio of NiO : ZnO = 17.5 : 33.2 (A), 24.9 : 24.9 (B), 31.7 : 16.5 (C), 39.0 : 9.4 (D), 48.2 : 0.7 (E), remainder Fe_2O_3.[21]

If, therefore, a high-frequency magnetic field with angular frequency given by (20.71) is applied, the magnetization rotation about the easy axis will resonate with the field, resulting in abrupt changes in μ' and μ''. If we take $K_1 = -5 \times 10^2\,\mathrm{J\,m^{-3}}$, $I_s = 0.3\,\mathrm{T}$ for Ni–Zn ferrite, we can calculate the anisotropy field from (12.56) as

$$H_a = \frac{4 \times 5 \times 10^2}{3 \times 3 \times 10^{-1}} \approx 2.2 \times 10^3\ (\mathrm{A\,m^{-1}}). \tag{20.73}$$

Using this value and assuming that $g = 2$, we have from (20.71),

$$\omega = (1.105 \times 10^5) \times 2 \times (2.2 \times 10^3) \approx 5 \times 10^8, \tag{20.74}$$

or

$$f = \frac{\omega}{2\pi} \approx 8 \times 10^7\ \mathrm{Hz} = 80\,\mathrm{MHz}. \tag{20.75}$$

This value is in the range of frequency where resonance occurs, as seen in Fig. 20.17. This phenomenon is called *natural resonance*.

We see in Fig. 20.17 that the higher the permeability, the lower the frequency where natural resonance occurs. This can be explained as follows: If we assume that $K_1 > 0$, the anisotropy field is given by (12.53) or

$$H_a = \frac{2K_1}{I_s}, \tag{20.76}$$

so that the resonance frequency increases with an increase of K_1 as

$$\omega = \frac{2\nu K_1}{I_s}. \tag{20.77}$$

On the other hand, the permeability decreases with an increase of K_1, as we know from relation (18.84), assuming the occurrence of magnetization rotation, to give

$$\bar{\mu} = \frac{I_s^2}{3K_1\mu_0}. \tag{20.78}$$

In order to eliminate K_1, we multiply (20.77) and (20.78) together, to obtain

$$\omega\bar{\mu} = \frac{2\nu I_s}{3\mu_0}. \tag{20.79}$$

The same relation holds for $K_1 < 0$. Assuming that $I_s = 0.3\,\mathrm{T}$, and using $\omega = 2\pi f$, and $\mu = 4\pi \times 10^{-7}\bar{\mu}$, we have

$$f\bar{\mu} \approx 5.6 \times 10^9\,\mathrm{Hz} = 5600\,\mathrm{MHz}. \tag{20.80}$$

The dashed line in Fig. 20.17 is drawn by connecting the points where μ' drops to one-half its maximum value. The condition expressed by (20.80) coincides approximately with this line: this line is called the *Snoek limit*. It is therefore predicted that no ferrite can have a permeability higher than the Snoek limit, as long as a cubic magnetocrystalline anisotropy is present.

It was discovered that this limit can be overcome by using a special magnetocrystalline anisotropy.[23] This anisotropy is a uniaxial anisotropy with $K_u < 0$, which exhibits an easy plane perpendicular to the c-axis. If the anisotropy in the c-plane is small, magnetization rotation in this plane can occur easily, so that the permeability may be high. Let the anisotropy field for this magnetization rotation be H_{a1}, while that for rotation out of this plane be H_{a2}. Then the resonance frequency is given, similar to (3.51), by

$$\omega = \nu\sqrt{H_{a1}H_{a2}}. \tag{20.81}$$

On the other hand, the permeability due to rotation in the plane is given, for random distribution of in-plane easy axes in a polycrystalline material, by

$$\mu_a \approx \chi_a = \frac{2I_s}{3H_{a1}}. \tag{20.82}$$

From (20.81) and (20.82), we have the relationship

$$\omega\bar{\mu} = \frac{2\nu I_s}{3\mu_0}\sqrt{\frac{H_{a2}}{H_{a1}}}. \tag{20.83}$$

If, therefore, $H_{a2} > H_{a1}$, this limit is higher than the Snoek limit given by (20.79). One of the materials that satisfies this condition is a magnetoplumbite-type hexagonal crystal (see Section 9.4) called *Ferroxplana*.[23] Table 20.1 lists various magnetic parameters for two Ferroxplanas. It was confirmed experimentally that the natural

Table 20.1. Various constants and resonance frequency of two types of ferroxplana

	$Co_2Ba_3Fe_{24}O_{41}$	$Mg_2Ba_2Fe_{12}O_{22}$
Saturation magnetization I_s	0.27 T (= 215 G)	0.09 T (= 72 G)
Anisotropy fields		
H_{a1}	1.0×10^4 A m^{-1} (= 1.3×10^2 Oe)	0.25×10^4 A m^{-1} (= 0.32×10^2 Oe)
H_{a2}	1.1×10^6 A m^{-1} (= 1.4×10^4 Oe)	1.2×10^6 Am^{-1} (= 1.5×10^4 Oe)
Resonance frequency f_r		
Theory	3700 MHz	2100 MHz
Experiment	2500 MHz	1000 MHz

resonance occurs at frequencies higher than the Snoek limit. Practically, however, it has been difficult to reduce the resistivity and accordingly the loss factor of these materials, so that they cannot be used at very high frequencies.

20.4 SPIN DYNAMICS

In this section, we discuss the dynamic character and switching mechanism of a spin system. As discussed in Section 3.3, a gyroscope performs a precession motion under the action of an external torque, but tends to rotate toward the external torque if its free precession is restricted by some boundary condition. This situation is described by the Landau–Lifshitz[24] equation

$$\frac{\mathrm{d}\boldsymbol{I}}{\mathrm{d}t} = -\nu(\boldsymbol{I} \times \boldsymbol{H}) - \frac{4\pi\mu_0\lambda}{I^2}(\boldsymbol{I} \times (\boldsymbol{I} \times \boldsymbol{H})), \tag{20.84}$$

where $\boldsymbol{I}$ is the magnetization vector and $\boldsymbol{H}$ is the magnetic field vector. The first term represents the precession motion of the magnetization (see (3.43); the magnetization moves in a direction perpendicular to both $\boldsymbol{I}$ and $\boldsymbol{H}$, or in the direction $-(\boldsymbol{I} \times \boldsymbol{H})$. The factor ν is the gyromagnetic constant given by (20.72). The $-$ sign comes from the fact that the angular momenta of electron spins are opposite to their magnetic moments, which are the origin of magnetization. The second term describes the damping of the precession motion in the direction $(\boldsymbol{I} \times \boldsymbol{H})$; the magnetization moves towards $-(\boldsymbol{I} \times (\boldsymbol{I} \times \boldsymbol{H}))$, thus approaching the axis about which the precession occurs. The factor λ, which has dimension s^{-1}, defines the degree of damping action, and is called the *relaxation frequency*. The equation of motion may also be written as

$$\frac{\mathrm{d}\boldsymbol{I}}{\mathrm{d}t} = -\nu(\boldsymbol{I} \times \boldsymbol{H}) + 4\pi\mu_0\lambda\left\{\boldsymbol{H} - \frac{((\mathbf{I} \cdot \mathbf{H})\boldsymbol{I})}{I^2}\right\}, \tag{20.85}$$

as shown by Kittel.[25] Since the second term in braces in (20.85) represents the component of $\boldsymbol{H}$ parallel to $\boldsymbol{I}$, the resultant vector given by both terms in braces represents the component of $\boldsymbol{H}$ perpendicular to $\boldsymbol{I}$. This component of $\boldsymbol{H}$ exerts a torque on the magnetization and causes a rotation of $\boldsymbol{I}$ towards the direction of $\boldsymbol{H}$ as a result of damping action on the precession motion. It is easily verified that (20.85) is mathematically equivalent to (20.84).

In equations (20.84) and (20.85), it is implicitly assumed that the first term (called the intertial term) is much larger than the second term (called the damping term). If we put

$$4\pi\mu_0\frac{\lambda}{\nu I}=\alpha, \tag{20.86}$$

this assumption is equivalent to $\alpha^2 \ll 1$. Strictly speaking, however, the damping should act not only on the precession motion, but also the motion induced by the second term in (20.84) or (20.85). In other words, the damping should act on the resultant motion of the magnetization $\mathrm{d}\boldsymbol{I}/\mathrm{d}t$. Thus the magnetization performs a precession motion under the action of the external force and the damping, so that the exact equation of motion should be

$$\frac{\mathrm{d}\boldsymbol{I}}{\mathrm{d}t}=-\nu\left\{\boldsymbol{I}\times\left(\boldsymbol{H}-\frac{\alpha}{\nu I}\frac{\mathrm{d}\boldsymbol{I}}{\mathrm{d}t}\right)\right\}. \tag{20.87}$$

This equation was first derived by Gilbert and Kelley.[26] Equations (20.84) and (20.85) can be derived from this equation by neglecting the higher-order terms in α^2.

First we consider the precession motion, neglecting the damping action. Suppose that a static magnetic field $\boldsymbol{H}$ is applied parallel to the $-z$-direction. The equation of motion is given by

$$\frac{\mathrm{d}\boldsymbol{I}}{\mathrm{d}t}=-\nu(\boldsymbol{I}\times\boldsymbol{H}), \tag{20.88}$$

which can be written for each component of Cartesian coordinates:

$$\left.\begin{aligned}\frac{\mathrm{d}I_x}{\mathrm{d}t}&=\nu I_y H\\ \frac{\mathrm{d}I_y}{\mathrm{d}t}&=-\nu I_x H\\ \frac{\mathrm{d}I_z}{\mathrm{d}t}&=0\end{aligned}\right\}. \tag{20.89}$$

On solving this equation, we have

$$\left.\begin{aligned}I_x&=I_s\sin\theta_0\mathrm{e}^{\mathrm{i}\omega_0 t}\\ I_y&=I_s\sin\theta_0\mathrm{e}^{\mathrm{i}\omega_0 t+\mathrm{i}\pi/2}\\ I_z&=I_s\cos\theta_0\end{aligned}\right\}, \tag{20.90}$$

where

$$\omega_0=\nu H. \tag{20.91}$$

This solution represents the precession motion of magnetization keeping a fixed angle

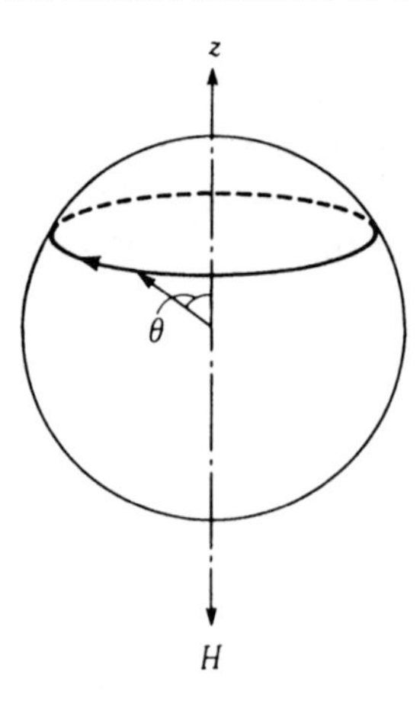

Fig. 20.18. Precession motion of magnetization in the absence of damping.

with respect to the z-axis (Fig. 20.18). It should be noted that without damping, an external field cannot rotate the magnetization towards the field direction.

If a non-zero damping acts on this precession motion, the precession motion will decay unless there is a source of external energy to maintain it.

We rewrite (20.87) for each component of the Cartesian coordinates as

$$\left.\begin{aligned}\frac{\mathrm{d}I_x}{\mathrm{d}t} &= \omega_0 I_y + \alpha\frac{I_y}{I_s}\frac{\mathrm{d}I_z}{\mathrm{d}t} - \alpha\frac{I_z}{I_s}\frac{\mathrm{d}I_y}{\mathrm{d}t}\\ \frac{\mathrm{d}I_y}{\mathrm{d}t} &= -\omega_0 I_x + \alpha\frac{I_z}{I_s}\frac{\mathrm{d}I_x}{\mathrm{d}t} - \alpha\frac{I_x}{I_s}\frac{\mathrm{d}I_z}{\mathrm{d}t}\\ \frac{\mathrm{d}I_z}{\mathrm{d}t} &= \alpha\frac{I_x}{I_s}\frac{\mathrm{d}I_y}{\mathrm{d}t} - \alpha\frac{I_y}{I_s}\frac{\mathrm{d}I_x}{\mathrm{d}t}\end{aligned}\right\}. \tag{20.92}$$

On solving these equations with respect to $\mathrm{d}I_x/\mathrm{d}t$, $\mathrm{d}I_y/\mathrm{d}t$ and $\mathrm{d}I_z/\mathrm{d}t$, we obtain

$$\left.\begin{aligned}\frac{\mathrm{d}I_x}{\mathrm{d}t} &= \frac{\omega_0}{1+\alpha^2}I_y + \frac{\omega_0\alpha}{1+\alpha^2}\frac{I_xI_z}{I_s}\\ \frac{\mathrm{d}I_y}{\mathrm{d}t} &= -\frac{\omega_0}{1+\alpha^2}I_x + \frac{\omega_0\alpha}{1+\alpha^2}\frac{I_yI_z}{I_s}\\ \frac{\mathrm{d}I_z}{\mathrm{d}t} &= -\frac{\omega_0\alpha}{1+\alpha^2}I_s + \frac{\omega_0\alpha}{1+\alpha^2}\frac{I_z^2}{I_s}\end{aligned}\right\}. \tag{20.93}$$

If we start from (20.84) or (20.85), we obtain a similar set of equations, with $(1+\alpha^2)$ everywhere replaced by 1. The two sets of solutions are the same when $\alpha^2 \ll 1$. In the

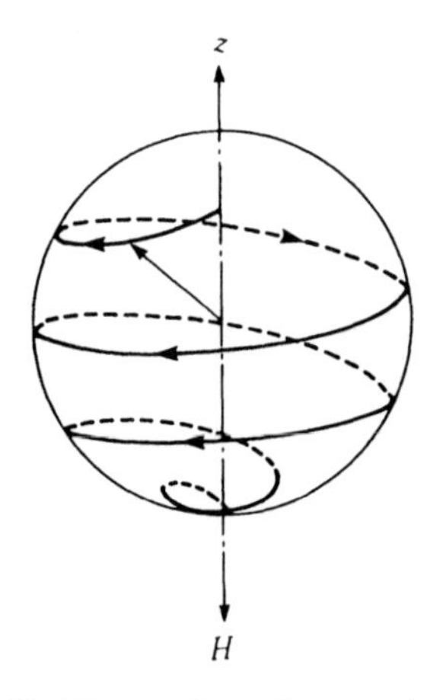

Fig. 20.19. The motion of magnetization for small damping.

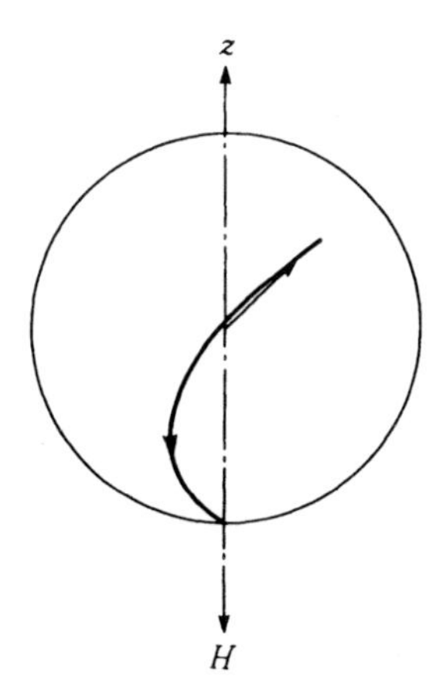

Fig. 20.20. The motion of magnetization for large damping.

general case including $\alpha^2 \gg 1$, however, we must use (20.93). Solving the differential equations (20.93) with respect to $I_x(t)$, $I_y(t)$ and $I_z(t)$, we obtain

$$\left.\begin{aligned} I_x &= I_s \sin\theta\, e^{i\omega t} \\ I_y &= I_s \sin\theta\, e^{i\omega t + i\pi/2} \\ I_z &= I_s \cos\theta \end{aligned}\right\}, \tag{20.94}$$

where θ is a function of time and is given by

$$\tan\frac{\theta}{2} = \tan\frac{\theta_0}{2}\, e^{-t/\tau}, \tag{20.95}$$

and θ_0 is the initial inclination of magnetization. The angular frequency ω and time constant τ in this equation are

$$\omega = \frac{\omega_0}{1+\alpha^2} = \frac{\omega_0}{1+(1/(\omega_0\tau_0))^2} \tag{20.96}$$

$$\tau = \tau_0(1+\alpha^2) = \tau_0\left(1+\left(\frac{1}{\omega_0\tau_0}\right)^2\right), \tag{20.97}$$

where ω_0 is the resonance frequency given by (20.91) and τ_0 is

$$\tau_0 = \frac{1}{\alpha\omega_0} = \frac{I_s}{4\pi\lambda\mu_0 H}. \tag{20.98}$$

If $\alpha^2 \ll 1$, we know from (20.96), (20.97) and (20.98) that $1/\omega = \alpha\tau$, so that $1/\omega \ll \tau$; hence the magnetization performs a number of precession rotations before it finally points to the $-z$-direction (Fig. 20.19). If $\alpha^2 \gg 1$, it turns out that $1/\omega \gg \tau$, and the magnetization rotates more directly towards the $-z$-direction without making many precession rotations (Fig. 20.20). This switching motion of magnetization becomes more viscous as the relaxation frequency becomes large. The switching time also

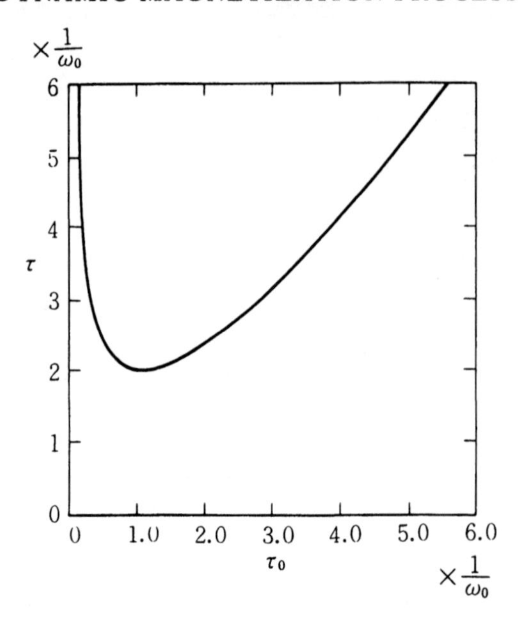

Fig. 20.21. The relaxation time for switching of magnetization as a function of the relaxation frequency.

becomes very large if the relaxation frequency is too small, because the magnetization performs too many precession rotations. The fastest switching is therefore attained for an intermediate value of the relaxation frequency. In Fig. 20.21, the relaxation time τ is plotted as a function of τ_0, which includes the relaxation frequency λ (see (20.98)). The graph shows that the relaxation time τ is minimum when

$$\tau_0 = \frac{1}{\omega_0}, \tag{20.99}$$

or

$$\lambda = \frac{\nu I_s}{4\pi\mu_0}. \tag{20.100}$$

This condition is called *critical damping*. The minimum value of τ is given by

$$\tau_{\min} = \frac{2}{\omega_0} = \frac{2}{\nu H}, \tag{20.101}$$

which can be estimated, taking a strong magnetic field $H_z = 1.6\,\mathrm{MA\,m^{-1}}$ (= 20000 Oe) and $g = 2$, as

$$\tau_{\min} = \frac{2}{2 \times 1.105 \times 10^5 \times 1.6 \times 10^3} \approx 5.7 \times 10^{-9}\,\mathrm{s}. \tag{20.102}$$

The condition for critical damping (16.51) is, for $I_s = 1\,\mathrm{T}$,

$$\lambda = 1.4 \times 10^{10}\,\mathrm{Hz}. \tag{20.103}$$

For instance, the relaxation frequency as determined for nickel ferrite from the width of the resonance line is 10^7–10^8 Hz, which is insufficient for the critical damping. One method to obtain critical damping is to use eddy current damping. Consider a thin metal wire. The demagnetizing field due to the total eddy currents is calculated from (20.50) as

$$H_d = \int_0^{r_0} i(r)\mathrm{d}r = -\frac{r_0^2}{4\rho}\frac{\mathrm{d}I}{\mathrm{d}t}. \tag{20.104}$$

On the other hand, it is easily seen in (20.87) that the damping action is equivalent to the presence of the demagnetizing field,

$$H_d = \frac{\alpha}{\nu I_s}\frac{\mathrm{d}I}{\mathrm{d}t} = \frac{4\pi\mu_o\lambda}{\nu^2 I_s^2}\frac{\mathrm{d}I}{\mathrm{d}t}. \tag{20.105}$$

On comparing (20.104) and (20.105), we see that the relaxation frequency due to eddy currents given by

$$\lambda = \frac{\nu^2 I_s^2 r_0^2}{16\pi\mu_0\,\rho}. \tag{20.106}$$

Using the values $\nu = 2.21 \times 10^5\ \mathrm{m\,A^{-1}\,s^{-1}}$, $\mu_0 = 4\pi \times 10^{-7}$, $I_s = 1\,\mathrm{T}$, $\rho = 1 \times 10^{-7}\ \Omega\,\mathrm{m}$, we obtain

$$\lambda = \frac{2.21^2 \times 10^{10} \times 1^2}{16\pi \times 4\pi \times 10^{-7} \times 1 \times 10^{-7}} r_0^2 = 7.73 \times 10^{21} r_0^2. \tag{20.107}$$

In order to attain critical damping, the relaxation frequency (20.107) must be equal to the value given by (20.103); thus the radius of the metal wire, r_0, must be

$$r_0 = \sqrt{\frac{1.4 \times 10^{10}}{7.73 \times 10^{21}}} = 1.35 \times 10^{-6}\ \mathrm{m} = 1.35\ \mu\mathrm{m}. \tag{20.108}$$

In the past, magnetic thin films and thin wires were investigated as possible fast switching devices. One of the reasons was to take advantage of their eddy currents to obtain critical damping.

20.5 FERRO-, FERRI-, AND ANTIFERROMAGNETIC RESONANCE

The fundamental concept of spin resonance has already been described in Section 3.3. There we assumed that all the spins forming the spontaneous magnetization maintain perfect parallelism during precession. We call this the *uniform mode* or *Kittel mode*. Starting from (20.88), we have the uniform mode of precession expressed by (20.90) with the angular frequency at resonance given by (20.91). If a damping force acts on the precession motion, the motion can be maintained only if energy is supplied to the system by an oscillating or rotating magnetic field. Let H_0 be the amplitude of the rotating field, and ω its angular frequency about the z-axis. In order to produce a

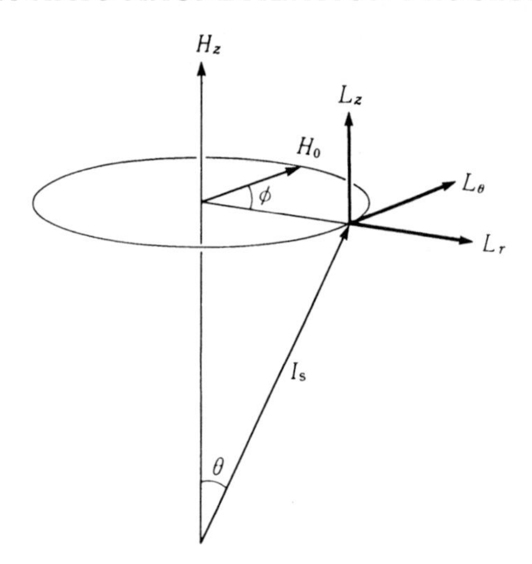

Fig. 20.22. Torque components exerted on magnetization I_s by the rotational field H_0.

non-zero torque on the magnetization vector, the rotating field vector H_0 must make a non-zero azimuthal angle ϕ with the component of the magnetization vector I_s normal to the z-axis (Fig. 20.22). The components of the torque acting on the magnetization vector are then expressed, in cylindrical coordinates, as

$$\left.\begin{aligned} L_r &= -I_s H_0 \cos\theta \sin\phi, \\ L_\theta &= I_s H_0 \cos\theta \cos\phi, \\ L_s &= I_s H_0 \sin\theta \sin\phi. \end{aligned}\right\}. \tag{20.109}$$

If we apply a static magnetic field parallel to the z-axis, the Landau–Lifshitz equation (20.84) becomes, for each component,

$$\left.\begin{aligned} \frac{dI_r}{dt} &= \nu I_s H_0 \cos\theta \sin\phi - 4\pi\lambda\mu_0 H_z \sin\theta\cos\theta, \\ \frac{dI_\theta}{dt} &= -\nu I_s H_0 \cos\phi + \nu I_s \sin\theta, \\ \frac{dI_s}{dt} &= -\nu I_s H_0 \sin\theta \sin\phi + 4\pi\lambda\mu_0 H_z \sin^2\theta. \end{aligned}\right\}. \tag{20.110}$$

In the stationary state,

$$\left.\begin{aligned} \frac{dI_r}{dt} &= 0 \\ \frac{dI_\theta}{dt} &= I_s \omega \sin\theta \\ \frac{dI_s}{dt} &= 0 \end{aligned}\right\}. \tag{20.111}$$

Comparing (20.111) with (20.110), we have

$$\left.\begin{aligned} \nu I_s H_0 \sin\phi - 4\pi\lambda\mu_0 H_z \sin\theta = 0 \\ H_0 \cos\theta\cos\phi - H_z \sin\theta = -\frac{\omega}{\nu}\sin\theta \end{aligned}\right\}, \tag{20.112}$$

from which we obtain

$$\sin\theta = \frac{\nu I_s}{4\pi\lambda\mu_0}\frac{H_0}{H_z}\sin\phi = \frac{1}{\alpha}\frac{H_0}{H_z}\sin\phi, \tag{20.113}$$

and

$$\tan\phi = \frac{4\pi\lambda\mu_0}{\nu I_s}\frac{H_z}{H_z - H_r} = \alpha\frac{H_z}{H_z - H_r} \tag{20.114}$$

for $\theta \ll \pi$, where H_r is the resonance field,

$$H_r = \frac{\omega}{\nu}. \tag{20.115}$$

The real and imaginary parts of the susceptibility are expressed as

$$\left.\begin{aligned} \chi' &= \frac{I_s}{H_0}\sin\theta\cos\phi \\ \chi'' &= \frac{I_s}{H_0}\sin\theta\sin\phi \end{aligned}\right\}, \tag{20.116}$$

or from (20.113),

$$\left.\begin{aligned} \chi' &= \frac{I_s}{\alpha H_z}\sin\phi\cos\phi \\ \chi'' &= \frac{I_s}{\alpha H_z}\sin^2\phi \end{aligned}\right\}. \tag{20.117}$$

These susceptibilities are plotted against the DC magnetic field H_z in Fig. 20.23 for various values of α. As H_z increases and becomes equal to H_r in (20.114), $\tan\phi$ becomes infinity, so that ϕ becomes 90°. As soon as H_z becomes larger than H_r, $\tan\phi$ changes to minus infinity, so that ϕ becomes $-90°$. Therefore χ' in (20.117) changes sign at $H_z = H_r$, while χ'' has a maximum at $H_z = H_r$, as shown in Fig. 20.23. As the parameter α or the relaxation frequency τ increases, the width of the absorption peak increases. The width of the absorption peak at half the height of the maximum value is called the *half-value width*. This value is calculated by putting $\phi = 45°$ in the equation for χ'' in (20.117), because at this value $\sin^2\phi$ becomes half its maximum value 1. Then $\tan\phi = 1$ in (20.114), so that the half-value width is given by

$$\Delta H = 2(H - H_r) = \frac{8\pi\lambda\mu_0}{\nu I_s - 4\pi\lambda\mu_0} H_r. \tag{20.118}$$

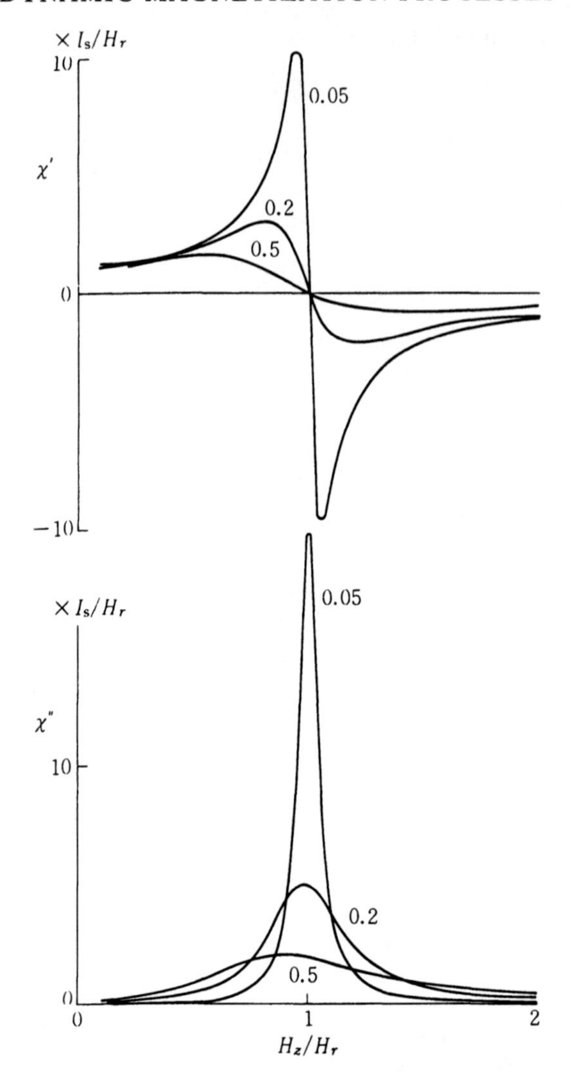

Fig. 20.23. Dependence of the real and imaginary parts of the rotational susceptibility, χ' and χ'', on the intensity of the DC field, H_z, near the resonance field H_r (numbers on the curves are values of α).

For Mn-Zn ferrite, $I_s = 0.25\,\mathrm{T}$, $H_r = 2.55 \times 10^5\,\mathrm{A\,m^{-1}}$, $\Delta H = 5.59 \times 10^3\,\mathrm{A\,m^{-1}}$, and $g = 2$, and we have

$$\lambda = 3.88 \times 10^7\ \mathrm{Hz}. \tag{20.119}$$

Thus we can determine the relaxation frequency, λ, from the half-value width. The reason why the line width is not zero is that the precession motion of the spin system is damped by various mechanisms such as eddy currents, electron hopping between Fe^{2+} and Fe^{3+} ions, generation of spin waves caused by inhomogeneity of the material, generation of lattice vibrations caused by magnetoelastic coupling, etc.

In the vibration of a string, various higher-order harmonic standing waves are

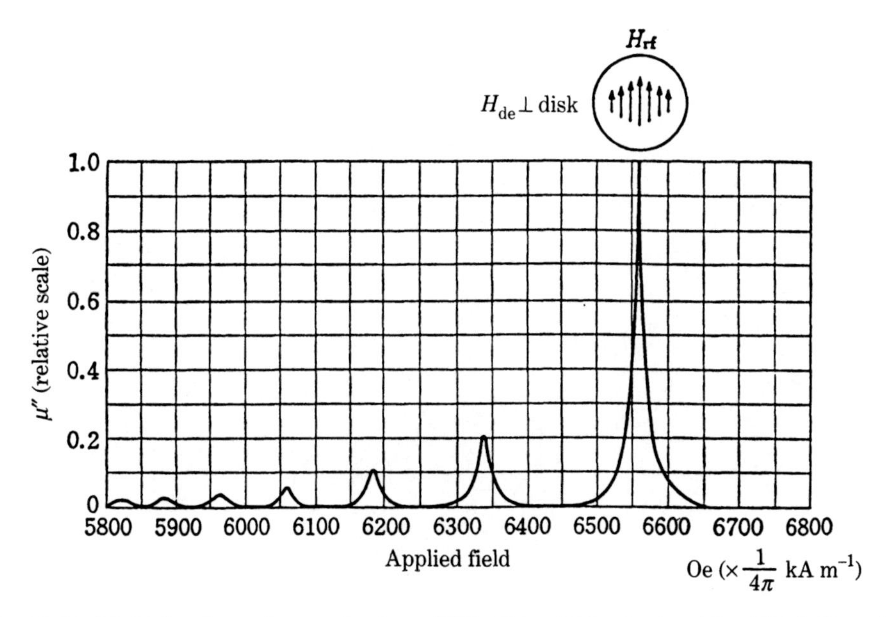

Fig. 20.24. Multiple absorption peaks in a (100) disk of Mn ferrite. The RF field variation across the disk is indicated. (Experiment by Dillon[27], after Walker[28])

generated in addition to the fundamental standing wave. Similarly, in ferromagnetic resonance, various higher-order harmonic modes can be generated in addition to the uniform mode. Dillon[27] observed many absorption peaks for a DC field less than the main resonance field H_r, as shown in Fig. 20.24, for a (100) disk of Mn ferrite placed in a static magnetic field parallel to [100] with an inhomogeneous RF field parallel to the disk as shown at the top of the figure. Walker[28] attributed this phenomenon to the excitation of non-uniform modes of precession, shown in Fig. 20.25. Assuming that the shape of the specimen is spherical, a possible higher precession mode is shown for each cross-section. Arrows in the figure represent the in-plane component of spins which rotate with the same rotational speed. In contrast to the uniform mode, which produces magnetic free poles on the side surface as shown in Fig. 3.17, thus storing a considerable magnetostatic energy, this mode produces distributed free poles, thus reducing the magnetostatic energy. Therefore this mode is referred to as a *magnetostatic mode* or *Walker mode*.

Many absorption lines were also observed for magnetic thin films to which a DC magnetic field is applied perpendicular to the surface.[29] The origin of these higher modes is the excitation of standing waves formed as a result of interference of spin waves propagating perpendicular to the surface.[30] Thus we call this resonance *spin-wave resonance*. Since the wavelength of such a spin wave is fairly short, it stores a considerable exchange energy. Therefore this mode is often referred to as the *exchange mode*.

Next we discuss *antiferromagnetic resonance*, which was first treated theoretically by Nagamiya[31] and independently by Kittel.[32] Suppose that the sublattice magnetizations I_A and I_B make different angles θ_A and θ_B with the easy axis (the z-axis), as shown in

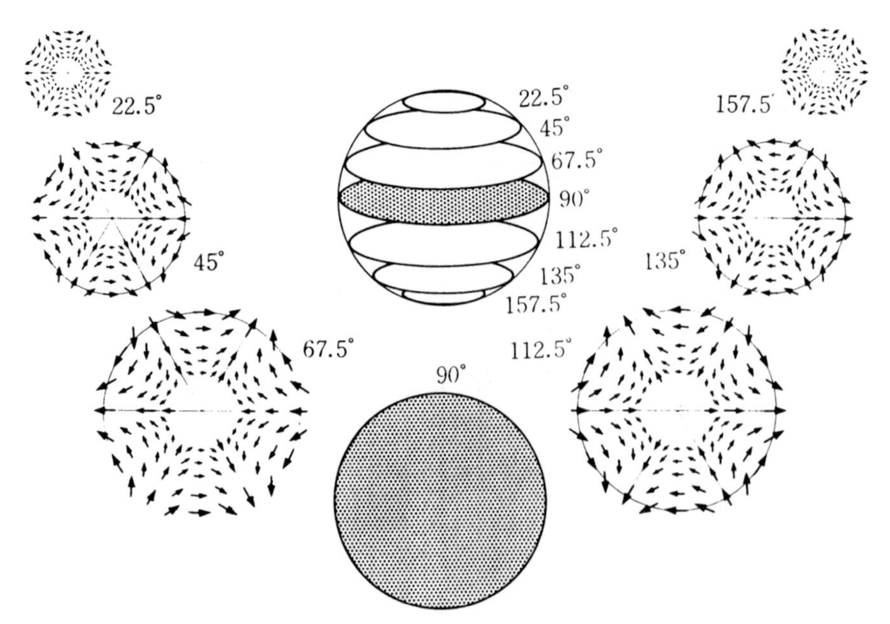

Fig. 20.25. The configuration of inplane component of magnetization in each cross section of a spherical specimen corresponding to the (4, 3, 0) mode. (After Walker[28])

Fig. 20.26, and a magnetic field $\boldsymbol{H}$ is applied parallel to the z-axis. The negative exchange interaction acting between I_A and I_B tends to align the two magnetization vectors antiparallel. These three effects, i.e., the external field, the magnetic anisotropy, and the exchange interaction, can combine to cause the same precessional speed of I_A and I_B about the z-axis keeping the angle between the two magnetizations constant.

Neglecting damping terms in (20.84), we have the equations of motion for two sublattice magnetizations:

$$\left.\begin{aligned} \frac{\mathrm{d}\boldsymbol{I}_A}{\mathrm{d}t} &= -\nu\{\boldsymbol{I}_A \times (\boldsymbol{H} + \boldsymbol{H}_{aA} + \boldsymbol{H}_{mA})\} \\ \frac{\mathrm{d}\boldsymbol{I}_B}{\mathrm{d}t} &= -\nu\{\boldsymbol{I}_B \times (\boldsymbol{H} + \boldsymbol{H}_{aB} + \boldsymbol{H}_{mB})\} \end{aligned}\right\}, \tag{20.120}$$

where $\boldsymbol{H}$ is the applied field, $\boldsymbol{H}_{aA}$ and $\boldsymbol{H}_{aB}$ the anisotropy fields acting on the A and B sublattices, respectively, and $\boldsymbol{H}_{mA}$ and $\boldsymbol{H}_{mB}$ the molecular fields, which are

$$\left.\begin{aligned} \boldsymbol{H}_{mA} &= w_{AA}\boldsymbol{I}_A + w_{AB}\boldsymbol{I}_B \\ \boldsymbol{H}_{mB} &= w_{BA}\boldsymbol{I}_A + w_{BB}\boldsymbol{I}_B \end{aligned}\right\}, \tag{20.121}$$

as shown in (7.1) and (7.2). For simplicity, we assume uniaxial anisotropy with its easy axis parallel to the z-axis, so that the anisotropy field is given by

$$\left.\begin{aligned} H_{aA} &= H_a \\ H_{aB} &= -H_a \end{aligned}\right\}, \tag{20.122}$$

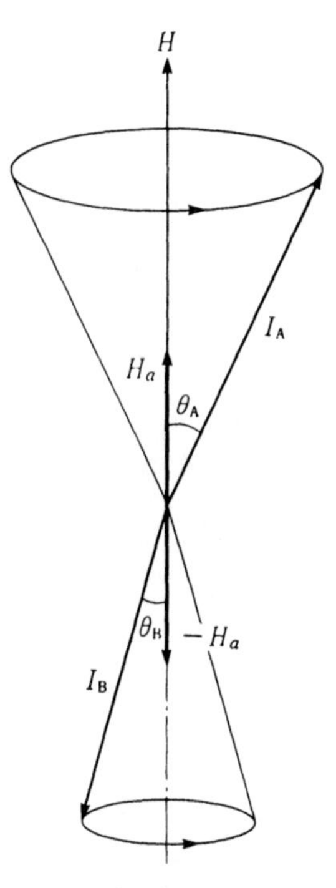

Fig. 20.26. Antiferromagnetic resonance.

where H_a is the absolute value of the anisotropy field. Let H_m be the absolute value of the exchange field acting between the two sublattices, or

$$H_m = |w_{AB} I_A| = |w_{BA} I_B|. \tag{20.123}$$

We assume that the external field $\boldsymbol{H}$ is applied in the + z-direction. Then the components of (20.120) become

$$\left.\begin{aligned}
\frac{dI_{Ax}}{dt} &= -\nu I_{Ay}(H_z + H_m + H_a) - \nu I_{By} H_m \\
\frac{dI_{Ay}}{dt} &= \nu I_{Bx} H_m + \nu I_{Ax}(H_z + H_m + H_a) \\
\frac{dI_{Az}}{dt} &= -\nu w_{AB}(I_{Ax} I_{By} - I_{Ay} I_{Bx}) \\
\frac{dI_{Bx}}{dt} &= -\nu I_{By}(H_z - H_m - H_a) + \nu I_{Ay} H_m \\
\frac{dI_{By}}{dt} &= -\nu I_{Ax} H_m + \nu I_{Bx}(H_z - H_m - H_a) \\
\frac{dI_{Bz}}{dt} &= -\nu w_{AB}(I_{Bx} I_{Ay} - I_{By} I_{Ax})
\end{aligned}\right\} \tag{20.124}$$

The components of the two sublattice magnetizations are given by

$$\left.\begin{aligned} I_{Ax} &= I \sin\theta_A \cos\omega t \\ I_{Ay} &= I \sin\theta_A \sin\omega t \\ I_{Az} &= I \cos\theta_A \\ I_{Bx} &= -I \sin\theta_B \cos\omega t \\ I_{By} &= -I \sin\theta_B \sin\omega t \\ I_{Bz} &= -I \cos\theta_B \end{aligned}\right\}, \tag{20.125}$$

where I is the absolute value of spontaneous magnetization. On putting these equations into (20.124), we have the relations

$$\left.\begin{aligned} \omega \sin\theta_A &= \nu\{\sin\theta_A(H_z + H_m + H_a) - \sin\theta_B H_m\} \\ \omega \sin\theta_B &= \nu\{\sin\theta_B(H_z - H_m - H_a) + \sin\theta_A H_m\} \end{aligned}\right\}, \tag{20.126}$$

from which we have

$$\omega = \nu\left\{H_z \pm \sqrt{H_a(H_a + 2H_m)}\right\}, \tag{20.127}$$

and for $H_m \gg H_a$ we obtain

$$\frac{\sin\theta_A}{\sin\theta_B} = 1 + \frac{H_a}{H_m} \pm \sqrt{2\frac{H_a}{H_m}}. \tag{20.128}$$

It is interesting to note that the solution for ω can exist even when $H_a = 0$, as is known from (20.127). However, in this case we find that $\theta_A = \theta_B$ from (20.128), so that such a resonance mode cannot be excited by an external RF field.

The theory was extended by Yosida[33] to a more general type of anisotropy. He treated the resonance of $CuCl_2 \cdot 2H_2O$, which has a fairly small exchange interaction that allows the observation of resonance in the microwave region, and obtained beautiful agreement with experiment.

Ferrimagnetic resonance[34,35] is essentially the same as ferromagnetic resonance, in the sense that the spontaneous magnetization precesses as a whole about the external field. In addition to this mode, however, another resonance is observed at higher frequency in the infrared region. The latter mode is caused by division of the spin axis between the two sublattice magnetizations, as is the case for antiferromagnetic resonance. Since this mode stores exchange energy, we call it the *exchange mode*. In the case of the N-type ferrimagnet, the angular momenta of the two sublattice magnetizations are almost compensated near the compensation point. In such a case the resonance frequency of the exchange mode drops down into the microwave region.[36,37] Observation of this phenomenon in GdIG was in good agreement with theory.[38]

20.6 EQUATION OF MOTION FOR DOMAIN WALLS

It was first pointed out by Döring[39] in 1948 that a moving domain wall exhibits

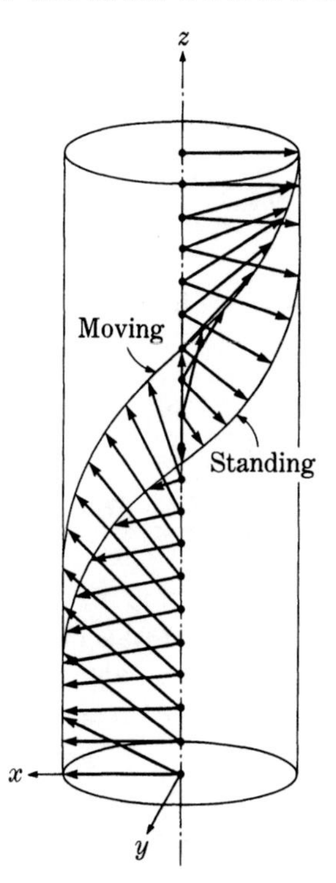

Fig. 20.27. Spin structure of standing and moving domain walls.

inertia, despite the absence of any mass displacement. In view of this property, the equation of motion for a 180° wall is

$$m\frac{d^2s}{dt^2}+\beta\frac{ds}{dt}+\alpha s=2I_sH, \tag{20.129}$$

where m is the mass of the wall per unit area, β is the damping coefficient, and α is the restoring coefficient as expressed by the second derivative of the wall energy (see (18.41)). The term on the right side of the equation represents the pressure acting on the 180° wall and should be replaced by $\sqrt{2}I_sH$ for a 90° wall.

The mass of a domain wall has its origin in the angular momenta of the spins forming the wall. In a Bloch wall, spins are confined in the plane (the x–y plane) perpendicular to the normal to the wall (the z-axis) (Fig. 20.27), if the wall is not moving. If a magnetic field is applied parallel to the x-direction, this field exerts a pressure on the wall in the z-direction and forces the spins in the wall to rotate clockwise (from the $+y$- to the $+x$-axis). This force, however, induces a precession motion of each spin, which results in a rotation of spins out of the x–y plane, and causes the appearance of magnetic free poles. The demagnetizing field produced by these free poles is given by

$$H_z=-\frac{I_z}{\mu_0}, \tag{20.130}$$

where I_z is the z-component of magnetization induced in the wall. This field acts on each spin so as to induce precession motion in the x–y plane, which results in a displacement of the wall in the z-direction. The rotational velocity of this precession is, from (20.91),

$$\frac{\mathrm{d}\varphi}{\mathrm{d}t} = \upsilon H_z = -\frac{\upsilon I_z}{\mu_0}. \tag{20.131}$$

This angular velocity can be related to the translational velocity of the wall, v, by

$$\frac{\mathrm{d}\varphi}{\mathrm{d}t} = \frac{\partial\varphi}{\partial z}\frac{\mathrm{d}z}{\mathrm{d}t} = -\frac{\partial\varphi}{\partial z}v. \tag{20.132}$$

Comparing these equations, we find

$$I_z = \frac{\mu_0}{\upsilon}\frac{\partial\varphi}{\partial z}v. \tag{20.133}$$

This means that the wall must have a z-component of magnetization in order to move with nonzero velocity. Furthermore, if the wall has a z-component of magnetization, the wall continues to move even without any external magnetic field. This phenomenon is described as the *inertia of the domain wall*. In this case the wall has an additional energy

$$\gamma_v = -\frac{1}{2}\int_{-\infty}^{\infty} I_z H_z \,\mathrm{d}z = \frac{\mu_0 v^2}{2\upsilon^2}\int_{-\infty}^{\infty}\left(\frac{\partial\varphi}{\partial z}\right)^2 \mathrm{d}z = \frac{\mu_0 v^2}{2\upsilon^2}\int_{-\pi/2}^{\pi/2}\left(\frac{\partial\varphi}{\partial z}\right)\mathrm{d}\varphi. \tag{20.134}$$

Since

$$\frac{\partial\varphi}{\partial z} = \frac{1}{\sqrt{A}}\sqrt{g(\varphi)}, \tag{20.135}$$

as shown in (16.39), (20.134) becomes

$$\gamma_v = \frac{\mu_0 v^2}{2\upsilon^2\sqrt{A}}\int_{-\pi/2}^{\pi/2}\sqrt{g(\varphi)}\,\mathrm{d}\varphi = \frac{\mu_0\gamma v^2}{4\upsilon^2 A}, \tag{20.136}$$

where γ is the wall energy per unit area (see (16.42)). This energy is proportional to v^2 and corresponds to the kinetic energy of the wall. We can express it as

$$\gamma_v = \tfrac{1}{2}mv^2, \tag{20.137}$$

where m is the *virtual mass of the wall* per unit area and is given, through comparison of (20.136) and (20.137), by

$$m = \frac{\mu_0\gamma}{2\upsilon^2 A}. \tag{20.138}$$

For a 180° wall in iron, with $\gamma = 1.6\times10^{-3}$, $A = 1.49\times10^{-11}$, $\upsilon = 2.21\times10^5$, and $\mu_0 = 4\pi\times10^{-7}$, we have

$$\begin{aligned} m &= \frac{(4\pi\times10^{-7})\times1.6\times10^{-3}}{2\times(2.21\times10^5)^2\times(1.49\times10^{-11})} \\ &= 1.38\times10^{-9}(\mathrm{kg\,m^{-2}})\ (=1.38\times10^{-10}\,\mathrm{g\,cm^{-2}}). \end{aligned} \tag{20.139}$$

Next we consider the damping term $\beta(\mathrm{d}s/\mathrm{d}t)$ in the equation of motion (20.129). If we neglect the first and third terms, (20.129) becomes

$$\frac{\mathrm{d}s}{\mathrm{d}t} = \frac{2I_s}{\beta} H. \tag{20.140}$$

For eddy current damping we can compare (20.140) with (20.63), and we have

$$\beta = \frac{4I_s^2 R \ln(r_0/R)}{\rho}. \tag{20.141}$$

For iron, with $I_s = 2.15\,(\mathrm{T})$, $\rho = 1 \times 10^{-7}$, and assuming $\ln(r_0/R) \approx 1$, we have

$$\beta = \frac{4 \times 2.15^2}{1 \times 10^{-7}} R \approx 1.8 \times 10^8 R. \tag{20.142}$$

Even if we assume that the radius is as small as $R = 1 \times 10^{-6}$ $(= 1\ \mu\mathrm{m})$, we have $\beta = 1.8 \times 10^2$.

For highly insulating materials such as ferrites, the origin of the wall damping is the same as the damping acting on the precession motion of the magnetization. Formally it can be expressed in terms of the relaxation frequency λ which appears in the Landau–Lifshitz equation (20.84). The z-component of (20.84) gives

$$\frac{\mathrm{d}I_z}{\mathrm{d}t} = 4\pi\lambda\mu_0 H_z. \tag{20.143}$$

Thus the internal magnetic field does, in one second, a quantity of work given by

$$H_z \frac{\mathrm{d}I_z}{\mathrm{d}t} = 4\pi\lambda\mu_0 H_z^2, \tag{20.144}$$

which is dissipated as heat. The power loss per unit area of the wall is then calculated to be

$$P_w = 4\pi\lambda\mu_0 \int_{-\infty}^{\infty} H_z^2\,\mathrm{d}z = \frac{4\pi\mu_0\lambda v^2}{\nu^2} \int_{-\infty}^{\infty} \left(\frac{\mathrm{d}\varphi}{\mathrm{d}z}\right)^2 \mathrm{d}z = \frac{2\pi\mu_0\lambda\gamma v^2}{\nu^2 A}. \tag{20.145}$$

The external field supplies the power to the travelling wall which is given, using (20.140), by

$$P_w = 2I_s H v = \beta v^2. \tag{20.146}$$

On comparing (20.145) and (20.146), we have

$$\beta = \frac{2\pi\mu_0\lambda\gamma}{\nu^2 A}. \tag{20.147}$$

For Ni-ferrite with $\lambda = 3.9 \times 10^7$ Hz, $\gamma = 4.75 \times 10^{-4}$, $A = 9.0 \times 10^{-12}$, $\nu = 2.21 \times 10^5$, and $\mu_0 = 4\pi \times 10^{-7}$, we have

$$\beta = \frac{2\pi(4\pi \times 10^{-7}) \times (3.9 \times 10^7) \times (4.75 \times 10^{-4})}{(2.21 \times 10^5)^2 \times (9.0 \times 10^{-12})} = 0.33. \tag{20.148}$$

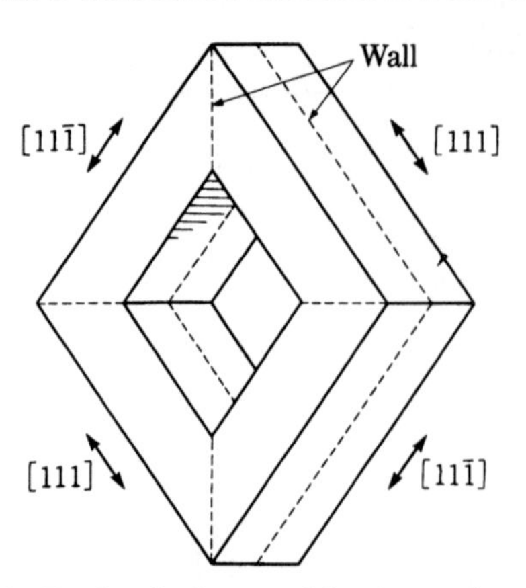

Fig. 20.28. Ferrite single crystal in picture-frame shape.

This value is much smaller than the value obtained for the eddy currents in a very thin iron wire. Galt[40] measured the velocity of a domain wall in single crystals of magnetite and Ni-ferrite, cut in a picture-frame shape as shown in Fig. 20.28, and confirmed the validity of (20.61). He found the values of c at room temperature to be $0.24\,\mathrm{m^2\,s^{-1}A^{-1}}$ ($=1900\ \mathrm{cm\,s^{-1}\,Oe^{-1}}$) for magnetite and $2.5\,\mathrm{m^2\,s^{-1}A^{-1}}$ ($=20\,000\,\mathrm{cm\,s^{-1}\,Oe^{-1}}$) for Ni-ferrite. On comparing (20.61) and (20.140), we have

$$c = \frac{2I_s}{\beta}. \tag{20.149}$$

Using this relationship, we can determine $\beta = 4.84$ for magnetite and $\beta = 0.26$ for Ni-ferrite. The latter value is of the same order of magnitude as the value of (20.148), which was deduced from the relaxation frequency. The value of β for magnetite is as large as 4.06 even after the effect of eddy currents is subtracted from the observed value. The reason for this large value of β is considered to be the hopping motion of electrons between Fe^{2+} and Fe^{3+} ions in the 16d sites associated with the wall displacement.

In the preceding discussion, the wall velocity is determined by the demagnetizing field inside the wall which is produced by the z-component of magnetization. Therefore, when the z-component of magnetization, I_z, reaches the saturation magnetization, I_s, the wall velocity is expected to saturate and then decrease with an increase in the external magnetic field.[41–43] Actually this characteristic was observed for a garnet crystal film with very low magnetic losses. Figure 20.29 shows the wall velocity as observed in the shrinking of bubble domains (Section 17.3). Curve (a), observed for a normal bubble, shows that the velocity exhibits a maximum and then decreases with increasing external magnetic field. On the other hand, curve (b), observed for a hard bubble in which spin rotation is not so easy as in a normal bubble, shows a monotonic increase in velocity with increasing field.

The third term in (20.129) signifies the restoring force acting on the wall. The origin

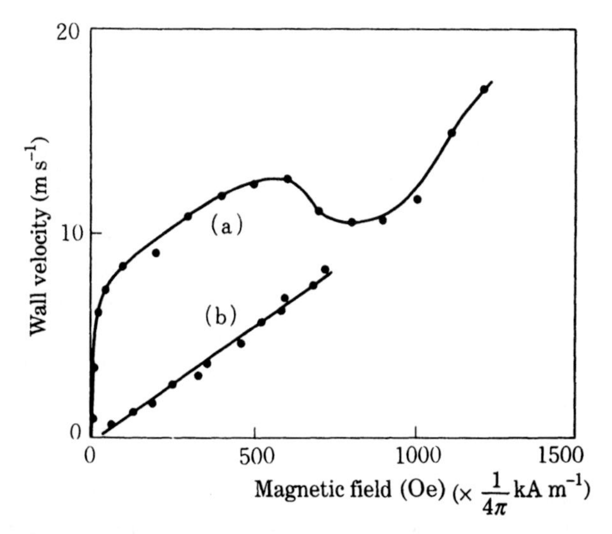

Fig. 20.29. Domain wall velocity of a shrinking bubble as a function of magnetic field, as observed for a garnet crystal film with very low magnetic losses: (a) normal bubble; (b) hard bubble.[44]

of this force is given in Section 18.2 for several examples. If we neglect the second term and the right-hand term, the equation of motion (20.129) becomes

$$m\frac{d^2s}{dt^2} = -\alpha s. \tag{20.150}$$

On solving this differential equation, we have

$$s = s_0 \sin \omega_0 t, \tag{20.151}$$

here

$$\omega_0 = \sqrt{\frac{\alpha}{m}}. \tag{20.152}$$

This means that the wall will oscillate about an equilibrium position. The factor α is related to the permeability μ by (18.36) ($\mu \approx \chi_a$) to give

$$\alpha = \frac{4I_s^2}{3\mu}S. \tag{20.153}$$

The mass m is given by (20.138), where γ is expressed, from the comparison of (16.46) and (16.47), as

$$\gamma = 4\pi\frac{A}{\delta}, \tag{20.154}$$

so that the mass becomes

$$m = \frac{2\pi\mu_0}{\nu^2\delta}. \tag{20.155}$$

On putting (20.153) and (20.155) into (20.152), we obtain

$$\omega_0\sqrt{\bar{\mu}} = 0.46\sqrt{S\delta}\,\nu I_s/\mu_0. \tag{20.156}$$

This relationship is similar to formula (20.79) for the natural resonance derived on the basis of magnetization rotation, except for the factor $\sqrt{S\delta}$. Since S is the total area and δ is the thickness, the quantity $S\delta$ is the fractional volume of wall per unit volume of the specimen. For instance, if the spacing of the walls is 0.1 mm, it turns out that $S\delta \approx 10^{-4}$ for $\delta = 10^{-5}$ mm; hence the wall resonance frequency is expected to be lower by a factor 10^2 than in the case of natural resonance. The reason is that for a given amplitude of magnetization, the angular velocity of spins in a wall is much larger than that for magnetization rotation. In order that the magnetization changes by I_s per second, magnetization rotation requires an angular velocity $(\pi/2)\,\mathrm{rad\,s^{-1}}$, while wall displacement requires $\pi/(2S\delta)\,\mathrm{rad\,s^{-1}}$. For the example mentioned above, the angular velocity is 10^4 times larger for wall displacement than for magnetization rotation.

By the same reasoning, the power loss should also be larger for wall displacement than for magnetization rotation. Since the power loss in the wall is proportional to ω^2 or $(S\delta)^{-2}$, the total loss is expected to be proportional to $(S\delta)^{-1}$. For a given amplitude of magnetic field, therefore, wall displacement is more rapidly damped than is magnetization rotation. Thus magnetization rotation becomes more important at high frequencies. However, note that these differences between wall displacement and rotation depend on the value of $S\delta$. If this value could be increased to approach 1, wall displacement could survive to high frequencies. This would require that the wall spacing be very small, so that the specimen would be filled with domain walls and the wall displacement would become practically equivalent to incoherent magnetization rotation. On the other hand, if $S\delta$ is very small, the difference between the two mechanisms would be large. It was observed by Galt[40] that the relative permeability of a frame-shaped single crystal of magnetite drops from 5000 to 1000 between 1 kHz and 10 kHz. One of the reasons for this is the high original value of permeability, but also it is partly because $S\delta$ is very small (10^{-6}), since the specimen contains only one domain wall.

PROBLEMS

20.1 After magnetizing a single crystal of carbon–iron parallel to [010] for a long time to its saturation, the magnetization is switched to [100]. Solve for the time change in the anisotropy energy, assuming diffusion after-effect and also $l_C \ll kT$.

20.2 Suppose that an aggregate of aligned, long, fine magnetic particles with distributed sizes is magnetized parallel to the aligned axis and is then subjected to a field which points in the negative direction. How fast is the magnetization change if the intensity of the field is maintained at 89% of the critical field, as compared to the case when the field is maintained at 90% of the critical field? Assume the occurrence of the thermal fluctuation after-effect.

20.3 Suppose that a plane 180° wall is travelling with velocity v from the top surface to the bottom surface of an infinitely wide plate of thickness d which is made from a ferromagnetic metal with resistivity ρ and saturation magnetization I_s. Calculate the eddy current loss per unit area of the plate as a function of the distance, z, of the wall from the top surface of the plate.

20.4 A spherical sample with saturation magnetization $I_s = 1\,\mathrm{T}$ and relaxation frequency $\lambda = 1 \times 10^8\,\mathrm{Hz}$ is magnetized along the z-axis ($\theta = 0$). A uniform magnetic field $H = -1 \times 10^2\,\mathrm{A\,m^{-1}}$ is applied in the $-z$-direction. Find the time required to rotate the magnetization from $\theta = 60°$ to $\theta = 120°$, and the number of precession rotations that occur in this time interval. Assume $g = 2$.

20.5 How far does a 180° wall, initially travelling at $10\,\mathrm{m\,s^{-1}}$ move after the magnetic field is switched off? Assume the relaxation frequency is 1×10^8 Hz.

REFERENCES

1. Y. Tomono, *J. Phys. Soc. Japan*, **7** (1952), 174, 180.
2. J. L. Snoek, *Physica*, **5** (1938), 663; *New development in ferromagnetic materials* (Elsevier, Amsterdam, 1949), §16, p. 46.
3. R. Becker and W. Döring, *Ferromagnetismus* (Springer, Berlin, 1939), p. 254.
4. G. Richter, *Ann. Physik*, **29** (1937), 605.
5. R. Street and J. C. Wooley, *Proc. Phys. Soc.*, **A62** (1949), 562.
6. L. Néel, *J. Phys. Radium*, **13** (1952), 249.
7. L. Néel, *Ann. Geophys.*, **5** (1949), 99; *Compt. Rend.*, **228** (1949), 664; *Rev. Mod. Phys.*, **25** (1953), 293.
8. C. P. Bean, *J. Appl. Phys.*, **26** (1955), 1381.
9. J. H. Phillips, J. C. Wooley, and R. Street, *Proc. Phys. Soc.*, **B68** (1955), 345.
10. L. Néel, *J. Phys. Radium*, **11** (1950), 49.
11. K. Ohashi, H. Tsuji, S. Tsunashima, and S. Uchiyama, *Jap. J. Appl. Phys.*, **19** (1980), 1333.
12. L. Néel, *J. Phys. Radium*, **12** (1951), 339.
13. J. L. Snoek, *New developments in ferromagnetic materials* (Elsevier, Amsterdam, 1949), §17, p. 54.
14. K. Ohta, *J. Phys. Soc. Japan*, **16** (1961), 250.
15. K. Ohta and T. Yamadaya, *J. Phys. Soc. Japan*, **17**, Suppl. B-I (1962), 291.
16. A. Yanase, *J. Phys. Soc. Japan*, **17** (1962), 1005.
17. R. M. Bozorth, *Ferromagnetism* (Van Nostrand, N.Y., 1951) Chap. 17.
18. H. J. Williams, W. Shockley and C. Kittel, *Phys. Rev.*, **80** (1950) 1090.
19. H. P. J. Wijn and H. van der Heide, *Rev. Mod. Phys.*, **25** (1953), 98.
20. F. G. Brockman, P. H. Dowling, and W. G. Steneck, *Phys. Rev.*, **77** (1950), 85.
21. E. W. Gorter, *Proc. IRE*, **43** (1955), 245.
22. J. L. Snoek, *Physica*, **14** (1948), 207.
23. G. H. Jonker, H. P. J. Wijn, and P. B. Brawn, *Philips Tech. Rev.*, **18** (1956–57), 145.
24. L. Landau and E. Lifshitz, *Phys. Z. Sowjetunion*, **8** (1935), 153; E. Lifshitz, *J. Phys. USSR*, **8** (1944), 337.
25. C. Kittel, *Phys. Rev.*, **80** (1950), 918.
26. T. L. Gilbert and J. M. Kelley, *Proc. 1st 3M Conf.* (1955), 253; T. L. Gilbert, *Phys. Rev.*, **100** (955), 1243.
27. J. F. Dillon, Jr, *Bull Am. Phys. Soc.*, Ser. II, **1** (1956), 125.

28. L. R. Walker, *Phys. Rev.*, **105** (1957), 390; *J. Appl. Phys.*, **29** (1958), 318.
29. M. H. Seavy, Jr., and P. E. Tannenwald, *Phys. Rev. Lett.*, **1** (1958), 168; *J. Appl. Phys.*, **30** (1959), 227S.
30. C. Kittel, *Phys. Rev.*, **110** (1958), 1295.
31. T. Nagamiya, *Progr. Theoret. Phys.* (Kyoto), **6** (1951), 342.
32. C. Kittel, *Phys. Rev.*, **82** (1951), 565.
33. K. Yosida, *Progr. Theoret. Phys.* (Kyoto), **7** (1952), 25, 425.
34. N. Tsuya, *Progr. Theoret. Phys.*, **7** (1953), 263.
35. R. K. Wangsness, *Phys. Rev.*, **93** (1954), 68.
36. R. K. Wangsness, *Phys. Rev.*, **97** (1955), 831.
37. T. R. McGuire, *Phys. Rev.*, **97** (1955), 831.
38. S. Gschwind and L. R. Walker, *J. Appl. Phys.*, **30** (1959), 163S.
39. W. Döring, *Z. Naturforsch.*, **3a** (1948), 373.
40. J. K. Galt, *Phys. Rev.*, **85** (1952), 664; *Rev. Mod. Phys.*, **25** (1953), 93; *Bell Sys. Tech. J.*, **33** (1954), 1023.
41. J. C. Slonczewski, *Int. J. Mag.*, **2** (1972), 85; *J. Appl. Phys.*, **45** (1974), 2705.
42. N. L. Schryer and L. R. Walker, *J. Appl. Phys.*, **45** (1974), 5406.
43. Y. Okabe, T. Toyooka, M. Takigawa, and T. Sugano, *Jap. J. Appl. Phys.*, **15 Suppl** (1976), 101.
44. S. Konoshi, K. Mizuno, F. Watanabe, and K. Narita, *AIP Conf. Proc.*, **No. 34** (1976), 145.

Part VIII

ASSOCIATED PHENOMENA AND ENGINEERING APPLICATIONS

In this part, we discuss various phenomena – thermal, electrical and optical – associated with magnetization. The engineering applications of magnetic materials are also summarized.

21

VARIOUS PHENOMENA ASSOCIATED WITH MAGNETIZATION

21.1 MAGNETOTHERMAL EFFECTS

In magnetic materials, the spins can undergo thermal motion. Therefore magnetic materials exhibit an additional or 'anomalous' specific heat in addition to the normal specific heat caused by the thermal motion of the crystal lattice (see Section 6.1). In other words, magnetic materials have additional entropy due to the presence of spins. This entropy is controllable by an external magnetic field.

Adiabatic demagnetization is a method to create a low temperature by manipulation of this extra entropy. Consider a paramagnetic material composed of a number of spins. For simplicity, we assume an *Ising model* (see Section 6.2) containing N spins in a unit volume. In this model, each spin magnetic moment (simply called 'spin' hereafter) can take either a + or − direction. When a magnetic field H is applied in the + direction, the number of + spins, N_+, increases, while the number of − spins, N_-, decreases, thus the magnetization

$$I = M(N_+ - N_-) \tag{21.1}$$

increases, where M is the magnetic moment of one spin and $N_+ + N_- = N$. The entropy S is generally given by $k \ln W$, where k is the Boltzmann factor and W is the number of different ways of realizing the state (in this case, magnetization). When N_+ and N_- are given, the number of ways of realizing this combination is expressed by

$$W = N!/(N_+!N_-!). \tag{21.2}$$

Using Stirling's formula $\ln(n!) = n \ln \mathrm{n} - n$, we have the entropy S for a given I,

$$\begin{aligned} S(I) &= k \ln W \\ &= k\{N \ln N - N_+ \ln N_+ - N_- \ln N_-\}. \end{aligned} \tag{21.3}$$

When $H = 0$, $I = 0$, or $N_+ = N_- = N/2$, so that the entropy becomes

$$S(0) = k\{N \ln N - N \ln(N/2)\} = kN \ln 2. \tag{21.4}$$

On the other hand, when H increases and the magnetization saturates at I_s, $N_- = 0$ and $N_+ = N$, so that the entropy becomes

$$S(I_s) = 0. \tag{21.5}$$

Thus the entropy is reduced by the application of a magnetic field.

First a strong magnetic field, H, is applied to a paramagnet, which magnetizes to the intensity, I, keeping the temperature constant at T (isothermal process). The work done by the field must be converted to heat, Q:

$$Q = \int_0^I H\,\mathrm{d}I, \tag{21.6}$$

which is transferred to the heat reservoir to keep the temperature constant. Then the paramagnet is thermally isolated from the heat reservoir, and the field is reduced (adiabatic process). In an adiabatic process, the heat change is zero, so

$$\delta Q = T\delta S = 0, \tag{21.7}$$

where δS is the entropy change. Therefore the entropy must be kept constant during adiabatic demagnetization. In order to keep the same entropy, the magnetization, I, must be held constant (see (21.3)). Since the magnetization of a paramagnet is a unique function of H/T (see Fig. 5.9 or Fig. 5.10), the temperature T decreases as the field H is reduced. This is the principle of adiabatic demagnetization.

Using a paramagnetic material containing a dilute concentration of magnetic ions, such as CrK alum, a temperature as low as 0.01 K can be attained by adiabatic demagnetization. The approach to low temperatures is limited by the interaction between magnetic ions. Super-low temperatures in the range of μK ($= 10^{-6}$ K) have been obtained by means of nuclear adiabatic demagnetization, using nuclear magnetic moments with very small interactions.

In the case of a ferromagnet, the spin entropy contributes to the *magnetocaloric effect*. At magnetization levels approaching saturation, the high field susceptibility, χ_0, is given by (18.133). Suppose that the magnetization is increased by δI under the application of a magnetic field H. The work δW done by the field (per unit volume) is given by

$$\delta W = H\,\delta I. \tag{21.8}$$

On the other hand, the internal energy is changed by

$$\delta E = -wI\,\delta I \tag{21.9}$$

(see (6.22)). Therefore the heat generated in a unit volume is given by

$$\delta Q = \delta W - \delta E = (H + wI)\,\delta I. \tag{21.10}$$

At $T \ll \Theta_f$, the molecular field wI is much larger than H, so that neglecting H in (21.10) we have

$$\delta Q = wI\,\delta I = \tfrac{1}{2}w\,\delta(I^2). \tag{21.11}$$

At $T > \Theta_f$, wI and H are comparable in magnitude, so that H cannot be neglected. In this temperature range the Curie–Weiss law (6.15) holds, so that we have

$$I = \chi H = \frac{NM^2H}{3k(T - \Theta_f)}. \tag{21.12}$$

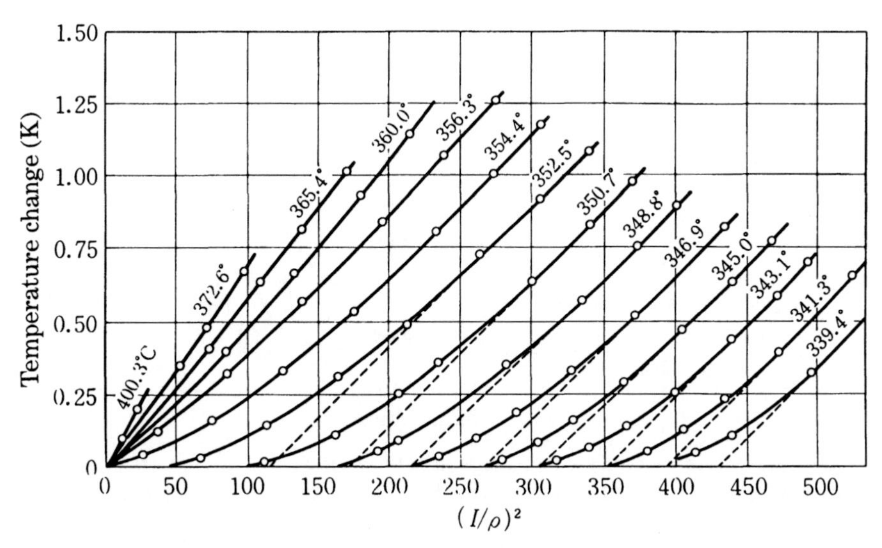

Fig. 21.1. Temperature change caused by magnetization in Ni at temperatures near the Curie point as a function of the square of magnetization/density (ρ: density). (Experiment by Weiss and Forrer,[1] after Becker and Döring[2])

Using the expression for the Curie point (6.8), (21.12) is converted to the form

$$H = \frac{T - \Theta_f}{\Theta_f} wI. \tag{21.13}$$

Using this in (21.10), we obtain

$$\delta Q = \frac{T}{\Theta_f} wI\,\delta I = \frac{T}{\Theta_f}\,\tfrac{1}{2} w\,\delta(I^2). \tag{21.14}$$

The temperature change is given by

$$\delta T = \frac{1}{C_I}\,\delta Q, \tag{21.15}$$

where C_I is the specific heat for constant I. It is predicted from (21.11) and (21.14) that the temperature change is proportional to I^2 below and above the Curie point. Figure 21.1 shows the magnetocaloric effect observed for Ni below and above its Curie point of 360°C. It is seen that the proportionality factor between δT and $\delta(I^2)$ is constant below 360°C, while it increases with increasing T, as expected from (21.11) and (21.14).

If we express (21.11) in terms of H, it becomes

$$\delta Q = w\chi_0 I\,\delta H = \frac{3\Theta_f L'(\alpha)}{T - 3\Theta_f L'(\alpha)}\, I\,\delta H \tag{21.16}$$

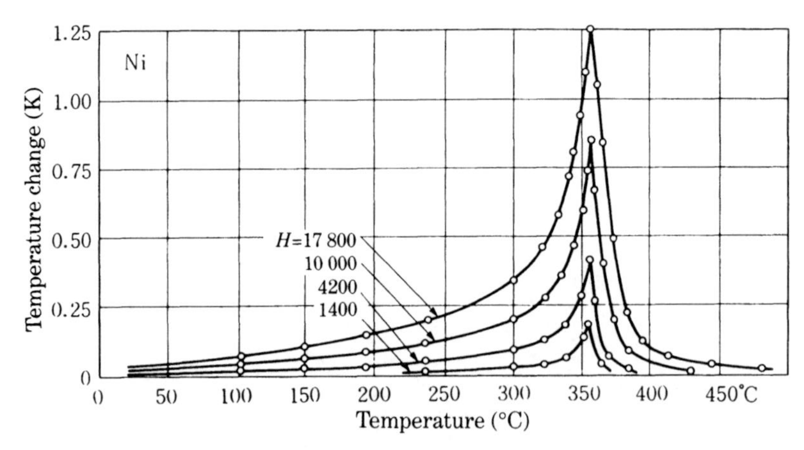

Fig. 21.2. Temperature dependence of magnetocaloric effect in Ni. Numerical values in the figure are fields in Oe or $(10^3/4\pi)$ $(\mathrm{A\,m^{-1}})$. (Experiment by Weiss and Forrer,[1] after Bozorth[3])

(see (6.15)). Since $L'(\alpha)$ approaches zero more rapidly than T does as $T \to 0$, δQ given by (21.16) becomes very small at low temperatures. On the other hand, at $T > \Theta_f$, (21.14) becomes

$$\delta Q = \frac{T}{\Theta_f} \tfrac{1}{2} w \chi_0^2 \, \delta(H^2) = \frac{T\Theta_f}{(T-\Theta_f)^2} \frac{1}{2w} \delta(H^2). \tag{21.17}$$

At very high temperatures, the denominator becomes larger more rapidly than the numerator as the temperature increases, so that δQ becomes very small again. The prediction is thus that δQ has a maximum at the Curie point and decreases with either decreasing or increasing temperature. Figure 21.2 shows the temperature dependence of the magnetocaloric effect as measured for Ni, showing the expected behavior. The curves are quite similar to the temperature dependence of the high field susceptibility (see Fig. 6.4). The temperature change at the Curie point is 0.85 K for Ni and 2 K for Fe at $H = 0.6\,\mathrm{MA\,m^{-1}}$ (= 8000 Oe).

The temperature of a ferromagnet is also changed by the generation of heat associated with technical magnetization processes. The temperature change due to this magnetothermal effect is the order of 0.001 K, much smaller than the magnetocaloric effect. The temperature change due to this magnetothermal effect is partly irreversible, so that the temperature will increase continuously if the material is cycled around the hysteresis loop. This corresponds to the heat generated by the hysteresis loss. This phenomenon has been investigated by a number of workers[4-6] for carbon steel, nickel and pure iron. Detailed investigations were made by Bates and his collaborators[7] for Fe, Co, and Ni. Investigations were also extended to ferrites[8] and permanent magnets.[9]

Figure 21.3 shows the temperature changes observed for Ni as it is cycled around the hysteresis loop.[7] In addition to the reversible change in temperature (magnetocaloric effect) occurring at high fields, an irreversible increase in temperature is

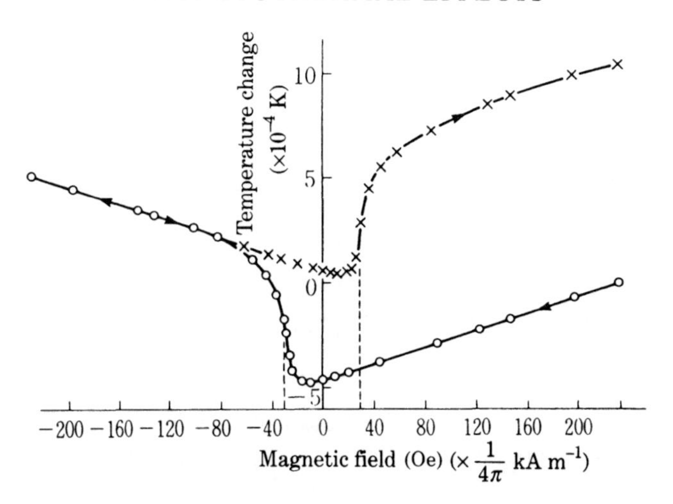

Fig. 21.3. Magnetothermal effect in Ni as a function of the applied field. (After Townsend[5])

observed at the coercive field. Figure 21.4 shows the magnetothermal effect observed for Fe. Reversible changes in temperature are observed not only at high fields but also in very low-field regions near the origin. This heat change is caused by magnetization rotation against the magnetocrystalline anisotropy. The work done by the magnetic field is mainly used to raise the anisotropy energy, and the rotation of magnetization results in a decrease in anisotropy field, which, like a decrease in molecular field, opens up the angular distribution of spins, increasing the entropy of the spin system and resulting in an absorption of heat. Since the anisotropy field is changed by $\Delta H = -10K_1/3I_s$ (see (12.57)), during the magnetization rotation from

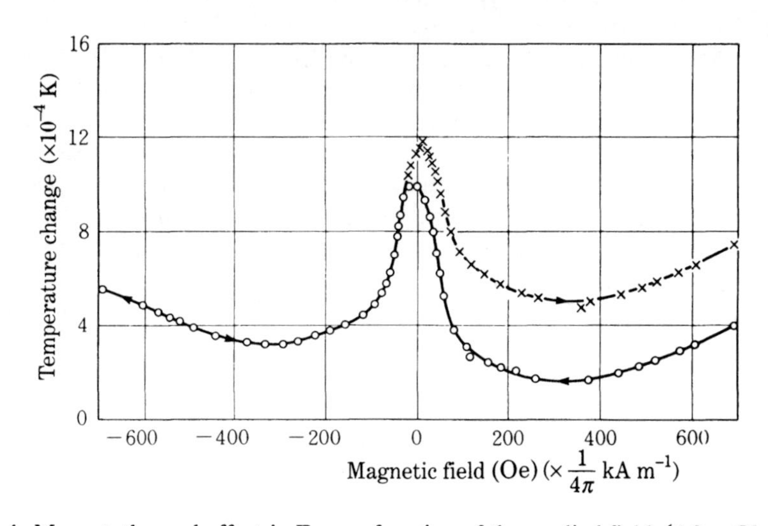

Fig. 21.4. Magnetothermal effect in Fe as a function of the applied field. (After Okamura[6])

$\langle 100 \rangle$ to $\langle 111 \rangle$, the heat generated during this process is calculated from (21.16) to be

$$\delta Q = -\frac{10\Theta_f L'(\alpha)}{T - 3\Theta_f L'(\alpha)} K_1. \tag{21.18}$$

When $K_1 < 0$, the magnetization rotates from $\langle 111 \rangle$ to $\langle 100 \rangle$, so that δQ is still negative. For iron, $K_1 = 4.2 \times 10^4$, $\Theta_f = 1043\,\mathrm{K}$, $T = 300\,\mathrm{K}$, $L'(\alpha) \approx 0.0025$ and $C = 3.5 \times 10^6\,\mathrm{J\,deg^{-1}\,m^{-3}}$, so that the temperature change is calculated to be

$$\begin{aligned}\delta T = \frac{\delta Q}{C} &= -\frac{10 \times 1043 \times 0.0025}{300 - 3 \times 1043 \times 0.0025} \times \frac{4.2 \times 10^4}{3.5 \times 10^6} \\ &= -1.07 \times 10^{-3}\,\mathrm{K},\end{aligned} \tag{21.19}$$

which is in good agreement with the heat absorption seen in Fig. 21.4. Experimental separation of the reversible and irreversible temperature changes was attempted by Bates and Sherry.[10]

21.2 MAGNETOELECTRIC EFFECTS

There are various magnetoelectric effects, including magnetoresistance, magnetic tunneling, the magnetothermoelectric effect, Hall effect, ME effect, etc.

There are two kinds of *magnetoresistance effect*: One is the dependence of resistivity on the magnitude of spontaneous magnetization, and the other is the dependence of resistivity on the orientation of the magnetization (*anisotropic magnetoresistance effect*). The first case causes a temperature dependence of resistivity in ferromagnetic metals resulting from the temperature change in spontaneous magnetization. Figure 21.5 compares the temperature dependence of the reduced resistivity of Ni and Pd. Pd has ten electrons outside the krypton shell, and in this sense is quite similar to Ni which has ten electrons outside the argon shell. In Fig. 21.5 the ordinate scales are adjusted so that the resistivities are equal at the Curie point of Ni. It is seen that the resistivity of Ni is considerably reduced by the appearance of spontaneous magnetization below its Curie point. The same effect is also observed in the forced increase of spontaneous magnetization by the application of high magnetic fields. Figure 21.6 shows the field dependence of resistivity of Ni parallel and perpendicular to the magnetic field, as measured by Englert.[12] It is seen that irrespective of the direction of the applied field with respect to the current, the resistivity always decreases with an increase in spontaneous magnetization.

Mott[13] interpreted these phenomena in terms of the scattering probability of the conduction electrons into $3d$ holes. If the material is ferromagnetic, half of the $3d$ shell is filled, so that the scattering of $4s$ electrons into the plus (magnetic) spin state is forbidden. This scattering is, however, permitted in a nonmagnetic metal, in which both the plus and minus spin states of the upper $3d$ levels are vacant. Mott explained the temperature variation of resistivity fairly well by this picture.

Kasuya[14] interpreted these phenomena by a picture quite different from the Mott theory. He considered that d electrons are localized at the lattice sites and interact

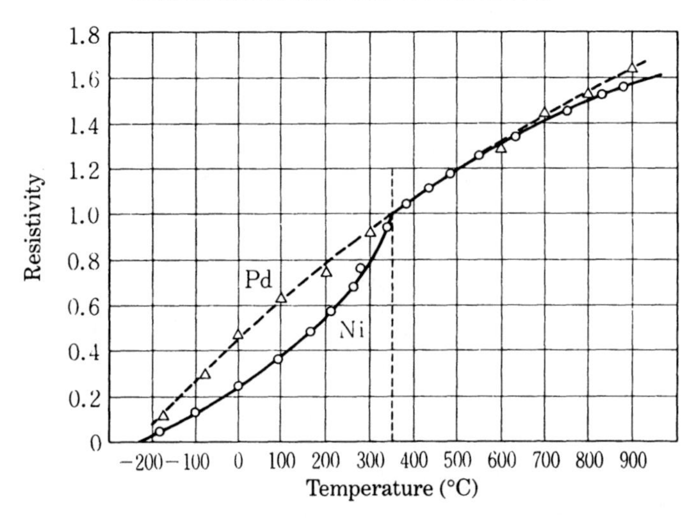

Fig. 21.5. Comparison of reduced resistivity of Ni and Pd. (After Becker and Döring[11])

with conduction electrons through the exchange interaction. At 0 K, the potential for conduction electrons is periodic, because the spins of $3d$ electrons on all the lattice sites point in the same direction. At nonzero temperatures, the spins of the $3d$ electrons are thermally agitated and the thermal motion breaks the periodicity of the potential. The $4s$ electrons are scattered by this irregularity of the periodic potential, resulting in an additional resistivity. Kasuya postulated that the temperature dependence of resistivity of ferromagnetic metals is composed of a monotonically increasing part due to lattice vibration and an anomalous part due to magnetic scattering, the magnitude of the latter being explained by his theory. In rare earth metals, in which the $4f$ electrons are responsible for the atomic magnetic moment, the conducting $6s$

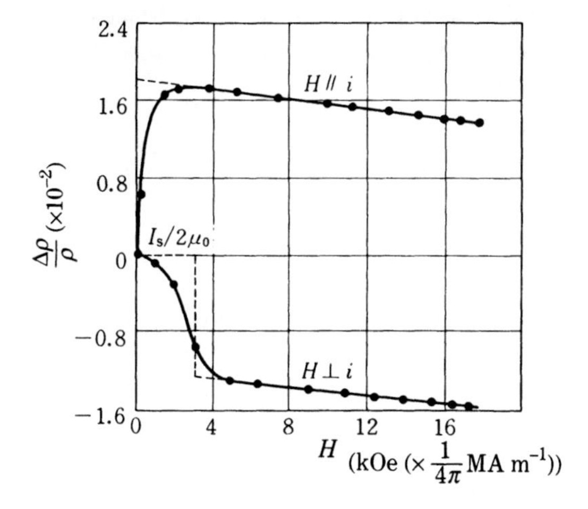

Fig. 21.6. Variation of resistivity of Ni as a function of magnetic field.[12]

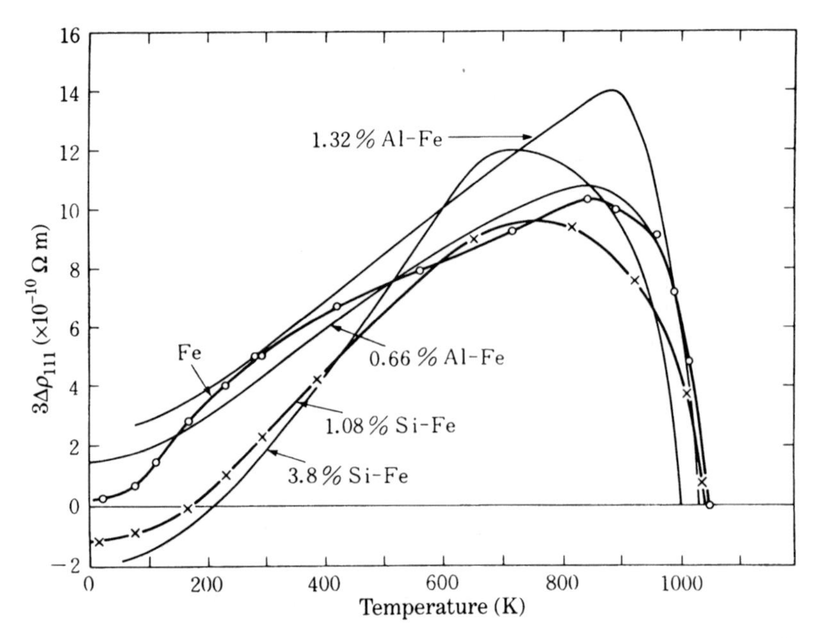

Fig. 21.7. Temperature dependence of $3\Delta\rho_{111}$ for Fe and Fe–Al alloys.[16]

electrons are considered to be scattered by the irregularity in the polarization of $4f$ electrons. Actually the anomalous part of the resistivity measured for a number of rare earth metals is proportional to the square of the magnitude of the spin magnetic moment,[15] in accordance with the Kasuya theory. Since the $4f$ electrons are completely localized, the Mott theory is considered to be invalid, at least for rare earth metals.

The second case of magnetoresistance, the anisotropic magnetoresistance effect, can be described in a similar way to magnetostriction, because both quantities depend not on the sense but only on the direction of spontaneous magnetization. The change in resistivity is described as

$$\frac{\delta\rho}{\rho} = \frac{3}{2}\frac{\Delta\rho_{100}}{\rho}\left(\alpha_1^2\beta_1^2 + \alpha_2^2\beta_2^2 + \alpha_3^2\beta_3^2 - \tfrac{1}{3}\right) + 3\frac{\Delta\rho_{111}}{\rho}(\alpha_1\alpha_2\beta_1\beta_2 + \alpha_2\alpha_3\beta_2\beta_3 + \alpha_3\alpha_1\beta_3\beta_1), \tag{21.20}$$

where $(\alpha_1, \alpha_2, \alpha_3)$ and $(\beta_1, \beta_2, \beta_3)$ are the direction cosines of spontaneous magnetization and electric current, respectively. These coefficients have been measured in Fe and Ni single crystals by a number of investigators. The numerical values are in the range 0.1–4%. Figure 21.7 shows the temperature dependence of $\Delta\rho_{111}$ as measured for Fe and Fe–Al alloys by Tatsumoto *et al.*[16]

The anisotropic magnetoresistance effect was explained by Kondo[17] in terms of the scattering of s electrons by a small unquenched orbital moment induced by $3d$ spin

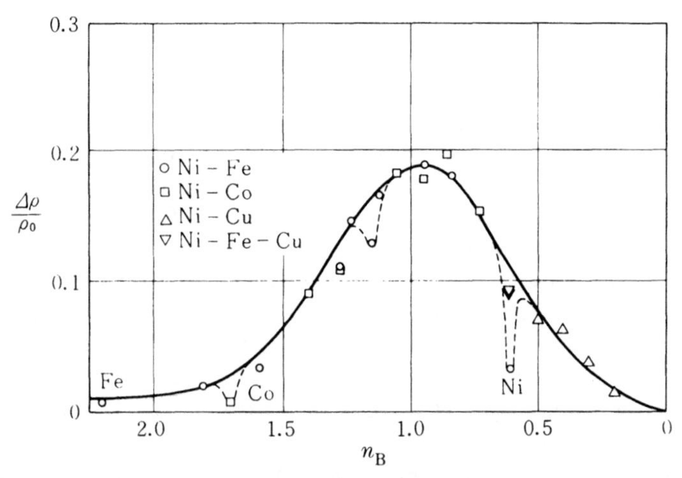

Fig. 21.8. Magnetoresistance effect as a function of the mean number of Bohr magnetons per atom observed for various magnetic $3d$ transition alloys. (Smit[18], after Jan[19])

magnetic moment. Figure 21.8 shows the average $\Delta\rho$ measured for various magnetic $3d$ alloys as a function of spontaneous magnetization. This curve is quite similar to the dependence of the gyromagnetic ratio on the number of $3d$ electrons measured for various magnetic $3d$ alloys.[20] Since the deviation of the gyromagnetic ratio from 2 is a good measure of the remaining orbital moment, this fact proves the validity of the Kondo theory. In fact, Kondo explained the magnitude and the temperature dependence of this effect satisfactorily.

The magnetoresistance effect is particularly large for magnetic semiconductors such as chalcogenide spinels (see Chapter 10). For instance, in In-doped $CdCr_2Se_4$[21] or $HgCr_2Se_4$,[22] the $\Delta\rho/\rho$ values amount to -100% near their Curie points (110–130 K). Another magnetic semiconductor, $Eu_{1-x}Gd_xSe$, exhibits very large temperature dependence of resistivity and also a very large magnetoresistance, the origin of which has been discussed by a number of investigators.[23-26]

Recently a *giant magnetoresistance effect* has been found in multilayer magnetic thin films. This phenomenon was first discovered by French scientists[27] in 1988 in multilayer Fe–Cr films, which are composed of alternating thin layers of Fe and Cr. They found that the resistance of the film drops to about half of its original value upon the application of a magnetic field, irrespective of the angle between the current and the field. The origin of this phenomenon is that the magnetizations of the Fe layers on opposite sides of a Cr layer are aligned antiparallel as a result of a negative exchange interaction through the Cr layer, and the spin-dependent magnetic scattering of conduction electrons is reduced by the application of a magnetic field which causes parallel alignment of the magnetization in the Fe layers. Similar phenomena have been observed for various combinations of ferromagnetic and antiferromagnetic metal multilayers.

Magnetic tunneling is observed for two ferromagnetic metal layers separated by a thin insulating layer about 1 nm thick. It was observed that the tunneling electric current between two metal layers depends on the relative angle between the magnetization in the two metal layers. This phenomenon was first observed by Julliere[28] in an Fe–Ge–Co junction in 1975. Maekawa and Garvert[29] observed hysteresis in tunneling resistance versus magnetic field for Ni–NiO–(Ni, Co, Fe) junctions. Slonczewski[30] showed theoretically that the tunneling electric conductance varies as $\cos\theta$, where θ is the angle between magnetizations in the metal layers. Miyazaki *et al.*[31] observed a change in tunneling resistance for an 82Ni–Fe/Al–Al_2O_3/Co junction as large as 3.5% at 77 K and 2.7% at room temperature.

The thermoelectric power of ferromagnetic metals also depends on magnetization. This is called the *magnetothermoelectric effect*. This effect is observed to exhibit an anomalous temperature dependence at the Curie point.[32] The anisotropic effect can be described in a similar way to the magnetostriction or magnetoresistance effect. The magnitude of this effect is of the order of $1\,\mu\mathrm{V\,K^{-1}}$ for Fe and Ni, and has a maximum at room temperature.

When a magnetic field H is applied to a ferromagnetic metal perpendicular to the electric current, an electric field is produced perpendicular to both the current and the magnetic field. This is the *Hall effect*. In the case of ferromagnetic metals, it was found[33] that the Hall electric field per unit current density depends not only on the magnetic field H but also on the magnetization I, so that it can be expressed by

$$\varepsilon = R_0 H + R_1 \frac{I}{\mu_0}, \tag{21.21}$$

where R_0 is the *ordinary Hall coefficient* and R_1 is the *extraordinary Hall coefficient*.[34] From the data of Smith,[35] the value of R_0 for Ni at room temperature was determined[36] to be

$$R_0 = -7.6 \times 10^{-17}\,\Omega\,\mathrm{m^2 A^{-1}} \qquad (= -6.1 \times 10^{-13}\,\Omega\,\mathrm{cm\,Oe^{-1}}). \tag{21.22}$$

This value is the same order of magnitude as that of nonmagnetic transition metals such as Mn or Cu. The temperature dependence is monotonic. The value of R_1 for Ni at room temperature is

$$R_1 = -7.49 \times 10^{-16}\,\Omega\,\mathrm{m^2 A^{-1}} \qquad (= -7.49 \times 10^{-11}\,\Omega\,\mathrm{cm\,G^{-1}}*). \tag{21.23}$$

This quantity increases with increasing temperature and exhibits an anomaly at the Curie point, as shown in Fig. 21.9. It should be noted that the value of R_1 is one order of magnitude larger than that of R_0. If the effect of magnetization is simply to apply an internal magnetic field $\mu_0 I$ to the conduction electrons, R_1 must be equal to R_0. Since this is not the case, we must look elsewhere for the origin of the extraordinary Hall effect.

Kaplus and Luttinger[38] considered that the extraordinary Hall effect can be separated into the two parts: One due to the internal field $\mu_0 I$ produced by the magnetization I, and the other due to the spin–orbit interaction between $3d$ spins and conduction electrons. Thus we can express

$$R_1 = R_0 + R_1', \tag{21.24}$$

* When this value is used, we must put $\mu_0 = 1$ in (21.21).

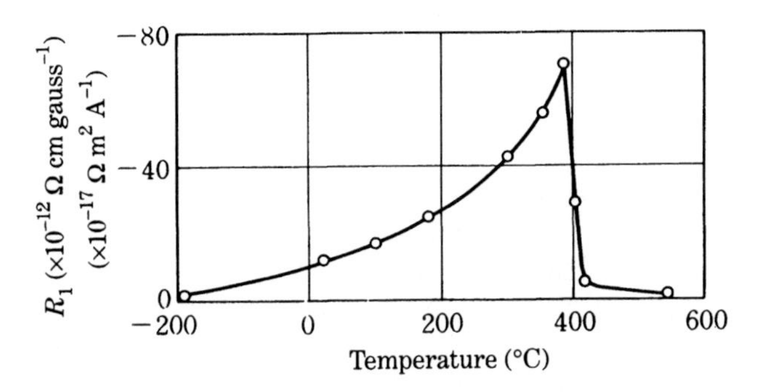

Fig. 21.9. Temperature dependence of extraordinary Hall coefficient of Ni.[37]

where R_0 is the ordinary Hall coefficient as given by (21.23) and R_1' is the extraordinary Hall coefficient due to spin–orbit interaction. Since $R_0 \ll R_1'$ except at very low temperatures, the main part of R_0 is R_1'. They considered that $3d$ electrons are conducting and their orbital motion is influenced by their own spin–orbit interaction. They showed that R_1' is related to the resistivity by

$$R_1' = A\rho^2. \tag{21.25}$$

This relation holds well for various kinds of Ni alloys. In contrast to this theory, Kondo[17] considered that $4s$ electrons are conducting and their orbital motions are influenced by the orbital motion of $3d$ electrons remaining unquenched. His calculation explains well the temperature dependence shown in Fig. 21.9. He extended his theory to rare earth metals and explained the extraordinary Hall effect of Gd metal.[39]

The *magnetoelectric polarization effect (ME effect)* is the phenomenon in which a magnetic field $\boldsymbol{H}$ produces an electric polarization $\boldsymbol{P}$ or an electric field $\boldsymbol{E}$ produces magnetization $\boldsymbol{I}$. When these quantities are small, the proportionality relationship holds between electric and magnetic quantities, but their directions are not parallel to each other. We can express these relationships as

$$\left.\begin{aligned} \boldsymbol{P} &= \alpha \boldsymbol{H} \\ \boldsymbol{I} &= \alpha' \boldsymbol{E} \end{aligned}\right\}. \tag{21.26}$$

where α and α' are the *ME tensor* and *EM tensor*, respectively. If we write (21.26) for each component in Cartesian coordinates, we have

$$\begin{pmatrix} P_x \\ P_y \\ P_z \end{pmatrix} = \begin{pmatrix} \alpha_{11} & \alpha_{12} & \alpha_{13} \\ \alpha_{21} & \alpha_{22} & \alpha_{23} \\ \alpha_{31} & \alpha_{32} & \alpha_{33} \end{pmatrix} \begin{pmatrix} H_x \\ H_y \\ H_z \end{pmatrix}, \tag{21.27}$$

and

$$\begin{pmatrix} I_x \\ I_y \\ I_z \end{pmatrix} = \begin{pmatrix} \alpha'_{11} & \alpha'_{12} & \alpha'_{13} \\ \alpha'_{21} & \alpha'_{22} & \alpha'_{23} \\ \alpha'_{31} & \alpha'_{32} & \alpha'_{33} \end{pmatrix} \begin{pmatrix} E_x \\ E_y \\ E_z \end{pmatrix}. \tag{21.28}$$

Between the two tensor components α and α' we have the relationship

$$\alpha'_{ij} = \alpha_{ji} \qquad (i = 1, 2, 3). \tag{21.29}$$

The unit of these tensor components is $\mathrm{s\,m^{-1}}$ or inverse velocity. In CGS Gaussian units, these tensor component are dimensionless, and the MKS (SI) values are $4\pi/c$ times the CGS values (c: velocity of light $= 3 \times 10^8\,\mathrm{m\,s^{-1}}$).

The ME effect is observed only for ferro- or ferrimagnetic materials with low crystal symmetry. In these materials one of the nondiagonal elements of the tensors in (21.27) or (21.28) is nonzero. ME effects are observed for boracite[40] $M_2B_7O_{13}X$ (M = Cr, Mn, Fe, Co, Ni, or Cu and X = Cl, Br, or I), which have canted spin systems (see Section 7.4), and magnetite[41,42] Fe_3O_4. In the latter material an interesting temperature dependence of the ME effect was observed near 10 K.[43,44]

21.3 MAGNETO-OPTICAL PHENOMENA

Magneto-optical phenomena can be classified into two categories: the *magneto-optical effect* in which the optical properties of a magnetic material can be altered by magnetic means, and the *photomagnetic effect* in which a magnetic property of a magnetic material is altered by optical means.

The magneto-optical effects include the *Faraday effect*, in which the plane of polarization of light transmitted through a magnetic material rotates in accordance with the sense of magnetization, and the *magnetic Kerr effect*, in which the plane of polarization of light reflected from a magnetic material rotates in accordance with the sense of magnetization. In both cases, only the component of magnetization parallel to the direction of propagation of the light is effective in rotating the plane of polarization. In the Faraday effect, the rotational angle, θ, of the plane of polarization is proportional to the path-length, l, and to the field, H, and is expressed as

$$\theta = VlH, \tag{21.30}$$

where V is the *Verdet constant*. The sign of θ is positive when θ has the same rotational sense as the electric current in the solenoid which produces the magnetic field. In this case the sign of the angle of rotation is independent of the sense of propagation of the light, because this phenomenon is caused by the difference in absorption between two oppositely circularly polarized light beams, due to the precession of spin, independently of the sense of propagation of the light. In spite of the fact that the frequency of precession motion of a spin is much smaller than the frequency of visible light, it causes a large difference in absorption between two circularly polarized light beams. This is due to the following mechanism: Suppose that a magnetic atom at an energy level A absorbs a photon $\hbar\omega$ and is excited to level B (see Fig. 21.10). If the atomic spin is $\frac{1}{2}$, each level splits into two levels $S = -\frac{1}{2}$ and $+\frac{1}{2}$ in the presence of a magnetic field H (*Zeeman splitting*). The allowed transition

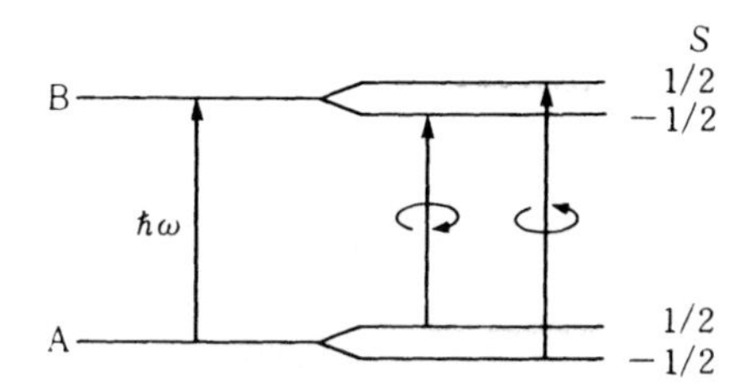

Fig. 21.10. Selective absorption of right-hand and left-hand circularly polarized light by a magnetic atom.

between the two levels is either from $S = \frac{1}{2}$ at level A to $S = -\frac{1}{2}$ at level B, or from $S = -\frac{1}{2}$ at level A to $S = \frac{1}{2}$ at level B. This is known as a *selection rule* in spectroscopy. Physically it corresponds to the *law of conservation of angular momentum*, because when circularly polarized light which has an angular momentum is absorbed by an atom, there must be a change in spin angular momentum. Therefore the two transitions mentioned above must be caused by the absorption of oppositely circularly polarized light as illustrated in Fig. 21.10. When a magnetic atom is magnetized, the population between the two spin states at the level A is different, so that the absorption of one circularly polarized light is greater than the other, thus resulting in a rotation of the plane of polarization of the light.

The mechanism of the magnetic Kerr effect is similar to the Faraday effect: when light is reflected at the surface of a magnetic material magnetized perpendicular to the surface, the plane of polarization will rotate in the same sense during penetration to the skin depth and back to the surface, thus resulting in a nonzero rotation of the plane of polarization.

The Verdet constants have been measured for a number of transparent magnetic materials and are found to be impurity-sensitive. For instance, the Verdet constant of YIG is appreciably increased by the introduction of a small amount of Bi^{3+} or Pb^{3+} ions. The reason is thought to be the transfer of some $2p$ electrons from O^{2-} to $6p$ orbits in Bi^{3+} ions, and this mixing of orbits enhances spin–orbit interactions.[46]

Faraday rotation has been used for measuring magnetization in ultra-high pulsed magnetic fields.[47] Faraday rotation and magnetic Kerr effect are also used for observing magnetic domain patterns (see Section 15.2) and for detecting the sense of magnetization of recorded magnetic patterns in magneto-optical memory devices (see Section 22.3).

Photo-induced magnetic anisotropy (Section 13.4.2) is an example of a photomagnetic effect. In a magneto-optical memory system, a magnetic signal is written using a laser beam (see Section 22.3). However, the role of light in this case is to locally heat the magnetic media above the Curie point so that the magnetization reverses on cooling in the demagnetizing field. Since this is not an optical function, we cannot regard this process as a photomagnetic effect.

The magnetic structure of Er orthochromite ($ErCrO_3$) was observed to change from antiferromagnetic to canted-spin upon illumination by light at 4.2 K. This is due to a change in superexchange interaction caused by light. Generally speaking, however, it has not been possible to change magnetic moments or exchange interactions by purely optical means, even if a powerful laser beam is used.

REFERENCES

1. R. Weiss and R. Forrer, *Ann. Physique*, **10** (1926), 153.
2. R. Becker and W. Döring, *Ferromagnetismus* (Springer, Berlin, 1939), p. 70.
3. R. M. Bozorth, *Ferromagnetism* (Van Nostrand, Princeton, N.J., 1951), p. 741, Fig. 15.9.
4. E. B. Ellwood, *Nature*, **123** (1929), 797; *Phys. Rev.*, **36** (1930), 1066.
5. A. Townsend, *Phys. Rev.*, **47** (1935), 306.
6. T. Okamura, *Sci. Rept. Tohoku Univ.*, **24** (1935), 745.
7. L. F. Bates and J. C. Weston, *Proc. Phys. Soc.* (London), **53** (1941), 5; L. F. Bates and D. R. Healey, *Proc. Phys. Soc.* (London), **55** (1943), 188; L. F. Bates and A. S. Edmondson, *Proc. Phys. Soc.* (London), **59** (1947), 329; L. F. Bates and E. G. Harrison, *Proc. Phys. Soc.* (London), **60** (1948), 213; L. F. Bates, *J. Phys. Radium*, **10** (1949), 353; **12** (1951), 459; L. F. Bates and G. Marshall, *Rev. Mod. Phys.*, **25** (1953), 17.
8. L. F. Bates and N. P. R. Sherry, *Proc. Phys. Soc.* (London), **68B** (1955), 304.
9. L. F. Bates and A. W. Simpson, *Proc. Phys. Soc.* (London), **68** (1955), 849.
10. L. F. Bates and N. P. R. Sherry, *Proc. Phys. Soc.* (London), **68** (1955), 642.
11. R. Becker and W. Döring, *Ferromagnetismus* (Springer, Berlin, 1939), p. 325.
12. E. Englert, *Ann. Physik*, **14** (1932), 589.
13. N. F. Mott, *Proc. Roy. Soc.* (London), **156** (1936), 368.
14. T. Kasuya, *Progr. Theor. Phys.* (Kyoto), **16** (1956), 58.
15. R. V. Colvin, S. Legvold, and F. H. Spedding, *Phys. Rev.*, **120** (1960), 741.
16. E. Tatsumoto, K. Kuwahara, and H. Kimura, *J. Sci. Hiroshima Univ.*, **24** (1960), 359.
17. J. Kondo, *Progr. Theor. Phys.* (Kyoto), **27** (1962), 772.
18. J. Smit, *Physica*, **17** (1952), 612.
19. J. P. Jan, *Solid State Physics* (Academic Press, New York, 1957), Vol. 5, p. 73.
20. R. M. Bozorth, *Ferromagnetism* (D. Van Nostrand Co., Princeton, N.J., 1951), p. 453, Fig. 10.18.
21. H. W. Lehmann and M. Robbins, *J. Appl. Phys.*, **37** (1966), 1389; H. W. Lehmann, *Phys. Rev.*, **163** (1967), 488; A. Amith and G. L. Gunsalus, *J. Appl. Phys.*, **40** (1969), 1020.
22. K. Miyatani, T. Takahashi, K. Minematsu, S. Osaka, and K. Yosida, *Proc. Int. Conf. on Ferrites* (Univ. Tokyo Press, 1971), p. 607.
23. T. Kasuya and A. Yanase, *Rev. Mod. Phys.*, **40** (1968), 684.
24. S. Methfessel and D. C. Mattis, *Handbuch der Physik*, **68** (1968), Vol. 18.
25. S. von Molner and T. Kasuya, *Proc. 10th Int. Conf. Semicon.* (1970), 233.
26. T. Kasuya, *Proc. 11th Int. Conf. Semicon.* (1972), 141.
27. M. N. Baibich, J. M. Broto, A. Fert, N. Nguyen, Van Dau, F. Pertroff, P. Eitenne, G. Creuzet, F. Friederich, and J. Chazelas, *Phys. Rev. Lett.*, **61** (1988), 2472.
28. M. Julliere, *Phys. Lett.*, **54A** (1975), 225.
29. S. Maekawa and U. Gafvert, *IEEE Trans. Mag.*, **Mag18** (1982), 707.
30. J. C. Slonczewski, *Phys. Rev.*, **B39** (1989), 6995.
31. T. Miyazaki, T. Yaoi, and S. Ishio, *J. Mag. Mag. Mat.*, **98** (1991), L7.
32. K. E. Grew, *Phys. Rev.*, **41** (1932), 356.
33. A. W. Smith and R. W. Sears, *Phys. Rev.*, **34** (1929), 1466.
34. E. W. Pugh, *Phys. Rev.*, **36** (1930), 1503.

35. A. W. Smith, *Phys. Rev.*, **30** (1910), 1.
36. E. M. Pugh, N. Rostoker, and A. I. Shindler, *Phys. Rev.*, **80** (1950), 688.
37. E. M. Pugh and N. Rostoker, *Rev. Mod. Phys.*, **25** (1953), 295.
38. R. Kaplus and J. M. Luttinger, *Phys. Rev.*, **95** (1954), 1154.
39. J. Kondo, *Progr. Theoret. Phys.*, **28** (1962), 846.
40. E. Asher, H. Rieder, and H. Schmid, *J. Appl. Phys.*, **37** (1966), 1404.
41. Y. Miyamoto, M. Ariga, A. Otuka, E. Morita, and S. Chikazumi, *J. Phys. Soc. Japan*, **46** (1979), 1947.
42. K. Siratori, E. Kita, G. Kaji, A. Tasaki, S. Kimura, I. Shindo, and K. Kohn, *J. Phys. Soc. Japan*, **47** (1979), 1779.
43. Y. Miyamoto, Y. Iwashita, Y. Egawa, T. Shirai, and S. Chikazumi, *J. Phys. Soc. Japan*, **48** (1980), 1389.
44. Y. Miyamoto, M. Kobayashi, and S. Chikazumi, *J. Phys. Soc. Japan*, **55** (1986), 660.
45. T. Teranishi, *Handbook on magnetic substances*, ed. by S. Chikazumi *et al.* (Asakura Publishing Company, Tokyo, 1975), No. 19.1.3.
46. K. Shinagawa, H. Takeuchi, and S. Taniguchi, *Jap. J. Appl. Phys.*, **12** (1973), 466.
47. K. Nakao, T. Goto, and N. Miura, *J. de Phys.*, **49** (1988), C8-953.
48. S. Kurita, K. Toyokawa, K. Tsushima, and S. Sugano, *Solid State Comm.*, **38** (1981), 235.

22

ENGINEERING APPLICATIONS OF MAGNETIC MATERIALS

Engineering applications of ferromagnetic materials are divided into three major categories: first is soft magnetic materials for AC applications in magnetic cores of transformers, motors, inductors and related devices. For this purpose, high permeability, low coercive field, and low hysteresis loss are required. Second is hard magnetic materials for permanent magnets. For this purpose, high coercive field, high residual magnetization, and accordingly high energy product (see Section 1.6) are required. Third is the semi-hard magnetic materials for magnetic recording. For this purpose, moderately high coercive field and high residual magnetization are required.

22.1 SOFT MAGNETIC MATERIALS

Soft magnetic materials are used for magnetic cores of transformers, motors, inductors, and generators. Desirable properties for core materials are high permeability, low coercive field and low magnetic losses. In addition, high induction and low cost are important factors. Since no material meets all these specifications, we must select suitable magnetic materials for particular applications based on a balance between competing requirements. From this point of view, we can classify soft magnetic materials in two groups: those for use in large machines where the cost of material is dominant, and those for use in small, low-power devices where material costs are less important than magnetic properties. In the former case, sizes of transformers, generators and motors are very large, up to many tons, so that magnetic cores must be magnetized to the maximum permeability point to make full use of the magnetization of the core material. Therefore high maximum permeability and low cost are necessary. In the latter case, excellent magnetic characteristics are more important than the cost of materials.

In the description of composition of practical alloys, it is customary to use wt%. For simplicity, we will omit 'wt' in the following description, so '%' will mean 'wt%.'

22.1.1 Soft magnetic alloys

(a) Silicon–iron alloys

The addition of a small amount of silicon to iron results in higher maximum permeability, lower magnetic loss and a substantial increase in electrical resistivity

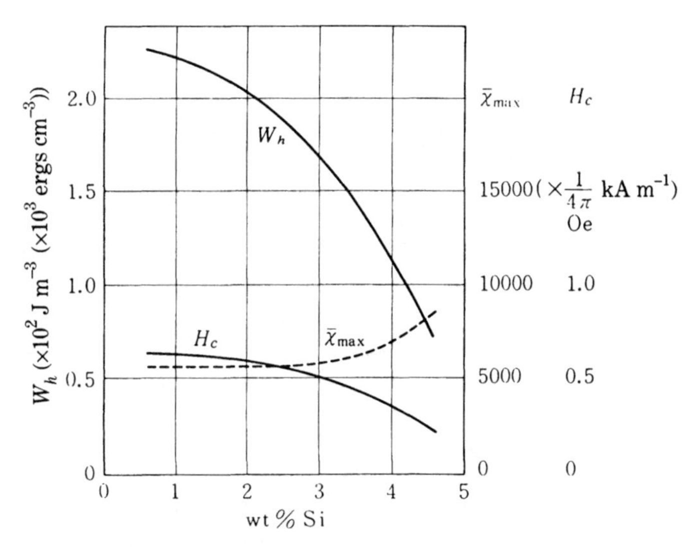

Fig. 22.1. Variation of magnetic properties of Si–Fe alloys with addition of silicon.[1]

which lowers eddy current losses. On the other hand, silicon causes a decrease in saturation magnetization and an increase in brittleness. Silicon is soluble in iron up to 15% Si. The Curie point decreases from 770°C to about 500°C as the silicon content increases from 0 to 15%.

Figure 22.1 shows composition dependence of various magnetic constants such as relative maximum susceptibility, $\bar{\chi}_{max}$, coercive field, H_c, and hysteresis loss, W_h, for Fe with small additions of Si. As seen in this figure, the soft magnetic properties are improved by the addition of Si. Increasing brittleness makes high silicon alloys difficult to produce in the form of thin sheets, so most commercial materials are made with Si contents from 1 to 3%.

The maximum permeability is greatly improved by orientating crystal grains so as to align the easy axis of each grain in a unique direction. Goss[2] first produced 'grain-oriented' silicon steel by a rather complicated process of hot-rolling, annealing, severe cold-rolling, and annealing for what is known as 'secondary recrystallization' to produce very large, well-oriented grains. It was later found that this process works because primary recrystallization is inhibited by the presence of naturally occurring MnS particles. The recrystallization texture is (110) [001], meaning that the grains have the (110) plane parallel to the plane of the sheet, and the [001] easy axis parallel to the rolling direction. This process was improved by Taguchi and Sakakura,[3] using AlN instead of MnS as an inhibitor against unwanted grain growth. The product developed by this technique is called *HIB (high bee)* steel, and has better performance than conventional *Goss steel*. Imanaka *et al.*[4] also developed a new product called *RGH* using SbSe as an inhibitor.

(b) Iron–nickel alloys

This alloy system is composed of two regions: the *'irreversible' alloys* containing 5% to 30% Ni–Fe, and the *'reversible' alloys* from 30% to 100% Ni–Fe. Irreversible alloys transform from fcc to bcc during cooling from high temperature. This transformation occurs over a certain temperature range, and the average transformation temperature differs considerably between cooling and heating. Because of this phenomenon, various magnetic properties exhibit thermal hysteresis, and this is the reason why these alloys are called irreversible alloys.

The reversible alloys are single phase fcc solid solutions. The saturation magnetization at 0 K as a function of composition shows a simple variation following the Slater–Pauling curve except near 30% Ni–Fe, where both the saturation magnetization and the Curie point of the fcc phase become very small and presumably drop off to zero.[5] The 36% Ni–Fe alloy where this anomalous behavior starts is called *Invar* and has nearly zero thermal expansion near room temperature. The alloys near 30% Ni–Fe show a more rapid decrease in saturation magnetization with temperature than other magnetic alloys, so they can be used as shunts in magnetic circuits to compensate for the normal temperature change of magnetic flux. Alloys of this kind are called *magnetic compensating alloys*.

The magnetocrystalline anisotropy and magnetostriction constants K and λ go through zero near 20% Fe–Ni (see Figs 12.35 and 14.11). Therefore the permeability has a maximum at 21.5 Fe–Ni. This alloy is called *Permalloy* (see Fig. 18.27), although the name is loosely applied to various Ni–Fe alloys. Very high permeability is obtained only when the alloy is quenched from 600°C to suppress the formation of directional order (see Section 13.1). The addition of a small amount of Mo, Cr, or Cu is also effective in suppressing the formation of directional order. Therefore *Supermalloy*, which contains 5% Mo in 79% Fe–Ni, attains very high permeability without quenching. Other effects of adding Mo are to attain the conditions $K = 0$ and $\lambda = 0$ simultaneously at the same composition, and also to increase electrical resistivity by a factor 4.

(c) Iron–cobalt alloys

Cobalt is soluble in iron up to 75% Co–Fe. The crystal structure is bcc and FeCo superlattice forms below about 730°C at compositions around 50% Co. As seen in the Slater–Pauling curve (Fig. 8.12), the saturation magnetization at 0 K has a maximum at about 35% Co–Fe. The maximum saturation value at room temperature occurs at about 40% Co–Fe, where K_1 goes through zero, so that we can expect high permeability. Actually, sharp maxima in χ_a and χ_{max} are observed to occur at about 50% Co–Fe. The alloy of this composition is made commercially under the name *Permendur*. Normally about 2% V is added to slow the ordering reaction and improve workability. Its high flux density is useful for pole pieces of high-power electromagnets, and in aircraft equipment where light weight is important.

(d) Iron–cobalt–nickel alloys

The best-known magnetic alloy which belongs to this system is *Perminvar*: 25% Co, 45% Ni, and 30% Fe. As reported by Elmen[6] and by Masumoto,[7] the unique feature of this alloy is its constant permeability over a wide range of magnetic induction. In order to attain this characteristic, the alloy must be annealed for a fairly long time

Table 22.1. Magnetic properties of soft magnetic alloys (mostly from Bozorth[9]).

Name	Content	Heat treatment (°C)	$\bar{\chi}_a$	$\bar{\chi}_{max}$	H_c (A m^{-1})	H_c (Oe)	I_s (T)	I_s (G)	Θ_f (°C)	ρ ($\times 10^{-8}$) (Ω m)
Iron	0.2imp	950	150	5 000	80	1.0	2.15	1710	770	10
Iron (pure)	0.05imp	1480 +880	10 000	200 000	4	0.05	2.15	1710	770	10
Si–Fe (random)	4Si	800	500	7 000	40	0.5	1.97	1570	690	60
Si–Fe (GOSS)	3Si	800	1 500	40 000	8	0.1	2.00	1590	740	47
78-Permalloy	78.5Ni	1050 600Q	8 000	100 000	4	0.05	1.08	860	600	16
Supermalloy	5Mo, 79Ni	1300	100 000	1 000 000	0.16	0.002	0.79	629	400	60
Permendur	50Co	800	800	5 000	160	2.0	2.45	1950	980	7
FeB amorphous[10]	8B6C	Q	680		8	0.1	1.73	1380	334	130

(for instance, 24 h) at 400–450°C. The necessity for this treatment was interpreted by Taniguchi[8] as follows: all the spins in the domain wall are stabilized by a uniaxial anisotropy induced by directional ordering during the anneal, which keeps the domain walls in their original positions. Since in this case the energy of the domain wall is a quadratic function of its displacement, the displacement, and accordingly the resultant magnetization, should be proportional to the applied field. When the alloy is magnetized by a strong field the domain walls may escape from their stabilized positions and thus become incapable of giving constant permeability. Then the hysteresis loop becomes a snake-shaped or constricted one as shown in Fig. 18.44. This behavior can be explained by the same model. It is also possible to make the hysteresis loop rectangular, because this alloy responds strongly to magnetic annealing (see Section 13.1).

(e) Amorphous magnetic alloys

The preparation procedure and magnetic properties of amorphous alloys are described in Chapter 11. The core materials are produced by continuous splat-cooling, or rapid solidification processing, in the form of long thin ribbons or sheets. Amorphous alloys have small anisotropy, so that they exhibit high permeability and low coercive field in addition to high electrical resistivity. They are mechanically strong, and some compositions have good corrosion resistance. For power machines, however, they have the disadvantage that the saturation magnetization is limited by the high content (about 25%) of non-magnetic metalloid elements. Also some compositions show relatively large aging and disaccommodation (Section 20.1) effects. Amorphous alloys are successfully used in power transformers up to several kVA capacity, and for small transformers and other devices operating at frequencies up to about 50 kHz.

Table 22.1 lists magnetic properties of various soft magnetic alloys.

22.1.2 Pressed powder and ferrite cores

For high-frequency applications, reduction of the various losses accompanying high-frequency magnetization is more important than static magnetic characteristics. The eddy current loss is of primary importance, and, in order to reduce it, magnetic metals and alloys for AC use are always in the form of thin sheets or fine particles. Ferrites and other nonmetallic compounds are particularly good for high-frequency applications because of their high resistivity (see Table 22.2).

(a) Sendust[11]

The alloy containing 5% Al, 10% Si and 85% Fe exhibits extremely high permeabilities $\bar{\chi}_a = 30\,000$ and $\bar{\chi}_{max} = 120\,000$ as cast from the melt, because this composition just meets the conditions $K_1 = 0$ and $\lambda = 0$.[12] Since this alloy is very brittle, it is ground into small particles about 10 μm in diameter, which, after annealing, are coated with a thin electrically insulating layer and compressed into final shape.

(b) Ferrites

As already discussed in Section 9.2, ferrites are electrically insulating, so that they are most suitable as high-frequency magnetic materials. Most commercial ferrites for this purpose contain Zn, for the following reasons: (1) The addition of Zn causes a decrease in the Curie point and brings the high permeability which is commonly attained just below the Curie point (Hopkinson effect, Section 18.2) closer to room temperature. (2) The addition of Zn causes an increase in saturation magnetization at 0 K (see Fig. 9.7), thus compensating a part of the decrease in saturation magnetization at room temperature caused by a decrease in Curie point.

Mn–Zn ferrites have relatively high saturation magnetization (see Table 9.2 or Fig. 9.7), so that they are extensively used as core materials. However, the resistivity is relatively low ($\rho = 0.1$ to $1\,\Omega\,\mathrm{m}$ ($= 10$–$100\,\Omega\,\mathrm{cm}$)), so that they are not suitable for extremely high-frequency use. Maximum permeability occurs at 50–75 mol% $MnFe_2O_4$, 50–25 mol% $ZnFe_2O_4$. After being sintered at 1400°C, they are usually quenched from 800°C or slowly cooled in a nitrogen atmosphere to prevent segregation of α-Fe_2O_3 and occurrence of disaccommodation (Section 20.1).

Ni–Zn ferrites have extremely high resistivity, so that they are suitable for extremely high-frequency uses (see Fig. 20.17). Initial permeability is maximum at 30 mol% $NiFe_2O_4$, 70 mol% $ZnFe_2O_4$. However, high-frequency properties are better at higher Ni content, because the 'natural resonance' shifts to higher frequency by an increase in K_1 (see Section 20.3). Since only the doubly charged state is stable for nickel ions, this ferrite can be sintered at high temperatures.

In Mg–Zn ferrites, high permeability is realized at 50 mol% $MgFe_2O_4$, 50 mol% $ZnFe_2O_4$.

Mg–Mn ferrite is used as a rectangular hysteresis material.

(c) Garnets

The crystal and magnetic properties of these oxides were discussed in Section 9.3. Since the metal ions are all triply charged, no hopping motion of electrons occurs, and the oxide has extremely high resistivity and exhibits extremely low magnetic losses. Single crystal garnets are used for microwave amplifiers. Thin films of this crystal are optically transparent, so that it is used for bubble domain devices (see Section 22.3).

Table 22.2 lists magnetic properties of pressed powder and ferrite cores.

22.2 HARD MAGNETIC MATERIALS

Another important engineering application of magnetic materials is as permanent magnets. The desirable properties for hard magnetic materials are high coercive field and high residual magnetization. A figure of merit to express the quality of permanent magnet materials is the maximum energy product $(BH)_{max}$, which is the maximum rectangular area under the B–H curve in the second quadrant of the hysteresis loop (see Fig. 1.27). For most efficient use of the magnetic material, the working point of the magnet should correspond to the maximum value of the energy product.

Table 22.2. Magnetic constants of pressed powder and ferrite cores.[13]

Name	Content	Heat treatment (°C)	$\bar{\chi}_a$	H_c ($\mathrm{A\,m^{-1}}$)	H_c (Oe)	I_s (T)	I_s (G)	Θ_f (°C)	ρ (Ω m)
Sendust powder	5Al,10Si	Press, 800	80	100	1.25	0.45	360	500	
MnZn ferrite	50Mn–50Zn	1150	2000	8	0.1	0.25	200	110	20
NiZn ferrite	30Ni70Zn	1050	80	240	3.0	0.40	320	130	5×10^4
Ca added YIG	0.2% Ca + $Y_3Fe_5O_{12}$		600			0.17	135	287	

In the history of permanent magnets, first iron and its alloys such as KS steel, MK steel, Alnico, etc., were used, and then compound magnets utilizing single domain characteristics such as Ba ferrites, MnAl, etc., were developed. These later materials have higher coercivity but lower residual magnetization as compared with the alloy magnets. Since the coercive field for B, ${}_BH_c$, cannot be greater than the value of the residual magnetization divided by μ_0 (see (1.120)), the aim in developing permanent magnet materials is to get the saturation magnetization as high as possible while maintaining high coercivity. Recently developed high-quality rare earth permanent magnets such as RCo_5 and $Nd_{15}Fe_{77}B_8$ meet both these requirements.

22.2.1 Permanent magnet alloys

As is well known in steel metallurgy, when a carbon iron is quenched from the fcc phase, it transforms completely or partially to a bct phase called *martensite*, which is mechanically and magnetically hard. W steel, Cr steel and Co steel are all old permanent magnets utilizing this phenomenon. Other examples are *KS magnet* (Fe–Co–W steel), invented by Honda and Saito,[14] and *MT magnet* (Fe–C–Al steel), invented by Mishima and Makino[15] which contain no expensive metals such as Ni and Co.

The *MK magnet* was invented by Mishima.[16] This alloy is also called *Alnico 5*. It has the approximate composition Fe_2NiAl, which decomposes into two bcc phases during slow cooling from the high-temperature ordered bcc phase. This process is known as *spinodal decomposition* (see Section 13.3.1) and it produces large internal stresses and at the same time finely divided elongated particles (see Fig. 13.15). Originally the high coercivity of this magnet was thought to be due to large internal stresses, but later Nesbitt[17] found that a high coercivity is also observed for an alloy of the same kind which has no magnetostriction. Accordingly, he suggested that a single-domain structure of fine precipitated particles must be the origin of the high coercivity. Cooling this alloy from 1300°C in a magnetic field results in a rectangular hysteresis loop and accordingly a large value of $(BH)_{max}$, when the hysteresis loop is measured in the direction of the cooling field. The origin of this magnetic annealing is the reorientation of the elongated precipitated particles (see Section 13.3.1).

Fe–Cr–Co alloy[18] is a deformable alloy magnet. The largest $(BH)_{max}$ is obtained at 15% Co, but even with 5% Co a usefully high value is obtained.

Fe–Pt and Co–Pt alloys form the superlattices FePt and CoPt, respectively, both of which have tetragonal crystal structures. Strong internal stresses produced by this superlattice formation are supposed to be the origin of their high coercivity.

22.2.2 Compound magnets

Ba ferrite is a hexagonal ferrite invented by Philips investigators.[19] It has a magnetoplumbite-type crystal structure with the chemical formula $BaO \cdot 6Fe_2O_3$ (see Section 9.4). The anisotropy is uniaxial and also very large (see Table 12.4). As discussed in Section 17.6, a sintered magnet can have a single domain structure, so that ferrite materials sintered into solid ceramic form can be used as permanent magnets. Moreover the crystal grains can be aligned by pressing in a magnetic field, which can increase $(BH)_{max}$ by a factor of 3.[20]

MnAl is a ferromagnetic compound which includes none of the ferromagnetic elements (see Section 10.1). This compound, containing 0.7% carbon,[21] is induction-melted and extruded at 700°C under a pressure of $80\,kg\,mm^{-2}$. In addition to excellent magnetic properties, it has good machinability and low density ($3.5\,g\,cm^{-3}$, less than that of Ba ferrite, $4.5\,g\,cm^{-3}$).

22.2.3 Rare earth compound magnets

RCo_5 and $R_2(Co,Fe)_{17}$ type compounds[22] have a strongly uniaxial crystal structure (see Fig. 8.26), and exhibit very high magnetocrystalline anisotropy. In the case of $R = Sm$, both compounds produce strong uniaxial anisotropies (see Table 12.3 and Strnat[22]). These compounds have fairly large saturation magnetization (see Table 12.3), so that the $(BH)_{max}$ values are one order of magnitude larger than those of previously available magnets.

The $Nd_{15}Fe_{77}B_8$ magnet includes grains of the phase $Nd_2Fe_{14}B$ with tetragonal crystal structure (Fig. 10.10), which contain no obstacles to domain wall displacement.[23] This phase has a very large uniaxial anisotropy. The grain size is much larger than the critical size for single-domain structure, so that once a domain wall is created, it will run across the grain, thus resulting in a very low coercive field. High coercivity is obtained because it is difficult to nucleate reverse domains. This kind of magnet is described as having *nucleation-controlled* coercivity. Necessary conditions for this are a defect-free structure of the crystals and also smooth grain boundaries containing Nd- and B-rich phases. RCo_5 and $R_2(Fe,Co)_{17}$ magnets belong to this category.

One of the features of this type of material is that once the magnet is magnetized by a strong field, the coercive field for I, ${}_IH_c$, increases with an increase of the maximum magnetizing field, H_m.[24] This phenomenon is considered to be due to the elimination of 'magnetic seeds' for the nucleation of reverse domains by the application of H_m.[25]

Nd–Fe–B and Pr–Fe–B alloys can be prepared by the melt-spinning technique[26] as well as by conventional sintering of powders. The melt-spun material is used in bonded magnets for accurately shaped parts.

Table 22.3. Magnetic properties of various permanent magnets.

Name	Composition	Heat treatment (°C)	B_r (T)	H_c (A m^{-1} $\times 10^2$)	H_c (Oe)	$(BH)_{max}$ J m^{-3}* $\times 10^3$	$(BH)_{max}$ (G Oe $\times 10^6$)	Density (g cm^{-3})
Alloy magnets								
W steel	0.7C, 0.3Cr, 6W	850Q	1.0	56	70	2.5	0.31	8.1
3.5%Cr steel	0.9C, 3.5Cr	830Q	0.98	48	60	2.1	0.27	7.7
15%Co steel	1.0C, 7Cr, 0.5Mo, 15Co	1150AQ, 780FC 1000Q	0.82	143	180	4.8	0.6	7.9
KS steel	0.9C, Cr, 4W, 35Co		0.90	200	250	8.0	1.0	
MT steel	2.0C, 8.0Al		0.60	160	200	3.6	0.45	6.9
MK magnet	16Ni, 10Al, 12Co, 6Cu		0.8	446	560	12.7	1.6	7.0
Orient. MK (Alnico5)	14Ni, 8Al 24Co, 3Cu	1300AF, 600B	1.2	438	550	40	5.0	7.3
Fe–Cr–Co	28Cr, 23Co, 1Si	600–540B	1.3	464	580	42	5.3	
Superlattice magnets								
Pt–Fe	78Pt		0.58	1250	1570	24	3.0	10
Pt–Co	23Co, 77Pt		0.45	2070	2600	30	3.8	11
Compound magnets								
Ba ferrite	$BaO \cdot 6Fe_2O_3$		0.20	1200	1500	8.0	1.0	4.5
MnAl	70Mn, 30Al, .7C		0.58	2560	3300	33.4	4.2	3.5
Rare earth compound magnets								
$SmCo_5$			0.93	7200	9000	166	20.9	8.3
$Sm_2(Co,Fe)_{17}$			1.5			477	60	
$Nd_{15}Fe_{77}B_8$			1.23	9600	12000	290	36	7.4

* The magnetic energy per unit volume is given by half of this value and is calculated as half of TA m^{-1} or $1/8\pi$ of G Oe.

Various constants are listed in Table 22.3 for a number of permanent magnet materials.

22.3 MAGNETIC MEMORY AND MEMORY MATERIALS

22.3.1 Magnetic tapes and disks

Magnetic tapes and *magnetic disks* are devices to record analog or digital signals magnetically by utilizing residual magnetization. The principle of these devices is

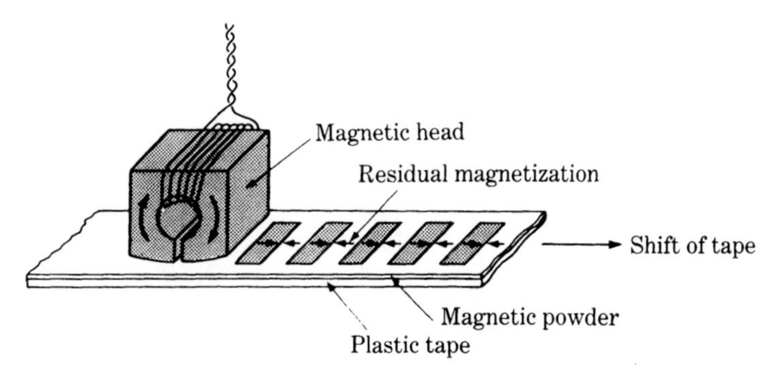

Fig. 22.2. The residual magnetization pattern recorded on a magnetic tape by a magnetic head.

shown in Fig. 22.2. Magnetic powder or an evaporated magnetic thin film carried on a nonmagnetic tape or disk is magnetized as it passes under a magnetic head, which in turn is magnetized by a signal current carrying the information to be recorded. Then a residual magnetization corresponding to the signal wave form is left on the tape or disk. In order to reproduce the signal, a magnetic reading head produces a voltage by electromagnetic induction when the recorded residual magnetization on the tape changes the magnetic flux in the head. The speed of magnetic tapes and disks relative to the magnetic head is becoming smaller and smaller, because of the increasing density of the recorded signal, so that the induction voltage of the signal becomes smaller. Therefore instead of the induction method a static detection of the recorded signal using the magnetoresistance effect has been developed (see Section 21.2). A magnetic head utilizing the magnetoresistance effect is called an *MR head*.

Magnetic tapes are used for recording audio signals (voice and music). They are also used for recording digital signals for large-scale data storage for computers. Magnetic tapes are also used for recording video signals. In this case, the running speed of the tape relative to the head must be high, so that usually both the tape and the magnetic head are moved. For example, a cylindrical head with multiple gaps is rotated, and the tape is wound around the rotating head cylinder and moved in the opposite direction to the head surface motion. Then the signals are recorded along many narrow parallel tracks tilted with respect to the edge of the tape. The signal from one track is converted to one horizontal visible line on the display tube, thus reproducing the recorded picture.

A magnetic disk consists of a rotating flexible plastic disk (floppy disk) or a rigid, relatively thick disk (hard disk) coated with a layer of magnetic medium. The advantage of magnetic disks is that a desired portion of the stored information can be quickly retrieved without spooling through an entire reel of tape.

The distribution of the recorded residual magnetization has been calculated by computer simulation, taking into consideration the coexistence of a magnetic head and magnetic medium whose hysteresis loop is assumed to be as shown in Fig. 22.3(a). The result is shown in Fig. 22.3(b). It is seen that the magnetization has a component perpendicular to the surface at all depths in the tape, tending to form closed magnetic flux paths. Such a magnetic structure stores some magnetostatic energy, and its

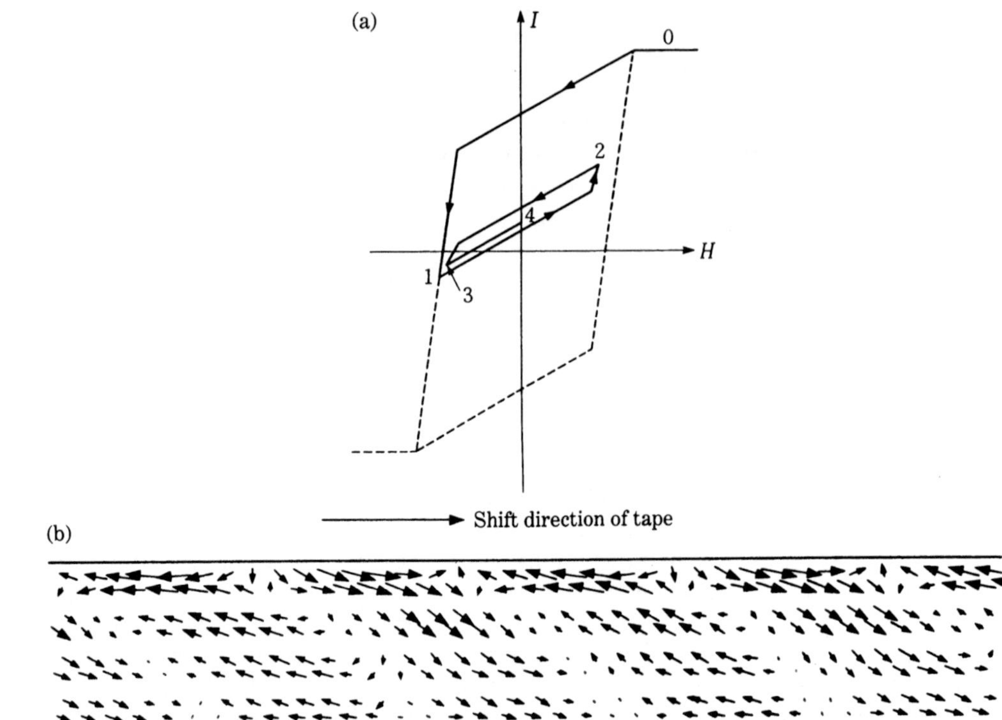

Fig. 22.3. Computer simulation of the distribution of residual magnetization in magnetic media: (a) assumed hysteresis loop (the dashed lines represent a complete hysteresis loop; the lines 1–4 are the magnetization process in a running tape); (b) the calculated distribution of residual magnetization in magnetic media. (After Shinagawa[27])

energy density becomes larger as the signal density becomes higher.[28] There is a physical limit to the minimum bit size, or the highest bit density. However, before reaching this physical limit, there are various technical limits in recording and reproducing magnetic signals, such as the miniaturization of the magnetic head and obtaining smooth contact between the magnetic head and the magnetic medium. One possible solution is a thin film magnetic head floating aerodynamically above a smooth magnetic medium, a system which has been developed for large-scale hard disks.

22.3.2 Magneto-optical recording

The magnetic media used for *magneto-optical recording*[29] are mostly R (rare earth)-Fe or R-Co sputtered amorphous thin films which have uniaxial magnetic anisotropy with the easy axis perpendicular to the film surface (see Section 13.4.5). The most promising material has been TbFeCo.[30] The recording is exclusively digital: either

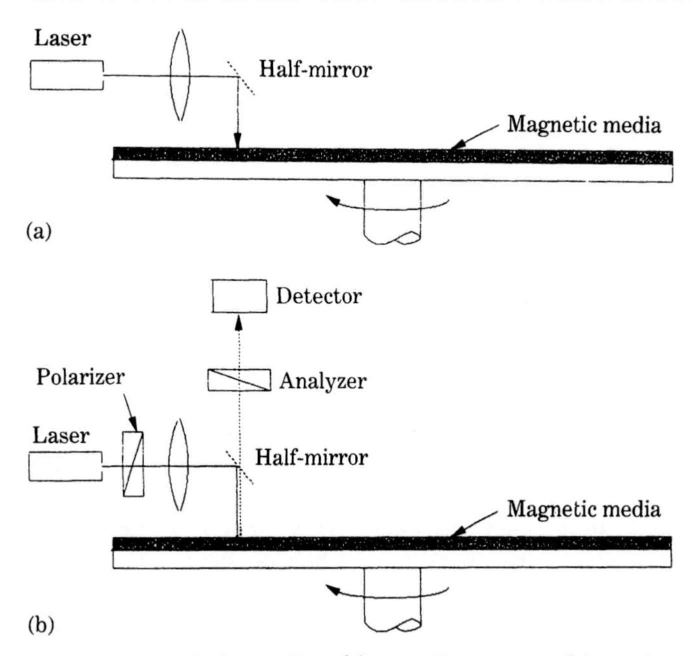

Fig. 22.4. Magneto-optical recording: (a) recording process; (b) reading process.

magnetization-up (+) or magnetization-down (−). In order to write a signal, first all the area is magnetized downward, and then a laser beam is focused on a narrow area to heat it above the Curie point (Fig. 22.4(a)). After switching off the beam, this area cools down and a region of upward-pointing magnetization appears, induced by the demagnetizing field of the surrounding material. Thus a + bit is written in.

In order to detect the signal, a polarized laser beam is directed onto the magnetic media, and the polarization of the reflected beam is rotated oppositely for + and − recorded bits (magnetic Kerr effect, see Section 21.3). This rotation of polarization is detected by an analyzer and detector (Fig. 22.4(b)).

The advantages of magneto-optical recording are high-density memory storage and greatly reduced friction and wear, because the writing and detecting of the signal

Table 22.4. Magnetic constants of magnetic recording materials.[28]

Materials	σ_s		I_s^*		H_c	
	$\mathrm{Wb\,m\,kg^{-1}}$	$\mathrm{emu\,g^{-1}}$	T	G	$\mathrm{kA\,m^{-1}}$	Oe
	$\times 10^{-4}$					
γ-Fe_2O_3 powder	1.0	80	0.1	80	20–31	250–390
Co-doped powder	1.0	80	0.1	80	20–80	250–1000
CrO_2 powder	1.1	90	0.15	120	16–64	200–800
Fe-Co powder	> 2.5	> 200	0.30	200	20–72	250–900
Co–Ni thin film	> 2.5	> 200	1.00	800	8–80	100–1000

* In the case of packing factor 0.4 for powders.

use highly focused laser beams and there is no physical contact with the magnetic medium.

Table 22.4 lists magnetic constants of various magnetic recording materials. Materials used for magnetic heads are Permalloy, Mn–Zn ferrite, Sendust, amorphous alloys, etc.

22.3.3 Bubble domain device

This is an all-magnetic memory system with no mechanically moving parts.[31] It is a digital memory using bubble domains (see Section 17.3), produced on a garnet single-crystal film which is grown on a gadolinium–gallium–garnet single-crystal plate. As shown in Fig. 22.5, a bubble domain can propagate along a path defined by Permalloy patterns magnetized by a rotating magnetic field. The bubble domain is attracted by the free pole appearing on a Permalloy pattern, and is shifted to the next pattern during one revolution of the rotating field. Such a shift of bubble domains can follow a rotational rate as high as 100 kHz. In a memory storage system, the digital number '1' is expressed as the presence of a bubble domain, while the number '0' is expressed as the absence of a bubble domain. An example of a memory storage system is shown in Fig. 22.6. First a digital number is converted to a bubble domain array by the bubble generator, and shifted to a major loop of Permalloy patterns. When the shift of a digital number is completed, the gates to minor loops are opened, and then each digital number is shifted to a minor loop. In this way many digital numbers can be stored in the minor loops. In reading a digital number stored in the minor loops, by opening all the gates to the major loops the digital number array is reproduced in the major loop, which is detected by a detector. For details on bubble domain devices the reader may refer to Mikami.[32]

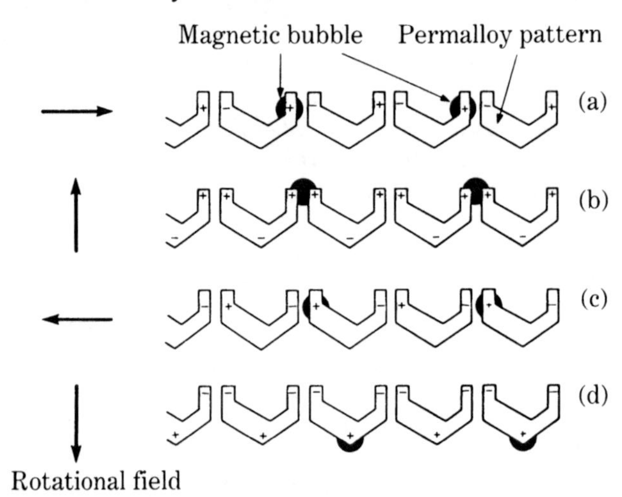

Fig. 22.5. Permalloy pattern for propagation of a bubble domain. (A bubble shifts to the next pattern by being attracted by a free pole appearing on the pattern during one revolution of a rotating magnetic field.)

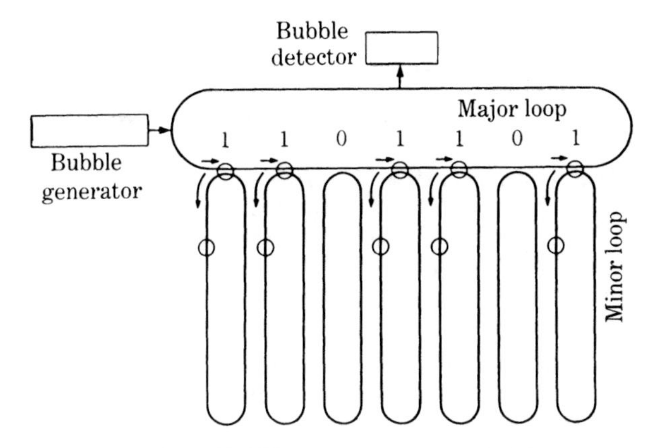

Fig. 22.6. Bubble memory storage circuit.

Bubble domain devices were for a time used for telephone exchange systems, numerically controlled machines, etc., but they have been largely replaced by semiconductor devices. Research on bubble devices, however, led to much useful understanding in solid-state physics and materials science, as discussed in Section 17.3, so that the topic should not be ignored.

REFERENCES

1. R. M. Bozorth, *Ferromagnetism* (Van Nostrand, Princeton, N.J., 1951) p. 80, Fig. 4.10.
2. N. P. Goss, *Trans. Am. Soc. Metals*, **23** (1935), 515.
3. S. Taguchi and A. Sakakura, US Pat. No. 3159511 (1964).
4. T. Imanaka, T. Kan, Y. Obata, and T. Sato, W. Germany Pat. OLS No. 2351142 (1974).
5. J. S. Kouvel and R. H. Wilson, *J. Appl. Phys.*, **32** (1961), 435.
6. G. W. Elmen, *J. Franklin Inst.*, **206** (1928), 317; **207** (1929), 583.
7. H. Masumoto, *Sci. Rept. Tohoku Imp. Univ.*, **18** (1929), 195.
8. S. Taniguchi, *Sci. Rept. Tohoku Univ.*, **8A** (1956), 173.
9. R. M. Bozorth, Ferromagnetism (Van Nostrand, Princeton, N.J., 1951) p. 870, Table 2.
10. S. Hatta, T. Egami, and C. D. Graham, Jr., *IEEE Trans.*, **MAG-1** (1978), 1013.
11. H. Masumoto, *Sci. Rept. Tohoku Imp. Univ.* (Honda) (1936), 388.
12. J. L. Snoek, *New developments in ferromagnetic materials* (Elsevier, Amsterdam, 1949).
13. T. Nagashima, *Handbook on magnetic substances* (ed. by Chikazumi *et al.*, Asakura Publishing Co., Tokyo, 1978) Sections 21, 22.
14. K. Honda and S. Saito, *Sci. Rept. Tohoku Imp. Univ.*, **9** (1920), 417.
15. T. Mishima and N. Makino, *Iron and Steel*, **43** (1956), 557, 647, 726.
16. T. Mishima, *Ohm*, **19** (1932), 353.
17. E. A. Nesbitt, *J. Appl. Phys.*, **21** (1950), 879.
18. H. Kaneko, M. Homma, and K. Nakamura, *AIP Conf. Proc.*, **5** (1971), 1088.
19. J. J. Went, G. W. Rathenau, E. W. Garter, and G. W. van Oosterhout, *Philips Tech. Rev*, **13** (1952), 194.
20. G. W. Rathenau, J. Smit, and A. L. Stuyts, *Z. Physik*, **133** (1952), 250.

21. T. Ohtani, M. Kato, S. Kojima, S. Sakamoto, I. Kuno, M. Tsukahara, and T. Kubo, *IEEE Trans. Mag.*, **MAG-13** (1977), 1328.
22. K. J. Strnat, *IEEE Trans. Mag.*, **MAG-8** (1972), 511.
23. M. Sagawa, S. Fujimura, N. Togawa, H. Yamamoto, and Y. Matsuura, *J. Appl. Phys.*, **55** (1984), 2083.
24. J. J. Becker, *J. Appl. Phys.*, **39** (1968), 1270; *IEEE Trans. Mag.*, **9** (1969), 214.
25. S. Chikazumi, *J. Mag. Mag. Mat.*, **54–7** (1986), 1551.
26. J. J. Croat, J. F. Herbst, R. W. Lee, and P. E. Pinkerton, *J. Appl. Phys.*, **55** (1984), 2078.
27. K. Shinagawa, Private communication.
28. S. Iwasaki, *Handbook on magnetic substances* (ed. by Chikazumi *et al.*, Asakura Publishing Co., Tokyo, 1975), Section 26.3.
29. N. Imamura and C. Ohta, *Jap. J. Appl. Phys.*, **19** (1980), L731.
30. N. Imamura, S. Tanaka, F. Tanaka, and Y. Nagao, *IEEE Trans. Mag.*, **MAG-21** (1985), 1607.
31. A. H. Bobeck, R. F. Fischer, A. J. Perneski, J. P. Remeika, and I. G. Van Uitert: *IEEE Trans. Mag.*, **MAG-5** (1969), 544.
32. I. Mikami, *Handbook on magnetic substances* (ed. by Chikazumi *et al.*, Asakura Publishing Co., Tokyo, 1975), Section 26.4.

SOLUTIONS TO PROBLEMS

CHAPTER 1

1.1 When the sphere is magnetized to the intensity I, the effective field becomes $H_{\rm eff} = H - (1/3\mu_0)I$. Since the magnetization I is attained by this effective field, we have the relationship $I = \chi H_{\rm eff} = \chi H - (\chi/3\mu_0)I$. Then we have

$$I = \frac{\chi}{1+(\bar{\chi}/3)} H.$$

The total magnetic moment of the sphere is

$$M = \frac{4\pi}{3} R^3 I = \frac{4\pi R^3 \chi}{3+\bar{\chi}} H.$$

When $\chi \to \infty$, the moment becomes $M = 4\pi R^3 \mu_0 H$.

1.2 The magnetic field produced in the air gap is given by $H = I/\mu_0$, which produces the Maxwell stress $T_{\parallel} = \frac{1}{2}\mu_0 H^2 = (1/2\mu_0)I^2$. Considering there are two gaps, we have the total force $F = (1/\mu_0)I^2 S$. From the energy point of view, this problem is interpreted as follows: Before the separation of the two halves, there are no free poles or there is no field in the space where I is present, so that the magnetostatic energy is zero. When the two halves are separated by the distance x, the air gaps store the magnetostatic energy $(1/\mu_0)I^2 Sx$. This expression can be deduced from (1.79) by considering that the magnetic potential difference between the magnetic poles $\pm IS$ is given by Hx, or from (1.105). Differentiating this expression with respect to x, we have the force acting between the two halves.

1.3 Using the magnetomotive force, $Ni = 200 \times 5 = 1 \times 10^3$ [A], and the reluctance,

$$R_{\rm m} = \frac{l_1}{\mu_0 \bar{\mu} S_1} + \frac{l_2}{\mu_0 S_2} = \frac{1}{4\pi \times 10^{-7}}\left[\frac{1}{500 \times 2.5 \times 10^{-3}} + \frac{0.01}{5 \times 10^{-4}}\right]$$
$$= \frac{0.8 + 20}{1.26 \times 10^{-6}} = 1.65 \times 10^7,$$

the field is calculated to be

$$H = \frac{\Phi}{\mu_0 S_2} = \frac{Ni}{V_{\rm m}\, \mu_0 S_2} = \frac{1 \times 10^3}{1.65 \times 10^7 \times 4\pi \times 10^{-7} \times 5 \times 10^{-4}} = 9.7 \times 10^4\,(\mathrm{A\,m^{-1}})$$
$$= 1200\,\mathrm{Oe}.$$

1.4 If there is no leakage of magnetic flux from the magnetic circuit, the cross-sectional area of the permanent magnet must be

$$S = \frac{\mu_0 \times 2.5 \times 10^5 \times 1.2 \times 10^{-3}}{0.15} = 2.5 \times 10^{-3}\,\mathrm{m}^2 = 25\,\mathrm{cm}^2.$$

From Ampère's theorem, the length of the permanent magnet parallel to the magnetization is given by

$$l = \frac{2.5\times 10^5 \times 1.4\times 10^{-2}}{6.0\times 10^4} = 0.058\,\text{m} = 58\,\text{mm}.$$

1.5 According to (1.99), we have

$$U = \frac{1}{2\mu_0}\,\tfrac{1}{2}I_r^2 \times \pi R^2 = \frac{\pi I_r^2 R^2}{4\mu_0}, \qquad \frac{\pi I_r^2 R^2}{4\mu_0\,\bar{\mu}}.$$

CHAPTER 2

2.1 By using $z = 0$, $l = \sqrt{3}\,r$, $R = r$ in (2.4), we have

$$C = n\frac{\sqrt{3}\,r}{\sqrt{r^2 + 3r^2}} = \frac{\sqrt{3}}{2}n.$$

(Remark: This value is equal to the cosine of the angle between the central axis and the line connecting the central point O and the edge of the solenoid in Fig. 2.1(a).)

2.2 From (2.7), we have $H = (Ni/l)\,(\text{A}\,\text{m}^{-1})$.

2.3 Refer to the text.

CHAPTER 3

3.1 Using $S = \frac{3}{2}$, $L = 6$, $J = 4\frac{1}{2}$, we have from (3.39)

$$g = 1 + \frac{\frac{9}{2}\times\frac{11}{2} + \frac{3}{2}\times\frac{5}{2} - 6\times 7}{2\times\frac{9}{2}\times\frac{11}{2}} = 0.73.$$

3.2 From (3.44), we have

$$\omega = \nu H = 2.21\times 10^5 \times 100 = 2.21\times 10^7\,\text{rad}\,\text{s}^{-1}.$$

3.3 Solving (3.47) with respect to H, we have

$$H = \frac{1}{2}\left[\sqrt{\left(\frac{I_s}{\mu_0}\right)^2 + 4\left(\frac{\omega}{\nu}\right)^2} - \frac{I_s}{\mu_0}\right].$$

Using $I_s/\mu_0 = 1.69\times 10^6$, $\omega/\nu = 2.71\times 10^5$, we find $H = 4.24\times 10^4\,\text{A}\,\text{m}^{-1}$.

3.4 Adding

$$\Theta_{20}^2 = \tfrac{5}{8}(9\cos^4\theta - 6\cos^2\theta + 1),$$

$$2\Theta_{2\pm 1}^2 = \tfrac{15}{2}(\cos^2\theta - \cos^4\theta),$$

and

$$2\Theta_{2\pm 2}^2 = \tfrac{15}{8}(1 - 2\cos^2\theta + \cos^4\theta),$$

we find that all the θ-dependent terms vanish.

CHAPTER 4

4.1 Using $g = 0.090/\frac{1}{2} = 0.180$, $H = -27 \times 10^6\,\mathrm{A\,m^{-1}}$, we have from (4.5)

$$f = \frac{0.180 \times (6.33 \times 10^{-33}) \times 2.7 \times 10^7}{2\pi \times (1.054 \times 10^{-34})} = 46.5\,\mathrm{MHz}.$$

4.2 The absorption lines in Fig. 4.5 from left to right correspond to the transitions ①, ②, ③, ④, ⑤, ⑥, in Fig. 4.3, respectively. Therefore the difference between ④ and ②, or ⑤ and ③, which is $0.39\,\mathrm{cm\,s^{-1}}$, must be equal to the Zeeman splitting of the ground state, $2MH$, from which we have $H = 26.2\,\mathrm{MA\,m^{-1}}$. The difference between ① and ②, or ⑤ and ⑥ corresponds to $4\Delta E$, so that we have $\Delta E = -0.12\,\mathrm{cm\,s^{-1}} = -9.2 \times 10^{-27}\,\mathrm{J}$. Moreover, taking the difference between ① and ⑥, and the difference between ② and ⑤ or ③ and ④, we have $\delta = 1.6\,\mathrm{mm\,s^{-1}}$.

4.3 From the relationship $\frac{1}{2}mv^2 = \frac{3}{2}kT$, we have $v^2 = 3kT/m$ or $v = \sqrt{3kT/m}$. Using this in (4.9), we have

$$\lambda = \frac{h}{m}\sqrt{\frac{m}{3kT}} = \frac{h}{\sqrt{3mkT}}.$$

Solving this equation, we have $T = h^2/3mk\lambda^2 = 40\,\mathrm{K}$.

CHAPTER 5

5.1 Using $N = 6.02 \times 10^{23} \times 8.94 \times 10^6/63.54 = 8.47 \times 10^{28}\,\mathrm{m}^3$, $a = 0.5 \times 10^{-10}\,\mathrm{m}$, $Z = 29$, $e = 1.60 \times 10^{-19}$, $m = 9.11 \times 10^{-31}$, we have from (5.8)

$$\bar{\chi} = -\frac{(8.47 \times 10^{28}) \times (4\pi \times 10^{-7}) \times (1.60 \times 10^{-19})^2 \times 29 \times (0.5 \times 10^{-10})^2}{6 \times (9.11 \times 10^{-31})}$$

$$= -\frac{8.47 \times 1.257 \times 1.60^2 \times 2.9 \times 0.25}{6 \times 9.1} \times 10^{28-6-38+1-20+31}$$

$$= 0.36 \times 10^{-4} = 3.6 \times 10^{-5}.$$

5.2 Using $N = 6.02 \times 10^{23}/22.4 \times 10^{-3} = 2.69 \times 10^{25}$, $g = 2$, $J = 1$, $M_B = 1.17 \times 10^{-29}$, $k = 1.38 \times 10^{-23}$, $T = 273$, $\mu_0 = 4\pi \times 10^{-7}$, we have from (5.26)

$$\bar{\chi} = \frac{(2.69 \times 10^{25}) \times 2^2 \times 1 \times 2 \times (1.17 \times 10^{-29})^2}{3 \times (1.38 \times 10^{-23}) \times 273 \times 4\pi \times 10^{-7}}$$

$$= \frac{2.69 \times 4 \times 2 \times 1.17^2}{3 \times 1.38 \times 2.73 \times 1.257} \times 10^{25-58+23-2+6}$$

$$= 2.07 \times 10^{-6}.$$

CHAPTER 6

6.1 From (6.9), we have

$$H_m = \frac{3Jk\Theta_f}{(J+1)M} = \frac{3\times 0.5\times(1.38\times 10^{-23})\times 628.3}{1.5\times 0.6\times(1.17\times 10^{-29})} = 1.24\times 10^9\,\mathrm{A\,m^{-1}}.$$

More accurate statistical treatment leads to a smaller value of $k\Theta_f/J$. Therefore the value of J or H_m deduced from the experimental value of Θ_f becomes larger.

6.2 At $T=\Theta_f$, (6.4) becomes $I = NM(\alpha/3) - H/w$. Comparing this with (6.2) and (5.22), we have $\alpha = (45)^{1/3}h^{1/3}$, where $h = H/NMw$. Using this in the above equation, we have the magnetization curve $I/NM = 1.19h^{1/3} - h$.

CHAPTER 7

7.1 We see in Fig. 7.7 that $\chi = \chi_0$ at $T=0$ and $T=\Theta_N/2$. Since $w_1=0$, we find that $\Theta_N = -\Theta_a$, so that $\chi = C/(T+\Theta_N)$. Therefore at $T=\Theta_N$, $\chi = C/(2\Theta_N) = \chi_0$. Thus we find that at $T = 2\Theta_N$, $\chi = \frac{2}{3}\chi_0 = 0.67\chi_0$.

7.2 Since $\alpha=\beta=0$, this ferrimagnet corresponds to the origin O in the α–β diagram (Fig. 7.11), which is in region II. Thus we find that $I_a = 2NM$, $I_b = -2NM$. Since $\lambda = \frac{2}{5}$ and $\mu = \frac{3}{5}$, we have $\Theta_f = -0.4899Cw$ from (7.41) and $\Theta_a = -0.4800Cw$ from (7.39) with (7.38), so that $\Theta_f/\Theta_a = 1.02$. Since the origin O is located in the region where $I_s < 0$ at $T=0$ as well as $T\sim\Theta_f$, no compensation point appears.

7.3 Since $J_2 = -(\sqrt{3}/6)J_1$, we have from (7.65) $\cos QC = \sqrt{3}/2$, from which we find that $QC = 2\pi/12$. Therefore we have from (7.62) $n = 2\pi/QC = 12$.

CHAPTER 8

8.1 From (8.29), we have

$$E_f(T) = E_f(0)\left[1 - \frac{\pi^2}{6}\frac{(kT)^2}{E_f(0)}\alpha\right].$$

Therefore the energy-dependence of the density of states becomes

$$g(E_f(T)) = g(E_f(0))\left(1 - \frac{\pi^2}{6}\alpha^2(kT)^2\right),$$

so that from (8.18) we have

$$\chi_p = 2M_B^2 g(E_f(0))\left(1 - \frac{\pi^2}{6}\alpha^2(kT)^2\right).$$

Thus we find that the susceptibility decreases with T, proportional to T^2.

8.2 $M = 1M_B$ is realized at electron numbers 27.6, 26.7 and 25.1 for Ni–Co, Co–Cr and Fe–Cr alloys, respectively. These numbers correspond to the alloy compositions 60 at% Ni–Co, 90 at% Co–Cr, and 55 at% Fe–Cr.

8.3, 8.4 Refer to the text.

CHAPTERS 9–11

Refer to the text.

CHAPTER 12

Table Sol. 12.1

	x	y	z
x'	$2/\sqrt{5}$	$1/\sqrt{5}$	0
y'	$1/\sqrt{5}$	$-2/\sqrt{5}$	0
z'	0	0	1

12.1 Set the new coordinates (x', y', z'), whose z'-axis is parallel to the z-axis and $(0, y', z')$ plane is parallel to the (210) plane. Then the normal to the $(0, y', z')$ plane, that is the x'-axis, has the direction cosines $(2/\sqrt{5}, 1/\sqrt{5}, 0)$ with respect to the cubic coordinates (x, y, z). The relationship between the cubic and new coordinates can be constructed as shown in Table Sol. 12.1 by using the relationship $\Sigma_i l_i^2 = \Sigma_i m_i^2 = \Sigma_i n_i^2 = 1$ and $\Sigma_i l_i m_i = \Sigma_i m_i n_i = \Sigma_i n_i l_i = 0$, where l_i, m_i and n_i represent the direction cosines of new coordinate axes with respect to the cubic coordinates. From Table Sol. 12.1, we have the relationships

$$\left.\begin{aligned} \alpha_1 &= (2/\sqrt{5})\alpha_1' + (1/\sqrt{5})\alpha_2' \\ \alpha_2 &= (1/\sqrt{5})\alpha_1' - (2/\sqrt{5})\alpha_2' \\ \alpha_3 &= \alpha_3'. \end{aligned}\right\}$$

Since $\alpha_1' = 0$, $\alpha_2' = \sin\theta$, and $\alpha_3' = \cos\theta$, we have

$$\left.\begin{aligned} \alpha_1 &= (1/\sqrt{5})\sin\theta \\ \alpha_2 &= -(2/\sqrt{5})\sin\theta \\ \alpha_3 &= \cos\theta. \end{aligned}\right\}$$

Then the K_1-term of the anisotropy energy becomes

$$\begin{aligned} E_a &= K_1(\sin^2\theta\cos^2\theta + \tfrac{4}{25}\sin^4\theta) \\ &= K_1(\sin^2\theta - \tfrac{21}{25}\sin^4\theta). \end{aligned}$$

12.2 Using (12.48), we have the anisotropy field

$$H_a = 2K_{u1}/I_s = 4.58 \times 10^5\,\mathrm{A\,m^{-1}} = 5750\,\mathrm{Oe}.$$

12.3 Since the number of atoms in a unit volume is calculated by

$$N = 2/a^3 = 2/(3\times 10^{-10})^3 = 7.41\times 10^{28}\ \mathrm{m}^{-3},$$

we have from (12.64)

$$K_1 = (16/9)Nq = (16/9)\times 7.41\times 10^{28}\times 3\times^{-25} = 3.95\times 10^4\ \mathrm{J\,m}^{-3}.$$

12.4 Assuming $n = 1$, we have from (12.72)

$$\langle K^{(1)}\rangle \propto I_s.$$

Therefore the anisotropy energy is given by

$$E_a = K\cos\varphi.$$

CHAPTER 13

13.1 Since $k_1 = 0$, the induced anisotropy (13.11) is expressed only by the second term. Let the angle between the spontaneous magnetization and the z-axis in the $(1\bar{1}0)$ plane be θ. Then we have $\alpha_1 = \alpha_2 = (1/\sqrt{2})\sin\theta$ and $\alpha_3 = \cos\theta$, so that Σ of the second term in (13.11) becomes

$$\Sigma = \alpha_1\alpha_2\beta_1\beta_2 + \alpha_2\alpha_3\beta_2\beta_3 + \alpha_3\alpha_1\beta_3\beta_1$$
$$= (1/2)\beta_1\beta_2\sin^2\theta + (1/\sqrt{2})\beta_2\beta_3\sin\theta\cos\theta + (1/\sqrt{2})\beta_3\beta_1\sin\theta\cos\theta.$$

Therefore for [001] annealing $\beta_1 = \beta_2 = 0$, so that $\Sigma = 0$. For [110] annealing $\beta_1 = \beta_2 = (1/\sqrt{2})$ and $\beta_3 = 0$, so that

$$\Sigma = \tfrac{1}{4}\sin^2\theta = -\tfrac{1}{4}\sin^2(\theta - \theta_0),$$

where θ_0 is the angle between [110] and the z-axis, that is $\pi/2$. For [111] annealing $\beta_1 = \beta_2 = \beta_3 = (1/\sqrt{3})$, so that

$$\Sigma = \tfrac{1}{6}\sin^2\theta + (\sqrt{2}/3)\sin\theta\cos\theta$$
$$= \tfrac{1}{4}\{-\tfrac{1}{3}\cos 2\theta + (\sqrt{2}/3)\sin 2\theta\} + \text{const.}$$
$$= \tfrac{1}{4}\{\cos 2\theta_0\cos 2\theta + \sin 2\theta_0\sin 2\theta\} + \text{const.}$$
$$= \tfrac{1}{4}\{\cos 2(\theta - \theta_0)\} + \text{const.}$$
$$= -\tfrac{1}{2}\sin^2(\theta - \theta_0) + \text{const.},$$

where θ_0 is the angle between the [111] and the z-axis. Note that $\sin\theta_0 = \sqrt{2}/\sqrt{3}$, $\cos\theta_0 = 1/\sqrt{3}$, $\sin 2\theta_0 = 2\sqrt{2}/3$, $\cos 2\theta_0 = -\tfrac{1}{3}$. Therefore the ratios of the induced anisotropies measured in the (110) plane for [001], [110] and [111] annealing are 0 : 1 : 2.

13.2 Let the angle between the x-axis and the spontaneous magnetization be θ. Then $\alpha_1 = \cos\theta$, $\alpha_2 = \sin\theta$, $\alpha_3 = 0$. For [110] magnetic annealing, $\beta_1 = \beta_2 = 1/\sqrt{2}$, $\beta_3 = 0$. Therefore the anisotropy energy (13.11) becomes

$$E_a = -\frac{N_{BB}l_0l_0'}{kT}\{k_1 + (k_2\sin\theta\cos\theta)/2\}$$
$$= -\frac{N_{BB}l_0l_0'}{kT}\{(k_2\cos^2(\theta - \pi/4)/2) + \text{const.}\}.$$

Comparing this with (13.20), we have for ⟨110⟩

$$K_u = \frac{N_{BB}l_0l_0'}{kT}k_2/2.$$

Comparing it with (13.22), the ratio of the induced anisotropies in ⟨110⟩ measured in the planes parallel to the (001) and (110) planes is given by $k_2/2:(k_1/2+k_2/4)$. Then using the values of k_1 and k_2 for simple cubic, body-centered cubic and face-centered cubic structures given in Table 13.1, i.e. $k_1=\frac{1}{3}$, $k_2=0$; $k_1=0$, $k_2=\frac{4}{5}$; and $k_1=\frac{1}{12}$, $k_2=\frac{4}{12}$, we have for sc $0:\frac{1}{6}=0:1$; for bcc $\frac{2}{5}:\frac{1}{5}=2:1$; and for fcc $\frac{1}{6}:\frac{1}{24}+\frac{1}{12}=4:3$.

13.3 Referring to (13.28) and (13.29) with slip system No. 3 in Table 13.3, we have Σ in (13.28)

$$\Sigma=s(\alpha_1^2/2+\alpha_2^2/2+\alpha_1\alpha_2),$$

where s is the slip density for slip system No. 3.

Table Sol. 13.3

	x	y	z
x'	$1/\sqrt{6}$	$1/\sqrt{6}$	$-\sqrt{2}/\sqrt{3}$
y'	$-1/\sqrt{2}$	$1/\sqrt{2}$	0
z'	$1/\sqrt{3}$	$1/\sqrt{3}$	$1/\sqrt{3}$

Let the angle between the spontaneous magnetization and $[1\bar{1}0]$ be θ, and set new coordinates (x', y', z'), in which y' is parallel to $[1\bar{1}0]$; z' is parallel to [111]. The relationships of the new coordinates and the cubic coordinates (x, y, z) are given by Table Sol. 13.3. Then we have

$$\left.\begin{aligned}\alpha_1&=1/\sqrt{6}\,\alpha_1'-1/\sqrt{2}\,\alpha_2'+1/\sqrt{3}\,\alpha_3'\\ \alpha_2&=1/\sqrt{6}\,\alpha_1'+1/\sqrt{2}\,\alpha_2'+1/\sqrt{3}\,\alpha_3'\\ \alpha_3&=-\sqrt{2}/\sqrt{3}\,\alpha_1'+1/\sqrt{3}\,\alpha_3'.\end{aligned}\right\}$$

Since $\alpha_1'=\cos\theta$, $\alpha_2'=\sin\theta$, $\alpha_3'=0$, we have

$$\left.\begin{aligned}\alpha_1&=1/\sqrt{6}\cos\theta-1/\sqrt{2}\sin\theta\\ \alpha_2&=1/\sqrt{6}\cos\theta+1/\sqrt{2}\sin\theta\\ \alpha_3&=-2/\sqrt{3}\cos\theta.\end{aligned}\right\}$$

Therefore

$$\Sigma=(s/3)\cos^2\theta,$$

so that we have the anisotropy energy

$$E_a=\tfrac{1}{24}Nl_0pS^2s\cos^2\theta.$$

CHAPTER 14

14.1 By applying the magnetic field parallel to the [010] direction, the domain magnetizations in [100] and $[\bar{1}00]$ change their directions to [010] by 90° wall displacement. Considering that the magnetostrictive elongation is proportional to the magnetization I, and the final elongation at saturation magnetization I_s is given by $\frac{3}{2}\lambda_{100}$, we have

$$\frac{\delta l}{l}=\tfrac{3}{2}\lambda_{100}\frac{I}{I_s}.$$

14.2 Referring to (14.33) and (14.24), we find

$$Nl \approx c\lambda = 10^{11} \times 10^{-5} = 10^{6}\,\mathrm{J\,m^{-3}}.$$

14.3 Putting $\alpha_1 = \cos\theta$, $\alpha_2 = \sin\theta$, we have from (14.43)

$$\frac{\delta l}{l} = h_1(\beta_1^2 \cos^2\theta + \beta_2^2 \sin^2\theta) + h_3(\sin^2\theta\cos^2\theta) + \text{const.}$$

For [100] elongation, $\beta_1 = 1$, $\beta_2 = 0$, so that we have

$$\left(\frac{\delta l}{l}\right)_{[100]} = h_1 \cos^2\theta + h_3 \sin^2\theta\cos^2\theta.$$

For [010] elongation, $\beta_1 = 0$, $\beta_2 = 1$, so that we have

$$\left(\frac{\delta l}{l}\right)_{[010]} = h_1 \sin^2\theta + h_3 \sin^2\theta\cos^2\theta.$$

14.4 Putting direction cosines of [123], or $\gamma_1 = 1/\sqrt{14}$, $\gamma_2 = 2/\sqrt{14}$, and $\gamma_3 = 3/\sqrt{14}$ in (14.89), we have

$$E_\sigma = -\tfrac{3}{28}\lambda_{100}\sigma(\alpha_1^2 + 4\alpha_2^2 + 3\alpha_3^2) - \tfrac{3}{14}\lambda_{111}\sigma(2\alpha_1\alpha_2 + 6\alpha_2\alpha_3 + 3\alpha_3\alpha_1).$$

Since $(\alpha_1, \alpha_2, \alpha_3)$ are $(1,0,0)$ for a [100] domain, we have $E_\sigma = -\tfrac{3}{28}\lambda_{100}\sigma$. Similarly, we have for a [010] domain $E_\sigma = -\tfrac{3}{7}\lambda_{100}\sigma$, and for a [001] domain $E_\sigma = -\tfrac{27}{28}\lambda_{100}\sigma$.

14.5 Using the values for Fe in Table 14.3, we have from (14.100)

$$\begin{aligned}\Delta K_1 &= -\tfrac{9}{4}\{(1.46 - 2.14)\times 10^{11} \times (20.7\times 10^{-6})^2 + 2\times 1.12\times 10^{11}\times(21.2\times 10^{-6})^2\}\\ &= -1.35\times 10^{2}\,\mathrm{J\,m^{-3}}.\end{aligned}$$

CHAPTER 16

16.1 Since the uniaxial anisotropy energy constant, K_u, in this case is given by $\tfrac{3}{2}\lambda\sigma$, the wall energy (16.47) becomes

$$\gamma = \sqrt{24A\lambda\sigma}.$$

16.2 Using (16.49) and assuming $\alpha_3' = 0$, we have for the second term of the anisotropy energy (12.5)

$$E_a = K_2 \sin^2\psi\cos^2\psi\,\alpha_1'^2\alpha_2'^4.$$

For (100), which is equivalent to (010), $\psi = 0$ (see Fig. 16.14), so that $\sin\psi = 0$. Therefore we have $E_a = 0$. Then the 180° wall energy (16.47) gives $\gamma_{100} = 0$. Similarly for (110), which is equivalent to (011), $\psi = \pi/4$, so that $\sin(\pi/4) = \cos(\pi/4) = 1/\sqrt{2}$. Therefore we have

$$E_a = (K_2/4)\alpha_1'^2\alpha_2'^4 = (K_2/4)\sin^4\varphi\cos^2\varphi.$$

Then the wall energy (16.42) becomes

$$\gamma_{110} = \sqrt{AK_2}\int_{-\pi/2}^{\pi/2} \sin^2\varphi\cos\varphi\,d\varphi.$$

If we put $\sin\varphi = x$, the above integral is converted to $\int_{-1}^{1} x^2\,dx$ and finally we have

$$\gamma_{110} = \sqrt{AK_2}\int_{-1}^{1} x^2\,dx = \sqrt{AK_2}\,|x^3/3|_{-1}^{1} = \tfrac{2}{3}\sqrt{AK_2}.$$

16.3 Putting $\Theta_f = 527 + 273 = 800$ in (16.5), we have

$$J = 0.34 \times 1.38 \times 10^{-23} \times 800 = 3.75 \times 10^{-21}\ \text{J}.$$

Using this value and $S = \frac{1}{2}$, $n = 2$ and $a = 3 \times 10^{-10}$ in (16.15), we have

$$A = 2 \times 3.75 \times 10^{-21} \times 0.5^2/3 \times 10^{-10} = 6.25 \times 10^{-12}.$$

Therefore we have the wall energy (16.55)

$$\gamma_{100} = 2 \times \sqrt{6.25 \times 10^{-12} \times 6 \times 10^4} = 1.22 \times 10^{-3}\ \text{J m}^{-2}.$$

16.4 Since the stable 90° walls make ±62° with the x–y plane (see Fig. 16.21 and equation (16.66)), the inner angles of the x-domain are $62° \times 2 = 124°$ and $180° - 124° = 56°$.

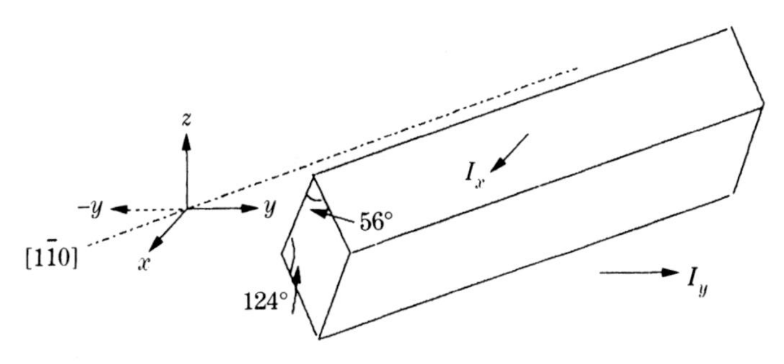

Fig. Sol. 16.4

CHAPTER 17

17.1 The magnetic pole density appearing on the 90° wall is given by

$$\omega = \pm I_s(\cos\phi - \sin\phi).$$

The magnetostatic energy per unit area of the 90° wall is given by (17.10) where I_s is replaced by ω, so that

$$\varepsilon_m = 5.40 \times 10^4 I_s^2 d(\cos\phi - \sin\phi)^2$$
$$= 5.40 \times 10^4 I_s^2 d(1 - \sin 2\phi).$$

In this case we must take into consideration the μ^* correction (17.31) where μ^* is given by (17.30). Therefore the magnetostatic energy per unit area of the 90° wall is given by

$$\varepsilon_m = 5.40 \times 10^4 I_s^2 d(1 - \sin 2\phi)/(1 + I_s^2/4\mu_0 K_1).$$

17.2 In case (i), the domain width d is given by (17.37), showing the dependence on the wall energy as $\sqrt{\gamma}$. Since the wall energy depends on K_u as $\sqrt{K_u}$, as shown by (16.47), we find that the domain width d depends on K_u as $K_u^{1/4}$.

In case (ii), we must take into consideration μ^* given by (17.29), where 1 can be ignored. Similarly in the μ^* correction A (17.26), 1 in the denominator can be ignored, so that A is proportional to K_u. Therefore we find in (17.37) that the domain width d depends on K_u as $K_u^{1/4}/K_u^{1/2} = K_u^{-1/4}$.

17.3 Using $I_s = 1.8$, $K_u = 4.1 \times 10^5$ in (17.39), we have

$$\chi = 1.8^2/(2 \times 4.1 \times 10^5) = 3.95 \times 10^{-6},$$

so that

$$\bar{\mu}^* = 1 + 3.95 \times 10^{-6}/4\pi \times 10^{-7} = 4.14.$$

The μ^* correction (17.26) becomes

$$A = 2/(1 + 4.14) = 0.389.$$

Using this correction in (17.36), we have the domain width

$$\begin{aligned} d &= \sqrt{\gamma l/1.08 \times 10^5 A}\,/(I_s \sin \theta_0) \\ &= \sqrt{(1.5 \times 10^{-2} \times 1 \times 10^{-2})/(1.08 \times 10^5 \times 0.389)}\,/(1.8 \times 0.1) \\ &= 3.31 \times 10^{-4} = 0.33\,\text{mm}. \end{aligned}$$

17.4 According to (17.86), the magnetostatic energy must be 0.20 times the value for an isolated single domain particle, so that the critical radius for single domain particle (17.75) must become $1/0.20 = 5.0$ times. Therefore the critical size for polycrystalline iron must be 5.0 times the value of 20 Å given by (17.76), which is equal to 100 Å or 0.01 μm.

CHAPTER 18

18.1 Since the residual magnetization remains in the positive direction in each particle, we have

$$I_r/I_s = \frac{1}{\pi}\int_{-\pi/2}^{\pi/2} \cos\theta \,\mathrm{d}\theta = \frac{1}{\pi}|\sin\theta|_{-\pi/2}^{\pi/2} = \frac{2}{\pi} = 0.637.$$

18.2 Since $l = a/2$ and $\theta = \pi/4$, so that $\cos\theta = 1/\sqrt{2}$, we have from (18.66)

$$H_0 = 2\sqrt{2}\,\gamma_{180°}/(I_s a).$$

18.3 Note that when H is parallel to the c-axis, only 180° wall displacement occurs, while when H is perpendicular to the c-axis, only magnetization rotation occurs. When the magnetic field makes an angle θ with the c-axis, the relative susceptibility is expressed by

$$\bar{\chi} = \bar{\chi}_{\text{wall}} \cos^2\theta + \bar{\chi}_{\text{rot}} \sin^2\theta,$$

where $\bar{\chi}_{\text{wall}} = 20$, $\bar{\chi}_{\text{rot}} = 5$. For $\theta = 30°$, $\cos^2\theta = \frac{3}{4}$, $\sin^2\theta = \frac{1}{4}$, so that $\bar{\chi} = (20 \times \frac{3}{4}) + \frac{5}{4} = 16.3$. For $\theta = 45°$, $\cos^2\theta = \sin^2\theta = \frac{1}{2}$, so that $\bar{\chi} = \frac{20}{2} + \frac{5}{2} = 12.5$. For $\theta = 60°$, $\cos^2\theta = \frac{1}{4}$, $\sin^2\theta = \frac{3}{4}$, so that $\bar{\chi} = \frac{20}{4} + (5 \times \frac{3}{4}) = 8.8$. For polycrystalline cobalt,

$$\overline{\cos^2\theta} = \int_0^{\pi/2} \cos^2\theta \sin\theta \,\mathrm{d}\theta = \tfrac{1}{3}, \qquad \overline{\sin^2\theta} = \int_0^{\pi/2} \sin^2\theta \sin\theta \,\mathrm{d}\theta = \tfrac{2}{3},$$

so that $\bar{\chi} = \frac{20}{3} + (5 \times \frac{2}{3}) = 10.$

18.4 Since $N_z = 0$, the shape anisotropy constant is deduced from (12.24) as

$$K_u = I_s^2/4\mu_0.$$

When a field slightly more than $I_s/4\mu_0 = K_u/I_s$ is applied in the x-direction, the magnetization which makes an angle $+45°$ with the $-y$-direction reverses its direction by irreversible magnetization rotation (see Fig. 18.29). After the field is rotated towards y, $-x$, $-y$, and finally comes back to the $+x$-direction, and decreases to zero, the magnetization which makes angles between $-135°$ to $45°$ to the $+x$-axis remain. Referring to the solution of (18.1), we can conclude that the residual magnetization of $2I_s/\pi$ is produced in the direction 45° from the $+x$-axis towards the $-y$-axis.

18.5 From (18.107), we have the average susceptibility

$$\chi = I_1/H_1 = \chi_a + \eta H_1.$$

Thus the susceptibility increases linearly with an increase of H_1.

CHAPTER 19

19.1 Since there is no magnetic anisotropy, the spin axis is always perpendicular to the field while the field is weak. As the field, H, is increased, spins on both the sites cant towards H, and finally become ferromagnetic at H_s. Since H_s is proportional to $I_s(= I_A = -|I_B|)$, the form of the H_s vs T curve is Q type (see Fig. Sol. 19.1).

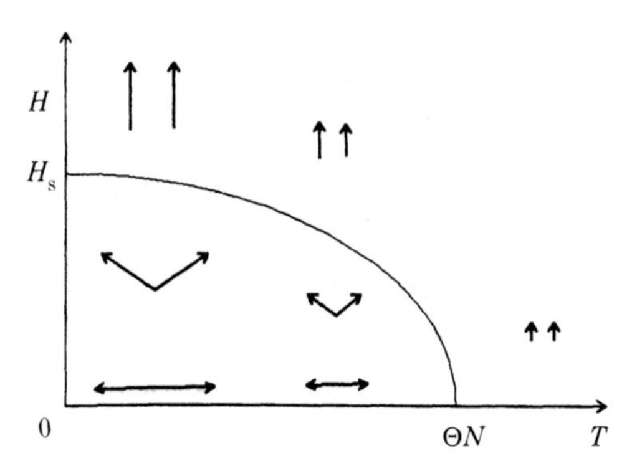

Fig. Sol. 19.1

19.2 The triangular arrangement starts at the field $H_{c1} = (w/3)NM$ and ends at $H_{c2} = wNM$, during which I increases from $NM/3$ to NM linearly with H. Since I_A and I_B are $2:1$ in magnitude, I_A makes the angle 30° with H and its magnitude is $(\sqrt{3}/2)wI_A$.

CHAPTER 20

20.1 Using the approximation for $l_c \ll kT$ in (20.22), we have

$$\text{at } t = 0 \qquad E_{a0} = (N_c l_c^2/3kT)(\tfrac{1}{3} - \alpha_2^2)$$

$$\text{at } t = \infty \qquad E_{a0} = (N_c l_c^2/3kT)(\tfrac{1}{3} - \alpha_1^2).$$

Using these relations in (20.26) we have

$$-\frac{N_c l_c^2}{3kT}\{\alpha_1^2 + (\alpha_2^2 - \alpha_1^2)e^{-t/\tau}\}.$$

20.2 The coefficient of $\ln t$ in (20.36) for $H = 0.98(2K_{umax}/I_s)$ is larger than that for $H = 0.90(2K_{umax}/I_s)$ by a factor

$$(1-0.90)^2/(1-0.98)^2 = 25.$$

20.3 Outside the moving wall, there is no flux change, so that the induced electric field must be constant. Let the induced field above and below the wall be E_a and E_b, respectively, then we have

$$E_a - E_b = 2I_s v.$$

Let the current density above and below the wall be i_a and i_b, respectively. By the condition for steady current, we have

$$i_a z = -i_b(d-z).$$

Since $i_a\rho = E_a$, and $i_b\rho = E_b$, we can solve for E_a and E_b as

$$E_a = 2I_s v(d-z)/d \quad \text{and} \quad E_b = -2I_s vz/d,$$

respectively. Therefore the total eddy current loss per unit area of the plate is given by

$$\begin{aligned} P &= E_a^2 z/\rho + E_b^2(d-z)/\rho \\ &= (2I_s v/d)^2\{(d-z)^2 z + z^2(d-z)\}/\rho \\ &= (2I_s v)^2(d-z)z/(\rho d). \end{aligned}$$

20.4 From (20.86), we have

$$\alpha = (4\pi)^2 \times 10^{-7} \times 1 \times 10^8/(1.105 \times 10^5 \times 2 \times 1) = 0.00715.$$

Since $\alpha \ll 1$, we have from (20.91) and (20.98)

$$\omega = \omega_0 = 1.105 \times 10^5 \times 2 \times (-1 \times 10^2) = -2.21 \times 10^7\,\text{s}^{-1},$$

and

$$\tau = \tau_0 = -1/(0.00715 \times 2.21 \times 10^7) = -6.33 \times 10^{-6}\,\text{s},$$

respectively. Since $\theta_0 = 60°$ and $\theta = 120°$, we have from (20.95)

$$\begin{aligned} t &= (\ln\tan(\theta_0/2) - \ln\tan(\theta/2)) \\ &= -6.33 \times 10^{-6}(\ln\tan 30° - \ln\tan 60°) \\ &= -6.33 \times 10^{-6} \times (-0.549 - 0.549) = 6.95 \times 10^{-6}\,\text{s}. \end{aligned}$$

Since the precession frequency $f = \omega/2\pi = 3.52 \times 10^6$, the number of turns during this time is given by

$$ft = 3.52 \times 10^6 \times 6.95 \times 10^{-6} = 24.4.$$

20.5 Putting $\alpha = H = 0$ in (20.129), we have the differential equation

$$m(\mathrm{d}^2 s/\mathrm{d}t^2) + \beta(\mathrm{d}s/\mathrm{d}t) = 0.$$

Integrating each term with respect to t, we have

$$m(\mathrm{d}s/\mathrm{d}t) + \beta s = 0,$$

from which we have

$$s = mv/\beta.$$

Using (20.138) and (20.147), this is converted and calculated to be

$$s = v/4\pi\lambda = 10/(4\pi \times 1 \times 10^8) = 8.0 \times 10^{-9}\ \mathrm{m} = 80\ \text{Å}.$$

Appendix 1

SYMBOLS USED IN THE TEXT

A	exchange stiffness constant
B	magnetic flux density
B_1, B_2	magnetoelastic constants
C	Curie constant; specific heat; nuclear scattering amplitude for neutrons
D	electric flux density; magnetic scattering amplitude for neutrons
E	energy density; electric field; Young's modulus
E_{H}	magnetic field energy density
E_{a}	magnetic anisotropy energy density
E_{mag}	magnetostatic energy density
E_{magel}	magnetoelastic energy density
E_{an}	magnetic after-effect anisotropy energy density
E_{σ}	magnetostrictive anisotropy energy density
F	force
H	magnetic field
H_{m}	molecular field
H_0	critical field
H_{c}	coercive field; critical field for superconductivity or spin flopping
H	anisotropy field
I	intensity of magnetization
I_{s}	saturation (spontaneous) magnetization
I_0	saturation magnetization at 0 K
I_{r}	residual magnetization
I_{rh}	rotational hysteresis integral
J	exchange integral; total angular momentum
K	magnetic anisotropy constant
K_1, K_2	cubic magnetocrystalline anisotropy constants
K_{u}	uniaxial anisotropy constant
K_{d}	unidirectional anisotropy constant
L	torque density; orbital angular momentum
M	magnetic moment
M_{B}	Bohr magneton
N	number in a unit volume; demagnetizing factor; total number of turns
N_{C}	number of carbon atoms in a unit volume
P	angular momentum; power
Q	activation energy; quadrupole moment
R	Hall coefficient; radius
R_{m}	reluctance
S	long-range order parameter; area; spin angular momentum
T	absolute temperature

U	energy
V	thermoelectric power
V_m	magnetomotive force
W	work
W_h	hysteresis loss
W_r	rotational hysteresis loss
Z	atomic number
a	lattice constant; coefficient of the shape effect (magnetostriction)
b	coefficient in law of approach to saturation
c	elastic modulus; compressibility; light velocity
c_{11}, c_{44}, c_{12}	elastic moduli
d	domain width
e	strain; electronic charge
f	frequency
g	g-factor; anisotropy energy function; exchange term
h	Planck's constant; magnetostriction constants for five constant expression
i	current; current density
i_{ma}	macroscopic current density
i_{mi}	microscopic current density
k	Klirr factor; dimensional ratio; Boltzmann factor
l	coefficient of dipole–dipole interaction; length; thickness; orbital angular momentum quantum number
m	magnetic pole; domain wall mass; magnetic quantum number
n	number of turns per unit length; principal quantum number
l, m, n	direction cosines
p	pressure; probability
q	coefficient of quadrupole interaction; electric field gradient
r	radius; distance; roll reduction
s	displacement; skin depth; slip density; spin quantum number
s_{11}, s_{44}, s_{12}	elastic constants
t	time
v	velocity; volume
w	molecular field coefficient; pair energy
x, y, z	Cartesian coordinates
z	number of nearest neighbors
α	curvature of potential valley; variables of Langevin and Brillouin functions; thermal expansion coefficient; damping factor
$(\alpha_1, \alpha_2, \alpha_3)$	direction cosines of magnetization
α, β	molecular field coefficients
β	damping coefficient of domain wall; packing factor
$(\beta_1, \beta_2, \beta_3)$	direction cosines of observation direction and the annealing field
γ	surface density of domain wall energy
$(\gamma_1, \gamma_2, \gamma_3)$	direction cosines of atomic pair
δ	thickness of domain wall; loss angle
ε	surface density of energy; ratio of orbit to spin
ε_m	magnetostatic energy per unit area

ε_w	wall energy per unit area
ζ	ratio of delayed to instantaneous magnetization
η	Rayleigh constant
θ	angle (particularly between magnetization and the field)
λ	magnetostriction constant; relaxation frequency; spin–orbit parameter
λ, μ	relative number of Fe^{3+} ions on A and B sites
μ	permeability
$\bar{\mu}$	relative permeability
μ_0	permeability of vacuum
μ_a	initial permeability
μ_{max}	maximum permeability
ν	gyromagnetic constant
ρ	resistivity
ρ_m	volume density of magnetic pole
σ	electrical conductivity; tension; short-range order parameter; mass magnetization
τ	relaxation time
φ, ϕ	angle (azimuthal); magnetic potential
χ	magnetic susceptibility
$\bar{\chi}$	relative magnetic susceptibility
ψ	angle
Θ	Curie point
Θ_f	ferromagnetic Curie point
Θ_N	Néel point
Θ_a	asymptotic Curie point
Φ	magnetic flux

Appendix 2

CONVERSION OF VARIOUS UNITS OF ENERGY*

eV	cm^{-1}	K	J	cal	MA/m†
1	$=0.80655 \times 10^{4}$	$=1.1604 \times 10^{4}$	$=1.60218 \times 10^{-19}$	$=3.8292 \times 10^{-20}$	$=1.37477 \times 10^{4}$
$1.23985 \times 10^{-4} =$	1	= 1.43872	$=1.98646 \times 10^{-23}$	$=4.7476 \times 10^{-24} =$	1.70450
$0.86177 \times 10^{-4} =$	0.69506	= 1	$=1.38071 \times 10^{-23}$	$=3.2999 \times 10^{-24} =$	1.18473
0.62415×10^{19}	$=0.50341 \times 10^{23}$	$=0.72426 \times 10^{23} =$	1	$=2.3900 \times 10^{-1}$	$=8.5806 \times 10^{22}$
2.61151×10^{19}	$=2.10631 \times 10^{23}$	$=3.03040 \times 10^{23} =$	4.1840	= 1	$=3.5901 \times 10^{23}$
$7.27396 \times 10^{-5} =$	0.58668	= 0.84407	$=1.16542 \times 10^{-23}$	$=2.7854 \times 10^{-24} =$	1

* The values in this table are based on the recommendations of the CODATA task group, 1973. Ref. *CODATA Bulletin* **11**, 7, Table IV (1973).

† This column expresses the value of the magnetic field H, which gives the corresponding energy $M_B H$, when it acts on 1 Bohr magneton (M_B).

Appendix 3

IMPORTANT PHYSICAL CONSTANTS*

Velocity of light	$c = 2.99792458 \times 10^{8}\ \mathrm{m\,s^{-1}}$
Acceleration due to gravity	$g = 9.80665\ \mathrm{m/s^{2}}$
Universal gravitation constant	$G = 6.67259 \times 10^{-11}\ \mathrm{N\,m^{2}\,kg^{-2}}$
Planck's constant	$h = 6.6260755 \times 10^{-34}\ \mathrm{J\,s}$
	$\hbar = h/2\pi = 1.05457266 \times 10^{-34}\ \mathrm{J\,s}$
Mechanical equivalent of heat	$J = 4.1840\ \mathrm{J\,(15°\,cal)^{-1}}$
Boltzmann's constant	$k = 1.380658 \times 10^{-23}\ \mathrm{J\,K^{-1}}$
Value of kT at 0°C	$kT_0 = 3.771 \times 10^{-21}\ \mathrm{J}$
Avogadro's number	$N = 6.0221367 \times 10^{23}\ \mathrm{mol^{-1}}$
Electronic mass	$m = 9.1093897 \times 10^{-31}\ \mathrm{kg}$
Electronic charge (absolute value)	$e = 1.60217733 \times 10^{-19}\ \mathrm{C}$
Ratio charge/mass	$e/m = 1.75881962 \times 10^{11}\ \mathrm{C\,kg^{-1}}$
Faraday constant	$F = Ne = 9.6485309 \times 10^{4}\ \mathrm{C\,mol^{-1}}$
Bohr magneton	$M_{\mathrm{B}} = 1.16540715 \times 10^{-29}\ \mathrm{Wb\,m}$
Gyromagnetic constant	$\nu = 1.10509896 \times 10^{5} g\ \mathrm{m\,A^{-1}\,s^{-1}}$
Flux quantum	$\Phi_0 = h/2e = 2.06783461 \times 10^{-15}\ \mathrm{Wb}$

* Mostly from CODATA 1986.

Appendix 4

PERIODIC TABLE OF ELEMENTS AND MAGNETIC ELEMENTS

Ia	IIa	IIIa	IVa
3 Li 1347 Lithium 180.54 ⊡ 2*s* −195 6.941 0.534 ⊙	**4 Be** 2970 Beryllium 1278 ⬡ $2s^2$ 500 9.01218 1.85 ⊙	**1 H** Hydrogen −252.87 −259.14 1*s* 1.0079 H_2 ⊙ 0.0763(−260)	**2 He** −268.93 Helium −272.2 26 MPa $1s^2$ ^3He ⊙ 4.00260 ^4He ⊙ 0.19(−273)
11 Na 882.9 Sodium 97.81 3*s* ⊡ 22.98977 0.9712 ✣	**12 Mg** 1090 Magnesium 648.8 $3s^2$ 24.305 ⊙ 1.74		
19 K 774 Potassium 63.65 4*s* 39.098 ⊡ 0.87	**20 Ca** 1484 Calcium 839 ⊡ $4s^2$ 464 40.08 ✣ 1.55	**21 Sc** 2832 Scandium 1539 $3d4s^2$ ✣ 44.9559 2.992 ⊙	**22 Ti** 3287 Titanium 1660 ⊡ $3d^24s^2$ 882 47.90 ⊙ 4.5
37 Rb 688 Rubidium 38.89 5*s* 85.4678 ⊡ 1.53	**38 Sr** 1384 Strontium 769 $5s^2$ ⊡ 87.62 605 ⊙ 2.60 215 ✣	**39 Y** 3337 Yttrium 1523 $4d^15s^2$ 88.9059 ⊙ 4.478	**40 Zr** 4377 Zirconium 1852 ⊡ $4d^25s^2$ 870 91.22 ⊙ 6.44
55 Cs 678.4 Cesium 28.40 6*s* 132.9054 1.873 ⊡	**56 Ba** 1640 Barium 725 $6s^2$ 137.34 ⊡ 3.5	57–71 (Lanthanides)	**72 Hf** 4602 Hafnium 2227 $5d^26s^2$ 178.49 ⊙ 13.3
87 Fr (667) Francium (27) 7*s* (223) —	**88 Ra** 1140 Radium 700 $7s^2$ 226.0254 5(?)	89–103 (Actinides)	**57 La** 3454 Lanthanum 920 ⊡ $5d6s^2$ 864 138.9055 β ✣ 6.174(α) 260 α ⬡
104 Ku — Kurchatovium — $6d^27s^2$ — —	**105 Ha** — Hahnium — — — —		**89 Ac** 3200 Actinium 1050 $6d7s^2$ (227) 10.07

Magnetic parameters

- θ_N : Néel pt. (K)
- θ_f : Curie pt. (K)
- θ_{fa} : ferro-antif. trans. pt. (K)
- M_{eff} : effect. mag. moment (M_{B})
- M_s : satur. mag. moment (M_{B})
- $M_{\pm}$: antif. subl. moment (M_{B})
- M_{sd} : amp. SDW (M_{B})
- σ : mass mag. moment (emu/g)
- I_s : satur. mag. (T)

Cr	Mn	Fe	
spin density wave $\theta_N = 311$ wavelength $\lambda_\perp$ = 27 lattice c. spin flip pt. = 123 K wavelength $\lambda_\parallel$ = 21 lattice c. $M_{sd} = 0.57 - 0.59$	γ Mn $\left\{ \begin{array}{l} \theta_N = 480\ ? \\ M_\pm = 2.25 \end{array} \right.$ βMn no magnetic order αMn $\theta_N = 95$ $M_{\mathrm{I}} = 1.9$, $M_{\mathrm{II}} = 1.7$ $M_{\mathrm{III}} = 0.6$, $M_{\mathrm{IV}} = 0.2$	bcc $M_{\mathrm{eff}} = 3.20$ $\theta_f = 1040.2$ $I_{s20} = 2.1580$ $\sigma_0 = 221.7$ $M_s = 2.216$ fcc $\theta_N = 67$ $M_\pm = 0.7?$	hex > 15 GPa nonferro-magnetic

Va	VIa	VIIa	
23 V 3380 **Vanadium** 1890 $3d^34s^2$ 50.9414 ⊡ 5.87	**24 Cr** 2672 **Chromium** 1857 $3d^54s$ 51.996 ⊡ 7.14	**25 Mn** 1962 **Manganese** 1244 □γ $3d^54s^2$ 1100 54.9380 □β 7.3 705 □α	**26 Fe** 2750 **Iron** 1535 ⊡ $3d^64s^2$ 1400 55.847 ✣ 7.86 910 ⊡
41 Nb 4742 **Niobium** 2468 $4d^45s$ 92.9064 ⊡ 8.4	**42 Mo** 4612 **Molybdenum** 2617 $4d^55s$ 95.94 ⊡ 9.01	**43 Tc** 4877 **Technetium** 2172 $4d^55s^2$ (97) ⊙ —	**44 Ru** 3900 **Ruthenium** 2310 $4d^75s$ 101.07 ⊙ 12.1
73 Ta 5425 **Tantalum** 2996 $5d^36s^2$ 180.9479 ⊡ 16.6	**74 W** 5660 **Tungsten** 3410 $5d^46s^2$ 183.85 ⊡ 19.3	**75 Re** 5627 **Rhenium** 3180 $5d^56s^2$ 186.207 ⊙ 20.53	**76 Os** 5027 **Osmium** 3045 $5d^66s$ 190.2 ⊙ 22.5

58 Ce 3257 **Cerium** 798 ⊡ $4f5d6s^2$ 730 140.12 γ ✣ 6.66(β), 6.768(γ) β ⊙	**59 Pr** 3212 **Praseodymium** 931 ✣ $4f^36s^2$ (800) 140.9077 α ⬡ 6.769(α)	**60 Nd** 3127 **Neodymium** 1010 ⊡ $4f^46s^2$ 862 144.24 α ⬡ 7.016(α)	**61 Pm** 2460(?) **Promethium** ~1080 $4f^56s^2$ (145) —
90 Th 4790 **Thorium** 1750 ⊡ $6d^27s^2$ 1400 232.0381 ✣ 11.00	**91 Pa** — **Protactinium** <1600 $5f^26d7s^2$ 231.0359 —	**92 U** 3818 **Uranium** 1132 $5f^36d7s^2$ 760 ⊡ 238.029 □ 18.7 660 ◫	**93 Np** 3902 **Neptunium** 640 $5f^46d7s^2$ ✣ 237.0482 577 □ — 280 ◫

Co		Ni
fcc M_{eff} = 3.15, θ_f = 1395		fcc M_{eff} = 1.61, θ_f = 628.3, I_{s15} = 0.6084, σ = 58.57, M_{s0} = 0.616
hex I_{s20} = 1.7870, σ = 161.9, M_s = 1.708		
fcc σ = 164.8, M_s = 1.739		

Contents in each frame

Atomic Number, Symbol	b.p. (°C)
	m.p. (°C)
electronic structure	
atomic weight	crystal type
density (g cm^{-3}) (°C, otherwise r.t.)	trans. p.t. (°C)

Symbols for Crystal types

□ cubic	diamond
fcc	△ rhombohedral
⊡ bcc	tetragonal
⬡ hex.	orth. rhomb.
c.p. hex.	▱ monoclinic

VIII		Ib	IIb
27 Co Cobalt 2870 / 1495; fcc (420) c.p. hex.; $3d^7 4s^2$; 58.9332; 8.71	**28 Ni** Nickel 2732 / 1453; $3d^8 4s^2$; 58.70; 8.8; fcc	**29 Cu** Copper 2567 / 1083; $3d^{10} 4s$; 63.546; 8.933; fcc	**30 Zn** Zinc 907 / 419.58; $3d^{10} 4s^2$; 65.38; 6.92; c.p. hex.
45 Rh Rhodium 3727 / 1966; $4d^8 5s$; 102.9055; 12.44; fcc	**46 Pd** Palladium 3140 / 1552; $4d^{10}$; 106.4; 12.16; fcc	**47 Ag** Silver 2212 / 961.93; $4d^{10} 5s$; 107.868; 10.492; fcc	**48 Cd** Cadmium 765 / 320.9; $4d^{10} 5s^2$; 112.40; 8.65; c.p. hex.
77 Ir Iridium 4130 / 2410; $5d^7 6s^2$; 192.22; 22.42; fcc	**78 Pt** Platinum 3827 / 1772; $5d^9 6s$; 195.09; 21.37; fcc	**79 Au** Gold 2807 / 1064; $5d^{10} 6s$; 196.9665; 18.88; fcc	**80 Hg** Mercury 356.58 / −38.87; $5d^{10} 6s^2$; 200.59; 14.193(−38.8); △ rhombohedral

62 Sm Samarium 1778 / 1072; $4f^6 6s^2$; 150.4; 7.536(α); α △	**63 Eu** Europium 1597 / 822; $4f^7 6s^2$; 151.96; 5.245; ⊡ bcc	**64 Gd** Gadolinium 3233 / 1311; $4f^7 5d 6s^2$; 157.25; 7.895(α); c.p. hex.	**65 Tb** Terbium 3041 / 1360; $4f^9 6s^2$; 158.9254; 8.253(α); c.p. hex.
94 Pu Plutonium 3232 / 641 ⊡; 476 tetragonal; 451 fcc; 319 orth. rhomb.; 206 ▱; 122 ▱; $5f^6 7s^2$; (244); —	**95 Am** Americium 2607 / 994; $5f^7 7s^2$; (243); —	**96 Cm** Curium — / 1340; $5f^7 6d 7s^2$; (247); —	**97 Bk** Berkelium — / —; $5f^9 7s^2$; (247); —

IIIb	IVb	Vb	VIb
5 B 2550 Boron 2330 $2s^22p$ 10.81 2.535 ▯	6 C 4827 Carbon 3550 $2s^22p^2$ 12.011 △ 3.52 (diamond) ⊙	7 N −195.8 Nitrogen −209.86 N_2 $2s^22p^3$ −210.1 14.0067 ⬡ 1.14(−273) −237.6 □	8 O −182.96 Oxygen −218.4 O_2 $2s^22p^4$ □ 15.9994 −229.8 1.568(−273) ◫ −249.7
13 Al 2467 Aluminum 660.37 $3s^23p$ 26.98154 2.70 ⊡	14 Si 2355 Silicon 1410 $3s^23p^2$ 28.086 □ 2.42 △	15 P 280 Phosphorus 44.1 red ▱ black ◫ $3s^23p^3$ 30.97376 2.20, 2.69 (red) (black)	16 S 444.67 Sulfur 119.0(α) 112.8(β) $3s^23p^4$ 32.06 ▱ 2.07(α),1.96(β) ◫
31 Ga 2403 Gallium 29.78 $4s^24p$ 69.72 5.93 ◫	32 Ge 2830 Germanium 937.4 $4s^24p^2$ 72.59 5.46 △	33 As Arsenic 613 sublimate $4s^24p^3$ 74.9216 5.73 △	34 Se 684.9 Selenium 217 $4s^24p^4$ ▱ 78.96 93 ▱ 4.82 91 ⬡
49 In 2080 Indium 156.61 $5s^25p$ 114.82 7.28 □	50 Sn 2270 Tin 231.97 $5s^25p^2$ β□ 118.69 α△ 7.30(β)	51 Sb 1750 Antimony 630.74 $5s^25p^3$ 121.75 △ 6.62	52 Te 989.8 Tellurium 449.5 $5s^25p^4$ 127.60 ⬡ 6.25
81 Tl 1457 Thallium 303.5 ⊡ $6s^26p$ 230 204.37 ⊙ 11.86	82 Pb 1740 Lead 327.50 $6s^26p^2$ 207.2 ⊡ 11.342	83 Bi 1560 Bismuth 271.3 $6s^26p^3$ 208.9804 △ 9.78	84 Po 962 Polonium 254 $6s^26p^4$ △ (209) □ —

66 Dy 2335 Dysprosium 1409 $4f^{10}6s^2$ 162.50 ⊙ 8.559	67 Ho 2720 Holmium 1470 $4f^{11}6s^2$ 164.9304 ⊙ 8.799	68 Er 2510 Erbium 1522 $4f^{12}6s^2$ 167.26 ⊙ 9.062	69 Tm 1727 Thulium 1545 $4f^{13}6s^2$ 168.9342 ⊙ 9.318

98 Cf — Californium — $5f^{10}7s^2$ (251) —	99 Es — Einsteinium — $5f^{11}7s^2$ (254) —	100 Fm — Fermium — $5f^{12}7s^2$ (257) —	101 Md — Mendelevium — $5f^{13}7s^2$ (258) —

VIIb	0
9 F −188.14 −219.62 Fluorine $2s^22p^5$ 18.99840 1.5(−273)	**10 Ne** −246.05 −248.67 Neon $2s^22p^6$ 20.179 1.204(−245) ⯐
17 Cl −34.6 −100.98 Cl_2 Chlorine $3s^23p^5$ 35.453 2.2(−273) □	**18 Ar** −185.7 −189.2 Argon $3s^23p^6$ 39.948 1.65(−233) ⯐
35 Br 58.78 −7.2 Br_2 Bromine $4s^24p^5$ 79.904 4.2(−273) ◫	**36 Kr** −152.30 −156.6 Krypton $4s^24p^6$ 83.80 3.4(−273) ⯐
53 I 184.35 113.5 I_2 Iodine $5s^25p^5$ 126.9045 4.94 ◫	**54 Xe** −107.1 −111.9 Xenon $5s^25p^6$ 131.30 — ⯐
85 At 337 302 Astatine $6s^26p^5$ (210) —	**86 Rn** −61.8 −71 Radon $6s^26p^6$ (222) —

70 Yb 1193 824 ⊡ Ytterbium $4f^{14}6s^2$ 173.04 660 ⯐ 6.959(α) 7 ~ 42 ⊙	**71 Lu** 3315 1656 Lutetium $4f^{14}5d6s^2$ 174.97 ⊙ 9.842
102 No — — Nobelium $5f^{14}7s^2$ (255) —	**103 Lr** — — Lawrencium $5f^{14}6d7s^2$ — —

Appendix 5

CONVERSION OF MAGNETIC QUANTITIES – MKSA AND CGS SYSTEMS

Quantity	Symbol	MKSA unit	Conversion ratio: $\frac{\text{MKSA value}}{\text{CGS value}}$	Conversion ratio: $\frac{\text{CGS value}}{\text{MKSA value}}$	CGS unit
Magnetic pole	m	Wb	1.257×10^{-7}	7.96×10^{6}	
Magnetic flux	Φ	Wb	1×10^{-8}	1×10^{8}	Mx
Magnetic moment	M	Wb m	1.257×10^{-9}	7.96×10^{8}	
Magnetization	I	T	1.257×10^{-3}	7.96×10^{2}	G
Magnetic flux density	B	T	1×10^{-4}	1×10^{4}	G
Magnetic field	H	$\mathrm{A\,m^{-1}}$	7.96×10	1.257×10^{-2}	Oe
Magnetic potential Magnetomotive force	ϕ_{m} V_{m}	A	7.96×10^{-1}	1.257	gilbert
Magnetic susceptibility	χ	$\mathrm{H\,m^{-1}}$	1.579×10^{-5}	6.33×10^{4}	
Relative susceptibility	$\bar{\chi}$		$= 4\pi\chi$ in CGS		
Permeability	μ	$\mathrm{H\,m^{-1}}$	1.257×10^{-6}	7.96×10^{5}	
Relative permeability	$\bar{\mu}$		$= \mu$ in CGS		
Permeability of vacuum	$\mu_0 = 4\pi \times 10^{-7}\,\mathrm{H\,m^{-1}}$				$= 1$
Demagnetizing factor	N		7.96×10^{-1}	1.257×10	
Rayleigh constant	η	H/A	1.579×10^{-8}	6.33×10^{7}	$\mathrm{Oe^{-1}}$
Reluctance	R_{m}	$\mathrm{H^{-1}}$	7.96×10^{7}	1.257×10^{-8}	gilbert $\mathrm{Mx^{-1}}$
Inductance	L	H	1×10^{-9}	1×10^{9}	abhenry
Anisotropy constant Energy density	K E_{m}	$\mathrm{J\,m^{-3}}$	1×10^{-1}	10	$\mathrm{erg\,cm^{-3}}$
Ordinary Hall coefficient	R	$\Omega\,\mathrm{m^2 A^{-1}}$	1.257×10^{-4}	7.96×10^{3}	$\Omega\,\mathrm{cm\,Oe^{-1}}$

Wb (weber), T (tesla), A (ampere), H (henry), J (joule), Ω (ohm), Mx (maxwell), G (gauss), Oe (oersted), $1.257 = 4\pi/10$, $7.96 = 10^2/4\pi$, $1.579 = (4\pi)^2/10^2$, $6.33 = 10^3/(4\pi)^2$.

Appendix 6

CONVERSION OF VARIOUS UNITS FOR MAGNETIC FIELD

$A\,m^{-1}$ (ampere per meter)	Oe (oersted)	T* (tesla)	Remarks
$1\,mA\,m^{-1}$	$= 1.26 \times 10^{-5}$ Oe	$= 1.26 \times 10^{-9}$ T	1×10^{-5} G $= 1\gamma$ (gamma)
$10\,mA\,m^{-1}$	$= 1.26 \times 10^{-4}$ Oe	$= 1.26 \times 10^{-8}$ T	
$100\,mA\,m^{-1}$	= 1.26 mOe	$= 1.26 \times 10^{-7}$ T	
$1\,A\,m^{-1}$	= 12.6 mOe	$= 1.26\,\mu T$	
$10\,A\,m^{-1}$	= 0.126 Oe	$= 12.6\,\mu T$	earth mag. field = 0.15–0.30 Oe
$100\,A\,m^{-1}$	= 1.26 Oe	= 0.126 mT	
$1\,kA\,m^{-1}$	= 12.6 Oe	= 1.26 mT	
$10\,kA\,m^{-1}$	= 126 Oe	= 12.6 mT	can be produced by permanent magnets
$100\,kA\,m^{-1}$	= 1.26 kOe	= 0.126 T	(can be produced by permanent magnets)
$1\,MA\,m^{-1}$	= 12.6 kOe	= 1.26 T	can be produced by electromagnets
$10\,MA\,m^{-1}$	= 126 kOe	= 12.6 T	can be produced by supercon. coil
$100\,MA\,m^{-1}$	= 1.26 MOe	= 126 T	can be produced by flux compression.

* T is a unit of magnetic flux density, but sometimes can be used as a unit of magnetic field. In this case, the symbol $\mu_0 H$ should be used and is preferably called an 'induction field'.

MATERIAL INDEX

Commercial, crystal, mineral, and chemical names are listed in the Subject Index. The general metallic elements are indicated by M, while the rare earth elements are indicated by R. The page number followed by 'f' or 't' indicate a reference to a figure and a table, respectively. In these cases, sometimes the material symbol may be found, not only in the figure or table, but also in the text. Bold page numbers indicate the pages where detailed explanations are given.

SUBJECT INDEX

Bold page numbers indicate the places where definitions or detailed explanations are given. The page number followed by 'f' or 't' indicate a reference to a figure or a table, respectively. In these cases, sometimes the word or term may be found not only in the figure or table, but also in the text.

Lightning Source UK Ltd.
Milton Keynes UK
UKOW021239300812

198250UK00002B/1/P

9 780199 564811